VOLUME ONE HUNDRED AND FORTY FOUR

Progress in Molecular Biology and Translational Science

Molecular and Cellular Changes in the Cancer Cell

VOLUME ONE HUNDRED AND FORTY FOUR

PROGRESS IN MOLECULAR BIOLOGY AND TRANSLATIONAL SCIENCE

Molecular and Cellular Changes in the Cancer Cell

Edited by

KEVIN PRUITT

Texas Tech University Health Sciences Center
Lubbock, TX, United States

ELSEVIER

AMSTERDAM • BOSTON • HEIDELBERG • LONDON
NEW YORK • OXFORD • PARIS • SAN DIEGO
SAN FRANCISCO • SINGAPORE • SYDNEY • TOKYO
Academic Press is an imprint of Elsevier

Academic Press is an imprint of Elsevier
50 Hampshire Street, 5th Floor, Cambridge, MA 02139, United States
525 B Street, Suite 1800, San Diego, CA 92101-4495, United States
The Boulevard, Langford Lane, Kidlington, Oxford OX5 1GB, United Kingdom
125 London Wall, London EC2Y 5AS, United Kingdom

First edition 2016

ISBN: 978-0-12-809328-3
ISSN: 1877-1173

For information on all Academic Press publications visit our website at https://www.elsevier.com/

Publisher: Zoe Kruze
Acquisition Editor: Alex White
Editorial Project Manager: Helene Kabes
Production Project Manager: Magesh Mahalingam
Cover Designer: Matthew Limbert

Typeset by SPi Global, India

CONTENTS

CONTRIBUTORS

A.U. Ahmed
Texas Tech University Health Sciences Center School of Medicine, Lubbock, TX, United States

A.T. Baines
Cancer Research Program, JLC-Biomedical/Biotechnology Research Institute, North Carolina Central University, Durham; School of Medicine, UNC-Chapel Hill, Chapel Hill, NC, United States

J. Bodily
Louisiana State University Health Sciences Center, Shreveport, LA, United States

M. Dancer
University of North Dakota School of Medicine and Health Sciences, Grand Forks, ND, United States

J. Deitrick
Texas Tech University Health Sciences Center School of Medicine, Lubbock, TX, United States

R.S. DeVaux
Cancer Research Center, University at Albany, Rensselaer, NY, United States

T.J. Grant
Boston University School of Medicine, Boston, MA, United States

P.W. Halcrow
University of North Dakota School of Medicine and Health Sciences, Grand Forks, ND, United States

J.I. Herschkowitz
Cancer Research Center, University at Albany, Rensselaer, NY, United States

K. Hua
Boston University School of Medicine, Boston, MA, United States

E. Karjalainen
Institute for Molecular Medicine Finland FIMM, Nordic EMBL Partnership for Molecular Medicine, University of Helsinki, Helsinki, Finland

M. Kumar
Cancer Research Center, University at Albany, Rensselaer, NY, United States

A.R. Maiuri
Medical Sciences, Indiana University School of Medicine, Bloomington, IN, United States

P.R. Manna
Texas Tech University Health Sciences Center School of Medicine, Lubbock, TX, United States

P.M. Martin
North Carolina Agricultural and Technical State University, Greensboro, NC, United States

D. Molehin
Texas Tech University Health Sciences Center School of Medicine, Lubbock, TX, United States

H.M. O'Hagan
Medical Sciences, Indiana University School of Medicine, Bloomington; Indiana University Melvin and Bren Simon Cancer Center, Indianapolis, IN, United States

J.E. Ohm
Roswell Park Cancer Institute, Buffalo, NY, United States

M. Panteah
University of North Dakota School of Medicine and Health Sciences, Grand Forks, ND, United States

K. Pruitt
Texas Tech University Health Sciences Center, Lubbock, TX, United States

W.M. Pruitt
Texas Tech University Health Sciences Center School of Medicine, Lubbock, TX, United States

S. Radhi
Texas Tech University Health Science Center, Lubbock, TX, United States

G.A. Repasky
Institute for Molecular Medicine Finland FIMM, Nordic EMBL Partnership for Molecular Medicine, University of Helsinki, Helsinki, Finland

G.W. Reuther
H. Lee Moffitt Cancer Center; University of South Florida, Tampa, FL, United States

C.J. Rorie
North Carolina Agricultural and Technical State University, Greensboro, NC, United States

M. Scott
Louisiana State University Health Sciences Center, Shreveport, LA, United States

A. Singh
Boston University School of Medicine, Boston, MA, United States

A.J. Smith
Texas Tech University Health Sciences Center, Lubbock, TX, United States

L.A. Smith
Texas Tech University Health Sciences Center, Lubbock, TX, United States

C. Walden
University of North Dakota School of Medicine and Health Sciences, Grand Forks, ND, United States

B. Woodby
Louisiana State University Health Sciences Center, Shreveport, LA, United States

PREFACE

Molecular and Cellular Changes in the Cancer Cell aims to provide a broad and comprehensive review of the prototypical changes that occur as cells progressively transition from a normal to a malignant cancerous state. As cancers evolve, the cells comprising the tumor mass undergo complex genomic and epigenomic alterations that lead to a rewriting of the intrinsic failsafe instructions that would otherwise prevent uncontrolled proliferation and migration. These molecular and cellular changes that accumulate over time lead to a crisis in the cell's identity and enable it to transition from normal to malignant. This volume examines many of the cellular alterations that enable tumor progression. The volume begins by discussing some of the most critical genomic and epigenomic changes across diverse tumor types and provides examples of key drivers of tumor progression that either modify the epigenome or activate signaling pathways found in the majority of select tumor types. Next, the interplay between inflammation and epigenetic changes are discussed as well as the role of viruses in promoting tumor initiation and progression. Additionally, this volume explores the molecular pathogenesis and some of the emerging molecular targeting therapies for leukemias and sarcomas as well as cancers of the colon, pancreas, breast, endometrium and ovary. The volume ends with a focus on new roles for noncoding RNAs in cancer initiation and progression, an increasingly important, but largely unexplored area where many of the discoveries are very recent. While each chapter provides a wealth of insight and is comprehensive, it also provides numerous key references should the reader want even more in-depth information. The editor wishes to express a deep gratitude to our distinguished chapter authors for their time, efforts, and incredible insight and to the senior team at Elsevier who greatly facilitated this endeavor.

Kevin Pruitt
Texas Tech University Health Sciences Center
Lubbock, Texas
October 2016

PART I

Overview: Genetic, Epigenetic and Cell Signaling Alterations

CHAPTER ONE

Molecular and Cellular Changes During Cancer Progression Resulting From Genetic and Epigenetic Alterations

K. Pruitt[1]
Texas Tech University Health Sciences Center, Lubbock, TX, United States
[1]Corresponding athor: e-mail address: kevin.pruitt@ttuhsc.edu

Contents

Progress in Molecular Biology and Translational Science, Volume 144
ISSN 1877-1173
http://dx.doi.org/10.1016/bs.pmbts.2016.09.001

Abstract

Tumorigenesis is a complex process that involves a persistent dismantling of cellular safeguards and checkpoints. These molecular and cellular changes that accumulate over months or decades lead to a change in the fundamental identity of a cell as it transitions from normal to malignant. In this chapter, we will examine some of the molecular changes in the evolving relationship between the genome and epigenome and highlight some of the key changes that occur as normal cells progress to tumor cells. For many years tumorigenesis was almost exclusively attributed to mutations in protein-coding genes. This notion that mutations in protein-coding genes were a fundamental driver of tumorigenesis enabled the development of several novel therapeutics that targeted the mutant protein or overactive pathway responsible for driving a significant portion of the tumor growth. However, because many therapeutic challenges remained in the face of these advances, it was clear that other pieces to the puzzle had yet to be discovered. Advances in molecular and genomics techniques continued and the study of epigenetics began to expand and helped reshape the view that drivers of tumorigenesis extended beyond mutations in protein-coding genes. Studies in the field of epigenetics began to identify aberrant epigenetic marks which created altered chromatin structures and enabled protein expression in tissues that defied rules governing tissue-specificity. Not only were epigenetic alterations found to enable overexpression of proto-oncogenes, they also led to the silencing of tumor suppressor genes. With these discoveries, it became clear that tumor growth could be stimulated by much more than mutations in protein-coding genes. In fact, it became increasingly clear that much of the human genome, while transcribed, did not lead to proteins. This discovery further led to studies that began to uncover the role of noncoding RNAs in regulating chromatin structure, gene transcription, and tumor biology. In this chapter, some of the key alterations in the genome and epigenome will be explored, and some of the cancer therapies that were developed as a result of these discoveries will be discussed.

1. INTRODUCTION

Over the last decade considerable progress has been made in defining the changes in cancer cells that enable them to become malignant. While the diversity and complexity of alterations in tumor cells that have been characterized are extensive, we will highlight examples of the changes that are generally considered to be the "drivers" of malignancy. Many mutations and driver genes have been identified across numerous cancer types and subtypes, but the primary focus here will be on genetic mutations or alterations in the four broad categories of enzymes or chromatin-binding proteins that influence epigenetic states.

1.1 Types of Chromatin

Remarkably, the billions of cells that comprise various tissues in organisms contain identical genetic material, yet each cell has the ability to carry out highly specific and unique biological functions directed by the same genetic material. This tissue-specificity is achieved by allowing or preventing transcription of clusters of genes via a mechanism that links transcription potential with specific chromatin structures and epigenetic states. Chromatin can be defined as the sum of DNA, histones, and associated RNA. The state of chromatin, whether it is open and accessible, or closed and inaccessible to transcription factors, will dictate gene expression in a cell-type and tissue-specific manner. Chromatin was once viewed as static and uninformative, but advances have now shown it to be dynamic and highly instructive with the capability of dictating cellular fates. In many respects, chromatin can be considered the depositor of cellular identity because although there is relative uniformity of genetic sequence in cells and tissues, different cells, and tissues display great phenotypic diversity. Chromatin not only regulates transcription but also serves to compact DNA and protect it from DNA damage. The general structural characteristics of chromatin are highly influenced by epigenetic states. In general, "epigenetics" refers to a heritable pattern of gene expression that is not the result of alterations in the primary nucleotide sequence of a gene. The specific epigenetic state is the result of dynamic and reversible covalent modifications to DNA or posttranslational modifications (PTMs) to histones, which together, create an epigenetic landscape that may look very different depending on the genomic locus being examined. Epigenetic marks play a major role in dictating the tightness of chromatin compaction, and this structural property will influence the ease with which RNA polymerase gains access of specific genes for transcription or whether DNA polymerases can access genomic regions for replication. In general, the more loosely packaged chromatin is referred to as euchromatin and the more tightly packaged chromatin is referred to as heterochromatin. Heterochromatin can be further subdivided into facultative heterochromatin which is infrequently transcribed or constitutive heterochromatin which is very tightly packaged and not transcribed at all.

1.2 The Nucleosome and the Core Epigenetic Modifications

Chromatin is structurally complex and is comprised of DNA that is wrapped around histone scaffolds which is referred to collectively as the nucleosome, or the basic building block of chromatin. The nucleosome is the basic unit of chromatin and provides a platform for wrapping approximately 147 bp of DNA around two copies of each of the "core" histones H2A, H2B, H3, and H4.[1] The remodeling of chromatin, which enables transcription factors to gain access to DNA, is regulated via diverse epigenetic mechanisms which lead to distinct patterns of gene expression.[2] To date there are at least four covalent modifications to DNA and 17 PTMs for histones that have been reasonably well characterized.[3] These PTMs to occurring on histone tails is critically important in determining whether the associated gene is expressed or repressed. While new epigenetic modifications have recently been discovered,[4,5] the most well-studied PTMs for histones implicated in cancer biology include acetylation, methylation, phosphorylation, and ubiquitination and the first two will largely be the focus of the following sections. For DNA, the most extensively studied modification is the covalent addition of a methyl group to the 5′ position of the cytosine base within the context of the cytosine-guanosine dinucleotide (CpG), which will be referred to as methyl-CpG (mCpG). More recently, three other covalent modifications to mCpG were identified. These modifications to methylcytosine result from active or passive DNA demethylation and include three products of oxidation of the 5′-methylcytosine, which are 5′-hydroxymethylcytosine, 5-formyl-carboxyl-methylcytosine, and 5-carboxylcytosine.[6,7] CpG methylation is found more frequently at small stretches of DNA called CpG islands. These islands are typically associated with the promoter regions of genes where the methylation status correlates with transcription.[8] Extensive investigation by a number of laboratories has shown that epigenetic regulation such as promoter hypermethylation can profoundly affect gene expression.[9] However, although CpG methylation is involved in the regulation of gene activity, it alone is insufficient to repress transcription and it often works in concert with other epigenetic modifications such as histone hypoacetylation.[10] In the following sections, how these modifications are altered in cancer cells will be examined in more depth.

1.3 DNase Hypersensitivity Sites Are Linked With the Epigenomic Landscape

DNA sequences that regulate transcription are characterized by the cooperative binding of sequence-specific transcription factors and other transcription regulators instead of the canonical nucleosome. Chromatin remodeling allows for a substitution of the histone proteins in the nucleosome for non-histone proteins that promote transcription, and this altered and more open chromatin state is characterized by a notable increase in accessibility to nucleases such as DNase I.[11] These so-called DNase I hypersensitive sites or DHSs in chromatin are flanked by nucleosomes with characteristic patterns of histone modifications that coincide with the functional role of the adjacent DHS region. DHS identification has been used to generate detailed maps of regulatory regions of the DNA such as active *cis*-regulatory elements including locus control regions, promoters, enhancers, silencers, and insulators.[12] Due to advances in methods of studying chromatin changes such as chromatin immunoprecipitation (ChIP) coupled with next-generation sequencing (ChIP-seq) and several other sophisticated molecular profiling approaches, genome-scale mapping of DHSs in mammalian cells have led to several breakthroughs in our understanding of the human epigenome. For example, 97.4% of a compilation of 1046 experimentally validated distal, nonpromoter *cis*-regulatory elements such as enhancers, insulators, and locus control regions were found to be encompassed within DHS chromatin.[13] Additional analysis revealed that nearly 250,000 CpGs falling within DHSs across 19 cell types showed increased DNA methylation was very consistently associated with less chromatin accessibility in 97% of the cases. When transcription factor transcript levels were compared to the average methylation at cognate recognition sites within DHSs, significant correlations emerged demonstrating a negative correlation between transcription factor expression and the methylation of its binding site for 70% of the transcription factors analyzed. Further analysis of large data sets suggested that DNA methylation patterns often result as a passive "filling in" of the voids left by vacating transcription factors, which is very reminiscent of a type of "use it or lose it" phenomenon.

1.4 Depositing, Removing, and Interpreting Epigenetic Marks

One important question to consider is why are epigenetic modifications so frequently targeted in cancer? First, these covalent changes to DNA and histone PTMs can alter chromatin structure by altering interactions within and between nucleosomes. Second, these marks provide platforms and docking sites for proteins with specific domains that anchor context-specific complexes of regulatory protein to genomic loci. Key protein families that specify epigenetic states have the ability to control both the content of the epigenetic information encoded within the epigenome and how this information gets interpreted. Just as the language of the genome is in the linear sequence of four distinct nucleotides, the language of the epigenome is largely in the form of PTMs to histones and covalent modifications to DNA. Early on, the term "histone code" was used to capture the importance of epigenetic modifications as combinations of histone PTMs appeared to codify specific molecular actions. This led to the "histone code hypothesis" which posited that multiple histone modifications, acting in a combinatorial or sequential fashion on one or multiple histone tails, specify unique downstream functions.[14] Revisions to this view have now lead to the consideration of epigenetics as a type of language which now include both modifications to DNA and histones as part of the "alphabet." To further describe the players involved in specifying epigenetic states, we will discuss the enzymes involved in depositing (writers), interpreting (readers), and removing (erasers) specific epigenetic marks and will begin to examine how the modifications and enzymes change during tumor progression.

2. DNA METHYLATION

In 1975, early studies first suggested that DNA methylation of cytosines within the context of CpG dinucleotides could potentially represent an epigenetic mark associated with gene silencing.[15,16] During this early stage it was suggested that mechanisms that may account for the stability of differentiation, or the ordered switching on or off of genes during development, may be linked with the enzymatic modification of specific bases in DNA. Additionally, a model based on DNA methylation was proposed to explain the initiation and maintenance of mammalian X inactivation and certain aspects of other permanent events in eukaryotic cell differentiation. These hypotheses came at a time when such enzymes had not yet been detected in eukaryotes. We now know the enzymes that deposit CpG methylation and remove this methylation. The most well-characterized mammalian DNA methyltransferases (DNMTs) include DNMT1, DNMT3a, and DNMT3b. DNMT3a and DNMT3b are "de novo" methyltransferases and can methylate DNA templates without any preexisting DNA methylation. However, DNMT1 is a maintenance methyltransferases that recognizes hemimethylated DNA, which may be generated during DNA replication where there will be the parental strand (with CpG methylation) and the daughter strand (lacking CpG methylation) that are intertwined. DNMT1 methylates newly synthesized CpGs when their complimentary parental strand already possesses CpG methylation. DNMT3a and DNMT3b do not require a hemimethylated DNA template but they are capable of methylating within this context as well.[17] Hemimethylation of DNA occurs when only one of the two complementary strands is methylated. For several decades, DNA methylation was considered to be such a stable covalent modification and therefore not amenable to active demethylation. With further investigation, this view began to conflict with evidence from the study of embryogenesis which revealed an active global loss of DNA methylation in the early zygote. However, it was not until 2009 that the real breakthrough occurred with two reports demonstrating the identification of 5-hydroxymethylcytosine (5hmC).[6,7] Continued investigation along this line of inquiry led to the identification and characterization of the ten-eleven translocation (TET1–3) family of proteins which act as a mammalian family of dioxygenases and serve as the central mediators of both passive and active DNA demethylation.[18] TET proteins perform a series of iterative oxidation reactions converting 5mC to 5hmC to 5-formylcytosine to 5-carboxycytosine which eventually gets converted to unmethylated cytosine. During the process of tumorigenesis both the writers (DNMTs) and erasers (TETs) of CpG methylation undergo alterations. In the following sections we will focus on significant advances that have occurred in the decades since the early conception of the importance of DNA methylation, and will examine how altered DNA methylation has now been firmly linked with tumor initiation, maintenance, and progression.

2.1 Genome-Wide DNA Hypomethylation

Eukaryotic cells maintain a form of inheritance based on DNA methylation patterns. The first described covalent modification of DNA which represents the most well-characterized epigenetic mark is the methylation of the 5-carbon on cytosine within the context of the CpG dinucleotide. This epigenetic mark strongly influences gene expression programs,[19,20] and changes in DNA methylation is one of the most notable features as cells progress from a normal to malignant state. Because DNA methylation influences active transcription,[21] it is frequently targeted for oncogene-induced changes and cancer cells universally show altered DNA methylation patterns. Global hypomethylation was first reported in primary human tumors in 1983 by two separate laboratories.[22,23] This pioneering work first established that substantial hypomethylation was found in the genomes of cancer cells compared with their normal counterparts in tumors representing multiple histological types of cancer. While the study of tumors from animal models, cancer and transformed cell lines provided increasing evidence, at that time that DNA methylation was important in gene expression, these early reports established that DNA methylation patterns are indeed perturbed in human tumors.

2.2 Tumor Suppressor Gene Hypermethylation

As discussed earlier, while global DNA methylation is often decreased during tumor progression,[24,25] the opposite frequently occurs for the promoter DNA associated with tumor suppressor genes (TSGs). CpG islands (CGIs) occur in about 70% of all mammalian promoters and are generally devoid of CpG methylation. During the process of tumorigenesis, unmethylated CGI of many TSGs become densely methylated.[26] Mammalian genome methylation normally occurs at a high frequency on CpGs throughout the genome, yet CGI are "protected" from methylation.[27] These aberrantly methylated CGIs are associated with the chromatin typical of nontranscribed DNA such as imprinted genes or silenced genes on the inactive X chromosome. Demonstrating the role of DNA hypermethylation, many studies have documented CpG-specific hypermethylation of promoter DNA of numerous TSGs across diverse tumor types. Almost half of the TSGs that cause familial cancers through germ-line mutations are inactivated in association with promoter DNA hypermethylation in sporadic cancers. Observations from several laboratories have demonstrated that, similar to genetic mutations, aberrant promoter methylation is associated with loss of gene function and provides a selective advantage to neoplastic cells.[20] For example, epigenetic silencing of potent TSGs in sporadic tumors occurs in the same tumor spectrum where germ-line mutations of genes such as VHL, BRCA1, and STK11 cause familial forms of renal, breast, and colon cancer, respectively. Additionally, promoter CGI methylation of E-cadherin in sporadic forms of diverse cancers is common. Another example involves the TSG transcription factor hypermethylated in cancer 1 (HIC1) which is often hypermethylated in human colon cancers and undergoes frequent loss of heterozygosity in tumors.[28] Finally, CGI methylation and silencing of

MLH-1 (mismatch repair gene) frequently occurs in sporadic tumors with microsatellite instability (MSI).[29] Importantly, SIRT1 localizes to the hypermethylated MLH-1 promoter[30] and represses HIC1.[31] Some clues into the mechanism by which SIRT1 is recruited to hypermethylated TSG promoters have been discovered,[32] but more investigation is needed to gain a deeper understanding of the mechanism. Because altered DNA methylation is linked with TSG silencing and genomic instability in diverse tumors, it is not surprising that the enzymes that deposit (DNMTs) or remove (TETs) cytosine methylation are either mutated or their expression level is altered in human tumors. For example, 22% of 280 patients with acute myeloid leukemia (AML) were shown to have DNMT3A mutations that affect protein levels.[33] Furthermore, DNMT mutations in 20% of 112 cases of acute monocytic leukemia,[34] the most common point mutation in these patients, drives cancer in vivo in mice by altering DNA methylation patterns and disrupting gene expression.[35] Moreover, DNMT alterations predispose stem cells to transformation[36] and defective differentiation.[37] When considering TET proteins the scenario is similar as they are mutated in diverse cancers including solid tumors.[38–40] In fact, the TET family derives its name from the initial discovery of the juxtaposition of the MLL (a lysine methyltransferase) and TET genes as part of recurrent chromosomal translocations in some AML patients.[41] In summary, altered expression or mutation of either the writer or eraser of DNA methylation may contribute to defective propagation of methylation patterns as cells transform from a normal to malignant state.

2.3 CpG Island Methylator Phenotype

Not only does DNA methylation occur at individual gene promoters, but this can also occur at clusters of loci. The term CpG island methylator phenotype (CIMP) was originally coined by Issa and colleagues[42] as a subset of colorectal cancers which consistently showed DNA methylation. This early study revealed that in colorectal cancer, there appears to be two types of methylation that are associated with cancer progression. The first type was viewed as age-related methylation and the second type as cancer-specific methylation. It was proposed that initially, age-related methylation arises as a function of age in normal colorectal epithelial cells which may lead to a predisposition state that precedes tumor formation in the colon. The second type, cancer-specific methylation was found exclusively in a subset of cancers, which display CIMP. This phenomenon of CIMP was proposed to be a novel molecular instability pathway responsible for most cases of aberrant TSG methylation in colorectal cancer. This study found that CIMP+ tumors account for the majority of sporadic colorectal cancers with MSI, through methylation of the mismatch repair gene hMLH1. Thus, CIMP has been associated with colorectal cancers having MSI. Years later, another link between CIMP status and the expression of the SIRT1 deacetylase, which will be discussed later in the chapter, was found in colorectal tumors. Upon analysis of 485 colorectal cancers, 37% showed nuclear overexpression of SIRT1 significantly associated with high tumor grade. SIRT1 expression was also significantly associated with CIMP-high and MSI-high phenotype.[43] Whether at a specific locus or gene or at a clustered set of promoters, aberrant DNA methylation contributes to tumor progression in diverse ways.

2.4 Viral Influences on Aberrant DNA Methylation

Erratic and abnormal DNA methylation is a well-documented hallmark of cancer and this epigenetic alteration promotes tumor initiation and progression. The cause of the misguided CpG methylation and errant histone modifications at transcription regulatory regions is not entirely clear but emerging evidence suggests that viral infection can contribute. The genomic regions that most alarmingly deviate from the norm tend to control the expression of growth control TSGs, and imbalances in the deposition or removal of epigenetic marks contribute to all stages of tumor progression.[44,45] As discussed in the previous section, virtually all human tumors exhibit abnormal DNA methylation resulting from the amplification or mutation in the enzymes that deposit or erase DNA methylation.[18,33,35,39,46] Interestingly, altered DNA methylation was also reported in chronic conditions that predispose cells to transformation. For example, chronic inflammation and viral infections that contribute to neoplastic transformation have been linked with aberrant DNA methylation.[47,48] One of the viruses which has been linked with aberrant DNA methylation is Epstein–Barr virus (EBV or HHV-4) which belongs to the gammaherpesvirus family.[49] Members of the gammaherpesvirus family are able to establish a persistent, lifelong infection in immunocompetent hosts.[50] Two of the types of gammaherpesviruses identified in humans include EBV and Kaposi's sarcoma-associated herpesvirus (KSHV). EBV has a well-documented association with Burkitt's lymphoma, Hodgkin's disease, and nasopharyngeal carcinoma,[50] and KSHV has been associated with AIDS-related lymphoproliferative disorders and Kaposi's sarcoma.

2.5 Therapeutic Reversal of Aberrant DNA Methylation

The contribution of aberrant DNA methylation to malignancy is without question and consequently there has been intense interest in understanding the factors that contribute to errant DNA methylation patterns and how it might be reversed. Because of intense interest and efforts from many labs, advances have been made in the treatment of some malignancies such as myelodysplastic syndromes (MDS). For decades, this group of disorders was refractory to various treatment strategies, but a series of studies demonstrated that demethylating agents (such as azacytidine and decitabine) could improve quality of life and survival time of even high-risk patients.[51,52] Interestingly, further investigation has demonstrated that administration of transient low doses of DNA-demethylating agents may also be very promising given the durable antitumor effects on hematological and epithelial tumor cells observed in experimental contexts.[53]

3. LYSINE ACETYLATION

Acetylation of the ε-amino group of lysine amino acids has been studied for many decades and was first reported to occur on histones[54] by Allfrey and colleagues. Since this discovery, the early studies focused on elucidating the mechanism and consequences of this acetylation within the context of regulating chromatin structure and gene regulation. While a few nonhistone substrates such as transcription factors and tubulin were studied early on, little was known about the role of acetylation in regulating nonhistone proteins in normal or pathological settings. After these initial pioneering studies, considerable attention began to be placed on nonhistone substrates especially following the seminal discovery of the acetylation of p53.[55] Since then, p53 acetylation on multiple lysine residues has been mapped and the role of these lysines in regulating p53 function is much clearer.[56–59] More recent advances using unbiased proteomics approaches have led to the discovery of "acetylomes" and have led to the identification of numerous proteins in every subcellular compartment across numerous species. We now know that acetylation occurs on nonnuclear proteins and can affect their structural properties, binding affinities, activity, stability, and localization. Depending on the specific protein target, this will affect diverse signaling pathways involved in a very wide range of cellular events such as cell survival, cell death, differentiation, motility, chromatin remodeling, DNA damage repair, metabolism, cell cycle control, nuclear transport, nuclear hormone signaling, cytoskeletal organization, nucleotide exchange, vesicular trafficking, and other processes.[60–62] Because histone-modifier gene expression profiles are associated with pathological and clinical outcomes of diverse cancers,[3,63] in the following sections, the writers and erasers of acetylation will be discussed. Table 1 provides a list of the writers and erasers of acetylation, and while several are known to participate in oncogenesis, only a few examples will be cited as representative examples of altering the expression or function of these enzymes in tumor progression.

3.1 Acetyltransferases

Histone acetyltransferases, also referred to as lysine acetyltransferases (KATs), are a subtype of transcription coactivators and enhance the ability of a transcription factor to promote transcription. KATs were the first enzymes shown to modify histones and have been extensively reported to be inactivated via mutations,[68,72] take part in recurrent chromosomal translocations,[70,72,75–77,83,111] undergo amplification or increased expression,[64,66,67,82] promote oncogenic signaling[78–81,84] or targeted therapeutically[65] in diverse tumors. Generally, while nucleosome acetylation increases DNA accessibility to transcription factors, transcription factor acetylation leads to more diverse outcomes such as altered DNA-binding ability, proteasomal degradation potential, protein–protein interaction, and nuclear vs cytosolic localization. The 17 human acetyltransferases can be divided into five families based on the degree of sequence similarity[112,113] (Table 1). These families consist of the GNAT, p300/CBP, MYST, SRC along with acetyltransferases that are not clearly categorized based on defining features of the first four classes. It is now well established that these enzymes acetylate diverse cellular proteins in every cellular compartment in response to upstream cues or changes in the cellular environment.[32,96,114–117] KATs are present across diverse organisms in multisubunit complexes and their catalytic activity and their recruitment to chromatin is guided by specific binding partners.[118] The availability of the acetyl group for acetylation is derived from acetyl-coenzyme A (acetyl-CoA) which is linked with the metabolic status of a cell. Therefore, the concentration of acetyl-CoA, the activity of an acetyltransferase and abundance of protein acetylation will be influenced by the cellular metabolic state.[119]

Table 1 Human Lysine Acetyltransferases and Their Cancer Association

HAT Family	Cancer Association	References
GNAT		
KAT1 (HAT1)	EC	64
KAT2A (GCN5)	Leukemias, HCC	65,66
KAT2B (P/CAF)	RMS	67
ATF2	Melanoma, renal	68,69
p300/CBP		
KAT3A (CBP, CREBBP)	DLCBL, AML, lung	70,71
KAT3B (p300, EP300)	FL, DLCBL	72
MYST		
KAT5 (Tip60)	CRC, melanoma, medulloblastoma	73,74
KAT6A (MOZ, MYST3)	AML	75–77
KAT6B (MORF, MYST4)	AML	76,77
KAT7 (HBO1, MYST2)	BC	78
KAT8 (MOF, MYST1)	BC, PC	63,79
SRC		
KAT13A (NCOA1, SRC-1)	BC	80,81
KAT13B (NCOA3, SRC-3, ACTR, AIB1)	BC	82
KAT13C (NCOA2, SRC-2, GRIP1, TIF2)	RMS	83
Other		
KAT4 (TAF1, TAF(II)250, DYT3, XDP)	–	–
KAT12 (TFIIIC90, GTF3C4, TF3C-δ)	–	–
Kat13D (CLOCK)	Colorectal	84

Continued

Table 1 Human Lysine Acetyltransferases and Their Cancer Association—cont'd

HDAC Family	Cancer Association	References
Class I		
HDAC1	BC, CRC, GC, lymphomas, HCC, PC	70,85,86
HDAC2	BC, CRC, HCC, PC	85–88
HDAC3	BC, HCC, Panc	85,86,89
HDAC8	Prostate	103
Class II		
HDAC4	EC	90
HDAC5	Medulloblastoma	91
HDAC6	BC, DLBCL	85
HDAC7	Panc, ALL, GB	89,92,93
HDAC9	ALL	89
HDAC10	HCC, lung	94,95
Class III		
SIRT1	BC, CRC, GC, HCC, PC, multiple cancers	30,96–101
SIRT2	HCC	102
SIRT3	Panc	103
SIRT4	CRC, GC	104,105
SIRT5	NSCLC	106
SIRT6	NSCLC, HCC, Panc	107–109
SIRT7	Panc	103
Class IV		
HDAC11	Carcinomas, lymphomas	74,110

ALL, acute lymphocytic leukemia; *AML*, acute myeloid leukemia; *BC*, breast; *CRC*, colorectal; *DLBCL*, diffuse large B-cell lymphoma; *EC*, esophageal; *GB*, glioblastoma; *GC*, gastric; *HCC*, hepatocellular carcinoma; *NSCLC*, nonsmall cell lung cancer; *RMS*, rhabdomyosarcomal; Panc, pancreatic; *PC*, prostate.

3.2 Class I, II, and IV Deacetylases

As their name implies, the erasers for the acetylation mark are the histone deacetylases (HDACs). While this name has been in place for quite some time, the term lysine deacetylase (KDACs) will sometimes be used interchangeably because this family deacetylates numerous nonhistone proteins. The mechanism of deacetylation differentiates class I/II/IV HDACs from sirtuins. Spanning about a 30 year period histone deacetylase activity could be demonstrated; however, it was not until 1996 that the first HDAC was identified. Using the HDAC inhibitor (HDI), trapoxin, and affinity purification coupled with peptide micro-sequencing and cDNA cloning,[120] this breakthrough lead to subsequent cloning and characterization of multiple HDAC family members which were demonstrated to act as corepressor molecules.[121] There are at least 18 distinct HDACs that target histone as well as nonhistone substrates and have been divided into four classes. Class I HDACs (HDAC1, 2, 3, and 8) are nuclear proteins that are associated with corepressors and sequence-specific transcriptional repressors and are defined by their homology to the yeast RPD3 protein.[122,123] Class II HDACs (HDAC4, 5, 6, 7, 9, and 10) shuttle between the nucleus and cytoplasm, are generally expressed in a tissue-specific manner and are defined by their homology to the yeast Hda1 protein.[124,125] HDAC11, possessing all of the necessary features to confer HDAC status, does not fit neatly into class I or II because of the low sequence similarity, so it is considered a class IV HDAC.[126] Class III HDACs comprise the seven sirtuin family members (SIRT1–7) and are related to yeast Sir2[127] and will be discussed in more detail later.

Early studies demonstrated that treatment of cultured mammalian cells with chemical agents such as a fungal antibiotic such as trichostatin A (TSA),[128] or a cyclic tetrapeptide fungal product such as trapoxin,[129] resulted in accumulation of acetylated histones in cells. Such accumulation was associated with an inhibition of histone deacetylase activity of these molecules and resulted in cell cycle arrest G1 and G2 phases, altered gene expression, reversal of morphology, and induction of differentiation of transformed cells. These studies showed early promise of the potential of HDAC inhibitors as potential anticancer agents. Trapoxin-based chemistry led to the identification of the first HDAC from bovine extracts which was an ortholog of yeast transcriptional regulator Rpd3p and the mammalian counterpart named HD1 (for histone deacetylase 1, later HDAC1) was cloned from Jurkat cells.[120] Mammalian HDACs (or KDACs) in the classes I, II, and IV show sequence similarity to yeast homologs are Zn^{2+}-dependent deacetylases with a highly similar deacetylase domain. However, the different N- and C-terminal domains distinguish their specificities and interactions with other proteins.

3.3 Alterations in Class I, II, and IV Deacetylases in Human Tumors

Several reports indicate that class I/II HDACs are altered in human tumors relative to normal tissue. In prostate cancer, HDAC1, 2, and 3 were shown to be highly expressed.[88] In this study, normal prostate tissue in the vicinity of prostate carcinomas was evaluated for 192 cases which showed varied levels of expression of HDAC1, HDAC2, and HDAC3. In tumors, strong expression was observed in the majority of the cases (HDAC1: 69.8%, HDAC2: 74%, HDAC3: 94.8%), and high rates of HDAC1 and HDAC2 expression were significantly associated with tumor dedifferentiation. Moreover, strong expression of these HDACs was accompanied by enhanced tumor cell proliferation, and HDAC2 was shown to be an independent prognostic marker in this particular prostate cancer cohort.[88] Other studies have demonstrated that specific class I/II deacetylases show altered expression in invasive ductal carcinomas (IDCs) of the breast,[85] prostate cancer[130] and serve as independent predictors of survival in hepatocellular carcinoma (HCC),[86,94] esophageal carcinoma,[90] medulloblastoma,[91] pancreatic adenocarcinomas,[92] glioblastomas,[93] lung cancers,[95] and acute lymphoblastic leukemia (ALL).[89]

3.4 Class I/II Deacetylase Inhibition

Almost two decades before HDACs were discovered and cloned, the first small molecule inhibitor of histone deacetylation was described. Early on, sodium butyrate was shown to induce hyperacetylation of histones H3 and H4 and have a wide range of morphological and physiological effects on cells. These studies provided clues to one of the mechanisms for altering chromatin structure at the nucleosomal level and explaining the selective DNase I sensitivity of transcriptionally active DNA across diverse cell types.[131,132] Since these first discoveries in 1978, significant progress has been made in efforts to target deacetylases therapeutically. While deacetylase inhibitors are now being considered for treatment of a diverse range of complications such as chronic pain, depression, and psychiatric disorders,[133,134] the primary focus here will be on advances related to cancer therapeutics. Because of the reversible nature of PTMs, such as acetylation or methylation, it makes them attractive targets for therapeutic development. Years of basic and clinical research and clinical trials have led to the approval of multiple inhibitors. There are four FDA-approved deacetylase inhibitors being used in the clinic to treat T-cell lymphoma and multiple myeloma and include Zolinza (vorinostat), Istodax (romidepsin), Beleodaq (belinostat), and Farydak (panobinostat lactate). Several others are also being tested in clinical trials for treatment of B-cell and T-cell malignancies along with some solid tumors.

3.5 Sirtuins

While the former are Zn^{2+}-dependent deacetylases, sirtuins catalyze the deacetylation reaction in a NAD^{+}-dependent manner. One NAD^{+} is hydrolyzed for every acetyl group removed from the substrate, resulting in a deacetylated protein and O-acetyl-ADP-ribose and nicotinamide as by-products. There are two distinguishing features of sirtuins relative to the other three classes. First, sirtuins require NAD^{+} for catalytic activity while the other deacetylase family members require Zn^{2+} as a cofactor. Second, the products of the deacetylation reaction between the sirtuins vs classes I, II, and IV deacetylases are different. As a product of the deacetylation reaction, sirtuins produce O-acetyl-ADP-ribose and nicotinamide along with the deacetylated lysine residue. Nicotinamide is the amide derivative of vitamin B3 and serves as an inhibitor of sirtuins as part of a negative feedback. However, the role of O-acetyl-ADP-ribose is less clear but some emerging reports suggest that it influences cellular metabolic pathways.[135] Sirtuin proteins were shown to play a critical role in regulating DNA accessibility early on. However, over the last several years our understanding of their contribution has increased and there is an almost exponential growth in research that rapidly expands the number of identified targets which are beyond well-established histone proteins. The balance of acetylation vs deacetylation can be affected by the metabolic status of a cell given the dependence of sirtuins on NAD^{+} which is a central component of metabolic pathways and is tightly regulated under normal physiological conditions. Sirtuins are evolutionary conserved enzymes whose presence can be traced all the way to archaea. The importance of cellular pathways regulated by sirtuin family members is very clear, whether it is from a single member in bacteria to seven in humans and other vertebrates.[136,137] Forming an extensive molecular network that responds to physiological stresses, human sirtuins are involved in a myriad of cellular processes such as cell proliferation, differentiation, DNA damage and stress genome stability, stress responses, and cell survival/death. As key regulators of homeostasis sirtuins' uncontrolled and/or ectopic activity is associated with wide spectrum of diseases. Even though our understanding of their role in aging, cancer, metabolic syndrome, neuropathologies, and other diseases is increasing, the role that each plays remains poorly understood. For example, the extent of redundancy of function among sirtuin family members is unclear. Also, because there is complex data with respect to their function, it is difficult to draw firm conclusions in different pathological conditions. However, there is a consensus in recognizing them as potential therapeutic targets. Since they regulate many cellular processes, it is reasonable to expect that their study will continue to garner considerable attention in the pursuit of targeted therapies. Of all of the newly identified and promising targets for novel therapeutics, this review will focus on SIRT1, which happens to be the most well-studied family member and has received considerable attention because of its regulation of key properties that are invariably deregulated in cancer and other diseases. Several excellent reviews have been written on the role of SIRT1 in aging,[138] adaptive cellular responses,[139] and endocrine signaling.[140]

3.6 Multiple SIRT1 Nonhistone Targets

One of the challenges of sorting out the impact of pharmacologic inhibition of histone-modifying enzymes is that nonhistone proteins are also substrates. For example, SIRT1 knockdown in multiple colon cancer cells and breast cancer cells was shown to induce p53 activation by acetylation in the absence of conventional stress and induced either apoptosis or growth arrest, depending on the cell type. Interestingly, apoptosis was induced with delayed kinetics in two different cancer cell lines lacking p53, while no apoptosis was observed in two different noncancer epithelial cell lines.[141] In a different study, SIRT1 was shown to bind to estrogen-related receptor alpha (ERRα) and deacetylate it[142] which resulted in enhancement of its DNA-binding potential. Interestingly, ERRα is one of the transcription factors that regulates transcription of the CYP19A1 gene which encodes for the aromatase protein. Aromatase is the enzyme that converts testosterone to estrogen in the estrogen biosynthesis pathway and aromatase inhibitors are used for treatment of breast cancer. Another report demonstrated that inhibition of SIRT1 reduced aromatase mRNA and protein levels in ER− breast cancer cells possibly due to loss of ERRα binding to the aromatase promoter and subsequent inhibition of aromatase transcription.[97]

While SIRT1 is shown to promote cancer phenotypes and acts as a positive regulator of many cancer hallmarks in cell lines[143] and is overexpressed in human cancer specimens compared to normal tissue (Table 1), some reports from transgenic or conditional overexpression mouse models have reported tumor suppressor properties of SIRT1.[144] Some of the controversy regarding whether SIRT1 acts in an oncogenic or tumor suppressive manner may be better reconciled as we consider the different experimental designs. For example, two studies came to different conclusions regarding SIRT1's in vivo influence on tumor biology. They both used $APC^{+/min}$ mouse models that enabled spontaneous adenomatous polyps and hyperplasia which results in colon cancer. Both studies investigated the role of SIRT1 in colon tumorigenesis. Firestein *et al.*[144] used conditional overexpression of SIRT1 in the intestine to show fewer polyps. However, Leko *et al.*[99] used a conditional enterocyte-specific SIRT1 knockout and observed reduced tumor size and number of polyps, no change in proliferation but an increase in apoptosis of tumor cells in knockouts vs wild-type $APC^{+/min}$ mice. Since the gain-of-function and loss-of-function phenotypes were similar, Leko and colleagues argued that super-physiological levels of SIRT1 due to overexpression might somehow be causing stoichiometry changes in protein complexes resulting in inactivation of overexpressed protein (as in the Firestein study). These types of differences in experimental design will need to be weighed as studies seeking to label chromatin regulators as tumor suppressers or oncogenes are performed.

3.7 Sirtuins in Cancer

Of the seven sirtuins, all of them have been implicated in some aspect of tumor biology. For example, SIRT1, SIRT2, and SIRT6 have been linked with HCC,[102,109,145] SIRT3 and SIRT7 have been linked with pancreatic cancer,[103] and SIRT1 and SIRT4 have been implicated in colorectal and gastric cancer.[98,99,104,105] Multiple sirtuins have been linked with NSCLC.[106–108] However, because of the complexity of the diverse target proteins of each of the sirtuins, some appear to exert more of an oncogenic influence while others more of a tumor suppressor influence. Also, of the sirtuins, SIRT1 has been studied most extensively and has been linked most consistently with tumor promotion across diverse tumor types.[30,31,96,141,146–148] Because mRNA abundance does not always reliably predict protein abundance between tumors,[149] several studies have assessed the expression of SIRT1 at the protein level in human tumors followed by studies involving animal models and in vitro analyses. First, SIRT1 has been linked with colon cancer. A study of 485 human colorectal cancers revealed SIRT1 overexpression in 37% of tumors. SIRT1 expression was associated with CGI methylator phenotype (CIMP) and MSI-high.[43] Another study reported that overexpression of SIRT1 was associated with a poor prognosis for colorectal cancer patients and promoted tumorigenesis.[150] Further, an analysis of SIRT1 protein expression in benign adenomas, liver metastases, and 120 paired colorectal cancer and normal mucosa tissues, demonstrated that SIRT1 overexpression was significantly correlated with poor overall survival and disease-free survival.[151] Because of the confounding effect of a whole-animal knockout, a deeper analysis of the role of SIRT1 using a tissue-specific transgenic model was performed. In this study, enterocyte-specific SIRT1-inactivation was found to cause a statistically significant reduction in total polyp surface, size, and number in an in vivo model of colorectal cancer.[99] This study concluded that SIRT1 acts as a tumor promoter in the $APC^{+/min}$ mouse model of intestinal tumorigenesis.[99]

Second, SIRT1 has been linked with breast cancer. In an analysis of SIRT1 protein levels across 90 IDC and 10 normal breast tissues adjacent to tumors, SIRT1 expression was significantly increased in IDC by approximately 2.6-fold across different grades and stages relative to normal adjacent tissue.[97] As with any cancer, the differences in subtypes can add to the complexity of identifying clinical correlates. In breast cancer there are multiple subtypes and SIRT1 has been linked both with ER+ tumors as well as triple-negative breast cancers. For example, in ER+ breast tumors, high levels of cytoplasmic SIRT1 was associated with increased tumor grade, size, and node positivity.[103] In triple-negative breast tumors, IHC analysis of SIRT1 expression was performed and increased SIRT1 expression was more frequently observed in lymph-node-metastasis-positive subgroup than negative subgroup,[152] but this trend was not observed in another subtype in the same study. Another analysis of a retrospective cohort of 460 breast cancer patients demonstrated that expression of SIRT1, but not of HDAC2, was significantly increased in tumor tissues compared to their normal counterparts.[87] In this study, multivariate survival analyses identified SIRT1 as independent prognostic factor for relapse-free survival. Additionally, when the combined expression levels of high expression of LSD1, HDAC2, and SIRT1 were grouped, it showed shorter patient survival time and shorter time to tumor relapse and correlated with poor tumor differentiation and high tumor cell proliferation.[87] Although SIRT1 contributes to mammary gland hyperplasia[153] and suppression of stress-induced apoptosis,[141,154] its mechanisms of involvement and overexpression in breast cancer initiation or progression remains unclear. To further address this issue, Shaker and colleagues examined associations between SIRT1 gene polymorphisms and breast cancer. They found that at least two polymorphisms of the

SIRT1 gene are associated with breast cancer risk and prognosis in a cohort of 541 breast cancer patients relative to 439 healthy individuals.[155] These findings and others demonstrated that in breast tumors SIRT1 expression is significantly higher in malignant tissue relative to their normal counterparts.[97,146,156]

Third, SIRT1 has been linked with leukemias and lymphomas. In a screen of 18 members of the HDAC family in primary AML samples, only SIRT1 was found to be upregulated relative to four different controls.[157] In other cancers, SIRT1 was found significantly increased in both mouse and human prostate cancer with progression toward malignancy.[158] In human breast and colon cancer cells, SIRT1 enables them to become more resistant to apoptotic signals[56,58] and its negative regulator, DBC1, was shown to mediate repression of SIRT1.[159,160] Finally, SIRT1 was shown to mediate heterochromatin formation and heritable silencing of endogenous mammalian TSGs. Inhibition of SIRT1 by pharmacologic, dominant negative, and siRNA-mediated inhibition in breast and colon cancer cells caused increased H4-K16 and H3-K9 acetylation at endogenous TSG promoters and gene reexpression.[30]

3.8 Sirtuin Inhibition

Several compounds are currently being tested in preclinical settings or are in various phases of clinical trials for cancers. Use of HDAC (also known as KDAC) inhibitors in combination with conventional chemotherapy or targeted therapy, such as receptor tyrosine kinase inhibitors, could be beneficial as combination therapies against certain tumors. Several sirtuin inhibitors belonging to different classes of chemical compounds have been and are being developed. So far, preclinical studies involving sirtuin inhibitors have shown promising anticancer effects in cell culture and mouse models.[96,97,99,100,141,147,161,162] Very early on, nicotinamide, the physiological product of the deacetylation reaction which acts as a noncompetitive inhibitor of sirtuin activity,[163,164] was first used to modulate its activity. Nicotinamide is the amide derivative of vitamin B3 and therefore has interesting implications for the role of diet in modulating sirtuin activity. It broadly inhibits multiple sirtuins with varying degrees to efficacy. Some of the early studies demonstrated that nicotinamide-induced apoptosis and exerted growth inhibitory effects on leukemia and prostate cancer cells.[165,166]

A β-naphthol compound (cambinol) was then developed that was more selective for inhibition of SIRT1 and SIRT2 relative to nicotinamide and was shown to cause increased apoptosis of BCL6-expressing Burkitt lymphoma cells and reduced tumor growth in xenograft models.[161] This reagent facilitated loss-of-function studies that could be combined corroborated with siRNA-mediated knockdown of SIRT1. Because SIRT1 regulates very diverse responses such as apoptosis, insulin sensitivity, autophagy, differentiation, and stem cell pluripotency, extensive efforts were underway to identify the mechanism by which SIRT1 orchestrates such pleiotropic cellular events. One of the key links that established a molecular basis for this came from the pioneering discovery that SIRT1 serves as a positive regulator of Wnt signaling.[100] SIRT1 loss of function via siRNA and pharmacological inhibition was shown to lead to a significant decrease in the levels of one or more of the three Disheveled (Dvl) proteins in various cellular contexts. Dvl proteins are critical mediators of both canonical (beta-catenin dependent) and noncanonical (beta-catenin independent) Wnt signaling. Furthermore, SIRT1 and Dvl proteins were shown to interact and inhibition of SIRT1 caused a reduction in Wnt target gene expression. Because the three mammalian Dvl proteins serve as key messengers for as many as 19 Wnt ligands, this report for the first time establishing a link between SIRT1-mediated regulation of Dvl proteins could help explain the diverse physiological responses observed in different cellular contexts.[100] Previously, SIRT1 had only been shown to mediate the epigenetic silencing of Wnt antagonists,[30] but the study by Holloway *et al.* demonstrated that SIRT1 regulates Dvl protein levels and Wnt signaling in several cellular contexts, but in particular in colorectal cancer and breast cancer. These findings demonstrated that SIRT1 acts as a positive regulator of transient and constitutive Wnt signaling and thereby may contribute to cancer cell signaling.

As the evidence mounted demonstrating that SIRT1 critically regulates cancer cell migration, this led to yet another question of how it regulated cell migration. Members of the Rho/Rac GTPase family were suspected to be involved because they had been shown to act as intermediary cellular switches and conduct transient and constitutive signals from upstream Wnt cues. Although SIRT1 was known to be overexpressed in several human cancers and was linked to cancer cell motility, its role in Rac1 activation had not been reported. To address this, a subsequent study probed the mechanism of sirtuin regulated motility and found that it positively regulated the levels of Rac1-GTP and the activity of the Rac1 activator, T-cell lymphoma invasion and metastasis 1 (TIAM1), a Rac guanine nucleotide exchange factor (GEF).[96] Transient inhibition of SIRT1 and SIRT2 was shown to increase acetylation of TIAM1 and cause disruption of the

DVL1–TIAM1 interaction. Hence, this study proposed a model for Rac activation where SIRT1/2 positively modulates the DVL/TIAM1/Rac axis and promotes sustained pathway activation.[96] This connection between SIRT1 and Wnt signaling was further strengthened with a study demonstrating that ectopic expression of SIRT1 markedly increased the migration and invasion of CRC cells and inhibition of SIRT1 decreased the aggressive CRC cells. Interestingly, tumor xenograft experiments revealed that knockdown of SIRT1 impaired CRC metastasis in vivo and induced a mesenchymal–epithelial transition. Moreover, SIRT1 expression was shown to correlate positively with Fra-1 expression, metastasis, and overall survival in patients with CRC, suggesting that SIRT1 may serve as a potential therapeutic target for metastatic CRC.[167]

Building on these previous connections, more screens led to the identification of other sirtuin inhibitors. Several of these were shown to inhibit proliferation or sensitize the antitumor activity of other natural compounds in various in vitro and in vivo model systems.[168] Another study using both Cambinol and other sirtuin inhibitors demonstrated a novel connection between SIRT1 and aromatase, the rate-limiting enzyme that converts androgens to estrogens.[97] Interestingly, this report assessed SIRT1 protein levels across 90 human breast IDC and found a significant increase of approximately 2.6-fold relative to normal adjacent tissue. This same report also demonstrated that SIRT1 inhibition in breast cancer cells caused an increase in ERRα acetylation and a decrease in aromatase mRNA and protein levels.[97] While cambinol inhibits deacetylase activity of both SIRT1 and SIRT2, the latest analogs of this compound have been shown to have more specific and potent inhibitory effects on SIRT1 or SIRT2 and showed cytotoxic effects in cell-based assays on epithelial cancer and lymphoma cell lines.[169] Moreover, use of different SIRT1 inhibitors in mouse tumor models revealed that pharmacologic SIRT1 inhibitors show antitumor activity across multiple in vivo tumor models of Burkitt lymphoma,[161] lung cancer,[170,171] AML,[172] CRC,[99,167] HCC,[145,173] and other diverse tumor types[174,175] as well as across multiple cancer cell lines.[96,97,100]

4. LYSINE AND ARGININE METHYLATION

4.1 Lysine Methyltransferases

Histone methylation is mediated by enzymes that belong to a superfamily of methyltransferases containing more than 100 members from bacteria to humans, and methylation can occur on different amino acids in both histone and nonhistone proteins.[176–178] First, lysine residues are subject to methylation by multiple families of methylating enzymes as shown in Table 2. In Table 2, several of the alternative names and aliases for the proteins are included for each family member. The lysine methyltransferases (KMTs) bring about mono-, di-, and tri-methylation of histone and nonhistone proteins. Virtually all of the enzymes of the KMT class share a conserved region of homology known as the SET domain which mediates methyltransferase catalytic activity. Because there are multiple lysine residues on each histone that can be methylated, each methyl mark provides a potential docking site for chromatin-associated proteins. For example, some lysine methylation (H3K4, H3K36, H3K79) tend to be associated with active transcription while other sites of lysine methylation (H3K9, H3K27, H4K20) are associated with gene repression.[3,178] Of the KMTs, the greatest number identified so far have been shown to methylate H3K4,[113,179–181] and some of these have been shown to have altered expression in colorectal and HCCs and regulate cancer cell proliferation.[181] Multiple KMTs have also been shown to methylate H3K9,[113,182,183] H3K36,[184–186] and H4K20.[183,187] However, considering the KMTs identified currently that methylate H3K27[188,189] and H3K79[190] or other lysines on histone H4[191] are not as abundant as those that target H3K4. Drug discovery efforts to identify inhibitors of lysine-modifying enzymes led to the identification of HDAC inhibitors earlier, but recent efforts have produced potent and increasingly more selective KMT inhibitors which have advanced to clinical trials against diverse cancer types.[192–195]

4.2 Arginine Methyltransferases

The second class of enzymes mediating histone and nonhistone methylation of arginine residues includes the histone/protein arginine methyltransferases (PRMTs) which can catalyze either mono- or di-methylation. We will primarily focus on the KMTs in this chapter, but there are several excellent reviews describing the characterization, function, and role of PRMTs in cancer biology.[196–198] Both KMTs and PRMTs are capable of transferring a methyl group from the cofactor *S*-adenosylmethionine to lysine or arginine residues, respectively, but the catalytic domains are structurally different.

Table 2 Human Lysine Methyltransferases and Demethylases and Their Substrates

KMT Family Member	Substrates
KMT2A (MLL, MLL1, HRX)	H3K4
KMT2B (MLL2, MLL4, HRX2)	H3K4
KMT2C (MLL3, HALR)	H3K4
KMT2D (MLL4, MLL2, ALR, KABUK1)	H3K4
KMT2E (MLL5)	H3K4
KMT2F (SETD1A, Set1, Set1A)	H3K4
KMT2G (SETD1B, Set1B)	H3K4
KMT2H (ASH1L)	H3K4, H3K36
ASH1 (ASCL1, HASH1)	H3K4
KMT3C (SMYD2, ZMYND14)	H3K4, H3K36
KMT3D (SMYD1, ZMYND18)	H3K4
KMT3E (SMYD3, ZMYND1)	H3K4, H4K5
KMT7 (SETD7, SET7, SET9)	H3K4
KMT1A (SUV39H1)	H3K9
KMT1B (SUV39H2)	H3K9
KMT1C (G9a, EHMT2)	H3K9
KMT1D (Eu-HMTase, GLP1, EHMT1)	H3K9
KMT1E (SETDB1, ESET)	H3K9
KMT1F (SETDB2, CLLD8)	H3K9
KMT8 (RIZ1, RIZ2, PRDM2)	H3K9
KMT6 (EZH2)	H3K27
KMT6B (EZH1)	H3K27
KMT3A (SETD2, SET2)	H3K36
KMT3B (NSD1)	H3K36
NSD2 (WHSC1, MMSET)	H3K36, H4K20
NSD3 (WHSC1L, WHISTLE)	H3K36
SETMAR (Metnase)	H3K36
KMT4 (DOT1L)+A3	H3K79

Continued

Table 2 Human Lysine Methyltransferases and Demethylases and Their Substrates—cont'd

KMT Family Member	Substrates
KMT5A (SETD8, PR-Set7, SET8)	H4K20
KMT5B (SUV4-20H1, CGI-85)	H4K20
KMT5C (SUV4-20H2)	H4K20
KDM Family	**Substrates**
KDM1 (LSD1, BHC110)	H3K4, H3K9
KDM5A (JARID1A, RBP2, RBBP2)	H3K4
KDM5B (JARID1B, PLU-1, RBP2-H1)	H3K4
KDM5C (JARID1C, SMCX)	H3K4
KDM5D (JARID1D, SMCY)	H3K4
KDM3A (JHDM2A, JMJD1, JMJD1A)	H3K9
KDM3B (JMJD1B, NET22)	H3K9
KDM4A (JHDM3A, JMJD2, JMJD2A)	H3K9, H3K36
KDM4B (JMJD2B, TDRD14B)	H3K9, H3K36
KDM4C (JHDM3C, JMJD2C, GASC1)	H3K9, H3K36
KDM4D (JMJD2D)	H3K9
KDM6A (UTX, KABUK2)	H3K27
KDM6B (JMJD3)	H3K27
KDM7A (JHDM1D)	H3K27, H3K9
KDM8 (JMJD5)	H3K36
KDM2A (FBXL11, JHDM1A)	H3K36
KDM2B (FBXL10, JHDM1B)	H3K36

4.3 Alterations in the MLL Lysine Methyltransferases in Human Tumors

Several critical lysine methyltransferases (KMTs) have been shown to be mutated across diverse tumors and contribute to multiple facets of tumor progression. For example, non-Hodgkin lymphomas (NHLs) are cancers of B, T or natural killer lymphocytes and harbor mutations in KMTs. In fact, the two most common types of NHL, follicular lymphoma (FL) and diffuse large B-cell lymphoma (DLBCL), show mutations in enzymes that target H3K4 for methylation. MLL2 mutations have been identified in 32% of DLBCL and 89% of FL.[199] MLL2 (also known as ALR or KMT2D) methylates H3K4 and creates an active mark and has been shown to undergo recurrent mutations in the two most common types of NHLs. These MLL2 mutations may contribute to genome instability in the lymphomas given reports demonstrating that MLL2 gene deletion in mouse and human cell culture induced elevated levels of sister chromatid exchange, gross chromosomal aberrations, and micronuclei. Additionally, MLL2 mutated cells were shown to display signs of substantial transcription stress and elevated levels of γH2AX, and suffer frequent mutation.[200] Another study involving targeted next-generation sequencing of 275 known and putative cancer genes from 38 treatment-naïve patient melanomas revealed that MLL2 was among the top three most frequently mutated.[201] This study found a high frequency of mutations identified in key epigenetic regulators and reported that 22.3% (165 of 740) of all nonsilent mutations occurred in genes impacting epigenetic regulators. Interestingly, mutations in genes encoding histone-modifying proteins were the most common of those involving epigenetic regulators. Another family member, MLL5 (KMT2E) has been implicated in patient responses to hypomethylating agents. Because hypomethylating agents such as decitabine are widely used in patients with MDS and some types of leukemia, it is important to understand why some patients are unresponsive to treatments. Efforts to investigate the role of MLL5 expression levels on the outcome of patients with AML who were treated with decitabine revealed that patients with MLL5 levels above the median expression level predicted longer overall survival independent of DNMT3A mutation status.[202] This study also found that in patients who received three or more courses decitabine, high MLL5 expression and wild-type DNMT3A independently predicted improved overall survival. Members of the SETD1A/B KMTs have also been shown to be altered in human tumors. Because H3K4 global and local methylation patterns are altered in human tumors and have been shown to be predictive for disease recurrence, there is considerable interest in identifying aberrant expression of the enzymes that deposit or remove this mark. SETD1A has been reported to be increased in primary colorectal tumors and cell lines[203] and its depletion has been shown to decrease breast cancer cell migration in cell culture and metastasis athymic nude mice.[204] Another KMT, EZH2, has been implicated extensively in other types of tumors, such as breast and prostate cancers. There are numerous studies demonstrating the prognostic and diagnostic utility of assessing not only EZH2 but H3K27 methylation levels in tumors.[178,205]

4.4 Lysine Demethylases

The identification of lysine demethylases (KDMs) in 2004 shattered decades of dogma which held that histone lysine methylation was thermodynamically stable and irreversible. It was viewed that histone demethylation, similar to DNA, was a passive process where histones would have to be replaced to remove methyl groups. This initial discovery directly addressed the long-standing question of whether histone demethylation was an active process.[206] The study from Shi and colleagues provided evidence that the first lysine demethylase identified, LSD1, a nuclear homolog of amine oxidases, functions as a histone demethylase and transcriptional corepressor. LSD1 was shown to specifically demethylate histone H3 lysine 4 (H3K4), which is linked to active transcription when H3K4 is trimethylated.[206] Following this initial breakthrough, using a biochemical assay coupled with chromatography, another family of lysine demethylases was identified.[207] This second KDM, named JHDM1, represented the second family of KDMs and was shown to have a novel Jumonji C (JmjC) domain and demethylate histone H3 at lysine 36 (H3K36).[207] Thus, the first class of KDMs, of which LSD1 (KDM1A) is a member, demethylate lysines via an amine oxidation reaction with flavin adenine dinucleotide as a cofactor.[208] The second class of KDMs is referred to as the Jumonji demethylases and have a conserved JmjC domain and utilize Fe(II) and α-ketoglutarate to mediate demethylation. Since this initial discovery several of these family members have been shown to play a key role in aspects of cellular transformation and tumorigenesis and efforts to identify small molecule inhibitors are actively underway.[209,210]

4.5 Alterations in Lysine Demethylases in Human Tumors

Interest in targeting lysine demethylases (KDMs) has increased considerably since their discovery in 2004, but less is known about their role in tumor promotion or suppression relative to other epigenetic regulators. Nonetheless, the prospect of targeting KDMs therapeutically became even more attractive when the expression of some family members were shown to be altered in various tumors relative to benign tissue.[211,212] This altered expression has been implicated in cancer cell proliferation and drug resistance[213] demethylation of tumor suppressors such as p53.[214] Because LSD1 was the first KDM to be identified, it has been most extensively studied. The link between LSD1 and pancreatic cancer was established with an analysis of 48 paired pancreatic cancer tissues and adjacent normal tissues by IHC. This study showed increased LSD1 expression in human tumors and found that elevated LSD1 expression correlated with poor prognosis for this cohort of pancreatic cancer patients. Further analysis of the potential mechanism of action of LSD1 was conducted in pancreatic cancer lines which revealed a synergy between LSD1 and hypoxia inducible factor-1α (HIF1α) in promoting cancer cell proliferation.[215] Another study of neuroblastoma demonstrated that LSD1 expression correlated with adverse outcome and was inversely correlated with differentiation in neuroblastomas. This study further found that differentiation of neuroblastoma cells resulted in downregulation of LSD1. Also, inhibition of LSD1 decreased cellular growth, induced expression of differentiation-associated genes, and increased target gene-specific and global H3K4 methylation. Moreover, targeting LSD1 reduced neuroblastoma xenograft growth in vivo. These data suggested that LSD1 is involved in maintaining the undifferentiated, malignant phenotype of neuroblastoma cells.[216] The same laboratory also assessed the connection between LSD1 and breast cancer. They found very high LSD1 levels in estrogen receptor (ER)-negative tumors and demonstrated that pharmacological LSD1 inhibition resulted in growth inhibition of breast cancer cells which was linked with proliferation-associated genes such as p21, ERBB2, and CCNA2.[217] Additionally, UTX (KDM6A), which demethylates H3K27, was shown to harbor 39 inactivating mutations in 1390 cancer samples across 12 histologically distinct cancers.[218] While significant progress has been made, much remains to be understood regarding the KDM family and whether specific members can be targeted therapeutically in specific cancer contexts.

5. NONCODING RNA

For many decades the driving force behind tumorigenesis was almost exclusively attributed to mutations and amplifications in protein-coding genes. This view that mutations in protein-coding genes was a fundamental driver of tumorigenesis enabled the development of small molecule inhibitors and immune-based therapeutics that essentially targeted the relentless rogue proteins responsible for driving much of the tumor growth. Years later with increased sequencing efforts and work for numerous labs, the study of epigenetics began to expand and views begin to develop which held that simply mutating protein-coding genes was not entirely sufficient for tumor initiation and progression. Studies in the field of epigenetics began to identify aberrant epigenetic marks which created chromatin structures allowing expression of proteins in tissues that essentially broke all of the rules governing tissue-specificity. Not only did epigenetic alterations lead to poorly timed or over production of stem cell factors, the epigenetic changes in regulatory sequences such as promoters, enhancers, and silencers were proving to play important roles. Thus, while the coding sequence of a TSG may have been wild-type, the epigenetic silencing of it rendered it nonfunctional. This next wave of discovery deepened the appreciation that cancer was the result of collaboration between mutations in protein-coding genes and aberrant epigenetic regulation of wild-type genes. At the same time, there were attempts to reconcile why such an enormous transcription potential of human genomes only led to a small number of protein-coding transcripts. Attempts to reconcile this peculiarity led to ground-breaking studies on the roles of noncoding RNAs (ncRNA). Studies began to uncover the role of ncRNAs in regulating epigenetic states and tumor biology. Recently, there have been several breakthroughs demonstrating the role of ncRNA in directing chromatin architecture and epigenetic marks based on their ability to recruit histone-modifying complexes or DNMTs to specific regions of the genome. Efforts from the ENCODE project revealed that at least 93% of the human genome was transcribed across a panel of cell lines,[173] yet only a very small fraction of less than 2% of these transcripts are translated into protein.[219] This suggests that there is a tremendous potential of aberrant ncRNA regulation in the transition of cells from normal to malignant states.

There are several excellent reviews on the classification and regulation of types of ncRNA,[220] but briefly, they can be characterized into long ncRNA (lncRNA) and small ncRNA which are more than or less than 200nts, respectively. The small ncRNAs include small nucleolar RNAs (snoRNAs),[221] transfer RNAs (tRNA),[222] PIWI-interacting RNAs (piRNAs),[223] small interfering RNAs (siRNAs), circular RNAs (circRNA),[224,225] and microRNAs (miRNA).[226] Several of the small RNAs were shown to be critically involved in regulating cellular transformation. For example, Guarnerio *et al.* reported a novel role for circRNAs associated with PML-RAR and MLL-AF9 translocations when expressed in mouse embryo fibroblasts (MEFs).[225] They found that MEFs in which specific circRNAs were expressed showed increased cellular proliferation and clonogenicity and decreased contact inhibition. Concerning lncRNAs, the number of genes encoding them is quite extensive and outnumbers the number of protein-coding genes.[227] With respect to chromatin structure, the lncRNAs appear to have a particularly important role in its regulation given their roles as molecular chaperones and scaffolds and given their link with malignant transformation.[228] The study of the role of ncRNA is rapidly expanding and will undoubtedly lead to significant and novel insights into their contribution to all stages of tumor progression.

6. EPIGENETIC READERS

6.1 Readers of DNA Methylation

MBD proteins such as MeCP2 have been shown to act as molecular scaffolds and can bridge DNA methylation with histone-modifying enzymes. Transcriptional silencing linked with DNA methylation is in part mediated by the binding of repressor proteins to methylated DNA.[229] The founding member of this protein family is MeCP2 and it contains multiple domains regulating gene silencing. The other MBD family members include MBD1–4 and Kaiso. Because multiple types of mutations in MeCP2 are causative for Rett syndrome, a neurological disorder, it has been studied the most extensively.[230] MeCP2 can bind to single symmetrically methylated CpG dinucleotides and repress transcription both in vitro and in vivo.[231] Furthermore, studies in mammalian and *Xenopus* systems have shown that transcription repression by MeCP2 is at least partially dependent on the recruitment of an HDAC complex.[232] An important question that remains an intense area of investigation is how promoter hypermethylation directs heritable silencing of genes. Regardless of whether methylation is the initial silencing event, or whether it is caused by earlier chromatin remodeling events,[233] the importance of MBD proteins in either scenario is very clear. Promoter methylation is an integral part of transcription repression because drugs that induce DNA demethylation can partially reactivate silenced genes in cancer cells and in patients.[26,234,235] Early studies demonstrated that the DNMTs collaborate with class I/II HDACs to silence several TSGs.[236] Due to the link between aberrant DNA methylation and cancer progression, it is clear that the role of methyl-CpG-binding domain (MBD) proteins in "reading" and mediating global or local repression will be critical. Without the MBD proteins serving as accomplices, aberrant methylation patterns will not have the same force in driving tumor progression. Given its role in propagating the effects of incorrect DNA methylation patterns, efforts to determine

MeCP2 expression across human tumors has increased. These efforts led to the discovery that MeCP2 is amplified in human tumors.[237] MeCP2 serves as a bridge for histone methyltransferases (HMTs),[238] deacetylases,[232] and other proteins that bind modified histones,[239] so this increased expression may be required for many of the malignant properties associated with these tumors. MeCP2 not only reads DNA methylation but also "integrates" the information encoded in CpG methylation, long-noncoding RNAs (lncRNA), and histone modifications. Remarkably, although MeCP2 binds methyl-CpG,[240] lncRNAs,[241,242] chromatin remodeling proteins,[232,238,243] is amplified in diverse human tumors,[237] and is causative for Rett syndrome,[230] and despite its central role in processes frequently targeted in tumorigenesis, its mechanism of regulation remains poorly defined. Some insights into the mechanism by which MeCP2 participates in a multiprotein complexes, such as those involving HDAC1, ATRX, and HP1, have been reported.[32] In a report by Pandey *et al.* inhibition of SIRT1 led to acetylation of novel sites on endogenous MeCP2 protein in breast and colon cancer cells. After mapping MeCP2 acetylation sites an analysis of lysine 171 (K171), which is situated in the uncharacterized intermediate domain between the MBD and the TRD, was performed. K171 is conserved in MeCP2 orthologs across species, suggesting it may be critical for some evolutionarily conserved function. To assess the role of K171, an acetylation mimetic mutant (K171Q) was generated and found to associate less with ATRX and HDAC1 relative to wild-type. This indicated that lysine 171 is important for MeCP2 interaction with at least two other chromatin remodeling enzymes. Thus, this site that is regulated by SIRT1 may serve as a regulatory switch that could potentially modulate protein–protein interaction and heterochromatin formation. To further address the significance of MeCP2 acetylation and its influence on interaction, pharmacologic SIRT1 inhibition was shown to decrease interaction between MeCP2 and ATRX/HDAC1 in a manner similar to that observed with the MeCP2 acetylation mimetic. These data support the idea that MeCP2 acetylation is important in modulating its interactions with ATRX, HDAC1, and other corepressors. Because MeCP2 is amplified in human tumors and globally alters the chromatin state in some cellular contexts, understanding this novel regulation helped explain the contribution of HDI therapy potentially not accounted for by simply assessing changes in histone acetylation.[32]

6.2 Readers of Lysine Acetylation and Methylation

Just as CpG methylation can be read by protein families, both acetyl-lysine and methyl-lysine motifs can be recognized by chromatin-binding proteins possessing various domains. It is important to note that these domains that confer the ability to read specific covalent modifications can be found in many chromatin modifiers that possess catalytic activity. For example, various acetyl-lysine motifs can be recognized and "read" by bromodomains and plant homeodomain (PHD) fingers. Methyl-lysine motifs can be recognized by multiple domains (such as chromodomains, Tudor domains, malignant brain tumor domains, proline-tryptophan-tryptophan-proline domains, PHD fingers, and WD40/β propeller). Because the role of nonhistone protein acetylation is increasingly being reported to regulate not only "readers," but also oncoproteins and tumor suppressor proteins, it will be important to continue to determine how PTMs or FDA-approved drugs impact their function.

7. CONCLUDING REMARKS

Over the course of this chapter, we have examined several aspects of the critical interplay between the cancer genome and epigenome and how their codependence leads to many of the hallmarks of cancer. While other factors that regulate the epigenetic landscape are certainly important, we focused primarily on the two classes of enzymes that regulate DNA methylation and four classes of histone-modifying enzymes (KATs,[69–71] HDACs,[88,100] KMTs,[113] and KDMs[188]) that target lysine residues in very diverse disease settings. From the involvement of KATs in colon cancer[244] to melanoma[73] to the role of HDACs in multiple cancers[74,96,101,109,110] or conditions that predispose to cancer,[245] clearly these enzymes represent very important therapeutic targets. For certain, understanding the role of each of these enzymes in regulating chromatin structure is critically important in identifying novel therapeutics; however, one area that will need to be addressed is the extent to which many of these enzymes modify lysines on nonhistone proteins. It will be important to continue to elucidate the mechanism of regulation and sort out the role of histone vs nonhistone PTMs in mediating the antitumor effects of many of the novel therapeutics targeting these enzymes.

REFERENCES

1. Kornberg RD, Lorch Y. Twenty-five years of the nucleosome, fundamental particle of the eukaryote chromosome. *Cell.* 1999;98:285–294.
2. Kingston RE, Narlikar GJ. ATP-dependent remodeling and acetylation as regulators of chromatin fluidity. *Genes Dev.* 1999;13:2339–2352.
3. Dawson MA, Kouzarides T. Cancer epigenetics: from mechanism to therapy. *Cell.* 2012;150:12–27.
4. Christophorou MA, Castelo-Branco G, Halley-Stott RP, et al. Citrullination regulates pluripotency and histone H1 binding to chromatin. *Nature.* 2014;507:104–108.
5. Tessarz P, Santos-Rosa H, Robson SC, et al. Glutamine methylation in histone H2A is an RNA-polymerase-I-dedicated modification. *Nature.* 2014;505:564–568.
6. Tahiliani M, Koh KP, Shen Y, et al. Conversion of 5-methylcytosine to 5-hydroxymethylcytosine in mammalian DNA by MLL partner TET1. *Science.* 2009;324:930–935.
7. Kriaucionis S, Heintz N. The nuclear DNA base 5-hydroxymethylcytosine is present in Purkinje neurons and the brain. *Science.* 2009;324:929–930.
8. Herman JG, Baylin SB. Gene silencing in cancer in association with promoter hypermethylation. *N Engl J Med.* 2003;349:2042–2054.
9. Esteller M. Epigenetics in cancer. *N Engl J Med.* 2008;358:1148–1159.
10. Ting AH, McGarvey KM, Baylin SB. The cancer epigenome—components and functional correlates. *Genes Dev.* 2006;20:3215–3231.
11. Gross DS, Garrard WT. Nuclease hypersensitive sites in chromatin. *Annu Rev Biochem.* 1988;57:159–197.
12. Felsenfeld G, Boyes J, Chung J, Clark D, Studitsky V. Chromatin structure and gene expression. *Proc Natl Acad Sci USA.* 1996;93:9384–9388.
13. Thurman RE, Rynes E, Humbert R, et al. The accessible chromatin landscape of the human genome. *Nature.* 2012;489:75–82.
14. Strahl BD, Allis CD. The language of covalent histone modifications. *Nature.* 2000;403:41–45.
15. Holliday R, Pugh JE. DNA modification mechanisms and gene activity during development. *Science.* 1975;187:226–232.
16. Riggs AD. X inactivation, differentiation, and DNA methylation. *Cytogenet Cell Genet.* 1975;14:9–25.
17. Chen T, Li E. Establishment and maintenance of DNA methylation patterns in mammals. *Curr Top Microbiol Immunol.* 2006;301:179–201.
18. Rasmussen KD, Helin K. Role of TET enzymes in DNA methylation, development, and cancer. *Genes Dev.* 2016;30:733–750.
19. Bird A. DNA methylation patterns and epigenetic memory. *Genes Dev.* 2002;16:6–21.
20. Jones PA, Baylin SB. The fundamental role of epigenetic events in cancer. *Nat Rev Genet.* 2002;3:415–428.
21. Surani MA. Imprinting and the initiation of gene silencing in the germ line. *Cell.* 1998;93:309–312.
22. Feinberg AP, Vogelstein B. Hypomethylation distinguishes genes of some human cancers from their normal counterparts. *Nature.* 1983;301:89–92.
23. Gama-Sosa MA, Slagel VA, Trewyn RW, et al. The 5-methylcytosine content of DNA from human tumors. *Nucleic Acids Res.* 1983;11:6883–6894.
24. Goelz SE, Vogelstein B, Hamilton SR, Feinberg AP. Hypomethylation of DNA from benign and malignant human colon neoplasms. *Science.* 1985;228:187–190.
25. Feinberg AP, Gehrke CW, Kuo KC, Ehrlich M. Reduced genomic 5-methylcytosine content in human colonic neoplasia. *Cancer Res.* 1988;48:1159–1161.

26. Baylin SB, Herman JG. DNA hypermethylation in tumorigenesis: epigenetics joins genetics. *Trends Genet.* 2000;16:168–174.
27. Ng HH, Bird A. DNA methylation and chromatin modification. *Curr Opin Genet Dev.* 1999;9:158–163.
28. Wales MM, Biel MA, el DW, et al. p53 activates expression of HIC-1, a new candidate tumour suppressor gene on 17p13.3. *Nat Med.* 1995;1:570–577.
29. Shields JM, Mehta H, Pruitt K, Der CJ. Opposing roles of the extracellular signal-regulated kinase and p38 mitogen-activated protein kinase cascades in Ras-mediated downregulation of tropomyosin. *Mol Cell Biol.* 2002;22:2304–2317.
30. Pruitt K, Zinn RL, Ohm JE, et al. Inhibition of SIRT1 reactivates silenced cancer genes without loss of promoter DNA hypermethylation. *PLoS Genet.* 2006;2:e40.
31. Chen WY, Wang DH, Yen RC, Luo J, Gu W, Baylin SB. Tumor suppressor HIC1 directly regulates SIRT1 to modulate p53-dependent DNA-damage responses. *Cell.* 2005;123:437–448.
32. Pandey S, Simmons Jr GE, Malyarchuk S, Calhoun TN, Pruitt K. A novel MeCP2 acetylation site regulates interaction with ATRX and HDAC1. *Genes Cancer.* 2015;6:408–421.
33. Ley TJ, Ding L, Walter MJ, et al. DNMT3A mutations in acute myeloid leukemia. *N Engl J Med.* 2010;363:2424–2433.
34. Yan XJ, Xu J, Gu ZH, et al. Exome sequencing identifies somatic mutations of DNA methyltransferase gene DNMT3A in acute monocytic leukemia. *Nat Genet.* 2011;43:309–315.
35. Xu J, Wang YY, Dai YJ, et al. DNMT3A Arg882 mutation drives chronic myelomonocytic leukemia through disturbing gene expression/DNA methylation in hematopoietic cells. *Proc Natl Acad Sci USA.* 2014;111:2620–2625.
36. Mayle A, Yang L, Rodriguez B, et al. Dnmt3a loss predisposes murine hematopoietic stem cells to malignant transformation. *Blood.* 2015;125:629–638.
37. Celik H, Mallaney C, Kothari A, et al. Enforced differentiation of Dnmt3a-null bone marrow leads to failure with c-Kit mutations driving leukemic transformation. *Blood.* 2015;125:619–628.
38. Delhommeau F, Dupont S, Della VV, et al. Mutation in TET2 in myeloid cancers. *N Engl J Med.* 2009;360:2289–2301.
39. Couronne L, Bastard C, Bernard OA. TET2 and DNMT3A mutations in human T-cell lymphoma. *N Engl J Med.* 2012;366:95–96.
40. Subramaniam D, Thombre R, Dhar A, Anant S. DNA methyltransferases: a novel target for prevention and therapy. *Front Oncol.* 2014;4:80.
41. Lorsbach RB, Moore J, Mathew S, Raimondi SC, Mukatira ST, Downing JR. TET1, a member of a novel protein family, is fused to MLL in acute myeloid leukemia containing the t(10;11)(q22;q23). *Leukemia.* 2003;17:637–641.
42. Toyota M, Issa JP. CpG island methylator phenotypes in aging and cancer. *Semin Cancer Biol.* 1999;9:349–357.
43. Nosho K, Shima K, Irahara N, et al. SIRT1 histone deacetylase expression is associated with microsatellite instability and CpG island methylator phenotype in colorectal cancer. *Mod Pathol.* 2009;22:922–932.
44. Baylin SB, Ohm JE. Epigenetic gene silencing in cancer—a mechanism for early oncogenic pathway addiction? *Nat Rev Cancer.* 2006;6:107–116.
45. Ballestar E, Paz MF, Valle L, et al. Methyl-CpG binding proteins identify novel sites of epigenetic inactivation in human cancer. *EMBO J.* 2003;22:6335–6345.
46. Munoz DP, Lee EL, Takayama S, et al. Activation-induced cytidine deaminase (AID) is necessary for the epithelial-mesenchymal transition in mammary epithelial cells. *Proc Natl Acad Sci USA.* 2013;110:E2977–E2986.

47. Okamoto Y, Shinjo K, Shimizu Y, et al. Hepatitis virus infection affects DNA methylation in mice with humanized livers. *Gastroenterology*. 2014;146:562–572.
48. Niwa T, Tsukamoto T, Toyoda T, et al. Inflammatory processes triggered by Helicobacter pylori infection cause aberrant DNA methylation in gastric epithelial cells. *Cancer Res*. 2010;70:1430–1440.
49. Jha HC, Banerjee S, Robertson ES. The role of gammaherpesviruses in cancer pathogenesis. *Pathogens*. 2016;5:E18.
50. Means RE, Lang SM, Jung JU. Human gammaherpesvirus immune evasion strategies. In: Arvin A, Campadelli-Fiume G, Mocarski E, Moore PS, Roizman B, Whitley R, Yamanishi K, eds. *Human Herpesviruses: Biology, Therapy, and Immunoprophylaxis*. Cambridge: Cambridge University Press; 2007:559–586.
51. Fenaux P, Mufti GJ, Hellstrom-Lindberg E, et al. Efficacy of azacitidine compared with that of conventional care regimens in the treatment of higher-risk myelodysplastic syndromes: a randomised, open-label, phase III study. *Lancet Oncol*. 2009;10:223–232.
52. Fenaux P, Mufti GJ, Hellstrom-Lindberg E, et al. Azacitidine prolongs overall survival compared with conventional care regimens in elderly patients with low bone marrow blast count acute myeloid leukemia. *J Clin Oncol*. 2010;28:562–569.
53. Tsai HC, Li H, Van NL, et al. Transient low doses of DNA-demethylating agents exert durable antitumor effects on hematological and epithelial tumor cells. *Cancer Cell*. 2012;21:430–446.
54. Allfrey VG, Faulkner R, Mirsky AE. Acetylation and methylation of histones and their possible role in the regulation of RNA synthesis. *Proc Natl Acad Sci USA*. 1964;51:786–794.
55. Gu W, Roeder RG. Activation of p53 sequence-specific DNA binding by acetylation of the p53 C-terminal domain. *Cell*. 1997;90:595–606.
56. Vaziri H, Dessain SK, Ng EE, et al. hSIR2(SIRT1) functions as an NAD-dependent p53 deacetylase. *Cell*. 2001;107:149–159.
57. Luo J, Su F, Chen D, Shiloh A, Gu W. Deacetylation of p53 modulates its effect on cell growth and apoptosis. *Nature*. 2000;408:377–381.
58. Luo J, Nikolaev AY, Imai S, et al. Negative control of p53 by Sir2alpha promotes cell survival under stress. *Cell*. 2001;107:137–148.
59. Tang Y, Zhao W, Chen Y, Zhao Y, Gu W. Acetylation is indispensable for p53 activation. *Cell*. 2008;133:612–626.
60. Kim GW, Yang XJ. Comprehensive lysine acetylomes emerging from bacteria to humans. *Trends Biochem Sci*. 2011;36:211–220.
61. Scott I. Regulation of cellular homoeostasis by reversible lysine acetylation. *Essays Biochem*. 2012;52:13–22.
62. Kaypee S, Sudarshan D, Shanmugam MK, Mukherjee D, Sethi G, Kundu TK. Aberrant lysine acetylation in tumorigenesis: implications in the development of therapeutics. *Pharmacol Ther*. 2016;162:98–119.
63. Patani N, Jiang WG, Newbold RF, Mokbel K. Histone-modifier gene expression profiles are associated with pathological and clinical outcomes in human breast cancer. *Anticancer Res*. 2011;31:4115–4125.
64. Xue L, Hou J, Wang Q, Yao L, Xu S, Ge D. RNAi screening identifies HAT1 as a potential drug target in esophageal squamous cell carcinoma. *Int J Clin Exp Pathol*. 2014;7:3898–3907.
65. Holmlund T, Lindberg MJ, Grander D, Wallberg AE. GCN5 acetylates and regulates the stability of the oncoprotein E2A-PBX1 in acute lymphoblastic leukemia. *Leukemia*. 2013;27:578–585.
66. Majaz S, Tong Z, Peng K, et al. Histone acetyl transferase GCN5 promotes human hepatocellular carcinoma progression by enhancing AIB1 expression. *Cell Biosci*. 2016;6:47.

67. Bharathy N, Suriyamurthy S, Rao VK, et al. P/CAF mediates PAX3-FOXO1 dependent oncogenesis in alveolar rhabdomyosarcoma. *J Pathol.* 2016;240:1–13.
68. Claps G, Cheli Y, Zhang T, et al. A transcriptionally inactive ATF2 variant drives melanomagenesis. *Cell Rep.* 2016;15:1884–1892.
69. Wu DS, Chen C, Wu ZJ, et al. ATF2 predicts poor prognosis and promotes malignant phenotypes in renal cell carcinoma. *J Exp Clin Cancer Res.* 2016;35:108.
70. Haery L, Thompson RC, Gilmore TD. Histone acetyltransferases and histone deacetylases in B- and T-cell development, physiology and malignancy. *Genes Cancer.* 2015;6:184–213.
71. Guo W, Lu J, Dai M, et al. Transcriptional coactivator CBP upregulates hTERT expression and tumor growth and predicts poor prognosis in human lung cancers. *Oncotarget.* 2014;5:9349–9361.
72. Pasqualucci L, Dominguez-Sola D, Chiarenza A, et al. Inactivating mutations of acetyltransferase genes in B-cell lymphoma. *Nature.* 2011;471:189–195.
73. Chen G, Cheng Y, Tang Y, Martinka M, Li G. Role of Tip60 in human melanoma cell migration, metastasis, and patient survival. *J Invest Dermatol.* 2012;132:2632–2641.
74. Deubzer HE, Schier MC, Oehme I, et al. HDAC11 is a novel drug target in carcinomas. *Int J Cancer.* 2013;132:2200–2208.
75. Murati A, Adelaide J, Quilichini B, et al. New types of MYST3-CBP and CBP-MYST3 fusion transcripts in t(8;16)(p11;p13) acute myeloid leukemias. *Haematologica.* 2007;92:262–263.
76. Yang XJ. MOZ and MORF acetyltransferases: molecular interaction, animal development and human disease. *Biochim Biophys Acta.* 1853;2015:1818–1826.
77. Dulak AM, Stojanov P, Peng S, et al. Exome and whole-genome sequencing of esophageal adenocarcinoma identifies recurrent driver events and mutational complexity. *Nat Genet.* 2013;45:478–486.
78. Iizuka M, Susa T, Takahashi Y, et al. Histone acetyltransferase Hbo1 destabilizes estrogen receptor alpha by ubiquitination and modulates proliferation of breast cancers. *Cancer Sci.* 2013;104:1647–1655.
79. Jaganathan A, Chaurasia P, Xiao GQ, et al. Coactivator MYST1 regulates nuclear factor-kappaB and androgen receptor functions during proliferation of prostate cancer cells. *Mol Endocrinol.* 2014;28:872–885.
80. Berns EM, van Staveren IL, Klijn JG, Foekens JA. Predictive value of SRC-1 for tamoxifen response of recurrent breast cancer. *Breast Cancer Res Treat.* 1998;48:87–92.
81. Fleming FJ, Myers E, Kelly G, et al. Expression of SRC-1, AIB1, and PEA3 in HER2 mediated endocrine resistant breast cancer; a predictive role for SRC-1. *J Clin Pathol.* 2004;57:1069–1074.
82. Burandt E, Jens G, Holst F, et al. Prognostic relevance of AIB1 (NCoA3) amplification and overexpression in breast cancer. *Breast Cancer Res Treat.* 2013;137:745–753.
83. Mosquera JM, Sboner A, Zhang L, et al. Recurrent NCOA2 gene rearrangements in congenital/infantile spindle cell rhabdomyosarcoma. *Genes Chromosomes Cancer.* 2013;52:538–550.
84. Karantanos T, Theodoropoulos G, Pektasides D, Gazouli M. Clock genes: their role in colorectal cancer. *World J Gastroenterol.* 2014;20:1986–1992.
85. Seo J, Min SK, Park HR, et al. Expression of histone deacetylases HDAC1, HDAC2, HDAC3, and HDAC6 in invasive ductal carcinomas of the breast. *J Breast Cancer.* 2014;17:323–331.
86. Quint K, Agaimy A, Di FP, et al. Clinical significance of histone deacetylases 1, 2, 3, and 7: HDAC2 is an independent predictor of survival in HCC. *Virchows Arch.* 2011;459:129–139.
87. Derr RS, van Hoesel AQ, Benard A, et al. High nuclear expression levels of histone-modifying enzymes LSD1, HDAC2 and SIRT1 in tumor cells correlate with decreased survival and increased relapse in breast cancer patients. *BMC Cancer.* 2014;14:604.

88. Weichert W, Roske A, Gekeler V, et al. Histone deacetylases 1, 2 and 3 are highly expressed in prostate cancer and HDAC2 expression is associated with shorter PSA relapse time after radical prostatectomy. *Br J Cancer.* 2008;98:604–610.
89. Moreno DA, Scrideli CA, Cortez MA, et al. Differential expression of HDAC3, HDAC7 and HDAC9 is associated with prognosis and survival in childhood acute lymphoblastic leukaemia. *Br J Haematol.* 2010;150:665–673.
90. Zeng LS, Yang XZ, Wen YF, et al. Overexpressed HDAC4 is associated with poor survival and promotes tumor progression in esophageal carcinoma. *Aging (Albany NY).* 2016;8:1236–1249.
91. Milde T, Oehme I, Korshunov A, et al. HDAC5 and HDAC9 in medulloblastoma: novel markers for risk stratification and role in tumor cell growth. *Clin Cancer Res.* 2010;16:3240–3252.
92. Ouaissi M, Sielezneff I, Silvestre R, et al. High histone deacetylase 7 (HDAC7) expression is significantly associated with adenocarcinomas of the pancreas. *Ann Surg Oncol.* 2008;15:2318–2328.
93. Peixoto P, Blomme A, Costanza B, et al. HDAC7 inhibition resets STAT3 tumorigenic activity in human glioblastoma independently of EGFR and PTEN: new opportunities for selected targeted therapies. *Oncogene.* 2016;35:4481–4494.
94. Park BL, Kim YJ, Cheong HS, et al. HDAC10 promoter polymorphism associated with development of HCC among chronic HBV patients. *Biochem Biophys Res Commun.* 2007;363:776–781.
95. Osada H, Tatematsu Y, Saito H, Yatabe Y, Mitsudomi T, Takahashi T. Reduced expression of class II histone deacetylase genes is associated with poor prognosis in lung cancer patients. *Int J Cancer.* 2004;112:26–32.
96. Saxena M, Dykes SS, Malyarchuk S, Wang AE, Cardelli JA, Pruitt K. The sirtuins promote Dishevelled-1 scaffolding of TIAM1, Rac activation and cell migration. *Oncogene.* 2013;34:188–198.
97. Holloway KR, Barbieri A, Malyarchuk S, et al. SIRT1 positively regulates breast cancer associated human aromatase (CYP19A1) expression. *Mol Endocrinol.* 2013;27:480–490.
98. Cha EJ, Noh SJ, Kwon KS, et al. Expression of DBC1 and SIRT1 is associated with poor prognosis of gastric carcinoma. *Clin Cancer Res.* 2009;15:4453–4459.
99. Leko V, Park GJ, Lao U, Simon JA, Bedalov A. Enterocyte-specific inactivation of SIRT1 reduces tumor load in the APC(+/min) mouse model. *PLoS One.* 2013;8: e66283.
100. Holloway KR, Calhoun TN, Saxena M, et al. SIRT1 regulates dishevelled proteins and promotes transient and constitutive Wnt signaling. *Proc Natl Acad Sci USA.* 2010;107:9216–9221.
101. Lin Z, Fang D. The roles of SIRT1 in cancer. *Genes Cancer.* 2013;4:97–104.
102. Chen J, Chan AW, To KF, et al. SIRT2 overexpression in hepatocellular carcinoma mediates epithelial to mesenchymal transition by protein kinase B/glycogen synthase kinase-3beta/beta-catenin signaling. *Hepatology.* 2013;57:2287–2298.
103. McGlynn LM, McCluney S, Jamieson NB, et al. SIRT3 & SIRT7: potential novel biomarkers for determining outcome in pancreatic cancer patients. *PLoS One.* 2015;10: e0131344.
104. Miyo M, Yamamoto H, Konno M, et al. Tumour-suppressive function of SIRT4 in human colorectal cancer. *Br J Cancer.* 2015;113:492–499.
105. Huang G, Cui F, Yu F, et al. Sirtuin-4 (SIRT4) is downregulated and associated with some clinicopathological features in gastric adenocarcinoma. *Biomed Pharmacother.* 2015;72:135–139.
106. Lu W, Zuo Y, Feng Y, Zhang M. SIRT5 facilitates cancer cell growth and drug resistance in non-small cell lung cancer. *Tumour Biol.* 2014;35:10699–10705.

107. Azuma Y, Yokobori T, Mogi A, et al. SIRT6 expression is associated with poor prognosis and chemosensitivity in patients with non-small cell lung cancer. *J Surg Oncol.* 2015;112:231–237.
108. Bai LH, Lin G, Sun L, et al. Upregulation of SIRT6 predicts poor prognosis and promotes metastasis of non-small cell lung cancer via the ERK1/2/MMP9 pathway. *Oncotarget.* 2016;7:40377–40386.
109. Lerrer B, Gertler AA, Cohen HY. The complex role of SIRT6 in carcinogenesis. *Carcinogenesis.* 2016;37:108–118.
110. Buglio D, Khaskhely NM, Voo KS, Martinez-Valdez H, Liu YJ, Younes A. HDAC11 plays an essential role in regulating OX40 ligand expression in Hodgkin lymphoma. *Blood.* 2011;117:2910–2917.
111. Ford DJ, Dingwall AK. The cancer COMPASS: navigating the functions of MLL complexes in cancer. *Cancer Genet.* 2015;208:178–191.
112. Marmorstein R. Structure and function of histone acetyltransferases. *Cell Mol Life Sci.* 2001;58:693–703.
113. Allis CD, Berger SL, Cote J, et al. New nomenclature for chromatin-modifying enzymes. *Cell.* 2007;131:633–636.
114. Choudhary C, Kumar C, Gnad F, et al. Lysine acetylation targets protein complexes and co-regulates major cellular functions. *Science.* 2009;325:834–840.
115. Bereshchenko OR, Gu W, la-Favera R. Acetylation inactivates the transcriptional repressor BCL6. *Nat Genet.* 2002;32:606–613.
116. Levy L, Wei Y, Labalette C, et al. Acetylation of beta-catenin by p300 regulates beta-catenin-Tcf4 interaction. *Mol Cell Biol.* 2004;24:3404–3414.
117. Yang XJ, Seto E. Lysine acetylation: codified crosstalk with other posttranslational modifications. *Mol Cell.* 2008;31:449–461.
118. Lee KK, Workman JL. Histone acetyltransferase complexes: one size doesn't fit all. *Nat Rev Mol Cell Biol.* 2007;8:284–295.
119. Peleg S, Feller C, Ladurner AG, Imhof A. The metabolic impact on histone acetylation and transcription in ageing. *Trends Biochem Sci.* 2016;41:700–711.
120. Taunton J, Hassig CA, Schreiber SL. A mammalian histone deacetylase related to the yeast transcriptional regulator Rpd3p. *Science.* 1996;272:408–411.
121. Ng HH, Bird A. Histone deacetylases: silencers for hire. *Trends Biochem Sci.* 2000;25:121–126.
122. Emiliani S, Fischle W, Van LC, Al-Abed Y, Verdin E. Characterization of a human RPD3 ortholog, HDAC3. *Proc Natl Acad Sci USA.* 1998;95:2795–2800.
123. Yang XJ, Seto E. The Rpd3/Hda1 family of lysine deacetylases: from bacteria and yeast to mice and men. *Nat Rev Mol Cell Biol.* 2008;9:206–218.
124. Grozinger CM, Hassig CA, Schreiber SL. Three proteins define a class of human histone deacetylases related to yeast Hda1p. *Proc Natl Acad Sci USA.* 1999;96:4868–4873.
125. Fischle W, Emiliani S, Hendzel MJ, et al. A new family of human histone deacetylases related to Saccharomyces cerevisiae HDA1p. *J Biol Chem.* 1999;274:11713–11720.
126. Gregoretti IV, Lee YM, Goodson HV. Molecular evolution of the histone deacetylase family: functional implications of phylogenetic analysis. *J Mol Biol.* 2004;338:17–31.
127. Blander G, Guarente L. The Sir2 family of protein deacetylases. *Annu Rev Biochem.* 2004;73:417–435.
128. Yoshida M, Kijima M, Akita M, Beppu T. Potent and specific inhibition of mammalian histone deacetylase both in vivo and in vitro by trichostatin A. *J Biol Chem.* 1990;265:17174–17179.
129. Kijima M, Yoshida M, Sugita K, Horinouchi S, Beppu T. Trapoxin, an antitumor cyclic tetrapeptide, is an irreversible inhibitor of mammalian histone deacetylase. *J Biol Chem.* 1993;268:22429–22435.

130. Waltregny D, North B, Van MF, De LJ, Verdin E, Castronovo V. Screening of histone deacetylases (HDAC) expression in human prostate cancer reveals distinct class I HDAC profiles between epithelial and stromal cells. *Eur J Histochem.* 2004;48:273–290.
131. Candido EP, Reeves R, Davie JR. Sodium butyrate inhibits histone deacetylation in cultured cells. *Cell.* 1978;14:105–113.
132. Vidali G, Boffa LC, Bradbury EM, ALLFREY VG. Butyrate suppression of histone deacetylation leads to accumulation of multiacetylated forms of histones H3 and H4 and increased DNase I sensitivity of the associated DNA sequences. *Proc Natl Acad Sci USA.* 1978;75:2239–2243.
133. Wang W, Cui SS, Lu R, Zhang H. Is there any therapeutic value for the use of histone deacetylase inhibitors for chronic pain? *Brain Res Bull.* 2016;125:44–52.
134. Fuchikami M, Yamamoto S, Morinobu S, Okada S, Yamawaki Y, Yamawaki S. The potential use of histone deacetylase inhibitors in the treatment of depression. *Prog Neuropsychopharmacol Biol Psychiatry.* 2016;64:320–324.
135. Pazienza V, Borghesan M, Mazza T, et al. SIRT1-metabolite binding histone macroH2A1.1 protects hepatocytes against lipid accumulation. *Aging (Albany NY).* 2014;6:35–47.
136. Frye RA. Phylogenetic classification of prokaryotic and eukaryotic Sir2-like proteins. *Biochem Biophys Res Commun.* 2000;273:793–798.
137. Greiss S, Gartner A. Sirtuin/Sir2 phylogeny, evolutionary considerations and structural conservation. *Mol Cells.* 2009;28:407–415.
138. Longo VD, Kennedy BK. Sirtuins in aging and age-related disease. *Cell.* 2006;126:257–268.
139. Anastasiou D, Krek W. SIRT1: linking adaptive cellular responses to aging-associated changes in organismal physiology. *Physiology (Bethesda).* 2006;21:404–410.
140. Yang T, Fu M, Pestell R, Sauve AA. SIRT1 and endocrine signaling. *Trends Endocrinol Metab.* 2006;17:186–191.
141. Ford J, Jiang M, Milner J. Cancer-specific functions of SIRT1 enable human epithelial cancer cell growth and survival. *Cancer Res.* 2005;65:10457–10463.
142. Wilson BJ, Tremblay AM, Deblois G, Sylvain-Drolet G, Giguere V. An acetylation switch modulates the transcriptional activity of estrogen-related receptor alpha. *Mol Endocrinol.* 2010;24:1349–1358.
143. Hanahan D, Weinberg RA. Hallmarks of cancer: the next generation. *Cell.* 2011;144:646–674.
144. Firestein R, Blander G, Michan S, et al. The SIRT1 deacetylase suppresses intestinal tumorigenesis and colon cancer growth. *PLoS One.* 2008;3:e2020.
145. Portmann S, Fahrner R, Lechleiter A, et al. Antitumor effect of SIRT1 inhibition in human HCC tumor models in vitro and in vivo. *Mol Cancer Ther.* 2013;12:499–508.
146. Elangovan S, Ramachandran S, Venkatesan N, et al. SIRT1 is essential for oncogenic signaling by estrogen/estrogen receptor alpha in breast cancer. *Cancer Res.* 2011;71:6654–6664.
147. Simmons Jr GE, Pandey S, Nedeljkovic-Kurepa A, Saxena M, Wang A, Pruitt K. Frizzled 7 expression is positively regulated by SIRT1 and beta-catenin in breast cancer cells. *PLoS One.* 2014;9:e98861.
148. Liu T, Liu PY, Marshall GM. The critical role of the class III histone deacetylase SIRT1 in cancer. *Cancer Res.* 2009;69:1702–1705.
149. Zhang B, Wang J, Wang X, et al. Proteogenomic characterization of human colon and rectal cancer. *Nature.* 2014;513:382–387.
150. Chen X, Sun K, Jiao S, et al. High levels of SIRT1 expression enhance tumorigenesis and associate with a poor prognosis of colorectal carcinoma patients. *Sci Rep.* 2014;4:7481.

151. Lv L, Shen Z, Zhang J, et al. Clinicopathological significance of SIRT1 expression in colorectal adenocarcinoma. *Med Oncol.* 2014;31:965.
152. Chung YR, Kim H, Park SY, et al. Distinctive role of SIRT1 expression on tumor invasion and metastasis in breast cancer by molecular subtype. *Hum Pathol.* 2015;46:1027–1035.
153. Li H, Rajendran GK, Liu N, Ware C, Rubin BP, Gu Y. SirT1 modulates the estrogen-insulin-like growth factor-1 signaling for postnatal development of mammary gland in mice. *Breast Cancer Res.* 2007;9:R1.
154. Zhao G, Cui J, Zhang JG, et al. SIRT1 RNAi knockdown induces apoptosis and senescence, inhibits invasion and enhances chemosensitivity in pancreatic cancer cells. *Gene Ther.* 2011;18:920–928.
155. Rizk SM, Shahin NN, Shaker OG. Association between SIRT1 gene polymorphisms and breast cancer in Egyptians. *PLoS One.* 2016;11:e0151901.
156. Jang KY, Hwang SH, Kwon KS, et al. SIRT1 expression is associated with poor prognosis of diffuse large B-cell lymphoma. *Am J Surg Pathol.* 2008;32:1523–1531.
157. Bradbury CA, Khanim FL, Hayden R, et al. Histone deacetylases in acute myeloid leukaemia show a distinctive pattern of expression that changes selectively in response to deacetylase inhibitors. *Leukemia.* 2005;19:1751–1759.
158. Huffman DM, Grizzle WE, Bamman MM, et al. SIRT1 is significantly elevated in mouse and human prostate cancer. *Cancer Res.* 2007;67:6612–6618.
159. Kim JE, Chen J, Lou Z. DBC1 is a negative regulator of SIRT1. *Nature.* 2008;451:583–586.
160. Zhao W, Kruse JP, Tang Y, Jung SY, Qin J, Gu W. Negative regulation of the deacetylase SIRT1 by DBC1. *Nature.* 2008;451:587–590.
161. Heltweg B, Gatbonton T, Schuler AD, et al. Antitumor activity of a small-molecule inhibitor of human silent information regulator 2 enzymes. *Cancer Res.* 2006;66:4368–4377.
162. Ota H, Tokunaga E, Chang K, et al. Sirt1 inhibitor, sirtinol, induces senescence-like growth arrest with attenuated Ras-MAPK signaling in human cancer cells. *Oncogene.* 2006;25:176–185.
163. Sauve AA, Moir RD, Schramm VL, Willis IM. Chemical activation of Sir2-dependent silencing by relief of nicotinamide inhibition. *Mol Cell.* 2005;17:595–601.
164. Avalos JL, Bever KM, Wolberger C. Mechanism of sirtuin inhibition by nicotinamide: altering the NAD(+) cosubstrate specificity of a Sir2 enzyme. *Mol Cell.* 2005;17:855–868.
165. Audrito V, Vaisitti T, Rossi D, et al. Nicotinamide blocks proliferation and induces apoptosis of chronic lymphocytic leukemia cells through activation of the p53/miR-34a/SIRT1 tumor suppressor network. *Cancer Res.* 2011;71:4473–4483.
166. Jung-Hynes B, Nihal M, Zhong W, Ahmad N. Role of sirtuin histone deacetylase SIRT1 in prostate cancer. A target for prostate cancer management via its inhibition? *J Biol Chem.* 2009;284:3823–3832.
167. Cheng F, Su L, Yao C, et al. SIRT1 promotes epithelial-mesenchymal transition and metastasis in colorectal cancer by regulating Fra-1 expression. *Cancer Lett.* 2016;375:274–283.
168. Liang Z, Yang Y, Wang H, et al. Inhibition of SIRT1 signaling sensitizes the antitumor activity of silybin against human lung adenocarcinoma cells in vitro and in vivo. *Mol Cancer Ther.* 2014;13:1860–1872.
169. Mahajan SS, Scian M, Sripathy S, et al. Development of pyrazolone and isoxazol-5-one cambinol analogues as sirtuin inhibitors. *J Med Chem.* 2014;57:3283–3294.
170. Zhu L, Chiao CY, Enzer KG, Stankiewicz AJ, Faller DV, Dai Y. SIRT1 inactivation evokes antitumor activities in NSCLC through the tumor suppressor p27. *Mol Cancer Res.* 2015;13:41–49.

171. Lu J, Zhang M, Huang Z, et al. SIRT1 in B[a]P-induced lung tumorigenesis. *Oncotarget*. 2015;6:27113–27129.
172. Sasca D, Hahnel PS, Szybinski J, et al. SIRT1 prevents genotoxic stress-induced p53 activation in acute myeloid leukemia. *Blood*. 2014;124:121–133.
173. Birney E, Stamatoyannopoulos JA, Dutta A, et al. Identification and analysis of functional elements in 1% of the human genome by the ENCODE pilot project. *Nature*. 2007;447:799–816.
174. Santolla MF, Avino S, Pellegrino M, et al. SIRT1 is involved in oncogenic signaling mediated by GPER in breast cancer. *Cell Death Dis*. 2015;6:e1834.
175. Roth M, Chen WY. Sorting out functions of sirtuins in cancer. *Oncogene*. 2013;33:1609–1620.
176. Dillon SC, Zhang X, Trievel RC, Cheng X. The SET-domain protein superfamily: protein lysine methyltransferases. *Genome Biol*. 2005;6:227.
177. Richon VM, Johnston D, Sneeringer CJ, et al. Chemogenetic analysis of human protein methyltransferases. *Chem Biol Drug Des*. 2011;78:199–210.
178. Song Y, Wu F, Wu J. Targeting histone methylation for cancer therapy: enzymes, inhibitors, biological activity and perspectives. *J Hematol Oncol*. 2016;9:49.
179. Abu-Farha M, Lambert JP, Al-Madhoun AS, Elisma F, Skerjanc IS, Figeys D. The tale of two domains: proteomics and genomics analysis of SMYD2, a new histone methyltransferase. *Mol Cell Proteomics*. 2008;7:560–572.
180. Sirinupong N, Brunzelle J, Ye J, Pirzada A, Nico L, Yang Z. Crystal structure of cardiac-specific histone methyltransferase SmyD1 reveals unusual active site architecture. *J Biol Chem*. 2010;285:40635–40644.
181. Hamamoto R, Furukawa Y, Morita M, et al. SMYD3 encodes a histone methyltransferase involved in the proliferation of cancer cells. *Nat Cell Biol*. 2004;6: 731–740.
182. Falandry C, Fourel G, Galy V, et al. CLLD8/KMT1F is a lysine methyltransferase that is important for chromosome segregation. *J Biol Chem*. 2010;285:20234–20241.
183. Schotta G, Lachner M, Sarma K, et al. A silencing pathway to induce H3-K9 and H4-K20 trimethylation at constitutive heterochromatin. *Genes Dev*. 2004;18:1251–1262.
184. Hsia DA, Tepper CG, Pochampalli MR, et al. KDM8, a H3K36me2 histone demethylase that acts in the cyclin A1 coding region to regulate cancer cell proliferation. *Proc Natl Acad Sci USA*. 2010;107:9671–9676.
185. He C, Li F, Zhang J, Wu J, Shi Y. The methyltransferase NSD3 has chromatin-binding motifs, PHD5-C5HCH, that are distinct from other NSD (nuclear receptor SET domain) family members in their histone H3 recognition. *J Biol Chem*. 2013;288:4692–4703.
186. Nimura K, Ura K, Shiratori H, et al. A histone H3 lysine 36 trimethyltransferase links Nkx2-5 to Wolf-Hirschhorn syndrome. *Nature*. 2009;460:287–291.
187. Pei H, Zhang L, Luo K, et al. MMSET regulates histone H4K20 methylation and 53BP1 accumulation at DNA damage sites. *Nature*. 2011;470:124–128.
188. Martin C, Zhang Y. The diverse functions of histone lysine methylation. *Nat Rev Mol Cell Biol*. 2005;6:838–849.
189. Shen X, Liu Y, Hsu YJ, et al. EZH1 mediates methylation on histone H3 lysine 27 and complements EZH2 in maintaining stem cell identity and executing pluripotency. *Mol Cell*. 2008;32:491–502.
190. Min J, Feng Q, Li Z, Zhang Y, Xu RM. Structure of the catalytic domain of human DOT1L, a non-SET domain nucleosomal histone methyltransferase. *Cell*. 2003;112:711–723.
191. Van Aller GS, Reynoird N, Barbash O, et al. Smyd3 regulates cancer cell phenotypes and catalyzes histone H4 lysine 5 methylation. *Epigenetics*. 2012;7:340–343.

192. Morera L, Lubbert M, Jung M. Targeting histone methyltransferases and demethylases in clinical trials for cancer therapy. *Clin Epigenetics*. 2016;8:57.
193. Gelato KA, Shaikhibrahim Z, Ocker M, Haendler B. Targeting epigenetic regulators for cancer therapy: modulation of bromodomain proteins, methyltransferases, demethylases, and microRNAs. *Expert Opin Ther Targets*. 2016;20:783–799.
194. Hamamoto R, Nakamura Y. Dysregulation of protein methyltransferases in human cancer: an emerging target class for anticancer therapy. *Cancer Sci*. 2016;107:377–384.
195. Simo-Riudalbas L, Esteller M. Targeting the histone orthography of cancer: drugs for writers, erasers and readers. *Br J Pharmacol*. 2015;172:2716–2732.
196. Wolf SS. The protein arginine methyltransferase family: an update about function, new perspectives and the physiological role in humans. *Cell Mol Life Sci*. 2009; 66:2109–2121.
197. Stopa N, Krebs JE, Shechter D. The PRMT5 arginine methyltransferase: many roles in development, cancer and beyond. *Cell Mol Life Sci*. 2015;72:2041–2059.
198. Lausen J. Contributions of the histone arginine methyltransferase PRMT6 to the epigenetic function of RUNX1. *Crit Rev Eukaryot Gene Expr*. 2013;23:265–274.
199. Morin RD, Mendez-Lago M, Mungall AJ, et al. Frequent mutation of histone-modifying genes in non-Hodgkin lymphoma. *Nature*. 2011;476:298–303.
200. Kantidakis T, Saponaro M, Mitter R, et al. Mutation of cancer driver MLL2 results in transcription stress and genome instability. *Genes Dev*. 2016;30:408–420.
201. Lee JJ, Sholl LM, Lindeman NI, et al. Targeted next-generation sequencing reveals high frequency of mutations in epigenetic regulators across treatment-naive patient melanomas. *Clin Epigenetics*. 2015;7:59.
202. Yun H, Damm F, Yap D, et al. Impact of MLL5 expression on decitabine efficacy and DNA methylation in acute myeloid leukemia. *Haematologica*. 2014;99:1456–1464.
203. Salz T, Li G, Kaye F, Zhou L, Qiu Y, Huang S. hSETD1A regulates Wnt target genes and controls tumor growth of colorectal cancer cells. *Cancer Res*. 2014;74:775–786.
204. Salz T, Deng C, Pampo C, et al. Histone methyltransferase hSETD1A is a novel regulator of metastasis in breast cancer. *Mol Cancer Res*. 2015;13:461–469.
205. Kleer CG, Cao Q, Varambally S, et al. EZH2 is a marker of aggressive breast cancer and promotes neoplastic transformation of breast epithelial cells. *Proc Natl Acad Sci USA*. 2003;100:11606–11611.
206. Shi Y, Lan F, Matson C, et al. Histone demethylation mediated by the nuclear amine oxidase homolog LSD1. *Cell*. 2004;119:941–953.
207. Tsukada Y, Fang J, Erdjument-Bromage H, et al. Histone demethylation by a family of JmjC domain-containing proteins. *Nature*. 2006;439:811–816.
208. Forneris F, Binda C, Vanoni MA, Mattevi A, Battaglioli E. Histone demethylation catalysed by LSD1 is a flavin-dependent oxidative process. *FEBS Lett*. 2005;579:2203–2207.
209. Suzuki T, Ozasa H, Itoh Y, et al. Identification of the KDM2/7 histone lysine demethylase subfamily inhibitor and its antiproliferative activity. *J Med Chem*. 2013;56:7222–7231.
210. Maes T, Carceller E, Salas J, Ortega A, Buesa C. Advances in the development of histone lysine demethylase inhibitors. *Curr Opin Pharmacol*. 2015;23:52–60.
211. Kauffman EC, Robinson BD, Downes MJ, et al. Role of androgen receptor and associated lysine-demethylase coregulators, LSD1 and JMJD2A, in localized and advanced human bladder cancer. *Mol Carcinog*. 2011;50:931–944.
212. Tokunaga R, Sakamoto Y, Nakagawa S, et al. The prognostic significance of histone lysine demethylase JMJD3/KDM6B in colorectal cancer. *Ann Surg Oncol*. 2016;23:678–685.
213. Gale M, Sayegh J, Cao J, et al. Screen-identified selective inhibitor of lysine demethylase 5A blocks cancer cell growth and drug resistance. *Oncotarget*. 2016;7:39931–39944.

214. Huang J, Sengupta R, Espejo AB, et al. p53 is regulated by the lysine demethylase LSD1. *Nature*. 2007;449:105–108.
215. Qin Y, Zhu W, Xu W, et al. LSD1 sustains pancreatic cancer growth via maintaining HIF1alpha-dependent glycolytic process. *Cancer Lett*. 2014;347:225–232.
216. Schulte JH, Lim S, Schramm A, et al. Lysine-specific demethylase 1 is strongly expressed in poorly differentiated neuroblastoma: implications for therapy. *Cancer Res*. 2009;69:2065–2071.
217. Lim S, Janzer A, Becker A, et al. Lysine-specific demethylase 1 (LSD1) is highly expressed in ER-negative breast cancers and a biomarker predicting aggressive biology. *Carcinogenesis*. 2010;31:512–520.
218. van HG, Dalgliesh GL, Davies H, et al. Somatic mutations of the histone H3K27 demethylase gene UTX in human cancer. *Nat Genet*. 2009;41:521–523.
219. Amaral PP, Dinger ME, Mercer TR, Mattick JS. The eukaryotic genome as an RNA machine. *Science*. 2008;319:1787–1789.
220. Beermann J, Piccoli MT, Viereck J, Thum T. Non-coding RNAs in development and disease: background, mechanisms, and therapeutic approaches. *Physiol Rev*. 2016;96:1297–1325.
221. Falaleeva M, Stamm S. Processing of snoRNAs as a new source of regulatory non-coding RNAs: snoRNA fragments form a new class of functional RNAs. *Bioessays*. 2013;35:46–54.
222. Green D, Fraser WD, Dalmay T. Transfer RNA-derived small RNAs in the cancer transcriptome. *Pflugers Arch*. 2016;468:1041–1047.
223. Watanabe T, Lin H. Posttranscriptional regulation of gene expression by Piwi proteins and piRNAs. *Mol Cell*. 2014;56:18–27.
224. Jeck WR, Sharpless NE. Detecting and characterizing circular RNAs. *Nat Biotechnol*. 2014;32:453–461.
225. Guarnerio J, Bezzi M, Jeong JC, et al. Oncogenic role of fusion-circRNAs derived from cancer-associated chromosomal translocations. *Cell*. 2016;165:289–302.
226. Di LG, Croce CM. miRNA profiling of cancer. *Curr Opin Genet Dev*. 2013;23:3–11.
227. Derrien T, Johnson R, Bussotti G, et al. The GENCODE v7 catalog of human long noncoding RNAs: analysis of their gene structure, evolution, and expression. *Genome Res*. 2012;22:1775–1789.
228. Schmitt AM, Chang HY. Long noncoding RNAs in cancer pathways. *Cancer Cell*. 2016;29:452–463.
229. Tate PH, Bird AP. Effects of DNA methylation on DNA-binding proteins and gene expression. *Curr Opin Genet Dev*. 1993;3:226–231.
230. Amir RE, Van den Veyver IB, Wan M, Tran CQ, Francke U, Zoghbi HY. Rett syndrome is caused by mutations in X-linked MECP2, encoding methyl-CpG-binding protein 2. *Nat Genet*. 1999;23:185–188.
231. Nan X, Campoy FJ, Bird A. MeCP2 is a transcriptional repressor with abundant binding sites in genomic chromatin. *Cell*. 1997;88:471–481.
232. Jones PL, Veenstra GJ, Wade PA, et al. Methylated DNA and MeCP2 recruit histone deacetylase to repress transcription. *Nat Genet*. 1998;19:187–191.
233. Bachman KE, Park BH, Rhee I, et al. Histone modifications and silencing prior to DNA methylation of a tumor suppressor gene. *Cancer Cell*. 2003;3:89–95.
234. Jones PA, Laird PW. Cancer epigenetics comes of age. *Nat Genet*. 1999;21:163–167.
235. Gore SD, Baylin S, Sugar E, et al. Combined DNA methyltransferase and histone deacetylase inhibition in the treatment of myeloid neoplasms. *Cancer Res*. 2006;66:6361–6369.
236. Cameron EE, Bachman KE, Myohanen S, Herman JG, Baylin SB. Synergy of demethylation and histone deacetylase inhibition in the re-expression of genes silenced in cancer. *Nat Genet*. 1999;21:103–107.

237. Neupane M, Clark AP, Landini S, et al. MECP2 is a frequently amplified oncogene with a novel epigenetic mechanism that mimics the role of activated RAS in malignancy. *Cancer Discov*. 2016;6:45–58.
238. Fuks F, Hurd PJ, Wolf D, Nan X, Bird AP, Kouzarides T. The methyl-CpG-binding protein MeCP2 links DNA methylation to histone methylation. *J Biol Chem*. 2003;278:4035–4040.
239. Agarwal N, Hardt T, Brero A, et al. MeCP2 interacts with HP1 and modulates its heterochromatin association during myogenic differentiation. *Nucleic Acids Res*. 2007;35:5402–5408.
240. Chahrour M, Jung SY, Shaw C, et al. MeCP2, a key contributor to neurological disease, activates and represses transcription. *Science*. 2008;320:1224–1229.
241. Filarsky M, Zillner K, Araya I, et al. The extended AT-hook is a novel RNA binding motif. *RNA Biol*. 2015;12:864–876.
242. Berghoff EG, Clark MF, Chen S, Cajigas I, Leib DE, Kohtz JD. Evf2 (Dlx6as) lncRNA regulates ultraconserved enhancer methylation and the differential transcriptional control of adjacent genes. *Development*. 2013;140:4407–4416.
243. Nan X, Hou J, Maclean A, et al. Interaction between chromatin proteins MECP2 and ATRX is disrupted by mutations that cause inherited mental retardation. *Proc Natl Acad Sci USA*. 2007;104:2709–2714.
244. Sakuraba K, Yasuda T, Sakata M, et al. Down-regulation of Tip60 gene as a potential marker for the malignancy of colorectal cancer. *Anticancer Res*. 2009;29:3953–3955.
245. Metoyer CF, Pruitt K. The role of sirtuin proteins in obesity. *Pathophysiology*. 2008;15:103–108.

CHAPTER TWO

Wnt/β Catenin-Mediated Signaling Commonly Altered in Colorectal Cancer

J. Deitrick, W.M. Pruitt[1]
Texas Tech University Health Sciences Center School of Medicine, Lubbock, TX, United States
[1]Corresponding author: e-mail address: wendy.pruitt@ttuhsc.edu

Contents

Progress in Molecular Biology and Translational Science, Volume 144
ISSN 1877-1173
http://dx.doi.org/10.1016/bs.pmbts.2016.09.010

Abstract

Colorectal cancer is the second most common cancer in females and the third most common cancer diagnosed in males (Torre *et al.*[1]). In 2012, there were about 1.4 million cases and 693,900 deaths due to colorectal cancer worldwide. It is more common in developed countries, and North America, Europe, and Australia have the highest incidence rates. In the United States, adults have a 5% chance of developing colorectal cancer (Cancer of the colon and rectum—SEER stat fact sheets[2]). Due to the high prevalence of colorectal cancer, understanding the mechanism underlying its initiation and progression in order to find better therapeutic agents will have a high impact in the field of oncology and may improve the treatment of other cancers with shared mechanistic properties. Aberrant Wnt/β-catenin signaling is a characteristic feature of colorectal cancer development and is the focus of this review.

1. WNT SIGNALING IN COLORECTAL CANCER

1.1 Introduction to Wnt Signaling

Wnt signaling plays an important role in proliferation, differentiation, motility, survival, apoptosis, embryonic development, and homeostasis.[3,4] Wnt is a highly insoluble protein ligand which binds to Frizzled (FZD), a cell membrane receptor with seven transmembrane domains.[5,6] The insolubility of Wnt is partially caused by the palmitoylation at a conserved cysteine residue, which is a critical posttranslational modification for its signaling ability.[5] There are 19 different Wnt ligands, each consisting of 350–400 amino acids, and 10 various Fz receptors.[6] This heterogeneity results in a variety of signaling pathways initiated by Wnt.

Wnt controls three major signaling pathways: the Wnt/calcium pathway, the planar cell polarity pathway, and the canonical Wnt pathway.[3] The combination of Wnt ligand and FZD receptor determines the pathway activated via Wnt. The Wnt/calcium pathway causes an increase in intracellular Ca^{2+}, which results in the activation of calmodulin-dependent protein kinase II and calcineurin. This activation results in the accumulation of NF-AT in the nucleus, where it acts as a transcription factor by increasing the expression of genes associated with cell adhesion and motility. The planar cell polarity pathway, which regulates cell polarity and morphogenic movements, utilizes small GTPases, such as Ras and RhoA to activate c-Jun N-terminal kinases. The canonical Wnt pathway, which results in the accumulation of β-catenin in the nucleus, is not only the most prevalent Wnt signaling pathway but also the most associated with the initiation and progression of cancer. Therefore, a more detailed inspection of the canonical Wnt pathway will provide insight into how to target this pathway to better treat and prevent recurrence of colorectal cancer.

The canonical Wnt Signaling commences with the binding of Wnt to FZD and co-receptor low-density lipoprotein receptor-related protein 5/6 (LRP5/6) and terminates with the transcription of Wnt target genes due to nuclear translocation of β-catenin. In the absence of Wnt signaling, β-catenin is constitutively degraded. A degradation complex consisting of Axin, adenomatous polyposis coli (APC), casein kinase 1 (CK1), and glycogen synthase kinase-3β (GSK-3β) targets β-catenin for ubiquination and subsequent proteasomal degradation. GSK-3β and CK1 both play a major role in phosphorylating β-catenin at specific serine and threonine residues to cause β-TrCP-induced polyubiquitination.

Wnt signaling initially causes the phosphorylation of disheveled (Dvl), which plays important roles in both the canonical and noncanonical Wnt signaling pathways. First, Dvl acts as a scaffolding protein to sequester glycogen synthase kinase-3β and axin, which are key members of the β-catenin degradation complex. This sequestration allows β-catenin to accumulate in the cytosol and subsequently translocate to the nucleus, where it carries out its function as a transcription coactivator. Second, Dvl clusters the FZD receptors and LRP6 co-receptors upon Wnt signaling stimulation to create a "signalsome".[6] This spatial concentration facilitates increased recruitment of the proteins that make up the degradation complex, thereby increasing the strength of the Wnt signaling. Last, Dvl may also play a significant role in the nucleus by associating with transcription factors to increase the expression of Wnt target genes.[7]

Once in the nucleus, β-catenin associates with T-cell factor (TCF) or lymphoid enhancer factor (LEF).[4] In the absence of Wnt signaling and nuclear β-catenin, these transcription factors associate with Groucho and other repressors to inhibit the transcription of Wnt target genes. The presence of nuclear β-catenin converts TCF and LEF from transcriptional repressors to activators. In addition to this association, β-catenin also associates with the coactivators p300 and CREB-binding protein (CBP).[3] This association may influence which Wnt target genes are expressed due to Wnt signaling: p300 is associated with cell differentiation, while CBP is associated with the maintenance of potency.[8] It is unclear how β-catenin enters the nucleus.[6] However, studies indicate that APC, in addition to contributing to the degradation complex, plays a role in exporting β-catenin from the nucleus after the Wnt signal has terminated.[4]

Wnt signaling directly regulates the transcription of a wide variety of genes. The Wnt target genes Oct4, Sox2, and Klf4 are known factors that induce pluripotency.[9] This association indicates the critical role Wnt plays in maintaining stem cell populations. Wnt signaling also induces the transcription of c-Myc, TCF1, and cyclin D1, which instigate cell growth and proliferation. When associated with p300, nuclear β-catenin activates transcription of c-Jun and Fra-1 to facilitate differentiation.[3] Furthermore, β-catenin regulates expression of components of the Wnt signaling pathway, including FZD, LRP6, Axin, Naked, TCF, and LEF.[6]

Due to the various roles of Wnt target genes, canonical Wnt signaling is critical for maintaining homeostasis within intestinal epithelium.[10] Expression of Wnt target genes drives proliferation and secretory cell differentiation within intestinal epithelium. Lack of Wnt signaling and downstream nuclear β-catenin greatly impedes these processes. Inhibition of β-catenin/TCF complex decreases expression of c-Myc, which is a repressor of p21.[11] This increased expression of p21 leads to G1 cell cycle arrest and subsequent differentiation, thwarting the uninhibited proliferation of cells with uninterrupted Wnt signaling.

1.2 Canonical Wnt Signaling and Colorectal Cancer

While Wnt signaling plays a critical role in numerous cellular and developmental processes in normal cells, aberrant Wnt signaling is highly associated with numerous cancers and may be responsible for drug resistance and recurrence of tumors. Overactive Wnt signaling can trigger tumorigenesis in skin, breast, bone marrow, and colon tissue.[12] Canonical Wnt signaling is upregulated in colon cancers, as supported by elevated levels of Wnt target genes, axin2 and human naked cuticle (hNKD), in colorectal cancer.[13]

In addition to facilitating initiation of tumor formation, Wnt signaling also induces resistance to conventional anticancer agents. Upregulation of the Wnt/β-catenin pathway stimulates resistance to the combination therapy of interferon-alpha and 5-fluorouracil in hepatocellular carcinomas.[14] Also, excessively active Wnt signaling induces radioresistance in head and neck cancers by upregulating Ku expression. This upregulation of β-catenin occurs via multiple mechanisms to confer resistance, including increased secretion of Wnt ligands and expression of Frizzled.[15] High levels of canonical Wnt signaling may initiate autocrine signaling that protects cancer cells from undergoing apoptosis due to exposure to cytotoxic anticancer agents.[16] Fortunately, inhibiting canonical Wnt signaling can induce sensitivity in drug-resistant cancer cells.[15] Therefore, the Wnt signaling pathway is a promising target not only for novel anticancer agents but also for inducing sensitivity to conventional therapies.

2. MUTATIONS IN THE WNT SIGNALING PATHWAY

Mutations in the Wnt pathway are linked to birth defects, cancer, and other diseases.[6] As aforementioned, deregulation of this pathway causes abnormal expression of Wnt target genes, which is highly implicated in tumorigenesis of colorectal cancer. Various mutations of the canonical Wnt pathway lead to deregulation resulting in constitutively increased expression of β-catenin in the cytosol and nucleus.

2.1 Adenomatous Polyposis Coli

APC mutations are the most common initiator of colorectal cancer.[17] About 85% of colorectal tumors have at least one mutation in the APC gene, while 60% have two mutations.[18] When both APC alleles are mutated, these mutations form interdependent of each other.[19] The 20R1 region of APC, which is required to bind to β-catenin, is necessary for development of colon polyps. Therefore, if one mutation truncates the protein to lose this region, the mutation on the following allele will always be one in which the region is conserved and vice versa. This interdependence is also demonstrated with respect to the β-*c*atenin *i*nhibitory *d*omain (CID) of APC, which is necessary to mediate degradation of β-catenin. The CID is often maintained in one allele and lost via truncation in the other allele. This truncated APC fragment is capable of inhibiting the functional CID to impede APC-mediated β-catenin degradation. The 95% of APC mutations resulted in a truncated protein, and 60% of the mutations were in exon 15, which constitutes only 10% of the total coding region.[18]

2.2 β-Catenin

While the majority of colorectal cancers are caused by APC mutations, those with wild-type APC often contain gain of function mutations in the *CTNNB1* gene, which codes for β-catenin.[20] About half of all colorectal cancers with wild-type APC have mutations in the amino-terminal regulatory domain of β-catenin. These mutant β-catenin proteins are six times as active as their wild-type counterparts. Therefore, most, if not all, colorectal cancers contain mutations in APC or β-catenin that result in drastic upregulation of the canonical Wnt signaling pathway.[20] Furthermore, β-catenin mutations are seen at a much higher frequency in later stage cancers.[21] Mutations in β-catenin are also very common in endometrioid ovarian cancer as well as pediatric hepatoblastoma and Wilms' pediatric kidney tumor.[17]

2.3 Axin and Other Wnt Signaling Components

While not as common, mutations in other components of canonical Wnt signaling are present in colorectal cancer. Axin is mutated in numerous cancers, including colorectal cancer, and its mutation is not mutually exclusive with APC or β-catenin.[17] TCF mutations have also been found in sporadic colorectal cancers, especially those with microsatellite instability.[22] Protein phosphatase-2A (PP2A), a phosphatase complex critical to the Wnt pathway, sometimes contains mutations within the regulatory A-subunit in various cancers.[17] Wnt inhibitory factor-1 (WIF1) is downregulated due to epigenetic silencing in breast cancer.[23] The WIF1 promoter is hypermethylated in 67% of breast tumors. While these mutations are not as common or highly associated with colorectal cancer as APC or β-catenin, their association with other cancers further demonstrates the imperative role of canonical Wnt signaling in tumorigenesis.

Mutations in proteins not directly involved in the Wnt signaling pathway may also facilitate upregulation of Wnt target genes and thus play a role in tumorigenesis. For instance, Kirsten rat sarcoma viral oncogene homolog (KRAS) regulates the nuclear localization of β-catenin.[24] KRAS triggers tyrosine phosphorylation of β-catenin, which allows it to be released from E-cadherin and transported to the nucleus.[12] Therefore, increased activation of KRAS results in increased nuclear β-catenin. Loss of APC function due to mutation often leads to KRAS oncogene activation.[12] This combination of loss of function of APC and gain of function of KRAS promotes β-catenin nuclear translocation. Other genes that influence nuclear localization of β-catenin may also have altered expression in cancers in order to fully activate constitutive Wnt signaling in colorectal cancer.

3. CANCER STEM CELLS

3.1 Introduction to Cancer Stem Cells

Human stem cells differ from most cells in the body in their ability of unlimited proliferation and self-renewal as well as differentiation into a wider variety of cell types. The potency and clonogenicity of this subset of cells in normal tissues make them unique and powerful cells in terms of development and maintaining homeostasis. Recent research indicates that while tumors used to be considered homogenous malignancies, they may also contain a subset of cells with similar properties to stem cells that have been termed cancer-initiating cells or cancer stem cells.

As with nonmalignant adult stem cells, cancer stem cells initiate the formation of a heterogeneous population of cells and replenish this population as well as self-renew to maintain their own subpopulation. Cancer stem cells are able to initiate the formation of tumors when transplanted into NOD/SCID mice, whereas other cancer cells from the same tumor are not.[25] It has been proposed that colorectal cancer initiation starts with a cancer stem cell that has been transformed from a nonmalignant intestinal stem cell.[26] In addition to being able to induce tumor formation and proliferate to replenish their own population, these multipotent tumorigenic cells can differentiate along multiple lineages to replenish differentiated cancer cells, and recent evidence demonstrates that phosphoinositide-3 kinase (PI3K) controls this lineage decision.[27] Anywhere from 1.8% to 24.5% of cells compromising tumors are cancer stem cells.[26] However, even a small number of cancer stem cells can initiate the formation of a new tumor.[28] The inability to eradicate these powerful cancer stem cells may be the cause of cancer relapse as well as resistance to current chemotherapy.[29]

Cancer stem cells are highly resistant to numerous conventional anticancer treatments, including platinum therapy.[30] The mechanism underlying this drug resistance has only begun to be elucidated. One possible mechanism: cancer stem cells secrete IL-4 in an autocrine manner in order to protect themselves against apoptosis, and blocking IL-4 or its receptor makes these cells more sensitive to the chemotherapeutic drugs oxaliplatin and 5-FU.[16] Further investigation may reveal all of the mechanisms underlying resistance in cancer stem cells.

While much has been learned about cancer stem cells in the last two decades, new studies are continuing to alter our view of how these unique cells promote tumor initiation and progression. First, all cancer stem cells may not be equal. Dieter *et al.* demonstrated that cancer stem cells have three distinct phenotypes with various functions in tumor initiation, progression, and metastasis.[31] They describe these subpopulations as: long-term TICs (LT-TIC), tumor transient-amplifying cells (T-TACs), and delay contributing TICs (DC-TIC). LT-TICs have the greatest self-renewal capability, and they drive metastasis in vivo. T-TACs play the predominant role in tumor formation, but they have negligible self-renewal capacity. DC-TICs are very rare, but they contribute to serial tumors.[31] Because LT-TICs play a major role in self-renewal and metastasis, they may be the prime target for therapeutic intervention.

Furthermore, cancer stem cells may have the ability to lose and regain their stemness. Cancer cells without stem cell properties are capable of generating cancer stem cells both in vitro and in vivo.[32] While cancer stem cells are more efficient at initiating and forming heterogeneous tumors, these results indicate that non-stem cells in tumors may be capable of promoting tumorigenicity and drug resistance as a means of replenishing the stem cell population of a tumor. Moreover, this conversion to stemness occurred not only spontaneously but also simultaneously in a large number of cells.[32] This indicates that cancer cells can replenish cancer stem cells in an organized event. Inhibiting the association of nuclear β-catenin with CBP, and the subsequent transcription of Wnt target genes, successfully thwarted this transition to stem cells.[33] Therefore, while this interconversion between non-stem and stem cell is still largely an enigma, canonical Wnt signaling must play a role in the acquisition of tumorigenicity, clonogenicity, and drug resistance.

3.2 Wnt: The Key Player in Cancer Cell Stemness

Due to Wnt's role in proliferation, potency, and differentiation, it has long been associated with stem cells. Canonical Wnt signaling regulates potency, proliferation, and differentiation in adult stem cell niches, including skin, intestinal crypts, hair follicles, mammary glands, and hematopoietic tissues.[12] In intestinal crypts, a gradient of Wnt signaling activity controls the transition of intestinal stem cells into differentiated cells of the intestinal epithelium.[26] Wnt signaling is greatest at the bottom of the crypt, where intestinal stem cells reside, and decreases moving upward toward the top of a villus. As the Wnt signal declines, the amount of potency of the cells also diminishes, resulting in more differentiated cells.[26] Therefore, Wnt signaling is positively correlated with potency and negatively correlated with differentiation in nonmalignant intestinal tissue.

In addition to maintaining stemness in adult stem cells, Wnt also plays a role in the maintenance of cancer stem cells. Aberrant Wnt signaling can transform intestinal stem cells into cancer stem cells.[34] Upregulation of Wnt signaling and increased transcription of Wnt target genes are highly associated with the cancer stem cell phenotype.[30] And, only cancer cells with high levels of canonical Wnt signaling express the characteristics of cancer stem cells, such as the capacity to self-renew and induce tumor formation.[35] Therefore, elevated Wnt signaling is not only sufficient but also necessary to maintain a population of cancer stem cells capable of clonogenesis and tumorigenesis.

Although all cells of a tumor may have the exactly same mutation in APC or β-catenin, there is a heterogeneous expression of nuclear β-catenin in colorectal tumors.[35] This difference in expression correlates with varying levels of clonogenicity and tumorigenicity. Cells with high levels of Wnt signaling have higher clonogenicity and express stem cell markers. Furthermore, these cells lose their stemness when grown in serum-containing medium, but this effect is easily reversed when grown with myofibroblasts.[35] This indicates that upregulation of Wnt signaling is not purely intrinsic but also depends on factors from the environment to increase nuclear β-catenin and cancer stemness. This should be no surprise as stem cell niches in normal intestinal crypts are maintained by surrounding epithelial cells as well as growth factors secreted by myofibroblasts.[36] Therefore, the environmental factors increasing Wnt signaling in adult stem cells may also be responsible for upregulation of Wnt signaling in cancer stem cells.

3.3 Cancer Stem Cell Markers

After isolating multipotent cancer stem cells, researchers have identified cellular markers common to these tumorigenic cells, and some of the cancer stem cell markers are direct transcriptional targets of the canonical Wnt signaling pathway. The Wnt target gene product leucine-rich repeat-containing G-protein-coupled receptor 5 (LGR5) is expressed in columnar stem cells at the bottom of intestinal crypts but not in neighboring intestinal cells higher up the crypts.[37,28] LGR5 is not only a marker of intestinal stem cells but also cancer stem cells. Its expression in colorectal cancer cells is highly associated with clonogenicity and tumorgenicity.[35] The Wnt target gene aldehyde dehydrogenase 1 (ALDH1) is also associated with the cancer stem cell phenotype.[30] Other cancer stem cell markers include CD133, CD44, CD24, CD29, CD166, Ascl2, Olfm4, EphB2, and Smoc2.[38,35,26,27] While not all of these stem cell markers are directly regulated by canonical Wnt signaling, they are all associated with increased accumulation of nuclear β-catenin, which indicates the critical role Wnt signaling plays in cancer stem cells.

4. TREATMENTS

4.1 Shortcomings of Current Treatment Regimes

Cetuximab and panitumumab are two current treatments for colorectal cancer. They are monoclonal antibody therapies that target the epidermal growth factor (EGF) receptor pathway.[39] However, these therapies are not effective in patients with KRAS or BRAF mutations.[40,41] Since KRAS mutations are typically induced after loss of function in APC mutations, as mentioned previously, it is imperative to find more effective treatments unaffected by these mutations. Due to the role of Wnt/β-catenin in cancer stem cells, tumor initiation, and drug resistance, this signaling pathway epitomizes a prime target for anticancer agents.

4.2 ICG-001: Specific β-Catenin Inhibitor

ICG-001 blocks transcription of Wnt target genes by interrupting the association between β-catenin and CBP, which is a crucial step in the β-catenin/TCF-stimulated transcription of many Wnt target genes, including cyclin D1 and survivin.[42] Even though CBP is 93% homologous to its related coactivator p300, ICG-001 does not interfere with the interaction between β-catenin and p300.[42,8] This reveals the specificity of inhibition of this powerful small-molecule inhibitor. Furthermore, ICG-001 is effective in inhibiting growth of colon cancer in vivo by increasing caspase activity in colon carcinoma cell lines.[42,43] These effects occur in colon cancer cells but not normal colonic epithelial cells.[42] The specificity and efficacy of this small-molecule inhibitor make it a promising anticancer agent for colorectal cancer.

4.3 Repurposed Drugs

While finding novel molecules as therapeutic anticancer agents remains a valuable source of new and more effective treatments, repurposing FDA-approved medications as anticancer therapies shows promising results. Furthermore, since these drugs are already FDA approved to treat other pathologies, the pharmokinetics and safety profiles are typically well known, which can expedite the transition from lab to clinical use.

4.4 Niclosamide: Antihelminth to Anticancer

Niclosamide is an oral antihelminth drug that has been used for nearly half a century to eradicate tapeworm infections.[44] Its proposed mechanism works by diminishing the potential of the inner mitochondrial membrane to inhibit oxidative phosphorylation. In addition to its recent use as a treatment for schistosomiasis, niclosamide has found a novel role in cancer therapy. Niclosamide inhibits canonical Wnt signaling via a variety of mechanisms. In colorectal cancer cells, niclosamide decreases the expression of Dvl-2 and β-catenin as well as prevents the association between β-catenin and TCF.[45,46] It also promotes endocytosis of the Wnt receptor FZD1 and decreases expression of the LRP 5/6 co-receptors.[47,48,29] This drug may also inhibit mTOR, Notch, and STAT3 pathways as well as activate the intrinsic mitochondrial apoptosis pathway in cancer cells.[29] Regardless of the exact mechanism, niclosamide inhibits proliferation of colorectal cancer cells and has little to no toxicity toward nonmalignant tissues.[45] Furthermore, niclosamide has the ability to induce apoptosis of cancer cells in both prostate and breast cancer cell lines.[49] These results indicate the potential anticancer therapeutic potential of this long-used medication.

4.5 Nitazoxanide: A Safer Relative of Niclosamide

Nitazoxanide is an FDA-approved antiprotozoal medication with a favorable pharmacokinetic and safety profile. It was discovered as an anticancer agent by the use of screening of the effects of 1600 various medications on a multicellular tumor spheroid (MCTS) 3D model that simulated the microenvironment of a tumor.[50] Of the drugs tested, promising anticancer results came from closantel, niclosamide, nitazoxanide, pyrvinium pamoate, and salinomycin. All of these compounds target mitochondria and oxidative phosphorylation. Closantel, niclosamide, and nitazoxanide are all uncouplers, which diminish the mitochondrial membrane potential. Nitazoxanide not only effectively inhibited tumor growth in vivo in colorectal cancer xenografts but also has a favorable safety profile to other medications, including niclosamide. Therefore, while niclosamide repeatedly proves to be an effective anticancer agent, its relative nitazoxanide may prove safer and more useful in clinical settings.

4.6 Silibinin

Silibinin is an active ingredient of milk thistle, which has historically been used in Chinese medicine and is now gaining recognition in Western medicine for treatment of liver disease and diabetes. In addition to decreasing the expression of β-catenin, silibinin decreases the nuclear localization of β-catenin in a dose-dependent manner.[51] It also has demonstrated the ability to inhibit tumor growth and decrease expression of Wnt target genes, cyclin D1, c-Myc, and CDK8, in colorectal cancer.[51] These effects were seen in colorectal cancer cells with APC mutations but not β-catenin mutations. Therefore, silibinin could be an effective agent against colorectal cancer in tumors with APC loss-of-function mutations and is known to be safe and well tolerated.

4.7 Monensin

Monensin is an ionophoric antibiotic used to treat bacterial, fungal, and parasitic infections. This pharmacological agent attenuates Wnt signaling, causing lower nuclear and cytoplasmic β-catenin as well as a significant decrease in the transcription of Wnt target genes.[52] Monensin inhibits the growth of colon cancer cells by inducing cell cycle arrest and apoptosis.[53] Furthermore, this drug suppressed colorectal tumor growth without negative effects on nonmalignant intestinal mucosa.[52] While studies on this antibiotic repurposed as an anticancer drug are still in early stages, this compound may quickly reach clinical trials as a repurposed medication.

4.8 Other Wnt Inhibitors

Numerous other molecules associated with inhibition of the canonical Wnt signaling pathway have demonstrated promising results in treating colorectal cancer. While these inhibitors target a wide variety of proteins in the pathway, they all display antiproliferative and antimetastatic effects in cancer cells.

Aeroplysinin-1 is a brominated tyrosine secondary metabolite derived from the marine sponge *Aplysina* species that exhibits cytotoxic effects against human cancer cells by facilitating the degradation of β-catenin.[54] Aeroplysinin-1 exhibits anticancer activity through multiple mechanisms of action. It suppresses the proliferation of EGF-dependent cancer cells by inhibiting the protein tyrosine kinase activity of the EGF receptor. This compound also decreases the transcription of Wnt target genes and the amount of cytosolic β-catenin independent of GSK-3β-mediated proteasomal degradation. While the exact mechanism is still unclear, it induces apoptosis in APC mutant colon cancer cells by increasing activity of caspases 3 and 7. However, this antiproliferative effect is not seen in colon cancers containing wild-type APC and mutant β-catenin. Future studies are required to understand the mechanism and possible shortcomings of this anticancer agent.

Glaucarubinone, a quassinoid derived from the Simaroubaceae plant family, suppresses colorectal cancer tumor growth and migration in a dose-dependent manner.[55] Ailanthinone, a derivative of glaucarubinone, also inhibits these tumor processes in colorectal cancer. Both of these compounds decrease protein levels of β-catenin, thereby decreasing canonical Wnt signaling. These derivatives also decrease the activity of hypoxia-inducible factor-1α and p21-activated kinase 1.[55] The mechanism behind this inhibition is not fully understood and requires further investigation, along with the pharmokinetics and safety profile of these compounds.

Histone deacetylase (HDAC) inhibitors have recently gained recognition for their anticancer properties, and a few are FDA approved to treat certain types of lymphomas.[56,57] The HDAC inhibitors SAHA and TSA cause downregulation of Wnt target genes and induce apoptosis in colorectal cancer cells.[58] However, the HDAC inhibitor, valproic acid, did not have the same results, indicating that selective HDAC inhibition is not as effective as universal HDAC inhibition. Inhibition of HDAC 6, 10, and 11 had the greatest impact on transcription of Wnt target genes by reducing the protein level of TCF7L2. As these potent HDAC inhibitors are studied further, they may prove to be a safe and effective therapy for colorectal cancer.[59,60]

Fentanyl, which has previously demonstrated the ability to inhibit gastric cancer progression, inhibits the growth and cell invasion of colorectal cancer cells.[61] The mechanism behind this therapeutic agent relies on its ability to decrease the expression of β-catenin. This results in decreased transcription of Wnt target genes indicated in the epithelial–mesenchymal transition that facilitates tumor invasion and metastasis of cancer cells. By inhibiting Wnt signaling, this compound is able to stop proliferation of colorectal cancer cells.

In addition to pharmacological agents, compounds found in food and drinks we normally consume may not only help treat colon cancer at a more concentrated level but also help prevent it with regular consumption. Resveratrol, which recently gained popularity due to its presence in red wine, successfully downregulates the transcription of Wnt target genes, such as c-Myc, and MMP-7, by inhibiting the nuclear localization of β-catenin.[62] It also inhibits proliferation, migration, and invasion of colorectal cells.[62] Furthermore, epigallocatechin-3-gallate (EGCG), which is a polyphenol found in green tea, inhibits the canonical Wnt signaling pathway and therefore downregulates transcription of target genes of β-catenin.[63] EGCG mediated the degradation of β-catenin via a beta-TrCP-dependent mechanism that relies on the phosphorylation of β-catenin at Ser45. However, this mechanism does not seem to depend on GSK-3β or PP2A. ECGC inhibited the growth of colorectal cells in a concentration-dependent manner.

5. MARKERS FOR EARLY DETECTION AND PROGNOSIS

5.1 Carcinoembryonic Antigen

Carcinoembryonic antigen (CEA) is currently the best-known marker in clinical use for monitoring efficacy of a patient's treatment.[39] After surgery, CEA levels should return to normal within 4–6 weeks. This marker is typically measured every 3 months for the first year postsurgery and every 6 months for the next few years. Any increase in CEA during this time may indicate metastases or infiltration.

5.2 APC and β-Catenin

Since altered expression of APC and β-catenin is common in sporadic colorectal cancer and it occurs early in the progression of cancer, the expression of APC and β-catenin in the large intestine may provide means of early detection of colon cancer. Ahearn *et al.* measured APC, β-catenin, and E-cadherin in intestinal crypts to find a correlation between their levels and risk for colorectal cancer.[64] Neither APC nor β-catenin measurements alone were statistically significant markers for colorectal cancer, but the combined APC/β-catenin score is inversely associated with the risk of colorectal cancer.[64] However, this score is also influenced by other factors, including NSAID use, physical activity, and dietary levels of folate and vitamin D. Therefore, while APC and β-catenin may be logical markers for early detection, they may not be the most sensitive or specific markers for colorectal cancer.

5.3 S100A4

S100A4 is a Wnt target gene that holds underutilized prognostic power in colorectal cancer. S100A4 expression is positively regulated by Wnt/β-catenin signaling.[65] It is a calcium-binding protein associated with metastasis in colon cancer.[46] And, it plays a critical role in cell migration and invasion of colorectal cancer cells.[65] Recent research has revealed the prognostic importance of S100A4. Patients with high S100A4 positivity survive, on average, half as long as those that are S100A4 negative.[66] In colorectal adenocarcinomas, nuclear S100A4 is inversely associated with metastasis free along with overall survival, and this prognostic significance is most pronounced in TNM stage II.[67] This correlation between S100A4 and poor prognosis is more sensitive and specific than previously used markers, such as p53 expression and tumor pT-stage.[66] These results reveal the clinical importance of S100A4 as a prognostic marker in colorectal cancer and will hopefully lead to more aggressive therapy in S100A4-positive patients.

5.4 Cancer Stem Cell Markers: CD133, CD44, ALDH1, and LGR5

Colon cancer stem cells are associated with high levels of nuclear β-catenin as well as the following markers: CD133, CD166, CD44, CD29, CD24, and Lgr5.[35] Because of the role cancer stem cells play in tumorigenesis, drug resistance, metastasis, and tumor recurrence, their markers may provide prognostic insight when treating colorectal cancer. As mentioned earlier, colon cancer-initiating cells, also referred to as cancer stem cells, contain the marker CD133.[25] CD133 is a reliable marker to identify tumorigenic and drug-resistant cancer cells.[25] While CD133 is a good marker of colorectal cancer stem cells, it is also a passive marker.[68] This means that it is present in all cancer stem cells, but its presence does not alter the function of stem cells. CD44, however, is a functional marker of cancer stem cells.[68] CD44 plays a pertinent role in tumorigenesis, and its presence is highly associated with drug resistance and poor prognosis in colorectal cancer. ALDH1 is a promising new marker for colorectal cancer stem cells. As a Wnt target gene, its expression correlates with the levels of canonical Wnt signaling. Cancer cells with ALDH1 are significantly more resistant to anticancer therapy than those lacking this marker.[30] LGR5, also a Wnt target gene, is another reliable cancer stem cell marker expressed in nonmalignant intestinal crypt cells as well as the subset of multipotent colorectal cancer cells capable of tumorigenesis.[11] Only cancer cells with the highest amount of nuclear β-catenin expressed these Wnt-dependent stem cell markers.[35] Colorectal cancers with higher expression of cancer stem cell markers were not only more aggressive but also more likely to recur after completion of treatment.[38] While only a few cancer stem cells are necessary for tumor formation, the presence of increased cancer stem cell markers is highly indicative of a poor prognosis.

6. SUMMARY

This review aims to explain key players and mechanisms involved in Wnt/β-catenin signaling and how regulation of this pathway influences cancer stem cell differentiation in colorectal cancer. Ongoing research in this field seeks to understand how aberrant β-catenin signaling associated with tumorigenesis and maintenance of stemness can be specifically targeted in the cancer cell and combined with other strategies aimed at obliterating metastatic disease and recurrence. An important question to address in identification or optimization of any new colorectal cancer treatments is "What makes a signaling molecule involved in cancer progression a good target for further therapeutic development?" The requirement in late-stage cancer progression that tumor cells must invade the local tissue and then go on to metastasize in distant sites is a coordinated process involving "outside-in" and "inside-out" signals to which the cancer cell responds.[69] β-Catenin is uniquely positioned as a cytoskeletal protein linked to membrane spanning receptors through E-cadherin, as well as a cytoplasmic transducer of Wnt signaling to the nucleus to modulate gene transcription. Ongoing efforts to elucidate the resulting genetic and epigenetic changes and their impact on molecular pathways will help tremendously in the development of more effective colorectal cancer treatments.

REFERENCES

1. Torre LA, Bray F, Siegel RL, Ferlay J, Lortet-Tieulent J, Jemal A. Global cancer statistics, 2012. *CA Cancer J Clin.* 2015;65(2):87–108. http://dx.doi.org/10.3322/caac.21262.
2. Cancer of the colon and rectum—SEER stat fact sheets. http://seer.cancer.gov/statfacts/html/colorect.html. Accessed June 20, 2016.
3. Takahashi-Yanaga F, Kahn M. Targeting Wnt signaling: can we safely eradicate cancer stem cells? *Clin Cancer Res.* 2010;16(12):3153–3162. http://dx.doi.org/10.1158/1078-0432.CCR-09-2943.
4. Van Es JH, Barker N, Clevers H. You Wnt some, you lose some: oncogenes in the Wnt signaling pathway. *Curr Opin Genet Dev.* 2003;13(1):28–33. http://dx.doi.org/10.1016/S0959-437X(02)00012-6.
5. Willert K, Brown JD, Danenberg E, et al. Wnt proteins are lipid-modified and can act as stem cell growth factors. *Nature.* 2003;423(6938):448–452. http://dx.doi.org/10.1038/nature01611.
6. MacDonald BT, Tamai K, He X. Wnt/β-catenin signaling: components, mechanisms, and diseases. *Dev Cell.* 2009;17(1):9–26. http://dx.doi.org/10.1016/j.devcel.2009.06.016.
7. Itoh K, Brott BK, Bae G-U, Ratcliffe MJ, Sokol SY. Nuclear localization is required for Dishevelled function in Wnt/beta-catenin signaling. *J Biol.* 2005;4(1):3. http://dx.doi.org/10.1186/jbiol20.
8. Teo J-L, Kahn M. The Wnt signaling pathway in cellular proliferation and differentiation: a tale of two coactivators. *Adv Drug Deliv Rev.* 2010;62(12):1149–1155. http://dx.doi.org/10.1016/j.addr.2010.09.012.
9. Marson A, Foreman R, Chevalier B, et al. Wnt signaling promotes reprogramming of somatic cells to pluripotency. *Cell Stem Cell.* 2008;3(2):132–135. http://dx.doi.org/10.1016/j.stem.2008.06.019.

10. Pinto D, Gregorieff A, Begthel H, Clevers H. Canonical Wnt signals are essential for homeostasis of the intestinal epithelium. *Genes Dev.* 2003;17(14):1709–1713. http://dx.doi.org/10.1101/gad.267103.
11. Van de Wetering M, Sancho E, Verweij C, et al. The beta-catenin/TCF-4 complex imposes a crypt progenitor phenotype on colorectal cancer cells. *Cell.* 2002;111(2): 241–250.
12. Fodde R, Brabletz T. Wnt/β-catenin signaling in cancer stemness and malignant behavior. *Curr Opin Cell Biol.* 2007;19(2):150–158. http://dx.doi.org/10.1016/j.ceb.2007.02.007.
13. Yan D, Wiesmann M, Rohan M, et al. Elevated expression of axin2 and hnkd mRNA provides evidence that Wnt/beta-catenin signaling is activated in human colon tumors. *Proc Natl Acad Sci USA.* 2001;98(26):14973–14978. http://dx.doi.org/10.1073/pnas.261574498.
14. Noda T, Nagano H, Takemasa I, et al. Activation of Wnt/β-catenin signalling pathway induces chemoresistance to interferon-α/5-fluorouracil combination therapy for hepatocellular carcinoma. *Br J Cancer.* 2009;100(10):1647–1658. http://dx.doi.org/10.1038/sj.bjc.6605064.
15. Mohammed MK, Shao C, Wang J, et al. Wnt/β-catenin signaling plays an ever-expanding role in stem cell self-renewal, tumorigenesis and cancer chemoresistance. *Genes Dis.* 2016;3(1):11–40. http://dx.doi.org/10.1016/j.gendis.2015.12.004.
16. Todaro M, Alea MP, Di Stefano AB, et al. Colon cancer stem cells dictate tumor growth and resist cell death by production of Interleukin-4. *Cell Stem Cell.* 2007;1(4):389–402. http://dx.doi.org/10.1016/j.stem.2007.08.001.
17. Polakis P. The many ways of Wnt in cancer. *Curr Opin Genet Dev.* 2007;17(1):45–51. http://dx.doi.org/10.1016/j.gde.2006.12.007.
18. Miyoshi Y, Nagase H, Ando H, et al. Somatic mutations of the APC gene in colorectal tumors: mutation cluster region in the APC gene. *Hum Mol Genet.* 1992;1(4): 229–233.
19. Kohler EM, Chandra SHV, Behrens J, Schneikert J. Catenin degradation mediated by the CID domain of APC provides a model for the selection of APC mutations in colorectal, desmoid and duodenal tumours. *Hum Mol Genet.* 2008;18(2):213–226. http://dx.doi.org/10.1093/hmg/ddn338.
20. Morin PJ, Sparks AB, Korinek V, et al. Activation of beta-catenin-Tcf signaling in colon cancer by mutations in beta-catenin or APC. *Science.* 1997;275(5307):1787–1790.
21. Johnson V. Exon 3-catenin mutations are specifically associated with colorectal carcinomas in hereditary non-polyposis colorectal cancer syndrome. *Gut.* 2005;54(2):264–267. http://dx.doi.org/10.1136/gut.2004.048132.
22. Duval A, Gayet J, Zhou XP, Iacopetta B, Thomas G, Hamelin R. Frequent frameshift mutations of the TCF-4 gene in colorectal cancers with microsatellite instability. *Cancer Res.* 1999;59(17):4213–4215.
23. Ai L. Inactivation of Wnt inhibitory factor-1 (WIF1) expression by epigenetic silencing is a common event in breast cancer. *Carcinogenesis.* 2006;27(7):1341–1348. http://dx.doi.org/10.1093/carcin/bgi379.
24. Phelps RA, Chidester S, Dehghanizadeh S, et al. A two-step model for colon adenoma initiation and progression caused by APC loss. *Cell.* 2009;137(4):623–634. http://dx.doi.org/10.1016/j.cell.2009.02.037.
25. O'Brien CA, Pollett A, Gallinger S, Dick JE. A human colon cancer cell capable of initiating tumor growth in immunodeficient mice. *Nature.* 2007;445(7123):106–110. http://dx.doi.org/10.1038/nature05372.
26. Valkenburg KC, Graveel CR, Zylstra-Diegel CR, Zhong Z, Williams BO. Wnt/β-catenin signaling in normal and cancer stem cells. *Cancers.* 2011;3(4):2050–2079. http://dx.doi.org/10.3390/cancers3022050.

27. Vermeulen L, Todaro M, de Sousa Mello F, et al. Single-cell cloning of colon cancer stem cells reveals a multi-lineage differentiation capacity. *Proc Natl Acad Sci USA*. 2008;105(36):13427–13432. http://dx.doi.org/10.1073/pnas.0805706105.
28. Barker N, Ridgway RA, van Es JH, et al. Crypt stem cells as the cells-of-origin of intestinal cancer. *Nature*. 2009;457(7229):608–611. http://dx.doi.org/10.1038/nature07602.
29. Li Y, Li P-K, Roberts MJ, Arend RC, Samant RS, Buchsbaum DJ. Multi-targeted therapy of cancer by niclosamide: a new application for an old drug. *Cancer Lett*. 2014;349(1):8–14. http://dx.doi.org/10.1016/j.canlet.2014.04.003.
30. Nagaraj AB, Joseph P, Kovalenko O, et al. Critical role of Wnt/β-catenin signaling in driving epithelial ovarian cancer platinum resistance. *Oncotarget*. 2015;6(27):23720–23734. http://dx.doi.org/10.18632/oncotarget.4690.
31. Dieter SM, Ball CR, Hoffmann CM, et al. Distinct types of tumor-initiating cells form human colon cancer tumors and metastases. *Cell Stem Cell*. 2011;9(4):357–365. http://dx.doi.org/10.1016/j.stem.2011.08.010.
32. He K, Xu T, Goldkorn A. Cancer cells cyclically lose and regain drug-resistant highly tumorigenic features characteristic of a cancer stem-like phenotype. *Mol Cancer Ther*. 2011;10(6):938–948. http://dx.doi.org/10.1158/1535-7163.MCT-10-1120.
33. He K, Xu T, Xu Y, Ring A, Kahn M, Goldkorn A. Cancer cells acquire a drug resistant, highly tumorigenic, cancer stem-like phenotype through modulation of the PI3K/Akt/β-catenin/CBP pathway: PI3K/Akt/β-catenin/CBP pathway regulates plasticity. *Int J Cancer*. 2014;134(1):43–54. http://dx.doi.org/10.1002/ijc.28341.
34. Wend P, Holland JD, Ziebold U, Birchmeier W. Wnt signaling in stem and cancer stem cells. *Semin Cell Dev Biol*. 2010;21(8):855–863. http://dx.doi.org/10.1016/j.semcdb.2010.09.004.
35. Vermeulen L, De Sousa E Melo F, van der Heijden M, et al. Wnt activity defines colon cancer stem cells and is regulated by the microenvironment. *Nat Cell Biol*. 2010;12(5):468–476. http://dx.doi.org/10.1038/ncb2048.
36. Fanali C. Cancer stem cells in colorectal cancer from pathogenesis to therapy: controversies and perspectives. *World J Gastroenterol*. 2014;20(4):923. http://dx.doi.org/10.3748/wjg.v20.i4.923.
37. Barker N, van Es JH, Kuipers J, et al. Identification of stem cells in small intestine and colon by marker gene Lgr5. *Nature*. 2007;449(7165):1003–1007. http://dx.doi.org/10.1038/nature06196.
38. Merlos-Suárez A, Barriga FM, Jung P, et al. The intestinal stem cell signature identifies colorectal cancer stem cells and predicts disease relapse. *Cell Stem Cell*. 2011;8(5):511–524. http://dx.doi.org/10.1016/j.stem.2011.02.020.
39. Lech G. Colorectal cancer tumor markers and biomarkers: recent therapeutic advances. *World J Gastroenterol*. 2016;22(5):1745. http://dx.doi.org/10.3748/wjg.v22.i5.1745.
40. Amado RG, Wolf M, Peeters M, et al. Wild-type KRAS is required for panitumumab efficacy in patients with metastatic colorectal cancer. *J Clin Oncol*. 2008;26(10):1626–1634. http://dx.doi.org/10.1200/JCO.2007.14.7116.
41. Di Nicolantonio F, Martini M, Molinari F, et al. Wild-type BRAF is required for response to panitumumab or cetuximab in metastatic colorectal cancer. *J Clin Oncol*. 2008;26(35):5705–5712. http://dx.doi.org/10.1200/JCO.2008.18.0786.
42. Emami KH, Nguyen C, Ma H, et al. A small molecule inhibitor of beta-catenin/CREB-binding protein transcription [corrected]. *Proc Natl Acad Sci USA*. 2004;101(34):12682–12687. http://dx.doi.org/10.1073/pnas.0404875101.
43. Arensman MD, Telesca D, Lay AR, et al. The CREB-binding protein inhibitor ICG-001 suppresses pancreatic cancer growth. *Mol Cancer Ther*. 2014;13(10):2303–2314. http://dx.doi.org/10.1158/1535-7163.MCT-13-1005.

44. Pan J-X, Ding K, Wang C-Y. Niclosamide, an old antihelminthic agent, demonstrates antitumor activity by blocking multiple signaling pathways of cancer stem cells. *Chin J Cancer*. 2012;31(4):178–184. http://dx.doi.org/10.5732/cjc.011.10290.
45. Osada T, Chen M, Yang XY, et al. Antihelminth compound niclosamide downregulates Wnt signaling and elicits antitumor responses in tumors with activating APC mutations. *Cancer Res*. 2011;71(12):4172–4182. http://dx.doi.org/10.1158/0008-5472.CAN-10-3978. 1.
46. Sack U, Walther W, Scudiero D, et al. Novel effect of antihelminthic niclosamide on S100A4-mediated metastatic progression in colon cancer. *J Natl Cancer Inst*. 2011;103(13):1018–1036. http://dx.doi.org/10.1093/jnci/djr190.
47. Arend RC, Londoño-Joshi AI, Samant RS, et al. Inhibition of Wnt/β-catenin pathway by niclosamide: a therapeutic target for ovarian cancer. *Gynecol Oncol*. 2014;134(1):112–120. http://dx.doi.org/10.1016/j.ygyno.2014.04.005.
48. Chen M, Wang J, Lu J, et al. The anti-helminthic niclosamide inhibits Wnt/Frizzled1 signaling. *Biochemistry*. 2009;48(43):10267–10274. http://dx.doi.org/10.1021/bi9009677.
49. Lu W, Lin C, Roberts MJ, Waud WR, Piazza GA, Li Y. Niclosamide suppresses cancer cell growth by inducing Wnt co-receptor LRP6 degradation and inhibiting the Wnt/β-catenin pathway. *PLoS One*. 2011;6(12):e29290. http://dx.doi.org/10.1371/journal.pone.0029290.
50. Senkowski W, Zhang X, Olofsson MH, et al. Three-dimensional cell culture-based screening identifies the anthelmintic drug nitazoxanide as a candidate for treatment of colorectal cancer. *Mol Cancer Ther*. 2015;14(6):1504–1516. http://dx.doi.org/10.1158/1535-7163.MCT-14-0792.
51. Kaur M, Velmurugan B, Tyagi A, Agarwal C, Singh RP, Agarwal R. Silibinin suppresses growth of human colorectal carcinoma SW480 cells in culture and xenograft through down-regulation of β-catenin-dependent signaling. *Neoplasia*. 2010;12(5):415–424. http://dx.doi.org/10.1593/neo.10188.
52. Tumova L, Pombinho AR, Vojtechova M, et al. Monensin inhibits canonical Wnt signaling in human colorectal cancer cells and suppresses tumor growth in multiple intestinal neoplasia mice. *Mol Cancer Ther*. 2014;13(4):812–822. http://dx.doi.org/10.1158/1535-7163.MCT-13-0625.
53. Park WH, Kim ES, Jung CW, Kim BK, Lee YY. Monensin-mediated growth inhibition of SNU-C1 colon cancer cells via cell cycle arrest and apoptosis. *Int J Oncol*. 2003;22(2):377–382.
54. Park S, Kim J-H, Kim JE, et al. Cytotoxic activity of aeroplysinin-1 against colon cancer cells by promoting β-catenin degradation. *Food Chem Toxicol*. 2016;93:66–72. http://dx.doi.org/10.1016/j.fct.2016.04.019.
55. Huynh N, Beutler JA, Shulkes A, Baldwin GS, He H. Glaucarubinone inhibits colorectal cancer growth by suppression of hypoxia-inducible factor 1α and β-catenin via a p-21 activated kinase 1-dependent pathway. *Biochim Biophys Acta*. 2015;1853(1):157–165. http://dx.doi.org/10.1016/j.bbamcr.2014.10.013.
56. Insinga A, Monestiroli S, Ronzoni S, et al. Inhibitors of histone deacetylases induce tumor-selective apoptosis through activation of the death receptor pathway. *Nat Med*. 2005;11(1):71–76. http://dx.doi.org/10.1038/nm1160.
57. Licciardi PV, Ververis K, Hiong A, Karagiannis TC. Histone deacetylase inhibitors (HDACIs): multitargeted anticancer agents. *Biologics*. 2013;7:47–60. http://dx.doi.org/10.2147/BTT.S29965.
58. Gotze S, Coersmeyer M, Muller O, Sievers S. Histone deacetylase inhibitors induce attenuation of Wnt signaling and TCF7L2 depletion in colorectal carcinoma cells. *Int J Oncol*. 2014;45:1715–1723. http://dx.doi.org/10.3892/ijo.2014.2550.

59. Holloway KR, Calhoun TN, Saxena M, et al. SIRT1 regulates Dishevelled proteins and promotes transient and constitutive Wnt signaling. *Proc Natl Acad Sci USA*. 2010;107(20):9216–9221. http://dx.doi.org/10.1073/pnas.0911325107.
60. Saxena M, Dykes SS, Malyarchuk S, Wang AE, Cardelli JA, Pruitt K. The sirtuins promote Dishevelled-1 scaffolding of TIAM1, Rac activation and cell migration. *Oncogene*. 2015;34(2):188–198. http://dx.doi.org/10.1038/onc.2013.549.
61. Zhang X-L, Chen M-L, Zhou S-L. Fentanyl inhibits proliferation and invasion of colorectal cancer via β-catenin. *Int J Clin Exp Pathol*. 2015;8(1):227–235.
62. Ji Q, Liu X, Fu X, et al. Resveratrol inhibits invasion and metastasis of colorectal cancer cells via MALAT1 mediated Wnt/β-catenin signal pathway. *PLoS One*. 2013;11: e78700. http://dx.doi.org/10.1371/journal.pone.0078700.
63. Oh S, Gwak J, Park S, Yang CS. Green tea polyphenol EGCG suppresses Wnt/β-catenin signaling by promoting GSK-3β- and PP2A-independent β-catenin phosphorylation/degradation: EGCG suppresses Wnt/β-catenin signaling. *Biofactors*. 2014;40(6): 586–595. http://dx.doi.org/10.1002/biof.1185.
64. Ahearn TU, Shaukat A, Flanders WD, Seabrook ME, Bostick RM. Markers of the APC/β-catenin signaling pathway as potential treatable, preneoplastic biomarkers of risk for colorectal neoplasms. *Cancer Epidemiol Biomarkers Prev*. 2012;21(6):969–979. http://dx.doi.org/10.1158/1055-9965.EPI-12-0126.
65. Stein U, Arlt F, Walther W, et al. The metastasis-associated gene S100A4 is a novel target of β-catenin/T-cell factor signaling in colon cancer. *Gastroenterology*. 2006;131(5): 1486–1500. http://dx.doi.org/10.1053/j.gastro.2006.08.041.
66. Gongoll S, Peters G, Mengel M, et al. Prognostic significance of calcium-binding protein S100A4 in colorectal cancer. *Gastroenterology*. 2002;123(5):1478–1484.
67. Boye K, Nesland JM, Sandstad B, Mælandsmo GM, Flatmark K. Nuclear S100A4 is a novel prognostic marker in colorectal cancer. *Eur J Cancer*. 2010;46(16):2919–2925. http://dx.doi.org/10.1016/j.ejca.2010.07.013.
68. Kemper K, Prasetyanti PR, De Lau W, Rodermond H, Clevers H, Medema JP. Monoclonal antibodies against Lgr5 identify human colorectal cancer stem cells. *Stem Cells*. 2012;30(11):2378–2386. http://dx.doi.org/10.1002/stem.1233.
69. McCrea PD, Maher MT, Gottardi CJ. Chapter five—nuclear signaling from cadherin adhesion complexes. In: Yap AS, ed. *Current Topics in Developmental Biology*. Academic Press; 2015:129–196. Cellular Adhesion in Development and Disease; vol. 112.

CHAPTER THREE

Interplay Between Inflammation and Epigenetic Changes in Cancer

A.R. Maiuri*, H.M. O'Hagan*,†,1
*Medical Sciences, Indiana University School of Medicine, Bloomington, IN, United States
†Indiana University Melvin and Bren Simon Cancer Center, Indianapolis, IN, United States
[1]Corresponding author: e-mail address: hmohagan@indiana.edu

Contents

Progress in Molecular Biology and Translational Science, Volume 144
ISSN 1877-1173
http://dx.doi.org/10.1016/bs.pmbts.2016.09.002

Abstract

Immune responses can suppress tumorigenesis, but also contribute to cancer initiation and progression suggesting a complex interaction between the immune system and cancer. Epigenetic alterations, which are heritable changes in gene expression without changes to the DNA sequence, also play a role in carcinogenesis through silencing expression of tumor suppressor genes and activating oncogenic signaling. Interestingly, epithelial cells at sites of chronic inflammation undergo DNA methylation alterations that are similar to those present in cancer cells, suggesting that inflammation may initiate cancer-specific epigenetic changes in epithelial cells. Furthermore, epigenetic changes occur during immune cell differentiation and participate in regulating the immune response, including the regulation of inflammatory cytokines. Cancer cells utilize epigenetic silencing of immune-related genes to evade the immune response. This chapter will detail the interactions between inflammation and epigenetics in tumor initiation, promotion, and immune evasion and how these connections are being leveraged in cancer prevention and treatment.

1. INTRODUCTION

Inflammation and cancer are intricately linked phenomena. People with chronic inflammatory diseases have an increased risk of developing cancer and interestingly, long-term consumption of drugs that suppress inflammation has been shown to reduce cancer incidence and mortality.[1,2] Although inflammation undeniably plays a critical role in cancer pathogenesis, the relationship between the immune system and cancer is very complex. Both arms of the immune system, innate and adaptive, play conflicting roles in the development of cancer and the specific roles that each arm plays are context dependent.

Genetic mutations and epigenetic alterations are essential to cancer initiation and progression. The genetic abnormalities leading to cancer have been well studied whereas the causal role of epigenetic changes in cancer is not as well proven. However, with increased identification of cancer-specific mutations in epigenetic modifiers and the effectiveness of epigenetic inhibitors in treating certain cancers, the role of epigenetic alterations in cancer is becoming more clearly defined.[3] In addition to their involvement in cancer, epigenetic alterations are also essential to normal human development. In particular, epigenetic events drive embryonic development and also contribute to immune cell differentiation. The mechanisms underlying how epigenetic alterations are initiated are not well characterized; however, several studies suggest that inflammation plays an important role.

The beginning of this chapter will provide an overview of the components (cell types and inflammatory mediators) of the innate and adaptive immune systems. Additionally, the dichotomous roles played by each arm of the immune system in the pathogenesis of cancer will be addressed. In the second half of this chapter, we provide an overview of epigenetics and discuss the involvement of epigenetic alterations in normal physiological processes and cancer. Studies that provide mechanistic insight concerning how inflammation causes epigenetic alterations are also discussed in detail. In addition, we review the current methods that physicians are using to target immune and epigenetic factors pharmacologically to treat cancer.

2. COMPLEX RELATIONSHIP BETWEEN THE IMMUNE SYSTEM AND CANCER

The immune system protects the human body against infection and responds to cellular damage. The goal of the immune response is to clear infection, heal the injury, and restore tissue homeostasis. However, chronic activation of the immune response can be protumorigenic with many chronic inflammatory diseases increasing cancer risk.[4] Because tumors share features with injured tissue, they also are characterized by chronic inflammation that can aid in the progression of cancer and is now included as a "hallmark of cancer."[5] As in other disease states, the immune system plays an initial role in decreasing tumorigenesis, but once the immune response becomes chronic it becomes detrimental. Epidemiological studies suggest that 15% of cancers are associated with chronic inflammation.[4] Here we will discuss the complex relationship between the immune system and cancer and how the immune response can be anti- or protumorigenic depending on the context.

2.1 Innate Immunity

The first response to an infection is from the innate immune system, which consists of the epithelial cell barrier and immune cells, including macrophages, neutrophils, dendritic cells (DCs), and natural killer (NK) cells, and causes inflammation. The jobs of the immune cells are to identify and eliminate the infection, but they do so in different ways. Macrophages, DCs, and mast cells reside in tissue normally and initiate the acute inflammatory response by producing chemoattractants, cytokines, and inflammatory mediators, which induce inflammation and recruit neutrophils, monocytes (precursors to macrophages and DCs), and DCs.[4]

Macrophages and neutrophils have cell surface pattern-recognition receptors called Toll-like receptors (TLRs) that bind conserved regions of pathogen-associated molecules.[6] These cells phagocytose the pathogen and kill it or when activated can release high levels of reactive oxygen species (ROS) and reactive nitrogen species (RNS) into the tissue surrounding the pathogen, which will kill the pathogens. Two macrophage phenotypes have been extensively characterized and this polarization is influenced by a combination of cytokine signaling and resulting transcription factor (TF) activation. M1 macrophages (classically activated) are induced by the cytokine interferon-gamma (IFN-γ), and TLR ligands and M2 macrophages (alternatively activated) are induced by Th2 cytokines.[7] These different subsets of macrophages are functionally distinct in that they release different factors (cytokines, chemokines, etc.).

Early activation of the innate immune response is also associated with inflammation, which increases blood flow to the site of infection increasing the amount of circulating leukocytes in the area. Vascular permeability is also increased to allow leukocytes to exit circulation and enter the inflamed tissue. DCs are antigen-presenting cells that stimulate priming of the adaptive immune response, including T and B cells. NK cells are innate lymphoid cells that recognize cellular stress through activation of the receptors NK group 2D (NKG2D) and DNAX accessory molecule-1.[8] NK cells contain small proteins in their cytoplasm that they release in proximity to the cell they need to kill. One of these proteins, perforin, forms pores in the cell membrane of the target cell allowing other proteases called granzymes to enter the cell, causing cell death. NK cells also directly kill bacteria by secreting antimicrobial proteins. γδ T cells are unique T cells that, in a naïve state, can be found in tissues such as the dermis, intestine, lungs, and uterus and do not require traditional antigen presentation to be activated.[9] They respond to infections in large numbers without extensive clonal expansion by recognizing key pathogen-associated molecules. Once activated they can act as antigen-presenting cells causing the activation of naïve CD4+ and CD8+ T cells and therefore are thought to bridge the innate and adaptive immune systems.

The inflammatory environment produced by the innate immune response also affects epithelial, endothelial, and mesenchymal cells in the surrounding environment.[4] Cytokines induce signaling in epithelial cells that results in the production of more cytokines and epithelial proliferation. ROS and RNS damage lipids, proteins, and DNA within epithelial cells causing the activation of various cellular stress response pathways within the epithelial cells.

Inflammasomes are large complexes of proteins that can sense pathogen- or damage-associated patterns and they are present in DCs, macrophages, and epithelial cells. When activated they initiate immune responses by promoting secretion of proinflammatory cytokines IL-1β, IL-18, IL-33, and FGF-2 and by promoting cell death.

The innate immune response activates TFs in immune cells and surrounding epithelial cells, including signal transducer and activator of transcription 3 (STAT3). STAT3 is activated in response to activation of cytokine or growth factor receptors or through activation of TLRs and then translocates into the nucleus where it acts as a TF. STAT3 induces further cytokine and growth factor expression crucial for the innate immune response.[10] STAT3 is often constitutively activated in cancer cells due to high levels of IL-6 signaling, abnormal signaling of various growth factor receptors, or to loss of negative regulators such as suppressor of cytokine signaling 3 (SOCS3).[10] Nuclear factor kappa-light-chain-enhancer of activated B cells (NF-κB) proteins function as dimeric TFs that regulate expression of genes involved in the innate and adaptive immune responses.[11] In unstimulated cells, NF-κB proteins are retained in the cytoplasm by inhibitor of κB (IκB). Ligand binding-mediated activation of receptors triggers activation of protein kinases that result in the degradation of IκB and subsequent nuclear localization of NF-κB. NF-κB then regulates the expression of many genes and induces proliferation, suppression of apoptosis, development of immune cells, and an increased innate immune response. NF-κB is also often constitutively active in cancer cells maintaining signals for proliferation, inhibition of apoptosis and angiogenesis.[11]

2.2 Innate Immune Response in Tumors

In tumors, even when there is no pathogen present, the innate immune response can limit cancer progression by responding to stress- or damage-associated molecular patterns present in the tumors. Macrophages and DC can eliminate apoptotic tumor cells by recognizing motifs that become present on these cells, but unlike in response to pathogens, they generally do so without generating proinflammatory cytokines. NK cells can be activated through the NKG2D receptor by cellular stress or DNA damage in tumor cells.[12] They can infiltrate solid tumors and in mouse models have been demonstrated to control tumor growth. Activated NK cells kill tumors in similar ways they kill pathogens, through the release of perforin and granzymes.[13] They also secrete IFN-γ (type II IFN), which can activate cell death signaling pathways and contribute to activation of cytotoxic T cells. The lysis of tumor cells can expose DCs to tumor antigens that will induce DC activation. γδ T cells are similarly involved in the antitumor immune response. Activated γδ T cells can be cytolytic, produce effector cytokines, and activate DCs.[12]

However, the innate immune response can also be protumorigenic. Tumor-associated macrophages (TAMs) are major contributors to chronic inflammation in the tumor microenvironment and they promote tumor growth by releasing factors such as RNS/ROS, growth factors, and cytokines IL-6 and IL-1β. Key chemokines expressed in the tumor microenvironment cause the infiltration and activation of TAMs.[13] TAMs also produce factors that increase angiogenesis, tumor invasion, and metastasis. The observation that macrophages can be polarized into two different states, M1 or M2, might explain why they behave differently in tumors compared to normal tissue.[13] M1 macrophages produce type I proinflammatory cytokines and are generally antitumorigenic. M2 macrophages produce type II antiinflammatory cytokines making them generally protumorigenic. However, protumorigenic M2 TAMs can be converted into M1 macrophages when exposed to IFN-γ. In tumors, the lack of expression of certain inflammatory cytokines such as type I IFNs causes DCs and NK cells present to remain in an inactive state thereby preventing the efficient activation of the adaptive immune response. For example, in CD8+ DC, uptake of antigens in the absence of certain inflammatory cytokines (type I IFNs) does not cause induction of adaptive immunity.[12]

Myeloid-derived suppressor cells (MDSCs) are precursors of DCs, macrophages, and granulocytes. Their normal role is to protect the host from harmful effects of excessive immune responses during infections and prevent autoimmunity. They are elevated in circulation in response to pathogen infection and tumors. Once at the tumor site, these cells suppress immune surveillance of cancer cells.[14] They inhibit proliferation and activation of CD4+ and CD8+ T cells, promote M2 polarization of TAMs, and inhibit the cytotoxicity of NK cells.

2.3 Adaptive Immunity

The adaptive immune response is highly specific and provides long-term protection against specific pathogens. Naïve lymphocytes with antigen receptors for unique antigens are located in lymph nodes. Antigen-presenting cells of the innate immune system activate lymphocytes specific for their antigen causing them to proliferate and then differentiate into effector cells. There are two broad classes of effector cells, B cells and T cells. B cells secrete antibodies, which circulate in the bloodstream. They bind specifically to the foreign antigen that stimulated their production, targeting the invading pathogen to be recognized more easily and to be destroyed by the cells of the innate immune system. T cell differentiation is initiated when the T cell receptor (TCR) and costimulatory molecules of naïve T cells become activated by antigen-presenting cells. This leads to activation of the TF nuclear factor of activated T cells (NFAT) and production of interleukin-2 (IL-2) by the naïve T cell. IL-2 induces activation of STAT5, which cooperates with additional TFs to facilitate T cell differentiation.[15] Naïve T cells recognize their specific antigen presented by antigen-presenting cells through major histocompatibility complex (MHC) II or normal cells through MHC I. Cytotoxic T cells are able to directly destroy infected cells, express CD8, and mainly recognize antigens displayed on MHC I molecules. Helper T cells activate other immune cells, express CD4, and mainly recognize antigens displayed by MHC II. There are distinct types of helper T cells and lineage commitment is ultimately determined by the extracellular cytokine milieu. For instance, the presence of IFN-γ promotes differentiation of naïve T cells into Th1 cells via activation of STAT1. STAT1 cooperates with the TCR-induced TFs NFAT and STAT5 to induce the production of IFN-γ by the T cell.[15] Additionally, other TFs are activated including STAT4, T-bet, H2.0-like homeobox (HLX), and runt-related TF 3 (RUNX3) which bind to the IFN-γ gene inducing its activation and further reinforcing Th1-lineage commitment.[16]

Alternatively, the presence of IL-4 in the extracellular milieu promotes Th2 cell lineage commitment by activating STAT6, which cooperates with NFAT and STAT5 to promote activation of the TF GATA3.[15] GATA3 induces activation of another TF, MAF, which cooperates with STAT6 to promote transcription of IL-4, IL-5, and IL-13 by the T cell further reinforcing its commitment to the Th2 lineage.[17] Differentiation of naïve T cells into Th17 cells follows a similar pattern, however, transforming growth factor beta (TGFβ) and IL-6 are responsible for initiating this process. TGFβ and IL-6 lead to activation of STAT3, which promotes an increase in expression of the TF RORγt which is critical to the maturation of Th17 cells.[18]

Th1 and Th17 subsets secrete factors that recruit and activate macrophages and B cells and promote inflammation. Th17 cells promote host defenses against infectious agents and are important in maintaining barrier immunity at mucosal surfaces of the gut and lung.[19] Th2 cells stimulate IgE production and eosinophils and are generally activated by an allergic response or parasitic infections. The process from activation of the naïve lymphocyte to active effector cells takes 4–5 days delaying the adaptive immune response. Regulatory Forkhead box protein P3 (FoxP3)+T cells (Tregs) repress the activity of the other T cells and B cells to turn off the immune response once the infection is under control and are required for immunological self-tolerance.[20]

2.4 Adaptive Immune Response in Tumors

The adaptive immune response plays a role in preventing tumors from developing. Mice deficient in recombination activating gene 2 are incapable of producing functional T, B, or NK cells and are less capable of preventing carcinogen-induced tumors than wild-type mice.[21] Furthermore, IFN-γ is produced mainly by CD4+ T cells and IFN-γ knockout mice are more susceptible to developing certain cancers.[22] The immune system can recognize aspects of tumor cells as nonself in similar ways as is done with pathogens. Most tumor antigens are presented by MHC class I molecules and therefore activate cytotoxic CD8+ T cells. Tumor-specific antigens can be neoantigens generated by point mutations in normal genes and therefore highly mutagenic tumors typically produce more neoantigens.[23] Products of oncogenes or genes that are normally silenced can also be tumor-specific antigens.

The tumors that manage to evade the immune system become cancer. A major subset of these cancers shows evidence of T cell priming and immune infiltration into tumor sites, including CD8 + T cells, macrophages, NK cells, and some B cells.[12] The tumor microenvironment contains corresponding expression of chemokines that likely recruit T cells, antigen-presenting cell markers, and a type I IFN signature. Some of the CD8 + cells recognize tumor antigens suggesting appropriate priming.[24] However, the immune failure is present at the effector phase possibly through a dominant-negative effect of immune inhibition in the tumor microenvironment. Immune inhibition is indicated by the presence of Tregs, ligands on the tumor cell surface for inhibitory receptors on activated T cells and indoleamine-2,3-dioxgenase (IDO), an enzyme associated with peripheral tolerance.[25] Another class of tumors seems to have exclusion of T cells from the tumor microenvironment possibly because of a lack of chemokines and other factors required for T cell migration or by the tumor producing high levels of inhibitory cytokines.[12]

Immune cells present in tumors produce cytokines that can be pro- or antitumorigenic. Th1 cells produce proinflammatory cytokines such IFN-γ, which is also produced by CD8+ T cells and NK cells and limits tumor growth. IFN-γ induces activation of antigen-presenting cells and CD8+ T cell cytotoxicity. Th2 cells secrete antiinflammatory cytokines and inhibit CD8+ T cell cytotoxicity making them protumorigenic and immunosuppressive. Th17 cells produce proinflammatory cytokines, including IL-17A, IL-17F, IL-21, IL-22, and depending on the experimental context, appear to be capable of being pro- or antitumorigenic. TAMs can induce Th17 cells in the tumor microenvironment because they express higher levels of IL-1β than normal macrophages and it is suggested that IL-1β plays a crucial role in promoting Th17 development.[26] Th17 cells express chemokines that recruit more DCs into the tumor microenvironment.[19] Th17 cells are also positively correlated with effector immune cells such as IFN-γ + T cells, cytotoxic T cells, and NK cells and this recruitment of other immune cells may be why in some cases Th17 cells are antitumorigenic. Additionally, Th17 cells can be plastic and gain features of Th1 cells, which are generally considered to be antitumorigenic. IL-17 deficiency has been shown to be protumorigenic, while enhanced Th17 improves antitumor immunity in some settings. Conversely, IL-17 has been shown to be protumorigenic in other mouse models.[27] IL-17 can induce IL-6 production in the tumor and microenvironment, which activates STAT3 and prosurvival and angiogenesis pathways.

3. CHRONIC INFLAMMATORY DISEASES PREDISPOSE INDIVIDUALS TO CANCER

While acute inflammation is needed to clear infections and repair tissue damage, the inability to turn off the inflammatory response or its repeated stimulation results in chronic deregulated inflammation that is detrimental to the organism. Chronic inflammation can be induced by environmental exposures such as asbestos, cigarette smoke, UV light; chronic infections such as *Helicobacter pylori* (*H. pylori*), gingivitis, or viral infections; or chronic inflammatory disease such as inflammatory bowel disease (IBD), gastroesophageal reflux disease, and chronic pancreatitis. The repetitive immune response causes immune cells to accumulate, epithelial cells to be exposed repeatedly to high levels of ROS/RNS and cytokines, and tissue repair functions to become excessive, creating a protumorigenic microenvironment.[1] Exposure of epithelial cells to IL-1β and tumor necrosis factor-alpha (TNF-α) induces the activation of NF-κB, a TF that induces epithelial cell proliferation, survival and cytokine, and ROS production. One of the induced cytokines, IL-6, is a potent activator of STAT3, which can be protumorigenic. These cytokines also activate other signaling pathways including the mitogen-activated protein kinases (MAPK) such as c-Jun N-terminal kinase. The inflammatory microenvironment also contributes to tumor progression by providing growth factors and cytokines that are protumorigenic.[4] Once tumors are formed most become associated with an inflammatory microenvironment, which can further promote tumor growth and cancer metastasis. The microenvironment contains inflammatory cells and their mediators, including chemokines, cytokines, and prostaglandins. Here we will briefly review some well-established chronic inflammatory diseases that predispose to cancer.

3.1 *H. pylori* Infection and Gastric Cancer

The gastrointestinal tract is inhabited by many microorganisms and is exposed to chemical agents directly through ingestion. Thus, there is constant immune surveillance of these tissues and many cancers in digestive organs are associated with inflammation. *H. pylori* are gram-negative bacteria that colonize the gastric epithelium. *H. pylori* infection can induce chronic gastritis and these patients have a higher risk of developing gastric cancer than patients who are *H. pylori* negative.[28] The mucosa in *H. pylori*-induced gastritis has increased levels of proinflammatory Th1 cells, neutrophils, and cytokines.[29] The cytokines IL-1β and TNF-α enhance NF-κB activation whereas IL-6 and IL-11 activate STAT3.[30] Further evidence that stomach inflammation increases cancer risk is from the observation that increased IL-1β expression in the stomachs of mice causes stomach inflammation and cancer.[31] Some strains of *H. pylori* contain the cytotoxin-associated gene A (*cagA*), which encodes for a toxin that is translocated into the host cell. Infection with *cagA*-positive *H. pylori* increases stomach cancer risk even further.[32] Transgenic expression of *cagA* in mice is sufficient to induce proliferation of epithelial cells and cancer formation.[33] *cagA*+ *H. pylori* may also promote gastric cancer by *cagA* directly affecting epithelial cell signaling and cell–cell adhesion, promoting cell proliferation, and cell migration.[34]

3.2 Hepatitis C Virus (HCV) and Liver Cancer

Hepatocellular carcinoma is highly associated with chronic hepatitis or cirrhosis and HCV infections, with the incidence of hepatocellular carcinoma increasing with the length of HCV infection. HCV is a single-stranded RNA virus that replicates in hepatocytes. Infection with the virus induces a CD8+ T cell response, but in many cases the response is not successful in clearing the virus, resulting in chronic inflammation.[35] The sites of chronic inflammation have high levels of ROS and RNS from the immune cells, but HCV proteins in infected cells also increase levels of ROS/RNS and induce activation of NF-κB and STAT3.[36] There is a high rate of hepatocellular turnover because the immune system or the virus induces apoptosis of infected cells, which are then replaced by increased hepatocellular proliferation.[37] In an HCV-infected liver, only a small number of hepatocytes contain detectable HCV. Therefore, even though HCV may have direct effects on DNA repair and proliferation in the cells it infects, the chronic hepatitis, cirrhosis, and hepatocellular carcinoma may be driven by the persistent immune response and not direct effects of the virus.[37]

3.3 Barrett's Esophagus and Esophageal Cancer

The major risk factors for esophageal adenocarcinoma are gastroesophageal acid reflux disease (GERD) and Barrett's esophagus.[38] These diseases all involve inflammation of the esophageal squamous epithelium and are thought to be progressive from one to the other. Barrett's esophagus occurs when GERD progresses to metaplasia, which can then progress to dysplasia and ultimately esophageal adenocarcinoma. Bile acids and acid in the reflux cause esophageal mucosal damage and epithelial injury. The cellular damage is repaired by proliferation of squamous cells or replacement with Barrett's cells, which are columnar cells that are more resistant to toxic agents.[39] The chronic inflammation in the esophagus is associated with increased levels of ROS, RNS, cytokines, and chemokines. The metaplasia of Barrett's esophagus contains Th2 and Th1 effector cells and elevated levels of IL-4, IL-8, IL-10, and IL-1β.[40] As in other tissue, this environment is associated with increased proliferation, invasion, and angiogenesis. NF-κB is rarely activated in GERD, but its activation is increased in Barrett's esophagus and esophageal adenocarcinoma along with activation of STAT3.[41,42]

3.4 IBD and Colorectal Cancer (CRC)

Crohn's disease (CD) and ulcerative colitis (UC) are two forms of IBD. The initial cause of IBD is not always clear, but IBD is associated with the intestinal microbiome composition, the immune system, genetics of the patient and environmental factors. Individuals develop colitis-associated colorectal cancer (CAC) at a younger age than sporadic CRC.[43] Like sporadic CRC, CACs develop from a dysplastic lesion. However, in CAC the lesion can be polypoid or flat, localized or multifocal, making surgical removal of the dysplastic lesion more difficult.[44] The sequence of molecular events leading to cancer appear to be different in CACs, with p53 mutations occurring early in the process and adenomatous polyposis coli (APC) mutations occurring later, the opposite of what tends to happen in sporadic CRC.[44] The risk of CAC increases with the length of time that a patient has UC and the extent of the disease suggesting that inflammation is the driver of CAC formation.[43] Furthermore, there is some evidence that antiinflammatory medications can reduce CAC.[45] Mouse models of colon inflammation-driven CRC have been useful in demonstrating the causality and understanding the role of the immune system and associated oxidative stress. A chemical irritant, dextran sulfate sodium (DSS) is an agent often used experimentally to induce colitis in mice that mimics IBD in humans. Repeated exposure to DSS induces colon inflammation and neoplasia in wild-type mice and mice predisposed to developing intestinal tumors.[46] Antioxidants in these mouse models reduce inflammation and tumor incidence. Mouse models of IBD have also been developed using knockout mice and toxigenic bacteria that further link the intestinal immune response to tumor formation.[27,47]

3.5 Ultraviolet (UV) Radiation and Skin Cancer

UV radiation is the main cause of skin cancer and causes injury to the epithelial layer of the skin, the epidermis.[48] This injury is followed by inflammation, proliferation, and wound healing. Exposure of skin to physical, chemical, or biological agents causes the infiltration of neutrophils into the injured epidermis. Neutrophils release ROS/RNS at the site. ROS/RNS can activate cell proliferation and survival signaling pathways involving NF-κB, activator protein-1, ERK/MAPK, PI3K/Akt, and STAT3. This environment may initiate cellular transformation and/or provide a microenvironment in which initiated cells can proliferate and lead to tumor promotion. While mouse models have demonstrated that inflammation can drive skin cancer, most skin cancer causing agents, including UV, damage DNA directly as well as induce inflammation, making it more difficult to directly demonstrate that inflammation plays a causative role in skin cancer as in some of the other diseases above.[49]

Chronic inflammation establishes an environment with many pro-tumorigenic cues including cell signaling activation, cell proliferation, migration, and angiogenesis. However, for normal cells to be transformed they have to gain genetic and/or epigenetic alterations. The relationship between epigenetic changes, inflammation and cancer are the focus of the remainder of this chapter.

4. OVERVIEW OF EPIGENETICS AND ITS ROLE IN NORMAL PHYSIOLOGICAL PROCESSES AND CANCER

4.1 Overview of Epigenetics

Epigenetics refers to the study of heritable factors, other than the underlying DNA sequence itself, that regulate gene expression. Several molecular factors have been identified that play an important role in regulating expression of genes, including noncoding RNAs (ncRNAs), histone modifications, and DNA methylation. The following sections will briefly discuss each of these factors.

4.1.1 Noncoding RNAs

A ncRNA is a RNA molecule that is not translated into protein. In the 1960s, it was hypothesized that ncRNAs might participate in regulating gene expression.[50,51] Several classes of ncRNAs have been defined including microRNAs (miRNAs), small interfering RNAs, long ncRNAs, ribosomal

RNA (rRNA), and others each with distinct functions.[52] ncRNAs can interact with a variety of molecules including RNA, DNA, and protein and these various RNA–RNA, RNA–DNA, and RNA–protein complexes play roles in regulating gene expression. For instance, ncRNAs can cooperate with histone- and DNA-modifying enzymes thereby influencing chromatin structure and consequently gene expression.[53] One classic example by which ncRNAs can influence chromatin structure and gene expression is the role of the ncRNA X-inactive specific transcript (*Xist*) in X-inactivation and dosage compensation. Female mammalian organisms only transcribe one X-chromosome in order to prevent an imbalance in the expression of hundreds of genes. This compensation is achieved by the transcription of the gene *Xist*. Upon its transcription, *Xist* coats one X-chromosome, thereby establishing chromosome silencing. This chromosome is subsequently hypoacetylated and hypermethylated by other epigenetic factors thereby permitting maintenance of the silent state.[54]

4.1.2 Histone Modifications

DNA is organized into structures known as nucleosomes, which comprise DNA wrapped around an octamer of histone proteins. Each histone protein has a long tail that can be posttranslationally modified and these modifications can influence many processes such as chromatin compaction and transcription. A variety of histone modifications have been observed, including acetylation, methylation, phosphorylation, sumoylation, ubiquitinylation, and others.[55] Acetylation at certain positions along the histone tail neutralizes the charge of these amino acids, reducing chromatin compaction, and increasing transcription.[56,57] Histone acetyltransferase (HAT) enzymes such as P300/CBP catalyze the acetylation of specific lysine residues on the histone tail. P300/CBP in particular participates with the trithorax group of proteins to acetylate H3K27 and this ultimately promotes activation of certain genes during development.[58] Histone deacetylase (HDAC) enzymes antagonize the effects of HATs by removing acetyl groups from lysine residues on the histone tail and this promotes repression of transcription. Two families of HDACs have been characterized and they include the classical family, which have a zinc-dependent active site and the silent information regulator 2-related protein (sirtuin) family,[59] which is NAD+-dependent. Methylation of histone tail residues can be repressive or activating depending on the residue methylated. Trimethylation of histone H3 at lysine 27 (H3K27me3) mediated by the histone methyltransferase (HMT) enhancer of zeste homolog 2 (EZH2), a member of the polycomb repressive complex (PRC2), is associated with transcriptional repression.[60] Importantly, these modifications can have direct and indirect effects on transcription by influencing the recruitment of proteins such as TFs and by interfering with the binding of protein complexes that can modulate chromatin compaction and accessibility.[61] As the name suggests, histone demethylases such as lysine-specific histone demethylase 1 (LSD1), remove methylation from histone lysine residues. LSD1 in particular is known to promote transcriptional repression by reducing H3K4me2 and H3K4me1 marks.[62] In addition to modifications of the histone tails, posttranslational modifications also occur on the lateral surface of the histone, which makes direct contact with the DNA, and have been shown to also play an important role in DNA-based processes including transcription.[63]

4.1.3 DNA Methylation

Methylation of DNA at cytosine residues that are adjacent to guanine residues (CpG sites) occurs frequently throughout the genome. This reaction is catalyzed by the DNA methyltransferase (DNMT) family of enzymes. DNMT3a and DNMT3b are de novo methyltransferases. They are responsible for newly methylating unmethylated DNA. Whereas DNMT1, a maintenance methyltransferase, methylates hemimethylated DNA in order to maintain the methylation pattern initiated by the de novo DNMTs. In humans, it is estimated that greater than 85% of CpG sites are methylated.[64] In contrast, a small portion of the genome, namely CpG islands, is largely unmethylated. CpG islands are regions with a high density of CpG sites and importantly, many promoter regions of genes contain CpG islands. This chemical mark on the DNA has functional significance because its presence in the promoter region of genes is strongly associated with permanent transcriptional silencing, presumably by interfering with the binding of transcriptional machinery. In addition to preventing TFs from binding to promoters, DNA methylation can lead to binding of DNA methyl-binding domain containing proteins (MBDs) which can further inhibit transcription by recruiting repressive complexes containing HDACs and HMTs.[65] Although CpG island DNA methylation has been shown to occur in other regions of genes, such as in the gene body or 3′ region of the gene, the functional role of CpG methylation in these nonpromoter regions is not as well understood. For instance, one study reported that the relationship between the level of gene–body methylation and the level of gene expression was bell shaped, meaning that genes with mid-level expression had the highest levels of methylation whereas genes with high-level or low-level expression had lower levels of methylation.[66] Furthermore, it was reported that non-promoter DNA methylation plays important roles in prostate tumorigenesis.[67] Interestingly, DNMT1 has other functions; it can modulate gene expression independently of its ability to methylate DNA. For instance, it can associate with histone modifying enzymes by acting as a scaffold protein.[68]

4.2 Importance of Epigenetics in Embryonic Development

DNA methylation plays an important role during embryonic development. In early embryonic development, beginning in the zygote, DNA methylation marks are almost completely erased, with the exception of a few regions.[69,70] During implantation, a new pattern of methylation is established throughout the genome. It has been suggested that the purpose of DNA methylation is to stabilize gene expression patterns throughout an organism's lifetime likely by cooperating with the factors mentioned in the previous sections (ncRNAs and histone modifiers).[64]

As discussed earlier, epigenetic alterations also play a critical role in X-inactivation during embryonic development. This process is complex and involves cooperation between various epigenetic factors including ncRNAs and histone modifying enzymes. Similar to X-chromosome inactivation, genomic imprinting is another process that is tightly regulated by epigenetic factors including ncRNAs, histone modifying enzymes, and DNMTs. Imprinting refers to the process whereby specific genes are silenced on one parent's chromosome but not on the other. Only a small subset of mammalian genes is imprinted.[71] The process of imprinting occurs during gametogenesis and involves DNA hypermethylation of specific genes. These are among the few genes that remain methylated in early embryonic development (preimplantation) when all other DNA methylation marks are erased from the genome.[64]

Polycomb group proteins and trithorax group proteins are two sets of epigenetic protein complexes that play a role in regulating gene expression during embryonic development. There are several known polycomb repressive complexes, PRC1 and PRC2 will be described in more detail later. PRC2 comprises four core subunits: the HMT EZH1/2, suppressor of zeste 12 (SUZ12; a zinc finger protein), embryonic ectoderm development, and retinoblastoma-associated protein 46/48 (RbAp46/48).[72] During development, PRC2 regulates expression of the homeobox (HOX) genes, which encode for a set of TFs that are responsible for anatomical patterning and differentiation. This repressive complex was first identified in studies involving *Drosophila melanogaster.*[73] Flies deficient in PRC2 exhibited ectopic expression of the HOX genes and also abnormal body patterning suggesting that PRC2 plays a critical role in fly development.[74,75] EZH2 in particular catalyzes the trimethylation of H3K27 leading to transcriptional repression of HOX genes in regions where they should not be expressed.[65] In mammals, PRC2 is recruited to specific regions in the genome, by mechanisms that are not yet well defined, where it trimethylates H3K27 which is recognized by the chromodomain of CBX-containing PRC1.[76,77] PRC1 contains the protein Ring1B which has E3 ubiquitin ligase activity and specifically catalyzes the monoubiquitination of histone 2A (H2A) at lysine 119.[78] It has been shown that monoubiquitination of H2A at lysine 119 mediated by PRC1 is indispensable for repression of genes, in particular HOX gene loci.[79] In addition, PRC1 might promote gene repression by other mechanisms including physically interfering with the binding of TFs and RNA polymerase and also by recruiting DNMT1 and promoting DNA methylation.[80,81] In addition to their involvement with HOX gene repression during development, the polycomb group proteins also play a critical role in mediating epigenetic silencing during X-chromosome inactivation and genomic imprinting.[82,83]

The trithorax group proteins antagonize the repressive function of the polycomb group proteins. More specifically, the trithorax group proteins promote active chromatin structure via their ability to catalyze the methylation of H3K4. In addition to histone acetylation, this complex has also been associated with promoting chromatin remodeling.[81]

4.3 Importance of Epigenetics in Immune Cell Differentiation and Cytokine Expression

Epigenetic processes are critical to the differentiation of immune cells particularly T helper cells. When naïve CD4+ T cells encounter antigen-presenting cells, a cascade of events is initiated including activation and relocalization of various TFs ultimately leading to changes in gene expression. These changes in gene expression determine the fate of the naïve T cell. More specifically, these changes determine whether a naïve T cell differentiates into a Th1, Th2, Th17, or Treg.

Although TFs are pivotal to determining T cell fate by modulating gene expression patterns, it has been suggested that more stable control of these gene expression changes is achieved via epigenetic mechanisms, including DNA methylation, histone modification, and chromatin remodeling.[15]

Previous work in vitro determined that treatment of T cell lines with a DNMT inhibitor, 5-azacytidine, led to production of IL-2 and IFN-γ.[84,85] Additionally, genetic ablation of DNMT1 or MBD2 led to increased expression of IFN-γ and Th2 cytokines. Specifically, ablation of DNMT1 or MBD2 interfered with the ability of Th1 and Th2 cells to silence the expression of the opposing cytokines; Th2 cytokines and Th1 cytokines, respectively.[86–88] These findings suggest that epigenetic factors, specifically DNA methylation, might play a role in suppressing transcription of these cytokine genes in a context-dependent fashion. Additional studies determined that the TFs involved in determining T cell fate recruit chromatin remodeling complexes to cytokine promoters.[15] In particular, STAT4 activation leads to recruitment of Brahma-related gene 1-containing chromatin remodeling complexes to the IFN-γ promoter in Th1 cells. This recruitment is required for enhanced expression of IFN-γ in these cells and contributes to Th1 lineage commitment.[89] Another study found that chromatin remodeling occurs in the Th2 cytokine locus when naïve T cells differentiate into Th1 or Th2 cells.[90] The Th2 cytokine locus contains the IL-4, IL-5, and IL-13 cytokine genes as well as other *cis*-regulatory elements. In naïve T cells, this particular locus has both permissive and repressive chromatin marks, allowing it to be poised for activation when the cell is exposed to the appropriate stimuli.[15] Activation of GATA3 in naïve T cells leads to recruitment of various epigenetic proteins such as HATs as well as displacement of HDACs and DNMTs at the Th2 cytokine locus, permitting activation of these cytokines, and ultimately Th2 lineage commitment.[17,87] As observed with the Th2 cytokine locus, the IFN-γ gene also contains both permissive and repressive chromatin marks in the naïve T cell suggesting that it is poised for activation. When STAT1 becomes activated in naïve T cells, this leads to activation of T-bet which has the ability to bind to the IFN-γ promoter.[91] Binding of T-bet to the IFN-γ promoter leads to displacement of HDACs and recruits HATs thereby permitting transcription of IFN-γ and contributing to Th1 lineage commitment.[92] With regard to Th17 lineage commitment, STAT3 has been shown to bind and to induce histone acetylation at the IL-17A and IL-17F promoters and it is thought this facilitates transcription of IL-17A and IL-17F.[93]

Epigenetic processes also play a role in the development of innate immune cells although knowledge concerning this role is limited. One example is the role of epigenetic factors in regulating macrophage polarization. Macrophages differentiate from monocytes and in their unstimulated state they play roles in maintaining tissue homeostasis. Following activation, macrophages become altered phenotypically and this polarization toward specific phenotypes is regulated by the extracellular cytokine milieu.[94,95] Active and repressive chromatin marks are known to regulate the expression of the cytokine genes relevant to macrophage polarization.[96] During monocyte differentiation, the genes responsible for M1 polarization are marked with both repressive and activating histone modifications ultimately allowing these genes to remain repressed in the unstimulated state but are poised for rapid activation upon stimulation. Another study suggested that polarization from the M2 to M1 phenotype is controlled by DNMT3b in obesity.[97] Specifically, in response to the fatty acid stearate, DNMT3b expression was induced in macrophages followed by binding of DNMT3b to the promoter of peroxisome proliferator-activated receptor gamma 1 (PPARγ1), a key regulator of macrophage polarization. This binding led to hypermethylation of the PPARγ1 promoter and likely contributed to the transition from the M2 to the M1 phenotype.[97]

4.4 Epigenetic Alterations and Their Importance in Cancer Initiation and Progression

Although epigenetic alterations clearly play an important role in normal physiology, it is widely known that aberrant epigenetic changes contribute to the pathogenesis of various diseases including cancer.[3] Aberrant DNA methylation in particular has been studied with regard to its involvement in the initiation and progression of cancer. Concerning DNA methylation and cancer, two prominent features have been observed across many types of cancer. In particular, there is a global loss of DNA methylation in cancer cells along with concurrent focal hypermethylation, particularly at promoter CpG islands.[98–100]

4.4.1 Silencing of Tumor Suppressor Genes and Activation of Oncogenes

Hypermethylation of the promoter CpG islands of tumor suppressor genes can lead to their silencing and ultimately contribute to the development of cancer. Roughly 60% of gene promoters contain CpG islands, and importantly, most of these CpG islands remain unmethylated throughout normal animal development. Interestingly, 5–10% of CpG islands are hypermethylated in cancer.[3] It is likely that not all of these hypermethylated CpG islands have functional significance with regard to cancer development. Identifying which of these methylation changes lead to functional consequences that promote cancer is a major challenge facing cancer researchers today.

In addition to being genetically mutated, many known tumor suppressor genes have been shown to be hypermethylated in cancer. Some examples include E-cadherin, MutL homolog 1 (MLH1), and APC.[3] It was hypothesized that following mutation, promoter hypermethylation of the second copy of a tumor suppressor gene might serve as a "second hit" thereby leading to more permanent loss of function of the gene.[101] This mechanism was observed with the E-cadherin gene CDH1 in sporadic diffuse gastric carcinoma.[102] More specifically, the CDH1 gene was found to be mutated in greater than 50% of diffuse gastric carcinomas, and furthermore, plasma membrane E-cadherin was lost in each of the samples containing mutated CDH1. Importantly, loss of heterozygosity was only observed in one case of diffuse gastric carcinoma suggesting that an alternative mechanism must explain the inactivation of the second CDH1 gene copy. Interestingly, 66.7% of the diffuse gastric carcinomas containing genetic mutations also exhibited promoter hypermethylation lending support to the idea that hypermethylation might act as a second hit to promote complete loss of function of CDH1 in diffuse gastric carcinoma.[102] These findings raise the possibility that a similar mechanism might underlie silencing of other tumor suppressor genes in cancer.

As mentioned earlier, there is a global loss of DNA methylation (hypomethylation) in cancer cells. It remains unclear the precise role of DNA hypomethylation in cancer initiation and progression. In fact, some reports suggest that global DNA hypomethylation plays a role in promoting tumorigenesis whereas other studies suggest that global DNA hypomethylation might suppress tumorigenesis.[103,104] DNA demethylation is thought to occur either passively by loss of DNMT1 or actively by the conversion of 5′ methylcytosine to 5′ hydroxymethylcytosine via the enzyme ten-eleven translocation methylcytosine dioxygenase 1 (TET1).[105] It was recently demonstrated that conditional ablation of DNMT1 specifically in the intestinal epithelium exacerbates microadenoma formation in multiple intestinal neoplasia (Min) mice, which are heterozygous for mutant APC.[104] Importantly, significant demethylation was observed at the oncogenes Dusp6 and c-Fos in the DNMT1 mutant adenomas relative to the control adenomas. Consistent with this observation, c-Fos expression is known to be negatively regulated by hypomethylation in cancer.[106,107] In addition to demethylation of oncogenes, conditional ablation of DNMT1 in the intestinal epithelium of Min mice leads to genomic instability and this likely contributed to enhanced loss of heterozygosity at the APC locus and ultimately increased microadenoma formation in Min mice.[104] These findings suggest that global DNA hypomethylation plays an important role in promoting the early stages of tumorigenesis. In contrast, another study reported that global DNA hypomethylation plays a dichotomous role in intestinal carcinogenesis.[108] In this study, Min mice with hypomethylated DNA exhibited increased microadenoma formation,[108] consistent with findings reported elsewhere.[104] However, in the same animals, despite increased microadenoma formation early on, macroscopic tumor formation, which occurs at a later time point, was significantly suppressed in Min mice with hypomethylated DNA. It was hypothesized that the initial increase in microadenoma formation in Min mice with hypomethylated DNA was due to increased chromosomal instability and loss of heterozygosity at the APC locus, whereas suppression of the macroscopic tumor formation later was due to reversal of epigenetic silencing of tumor suppressor genes.[108] Importantly, findings from this study demonstrate that global DNA hypomethylation plays a dual role in intestinal carcinogenesis and it is possible that this holds true in other cancer types as well.

4.4.2 Epigenetic Alterations That Allow Cancer Cells to Evade the Immune System

The ability to evade destruction by the immune system is a hallmark of cancer cells.[5] Both innate and adaptive immune cells can engage in direct killing of cancer cells. This process, i.e., immune cell-mediated killing of cancer cells, requires that cancer cells express certain molecular features including death receptors, stress-induced ligands, and tumor-associated antigens on the surface of their plasma membrane. With regard to immune cell-mediated killing of cancer cells, there are antigen-specific processes and nonantigen-specific processes. The former are initiated primarily by CD8+ and CD4+ T cells whereas the latter are initiated primarily by NK cells.[109] In order for T cells to effectively kill cancer cells, cancer cells must process and present tumor-associated antigen peptides on MHC molecules located on the surface of their plasma membrane. To do this, cancer cells must express antigen processing and presentation machinery.[109] Cancer cell killing mediated by NK cells requires that cancer cells express death receptors such as the Fas ligand receptor and TNF-α receptor (TNFR) as well as various stress-induced ligands on their plasma membrane.[110]

Epigenetic processes play an extremely important role in a cancer cell's ability to evade the immune system. For instance, DNA methylation and certain histone posttranslational modifications are known to contribute to silencing of various genes involved in antigen processing and presentation including, but not limited to, tumor-associated antigen genes, MHC class I and II genes and costimulatory molecule genes.[109,111] Further supporting this notion is the observation that treatment of cancer cells with epigenetic therapies such as HDAC inhibitors (HDACi) leads to their upregulation of NK cell ligands[112–115] and antigen presentation machinery including tumor-associated antigens, cancer testis antigen, proteasome subunits, and MHC class I and II molecules.[109] Furthermore, CRC cells deficient in DNMTs had decreased methylation and increased expression of MHC class I genes and NK cell ligands.[116] Collectively, these studies and others lend credence to the notion that epigenetic processes play a pivotal role in the immune evasion capabilities of cancer cells. In light of this, recent efforts have been made to determine if epigenetic therapies could be used to enhance the efficacy of immunotherapies in the treatment of cancer and this topic will be visited in more depth later in this chapter.

5. THE ROLE OF INFLAMMATION IN INITIATING EPIGENETIC ALTERATIONS

5.1 Epidemiological Evidence Supporting a Connection Between Inflammation and Aberrant Epigenetic Alterations

There is a considerable amount of epidemiological data supporting a connection between inflammation and epigenetic alterations. As mentioned earlier, *H. pylori* infection is strongly associated with gastric cancer incidence. Incidentally, epigenetic alterations play a critical role in the pathogenesis of gastric cancer.[117] One study demonstrated that CDH1 promoter methylation occurs at a higher frequency in the gastric mucosa of individuals infected with *H. pylori* in comparison to individuals that are not infected with *H. pylori*, suggesting an association between inflammation and aberrant epigenetic alterations.[118] Similarly, another study examined a panel of eight CpG island containing genes and reported that patients infected with *H. pylori* have much higher levels of methylation in these genes, specifically in their gastric mucosa, relative to uninfected individuals.[119] This association between inflammation and aberrant epigenetic alterations is not unique to *H. pylori* infection and gastric cancer.[120] For instance, HCV infection and hepatocellular carcinoma are also strongly linked, and it is possible that aberrant epigenetic alterations present in

HCV-infected individuals might predispose them to developing hepatocellular carcinoma. In support of this notion is the observation that noncancerous tissue from HCV-infected individuals had increased methylation levels at several important tumor suppressor gene loci including APC, SOCS1, RASSF1, RUNX3, SFRP2, and CDH1 compared to normal tissue from HCV-negative individuals.[121] As previously mentioned, patients with Barrett's esophagus, a condition characterized by high levels of inflammation in the esophagus, are at an increased risk of developing esophageal cancer. Interestingly, patients with Barrett's esophagus were shown to have increased methylation in the promoters of the tumor suppressor genes APC and CDKN2A.[122] IBD and UC are chronic inflammatory diseases highly associated with an increased risk of CRC. Importantly, patients with extensive UC were reported to have increased methylation in the tumor suppressor gene CDKN2A.[123] Additionally, many genes (several with tumor suppressive function) have been shown to be differentially methylated in IBD patients compared to control patients.[124] UV exposure, which causes skin inflammation, is strongly associated with skin cancer and therefore it stands to reason that UV exposure-induced inflammation might be associated with epigenetic alterations as well. Interestingly, malignant and nonmalignant skin lesion samples from sun-exposed areas of the human body had very high levels of methylation in the promoters of several important tumor suppressor genes such as CDH1 and RASSF1A, whereas samples taken from sun protected areas did not exhibit these changes.[125] Collectively, these findings in humans suggest a strong association between chronic inflammation and aberrant epigenetic alterations. That being said, based on these human studies, it remains unclear whether chronic inflammation is causally related to epigenetic alterations.

5.2 In Vivo Studies Demonstrating a Relationship Between Inflammation and Epigenetic Alterations

Although the human studies described earlier demonstrate an association between inflammation and epigenetic alterations, more evidence is needed to establish cause and effect. Importantly, animal models have provided substantial evidence indicating that inflammation can cause aberrant epigenetic alterations.[1] One study in vivo reported that suppression of *H. pylori*-mediated gastric inflammation using cyclosporine A led to suppression of *H. pylori*-induced aberrant DNA methylation changes in the gastric mucosa of Mongolian gerbils.[126] Another study reported that exposure of mice to

UVB radiation caused a distinct pattern of DNA hypermethylation in UVB-exposed skin as well as UVB-induced tumors. In particular, the tumor suppressor genes RASSF1A and CDKN2A were hypermethylated and silenced in both the UVB-exposed skin and UVB-induced tumor tissue.[127] In a study of inflammation in the colon, treatment of mice with DSS led to aberrant DNA methylation changes in multiple gene loci.[128] These DNA methylation changes occurred in the absence of T and B cells, suggesting that innate immune-mediated inflammation is likely the culprit in initiating DNA methylation changes in response to treatment with DSS. Treatment of mice with DSS in combination with the mutagen azoxymethane (AOM) induces colitis followed by tumorigenesis in the colon.[129] In this model, it is suspected that AOM initiates epithelial transformation and DSS acts as a tumor promoting agent. Interestingly, one study reported that HDAC levels and activity are elevated in precancerous colonic tissue from mice treated with AOM/DSS.[130] Moreover, levels of acetylated H3K27 (H3K27ac) were depleted in AOM/DSS-treated mice compared to control mice. Hypoacetylation of chromatin can alter the expression of important tumor suppressor genes and contribute to the development of cancer.[131] Interestingly, the AOM/DSS-induced epigenetic alteration (i.e., histone hypoacetylation) was reversed by treatment with the nonsteroidal antiinflammatory agent aspirin, supporting the concept that inflammation causes epigenetic alterations.[130] Using the AOM/DSS mouse model of inflammation-induced tumorigenesis, another study revealed significant hypermethylation of DNA methylation valleys (DMVs) in inflamed colon tissue and tumor tissue compared to normal tissue.[132] Furthermore, many of the hypermethylated DMV-associated genes were silenced in inflamed colon and tumor tissue and, interestingly, this pattern was conserved in human colon tissue. Another mouse model of inflammation-induced carcinogenesis revealed that inflammation promotes aberrant DNA methylation early and this signature is conserved in malignant tissue.[133] Specifically, glutathione double knockout mice, which have high levels of inflammation in the ileum and are predisposed to tumorigenesis in this region, exhibited DNA hypermethylation of many polycomb group target genes in both inflamed and malignant tissue and the methylation gains in these regions were correlated with loss of H3K27me3.[133] These in vivo studies and many others not described herein provide considerable evidence that inflammation does indeed cause epigenetic alterations.

5.3 In Vitro Studies Demonstrating a Relationship Between Inflammation and Epigenetic Alterations

In addition to the in vivo studies described earlier, studies in vitro have greatly contributed to our understanding of the relationship between inflammation and epigenetic alterations. In particular, several studies in vitro have identified some of the specific inflammatory mediators responsible for inducing epigenetic alterations.[1] Several immune mediators including ROS/RNS and cytokines are known to modulate the activity of epigenetic proteins. Treatment of several different cell lines with nitric oxide or the cytokine IL-1β, two proinflammatory mediators, resulted in increased DNMT activity, increased methylation, and decreased expression of several CpG island containing genes.[134] Another study revealed that treatment of neuronal cells with the RNS nitric oxide led to S-nitrosylation of HDAC2 resulting in histone modifications and altered transcription of genes.[135] Interestingly, S-nitrosylation of HDAC2 did not affect its deacetylase activity but it did cause HDAC2 to release from chromatin thereby leading to increased chromatin acetylation and activation of genes. Another study reported that the proinflammatory cytokine IL-6 induces DNMT1 transcription by regulating the TF Fli1.[136] Furthermore, IL-6 regulates DNMT1 expression by suppressing two miRNAs known to target and silence DNMT1.[137] Treatment of cancer cells with the inflammatory mediator prostaglandin E2 (PGE2) caused increases in DNMT1 and DNMT3b expression and induced DNA methylation of several genes.[138] Although the studies discussed earlier provide some mechanistic insight concerning the specific inflammatory mediators that contribute to inflammation-induced epigenetic alterations they do not fill in the gaps needed to explain the molecular progression from inflammation to resulting epigenetic alterations.

5.4 Molecular Mechanisms Underlying Inflammation-Induced Epigenetic Alterations

From the studies described earlier, we learned that inflammation causes epigenetic alterations and also gained insight concerning the specific inflammatory mediators that play a role in this process. However, what remains to be determined are the molecular mechanisms driving inflammation-induced epigenetic alterations. This section will discuss studies that have provided a deeper understanding of these mechanisms with a particular focus on the role of cellular metabolism and oxidative stress-induced DNA damage in initiating epigenetic alterations.

5.4.1 Alteration of Cellular Metabolism and Epigenetic Protein Cofactors

A mechanism by which inflammation can alter epigenetics is through altering the metabolic state of the cell. The metabolism of immune cells is altered when they are activated and activated immune cells alter metabolic processes in surrounding tissue.[139] Furthermore, activated STAT3 can regulate metabolism by inducing aerobic glycolysis and decreasing mitochondrial activity.[140] Many epigenetic enzymes require cellular metabolism intermediates for their activity.[141] S-adenosyl methinonine (SAM) is a required cofactor for DNMTs and HMTs, whereas S-adenosyl homocysteine is an inhibitor. SAM is produced by 1-carbon metabolism and through dietary consumption of methyl group donors (i.e., folate). HDMs and TET proteins use alpha-ketoglutarate produced in the TCA cycle as a cofactor. Acetyl–CoA is produced from glucose metabolism and is a cofactor required by HATs. NAD+ is an essential cofactor for the HDAC SIRT1 and SIRT1 has been linked to metabolic disorders.

5.4.2 The Role of DNA Damage in Inflammation-Induced Epigenetic Alterations

As mentioned earlier, inflammatory cells such as neutrophils and macrophages release many factors such as cytokines, chemokines, and ROS including hydrogen peroxide. There are many examples in the literature demonstrating that ROS can induce epigenetic alterations. In a CRC cell line, CDX1, a suspected tumor suppressor gene, became methylated and silenced in response to hydrogen peroxide treatment.[142] Additionally, treatment with either the free radical scavenger N-acetyl cysteine or a DNMT1 inhibitor (5-azacytidine) reversed these effects. Another study reported similar findings in CRC cells with the gene RUNX3, another putative tumor suppressor gene.[143] Specifically, hydrogen peroxide treatment led to increased methylation of the RUNX3 promoter and silencing of the gene, and these effects were reversed upon treatment with either N-acetyl cysteine or 5-azacytidine.[143] These studies raise the question: how exactly does oxidative stress (ROS) lead to epigenetic alterations in cancer cells? Studies discussed later suggest that DNA damage induced by ROS plays a particularly important role in initiating epigenetic alterations. DNA molecules are prone to damage mediated by ROS particularly at guanine residues, which is the most easily oxidized base.[144] Hydroxyl radicals, a particularly pernicious ROS that can form from hydrogen peroxide when in the presence of iron (Fenton reaction), can induce several types of DNA damage including DNA strand breakage, base modifications, deletions, and chromosomal rearrangements.[145] DNA damage induced by exposure to ROS has been reported to cause either DNA hypomethylation or DNA hypermethylation depending on the context.[146] For instance, certain DNA lesions induced by hydroxyl radicals can interfere with the ability of DNMTs to bind to hemimethylated DNA and therefore promote global DNA hypomethylation.[145] Conversely, exposure of hepatocellular carcinoma cells to hydrogen peroxide led to hypermethylation of the E-cadherin promoter. Specifically, hydrogen peroxide treatment led to increased expression of the TF Snail, a negative regulator of E-cadherin expression, which caused E-cadherin promoter methylation by recruiting HDAC1 and DNMT1 to the E-cadherin promoter.[147]

Repair of DNA damage utilizes chromatin remodelers as has been explained by the access, repair, restore model postulated by Polo and Almouzni.[148] One hypothesis linking DNA damage to changes in DNA methylation is that these transient chromatin changes may occasionally lead to heritable epigenetic alterations in a similar fashion as to how rare mistakes in DNA repair leads to genetic mutations.[149] DNA double-strand breaks cause transient silencing of transcription.[150,151] It was reported that induction of a DNA double-strand break in the E-cadherin promoter CpG island can lead to heritable silencing of the gene, and ultimately DNA hypermethylation of the CpG island.[152] Interestingly, introduction of the DNA double-strand break led to recruitment of the epigenetic proteins SIRT1, EZH2, DNMT1, and DNMT3b to the site of DNA damage. Importantly, most of the cells harboring the DNA double-strand break exhibited normal repair of the break; however, a minority of cells maintained silencing of the E-cadherin promoter and passed these marks onto subsequent generations. Another study demonstrated that treatment of colon cancer cells with hydrogen peroxide induces relocalization of an epigenetic silencing complex containing DNMT1, DNMT3b, and polycomb group proteins from non-GC-rich regions to GC-rich regions, including promoter CpG islands of tumor suppressor genes.[153] Importantly, some of these findings were recapitulated in a mouse model of colitis.[153] A related study revealed that the DNA repair process induced in response to DNA damage is critical to the initiation of the epigenetic alterations observed in cancer cells treated with hydrogen peroxide.[154] In particular, this study found that hydrogen peroxide treatment promoted recruitment of the mismatch repair machinery to sites of DNA damage and these repair proteins, MSH2 and MSH6 in particular, were essential to the recruitment of DNMT1 to chromatin. Platinum-derived drugs can induce epigenetic alterations including DNA methylation changes and these alterations in DNA methylation are thought to partially underlie the acquisition of platinum-derived drug resistance.[155,156] Platinum-derived compounds such as cisplatin are common anticancer therapies and they kill cancer cells by inducing crosslinking of DNA and ultimately DNA damage. The observation that exposure to platinum-containing compounds induces DNA methylation changes strengthens the concept that DNA damage leads to epigenetic alterations. Although the picture is not entirely complete yet, these studies provide tremendous insight concerning the mechanisms underlying inflammation-induced epigenetic alterations particularly with regard to the role of DNA damage.

6. CANCER PREVENTION AND TREATMENT

Cancers often have overexpressed epigenetic modifiers including DNMTs, EZH2, LSD1, SIRT1, HDACs, and G9a. Furthermore, recurrent mutations of epigenetic modifiers, including IDH1/2, EZH2, DNMT3A, KDM6A, and SMARCB1 have been demonstrated in many cancer types.[157] Since both inflammation and epigenetic alterations can drive tumor formation, both the immune system and epigenetic proteins have been targeted to prevent and treat cancer. Interestingly, many drugs and naturally occurring dietary molecules affect both of these pathways at the same time.

6.1 Antiinflammatory Agents

Epidemiological studies have demonstrated reduced incidence of cancers in those with long-term use of nonsteroidal antiinflammatory drugs (NSAIDs) and cyclooxygenase-2 (COX-2) selective antiinflammatory drugs.[158] NSAIDs, including aspirin, primarily inhibit COX enzymes, which produce prostaglandins, and have also been associated with decreased metastasis. In a mouse model of inflammation-induced CRC, aspirin treatment was associated with reduced HDAC expression and activity and resulted in increased H3K27ac levels.[130] Additional studies suggest that aspirin and other NSAIDs may reverse tumor suppressor gene hypermethylation in cancer cells and alter miRNA expression.[159] Most of these studies do not establish a mechanism so it is presently unclear whether NSAIDs directly regulate epigenetics or if they regulate inflammation, which in turn reduces epigenetic alterations. Some NSAIDs have adverse side effects including stomach ulcers and bleeding, namely the nonselective COX inhibitors such as aspirin, and therefore, may not be useful as cancer preventative agents.

As discussed earlier, cytokines and chemokines are elevated at sites of inflammation and play a key role in the tumor microenvironment. Inhibition of key cytokines may be able to disrupt the larger cytokine network and influence several tumor promoting pathways at once. However, in certain instances cytokines can also be antitumorigenic. Biological inhibitors of cytokines, including TNF-α and IL-6, were developed for the treatment of chronic inflammatory disease such as rheumatoid arthritis. Some of these inhibitors have shown early promise in groups of cancer patients, but clinical trials are still ongoing.[160] Since the inflammatory environment can induce epigenetic changes, both NSAIDs and biological inhibitors of cytokines may be effective in reducing epigenetic changes in individuals with chronic inflammatory diseases and therefore reduce cancer risk.

6.2 HDAC Inhibitors

HDACs are often overexpressed or dysregulated in cancer. HDAC dysregulation can result in aberrant gene transcription associated with increased tumor cell proliferation and proinflammatory responses. HDACi induce increases in gene expression and alteration of protein acetylation. They have been used to treat cancer because of their antiproliferative activity and activation of the apoptotic pathway.[161] HDACi also have antiangiogenic and antimetastatic effects.[162] Many of the current HDACis affect multiple HDACs (panHDACi) and therefore affect many different cellular pathways. An effort is being made to produce more specific HDACis as each HDAC likely has specific roles and targeting only one would improve the effectiveness of the HDACi.[163]

HDACi can both stimulate and repress the immune system depending on the cell type, activation status, and type of inhibitor. HDAC1 and 2 have been shown to bind with the NF-κB family corepressor p65 and downregulate NF-κB transcription.[164] When activated, this repressive complex is replaced by an activating complex, increasing gene transcription.

HDAC3 also regulates NF-κB activation by regulating acetylation of p65 and histone acetylation at NF-κB target promoters.[165] Therefore, HDAC enzymes play a role in maintaining a balanced inflammatory response by regulating NF-κB and downstream cytokine expression. HDACs also decrease acetylation of STAT3 and in doing so inhibit DCs.[166] Because the HDACs are involved in regulating proinflammatory responses normally, HDACis can also modulate the immune response and animal studies have demonstrated that HDACis reduce inflammatory diseases, enhance allograft survival, and induce immune tolerance in graft-vs-host disease.[167] In transplant patients, vorinostat, a panHDACi, treatment reduced proinflammatory cytokines after transplant and increased Tregs, but did not alter other T cell responses.[168] It appears to do so through increasing the acetylation of STAT3, which results in increased IDO expression.[166,169] Vorinostat has also been effective in inducing immunogenic cell death in tumor cells where dying tumor cells increase the activation of antigen-presenting cells and induce antitumor T cell responses.[170] The panHDAC valproic acid increases the expression of activating NKG2KD ligands for NK cells on tumor cells, but not on normal cells, leading to increased killing of tumor cells.[115] However, other studies of panHDACis have shown that these inhibitors reduce expression of costimulatory molecules and proinflammatory cytokines.[171] Interestingly, the class I specific HDACi entinostat does not downregulate proinflammatory cytokines, suggesting that using more targeted HDACis can selectively regulate the immune response.[172] Overall it seems that HDACis may be able to increase recognition of tumor cells by immune cells, but this effect may depend on the tumor type and specific HDACi.

Butyrate is an example of a short chain fatty acid panHDACi that is produced by microbes in the colon. Butyrate treatment of cancer cells causes derepression of epigenetically silenced tumor suppressor genes.[173,174] Butyrate also affects immune cell migration, adhesion, and cytokine expression. Butyrate suppresses NF-κB and the inhibition of INF-γ production, which may be mediated by its inhibition of HDACs.[175] In patients with IBD, butyrate reduced inflammation and symptoms.[176]

Sirtuins are class III HDACs, but unlike the other HDACs sirtuins require NAD+ to deacetylate histones and proteins. In humans, there are seven family members (SIRT1–7) that have different cellular localization and targets. SIRT1 deacetylates H4K16, making chromatin more repressive and participates in the silencing of tumor suppressor genes in cancer cells.[177] SIRT1 can also deacetylate nonhistone proteins, suppressing the function of tumor suppressors, including p53, p73, and HIC1 and tumor promoters, including NF-κB.[178–181] Therefore, SIRT1 can be tumor suppressive or promoting depending on the setting. There are several inhibitors of SIRT1, including nicotinamide, sirtinol, and EX-527 as well as activators. SIRT1/2 inhibitors can induce apoptosis in cancer cells likely through enhanced p53 acetylation and stabilization.[178] However, most tumor studies in mouse models have had mixed results as to whether SIRT1 inhibitors reduce or promote tumorigenesis.[182] SIRT1 activators and inhibitors also affect the inflammatory response.[183] Activators can suppress the inflammatory response by enhancing the deacetylation of p65 and inhibition of NF-κB.[184] SIRT1 is generally thought to be beneficial in aging and metabolic and inflammatory diseases because of its antiinflammatory role.[182]

6.3 Histone Lysine Demethylase Inhibitors

LSD1 is required for normal differentiation and stem cell maintenance. However, LSD1 is overexpressed in several types of cancer and thought to be tumor promoting.[185] Studies involving the knockdown of LSD1 suggest that reduced LSD1 decreases cancer cell growth, migration, and invasion.[186] First generation LSD1 inhibitors had many of target effects making them fairly toxic and difficult to study, but they were used to demonstrate a sensitivity of AML to LSD1 inhibition.[187,188] Recently more specific LSD1 catalytic inhibitors have been developed, which inhibit the growth of small cell lung cancer cells and xenografts through an alteration of cell state.[189] In normal cells, LSD1 inhibition or knockdown induces expression of proinflammatory cytokines suggesting that LSD1 normally helps repress the expression of these genes.[190,191]

6.4 BET Inhibitors

Epigenetic readers are proteins that recognize posttranslational modifications of histones. The bromodomain and extraterminal domain (BET) family of proteins are proteins that "read" acetylated lysines and interact with other proteins involved in transcription. BET proteins were originally connected to cancer because in NUT midline carcinomas there is a genetic driving event involving a translocation between the nuclear protein of testis (NUT) gene and BRD4, a BET protein, creating a BRD4-NUT fusion gene.[192] Targeting of these fusion proteins and other BET proteins using inhibitors has been successful in some cancer types, including hematological malignancies.[193] In cancer cells, it appears that BET inhibitors are successful because they reduce the localization of BET proteins at super-enhancers that drive high level of expression of tumor promoting genes, including *MYC*.[194,195] BET proteins are required for transcription of inflammatory genes[196] and BET inhibitors were successfully used to suppress inflammation in mouse models of sepsis, HIV-induced nephropathy, and IBD.[197,198]

6.5 Histone Lysine Methyltransferase Inhibitors

The H3K27 methyltransferase EZH2 is overexpressed and/or mutated in many cancer types.[157] It is involved in cancer initiation, progression, metastasis, and drug resistance. In diffuse large B-cell lymphoma and follicular lymphoma, there are recurrent somatic mutations in the catalytic SET domain of EZH2 that alter the enzyme's activity resulting in higher H3K27me3 levels.[199] EZH2 also methylates nonhistone substrates including STAT3, increasing its activation, and retinoic acid-related orphan nuclear receptor alpha (RORα) targeting it for degradation.[200,201] Several highly selective inhibitors of EZH2 have been developed. GSK126 is a S-adenosylmethionine-competitive, small-molecule inhibitor of EZH2 that reduces global H3K27me3 levels and induces the repression of PRC2 target genes.[199] This inhibitor reduced proliferation of B cell lymphoma cell lines and xenografts with mutant EZH2. Using a less specific inhibitor of EZH2 results in transcriptional activation of immune response genes that is cancer specific, again demonstrating the crosstalk between epigenetic regulation and the immune response.[202]

G9a is a HMT for H3K9 that is generally involved in gene repression within euchromatin. G9a plays a critical role in development, the immune response, and tumor cell growth.[203] Loss of G9a impairs the lineage differentiation of CD4+ cells into Th2, Th17, or Treg cells after immune challenges.[204,205] In cancer, elevated G9a levels are associated with higher DNA methylation levels and transcriptional repression of tumor suppressor genes.[203] Inhibition of G9a induces a reduction in cell proliferation and an induction of autophagy in cancer cells.[203] Chemical inhibitors of G9a work well in vitro but have poor pharmacokinetics in vivo requiring better inhibitors to be produced before the in vivo effects of inhibiting G9a can be tested.

6.6 DNA Methyltransferase Inhibitors

Since cancer cells have aberrant DNA methylation changes and often overexpress DNMTs, DNMTs are a target for cancer therapy. There are several drugs that inhibit the DNMTs that are in clinical trials and approved for use in solid and hematological tumors.[163] One class of these inhibitors is nucleoside analogs. Decitabine (5-aza-2′-deoxycytidine) is incorporated into replicating DNA and azacitidine (5-azacytidine) is incorporated into RNA with incorporation into DNA as 5-aza-deoxyribonucleotides.[206] When the DNMTs bind to the analogs in DNA a covalent bond is formed, trapping the DNMT. This process results in a decrease in DNA methylation that is cell cycle dependent and a decrease in total protein levels of DNMT1. These inhibitors cause reexpression of genes silenced by DNA methylation and induce cellular differentiation in cancer cells.[207] However, these nucleoside analogs are quickly deaminated by cytidine deaminase giving them a short half-life.[208] SGI-100 (guadecitabine) is a more stable inhibitor because it consists of decitabine linked to guanosine, making it resistant to cytidine deaminase, and prolonging the exposure time.[209] Catalytic inhibitors of DNMTs have not been as successful and this may be in part because nucleoside analogs cause a loss of the DNMT protein whereas catalytic inhibitors do not. As mentioned earlier, DNMT1 has roles in addition to its methyltransferase activity, which may explain the increased effectiveness of inhibitors that reduce DNMT protein levels over those that do not.

DNMT1 is altered by inflammation and DNMT inhibitors regulate the immune system. DNMT1 activity is unregulated in IBD, especially in actively inflamed colonic mucosa.[210] IL-6 expression has been shown to increase and stabilize DNMT1 and further activate STAT3.[211,212] Increased STAT3 activation is also associated with induction of type I IFN genes after DNMT inhibitor treatment.[213] Additionally, STAT3 has been shown to directly induce DNMT1 expression.[214] DNMT inhibitors also upregulate dsRNA and trigger the viral defense system.[215,216] DNMT inhibitors have been shown to have immune suppressive effects in murine graft-vs-host disease by increasing the levels of FOXP3+ Tregs.[217,218] Resolution of lung inflammation in mice was accelerated with decitabine treatment, which was associated with an increased number of Foxp3+ Tregs.[219] Decitabine treatment also reduced atherosclerosis development in mice by decreasing proinflammatory gene expression in macrophages, reducing their infiltration into atherosclerotic plaques.[220]

6.7 Dietary Compounds

Dietary compounds are studied in cancer chemoprevention because they are readily available and have been found to target multiple pathways. Dietary folate, B-vitamins, choline, and betaine may influence the methyl donor pool and therefore levels of DNA and histone methylation. Additionally, other dietary molecules have been shown to modulate epigenetic changes. (−)-Epigallocatechin-3-gallate (EGCG), curcumin, retinoic acid, genistein, and isothiocyanates have been shown in various studies to modulate DNA methylation, histone methylation, and/or miRNAs.[221] Polyphenols are naturally occurring in plants. EGCG is a polyphenol in green tea and a strong antioxidant. Some studies have demonstrated that EGCG can inhibit DNMT activity.[222,223] Resveratrol and pterostilbene, polyphenols found in grapes and blueberries, respectively, have been shown to be antioxidants and regulate inflammation and the activity of SIRT1 and DNMTs.[221,224] Curcumin is from the spice turmeric and has been used to treat a wide variety of illnesses, including those with an inflammatory component such as asthma, COPD, IBD, and psoriasis. It is a cell permeable antioxidant and can inhibit the HAT P300/CBP and DNMTs.[225–227] It has been used to treat several forms of cancer, including breast, pancreatic, and CRC.[228] Therefore, the effectiveness of these dietary compounds in chemoprevention may be explained by their ability to affect both inflammation and the activity of epigenetic proteins.

6.8 Immunotherapy

The goal of immunotherapy is to activate the body's immune response to target the tumor. As mentioned earlier, tumors have developed ways to evade the immune response by downregulating MHC molecules and tumor antigens or by actively suppressing the immune response.[109] One concept of immunotherapy is checkpoint blockade, which uses agents to target the ligand receptor interactions that control the activity of T cells.[229] Binding of ligands and receptors on the tumor and immune cells causes the immune cells to be tolerant of the tumor. Antibodies can block these interactions and therefore increase the activity of the immune cells. Cytotoxic T-lymphocyte-associated antigen 4 (CTLA-4), is an inhibitory molecule on T cells that counteracts costimulatory receptor CD28, likely by out competing CD28 for CD80 and CD86 binding.[229] Programmed death-1 (PD-1) is a receptor on T cells, which binds to the inhibitory ligands PD-L1 and PD-L2 ligands. PD-1's main role is to limit activity of T cells at the site of inflammation.[229] These inhibitory pathways play normal roles during inflammation to keep the inflammatory response in check and limit autoimmunity. However, in cancer these pathways can be utilized by the tumor to evade the immune response by, for example, the cancer cells upregulating PD-L1 expression on their cell surfaces. Antibodies against CTLA-4, PD-1, and PD-L1 have been effective in treating melanoma, neuroblastoma, and nonsmall cell lung cancer; however, only a minority of patients responds to these treatments, suggesting that complimentary cotherapies may improve the magnitude and frequency of response to immunotherapy.[109]

6.9 Combination Therapy

Epigenetic silencing can alter antigen processing and presentation and expression of costimulatory molecules and of stress-induced ligands. Therefore, epigenetic inhibitors will likely improve the immunogenicity of tumors and are being tested in conjunction with immunotherapy.

As mentioned earlier, HDACis can increase tumor-associated antigen expression by tumor cells and upregulate genes involved in antigen presentation and costimulatory molecules.[161] This altered expression would enhance the ability of T cells to recognize and kill tumor cells. Combining immunotherapy with HDACis may further activate the responding immune cells. For example, combining HDACis with antibodies against activating immune checkpoint targets (CD40 and CD137) increases DC activation and CD8+ cytotoxic T cell proliferation resulting in tumor killing by CD8+ T cells and NK cells.[230] The HDACi MS-275 can also enhance the ability of oncolytic viruses to kill tumor cells by modulating the innate antiviral immune response and enhancing T cell tumor response.[167]

DNMTis have been shown to upregulate normally silenced cancer testis antigens, tumor-associated antigens, and MHC class I and II molecules.[231,232] DNMTi also induce long-lasting expression of tumor-associated antigens.[233–235] Furthermore, they improve costimulatory properties of tumor cells by upregulating surface expression of CD40, CD80, CD86, and ICAM1 and restore sensitivity to immune cell triggered apoptosis. When DNMTi treatment induces upregulation of cancer testis antigens such as

NY-ESO-1, the tumor cells can then be targeted with vaccines against these antigens.[236,237] Additionally, combining treatment with CTLA-4 blocking antibodies and decitabine- or guadecitabine-induced expression of tumor associated and MHC class I antigens, increased tumor infiltration of CD3 + T cells, and reduced syngeneic tumor growth in mouse models more than the single agents.[238] Furthermore, combining azacytidine with an HDACi and anti-PD-1 and anti-CTLA-4 antibodies reduced mammary and colon syngeneic tumor growth and this occurred by the epigenetic inhibitors depleting MDSCs.[239] Based on the in vitro and in vivo preclinical data, clinical trials have been initiated to test the combination of epigenetic and immunotherapy in patients. Decitabine was combined with NY-ESO-1 whole protein vaccination and chemotherapy in ovarian cancer patients. Increased tumor-associated antigen expression and T cell immune response against NY-ESO-1 were observed along with some clinical response.[240] Trials are also underway that combine anti-CTLA-4 antibodies with guadecitabine in patients with metastatic melanoma and azacytidine with an HDACi and anti-PD-1 immunotherapy in patients with nonsmall cell lung cancer.

7. CONCLUDING REMARKS

Inflammation and epigenetic alterations both play important roles in the initiation and development of cancer (Fig. 1). The relationship between inflammation and epigenetic alterations in the context of cancer has recently garnered high levels of interest. Based on the work described earlier, we now know that these two phenomena are intricately connected. More research is needed to improve our understanding of the mechanisms underlying how inflammation drives epigenetic alterations that are specific to cancer. Focusing future research on the role of DNA damage and repair might provide illuminating details with regard to the mechanisms of inflammation-induced epigenetic changes and lead to informed chemopreventive treatments for individuals with chronic inflammatory diseases. A deeper understanding of the interplay between inflammation and epigenetic alterations will contribute to the development of more efficacious cancer treatment modalities that preserve as much of the normal function of the immune system as possible.

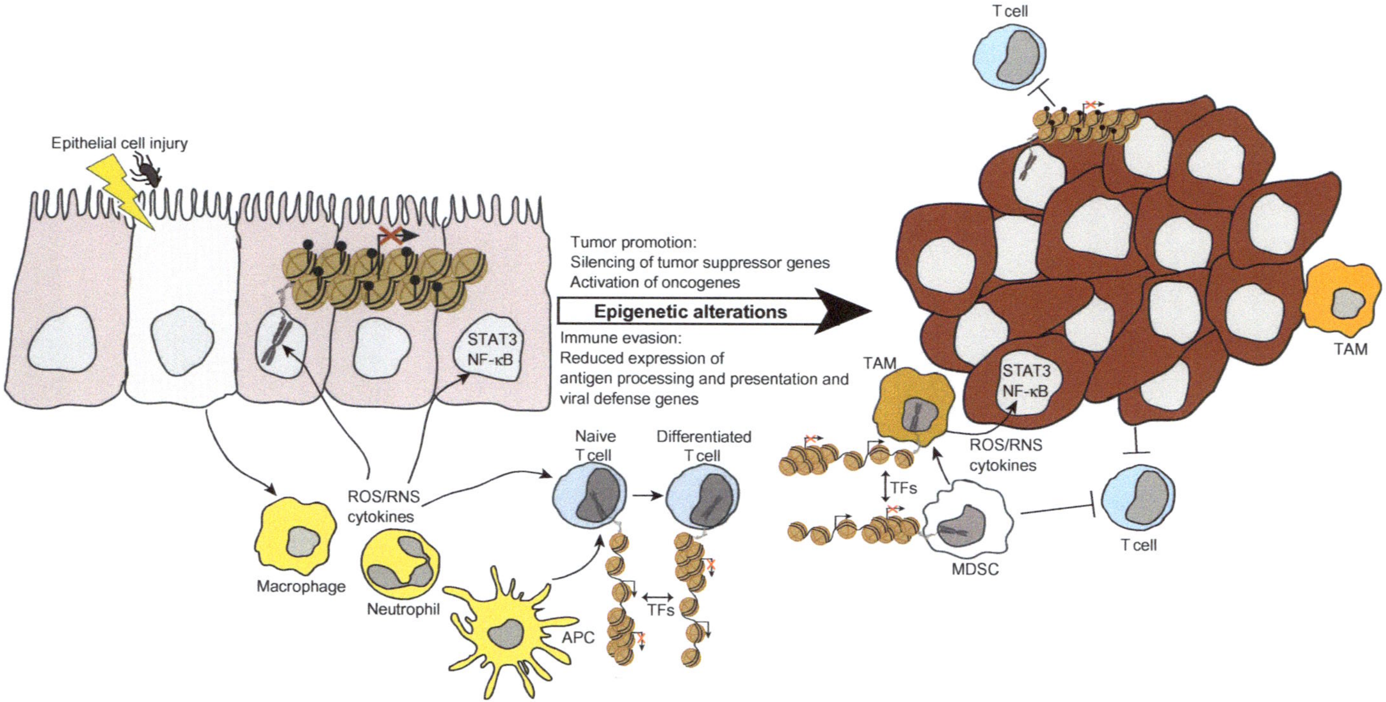

Fig. 1 Inflammation induces epigenetic changes in epithelial cells that promote tumorigenesis and immune avoidance. Innate immune cells are activated in response to epithelial injury. The inflammatory environment contains reactive oxygen and nitrogen species (ROS/RNS) and cytokines. Antigen-presenting cells (APCs) and cytokines induce activation of T cells, which in part involves transcription factors (TFs) inducing transcriptional reprogramming that results in epigenetic changes. Epithelial cells in this environment activate STAT3 and NF-κB transcription, which is antiapoptotic and proproliferative. The ROS/RNS and cytokine exposures induce epigenetic changes in the epithelial cells. Epigenetic changes can be tumor promoting by silencing tumor suppressor genes and activating oncogenes. Epigenetic changes can also help the tumor evade the immune response by reducing the expression of genes involved in antigen processing and presentation and viral defense. Tumor-associated macrophages (TAMs) contribute to the proinflammatory tumor microenvironment by release of ROS/RNS and cytokines, which contribute to continued activation of STAT3 and NF-κB in the tumors. Myeloid-derived suppressor cells (MDSCs) inhibit T cell activation and can differentiate into TAMs as TFs induce alternate transcriptional programs that result in epigenetic changes.

REFERENCES

1. Chiba T, Marusawa H, Ushijima T. Inflammation-associated cancer development in digestive organs: mechanisms and roles for genetic and epigenetic modulation. *Gastroenterology*. 2012;143(3):550–563.
2. Rothwell PM, Wilson M, Elwin CE, et al. Long-term effect of aspirin on colorectal cancer incidence and mortality: 20-year follow-up of five randomised trials. *Lancet*. 2010;376(9754):1741–1750.
3. Baylin SB, Jones PA. Epigenetic determinants of cancer. *Cold Spring Harb Perspect Biol*. 2016;8(9):a019505.
4. Coussens LM, Werb Z. Inflammation and cancer. *Nature*. 2002;420(6917):860–867.
5. Hanahan D, Weinberg RA. Hallmarks of cancer: the next generation. *Cell*. 2011;144(5):646–674.
6. Nowarski R, Gagliani N, Huber S, Flavell RA. Innate immune cells in inflammation and cancer. *Cancer Immunol Res*. 2013;1(2):77–84.
7. Ivashkiv LB. Epigenetic regulation of macrophage polarization and function. *Trends Immunol*. 2013;34(5):216–223.
8. Chan CJ, Smyth MJ, Martinet L. Molecular mechanisms of natural killer cell activation in response to cellular stress. *Cell Death Differ*. 2014;21(1):5–14.
9. Vantourout P, Hayday A. Six-of-the-best: unique contributions of gammadelta T cells to immunology. *Nat Rev Immunol*. 2013;13(2):88–100.
10. Yu H, Pardoll D, Jove R. STATs in cancer inflammation and immunity: a leading role for STAT3. *Nat Rev Cancer*. 2009;9(11):798–809.
11. Karin M, Cao Y, Greten FR, Li ZW. NF-kappaB in cancer: from innocent bystander to major culprit. *Nat Rev Cancer*. 2002;2(4):301–310.
12. Gajewski TF, Schreiber H, Fu YX. Innate and adaptive immune cells in the tumor microenvironment. *Nat Immunol*. 2013;14(10):1014–1022.
13. Woo SR, Corrales L, Gajewski TF. Innate immune recognition of cancer. *Annu Rev Immunol*. 2015;33:445–474.
14. Marvel D, Gabrilovich DI. Myeloid-derived suppressor cells in the tumor microenvironment: expect the unexpected. *J Clin Invest*. 2015;125(9):3356–3364.
15. Wilson CB, Rowell E, Sekimata M. Epigenetic control of T-helper-cell differentiation. *Nat Rev Immunol*. 2009;9(2):91–105.
16. Djuretic IM, Levanon D, Negreanu V, Groner Y, Rao A, Ansel KM. Transcription factors T-bet and Runx3 cooperate to activate Ifng and silence Il4 in T helper type 1 cells. *Nat Immunol*. 2007;8(2):145–153.
17. Ansel KM, Djuretic I, Tanasa B, Rao A. Regulation of Th2 differentiation and Il4 locus accessibility. *Annu Rev Immunol*. 2006;24:607–656.
18. McGeachy MJ, Cua DJ. Th17 cell differentiation: the long and winding road. *Immunity*. 2008;28(4):445–453.
19. Miossec P, Kolls JK. Targeting IL-17 and TH17 cells in chronic inflammation. *Nat Rev Drug Discov*. 2012;11(10):763–776.
20. Nishikawa H, Sakaguchi S. Regulatory T cells in cancer immunotherapy. *Curr Opin Immunol*. 2014;27:1–7.
21. Shankaran V, Ikeda H, Bruce AT, et al. IFNgamma and lymphocytes prevent primary tumour development and shape tumour immunogenicity. *Nature*. 2001;410(6832):1107–1111.
22. Kaplan DH, Shankaran V, Dighe AS, et al. Demonstration of an interferon gamma-dependent tumor surveillance system in immunocompetent mice. *Proc Natl Acad Sci U S A*. 1998;95(13):7556–7561.
23. Schumacher TN, Schreiber RD. Neoantigens in cancer immunotherapy. *Science*. 2015;348(6230):69–74.

24. Mortarini R, Piris A, Maurichi A, et al. Lack of terminally differentiated tumor-specific CD8+ T cells at tumor site in spite of antitumor immunity to self-antigens in human metastatic melanoma. *Cancer Res.* 2003;63(10):2535–2545.
25. Gajewski TF. Failure at the effector phase: immune barriers at the level of the melanoma tumor microenvironment. *Clin Cancer Res.* 2007;13(18 Pt 1):5256–5261.
26. Kryczek I, Banerjee M, Cheng P, et al. Phenotype, distribution, generation, and functional and clinical relevance of Th17 cells in the human tumor environments. *Blood.* 2009;114(6):1141–1149.
27. Wu S, Rhee KJ, Albesiano E, et al. A human colonic commensal promotes colon tumorigenesis via activation of T helper type 17T cell responses. *Nat Med.* 2009;15(9): 1016–1022.
28. Uemura N, Okamoto S, Yamamoto S, et al. Helicobacter pylori infection and the development of gastric cancer. *N Engl J Med.* 2001;345(11):784–789.
29. Smythies LE, Waites KB, Lindsey JR, Harris PR, Ghiara P, Smith PD. Helicobacter pylori-induced mucosal inflammation is Th1 mediated and exacerbated in IL-4, but not IFN-gamma, gene-deficient mice. *J Immunol.* 2000;165(2):1022–1029.
30. Sue S, Shibata W, Maeda S. Helicobacter pylori-induced signaling pathways contribute to intestinal metaplasia and gastric carcinogenesis. *Biomed Res Int.* 2015;2015:737621.
31. Tu S, Bhagat G, Cui G, et al. Overexpression of interleukin-1beta induces gastric inflammation and cancer and mobilizes myeloid-derived suppressor cells in mice. *Cancer Cell.* 2008;14(5):408–419.
32. Huang JQ, Zheng GF, Sumanac K, Irvine EJ, Hunt RH. Meta-analysis of the relationship between cagA seropositivity and gastric cancer. *Gastroenterology.* 2003;125(6): 1636–1644.
33. Ohnishi N, Yuasa H, Tanaka S, et al. Transgenic expression of Helicobacter pylori CagA induces gastrointestinal and hematopoietic neoplasms in mouse. *Proc Natl Acad Sci U S A.* 2008;105(3):1003–1008.
34. Yong X, Tang B, Li BS, et al. Helicobacter pylori virulence factor CagA promotes tumorigenesis of gastric cancer via multiple signaling pathways. *Cell Commun Signal.* 2015;13:30.
35. Hajarizadeh B, Grebely J, Dore GJ. Epidemiology and natural history of HCV infection. *Nat Rev Gastroenterol Hepatol.* 2013;10(9):553–562.
36. Gong G, Waris G, Tanveer R, Siddiqui A. Human hepatitis C virus NS5A protein alters intracellular calcium levels, induces oxidative stress, and activates STAT-3 and NF-kappa B. *Proc Natl Acad Sci U S A.* 2001;98(17):9599–9604.
37. Mitchell JK, Lemon SM, McGivern DR. How do persistent infections with hepatitis C virus cause liver cancer? *Curr Opin Virol.* 2015;14:101–108.
38. Wild CP, Hardie LJ. Reflux, Barrett's oesophagus and adenocarcinoma: burning questions. *Nat Rev Cancer.* 2003;3(9):676–684.
39. Fitzgerald RC, Omary MB, Triadafilopoulos G. Dynamic effects of acid on Barrett's esophagus. An ex vivo proliferation and differentiation model. *J Clin Invest.* 1996;98(9): 2120–2128.
40. Abdel-Latif MM, Duggan S, Reynolds JV, Kelleher D. Inflammation and esophageal carcinogenesis. *Curr Opin Pharmacol.* 2009;9(4):396–404.
41. O'Riordan JM, Abdel-latif MM, Ravi N, et al. Proinflammatory cytokine and nuclear factor kappa-B expression along the inflammation-metaplasia-dysplasia-adenocarcinoma sequence in the esophagus. *Am J Gastroenterol.* 2005;100(6):1257–1264.
42. Dvorak K, Chavarria M, Payne CM, et al. Activation of the interleukin-6/STAT3 anti-apoptotic pathway in esophageal cells by bile acids and low pH: relevance to Barrett's esophagus. *Clin Cancer Res.* 2007;13(18 Pt 1):5305–5313.
43. Eaden JA, Abrams KR, Mayberry JF. The risk of colorectal cancer in ulcerative colitis: a meta-analysis. *Gut.* 2001;48(4):526–535.

44. Potack J, Itzkowitz SH. Colorectal cancer in inflammatory bowel disease. *Gut Liver.* 2008;2(2):61–73.
45. Rubin DT, Cruz-Correa MR, Gasche C, et al. Colorectal cancer prevention in inflammatory bowel disease and the role of 5-aminosalicylic acid: a clinical review and update. *Inflamm Bowel Dis.* 2008;14(2):265–274.
46. Cooper HS, Everley L, Chang WC, et al. The role of mutant Apc in the development of dysplasia and cancer in the mouse model of dextran sulfate sodium-induced colitis. *Gastroenterology.* 2001;121(6):1407–1416.
47. Itzkowitz SH, Yio X. Inflammation and cancer IV. Colorectal cancer in inflammatory bowel disease: the role of inflammation. *Am J Physiol Gastrointest Liver Physiol.* 2004;287(1):G7–G17.
48. Madan V, Lear JT, Szeimies RM. Non-melanoma skin cancer. *Lancet.* 2010;375(9715): 673–685.
49. Tang L, Wang K. Chronic inflammation in skin malignancies. *J Mol Signal.* 2016;11(1):2.
50. Jacob F, Monod J. Genetic regulatory mechanisms in the synthesis of proteins. *J Mol Biol.* 1961;3:318–356.
51. Britten RJ, Davidson EH. Gene regulation for higher cells: a theory. *Science.* 1969;165(3891):349–357.
52. Eddy SR. Non-coding RNA, genes and the modern RNA world. *Nat Rev Genet.* 2001;2(12):919–929.
53. Jain S, Thakkar N, Chhatai J, Bhadra MP, Bhadra U. Long non-coding RNA: functional agent for disease traits. *RNA Biol.* 2016;1–14.
54. Kelley RL, Kuroda MI. Noncoding RNA genes in dosage compensation and imprinting. *Cell.* 2000;103(1):9–12.
55. Cosgrove MS, Boeke JD, Wolberger C. Regulated nucleosome mobility and the histone code. *Nat Struct Mol Biol.* 2004;11(11):1037–1043.
56. Akhtar A, Becker PB. Activation of transcription through histone H4 acetylation by MOF, an acetyltransferase essential for dosage compensation in Drosophila. *Mol Cell* 2000;5(2):367–375.
57. Ren Q, Gorovsky MA. Histone H2A.Z acetylation modulates an essential charge patch. *Mol Cell.* 2001;7(6):1329–1335.
58. Tie F, Banerjee R, Stratton CA, et al. CBP-mediated acetylation of histone H3 lysine 27 antagonizes Drosophila Polycomb silencing. *Development.* 2009;136(18):3131–3141.
59. Yang XJ, Seto E. HATs and HDACs: from structure, function and regulation to novel strategies for therapy and prevention. *Oncogene.* 2007;26(37):5310–5318.
60. Cao R, Wang L, Wang H, et al. Role of histone H3 lysine 27 methylation in Polycomb-group silencing. *Science.* 2002;298(5595):1039–1043.
61. Tessarz P, Kouzarides T. Histone core modifications regulating nucleosome structure and dynamics. *Nat Rev Mol Cell Biol.* 2014;15(11):703–708.
62. Forneris F, Binda C, Battaglioli E, Mattevi A. LSD1: oxidative chemistry for multifaceted functions in chromatin regulation. *Trends Biochem Sci.* 2008;33(4):181–189.
63. Lawrence M, Daujat S, Schneider R. Lateral thinking: how histone modifications regulate gene expression. *Trends Genet.* 2016;32(1):42–56.
64. Bergman Y, Cedar H. DNA methylation dynamics in health and disease. *Nat Struct Mol Biol.* 2013;20(3):274–281.
65. Kiefer JC. Epigenetics in development. *Dev Dyn.* 2007;236(4):1144–1156.
66. Jjingo D, Conley AB, Yi SV, Lunyak VV, Jordan IK. On the presence and role of human gene-body DNA methylation. *Oncotarget.* 2012;3(4):462–474.
67. Wang Y, Jadhav RR, Liu J, et al. Roles of distal and genic methylation in the development of prostate tumorigenesis revealed by genome-wide DNA methylation analysis. *Sci Rep.* 2016;6:22051.

68. Clements EG, Mohammad HP, Leadem BR, et al. DNMT1 modulates gene expression without its catalytic activity partially through its interactions with histone-modifying enzymes. *Nucleic Acids Res.* 2012;40(10):4334–4346.
69. Oswald J, Engemann S, Lane N, et al. Active demethylation of the paternal genome in the mouse zygote. *Curr Biol.* 2000;10(8):475–478.
70. Mayer W, Niveleau A, Walter J, Fundele R, Haaf T. Demethylation of the zygotic paternal genome. *Nature.* 2000;403(6769):501–502.
71. Butler MG. Genomic imprinting disorders in humans: a mini-review. *J Assist Reprod Genet.* 2009;26(9–10):477–486.
72. Margueron R, Reinberg D. The Polycomb complex PRC2 and its mark in life. *Nature.* 2011;469(7330):343–349.
73. Kennison JA. The Polycomb and trithorax group proteins of Drosophila: trans-regulators of homeotic gene function. *Annu Rev Genet.* 1995;29:289–303.
74. Lewis EB. A gene complex controlling segmentation in Drosophila. *Nature.* 1978;276(5688):565–570.
75. Struhl G. A gene product required for correct initiation of segmental determination in Drosophila. *Nature.* 1981;293(5827):36–41.
76. Simon JA, Kingston RE. Occupying chromatin: polycomb mechanisms for getting to genomic targets, stopping transcriptional traffic, and staying put. *Mol Cell.* 2013;49(5):808–824.
77. Gil J, O'Loghlen A. PRC1 complex diversity: where is it taking us? *Trends Cell Biol.* 2014;24(11):632–641.
78. Wang H, Wang L, Erdjument-Bromage H, et al. Role of histone H2A ubiquitination in Polycomb silencing. *Nature.* 2004;431(7010):873–878.
79. Endoh M, Endo TA, Endoh T, et al. Histone H2A mono-ubiquitination is a crucial step to mediate PRC1-dependent repression of developmental genes to maintain ES cell identity. *PLoS Genet.* 2012;8(7):e1002774.
80. Wu X, Gong Y, Yue J, Qiang B, Yuan J, Peng X. Cooperation between EZH2, NSPc1-mediated histone H2A ubiquitination and Dnmt1 in HOX gene silencing. *Nucleic Acids Res.* 2008;36(11):3590–3599.
81. Simon JA, Kingston RE. Mechanisms of polycomb gene silencing: knowns and unknowns. *Nat Rev Mol Cell Biol.* 2009;10(10):697–708.
82. Payer B, Lee JT. X chromosome dosage compensation: how mammals keep the balance. *Annu Rev Genet.* 2008;42:733–772.
83. Wu HA, Bernstein E. Partners in imprinting: noncoding RNA and polycomb group proteins. *Dev Cell.* 2008;15(5):637–638.
84. Ballas ZK. The use of 5-azacytidine to establish constitutive interleukin 2-producing clones of the EL4 thymoma. *J Immunol.* 1984;133(1):7–9.
85. Young HA, Ghosh P, Ye J, et al. Differentiation of the T helper phenotypes by analysis of the methylation state of the IFN-gamma gene. *J Immunol.* 1994;153(8):3603–3610.
86. Makar KW, Wilson CB. DNA methylation is a nonredundant repressor of the Th2 effector program. *J Immunol.* 2004;173(7):4402–4406.
87. Hutchins AS, Mullen AC, Lee HW, et al. Gene silencing quantitatively controls the function of a developmental trans-activator. *Mol Cell.* 2002;10(1):81–91.
88. Lee PP, Fitzpatrick DR, Beard C, et al. A critical role for Dnmt1 and DNA methylation in T cell development, function, and survival. *Immunity.* 2001;15(5):763–774.
89. Zhang F, Boothby M. T helper type 1-specific Brg1 recruitment and remodeling of nucleosomes positioned at the IFN-gamma promoter are Stat4 dependent. *J Exp Med.* 2006;203(6):1493–1505.
90. Agarwal S, Rao A. Modulation of chromatin structure regulates cytokine gene expression during T cell differentiation. *Immunity.* 1998;9(6):765–775.

91. Tong Y, Aune T, Boothby M. T-bet antagonizes mSin3a recruitment and transactivates a fully methylated IFN-gamma promoter via a conserved T-box half-site. *Proc Natl Acad Sci U S A*. 2005;102(6):2034–2039.
92. Chen GY, Osada H, Santamaria-Babi LF, Kannagi R. Interaction of GATA-3/T-bet transcription factors regulates expression of sialyl Lewis X homing receptors on Th1/ Th2 lymphocytes. *Proc Natl Acad Sci U S A*. 2006;103(45):16894–16899.
93. Ivanov II, Zhou L, Littman DR. Transcriptional regulation of Th17 cell differentiation. *Semin Immunol*. 2007;19(6):409–417.
94. Mosser DM, Edwards JP. Exploring the full spectrum of macrophage activation. *Nat Rev Immunol*. 2008;8(12):958–969.
95. Gordon S, Martinez FO. Alternative activation of macrophages: mechanism and functions. *Immunity*. 2010;32(5):593–604.
96. Dabritz J, Menheniott TR. Linking immunity, epigenetics, and cancer in inflammatory bowel disease. *Inflamm Bowel Dis*. 2014;20(9):1638–1654.
97. Yang X, Wang X, Liu D, Yu L, Xue B, Shi H. Epigenetic regulation of macrophage polarization by DNA methyltransferase 3b. *Mol Endocrinol*. 2014;28(4):565–574.
98. Berman BP, Weisenberger DJ, Aman JF, et al. Regions of focal DNA hypermethylation and long-range hypomethylation in colorectal cancer coincide with nuclear lamina-associated domains. *Nat Genet*. 2012;44(1):40–46.
99. Hansen KD, Timp W, Bravo HC, et al. Increased methylation variation in epigenetic domains across cancer types. *Nat Genet*. 2011;43(8):768–775.
100. Zelic R, Fiano V, Zugna D, et al. Global hypomethylation (LINE-1) and gene-specific hypermethylation (GSTP1) on initial negative prostate biopsy as markers of prostate cancer on a rebiopsy. *Clin Cancer Res*. 2016;22(4):984–992.
101. Grady WM, Willis J, Guilford PJ, et al. Methylation of the CDH1 promoter as the second genetic hit in hereditary diffuse gastric cancer. *Nat Genet*. 2000;26(1):16–17.
102. Machado JC, Oliveira C, Carvalho R, et al. E-cadherin gene (CDH1) promoter methylation as the second hit in sporadic diffuse gastric carcinoma. *Oncogene*. 2001;20(12):1525–1528.
103. Baba S, Yamada Y, Hatano Y, et al. Global DNA hypomethylation suppresses squamous carcinogenesis in the tongue and esophagus. *Cancer Sci*. 2009;100(7): 1186–1191.
104. Sheaffer KL, Elliott EN, Kaestner KH. DNA hypomethylation contributes to genomic instability and intestinal cancer initiation. *Cancer Prev Res (Phila)*. 2016;9(7):534–546.
105. Tahiliani M, Koh KP, Shen Y, et al. Conversion of 5-methylcytosine to 5-hydroxymethylcytosine in mammalian DNA by MLL partner TET1. *Science*. 2009;324(5929):930–935.
106. Shukla V, Coumoul X, Lahusen T, et al. BRCA1 affects global DNA methylation through regulation of DNMT1. *Cell Res*. 2010;20(11):1201–1215.
107. Dizik M, Christman JK, Wainfan E. Alterations in expression and methylation of specific genes in livers of rats fed a cancer promoting methyl-deficient diet. *Carcinogenesis*. 1991;12(7):1307–1312.
108. Yamada Y, Jackson-Grusby L, Linhart H, et al. Opposing effects of DNA hypomethylation on intestinal and liver carcinogenesis. *Proc Natl Acad Sci U S A*. 2005;102(38): 13580–13585.
109. Maio M, Covre A, Fratta E, et al. Molecular pathways: at the crossroads of cancer epigenetics and immunotherapy. *Clin Cancer Res*. 2015;21(18):4040–4047.
110. Waldhauer I, Steinle A. NK cells and cancer immunosurveillance. *Oncogene*. 2008;27(45): 5932–5943.
111. Sigalotti L, Fratta E, Coral S, Maio M. Epigenetic drugs as immunomodulators for combination therapies in solid tumors. *Pharmacol Ther*. 2014;142(3):339–350.

112. Diermayr S, Himmelreich H, Durovic B, et al. NKG2D ligand expression in AML increases in response to HDAC inhibitor valproic acid and contributes to allorecognition by NK-cell lines with single KIR-HLA class I specificities. *Blood*. 2008;111(3):1428–1436.
113. Yamanegi K, Yamane J, Kobayashi K, et al. Sodium valproate, a histone deacetylase inhibitor, augments the expression of cell-surface NKG2D ligands, MICA/B, without increasing their soluble forms to enhance susceptibility of human osteosarcoma cells to NK cell-mediated cytotoxicity. *Oncol Rep*. 2010;24(6):1621–1627.
114. Zhang C, Wang Y, Zhou Z, Zhang J, Tian Z. Sodium butyrate upregulates expression of NKG2D ligand MICA/B in HeLa and HepG2 cell lines and increases their susceptibility to NK lysis. *Cancer Immunol Immunother*. 2009;58(8):1275–1285.
115. Armeanu S, Bitzer M, Lauer UM, et al. Natural killer cell-mediated lysis of hepatoma cells via specific induction of NKG2D ligands by the histone deacetylase inhibitor sodium valproate. *Cancer Res*. 2005;65(14):6321–6329.
116. Sers C, Kuner R, Falk CS, et al. Down-regulation of HLA Class I and NKG2D ligands through a concerted action of MAPK and DNA methyltransferases in colorectal cancer cells. *Int J Cancer*. 2009;125(7):1626–1639.
117. Ushijima T, Sasako M. Focus on gastric cancer. *Cancer Cell*. 2004;5(2):121–125.
118. Chan AO, Lam SK, Wong BC, et al. Promoter methylation of E-cadherin gene in gastric mucosa associated with Helicobacter pylori infection and in gastric cancer. *Gut*. 2003;52(4):502–506.
119. Maekita T, Nakazawa K, Mihara M, et al. High levels of aberrant DNA methylation in Helicobacter pylori-infected gastric mucosae and its possible association with gastric cancer risk. *Clin Cancer Res*. 2006;12(3 Pt 1):989–995.
120. Ushijima T, Hattori N. Molecular pathways: involvement of Helicobacter pylori-triggered inflammation in the formation of an epigenetic field defect, and its usefulness as cancer risk and exposure markers. *Clin Cancer Res*. 2012;18(4):923–929.
121. Nishida N, Nagasaka T, Nishimura T, Ikai I, Boland CR, Goel A. Aberrant methylation of multiple tumor suppressor genes in aging liver, chronic hepatitis, and hepatocellular carcinoma. *Hepatology*. 2008;47(3):908–918.
122. Eads CA, Lord RV, Kurumboor SK, et al. Fields of aberrant CpG island hypermethylation in Barrett's esophagus and associated adenocarcinoma. *Cancer Res*. 2000;60(18):5021–5026.
123. Hsieh CJ, Klump B, Holzmann K, Borchard F, Gregor M, Porschen R. Hypermethylation of the p16INK4a promoter in colectomy specimens of patients with long-standing and extensive ulcerative colitis. *Cancer Res*. 1998;58(17):3942–3945.
124. Yi JM, Kim TO. Epigenetic alterations in inflammatory bowel disease and cancer. *Intest Res*. 2015;13(2):112–121.
125. Sathyanarayana UG, Moore AY, Li L, et al. Sun exposure related methylation in malignant and non-malignant skin lesions. *Cancer Lett*. 2007;245(1–2):112–120.
126. Niwa T, Tsukamoto T, Toyoda T, et al. Inflammatory processes triggered by Helicobacter pylori infection cause aberrant DNA methylation in gastric epithelial cells. *Cancer Res*. 2010;70(4):1430–1440.
127. Nandakumar V, Vaid M, Tollefsbol TO, Katiyar SK. Aberrant DNA hypermethylation patterns lead to transcriptional silencing of tumor suppressor genes in UVB-exposed skin and UVB-induced skin tumors of mice. *Carcinogenesis*. 2011;32(4):597–604.
128. Katsurano M, Niwa T, Yasui Y, et al. Early-stage formation of an epigenetic field defect in a mouse colitis model, and non-essential roles of T- and B-cells in DNA methylation induction. *Oncogene*. 2012;31(3):342–351.
129. De Robertis M, Massi E, Poeta ML, et al. The AOM/DSS murine model for the study of colon carcinogenesis: from pathways to diagnosis and therapy studies. *J Carcinog*. 2011;10:9.

130. Guo Y, Liu Y, Zhang C, et al. The epigenetic effects of aspirin: the modification of histone H3 lysine 27 acetylation in the prevention of colon carcinogenesis in azoxymethane- and dextran sulfate sodium-treated CF-1 mice. *Carcinogenesis*. 2016;37(6):616–624.
131. Villagra A, Sotomayor EM, Seto E. Histone deacetylases and the immunological network: implications in cancer and inflammation. *Oncogene*. 2010;29(2):157–173.
132. Abu-Remaileh M, Bender S, Raddatz G, et al. Chronic inflammation induces a novel epigenetic program that is conserved in intestinal adenomas and in colorectal cancer. *Cancer Res*. 2015;75(10):2120–2130.
133. Hahn MA, Hahn T, Lee DH, et al. Methylation of polycomb target genes in intestinal cancer is mediated by inflammation. *Cancer Res*. 2008;68(24):10280–10289.
134. Hmadcha A, Bedoya FJ, Sobrino F, Pintado E. Methylation-dependent gene silencing induced by interleukin 1beta via nitric oxide production. *J Exp Med*. 1999;190(11):1595–1604.
135. Nott A, Watson PM, Robinson JD, Crepaldi L, Riccio A. S-nitrosylation of histone deacetylase 2 induces chromatin remodelling in neurons. *Nature*. 2008;455(7211):411–415.
136. Hodge DR, Xiao W, Clausen PA, Heidecker G, Szyf M, Farrar WL. Interleukin-6 regulation of the human DNA methyltransferase (HDNMT) gene in human erythroleukemia cells. *J Biol Chem*. 2001;276(43):39508–39511.
137. Braconi C, Huang N, Patel T. MicroRNA-dependent regulation of DNA methyltransferase-1 and tumor suppressor gene expression by interleukin-6 in human malignant cholangiocytes. *Hepatology*. 2010;51(3):881–890.
138. Xia D, Wang D, Kim SH, Katoh H, DuBois RN. Prostaglandin E2 promotes intestinal tumor growth via DNA methylation. *Nat Med*. 2012;18(2):224–226.
139. Kominsky DJ, Campbell EL, Colgan SP. Metabolic shifts in immunity and inflammation. *J Immunol*. 2010;184(8):4062–4068.
140. Demaria M, Giorgi C, Lebiedzinska M, et al. A STAT3-mediated metabolic switch is involved in tumour transformation and STAT3 addiction. *Aging*. 2010;2(11):823–842.
141. Johnson C, Warmoes MO, Shen X, Locasale JW. Epigenetics and cancer metabolism. *Cancer Lett*. 2015;356(2 Pt A):309–314.
142. Zhang R, Kang KA, Kim KC, et al. Oxidative stress causes epigenetic alteration of CDX1 expression in colorectal cancer cells. *Gene*. 2013;524(2):214–219.
143. Kang KA, Zhang R, Kim GY, Bae SC, Hyun JW. Epigenetic changes induced by oxidative stress in colorectal cancer cells: methylation of tumor suppressor RUNX3. *Tumour Biol*. 2012;33(2):403–412.
144. Steenken S, Jovanovic SV. How easily oxidizable is DNA? One-electron reduction potentials of adenosine and guanosine radicals in aqueous solution. *J Am Chem Soc*. 1997;119(3):617–618.
145. Franco R, Schoneveld O, Georgakilas AG, Panayiotidis MI. Oxidative stress, DNA methylation and carcinogenesis. *Cancer Lett*. 2008;266(1):6–11.
146. Ziech D, Franco R, Pappa A, Panayiotidis MI. Reactive oxygen species (ROS)-induced genetic and epigenetic alterations in human carcinogenesis. *Mutat Res*. 2011;711(1–2):167–173.
147. Lim SO, Gu JM, Kim MS, et al. Epigenetic changes induced by reactive oxygen species in hepatocellular carcinoma: methylation of the E-cadherin promoter. *Gastroenterology*. 2008;135(6):2128–2140. 2140.e2121–2128.
148. Polo SE, Almouzni G. Chromatin dynamics after DNA damage: the legacy of the access-repair-restore model. *DNA Repair*. 2015;36:114–121.
149. O'Hagan HM. Chromatin modifications during repair of environmental exposure-induced DNA damage: a potential mechanism for stable epigenetic alterations. *Environ Mol Mutagen*. 2014;55(3):278–291.

150. Shanbhag NM, Greenberg RA. Neighborly DISCourse: DNA double strand breaks silence transcription. *Cell Cycle*. 2010;9(18):3635–3636.
151. Pankotai T, Bonhomme C, Chen D, Soutoglou E. DNAPKcs-dependent arrest of RNA polymerase II transcription in the presence of DNA breaks. *Nat Struct Mol Biol* 2012;19(3):276–282.
152. O'Hagan HM, Mohammad HP, Baylin SB. Double strand breaks can initiate gene silencing and SIRT1-dependent onset of DNA methylation in an exogenous promoter CpG island. *PLoS Genet*. 2008;4(8):e1000155.
153. O'Hagan HM, Wang W, Sen S, et al. Oxidative damage targets complexes containing DNA methyltransferases, SIRT1, and polycomb members to promoter CpG Islands. *Cancer Cell*. 2011;20(5):606–619.
154. Ding N, Bonham EM, Hannon BE, Amick TR, Baylin SB, O'Hagan HM. Mismatch repair proteins recruit DNA methyltransferase 1 to sites of oxidative DNA damage. *J Mol Cell Biol*. 2016;8(3):244–254.
155. Zeller C, Dai W, Steele NL, et al. Candidate DNA methylation drivers of acquired cisplatin resistance in ovarian cancer identified by methylome and expression profiling. *Oncogene*. 2012;31(42):4567–4576.
156. Nogales V, Reinhold WC, Varma S, et al. Epigenetic inactivation of the putative DNA/RNA helicase SLFN11 in human cancer confers resistance to platinum drugs. *Oncotarget*. 2016;7(3):3084–3097.
157. Feinberg AP, Koldobskiy MA, Gondor A. Epigenetic modulators, modifiers and mediators in cancer aetiology and progression. *Nat Rev Genet*. 2016;17(5):284–299.
158. Rothwell PM, Fowkes FG, Belch JF, Ogawa H, Warlow CP, Meade TW. Effect of daily aspirin on long-term risk of death due to cancer: analysis of individual patient data from randomised trials. *Lancet*. 2011;377(9759):31–41.
159. Yiannakopoulou E. Targeting epigenetic mechanisms and microRNAs by aspirin and other non steroidal anti-inflammatory agents—implications for cancer treatment and chemoprevention. *Cell Oncol (Dordr)*. 2014;37(3):167–178.
160. Crusz SM, Balkwill FR. Inflammation and cancer: advances and new agents. *Nat Rev Clin Oncol*. 2015;12(10):584–596.
161. Kroesen M, Gielen P, Brok IC, Armandari I, Hoogerbrugge PM, Adema GJ. HDAC inhibitors and immunotherapy; a double edged sword? *Oncotarget*. 2014;5(16):6558–6572.
162. Berni Canani R, Di Costanzo M, Leone L. The epigenetic effects of butyrate: potential therapeutic implications for clinical practice. *Clin Epigenetics*. 2012;4(1):4.
163. Nebbioso A, Carafa V, Benedetti R, Altucci L. Trials with 'epigenetic' drugs: an update. *Mol Oncol*. 2012;6(6):657–682.
164. Ashburner BP, Westerheide SD, Baldwin Jr AS. The p65 (RelA) subunit of NF-kappaB interacts with the histone deacetylase (HDAC) corepressors HDAC1 and HDAC2 to negatively regulate gene expression. *Mol Cell Biol*. 2001;21(20): 7065–7077.
165. Ziesche E, Kettner-Buhrow D, Weber A, et al. The coactivator role of histone deacetylase 3 in IL-1-signaling involves deacetylation of p65 NF-kappaB. *Nucleic Acids Res*. 2013;41(1):90–109.
166. Sun Y, Chin YE, Weisiger E, et al. Cutting edge: negative regulation of dendritic cells through acetylation of the nonhistone protein STAT-3. *J Immunol*. 2009;182(10): 5899–5903.
167. Bridle BW, Chen L, Lemay CG, et al. HDAC inhibition suppresses primary immune responses, enhances secondary immune responses, and abrogates autoimmunity during tumor immunotherapy. *Mol Ther*. 2013;21(4):887–894.
168. Choi SW, Gatza E, Hou G, et al. Histone deacetylase inhibition regulates inflammation and enhances Tregs after allogeneic hematopoietic cell transplantation in humans. *Blood*. 2015;125(5):815–819.

169. Reddy P, Sun Y, Toubai T, et al. Histone deacetylase inhibition modulates indoleamine 2, 3-dioxygenase-dependent DC functions and regulates experimental graft-versus-host disease in mice. *J Clin Invest.* 2008;118(7):2562–2573.
170. Sonnemann J, Greßmann S, Becker S, Wittig S, Schmudde M, Beck JF. The histone deacetylase inhibitor vorinostat induces calreticulin exposure in childhood brain tumour cells in vitro. *Cancer Chemother Pharmacol.* 2010;66(3):611–616.
171. Fiegler N, Textor S, Arnold A, et al. Downregulation of the activating NKp30 ligand B7-H6 by HDAC inhibitors impairs tumor cell recognition by NK cells. *Blood.* 2013;122(5):684–693.
172. Halili MA, Andrews MR, Labzin LI, et al. Differential effects of selective HDAC inhibitors on macrophage inflammatory responses to the Toll-like receptor 4 agonist LPS. *J Leukoc Biol.* 2010;87(6):1103–1114.
173. Davie JR. Inhibition of histone deacetylase activity by butyrate. *J Nutr.* 2003;133(7): 2485S–2493S.
174. Spurling CC, Suhl JA, Boucher N, Nelson CE, Rosenberg DW, Giardina C. The short chain fatty acid butyrate induces promoter demethylation and reactivation of RARβ2 in colon cancer cells. *Nutr Cancer.* 2008;60(5):692–702.
175. Inan MS, Rasoulpour RJ, Yin L, Hubbard AK, Rosenberg DW, Giardina C. The luminal short-chain fatty acid butyrate modulates NF-κB activity in a human colonic epithelial cell line. *Gastroenterology.* 2000;118(4):724–734.
176. Hallert C, Björck I, Nyman M, Pousette A, Grännö C, Svensson H. Increasing fecal butyrate in ulcerative colitis patients by diet: controlled pilot study. *Inflamm Bowel Dis.* 2003;9(2):116–121.
177. Pruitt K, Zinn RL, Ohm JE, et al. Inhibition of SIRT1 reactivates silenced cancer genes without loss of promoter DNA hypermethylation. *PLoS Genet.* 2006;2(3):e40.
178. Vaziri H, Dessain SK, Eaton EN, et al. hSIR2SIRT1 functions as an NAD-dependent p53 deacetylase. *Cell.* 2001;107(2):149–159.
179. Dai JM, Wang ZY, Sun DC, Lin RX, Wang SQ. SIRT1 interacts with p73 and suppresses p73-dependent transcriptional activity. *J Cell Physiol.* 2007;210(1):161–166.
180. Chen WY, Wang DH, Yen RC, Luo J, Gu W, Baylin SB. Tumor suppressor HIC1 directly regulates SIRT1 to modulate p53-dependent DNA-damage responses. *Cell.* 2005;123(3):437–448.
181. Yeung F, Hoberg JE, Ramsey CS, et al. Modulation of NF-κB-dependent transcription and cell survival by the SIRT1 deacetylase. *EMBO J.* 2004;23(12): 2369–2380.
182. Liu TF, McCall CE. Deacetylation by SIRT1 reprograms inflammation and cancer. *Genes Cancer.* 2013;4(3–4):135–147.
183. Baur JA, Sinclair DA. Therapeutic potential of resveratrol: the in vivo evidence. *Nat Rev Drug Discov.* 2006;5(6):493–506.
184. Yang H, Zhang W, Pan H, et al. SIRT1 activators suppress inflammatory responses through promotion of p65 deacetylation and inhibition of NF-κB activity. *PLoS One.* 2012;7(9):e46364.
185. Lynch JT, Harris WJ, Somervaille TC. LSD1 inhibition: a therapeutic strategy in cancer? *Expert Opin Ther Targets.* 2012;16(12):1239–1249.
186. Lv T, Yuan D, Miao X, et al. Over-expression of LSD1 promotes proliferation, migration and invasion in non-small cell lung cancer. *PLoS One.* 2012;7(4):e35065.
187. Schenk T, Chen WC, Göllner S, et al. Inhibition of the LSD1 (KDM1A) demethylase reactivates the all-trans-retinoic acid differentiation pathway in acute myeloid leukemia. *Nat Med.* 2012;18(4):605–611.
188. Harris WJ, Huang X, Lynch JT, et al. The histone demethylase KDM1A sustains the oncogenic potential of MLL-AF9 leukemia stem cells. *Cancer Cell.* 2012;21(4): 473–487.

189. Mohammad HP, Smitheman KN, Kamat CD, et al. A DNA hypomethylation signature predicts antitumor activity of LSD1 inhibitors in SCLC. *Cancer Cell.* 2015;28(1):57–69.
190. Liu D, Zempleni J. Low activity of LSD1 elicits a pro-inflammatory gene expression profile in riboflavin-deficient human T Lymphoma Jurkat cells. *Genes Nutr.* 2014;9(5):1–8.
191. Reddy MA, Villeneuve LM, Wang M, Lanting L, Natarajan R. Role of the lysine-specific demethylase 1 in the proinflammatory phenotype of vascular smooth muscle cells of diabetic mice. *Circ Res.* 2008;103(6):615–623.
192. French CA, Miyoshi I, Aster JC, et al. BRD4 bromodomain gene rearrangement in aggressive carcinoma with translocation t (15; 19). *Am J Pathol.* 2001;159(6):1987–1992.
193. Chaidos A, Caputo V, Karadimitris A. Inhibition of bromodomain and extra-terminal proteins (BET) as a potential therapeutic approach in haematological malignancies: emerging preclinical and clinical evidence. *Ther Adv Hematol.* 2015;6(3):128–141.
194. Chapuy B, McKeown MR, Lin CY, et al. Discovery and characterization of super-enhancer-associated dependencies in diffuse large B cell lymphoma. *Cancer Cell.* 2013;24(6):777–790.
195. Lovén J, Hoke HA, Lin CY, et al. Selective inhibition of tumor oncogenes by disruption of super-enhancers. *Cell.* 2013;153(2):320–334.
196. Hargreaves DC, Horng T, Medzhitov R. Control of inducible gene expression by signal-dependent transcriptional elongation. *Cell.* 2009;138(1):129–145.
197. Nicodeme E, Jeffrey KL, Schaefer U, et al. Suppression of inflammation by a synthetic histone mimic. *Nature.* 2010;468(7327):1119–1123.
198. Zhang G, Liu R, Zhong Y, et al. Down-regulation of NF-κB transcriptional activity in HIV-associated kidney disease by BRD4 inhibition. *J Biol Chem.* 2012;287(34): 28840–28851.
199. McCabe MT, Ott HM, Ganji G, et al. EZH2 inhibition as a therapeutic strategy for lymphoma with EZH2-activating mutations. *Nature.* 2012;492(7427):108–112.
200. Kim E, Kim M, Woo D-H, et al. Phosphorylation of EZH2 activates STAT3 signaling via STAT3 methylation and promotes tumorigenicity of glioblastoma stem-like cells. *Cancer Cell.* 2013;23(6):839–852.
201. Lee JM, Lee JS, Kim H, et al. EZH2 generates a methyl degron that is recognized by the DCAF1/DDB1/CUL4 E3 ubiquitin ligase complex. *Mol Cell.* 2012;48(4):572–586.
202. Sun F, Chan E, Wu Z, Yang X, Marquez VE, Yu Q. Combinatorial pharmacologic approaches target EZH2-mediated gene repression in breast cancer cells. *Mol Cancer Ther.* 2009;8(12):3191–3202.
203. Casciello F, Windloch K, Gannon F, Lee JS. Functional role of G9a histone methyltransferase in cancer. *Front Immunol.* 2015;6:487.
204. Lehnertz B, Northrop JP, Antignano F, et al. Activating and inhibitory functions for the histone lysine methyltransferase G9a in T helper cell differentiation and function. *J Exp Med.* 2010;207(5):915–922.
205. Antignano F, Burrows K, Hughes MR, et al. Methyltransferase G9A regulates T cell differentiation during murine intestinal inflammation. *J Clin Invest.* 2014;124(5):1945.
206. Leone G, Teofili L, Voso MT, Lubbert M. DNA methylation and demethylating drugs in myelodysplastic syndromes and secondary leukemias. *Haematologica.* 2002;87(12): 1324–1341.
207. Baylin SB, Jones PA. A decade of exploring the cancer epigenome—biological and translational implications. *Nat Rev Cancer.* 2011;11(10):726–734.
208. Rius M, Lyko F. Epigenetic cancer therapy: rationales, targets and drugs. *Oncogene.* 2012;31(39):4257–4265.

209. Yoo CB, Jeong S, Egger G, et al. Delivery of 5-aza-2′-deoxycytidine to cells using oligodeoxynucleotides. *Cancer Res.* 2007;67(13):6400–6408.
210. Saito S, Kato J, Hiraoka S, et al. DNA methylation of colon mucosa in ulcerative colitis patients: correlation with inflammatory status. *Inflamm Bowel Dis.* 2011;17(9):1955–1965.
211. Foran E, Garrity-Park MM, Mureau C, et al. Upregulation of DNA methyltransferase-mediated gene silencing, anchorage-independent growth, and migration of colon cancer cells by interleukin-6. *Mol Cancer Res.* 2010;8(4):471–481.
212. Li Y, Deuring J, Peppelenbosch MP, Kuipers EJ, de Haar C, van der Woude CJ. IL-6-induced DNMT1 activity mediates SOCS3 promoter hypermethylation in ulcerative colitis-related colorectal cancer. *Carcinogenesis.* 2012;33(10):1889–1896.
213. Karpf AR, Peterson PW, Rawlins JT, et al. Inhibition of DNA methyltransferase stimulates the expression of signal transducer and activator of transcription 1, 2, and 3 genes in colon tumor cells. *Proc Natl Acad Sci.* 1999;96(24):14007–14012.
214. Zhang Q, Wang HY, Woetmann A, Raghunath PN, Odum N, Wasik MA. STAT3 induces transcription of the DNA methyltransferase 1 gene (DNMT1) in malignant T lymphocytes. *Blood.* 2006;108(3):1058–1064.
215. Roulois D, Yau HL, Singhania R, et al. DNA-demethylating agents target colorectal cancer cells by inducing viral mimicry by endogenous transcripts. *Cell.* 2015;162(5):961–973.
216. Chiappinelli KB, Strissel PL, Desrichard A, et al. Inhibiting DNA methylation causes an interferon response in cancer via dsRNA including endogenous retroviruses. *Cell.* 2015;162(5):974–986.
217. Choi J, Ritchey J, Prior JL, et al. In vivo administration of hypomethylating agents mitigate graft-versus-host disease without sacrificing graft-versus-leukemia. *Blood.* 2010;116(1):129–139.
218. Hu Y, Cui Q, Gu Y, et al. Decitabine facilitates the generation and immunosuppressive function of regulatory γδT cells derived from human peripheral blood mononuclear cells. *Leukemia.* 2013;27(7):1580–1585.
219. Singer BD, Mock JR, Aggarwal NR, et al. Regulatory T cell DNA methyltransferase inhibition accelerates resolution of lung inflammation. *Am J Respir Cell Mol Biol.* 2015;52(5):641–652.
220. Cao Q, Wang X, Jia L, et al. Inhibiting DNA Methylation by 5-aza-2′-deoxycytidine ameliorates atherosclerosis through suppressing macrophage inflammation. *Endocrinology.* 2014;155(12):4925–4938.
221. Stefanska B, Karlic H, Varga F, Fabianowska-Majewska K, Haslberger A. Epigenetic mechanisms in anti-cancer actions of bioactive food components—the implications in cancer prevention. *Br J Pharmacol.* 2012;167(2):279–297.
222. Nandakumar V, Vaid M, Katiyar SK. (−)-Epigallocatechin-3-gallate reactivates silenced tumor suppressor genes, Cip1/p21 and p16INK4a, by reducing DNA methylation and increasing histones acetylation in human skin cancer cells. *Carcinogenesis.* 2011;32(4):537–544.
223. Gu B, Ding Q, Xia G, Fang Z. EGCG inhibits growth and induces apoptosis in renal cell carcinoma through TFPI-2 overexpression. *Oncol Rep.* 2009;21(3):635.
224. Kala R, Shah HN, Martin SL, Tollefsbol TO. Epigenetic-based combinatorial resveratrol and pterostilbene alters DNA damage response by affecting SIRT1 and DNMT enzyme expression, including SIRT1-dependent γ-H2AX and telomerase regulation in triple-negative breast cancer. *BMC Cancer.* 2015;15(1):1.
225. Morimoto T, Sunagawa Y, Kawamura T, et al. The dietary compound curcumin inhibits p300 histone acetyltransferase activity and prevents heart failure in rats. *J Clin Invest.* 2008;118(3):868–878.
226. Balasubramanyam K, Varier RA, Altaf M, et al. Curcumin, a novel p300/CREB-binding protein-specific inhibitor of acetyltransferase, represses the acetylation of

histone/nonhistone proteins and histone acetyltransferase-dependent chromatin transcription. *J Biol Chem.* 2004;279(49):51163–51171.
227. Liu Z, Xie Z, Jones W, et al. Curcumin is a potent DNA hypomethylation agent. *Bioorg Med Chem Lett.* 2009;19(3):706–709.
228. Gupta SC, Kismali G, Aggarwal BB. Curcumin, a component of turmeric: from farm to pharmacy. *Biofactors.* 2013;39(1):2–13.
229. Pardoll DM. The blockade of immune checkpoints in cancer immunotherapy. *Nat Rev Cancer.* 2012;12(4):252–264.
230. Christiansen AJ, West A, Banks K-M, et al. Eradication of solid tumors using histone deacetylase inhibitors combined with immune-stimulating antibodies. *Proc Natl Acad Sci.* 2011;108(10):4141–4146.
231. Karpf AR, Lasek AW, Ririe TO, Hanks AN, Grossman D, Jones DA. Limited gene activation in tumor and normal epithelial cells treated with the DNA methyltransferase inhibitor 5-aza-2′-deoxycytidine. *Mol Pharmacol.* 2004;65(1):18–27.
232. James S, Link P, Karpf AR. Epigenetic regulation of X-linked cancer/germline antigen genes by DNMT1 and DNMT3b. *Oncogene.* 2006;25(52):6975–6985.
233. Goodyear O, Agathanggelou A, Novitzky-Basso I, et al. Induction of a CD8+ T-cell response to the MAGE cancer testis antigen by combined treatment with azacitidine and sodium valproate in patients with acute myeloid leukemia and myelodysplasia. *Blood.* 2010;116(11):1908–1918.
234. Guo ZS, Hong JA, Irvine KR, et al. De novo induction of a cancer/testis antigen by 5-aza-2′-deoxycytidine augments adoptive immunotherapy in a murine tumor model. *Cancer Res.* 2006;66(2):1105–1113.
235. Li H, Chiappinelli KB, Guzzetta AA, et al. Immune regulation by low doses of the DNA methyltransferase inhibitor 5-azacitidine in common human epithelial cancers. *Oncotarget.* 2014;5(3):587–598.
236. Srivastava P, Paluch BE, Matsuzaki J, et al. Immunomodulatory action of SGI-110, a hypomethylating agent, in acute myeloid leukemia cells and xenografts. *Leuk Res.* 2014;38(11):1332–1341.
237. Srivastava P, Paluch BE, Matsuzaki J, et al. Immunomodulatory action of the DNA methyltransferase inhibitor SGI-110 in epithelial ovarian cancer cells and xenografts. *Epigenetics.* 2015;10(3):237–246.
238. Covre A, Coral S, Nicolay H, et al. Antitumor activity of epigenetic immunomodulation combined with CTLA-4 blockade in syngeneic mouse models. *Oncoimmunology.* 2015;4(8):e1019978.
239. Kim K, Skora AD, Li Z, et al. Eradication of metastatic mouse cancers resistant to immune checkpoint blockade by suppression of myeloid-derived cells. *Proc Natl Acad Sci U S A.* 2014;111(32):11774–11779.
240. Odunsi K, Matsuzaki J, James SR, et al. Epigenetic potentiation of NY-ESO-1 vaccine therapy in human ovarian cancer. *Cancer Immunol Res.* 2014;2(1):37–49.

PART II

Viral Influences in Cancer Initiation and Progression

CHAPTER FOUR

Viral Carcinogenesis

A.J. Smith, L.A. Smith[1]
Texas Tech University Health Sciences Center, Lubbock, TX, United States
[1]Corresponding author: e-mail address: lisa.ann.smith@ttuhsc.edu

Contents

Abstract

Cancer has been recognized for thousands of years. Egyptians believed that cancer occurred at the will of the gods. Hippocrates believed human disease resulted from an imbalance of the four humors: blood, phlegm, yellow bile, and black bile with cancer being caused by excess black bile. The lymph theory of cancer replaced the humoral theory and the blastema theory replaced the lymph theory. Rudolph Virchow was the first to recognize that cancer cells like all cells came from other cells and believed chronic irritation caused cancer. At the same time there was a belief that trauma caused cancer, though it never evolved after many experiments inducing trauma. The birth of virology occurred in 1892 when Dimitri Ivanofsky demonstrated that diseased tobacco plants remained infective after filtering their sap through a filter that trapped bacteria. Martinus Beijerinck would call the tiny infective agent a virus and both Dimitri Ivanofsky and Marinus Beijerinck would become the fathers of virology. Not to long thereafter, Payton Rous founded the field of tumor virology in 1911 with his discovery of a

Progress in Molecular Biology and Translational Science, Volume 144
ISSN 1877-1173
http://dx.doi.org/10.1016/bs.pmbts.2016.09.007

transmittable sarcoma of chickens by what would come to be called Rous sarcoma virus or RSV for short. The first identified human tumor virus was the Epstein–Barr virus (EBV), named after Tony Epstein and Yvonne Barr who visualized the virus particles in Burkitt's lymphoma cells by electron microscopy in 1965. Since that time, many viruses have been associated with carcinogenesis including the most studied, human papilloma virus associated with cervical carcinoma, many other anogenital carcinomas, and oropharyngeal carcinoma. The World Health Organization currently estimates that approximately 22% of worldwide cancers are attributable to infectious etiologies, of which viral etiologies is estimated at 15–20%. The field of tumor virology/viral carcinogenesis has not only identified viruses as etiologic agents of human cancers, but has also given molecular insights to all human cancers including the oncogene activation and tumor suppressor gene inactivation.

1. RNA RETROVIRUSES

1.1 History

Cancer has been recognized for thousands of years. Egyptians believed cancer occurred at the will of the gods. Hippocrates believed human disease resulted from an imbalance of the four humors: blood, phlegm, yellow bile, and black bile with cancer caused by excess black bile. The lymph theory of cancer replaced the humoral theory and the blastema theory replaced the lymph theory. Rudolph Virchow was the first to recognize that cancer cells, like all cells, came from other cells. Additionally, he believed chronic irritation caused cancer. At the same time there was a belief that trauma caused cancer. That belief quickly faded after many experiments inducing trauma never produced cancer. The birth of virology occurred in 1892 when Dimitri Ivanofsky demonstrated that diseased tobacco plants remained infective after passing their sap through a filter that trapped bacteria. Martinus Beijerinck would call the tiny infective agent "a virus" and both Dimitri Ivanofsky and Marinus Beijerinck became the fathers of virology. The first human virus, yellow fever, was discovered in 1901. In 1907, Giuseppe Ciuffo published "Innesto positivo con filtrato de verruca volgare" where he substantiated a viral etiology for warts.[1] In 1908, Vilhelm Ellerman and Olaf Bang reported transferring a filterable virus from a chicken with leukocytosis to healthy chickens by injecting the healthy chicken with a cell-free filtrate derived from the chicken with leukocytosis. The virus would be identified as avian leukosis virus (ALV), however, as leukemia was not identified as a cancer, the paper went largely unnoticed.[2,3]

The transmission of tumor from one animal to a different healthy animal was being performed as early as 1876. From an animal with a tumor, a small fragment would be taken and implanted into a second healthy animal where it subsequently grew. However, progress beyond simple transference did not occur until later. In 1909, Peyton Rous began to study a chicken sarcoma. He had received a 15-month-old Plymouth Rock hen with a sarcoma of its right breast from a local farmer along with additional fowl from the farmer's inbred flock. Rous started his investigation by transferring small bits of tumor to the opposite breast and peritoneal cavity of the same hen and to other young hens from the same flock. The tumors that grew were morphologically similar to the original sarcoma. At necropsy, the mitotically active tumors showed central necrosis surrounded by a rim of multinucleated cells with peripheral spindled cells. In a follow-up study, Rous ground up tumor, filtered it through a Berkefeld filter with pores that retained bacteria and larger cell particles, and injected the filtrate into a closely related hen while at the same time transferring a small bit of tumor to a second closely related hen. Both hens developed the sarcoma, but the hen injected with the filtrate developed the sarcoma at a slower rate.[4] He published his findings in 1910 and 1911 and the virus came be known as the RSV.[5–7]

In 1933, Richard Shope and E. Weston Hurst investigated the cause of keratin horn like structures that had plagued some cottontail rabbits. They discovered a filterable infectious papillomatosis.[8] In 1935, when Peyton Rous and Joseph Beard placed that filterable infectious papillomatosis in domestic rabbits, it resulted not in a benign tumor (wart), but in a skin carcinoma.[3,9,10] Still, in the early part of the 1900s the scientific tools to study a transmissible cancer and the intellectual atmosphere was not right for further investigation of a possible infectious etiology for cancer of a chicken, a rabbit, or other nonhuman animal.

The subsequent era of bacteriophage research and Max Delbruck inspired Renato Dulbecco and Marguerite Vogt to develop the plaque assay for cytocidal viruses, introduced in 1952–54. A Petri dish with a susceptible monolayer of host cells would be inoculated with a virus. The virus would infect a cell, multiply, lyse/kill the cell, and infect neighboring cells. After an incubation period, the monolayer would be covered with agar, which would halt progression of the virus and within a short period of time, a visible plaque would form where a virus had parasitized an individual cell, allowing for viral quantitation. The problem, however, was that RSV was not a cytocidal animal virus.[11,12] In 1956, Manaker and Groupe in the journal, *Virology*, published "Discrete foci of altered chicken embryo cells associated with Rous sarcoma virus in tissue culture," suggesting that oncogenic transformation might be possible in a cell culture-based assay.[10,13,14] In the laboratory of Renato Dulbecco, Howard Temin, and Harry Rubin developed a variant of the plaque assay using a chick fibroblast monolayer inoculated with RSV. Instead of plaques forming, the cells lived and were transformed into cells resembling in both morphology and metabolism the sarcoma cells derived from chickens. The focus assay was the first oncogenic cell culture, it was a quantitative assay, it was the first to demonstrate that oncogenic transformation could occur outside an organism, and it made it possible to study the interaction of one virus with one cell. As they continued to study this phenomenon, other observations emerged. When normal chick fibroblasts were introduced to a Petri dish, they would proliferate until they filled the Petri dish in a monolayer, then they would become senescent. This phenomenon is known as contact inhibition. However, the cells infected with the RSV did not exhibit or had lost contact inhibition. They would continue to proliferate forming small piles of tumor cells.[4,10,15]

Since its discovery, many RSV variants and mutations have been developed, observed, and studied. In 1960, Howard Temin published, "The control of cellular morphology in embryonic cells infected with Rous sarcoma virus in vitro," where he reported the first morphological mutation of transformed cells from RSV.[16] In 1963, Hidesaburo Hanafusa and Harry Rubin published "The defectiveness of Rous sarcoma virus" and its analysis was published in 1964. In those publications, they describe a mutant RSV that required a helper virus for replication. The helper virus was named Rous associated virus (RAV) and it was indistinguishable from RSV. The defective RSV mutant could induce malignant transformation without help and the transformed cells would maintain their transformed morphology for many generations, but they could not produce infectious progeny. However, when the helper virus was introduced, the transformed cells would produce abundant infectious viral particles.[17,18] It was concluded that the ability to transform and the ability to infect were independent of each other. In 1970, G. Steven Martin published, "Rous sarcoma virus: a function required for the maintenance of the transformed state," where Martin describes a temperature sensitive (ts) mutant he isolated. The RSV mutant would transform cells when cultured at 37°C (human body temperature), but not at 41°C (chicken body temperature). When a cell culture was infected with tsRSV at 37°C, the permissive temperature, the cells would transform. After a period of time the cell culture would be moved to 41°C, the nonpermissive temperature, and the cells would revert to an uninfected morphology. This finding suggested that an RSV protein was required not only to initiate tumor transformation, but also to maintain tumor transformation. When the temperature was again shifted to the permissive temperature of 37 degree, the cells again appeared transformed.[19] This finding suggested that the RSV infection persisted despite the change in temperature and was passed on to its progeny.[5,20] The many studies of RSV had produced infectious mutants (RAV) and oncogenic transformation mutants (tsRSV). The genomes of a defective mutant could then be compared with the genome of a competent mutant; the difference may be the gene or genes responsible for the loss or gain of function. In 1970, Peter Duesberg and Peter Vogt published, "Differences between the ribonucleic acids of transforming and non-transforming avian tumor viruses" where they noted a difference in the size of the genomes.[21] That genomic sequence difference would eventually be named "src" short for sarcoma.

In 1976, Temin published "The DNA provirus hypothesis."[22] He was the first to suggest RSV, a single-stranded RNA virus, may be able to make a double-stranded DNA copy of its genome that could insert itself into the host DNA as a provirus that would later be replicated concurrently with the host DNA during cellular replication. This flew in the face of the central dogma of biology, first published by Francis Crick in 1956, which states in essence, DNA is translated to RNA, which is synthesized to protein.[4] Under great scrutiny he remained confident largely due to previous studies he had performed. In 1963, he published, "The effects of actinomycin D on growth of Rous sarcoma virus in vitro."[23] Actinomycin D is a strong inhibitor of DNA synthesis. Because of the inhibitory effect it had on RSV, Temin believed there was a DNA link to this RNA virus. Concurrently in 1970, David Baltimore published "RNA-dependent DNA polymerase in virions of RNA tumour viruses"[24] and Howard Temin and Satoshi Muzutani published, "RNA-dependent DNA polymerase in virions of Rous sarcoma virus."[25] These RNA-dependent DNA polymerases would later be called reverse transcriptase. They combined efforts and in 1972 together publish "RNA-directed DNA synthesis and RNA tumor viruses."[26] The RNA viruses with reverse transcriptase would come to be called retroviruses. Oncogenic retroviruses, like DNA viruses to be discussed later, rely on integration of their genome into the host genome to facilitate stable retention and replication of their own genome.[5]

Reverse transcriptase, discovered in 1970 changed the study of src from RNA-based techniques to DNA-based techniques, of which there existed more variation and more power. In 1974, reverse transcriptase was used to generate a DNA probe of the viral src (v-src) sequence. The probe was used to track the v-src sequence during viral interaction with a host cell. In 1976, D. Stehelin, H.E. Varmus, J.M. Bishop, and Peter Vogt published, "DNA related to the transforming gene(s) of avian sarcoma viruses is present in normal avian DNA," whereby they document the presence of at least part of the v-src sequence in uninfected chicken cells and in uninfected cells of other avian species.[27] They concluded that the origin of the v-src gene was cellular and not viral.[20] This finding, the first oncogene, solidified the concept first hypothesized in 1969 by Robert Huebner and George Todaro in their paper, "Oncogenes of RNA tumor viruses as determinants of cancer," that the genome of most species contain oncogenes that can transform a normal benign cell to cancer.[28] How retroviruses acquire cellular oncogenes is still debatable. Nonetheless, the observation that cellular oncogenes do not regularly cause cancer and viral oncogenes do suggests that cellular oncogenes require a mutation before they can become an element of transformation, thus they were called protooncogenes.[3] Furthermore, if cellular protooncogenes required a mutation for oncogenic transformation, then the origin of the mutation may not require a virus. This discovery and its implications set in motion a frenzy to discover new oncogenes.

Now that a genetic basis for oncogenic transformation had been established, the question remained; what was the protein product and how did it transform the cell? In 1977, Joan Brugge and Ray Erikson identified the protein product of the src oncogene using an antibody developed in a rabbit.[29] In 1978, two independent groups identified the protein as a phosphokinase. In 1979, Tony Hunter and Bartholomew Sefton identified it as a tyrosine kinase.[30] Tyrosine kinases are now known to be signal transducers and are identified as a class of oncogenes from which over 40 targeted tyrosine kinase inhibitor therapeutic agents have been developed.[14,20]

By definition, all retroviruses are RNA-based viruses that contain reverse transcriptase, which they use to make a ds-DNA copy of their RNA genome. The DNA copy, termed a provirus, is thereafter inserted into the host DNA. Oncogenic retroviruses that rapidly transform cell cultures (days to weeks) to tumor were found to contain an oncogene within their genome and are known as acute transforming retroviruses. However, there are some oncogenic retroviruses which transformed host cells much more slowly (months), notably ALV and murine leukemia virus (MLV). Their genomes contain no oncogenes and they are known as slow transforming retroviruses. Transformation by these viruses is induced by insertion of the provirus into the host genome resulting in altered gene expression,[3,31] termed insertional mutagenesis. Retroviruses that lack specific oncogenes can, by insertional mutagenesis, activate host cellular protooncogenes in this way,[20] but because insertion is random the process takes much longer. In 1981, William Hayward, Benajamin Neel, and Susan Astrin in their paper "Activation of a cellular onc gene by promoter insertion in ALV-induced lymphoid leukosis," showed that in most, not all, cases of ALV-induced lymphomas, the cellular myc gene was activated by adjacent ALV proviral insertion thus utilizing the viral promotor instead of its own, and enhancing its expression resulting in neoplastic transformation.[32]

1.2 Human T Cell Lymphotropic Virus 1 (HTLV-1)

Thus far, the story of oncogenic retroviruses has included acute transforming retroviruses and slow transforming retroviruses. Both are classified as simple retroviruses. Their genome contains only gag (group antigen/core protein), pol (polymerase/reverse transcriptase), and env (envelope) genes. However, there is one more class of retroviruses to consider; the complex oncogenic retroviruses. Their genome contains additional nonstructural genes having regulatory or supplementary functions. In 1977, Takashi Uchiyama and colleagues discovered adult T-cell leukemia (ATL) and in their paper, "Adult T-cell leukemia: clinical and hematologic features of 16 cases," they described the clinical and hematologic characteristics including the unique morphology of lobulated nuclei, coarsely clumped nuclear chromatin, and scant cytoplasm of this new entity.[33] Of interest, they noted that most of the cases were born or raised near Kyushu, Japan. Epidemiologic studies would later support the clustering of ATL and suggest a distinct etiology. In 1980, Bernard Poiesz, Robert Gallo and colleagues would identify retroviral particles in lymphocytes from a patient with ATL.[34] In 1982, Motoharu Seiki, Seisuke Hattori, Yoko Hirayama, and Matsuaki Yoshida had sequenced the provirus genome and found that the genome contained the gag, pol, and env genes typical of known retroviruses, but in addition, it had four open frames that they hypothesized might code for transforming genes.[35] To determine if the virus was truly oncogenic or just an observing passenger in ATL, Yoshida, Seiki, Yamaguchi, and Takatsuki studied the location of provirus insertion into host DNA. They asserted that proviral insertion occurs randomly, but tumors proliferate clonally. If the provirus were a passenger in the transformation process, then they would find proviral DNA integrated at different host sites within the tumor cells. If the provirus were the cause of the transformation, then it would be found at one site within the tumor cells.

They published their findings, "Monoclonal integration of human T-cell leukemia provirus in all primary tumors of adult T-cell leukemia suggests causative role of human T-cell leukemia virus in the disease."[36] This new genome which contained no oncogenes and did not utilize insertional mutagenesis, constituted a new class of oncogenic retrovirus. The open reading frames were found to be overlapping and encode three proteins, Tax, Rex, and p21. Tax and Rex were found to work together to regulate viral expression. Additionally, in vitro studies found Tax involved in many cellular functions. It binds enhancer binding proteins nuclear factor-kB (NF-kB) and CREB; binds transcriptional cofactor CBP; targets tumor suppressor proteins p16INK4, p15INK4, and hDlg for inactivation; suppresses apoptosis by activation of NF-kB, activation of Bcl-X, and repression of Bax; suppresses DNA repair by topoisomerase I, and suppresses DNA checkpoint activity by polymerase B. Overall, Tax drives cellular proliferation allowing for accumulation of mutations while suppressing repair and apoptotic mechanisms, ultimately resulting in transformation of HTLV-I infected lymphocytes.[33] More recently, there is growing evidence that bZIP factor (HBZ) may be important for proliferation and oncogenic transformation as well. HBZ regulates and modulates multiple cellular signaling pathways related to cell growth, apoptosis, immune escape, and T-cell differentiation among others. Whereas HBZ is constitutively expressed, Tax is often lost. It has been suggested that Tax contributes to pathogenesis and oncogenesis in the early stages of infection, but after being targeted by cytotoxic T cells, transformed cells utilizing NF-kB inducing kinase and HBZ for proliferation in the later stages of infection are selected for. It has therefore been proposed that Tax and HBZ may be cooperated for maintenance transformation.[37]

HTLV is a group of genetically similar viruses. There are four human T-cell lymphotrophic viruses: I, II, III, and IV. HTLV-I is composed of subtypes A–F with the majority of infections caused by subtype A. HTLV-II was discovered in a patient with a T-cell variant of hairy cell leukemia by Robert Gallo and colleagues.[38] HTLV-III and -IV were discovered in Camaroon in 2005. Because HTLV-III and -IV were found to be closely related to simian T-lymphotropic viruses (STLVs), all four HTLVs are believed to have originated as a result of zoonotic trans-species transmission. Primate T-lymphotropic viruses include HTLVs and STLVs.[39,40]

Current estimates of worldwide HTLV-1 infection are believed to be underestimated at approximately 5–10 millions individuals.[41] Epidemiologic studies have found that less than 5% of human T-cell leukemia virus

(HTLV-I) carriers developed ATL; that is most infected individual are asymptomatic carriers.[40] ATL is endemic in southwestern Japan, the Carribean, and Africa. Four clinical variants have been identified: smoldering, chronic, lymphoma, and acute. Acute and lymphoma variants are the most aggressive. The acute variant is the most common and is characterized by a leukemic phase, diffuse lymphadenopathy, and skin rash. Additionally, many will also have an associated T-cell immunodeficiency and opportunistic infections, most commonly with pneumocystis carinii and/or Strongyloides. The lymphomatous variant is characterized by lymphadenopathy and skin rash. The chronic variant is associated with a skin rash. The smoldering variant has greater than 5% abnormal circulating lymphocytes. The neoplastic lymphocytes are typically medium to large in size, with marked nuclear pleomorphism, coarsely clumped chromatin, and prominent nucleoli, often described as "flower cells." That said, there are many morphologic variants ranging in size from small to giant and in morphology from having irregular nuclear contours to being anaplastic.[42]

The virus is transmitted as a result of exposure to contaminated blood. The route of transmission includes transfusion of blood products and tissue donation, sexual intercourse, needle sharing, accidental or purposeful, and birthing and breast feeding. The latent period between infection and tumor development is estimated at 20–30 years. Because there are no effective therapies available, social strategies have been instituted to limit the spread of HTLV-1. Blood banks test for HTLV-1 antibodies and exclude HTLV-1 positive blood and donor from the blood supply. Likewise, HTLV-1 positive mothers are encouraged to bottle feed their babies instead of breast feeding. Though a vaccine has been sought, none as yet have been found. Of interest, HTLV-1 also causes tropical spastic paraparesis/HTLV-1-associated myelopathy and HTLV-1 uveitis.[40,43]

1.3 Human Immunodeficiency Virus (HIV)

HIV is also considered a complex retrovirus. It is believed to have been transmitted to humans through multiple zoonotic infections of simian immunodeficiency virus (SIV). SIV, like HIV, are groups of viruses found in various nonhuman primates from sub-Sahara Africa.[44] Most SIVs are not pathologic in their host; however, the link was made in 1986 when the newly discovered HIV type 2 was found to be related to HIV1, but more closely related to the SIV that caused immunodeficiency in captive macaques.[45]

HIV1 is made up of groups M, N, O, and P. Group M is further divided into subtypes A–K based on similar genetic grouping. Group M, subgroup C is the dominant form worldwide, producing 55–60% of HIV1 infections. There are also many recombinant forms. The HIV viral genome has two copies of single stranded, positive sense, RNA (not ds-RNA). If a host cell is infected with two different viral genomes and one copy from each of the viral genomes is packaged into a virion, then a recombinant form can be produced.[46] HIV2 is made up of groups A–H with groups A and B representing more than an individual case. Each group resulted from an independent cross-species zoonotic transmission.[44] HIV1 group M (main group) was the first to be discovered in 1983 and is the pandemic form.[47] Group N (non-M, non-O) was discovered in 1998[48] and is responsible for a handful of cases of HIV1 infections. Group O (outlier) was discovered in 1990[49] and is responsible for <1% of HIV1 infections. Group P was discovered in 2009[50] and is responsible for only a couple cases of HIV1 infection. It is believed, groups M and N are derived from chimpanzees, P from apes, and O remains unknown. Though the actual time and place of transmission cannot be definitively known, it is believed that HIV1 groups M and O were transmitted in the early part of the 20th century and spread among humans for most of the century before being discovered. It is believed HIV1 group M transmission occurred near Leopoldville/Kinshasa, within the natural habitat of its reservoir.[44] Furthermore, it is believed transmission occurred by cutaneous or mucosal exposure to infected blood and/or body fluids during the butchering of chimpanzees (HIV1), sooty mangabeys (HIV2), and other nonhuman primates.[51]

Greater than 70 million people have been infected with the HIV virus since the World Health Organization began keeping track and approximately 35 million people have died of it. It is estimated at the end of 2015, 36.7 million people worldwide have HIV and 1.1 million will have died of it in 2015. Most of the cases, almost 70%, are in sub-Saharan Africa with approximately 1 in 25 adults infected. The dominant mode of transmission is heterosexual transmission.[52] In the United States, greater than 1.2 million people have HIV with approximately 44,000 new diagnoses made in 2014. The dominant mode of transmission in the United States is homosexual transmission and to a lesser degree intravenous drug use.[53]

The HIV genome is made up of three structural proteins: gag, pol, and env. Env codes for surface glycoprotein gp120 and it is associated with transmembrane protein gp41. Gp120 attaches to a host CD4 receptor. Binding causes a transformational change in gp120 that allows gp120 to also bind a chemokine coreceptor, CCR5 or CXCR4. Gp41 subsequently facilitates fusion of the viral envelope with the host plasma membrane and entry of the viral core into the host cytoplasm. Pol codes for reverse transcriptase, RNase, integrase, and HIV protease. After entry of the viral core into the host cytoplasm, reverse transcriptase produces a ds-DNA copy from the viral RNA genome template. Reverse transcriptase is error prone and has no proof reading ability accounting for HIV's genomic variability. Integrase inserts the DNA copy into the host DNA as a provirus. The HIV genome has two essential regulatory proteins, tat and rev, and four accessory regulatory proteins, nef, vpr, vif, and vpu. Rather than vpu, HIV2 has vpx. Tat is a transcriptional activator of HIV RNA. Rev is an RNA-binding protein that induces the switch from early to late gene expression and the export of RNA from the nucleus to the cytoplasm. Tat and rev are required for replication. The accessory proteins are considered virulence factors. None are required for replication. Except for nef, all are rev-dependent late phase proteins. All accessory proteins have multiple functions. Nef is expressed early and is the first measurable protein after HIV entry into the host cell. Nef downregulates CD4 cell surface expression, disrupts T-cell activation, and stimulates HIV infectivity. Vpr facilitates the transportation of a pre-integration complex composed of viral genome, viral integration proteins, and host proteins from the cytoplasm through a nuclear pore into the nucleus. This allows integration of a provirus without disrupting the nuclear membrane; significant because it can be done in the cellular resting state rather than during division. Vpu facilitates virion release by budding. Vif is required for virion assembly and is included in the virion. Vif also counteracts intrinsic immunity, that is, host constitutively expressed cellular defense mechanisms against viral infection (different from innate or adaptive immunity). HIV transcription is initiated at one promoter. The transcript can then be spliced and modified into greater than 30 mRNAs or packaged as viral RNA genome into virion particles.[54]

Transmission generally occurs as the result of mucus membrane or broken skin exposure to infected body fluids including blood, semen, vaginal fluid, rectal fluid, and breast milk. The route of transmission includes transfusion of blood products and tissue donation; sexual including oral, vaginal, and/or anal sex; needle sharing, accidental or purposeful; and birthing and breast feeding.[55] The risk of infection is associated with the route of transmission (anal sex has a high risk than vaginal and vaginal has a higher risk than oral), the virulence of the virus, the viral load (HIV2 infected individuals typically have a lower viral load than HIV1 infected individuals and resulting decreased transmission rates), and host factors (chemokine receptor 5 mutation (CCR5) is protective). Initial transmission results in the infection of a small number of CD4 positive T helper lymphocytes, macrophages, and/or dendritic cells. The virus is transported to the draining lymph node where initial immune activation occurs and initial viral replication begins. A systemic viraema, the acute infection, subsequently occurs with viral replication occurring in the blood stream, bone marrow, gastrointestinal lymphoid tissue, spleen, central nervous system, and other sites. The high concentrations of virus fall with the appearance of CD8 positive HIV-specific cytotoxic T cells. Within a few months a steady state of viral replication/levels occurs that will remain stable for years.[46] Overall, HIV induces a hyperactive immune state that is dysfunctional, hyporesponsive, depleted, and eventually deficient/deteriorated.[56] At the same time, faulty reverse transcriptase produces many HIV quasi species. Some of the mutants will have survival advantages and may be harbored in reservoir cells such as resting T cells, resting central nervous system macrophages, and other resting tissue macrophages. Plasma HIV1 RNA viral loads and CD4 T lymphocyte counts are used to monitor disease progression. The CD4 T lymphocyte counts is the most important laboratory test of patient immune function and is usually performed by flow cytometry.[46] Immune deficiency, characterized by decreased circulating CD4 T lymphocyte counts, results from cellular destruction as well as diminished cellular production and cellular sequestration.[56] Acquired immune deficiency syndrome (AIDS) is the last disease stage characterized by circulating CD4 T lymphocytes counts of less than 200 per microliter ($<200/\mu L$), AIDS defining opportunistic infection or malignancies including invasive cervical cancer, Kaposi's sarcoma, Burkitt lymphoma, immunoblastic lymphoma, or primary brain lymphoma.[57]

The first published event of the current HIV pandemic was June 5, 1981. It documented a cluster of five young homosexual males with pneumocystis carinii, CMV, and mucosal candidiasis. Two of the five young men died.[58] The CDC Morbidity and Mortality Weekly Report of July 3, 1981 reported on Kaposi's sarcoma and Pneumocystis pneumonia among homosexual men—New York City and California.[59] Subsequently, the CDC initiated a task force whose first purpose was to develop a case definition for surveillance and investigation of outbreaks of the opportunistic infections and Kaposi's sarcoma. The original CDC case definition was biopsy-proven Kaposi's sarcoma in persons under 60 years old or biopsy/culture-proven life-threatening/fatal opportunistic infections in persons with no known cancer, underlying immune deficiency, or history of immunosuppression. By the end of 1981, they identified 159 cases of Kaposi's sarcoma and/or opportunistic infections in the United States dating back to 1978. There had been a steady increase in cases over time and about half were Kaposi's sarcoma, 40% Pneumocystis carinii, and 10% of cases had both.[60] HIV was isolated by Luc Montagnier and colleagues in 1983.[47,60] Robert Gallo and colleagues further substantiated HIV as the causative agent of AIDS in 1984.[61] HIV is not included in this chapter because it can directly transform host cells into malignant cells. Rather by causing a dysfunctional, depleted, and eventually deficient immune system, HIV is included in this chapter because it predisposes to cancers including those caused by other viruses. That said, a more direct process of transformation has been proposed. Robert Gallo and colleagues found when HHV8 infected vascular cells were exposed to inflammatory cytokines induced by HIV, the vascular cells developed a spindle shaped morphology and respond to extracellular HIV Tat protein with proliferative cell growth.[62]

Overall, HIV and AIDS are associated with an increased risk of cancer.[63–65] Likewise, HIV and AIDS are associated with an increased risk and progression of viral-induced cancers including Kaposi's sarcoma caused by HHV8; Hodgkin's lymphoma caused by EBV (nearly all); non-Hodgkin's lymphomas including primary central nervous system lymphoma caused by EBV (80–100%); primary effusion lymphoma caused by HHV8 and/or EBV, and Burkitt's lymphoma caused by EBV (30–50%); diffuse large B cell lymphoma with immunoblastic features by EBV (80%); plasmablastic lymphoma caused by EBV; Hodgkin's disease caused by EBV; and cervical squamous cell carcinoma caused by HPV.[65–67]

2. DNA VIRUSES

2.1 History

Though the discussion thus far has been limited to RNA viruses, specifically retroviruses, it is now time to take a close look at DNA viruses. The discovery of DNA viruses lagged behind the discovery of RNA viruses. Likewise, the discoveries made with DNA viruses somewhat lagged behind the discoveries made with RNA viruses. In part because the discoveries made with DNA viruses were often misinterpreted secondary to preconceived notions forged by the discoveries made with RNA viruses. Nonetheless, the small genomes of some DNA viruses proved beneficial for study.

2.2 Simian Vacuolating Virus (SV40)

The 1950s and 1960s saw the discovery of DNA viruses such as the hepatitis B virus, SV40/polyoma virus, human adenovirus, and human herpesvirus (HHV) among others. Benjamin Sweet and Maurice Hilleman originally discovered the SV40 (simian vacuolating virus 40/simian virus 40) in 1960 as a contaminant of the rhesus monkey kidney cultures used to propagate poliovirus for the polio vaccine (both salk and sabin vaccines).[68] The rhesus monkey is an asymptomatic carrier of SV40. When the rhesus monkey cultures were used to infect green monkey kidney cells with the attenuated poliovirus, the SV40 virus revealed itself by produced abundant vacuoles and in 3–4 days lysing/killing the cell and releasing hundreds of SV40 virus particles, much more than polio virus particles, to infect surrounding cells. In 1960 and 1961, Bernice Eddy and colleagues found when SV40 infected monkey kidney cells were injected into a nonpermissive host such as a hamster, tumors were produced.[3,69] The finding of contaminated polio vaccine and the resulting tumor production in nonpermissive hosts has resulted in a continuing search for human cancers caused by SV40.

Early experiments identified a 90–100 kDa protein that was expressed within a few hours of infection and was active during infection and tumor transformation. Proteins involved with transformation were commonly named tumor antigens or T antigens. The origin of the protein, whether cellular or viral, was not known.[70] Additional early experiments found the T antigen to be required for viral replication and the stimulation of cellular replication.

By 1968, Joe Sambook, Heiner Westphal, P.R. Srinivasan, and Renalto Dulbecco in two studies found that transformed cells contained integrated SV40 DNA.[71,72] In 1976, Michael Botchan, William Topp, and Joe Sambrook found the integrated SV40 DNA in transformed cells randomly integrated at different sites.[73] This finding eliminated the proposition that insertion mutagenesis of cellular protooncogenes might be the cause of transformation. Always however, the early promotor and the T antigen were preserved resulting in uninterrupted expression of T antigen.[70]

The use of restriction endonucleases combined with polyacrylamide gel electrophoresis performed by Kathleen Danna and Daniel Nathans in 1971[74] allowed for the eventual mapping of the circular ds-DNA SV40 genome by George Khoury, Malcom Marin, Theresa Lee and Daniel Nathans in 1975.[75] In the same year, using endonucleases, Abraham and colleagues isolated the transforming property to the early region of the SV40 genome.[76]

They also noted that the T antigen was present in nearly all transformed cells. In 1977, Kathleen Rundell and colleagues identified the 100,000-Da protein observed in SV40-infected cells as the virus T antigen, a product of the A gene.[77] Then, in May 1978 two separate studies, one by Fiers and his colleagues, the second by Reddy and his colleagues reported they had sequenced the SV40 genome.[78,79] At the same time, David Shortle and Daniel Nathans published their paper, "Local mutagenesis: a method for generating viral mutants with base substitutions in preselected regions of the viral genome," where they described how they were able to develop and then functionally test preselected site-specific SV40 mutants using different endonucleases.[80] This prompted the creation of countless mutants and allowed for the functional evaluation of protein products. The T antigen was the obvious primary target of these studies. In 1980, SV40's small DNA genome was inserted into a bacterial plasmid and cloned in *Escherichia coli* where it caused morphological transformation of mouse cells in culture.[81] Shortly thereafter in 1981, Ed Harlow, Lionel Crawford, David Pim and Nicola Williamson created monoclonal antibodies against SV40 tumor antigens, allowing for the identification of its cellular associations.[82] Science now had the necessary tools to detail the T antigen's various associations and functions with the goal of determining how it transforms cells.[70]

2.3 SV40 and Rb

In April 1971, Alfred Knudson in his study, "Mutation and cancer: statistical study of retinoblastoma," presented the two hit hypothesis wherein he claims retinoblastoma, a tumor of the eye most often affecting infants and young children, is caused by two mutational events. In the dominant hereditary form, one hit/mutation is inherited and the other occurs later with an average development of tumor or tumors by 15 months. In the sporadic form, two hits/mutations are required and often occur later than the hereditary form. There is usually development of a single tumor by 24 months on average.[83] David Comings in 1973 suggested in his paper "A general theory of carcinogenesis," that the two hits/mutations occur in two alleles of the same gene.[84] The first is an inherited germ-line inactivation of a regulatory gene with subsequent somatic mutation of the other regulatory gene. Several studies of the 1970s and 1980s injected adenovirus 12 into the eyes of newborn lab animals resulting in tumors that histologically and by electron microscopy were considered retinoblastoma. Subsequent searches for adenoviruses in the tumors of human retinoblastoma failed to reveal its presence. Ed Harlow, Peter Whyte Robert Franza and Carol Schley, in 1986, identified a 110 K protein associating with adenovirus early region 1A protein they believed was responsible for transformation caused by adenovirus.[85] The adenovirus early region 1A protein is analogous to the SV40 T antigen. In October 1987, Wen-Hwa Lee and associated created an antibody to the Rb gene product and discovered that it was a nuclear phosphoprotein that they hypothesized might regulate other cellular genes.[86] Then in July 1988, Peter Whyte, Karen Buchkovich, Johathan Horowitz, Stephen Friends, Margaret Raybuck, Robert Weinberg and Ed Harlow in their publication, "Association between an oncogene and an anti-oncogene: the adenovirus E1A proteins bind to the retinoblastoma gene product," reveal that the 110 K protein that associates with the adenovirus early region 1A protein is the protein product of the retinoblastoma gene and is the first oncogene, antioncogene interaction identified.[87] Soon thereafter, James DeCaprio and colleagues using antibodies to both the SV40 T antigen and to the 110 K protein product of Rb found that when they immunoprecipitated lysed extracts of monkey cells synthesizing SV40 large T antigen, the two antibodies coprecipitated together demonstrating they form a complex. When performed on transforming mutants, the antibodies again coprecipitated. When performed on nontransforming mutants, the antibodies did not coprecipitate. They suggested that the complex of the SV40 large T antigen to the cellular RB gene protein product with disruption of Rb protein function and/or T antigen function had a part in cellular transformation.[88] Then in 1990, Jolene Windle and colleagues in their paper, "Retinoblastoma in transgenic mice," report that retinoblastoma is caused by a double hit, two mutations, one to each of the two alleles of the Rb gene resulting in inactivate or eliminate of its protein product, p105Rb.[89] They demonstrated it by using simian virus 40T antigen in the retina of transgenic mice and showing it produces heritable ocular tumors which are histologically, ultrastructurally, and immunohistochemically identical to human retinoblastoma. They also suggested using the transgenic mouse with SV40 T antigen expression in its retina as an animal model for hereditary retinoblastoma. Many studies of the late 1980s and 1990s isolated the Rb gene from retinoblastoma tumors where they demonstrated multiple different mutations including premature stop codons, deletions, point substitutions, all ultimately inactivating, or eliminating Rb gene protein products.[90]

So, what is the function of Rb so that when inactive or absent, malignant tumors form? In the September issue of the journal *Cell*, three separate studies found the Rb protein was phosphorylated differently depending on the phase of the cell cycle.[91–93] In G0, resting senescent nonreplicating cells, the Rb protein is unphosphorylated. In the G1 phase of replicating cells, the Rb protein remains unphosphorylated. The Rb protein becomes phosphorylated at the G1/S transition. The highest percent of Rb protein phosphorylation occurs during the S or synthesis phase when DNA replication is occurring. The Rb protein product remains phosphorylated during the G2 and M or mitotic phase when cell division occurs. Rb had been previously speculated to inhibit cell growth, combined now with the findings of Rb modulation by phosphorylation, suggested that Rb might regulate the cell cycle. That is, unphosphorylated Rb gene product maintained senescence and inhibited cell division whereas when phosphorylated, it allowed cell division to progress. It was previously recognized that the SV40 T antigen binds to the unphosphorylated Rb gene product[94] suggesting that the T antigen acts similarly to phosphorylation, downregulating growth suppression, or alternatively allowing replication.[90]

Three studies in 1990 mapped the Rb protein to characterize the regions that interact with the SV40 T antigen.[95–97] Those studies found that two noncontiguous regions are necessary for binding to the SV40 T antigen. If a mutation occurs in either of the two regions, the Rb protein–SV40 T antigen complex will not form. It was noted that an inactivating mutation of the RB gene also occur within the same two region suggesting that those two regions are essential for normal RB protein function and that viral oncoproteins specifically target those two regions to inactivate or eliminate RB function. Also, this finding was consistent with previous studies that showed mutant Rb proteins did not complex with the SV40 T antigen.[90]

Srilata Bagchi, Pradip Raychaudhuri, and Joseph R. Nevins in their 1990 paper, "Adenovirus E1A proteins can dissociate heteromeric complexes involving the E2F transcription factor: a novel mechanism for E1A transactivation," showed that cellular factors complexed with cellular E2F transcription factor can be dissociated by the E1A protein of adenovirus.[98] The free cellular E2F transcription factor is then able to interact with adenovirus early E4 protein generating a stable E2 promoter complex and stimulating transcription. Shortly thereafter, it was found Rb protein complexes with E2F transcription factor and specifically, the unphosphorylated Rb protein. Likewise, adenovirus E1a protein can dissociated the E2F–Rb complex.[99] It was known that the adenovirus E1A protein, the SV40 T antigen and the human papilloma virus (HPV) E7 protein have a shared domain that binds Rb protein and is used for transformation. It was subsequently shown in 1992, that the SV40 T antigen and the HPV E7 protein can both dissociate the E2F–Rb protein complex just as the adenovirus E1A protein had. Furthermore, it was found that the E2F–RB protein complex was absent in cervical carcinoma cell line with expression of E7 protein or RB mutation. The study suggested adenovirus E1A, SV40 T antigen, and HPV E7 may act to incite a cell to proliferate so that the integrated virus can proliferate.[100]

In 1995, Rb-like proteins were discovered to also interact with SV40 T antigen and E2F transcription factor. Zalvide and DeCaprio, in their study, "Role of pRb-related proteins in simian virus 40 large-T-antigen-mediated transformation," found a particular SV40 T antigen domain from residues 102–114 that included the LXCXE motif was required for binding to the Rb protein and ultimately to transformation.[101] Likewise, mutations within the LXCXE motif inhibited binding to the Rb protein and transformation did not occur. In addition to the Rb protein (RB1, also known as p105), the SV40 T antigen was found to bind to p107 (also known as retinoblastoma like protein or RBL1) and p130 (also known as retinoblastoma like protein 2 or RBL2). It was found that the three proteins making up the Rb protein family shared sequence homology, had the ability to bind E2F, and their complexes with E2F could be disrupted by the SV40 T antigen, suggesting they may be tumor suppressor proteins as well. By 1998, three SV40 T antigen domains had been identified mediating cellular transformation; the p53 binding domain, the LXCXE or Rb protein-binding motif, and the J domain. Zalvide, Stubdal, and DeCaprio showed not only was the LXCXE motif required for binding to the Rb protein family, but also the J domain was required for inactivation of the Rb protein family.[102] They work together to disrupt Rb family E2F complexes, inhibit suppression of E2F-dependent transactivation by RB1 and p130, and inactivate growth suppression by Rb protein, p107, and p105.

In 2000, Christopher Sullivan, Paul Cantalupo, and James Pipas in their publication, "The molecular chaperone activity of simian virus 40 large T antigen is required to disrupt Rb-E2F family complexes by an ATP-dependent mechanism," describe the two hypotheses forwarded to explain the requirement of the J domain for disruption of Rb–E2F complexes; the affinity displacement model and the chaperone model, and how they intended to determine which was in play.[103] First, they added SV40 T antigen to preformed p130-E2F-4-DP-1 complexes. In the affinity displacement model, when the T antigen was added to the complexes, the T antigen should, because of a higher affinity for the site, out compete E2F for the Rb protein-binding site, ultimately releasing E2F, which can be measured. However, Sullivan and colleagues found that E2F was not released and instead the T antigen formed a stable association with the preformed complex. They concluded the T antigen and the E2F do not share the same Rb protein-binding site. Previously, it had been shown the SV40 T antigen J domain binds Hsc70.[104] The chaperone model predicted that Rb–E2F complex disruption requires energy supplied by ATP hydrolysis mediated by Hsc70 binding to the J domain of SV40 T antigen. When the study added SV40 T antigen to p130-E2F-4-DP-1 complexes with exogenous Hsc70 and ATP, Rb-E2F complexes were disrupted. When they did the same thing using an Hsc70 mutant, the complexes were not disrupted. Since both Hsc70 and T antigen have ATPase activity, they added ATPase defective T antigen mutants to the p130–E2F complexes with exogenous Hsc70 and ATP, which disrupted the Rb–E2F complexes. Therefore, they concluded the chaperone model where the SV40 T antigen binds to the p130–E2F complex through an LXCXE motif with cooperation by the J domain which recruits Hsc70 to the complex stimulating its ATPase activity releasing the SV40 T antigen, p130, and E2F from the complex.

In summary, unphosphorylated Rb inhibits G1 to S phase progression through the cell cycle by complexing with E2F transcription factor, inhibiting activation of most genes involved in synthesis of cellular DNA. G1 to S phase progression is dependent on the accumulation of G1 cyclin-dependent kinases, which phosphorylate Rb protein, inactivating it and causing it to release E2F to induce activation of genes involved in DNA synthesis. Adenovirus E1A, SV40 T antigen, and HPV E7 can disrupt Rb–E2F complexes and bind to E2F forming a stable E2 promotor DNA complex that stimulates transcription. As historically presented, it has been grossly oversimplified. For example, each Rb family protein has different levels of expression in the different phases of the cell cycle and the levels are different depending on the cell type. Like the Rb family, E2F transcription factor also forms a family of related molecules that function just slightly different from one another. Nonetheless, the study of tumor viruses was pivotal in working out the mechanisms involved in cell cycle progression, as well as the structure and function of Rb, the first discovered tumor suppressor protein.[90,105]

2.4 SV40 and p53

David Lane and Lionel Crawford in their 1979 paper, "T antigen is bound to a host protein in SV40-transformed cells," were the first to identify a particular cellular protein complexed with the SV40 T antigen in transformed cells.[106] They used a monoclonal antibody to the large T antigen and mixed it with a mixture of mouse cells infected with SV40 virus. They then exposed the mixture to electrophoresis, which sorted the proteins by size and found a nonviral 53 kDa protein complexed to the large T antigen. Daniel Linzer and Arnold Levine, using a similar design found essentially the same thing; but in addition, found the 53 kDa protein in embryonal carcinoma cells that had not been infected with SV40.[107] Albert DeLeo and colleagues subsequently found the 53 kDa protein in all transformed mouse cells he tested, including chemically induced sarcomas, leukemias, spontaneously transformed fibroblasts, and cells transformed by simian virus 40 and murine sarcoma virus.[108] Because the 53 kDa protein was found in transformed cell of other than viral infection, the protein must be cellular in origin. Daniel Linzer, Warren Maltzman, and Arnold J. Levine monitored the amount of T antigen and the 53 kDa protein in mouse cells prior to infection, following infection, following transformation, and using a temperature sensitive mutant at permissive and nonpermissive temperatures found that the T antigen is required to initiate and possibly maintain the high concentration of the 53 kDa protein in infected and transformed cells.[109] Additionally, Moshe Oren and colleagues found the concentration of the p53 kDa was approximately 100-fold higher in transformed cells than in nontransformed cell, and the half-life of p53 was much longer.[110] These findings and the general scientific attitudes at the time were convinced that the 53 kDa protein was a protooncogene.

The 53 kDa protein was officially named p53 in 1983 during the first p53 workshop in London. Of course it was named after its size, except that it was actually found later to weigh 43.7 kDa. The difference is speculated to be due to its proline-rich regions that would slow migration during electrophoresis.[111]

In1989, Baker and colleagues, studying colorectal carcinoma noted allelic deletions of the short arm of chromosome 17 occurring in over 75% of colorectal carcinomas, set out to determine the identity of the deleted segment. The deleted segment corresponded to p53. Inspection of the nondeleted somatic p53 gene revealed mutations. The data, Baker argued, suggested that p53 gene mutations might be involved in colorectal carcinoma through inactivation of a tumor suppressor function of wild-type p53 gene.[112] Also in 1989, Cathy Finlay, Philip Hinds, and Arnold J. Levine published, "The p53 proto-oncogene can act as a suppressor of transformation," where they documented the wild-type p53 suppresses mutant p53 in transformation and suggests a reinterpretation of cellular p53 function.[113] Furthermore, they propose that the T-ag may bind to p53 so as to inactivate it predispose the cell to transformation. This was the first suggestion that p53 was not an oncogene, but rather a tumor suppressor gene. As continuing evidence mounted, it became clear that the wild-type p53 suppressed transformation and mutated p53 allowed transformation.

In 1990, two studies reported on the dominantly inherited Li–Fraumeni syndrome, a rare autosomal dominant syndrome characterized by the development of a variety of tumors at various sites, and its association with p53. In the study by Malkin and associates, p53 was studied within the familial cancer syndrome because the sporadic nonfamilial forms of most of the cancers associated with the disease showed inactivation of p53. All five Li–Fraumeni families analyzed in the study had germ-line p53 mutations.[114] In the study by Srivastava and associates, different members of a Li–Fraumeni family were found to have the same point mutation of one of their p53 genes, also retaining a normal p53 allele.[115] Srivastava and associates suggested the inherited p53 mutation might increase susceptibility to cancer in the members of the Li–Fraumeni family. As a result of the overwhelming evidence, p53 has been reclassified as a tumor suppressor gene.[111]

Many mutants of p53 were subsequently developed including temperature sensitive mutants. It was known that wild-type p53 inhibits transformation, while mutant p53 contributed to transformation. Dan Michalovitz, Orna Halevy, and Moshe Oren in their 1990 study, "Conditional inhibition of transformation and of cell proliferation by a temperature-sensitive mutant of p53," used a temperature sensitive mutant that allowed transformation at 37.5°C, but acted like wild-type p53 at 32.5°C, suppressing transformation.[116] Additionally, when transformed cells are transferred from 37.5 to 32.5°C, both transformation and proliferation are reversed suggesting that wild-type p53 acts to arrest growth. In 1991, Yonish-Rouach and colleagues used a p53 deficient murine myeloid leukemia cell line and introduced wild-type p53, resulting in apoptosis.[117] Since the function of tumor suppressor genes was largely unknown at the time, it was suggested tumor suppressor genes might work to inhibit cellular transformation by inducing apoptosis. In 1998, Wang and colleagues, using p53 deficient human lung cancer cells introduced wild-type p53 resulting in irreversible cell growth arrest most like senescence.[118] They then suggested that p53 might prevent immortalization by inducing senescence.

p53 is now known to also induce transcription. In the August 1990 issue of *Science*, two separate studies found that p53 bound DNA and activated transcription. Field and Jang found the 72 NH2 terminal residues of p53 activated transcription as strong as one of the strongest known.[119] They surmised that since p53 was previously localized to the nucleus and that it can bind DNA, then it might possibly activate transcription of genes responsible for inhibiting cell proliferation. Raycroft, Wu, and Lozano, using a fusion protein anchored p53 to a DNA target sequence, measured transcriptional activation.[120] They found wild-type p53 stimulated transcription, whereas two p53 transforming mutants did not. They suggested the ability to activate transcription was not only the difference between wild-type and mutant p53, but also the inability to activate transcription is what made p53 mutants transformation competent. Soon thereafter, Bargonetti and colleagues evaluated both wild-type and tumor-derived mutant p53 interactions with SV40 DNA.[121] They found wild-type p53, both human and murine, bound to specific sequence next to the late border of viral replication origin, whereas, mutant p53 did not. Additionally, they found that the SV40 T antigen prevented wild-type p53 binding. p53 binds to a specific DNA sequence named the p53 response element. p53 activates transcription of genes containing its site-specific response element and repress transcription of other genes that do not contain its response element. Additionally, p53 has biochemical activity of its own. Through a web of interactions, it has the ability to alter the expression of multitudes of genes. But, in the early 1990s when its function was largely unknown, discovering its function meant discovering what it complexed with. In 1991, Jamil Momand and colleagues found the oncogene murine double minute 2 gene (mdm-2) product complexed to p53.[122] The mdm-2 gene product is a phosphoprotein and was found to inhibit p53 gene transactivation. The human counterpart to the murine mdm-2 is the hdm-2. In "Mapping of the p53 and mdm-2 interaction domains," Chen and colleagues demonstrated that mdm-2 protein product binds to the transactivation domain of p53 blocking transactivation by p53.[123] By 1993, Xiangwei Wu and colleagues determined mdm-2 and p53 formed an autoregulatory feedback loop.[124] That is, the mdm-2 gene contains a p53 binding site/response element that induced transcription of mdm-2. In return, mdm-2 can bind p53 inhibiting it's binding to the mdm-2 binding site/response element, inhibiting its own transcription. Wu and colleague point out that mdm-2 is self-regulated and the feedback loop regulates p53 activity as well as mdm-2 concentration/expression. Michael Kubbutat, Stephen Jones, and Karen Vousden in their 1997 paper, "Regulation of p53 stability by Mdm2," demonstrated that mdm-2 can enhance proteasome-dependent p53 degradation.[125] They explain that p53 is maintained at low levels in normal cells by the negative feedback loop where mdm-2 inhibits p53 cell growth arrest and apoptosis. However, when DNA damage occurs there is a rise in p53 levels; and, increased levels of p53 transcriptionally activate mdm2. The increased levels of mdm-2 can subsequently reduce the levels of p53 by the degradation pathway, possibly controlling the extent and duration of the p53 response. The p53 family has subsequently grown to include p63 and p73, which in addition to tumor suppressor activity also function in embryonic development and differentiation.[126] Likewise, p53 has many isoforms functioning in different tissues and at different times of development.[111]

Inactivation of cellular tumor suppressor genes by viral oncoproteins is paramount to the survival and replication of small DNA tumor viruses. In short, these small DNA tumor viruses have oncogenic proteins such as SV40 T antigen, adenovirus E1A, and HPV E7 that associate with the Rb protein–E2F complex, disassociating E2F transcription factor initiating S phase. By initiating cell replication, they are initiating their own replication. Cellular stressors including inappropriate growth signaling by viruses activate p53. Though p53 has many functions, it is most notable for its ability to suspend the cell cycle at the G1/S checkpoint, initiate DNA repair, and/or induce apoptosis when DNA damage cannot be resolved. The p53 inhibition of cellular replication and apoptosis is inhibition of viral replication and viral termination. Therefore, inactivation of the p53 pathway by virus and tumor is required for their subsequent survival and growth. Nearly 50% of all human tumors contain p53 mutations. Likewise, DNA tumor viruses including HPV, adenovirus, and SV40, produced viral oncoproteins that complex with the p53, inactivating it and causing accumulation of intracellular p53 and cellular transformation. It is no wonder p53 is called the guardian of the genome.

SV40 has been classified as an oncogenic virus. SV40 is known to cause cancer is animals, but is still debatable in humans. *Pa*pilloma viruses, *po*lyoma viruses, and *va*cuolating SV40 virus have been group as papovavirus class of DNA viruses.[5]

2.5 Human Papilloma Virus (HPV)

The obvious verrucous lesions of skin and genital warts theoretically should have been recognized for all of human history. Certainly the Greek and Romans knew of them.[127] Certainly, there was a clue to the infectious nature. In 1842, Dr. Domenico Rigoni-Stern, after analyzing the death certificates of women in Verona, Italy from 1760 to 1839 observed that women of the city, namely married women, widows, and prostitutes had a higher rate of cervical cancer that the virgin nuns who lived in the country. He summarized that cervical cancer must be sexually transmitted.[128] In 1891, J. Payne published, "On the contagious rise of common warts."[129] Then in 1907, Giuseppe Ciuffo using filtration and common warts substantiated a viral etiology.[1] Richard Shope discovered the first papilloma virus, the Shope papilloma virus or cottontail rabbit papilloma virus, in 1933.[8] In 1935, when Peyton Rous and Joseph Beard placed the virus in domestic rabbits it resulted not in a benign tumor (wart), but in a skin carcinoma.[3,9,10] Electron microscopy of skin warts had demonstrated viral intranuclear inclusions and cytoplasmic masses in 1949.[130] By the 1950s, it was known that the papilloma virus contained a DNA genome.[10] In 1959, bovine urothelial carcinoma was induced by the injection of cutaneous papillomas/warts from other cattle. Additionally, it was noted that the bovine cutaneous papillomas induced cancer in other animals.[131] Of note, George Otto Gey derived the HeLa cell line from cervical cancer cells taken on February 8, 1951 from Henrietta Lacks.[132] The HeLa immortal cell line quickly spread to research labs throughout the World.

The belief in the early 1970s was that Herpes simplex virus type 2 was the likely cause of cervical and other anogenital carcinomas. Genital herpes, with its characteristic blistering, could be easily diagnosed by a Tzanck smear. The Tzanck smear was first described in 1947 by Arnault Tzanck as a tool to differentiate various skin blistering conditions. To perform a Tzanck smear, the clinician would remove any crusting over the ulcer, scrape the edge of the ulcer, smear the scraping on a slide, allow the slide to air dry, and stain the slide with Geimsa stain.[133] Under a microscope, herpes could be diagnosed when cells displayed the distinctive viral cytopathic effect; namely, nuclear changes characterized by multinucleation, nuclear molding, and chromatin margination caused by a nuclear inclusion.

Then, in 1972, Stefania Jablonska, Jan Dabrowski, and Kazimierz Jakubowicza reported on a family with epidermodysplasia verruciformis, a hereditary skin disorder characterized by vulnerability to verrucous lesions (warts), which, over time may progress to carcinoma. Verrucous lesions with histologic evidence of papovavirus were harvested from one family member and successfully transmitted to a healthy volunteer. Additionally, the presence of the virus in the verrucous lesions was demonstrated by electron microscopy. Papilloma virus by electron microscopy looks similar to a golf ball. Together the evidence was sufficient to recognize the virus as the causative agent of the verruca/warts and the initiator of the cancerous process.[134]

Harald zur Hausen, in 1974 and 1976 was the first to suggest that perhaps a papilloma virus was the cause of cervical and other anogenital carcinomas.[135] In 1974, zur Hausen using the DNA of various herpes viruses (EBV, CMV, HSV, and VZV) made complementary RNA and attempted to hybridize it to various human malignant and nonmalignant tissues. EBV was found to anneal to nasopharyngeal carcinoma, and multiple tissue from patients with infectious mononucleosis. Herpes simplex virus type 1 and type 2 (HSV) did not hybridize/anneal to any of the cervical carcinoma biopsies.[136,137] In 1983 and 1984, zur Hausen and colleagues were able to identified HPV 16 and 18 in cervical and anogenital carcinomas, as well as, HPV 6 and 11 in condylomas.[138–140]

Papilloma viruses are ubiquitous. They infect humans and animals. However, they are very specific for not only the host, but also the tissue they infect. There are greater than 200 different types of HPV, 85 are well characterized, the others only partially.[141] HPVs are small, nonenveloped DNA viruses. The HPVs are grouped as alpha, beta, and gamma genera. Alpha HPV infects the genital and oropharnygeal mucosa. Beta and gamma HPV infect skin.[142] The discussion will now be limited to the alpha HPV types that infect mucosa, more specifically those that infect genital mucosa, and even more specifically those that cause cervical carcinoma, unless stated otherwise. The HPV types that infect the genital tract are classified as "low risk" or "high risk" types based on their oncogenic potential. Low risk types generally cause genital warts; "condyloma accuminata." High risk types have a greater potential for development of carcinoma. Low risk HPV types include: 6, 11, 40, 42, 43, 44, 54, 61, 72, and 81. HPV 6 and 11 account for approximately 90% of genital warts. High risk HPV types include: 16, 18, 31, 33, 35, 39, 45, 51, 52, 56, 68, 59, 68, and 82. Of invasive carcinomas of the cervix, squamous cell carcinomas account for 95% and adenocarcinomas account for 5%. HPV 16 and 18 cause greater than 95% of invasive squamous cell carcinomas of the cervix and HPV 16, 18, and 45 are associated with adenocarcinomas of the cervix. Additionally, HPV types have variants. For example, HPV 16 has four variants: European (E), African (AF-1, AF-2), and Asian American (AA). The various types have at least a 10% difference in the nucleotide sequence of their L1 open reading frame, whereas, variants have <3%.[143]

HPV is transmitted by skin and/or mucosal contact. Most sexually active women and men, at some point in their lives, usually shortly after becoming sexually active, will be infected with HPV.[144] Acquiring one HPV infection does not grant immunity to other types of HPV infection, so repeated HPV infections can occur. Most HPV infections clear within a couple months and about 90% clear within 2 years. Whether the virus is actually cleared or just undetectable is debatable. Development of carcinoma occurs with persistent infections over 10–20 years through a series of distinct steps.[143]

The endocervix histologically is a mucin secreting simple columnar epithelium. The ectocervix histologically is nonkeratinized stratified squamous epithelium that sits on a basement membrane with underlying connective tissue. The junction of the endocervix and the ectocervix is called the squamocolumnar junction. The squamocolumnar junction is within the cervical canal in young girls and migrates out during puberty and pregnancy. It eventually migrates back into the cervical canal after menopause. When the squamocolumnar junction migrates out of the cervical canal, the epithelium of the old squamocolumnar junction undergoes metaplastic changes, usually squamous metaplasia, as a protective mechanism against the acidic environment of the vagina while forming a new squamocolumnar junction within the cervical canal. The difference between the old and the new squamocolumnar junction is called the transformation zone and it is the location of HPV infection. Stratified squamous epithelium is arranged in layers, attached to the basement membrane and matures/differentiates as it progresses superficially and is eventually shed from the surface. The basal cell layer is one cell thick and sits on the basement membrane. It is composed of round cells with a high nuclear to cytoplasmic ratio. The parabasal layer is a couple cells thick and the cells are round like basal cells, but have slightly more cytoplasm. Mitoses are limited to the basal and occasionally the parabasal layers. In normal epithelium, mitoses do not occur in the intermediate or superficial layers. The intermediate layer is the thickest layer. It is composed of maturing cells with progressively more cytoplasm and smaller nuclei. By pap smear/cytology, the intermediate cell is the standard cell for comparing all the other cells. The superficial layer is the mature layer composed of cells with abundant flat cytoplasm and small, dark, pyknotic nuclei. These cells appear like fried eggs and are eventually shed from the surface.

As mentioned earlier, each HPV type is tissue tropic and the HPVs that infect stratified squamous epithelium mature as the squamous epithelium matures. Likewise, HPV cytopathic effects progress from basal cells to superficial cells. As well, HPV lesions progress from basement membrane to surface. The pathognomonic cytopathic effect of HPV is koilocytosis.

A koilocyte displays large (>2–3 ×), hyperchromatic nuclei of varying sizes with irregular nuclear membranes and a large, clear perinuclear halo that pushes and compresses cytoplasmic elements to the periphery. Multinucleation is often seen and polarity is often lost. Polarity describes the orientation of the squamous cell in relation to the basement membrane. The earliest lesion is low grade squamous intraepithelial lesion. The term describes a state of mild dysplasia in which infected squamous cells with a basal cell appearance proliferate up to a third the thickness of the total squamous epithelium. It encompasses both HPV atypia and cervical intraepithelial neoplasia I (CIN I). High grade squamous intraepithelial lesion is the term used for a state of moderate and marked dysplasia in which infected squamous cells with a basal cell appearance proliferate up to two-third and beyond two-thirds the thickness of the total squamous epithelium. It encompasses cervical intraepithelial lesion II and III (CIN II and CIN III). CIN III is equivalent to squamous cell carcinoma in situ in which the entire thickness appears like immature basal cells with an intact basement membrane. If the infected cells have breached the basement membrane, the squamous cell carcinoma is invasive.

Cervical microtrauma, as occurs during intercourse, exposes basal cells to HPV. HPV entry into a basal cell is believed mediated by heparan sulfate receptor. The basal cell divides forming two daughter cells, one remains in the basal layer as a stem cell and the other continues upward. After entry into the basal cells, HPV replicates up to approximately 100 episomal copies per basal cell. Because HPV requires the cellular machinery for replication, the virus keeps the cervical cell in a constant state of division. The HPV genome is circular double-stranded DNA. It is composed of five early genes (E1, E2, E5, E6, and E7) and three late genes (E4, L1, and L2). There are two promoters, one directing the expression of early genes and another one directing the expression of late genes. After HPV entry to the nucleus, E1 and E2 induce viral replication. E2 is required for viral DNA replication and regulation of transcription. E2 has a DNA-binding domain and a transactivation domain. Alternatively, it can be spliced to a truncated variant lacking the transactivation domain. E2 proteins function as dimmers. At low E2 concentrations, E2 activates early gene expression by encouraging viral genome transcription. At high E2 concentration, E2 interferes with binding of transcriptional factors to their recognition site, which overlaps with the early E2 binding site. Over expression of E2 protein can induce apoptosis, independent of p53 and beyond rescue by E6. Generally, E2 binds proximal to the HPV origin of replication/promoter at the early E2 binding site.

It subsequently recruits and loads E1 on to the origin. E1 unwinds the DNA and recruits polymerase to the viral origin of replication to begin transcription. As the squamous epithelial cell matures and progresses superficially, E1 and E2 switch to the late promoter, which is not regulated by E2. The viral genomic copies quickly rise to 1000 per maturing cervical squamous cell. E2 performs many other function, but an important one to note is that E2 downregulates E6 and E7. E2 expression is often lost as well as its influence on E6 and E7 upon viral genome integration into host DNA. The viral genome is typically maintained as an episome. Rarely and by accident, the viral genome is integrated into the host cell DNA. Integration is the initiating step in cellular transformation. E4 and E5 are synthesized as early and late genes, but primarily expressed as a late transcription. Many of the specific functions of E5 remain unclear. E5 is expressed with both early gene and late gene expression, but not in transformed malignant cells. It is a membrane spanning protein located within the endosomal membranes, Golgi, and occasional in the cellular membranes. It complexes with platelet-derived growth factor receptor and epidermal growth factor receptor. Additionally, E5 downregulates MHC1 expression, possibly evading immune surveillance and response. E6 is present in both the nucleus and the cytoplasm and lacks enzymatic activity. In all, E6 binds over 12 proteins. It's most noted function is the clearance of p53 allowing for uncontrolled cell proliferation. The ability of E6 to dispose of p53 is not present in the low risk HPV types. E6 binds cellular E6AP, a ubiquitin ligase, then it binds p53 forming a trimeric complex. E6AP tags p53 with ubiquitin which when recognized by a proteasome, digests the p53. Furthermore, E6 downregulates p53 through its association with p300/CBP. Another important function of E6 is reactivation of hTert (human telomerase reverse transcriptase) transcription, a function performs with the help of cMyc. hTert is the catalytic subunit of telomerase. E7 binds Rb protein, p107, and p130, resulting is their proteasomal degradation, unrestricted E2F activity, drives cellular proliferation. The E7 of low risk HPVs also bind the Rb protein, but with a much lower affinity. E7 also binds histone deacetylases resulting in increased E2F-driven transcription and cellular proliferation. Additionally, E7 has an interesting ability to induce genomic instability. E6 and E7 are considered viral oncogenes. Genomic proliferation switches to genomic packaging with the help of high E2 levels. E2 recruits L2, the minor coat protein, to areas of replication prior to L1 expression. The dying squamous cell of the most superficial levels loose mitochondrial oxidative phosphorylation creating an environment suitable for capsid self-assembly by L1 and L2 proteins.

E4 lacks an initiator codon and thus uses the E1 codon for translational initiation resulting in E1–E4 fusion proteins. The E1–E4 fusion protein associates with keratin networks and when overexpressed can induce keratin network collapse. It is therefore believed that E4 facilitates the release of viral particles.[143,145] The intracellular replication of the virus, absence of a blood borne acute phase virema, and nonlytic viral replication with release of viral particle from superficial epithelial cells upon their cellular sloughing may function in viral evasion of immune surveillance and response.[146]

Cervical carcinoma is the most common and the best characterized HPV-related disease/carcinoma with nearly all cases caused by HPV infection. However, HPV also causes 65% of vaginal, 50% of vulvar, 95% of anal (most HPV 16), 35% of penile, and 70% oropharyngeal carcinomas (>50% HPV 16).[146] According to the World Health Organization, cervical cancer is the second most common cancer of women living in less developed areas. In those less developed areas, there were approximately 445,000 new cases, accounting for 84% of new cases worldwide in 2012. In that same year, approximately 27,000 women died from cervical cancer with greater than 85% occurring in low- to middle-income countries.[144] In 2013, in the United States, nearly 12,000 women were diagnosed with cervical carcinoma and slightly more than 4000 died from it.[147] Because of the implementation of the pap smear as a screening method to detect early precursor lesions, the incidence of cervical cancer in the United States has dropped approximately 90%. There are currently two HPV vaccines available. Gardasil is a quadrivalent HPV vaccine composed of recombinant L1 protein viral like particles from HPV 6, 11, 16, and 18 for use in young ladies 9–26 years old. Cervarix is a bivalent HPV vaccine is composed of recombinant L1 protein viral like particle from HPV 16 and 18 for use in young ladies 10–25 years old. Both are given in three doses before the onset of sexual activity.[143]

2.6 Human Polyoma Virus

In New York, 1972, Cyril Toker first described Merkel cell carcinoma (MCC) as trabecular cell carcinoma.[148] It is interesting to note, that despite the namesake, it is very controversial as to whether MCC actually derives from Merkel cells. In 1978, the histology of MCC led Tang and Toker to conclude this new malignancy was from Merkel cells.[149] However, later advances in science, in particular tumor markers, have pointed to a possible stem cell origin.[150,151] The clinical presentation of MCC is generally asymptomatic reddish–blue nodules on sun-exposed areas that develop rapidly and are commonly seen in older males.[152] Clinically, MCC is often first suspected to be basal cell carcinoma or perhaps a epidermoid cyst due to its rarity.[153] Histologically, MCC is made of small, ill-defined blue cells with a salt and pepper nucleus arranged in cords, bands, and clusters[153] similar to that of small cell carcinoma of the lung or melanoma. As with many viral-associated malignancies, MCC is significantly more common among the immunocompromised like organ transplant and AIDS patients.[154] The cause of MCC was not known for over 30 years until 2008 when Feng *et al.* detected Merkel cell polyoma virus (MCPyV) DNA in 80% of MCC tumors. This number is staggering compared to the 16% normally found in control tissue samples.[155] MCPyV, like all polyoma viruses, is a naked circular double-stranded DNA virus. MCPyV codes for early T antigen genes and late capsid proteins.[153] Unlike most polyomaviruses, MCPyV has not been identified to have a middle T antigen protein, but rather only small and large T antigen proteins.[156]

The work continued for Feng and his colleagues who found a mutation in MCPyV from MCC positive samples that was not present in MCC negative samples of MCPyV.[157] Specifically, the LT antigen protein of MCPyV was truncated leading to a loss of helicase function.[157] Normally, the LT antigen protein interacts with helicase and RB1 tumor suppressor protein among others. The interaction with helicase will lead to autonomous viral replication. The LT antigen's interaction with Rb1 inhibits tumor suppression via deregulating E2F transcription.[156] With the LT antigen protein truncated, viral replication is lost but unregulated cell division occurs.[153] In total, researchers concluded that patients required the integration of MCPyV DNA and then the LT antigen mutation to develop MCC.[157] However, there are other proteins at play that are key to carcinogenesis. LT or large T antigen protein works in coordinance with ST or small-T antigen protein in the pathogenesis of MCPyV. ST antigen redundantly stimulates cell proliferation via different cellular pathways. However, a knockout study by Shuda *et al.* demonstrated the redundancy was necessary for MCC tumor proliferation and survival.[158]

Though the mechanism for MCC from MCPyV is becoming more and more clear every year with research, it does not take into account those MCC cases that do not have a direct link to MCPyV. It has been demonstrated more than once that MCC in the Australian population may have a different pathophysiology than that of North America.[159,160] In one study, a significantly lower percentage of patients with MCC had MCPyV DNA in their tumor site genome than the original 80% seen in Feng *et al.*'s research in 2008.[159] This was later confirmed in another study that demonstrated the usual MCPyV immunohistochemistry was not near as sensitive in Australian populations as compared to the control.[160] Both of these studies support alternate causes to MCC. Those alternate causes are still unknown, and there is still much to be discovered about the causes of MCC and the mechanism and treatment of MCPyV linked MCC.

2.7 Epstein–Barr Virus (EBV, HHV-4)

London 1961, Anthony Epstein went to a lecture being given by a British surgeon working in east Africa. Dr. Denis Burkitt described a tumor of young children with distinct geography determined by temperature and rainfall. Epstein, at the time, had been working on oncogenic viruses and the lecture immediately brought to mind the possibility of a vector borne virus. Epstein immediate stopped his current work and decided to investigate the etiology of Burkitt's lymphoma. Early attempts at viral isolation and cell culture development failed. In December 1963, there was a delay in tumor transportation from Africa. By the time the tumor sample had arrived, the suspension was cloudy. Microscopic evaluation revealed a cellular suspension instead of bacterial contamination as had been suspected. The cellular suspension, it turned out, was more likely to yield a viable cell culture. The delayed suspension grew the first lymphoma cell culture. Methods at the time could not isolate the virus. Epstein, convinced that the virus could be identified by electron microscopy much like bacteria could by light microscopy, prepared his first sample in February 1964. Immediately, the pictures showed a herpesvirus, but unlike other known herpesviruses. The new herpesvirus did not kill the host cell. The peculiarities of the virus caused Epstein to send the virus to Werner and Gertrude Henle who confirmed Epstein's doubtful findings. His virus was indeed a new herpesvirus. It was the Henle's that would give the EBV its name.[3,161] Some original arguments have been overcome in assigning EBV as a cause for Burkitt's lymphoma. First, the virus is ubiquitous with EBV antibodies found in greater than 90% of people everywhere. Second, Burkitt's lymphoma is identified by one of the three translocations, all involving c-myc on chromosome 8. While that translocation was present is Burkitt's lymphoma cases in developed countries, the EBV virus were not. Nonetheless, it is now generally accepted that EBV is the causative agent of African Burkitt's lymphoma along with AIDS-associated lymphoma, posttransplant lymphoma, nasopharyngeal carcinoma, nasal type T/NK lymphoma, Hodgkin lymphoma, gastric adenocarcinoma, lymphoepithelioma like carcinoma, and immune deficiency-related leiomyosarcoma.[3] Additionally, EBV, also known as HHV4, causes infectious mononucleosis. EBV usually spreads through body fluids and is usually contracted in childhood. The symptoms range from asymptomatic to mild flu like illness. Once contracted, the virus remains latent for the life of the host/person.

2.8 Kaposi's Sarcoma-Associated Herpesvirus (KSHV, HHV-8)

Moritz Kaposi first described the dermatological presentation of this disease in 1872 in his paper "Idiopathic multiple pigmented sarcoma of the skin." In what would later be subtyped as Western or European Kaposi's sarcoma, Kaposi described the violet colored nodules that are commonly found on patient's extremities, head, and mucus membranes when affected.[162] This etiological subtype was more common in elderly men, restricted to skin, and rarely lethal.[163] Over the next 100 years, three other etiological subtypes would be discovered; all presenting with similar stereotypical findings. During the 1950s, in Africa, reports of a more fulminant KS surfaced.[163] However, unlike European KS, this African variant was found in children as well as adults. More than 50 years later, studies have shown though an age-independent prevalence is often observed in these African KS populations, there are some countries, such as Uganda, that demonstrate a clear association with age and HHV-8 infection. This leads researchers to believe there is more than one transmission pattern in Africa.[164] In the 1970s, just 20 years after the invention of immunosuppressive drugs, KS was seen again in transplant patients.[165,166] In this setting, regression of KS is often seen with immunosuppression discontinuation.[166] Finally, KS became the most common neoplasm of AIDS patients in the 1980s; nearly 40% of people with HIV, concurrently had KS.[167] Mortality among HIV-related KS is significantly higher than all other forms. This is most likely due to visceral involvement of the gastrointestinal tract as well as bronchi mucosa.[163] However, with the arrival of antiretroviral therapy, the rate of KS among HIV infected individuals saw a dramatic reduction.[168,169] The surge in research following the AIDS epidemic would later link KS with HHV-8 in 1994 via the relatively new method of PCR testing and serological assays. This led many researchers to assume a direct association between HIV and HHV-8. However, that is not necessarily true. Trends in the mode of HIV infection, as well as various sexual behaviors weighed greatly on a patient's risk of KS.

For instance, KS is significantly less likely in vertical and blood borne HIV transmission. Rather, it is seen in men who have sex with men and those who have more partners carry a higher risk.[170,171] Even in the African variant of KS where children can be infected, studies have shown transmission to be horizontal rather than vertical. In addition to KS, HHV-8 has been linked to multicentric Castleman's disease and primary effusion lymphoma, among others.

As with all herpes viruses, HHV-8 has a biphasic life cycle with both latent and lytic phases. HHV-8 defaults to the latent phase in KS and primary effusion lymphoma. It is interesting to note that this is not the case with multicentric Castleman's disease, where the lytic phase is associated with malignancy.[172] Each phase has associated key viral proteins. The latent phase key proteins are viral cyclin D (v-cyc), latency-associated nuclear antigen (LANA1), and viral Fas-associated death domain interleukin-1B converting enzyme inhibitory protein (vFLIP).[173,174] V-cyc, transcribed from the gene orf72, causes genomic instability via overriding host growth arrest by cyclin-dependent kinases.[175] LANA1, transcribed from orf73, is perhaps the biggest and most researched protein of this particular viral caused malignancy due to the multiple functions it performs. This protein is essential to maintaining latency by attaching to the H1 histone on host chromatin.[176] It also exercises its E3 ubiquitin ligase activity on p53 and von Hippel–Lindau, thereby effectively impairing apoptosis and inducing angiogenesis and cell proliferation.[177] vFLIP redundantly protects the infected cell from apoptosis by inhibiting the Fas death receptor pathway.[178,179] vFLIP has also been indicated in maintaining the latent phase via activation of NF-kB. The NF-kB then inhibits reactivation transcription activation expression commonly seen during the HHV-8 lytic phase.[180]

In addition to the latent phase proteins, HHV-8 also produces viral G protein-coupled receptor (vGPCR), viral BCL-2 (vBCL-2), viral IL-6 (vIL-6), and K1 protein during the lytic phase.[173,174] Viral GPCR, transcribed by orf74, causes an increase in VEG-F via the mTOR signaling pathway.[181] This increase in VEG-F causes the KS stereotypical angiogenesis and, in turn, violet colored nodules.[182,183] With a 60% homology to human BCL-2, it is easy to understand how vBCL-2 lends itself to protecting cells from apoptosis.[184,185] Unlike vBCL-2, vIL-6 does not share such a close relation to its human homolog.[186] However, vIL-6 does induce the production of natural IL-6 leading to proliferation of myeloma cells.[186,187] This viral protein does not play a big role in KS. However, it is a key player in the pathogenesis of other HHV-8 linked illnesses.[188] The final lytic phase

HHV-8 protein produced is K1. As a highly variable transmembrane glycoprotein, K1 is very useful for geographic mapping of the different subtypes of KS.[189,190] The pathogenetic significant of the protein is not well known as of yet.

2.9 Hepatitis B Virus (HBV)

Hepatitis, with its resulting jaundice, has been easily recognized for thousands of years. However, serum hepatitis was perhaps first recognized in Germany in 1883 after shipyard workers had been administered a small pox vaccine containing human lymph different from another vaccine administered at the shipyard.[191] In 1966, Harvey Alter and Baruch Blumberg in their paper, "Further studies on a "new" human isoprecipitin system (Australia antigen)," describe a new antigen; the Australia antigen, in the blood of an Australian aborigine from which it got its name.[192] The paper goes on to tell that the Australia antigen is more commonly found in foreign populations than American populations. It is relatively common in US patients with leukemia, and the anti-Australia antibody is relatively common in patients with hemophilia, but rare in transfused patients with other diseases. Although Blumberg in 1967 associated the Australia antigen with hepatitis, it was Alfred Prince in 1968 that associated the Australia antigen specifically with serum hepatitis.[193] The Australia antigen was eventually found to be the surface antigen of the hepatitis B virus, the cause of serum hepatitis. In 1975, it was Blumberg and his colleagues that first linked hepatocellular carcinoma with chronic hepatitis B infections in their paper, "The relation of infection with the hepatitis B agent to primary hepatic carcinoma."[194]

In the United States, there are an estimated 19,000 new cases of hepatitis C a year with an estimated 850,000—2.2 million people with chronic HBV infection. Less than 1% of infants under a year develop acute symptoms, 5–15% of children ages 1–5 years develop acute symptoms, 30–50% of people greater than 5 years of age develop acute symptoms, and 5–15% of immunosuppressed adults develop acute symptoms. Greater than 90% of infants will develop chronic infection, 25–50% of children age 1–5 years will develop chronic infection, and 6–10% of people over 5 years of age will develop chronic infection. While acute illness is rarely fatal, it generally has no lasting liver damage. Of chronically infected people, 15–25% will develop chronic liver disease, cirrhosis, liver failure, or hepatocellular carcinoma. In 2014, that accounted for approximately 1800 deaths.[195]

The first "Vaccine against human hepatitis B" was developed in 1976 by Eugene Buynak, Robert Roehm, Alfred Tytell, Alexander Bertland, George Lampson, and Maurice Hilleman.[196] That vaccine was replaced in 1980 by a recombinant HBV surface antigen vaccine that is given in three doses starting at birth if possible. The hepatitis B vaccine was the first vaccine developed that has the potential to prevent cancer.

3. RNA VIRUSES

3.1 Hepatitis C Virus (HCV)

In April 1975, Stephen Feinstone, Albert Kapikian, Robert Purcell, Harvey Alter, and Paul Holland, in their article, "Transfusion-associated hepatitis not due to viral hepatitis type A or B," examined 22 patients with transfusion-associated hepatitis for new development of antibodies to hepatitis A, hepatitis B, cytomegalovirus, and EBV. Though nine had developed antibodies to cytomegalovirus, CMV could not account for all 20 cases. They therefore proposed a not yet identified infectious agent as the likely cause of transfusion-associated non-A, non-B hepatitis.[197] In November of the same year, in a prospective study, they suggest the existence of previously unrecognized human hepatitis virus is not only likely, it is probable.[198] The transfusion-transmitted virus study found a 10% attack rate for transfusion recipients.[199] In 1978, Harvey Alter and colleagues inoculated five chimpanzees with the plasma or serum from four patients with acute or chronic non-A, non-B posttransfusion hepatitis, and from a blood-donor suspected to be the likely cause of two of the posttransfusion-associated hepatitis cases. The five chimpanzees developed elevated ALT and histological evidence of hepatitis. The severity of the histologic hepatitis correlated with the elevation in ALT. All animals completely recovered and the chimpanzee became an animal model for non-A, non-B hepatitis. The finding that chronic hepatitis could cause posttransfusion non-A, non-B hepatitis suggested a possible carrier state.[200] Using a similar experimental design in 1979, two groups of chimpanzees were inoculated with plasma from two different patients with non-A, non-B hepatitis. The hepatocytes in the first group, harvested during the acute phase, showed cytoplasmic tubules not seen in the second group. The second groups showed nuclear changes, not seen in the first group. The nuclear changes were associated with aggregates of 20–27 nm particles resembling viruses, but they did not appear as uniform as small virus particles frequently seem.[201] The non-A, non-B tubule forming agent could pass through an 80 nm membrane filter and had a known sensitivity to chloroform,[202] a lipid solvent, suggesting that it may be a small, enveloped RNA virus.[203] For a decade, various molecular methods, serologic methods, and cell culturing attempts yielded little to the identification of the non-A, non-B hepatitis virus.[204]

Though Michael Houghton had considered using a serum samples from non-A, non-B diagnosed patients against cDNA expression libraries from infectious liver and plasma samples, the method was not consistently reliable in identifying targets. In a casual conversation with George Kuo in 1985, Kuo strongly argued for a blind immunoscreening cDNA cloning strategy since in his belief the antigen concentrations in infected livers and plasma are low and undetectably using conventional methods. Shortly thereafter, Dan Bradley also discussed with Houghton the possibility of using a blind immunoscreening method. Houghton began the study. George Kuo suggested using radioactive labeled I^{125} antihuman IgG for detecting human Ig binding to cDNA clones. For months Qui-Lim Choo screened millions of cDNA clones from libraries that Houghton prepared. Success final came in 1988 and 1989 using high infectious titer chimpanzee plasma, produced by Dan Bradley, from chimpanzee 910 and screening with serum from a chronic non-A, non-B hepatitis patient with high ALT. Clone 5-1-1 proved to be from the genome of non-A, non-B hepatitis virus. It was a positive stranded RNA molecule of at least 10,000 nucleotides, similar to togaviridae or flaviviridae.[205] Of course, togaviridae and flaviviridae were once one in the family togaviridae. Then in 1984, because of reproductive differences, the togaviridae family was split and flaviviridae became its own family, though still somewhat closely related. This newly discovered flavivirus got its name, hepatitis C virus or HCV for short. Houghton and his team quickly developed c100-3 antigen in recombinant yeast from 5-1-1 antigen, as well as an enzyme immunoassay for circulating antibodies to c100-3.[206]

Using radioimmunoassay, they followed and measured antibody to HCV (non-A, non-B hepatitis) in transfusion donors and recipients documenting the progression of virus and disease from donor to recipients. They concluded that HCV was the primary etiologic agent of non-A, non-B hepatitis. Furthermore, screening donors would effectively prevent the majority of transmission by transfusion.[207] This opened up HCV for our discovery. Quickly, blood tests used to test the blood supply preventing transfusion-associated hepatitis C, methods to monitor progression and therapy of HCV, and generation of cell lines were developed.[204]

HCV is a positive RNA virus with one open reading frame. The open reading frame codes a single polyprotein, which can be cleaved by viral or cellular proteases into 10 proteins. The structural proteins consist of core, E1, E2/p7. The nonstructural/enzymatic proteins consist of NS2, NS3, NS4A, NS4B, NS5A, and NS5B.

In the United States, there are an estimated 30,000 new cases of hepatitis C a year with an estimated 2.7–3.9 million people with chronic HCV infection. Only 20–30% of newly infected people develop acute symptoms whereas 75–85% will develop chronic infection. While acute illness has no lasting liver damage, 60–70% of chronically infected will develop chronic liver disease, 5–20% will develop cirrhoses over 20–30 years, and 1–5% will die more commonly from decompensated cirrhosis or resulting hepatocellular carcinoma. In 2014, that accounted for approximately 19,600 deaths. Though there have been continued efforts at a hepatitis C viral vaccine, none as yet are available.[195]

Because of the association of hepatitis B virus with chronic hepatitis, cirrhosis, and hepatocellular carcinoma, HCV was suspected of the same. In 1989, the same year HCV was discovered and named, Houghton and colleagues using radioimmunoassay found HCV to be an important factor associated with HCC.[208] Likewise, epidemiologic studies showed HCC is significantly higher in people with HCV infection.[209] HCV is thought to be an indirect cause of HCC, causing chronic inflammation, cell death, and proliferation through an adenoma–HCC pathway. Yet more recently, some argue HCV is a direct cause of HCC. As an RNA virus, HCV does not integrate into the host genome. HCV core protein, NS3, and NS5A have been implicated in the process of transformation. The HCV core protein is a structural protein. It has been shown to induce HCC in two independent lines of transgenic mice[210] and cholangiocarcinoma in human bile duct cells.[211] Likewise, the HCV serine protease enzyme NS3 transforms transfected rat fibroblasts cells in culture.[212] The core protein, NS5A, and NS3 inhibit p53; the core protein and NS5B inhibit RB1; and the core protein inhibits p73. The core protein, E2, NS5A, and NS5B activate cellular proliferation through the RAF/MAPK/ERK kinase and E2F1 pathways. NS5A is involved in activation of PI3K/ATK and B-catenin/WNT pathways.[213] Many additional published associations resulting in activations and inhibitions are plentiful implicating HCV in transformation, both directly and indirectly. The underlying mechanism of malignant transformation of infected hepatocytes has not been fully worked out and is a continuing area of investigation.

4. CONCLUSION

The study of oncogenic viruses has contributed greatly to the scientific advances made during this molecular age in the field of cancer biology. They have been the key to unlocking cell cycle function and control. Research with acute transforming retroviruses led to the discovery of reverse transcriptase and oncogenes. Slow transforming retroviruses integrate their genome into the host cell genome possibly disrupting a protooncogene in a process called insertional mutagenesis. Complex oncogenic retroviruses contain in their genome not only the basic genes of the aforementioned simple retroviruses, but also regulatory and supplemental genes in which they manipulate the host cell and its machinery. The studies of DNA oncogenic viruses lead to the discovery of tumor suppressor genes including RB and p53. And obviously, oncogenic viruses provided a viral etiology and genomic basis for cancer.

Oncogenic viruses infect, but do not kill their host cell, instead they establishing long-term persistent infections that after many years may produce cancer. Vaccines for hepatitis B virus, an etiologic agent of hepatocellular carcinoma, and HPVs, the etiologic agent for nearly all cervical carcinomas, are now available. The goal of everyone who has ever entered medicine is to save lives. Oncoviruses, or rather oncoviral vaccinations, give all of us involved in medicine an opportunity to cure cancer before it even happens.

REFERENCES

[1] Ciuffo G. Innesto positivo con filtrato di verruca volgare. *G Ital Mal Ven Pelle.* 1907;48:12–17.

[2] Ellerman V, Bang O. Experimentelle Leukämie bei Hühnern. *Zentralbl Bakteriol Parasitenkd Infectionskr Hyg Abt I.* 1908;46:595–609.

[3] Javier RT, Butel JS. The history of tumor virology. *Cancer Res.* 2008;68(19):7693–7706.

[4] Rubin H. The early history of tumor virology: Rous, RIF, and RAV. *Proc Natl Acad Sci U S A.* 2011;108(37):15534.

[5] Weinberg R. Tumor viruses. In: *The Biology of Cancer.* 2nd ed. New York, NY: Garland Science; 2006:57–90.

[6] Rous P. A sarcoma of the fowl transmissible by an agent separable from the tumor cells. *J Exp Med.* 1911;13(4):397–411.

[7] Rous P. A transmissible avian neoplasm (sarcoma of the common fowl). *J Exp Med.* 1910;12(5):696–705.

[8] Shope RE, Hurst EW. Infectious papillomatosis of rabbits: with a note on the histopathology. *J Exp Med.* 1933;58(5):607–624.

[9] Rous P, Beard JW. The progression to carcinoma of virus-induced rabbit papillomas (Shope). *J Exp Med.* 1935;62(4):523–548.

[10] Vogt PK. Oncogenes and the revolution in cancer research: homage to hidesaburo hanafusa (1929–2009). *Genes Cancer*. 2010;1(1):6–11.
[11] Dulbecco R. Production of plaques in monolayer tissue cultures by single particles of an animal virus. *Proc Natl Acad Sci U S A*. 1952;38(8):747–752.
[12] Dulbecco R, Vogt M. Plaque formation and isolation of pure lines with poliomyelitis viruses. *J Exp Med*. 1954;99(2):167–182.
[13] Groupe V, Manaker RA. Discrete foci of altered chicken embryo cells associated with Rous sarcoma virus in tissue culture. *Virology*. 1956;2(6):838–840.
[14] Martin GS. The road to Src. *Oncogene*. 2004;23:7910–7917.
[15] Temin HM, Rubin H. Characteristics of an assay for Rous sarcoma virus and Rous sarcoma cells in tissue culture. *Virology*. 1958;6:669–688.
[16] Temin HM. The control of cellular morphology in embryonic cells infected with Rous sarcoma virus in vitro. *Virology*. 1960;10:182–197.
[17] Hanafusa H, Hanafusa T, Rubin H. The defectiveness of Rous sarcoma virus. *Proc Natl Acad Sci U S A*. 1963;49(4):572–580.
[18] Hanafusa H, Hanafusa T, Rubin H. Analysis of the defectiveness of Rous sarcoma virus: I. Characterization of the helper virus. *Virology*. 1964;22(4):591–601.
[19] Martin GS. Rous sarcoma virus: a function required for the maintenance of the transformed state. *Nature*. 1970;227:1021–1023.
[20] Vogt PK. Retroviral oncogenes: a historical primer. *Nat Rev Cancer*. 2012;12(9): 639–648.
[21] Duesberg PH, Vogt PK. Differences between the ribonucleic acids of transforming and nontransforming avian tumor viruses. *Proc Natl Acad Sci U S A*. 1970;67(4): 1673–1680.
[22] Temin HM. The DNA, provirus hypothesis. *Science*. 1976;192(4244):1075–1080.
[23] Temin HM. The effects of actinomycin D on growth of Rous sarcoma virus in vitro. *Virology*. 1963;20:577–582.
[24] Baltimore D. RNA-dependent DNA, polymerase in virions of RNA tumour viruses. *Nature*. 1970;226(5252):1209–1211.
[25] Temin HM, Mizutani S. RNA-dependent DNA polymerase in virions of Rous sarcoma virus. *Nature*. 1970;226(5252):1211–1213.
[26] Temin HM, Baltimore D. RNA-directed DNA synthesis and RNA tumor viruses. *Adv Virus Res*. 1972;17:129–186.
[27] Stehelin D, Varmus HE, Bishop JM, Vogt PK. DNA related to the transforming gene(s) of avian sarcoma viruses is present in normal avian DNA. *Nature*. 1976;260(5547): 170–173.
[28] Huebner RJ, Todaro GJ. Oncogenes of RNA tumor viruses as determinants of cancer. *Proc Natl Acad Sci U S A*. 1969;64(3):1087–1094.
[29] Brugge JS, Erikson RL. Identification of a transformation-specific antigen induced by an avian sarcoma virus. *Nature*. 1977;269(5626):346–348.
[30] Hunter T, Sefton BM. Transforming gene product of Rous sarcoma virus phosphorylates tyrosine. *Proc Natl Acad Sci U S A*. 1980;77(3):1311–1315.
[31] Uren AG, Kool J, Berns A, van Lohuizen M. Retroviral insertional mutagenesis: past, present and future. *Oncogene*. 2005;24(52):7656–7672.
[32] Hayward WS, Neel BG, Astrin SM. Activation of a cellular onc gene by promoter insertion in ALV-induced lymphoid leukosis. *Nature*. 1981;290:475–480.
[33] Uchiyama T, Yodoi J, Sagawa K, Takatsuki K, Uchino H. Adult T-cell leukemia: clinical and hematologic features of 16 cases. *Blood*. 1977;50(3):481–492.
[34] Poiesz BJ, Ruscetti FW, Gazdar AF, Bunn PA, Minna JD, Gallo RC. Detection and isolation of type C retrovirus particles from fresh and cultured lymphocytes of a patient with cutaneous T-cell lymphoma. *Proc Natl Acad Sci U S A*. 1980;77(12): 7415–7419.

[35] Seiki M, Hattori S, Hirayama Y, Yoshida M. Human adult T-cell leukemia virus: complete nucleotide sequence of the provirus genome integrated in leukemia cell DNA. *Proc Natl Acad Sci U S A*. 1983;80(12):3618–3622.
[36] Yoshida M, Seiki M, Yamaguchi K, Takatsuki K. Monoclonal integration of human T-cell leukemia provirus in all primary tumors of adult T-cell leukemia suggests causative role of human T-cell leukemia virus in the disease. *Proc Natl Acad Sci U S A*. 1984;81(8):2534–2537.
[37] Zhao T. The role of HBZ in HTLV-1-induced oncogenesis. *Viruses*. 2016;8(2):34.
[38] Kalyanaraman VS, Sarngadharan MG, Robert-Guroff M, Miyoshi I, Golde D, Gallo RC. A new subtype of human T-cell leukemia virus (HTLV-II) associated with a T-cell variant of hairy cell leukemia. *Science*. 1982;218(4572):571–573.
[39] Zheng H, Wolfe ND, Sintasath DM, et al. Emergence of a novel and highly divergent HTLV-3 in a primate hunter in Cameroon. *Virology*. 2010;401(2):137–145.
[40] Gonçalves DU, Proietti FA, Ribas JGR, et al. Epidemiology, treatment, and prevention of human T-cell leukemia virus type 1-associated diseases. *Clin Microbiol Rev*. 2010;23(3):577–589.
[41] Gessain A, Cassar O. Epidemiological aspects and world distribution of HTLV-1 infection. *Front Microbiol*. 2012;3:388.
[42] Ohshima K, Jaffe ES, Kikuchi M. Adult T-cell leukaemia/lymphoma. In: Swerdlow S, Campo E, Harris NL, Jaffe ES, Pileri SA, Stein H, et al., eds. *WHO Classification of Tumours of Haematopoietic and Lymphoid Tissues*. Lyon: IARC Press; 2008.
[43] Yoshida M. Discovery of HTLV-1, the first human retrovirus, its unique regulatory mechanisms, and insights into pathogenesis. *Oncogene*. 2005;24:5931–5937.
[44] Sharp PM, Hahn BH. Origins of HIV and the AIDS pandemic. *Cold Spring Harb Perspect Med*. 2011;1(1):a006841.
[45] Clavel F, Guétard D, Brun-Vézinet F, et al. Isolation of a new human retrovirus from West African patients with AIDS. *Science*. 1986;233(4761):343–346.
[46] Simon V, Ho DD, Karim QA. HIV/AIDS epidemiology, pathogenesis, prevention, and treatment. *Lancet*. 2006;368(9534):489–504.
[47] Barré-Sinoussi F, Chermann JC, Rey F, et al. Isolation of a T-lymphotropic retrovirus from a patient at risk for acquired immune deficiency syndrome (AIDS). *Science*. 1983;220(4599):868–871.
[48] Simon F, Mauclère P, Roques P, et al. Identification of a new human immunodeficiency virus type 1 distinct from group M and group O. *Nat Med*. 1998;4(9): 1032–1037.
[49] De Leys R, Vanderborght B, Vanden Haesevelde M, et al. Isolation and partial characterization of an unusual human immunodeficiency retrovirus from two persons of west-central African origin. *J Virol*. 1990;64(3):1207–1216.
[50] Plantier JC, Leoz M, Dickerson JE, et al. A new human immunodeficiency virus derived from gorillas. *Nat Med*. 2009;15:871–872.
[51] Peeters M, Courgnaud V, Abela B, et al. Risk to human health from a plethora of simian immunodeficiency viruses in primate bushmeat. *Emerg Infect Dis*. 2002;8: 451–457.
[52] Global health observatory data: HIV/AIDS. World Health Organization web site. http://www.who.int/gho/hiv/en/. Accessed September 3, 2016.
[53] HIV in the United States: at a glance. Centers for disease control and prevention. http://www.cdc.gov/hiv/statistics/overview/ataglance.html. Accessed September 3, 2016.
[54] Hope TJ, Trono D. *Structure, Expression, and Regulation of the HIV Genome. HIV InSite*. San Francisco: University of California; 2000. http://hivinsite.ucsf.edu/InSite?page=kb-02-01-02#S2X. Accessed September 3, 2016.

[55] HIV transmission. Centers for Disease Control and Prevention. http://www.cdc.gov/hiv/basics/transmission.html. Accessed September 3, 2016.
[56] Lederman MM, Rodriguez B, Sieg S. *Immunopathogenesis of HIV infection. HIV InSite*. San Francisco: University of California; October 2004. content reviewed January 2006. Accessed September 3, 2016. *http://hivinsite.ucsf.edu/InSite?page=kb-02-01-04#S4X*.
[57] Selik RM, Mokotoff ED, Branson B, Owen SM, Whitmore S, Hall HI. *Revised Surveillance Case Definition for HIV Infection—United States, 2014*. Recommendations and Reports. Morbidity and Mortality Weekly Report, Atlanta, GA: Centers for Disease Control and Prevention; 2014. http://www.cdc.gov/mmwr/preview/mmwrhtml/rr6303a1.htm?s_cid=rr6303a1_e#Appendix. Accessed September 3, 2016.
[58] CDC. *Pneumocystis* pneumonia—Los Angeles. *MMWR*. 1981;30:1–3. Accessed September 3, 2016.
[59] CDC. Kaposi's sarcoma and Pneumocystis pneumonia among homosexual men—New York City and California. *MMWR*. 1981;30:305–308. Accessed September 3, 2016.
[60] Curran JW, Jaffe HW. AIDS: the early years and CDC's response, supplements. *MMWR*. 2011;60(04):64–69.
[61] Gallo RC, Salahuddin SZ, Popovic M, et al. Frequent detection and isolation of cytopathic retroviruses (HTLV-III) from patients with AIDS and at risk for AIDS. *Science*. 1984;224(4648):500–503.
[62] Barillari G, Gendelman R, Gallo RC, Ensoli B. The Tat protein of human immunodeficiency virus type 1, a growth factor for AIDS Kaposi sarcoma and cytokine-activated vascular cells, induces adhesion of the same cell types by using integrin receptors recognizing the RGD amino acid sequence. *Proc Natl Acad Sci U S A*. 1993;90(17):7941–7945.
[63] Goedert JJ, Coté TR, Virgo P, et al. Spectrum of AIDS-associated malignant disorders. *Lancet*. 1998;351(9119):1833–1839.
[64] Grulich AE, Li Y, McDonald A, Correll PK, Law MG, Kaldor JM. Rates of non-AIDS-defining cancers in people with HIV infection before and after AIDS diagnosis. *AIDS*. 2002;16(8):1155–1161.
[65] Frisch M, Biggar RJ, Engels EA, Goedert JJ. AIDS-cancer match registry study group. Association of cancer with AIDS-related immunosuppression in adults. *JAMA*. 2001;285(13):1736–1745.
[66] Raphaël M, Said J, Borisch B, Cesarman E. Lymphomas associated with HIV infection. In: Swerdlow S, Campo E, Harris NL, Jaffe ES, Pileri SA, Stein H, et al., eds. *WHO Classification of Tumours of Haematopoietic and Lymphoid Tissues*. Lyon: IARC Press; 2008:340–342.
[67] Wood Charles. *AIDS-Associated Viral Oncogenesis: Overview*. Virology Papers. Paper 187, 2007. http://digitalcommons.unl.edu/virologypub/187. Accessed September 3, 2016.
[68] Sweet BH, Hilleman MR. The vacuolating virus, S.V. 40. *Proc Soc Exp Biol Med*. 1960;105:420–427.
[69] Eddy BE, Borman GS, Grubbs GE, Young RD. Identification of the oncogenic substance in rhesus monkey kidney cell culture as simian virus 40. *Virology*. 1962;17:65–75.
[70] Pipas JM. SV40: cell transformation and tumorigenesis. *Virology*. 2009;384(2):294–303.
[71] Westphal H, Dulbecco R. Viral DNA in polyoma- and SV40-transformed cell lines. *Proc Natl Acad Sci U S A*. 1968;59(4):1158–1165.
[72] Sambrook J, Westphal H, Srinivasan PR, Dulbecco R. The integrated state of viral DNA in SV40-transformed cells. *Proc Natl Acad Sci U S A*. 1968;60(4):1288–1295.

[73] Botchan M, Topp W, Sambrook J. The arrangement of simian virus 40 sequences in the DNA of transformed cells. *Cell.* 1976;9(2):269–287.
[74] Danna K, Nathans D. Specific cleavage of simian virus 40 DNA by restriction endonuclease of Hemophilus influenzae. *Proc Natl Acad Sci U S A.* 1971;68(12): 2913–2917.
[75] Khoury G, Martin MA, Lee TNH, Nathans D. A transcriptional map of the SV40 genome in transformed cell lines. *Virology.* 1975;63:263–272.
[76] Abrahams PJ, Mulder C, Van De Voorde A, Warnaar SO, Van der Eb AJ. Transformation of primary rat kidney cells by fragments of simian virus 40 DNA. *J Virol.* 1975;16(4):818–823.
[77] Rundell K, Collins JK, Tegtmeyer P, Ozer HL, Lai CJ, Nathans D. Identification of simian virus 40 protein A. *J Virol.* 1977;21(2):636–646.
[78] Reddy VB, Thimmappaya B, Dhar R, et al. The genome of simian virus 40. *Science.* 1978;200(4341):494–502.
[79] Fiers W, Contreras R, Haegeman G, et al. Complete nucleotide sequence of SV40 DNA. *Nature.* 1978;273:113–120.
[80] Shortle D, Nathans D. Local mutagenesis: a method for generating viral mutants with base substitutions in preselected regions of the viral genome. *Proc Natl Acad Sci U S A.* 1978;75(5):2170–2174.
[81] Peden KW, Pipas JM, Pearson-White S, Nathans D. Isolation of mutants of an animal virus in bacteria. *Science.* 1980;209(4463):1392–1396.
[82] Harlow E, Crawford LV, Pim DC, Williamson NM. Monoclonal antibodies specific for simian virus 40 tumor antigens. *J Virol.* 1981;39(3):861–869.
[83] Knudson A. Mutation and cancer: statistical study of retinoblastoma. *Proc Natl Acad Sci U S A.* 1971;68(4):820–823.
[84] Comings DE. A general theory of carcinogenesis. *Proc Natl Acad Sci U S A.* 1973;70(12):3324–3328.
[85] Harlow E, Whyte P, Franza Jr BR, Schley C. Association of adenovirus early-region 1A proteins with cellular polypeptides. *Mol Cell Biol.* 1986;6(5):1579–1589.
[86] Lee WH, Shew JY, Hong FD, et al. The retinoblastoma susceptibility gene encodes a nuclear phosphoprotein associated with DNA binding activity. *Nature.* 1987; 329(6140):642–645.
[87] Whyte P, Buchkovich KJ, Horowitz JM, et al. Association between an oncogene and an anti-oncogene: the adenovirus E1A proteins bind to the retinoblastoma gene product. *Nature.* 1988;334(6178):124–129.
[88] DeCaprio JA, Ludlow JW, Figge J, et al. SV40 large tumor antigen forms a specific complex with the product of the retinoblastoma susceptibility gene. *Cell.* 1988;54(2): 275–283.
[89] Windle JJ, Albert DM, O'Brien JM, et al. Retinoblastoma in transgenic mice. *Nature.* 1990;343(6259):665–669.
[90] DeCaprio JA. How the Rb tumor suppressor structure and function was revealed by the study of Adenovirus and SV40. *Virology.* 2009;384(2):274–284.
[91] Buchkovich K, Duffy LA, Harlow E. The retinoblastoma protein is phosphorylated during specific phases of the cell cycle. *Cell.* 1989;58(6):1097–1105.
[92] Chen PL, Scully P, Shew JY, Wang JY, Lee WH. Phosphorylation of the retinoblastoma gene product is modulated during the cell cycle and cellular differentiation. *Cell.* 1989;58(6):1193–1198.
[93] DeCaprio JA, Ludlow JW, Lynch D, et al. The product of the retinoblastoma susceptibility gene has properties of a cell cycle regulatory element. *Cell.* 1989;58(6):1085–1095.
[94] Ludlow JW, DeCaprio JA, Huang CM, Lee WH, Paucha E, Livingston DM. SV40 large T antigen binds preferentially to an underphosphorylated member of the retinoblastoma susceptibility gene product family. *Cell.* 1989;56(1):57–65.

[95] Hu QJ, Dyson N, Harlow E. The regions of the retinoblastoma protein needed for binding to adenovirus E1A or SV40 large T antigen are common sites for mutations. *EMBO J.* 1990;9(4):1147–1155.
[96] Huang S, Wang NP, Tseng BY, Lee WH, Lee EH. Two distinct and frequently mutated regions of retinoblastoma protein are required for binding to SV40 T antigen. *EMBO J.* 1990;9(6):1815–1822.
[97] Kaelin Jr WG, Ewen ME, Livingston DM. Definition of the minimal simian virus 40 large T antigen- and adenovirus E1A-binding domain in the retinoblastoma gene product. *Mol Cell Biol.* 1990;10(7):3761–3769.
[98] Bagchi S, Raychaudhuri P, Nevins JR. Adenovirus E1A proteins can dissociate heteromeric complexes involving the E2F transcription factor: a novel mechanism for E1A trans-activation. *Cell.* 1990;62(4):659–669.
[99] Chellappan SP, Hiebert S, Mudryj M, Horowitz JM, Nevins JR. The E2F transcription factor is a cellular target for the RB protein. *Cell.* 1991;65(6):1053–1061.
[100] Chellappan S, Kraus VB, Kroger B, et al. Adenovirus E1A, simian virus 40 tumor antigen, and human papillomavirus E7 protein share the capacity to disrupt the interaction between transcription factor E2F and the retinoblastoma gene product. *Proc Natl Acad Sci U S A.* 1992;89(10):4549–4553.
[101] Zalvide J, DeCaprio JA. Role of pRb-related proteins in simian virus 40 large-T-antigen-mediated transformation. *Mol Cell Biol.* 1995;15(10):5800–5810.
[102] Zalvide J, Stubdal H, DeCaprio JA. The J domain of simian virus 40 large T antigen is required to functionally inactivate RB family proteins. *Mol Cell Biol.* 1998;18(3):1408–1415.
[103] Sullivan CS, Cantalupo P, Pipas JM. The molecular chaperone activity of simian virus 40 large T antigen is required to disrupt Rb-E2F family complexes by an ATP-dependent mechanism. *Mol Cell Biol.* 2000;20(17):6233–6243.
[104] Campbell KS, Mullane KP, Aksoy IA, et al. DnaJ/hsp40 chaperone domain of SV40 large T antigen promotes efficient viral DNA replication. *Genes Dev.* 1997;11:1098–1110.
[105] Lapenna S, Giordano A. Cell cycle kinases as therapeutic targets for cancer. *Nat Rev Drug Discov.* 2009;8:547–566.
[106] Lane DP, Crawford LV. T antigen is bound to a host protein in SY40-transformed cells. *Nature.* 1979;278:261–263.
[107] Linzer DI, Levine AJ. Characterization of a 54k Dalton cellular SV40 tumor antigen present in SV40-transformed cells and uninfected embryonal carcinoma cells. *Cell.* 1979;17(1):43–52.
[108] DeLeo AB, Jay G, Appella E, Dubois GC, Law LW, Old LJ. Detection of a transformation-related antigen in chemically induced sarcomas and other transformed cells of the mouse. *Proc Natl Acad Sci U S A.* 1979;76(5):2420–2424.
[109] Linzer D, Maltzman W, Levine AJ. The SV40 a gene product is required for the production of a 54,000 MW cellular tumor antigen. *Virology.* 1979;98(2):308–318.
[110] Oren M, Maltzman W, Levine AJ. Post-translational regulation of the 54K cellular tumor antigen in normal and transformed cells. *Mol Cell Biol.* 1981;1(2):101–110.
[111] Levine AJ, Oren M. The first 30 years of p53: growing ever more complex. *Nat Rev Cancer.* 2009;9(10):749–758.
[112] Baker SJ, Fearon ER, Nigro JM, et al. Chromosome 17 deletions and p53 gene mutations in colorectal carcinomas. *Science.* 1989;244(4901):217–221.
[113] Finlay CA, Hinds PW, Levine AJ. The p53 proto-oncogene can act as a suppressor of transformation. *Cell.* 1989;57(7):1083–1093.
[114] Malkin D, Li FP, Strong LC, et al. Germ line p53 mutations in a familial syndrome of breast cancer, sarcomas, and other neoplasms. *Science.* 1990;250(4985):1233–1238.

[115] Srivastava S, Zou ZQ, Pirollo K, Blattner W, Chang EH. Germ-line transmission of a mutated p53 gene in a cancer-prone family with Li–Fraumeni syndrome. *Nature.* 1990;348(6303):747–749.

[116] Michalovitz D, Halevy O, Oren M. Conditional inhibition of transformation and of cell proliferation by a temperature-sensitive mutant of p53. *Cell.* 1990;62(4):671–680.

[117] Yonish-Rouach E, Resnitzky D, Lotem J, Sachs L, Kimchi A, Oren M. Wild-type p53 induces apoptosis of myeloid leukaemic cells that is inhibited by interleukin-6. *Nature.* 1991;352(6333):345–347.

[118] Wang Y, Blandino G, Oren M, Givol D. Induced p53 expression in lung cancer cell line promotes cell senescence and differentially modifies the cytotoxicity of anti-cancer drugs. *Oncogene.* 1998;17(15):1923–1930.

[119] Fields S, Jang SK. Presence of a potent transcription activating sequence in the p53 protein. *Science.* 1990;249(4972):1046–1049.

[120] Raycroft L, Wu HY, Lozano G. Transcriptional activation by wild-type but not transforming mutants of the p53 anti-oncogene. *Science.* 1990;249(4972):1049–1051.

[121] Bargonetti J, Friedman PN, Kern SE, Vogelstein B, Prives C. Wild-type but not mutant p53 immunopurified proteins bind to sequences adjacent to the SV40 origin of replication. *Cell.* 1991;65(6):1083–1091.

[122] Momand J, Zambetti GP, Olson DC, George D, Levine AJ. The mdm-2 oncogene product forms a complex with the p53 protein and inhibits p53-mediated transactivation. *Cell.* 1992;69(7):1237–1245.

[123] Chen J, Marechal V, Levine AJ. Mapping of the p53 and mdm-2 interaction domains. *Mol Cell Biol.* 1993;13(7):4107–4114.

[124] Wu X, Bayle JH, Olson D, Levine AJ. The p53-mdm-2 autoregulatory feedback loop. *Genes Dev.* 1993;7(7A):1126–1132.

[125] Kubbutat MH, Jones SN, Vousden KH. Regulation of p53 stability by Mdm2. *Nature.* 1997;387(6630):299–303.

[126] Stiewe T. The p53 family in differentiation and tumorigenesis. *Nat Rev Cancer.* 2007;7(3):165–168.

[127] Bäfverstedt B. Condylomata acuminata past and present. *Acta Derm Venereol.* 1967;47:376–381.

[128] Rigoni-Stern DA. Fatti Statistici relativi alle malattie cancerose che Servirono di Base Alle Poche Cose Dette dal Dott. *G Servire Progr Path Tera.* 1842;2:507–517.

[129] Payne J. On the contagious rise of common warts. *Br J Dermatol.* 1891;3:185–189.

[130] Strauss MJ, Shaw EW, Bunting H, Melnick JL. Crystalline virus-like particles from skin papillomas characterized by intranuclear inclusion bodies. *Proc Soc Exp Biol Med.* 1949;72:46–50.

[131] Olson C, Pamukcu AM, Brobst DF, Kowalczyk T, Satter EJ, Price JM. A urinary bladder tumor induced by a bovine cutaneous papilloma agent. *Cancer Res.* 1959;19: 779–782.

[132] Masters JR. HeLa cells 50 years on: the good, the bad and the ugly. *Nat Rev Cancer.* 2002;2(4):315–319.

[133] Kelly B, Shimoni T. Reintroducing the Tzanck smear. *Am J Clin Dermatol.* 2009;10(3):141–152.

[134] Jablonska S, Dabrowski J, Jakubowicz K. Epidermodysplasia verruciformis as a model in studies on the role of papovaviruses in oncogenesis. *Cancer Res.* 1972;32(3): 583–589.

[135] zur Hausen H. Condylomata acuminata and human genital cancer. *Cancer Res.* 1976;36(2 pt 2):794.

[136] zur Hausen H, Meinhof W, Scheiber W, Bornkamm GW. Attempts to detect virus-specific DNA in human tumors. I. Nucleic acid hybridizations with complementary RNA of human wart virus. *Int J Cancer.* 1974;13:650–656.

[137] zur Hausen H, Schulte-Holthausen H, Wolf H, Dörries K, Egger H. Attempts to detect virus-specific DNA in human tumors. II. Nucleic acid hybridizations with complementary RNA of human herpes group viruses. *Int J Cancer*. 1974;13: 657–664.
[138] Dürst M, Gissmann L, Ikenberg H, zur Hausen H. A papillomavirus DNA from a cervical carcinoma and its prevalence in cancer biopsy samples from different geographic regions. *Proc Natl Acad Sci U S A*. 1983;80(12):3812–3815.
[139] Boshart M, Gissmann L, Ikenberg H, Kleinheinz A, Scheurlen W, zur Hausen H. A new type of papillomavirus DNA, its presence in genital cancer biopsies and in cell lines derived from cervical cancer. *EMBO J*. 1984;3(5):1151–1157.
[140] zur Hausen H. Papillomaviruses in the causation of human cancers—a brief historical account. *Virology*. 2009;384(2):260–265. http://dx.doi.org/10.1016/j.virol.2008.11.046. Epub 2009 Jan 8.
[141] Burd EM. Human papillomavirus and cervical cancer. *Clin Microbiol Rev*. 2003;16(1): 1–17.
[142] de Villiers EM, Fauquet C, Broker TR, Bernard HU, zur Hausen H. Classification of papillomaviruses. *Virology*. 2004;324(1):17–27.
[143] Ibeanu OA. Molecular pathogenesis of cervical cancer. *Cancer Biol Ther*. 2011;11(3):295–306. Epub 2011 Feb 1.
[144] World Health Organization. Human Papillomavirus (HPV) and Cervical Cancer Fact Sheet. http://www.who.int/mediacentre/factsheets/fs380/en/. Accessed September 5, 2016.
[145] Longworth MS, Laimins LA. Pathogenesis of human papillomaviruses in differentiating epithelia. *Microbiol Mol Biol Rev*. 2004;68(2):362–372.
[146] National Institute of Health. National Cancer Institute. HPV and Cancer. http://www.cancer.gov/about-cancer/causes-prevention/risk/infectious-agents/hpv-fact-sheet. Accessed September 5, 2016.
[147] Centers for Disease Control and Prevention. Cervical Cancer Statistics. https://www.cdc.gov/cancer/cervical/statistics/. Accessed September 5, 2016.
[148] Toker C. Trabecular carcinoma of the skin. *Arch Dermatol*. 1972;105(1):107–110.
[149] Tang CK, Toker C. Trabecular carcinoma of the skin: an ultrastructural study. *Cancer*. 1978;42(5):2311–2321.
[150] Brunner M, Thurnher D, Pammer J, et al. Expression of VEGF-A/C, VEGF-R2, PDGF-alpha/beta, c-kit, EGFR, Her-2/Neu, Mcl-1 and Bmi-1 in Merkel cell carcinoma. *Mod Pathol*. 2008;21(7):876–884. http://dx.doi.org/10.1038/modpathol.2008.63. Epub 2008 Apr 11.
[151] Erovic I, Erovic BM. Merkel cell carcinoma: the past, the present, and the future. *J Skin Cancer*. 2013;2013:929364. http://dx.doi.org/10.1155/2013/929364.
[152] Heath M, Jaimes N, Lemos B, et al. Clinical characteristics of Merkel cell carcinoma at diagnosis in 195 patients: the AEIOU features. *J Am Acad Dermatol*. 2008;58(3): 375–381.
[153] Wang TS, Byrne PJ, Jacobs LK, Taube JM. Merkel cell carcinoma: update and review. *Semin Cutan Med Surg*. 2011;30(1):48–56.
[154] Engels EA, Frisch M, Goedert JJ, Biggar RJ, Miller RW. Merkel cell carcinoma and HIV infection. *Lancet*. 2002;359(9305):497–498.
[155] Feng H, Shuda M, Chang Y, Moore PS. Clonal integration of a polyomavirus in human Merkel cell carcinoma. *Science*. 2008;319(5866):1096–1100.
[156] Arora R, Chang Y, Moore PS. MCV and Merkel cell carcinoma: a molecular success story. *Curr Opin Virol*. 2012;2(4):489–498.
[157] Shuda M, Feng H, Kwun HJ, et al. T antigen mutations are a human tumor-specific signature for Merkel cell polyomavirus. *Proc Natl Acad Sci U S A*. 2008;105(42): 16272–16277.

[158] Shuda M, Kwun HJ, Feng H, Chang Y, Moore PS. Human Merkel cell polyomavirus small T antigen is an oncoprotein targeting the 4E-BP1 translation regulator. *J Clin Invest.* 2011;121(9):3623–3634.
[159] Garneski KM, Warcola AH, Feng Q, Kiviat NB, Leonard JH, Nghiem P. Merkel cell polyomavirus is more frequently present in North American than Australian Merkel cell carcinoma tumors. *J Invest Dermatol.* 2009;129(1):246–248. http://dx.doi.org/10.1038/jid.2008.229. Epub 2008 Jul 24.
[160] Paik JY, Hall G, Clarkson A, et al. Immunohistochemistry for Merkel cell polyomavirus is highly specific but not sensitive for the diagnosis of Merkel cell carcinoma in the Australian population. *Hum Pathol.* 2011;42(10):1385–1390. http://dx.doi.org/10.1016/j.humpath.2010.12.013. Epub 2011 Mar 30.
[161] Epstein A. Burkitt lymphoma and the discovery of Epstein–Barr virus. *Br J Haematol.* 2012;156(6):777–779.
[162] Kaposi M. Idiopathisches multiples pigmentsarkom der Haut. *Arch Dermatol Syph.* 1872;4:265–273.
[163] Shiels RA. A history of Kaposi's sarcoma. *J R Soc Med.* 1986;79(9):532–534.
[164] Butler LM, Dorsey G, Hladik W, et al. Kaposi sarcoma-associated herpesvirus (KSHV) seroprevalence in population-based samples of African children: evidence for at least 2 patterns of KSHV transmission. *J Infect Dis.* 2009;200:430–438.
[165] Ballestas ME, Chatis PA, Kaye KM. Efficient persistence of extrachromosomal KSHV DNA mediated by latency-associated nuclear antigen. *Science.* 1999;284:641–644.
[166] Einollahi B, Lessan-Pezeshki M, Nourbala MH, et al. Kaposi's sarcoma following living donor kidney transplantation: review of 7, 939 recipients. *Int Urol Nephrol.* 2009;41:679–685.
[167] Beral V, Peterman TA, Berkelman RL, Jaffe HW. Kaposi's sarcoma among persons with AIDS: a sexually transmitted infection? *Lancet.* 1990;335(8682):123–128.
[168] Dore GJ, Li Y, Grulich AE, et al. Declining incidence and later occurrence of Kaposi's sarcoma among persons with AIDS in Australia: the Australian AIDS cohort. *AIDS.* 1996;10(12):1401–1406.
[169] Joffe MM, Hoover DR, Jacobson LP, Kingsley L, Chmiel JS, Visscher BR. Effect of treatment with zidovudine on subsequent incidence of Kaposi's sarcoma. *Clin Infect Dis.* 1997;25(5):1125–1133.
[170] Armenian HK, Hoover DR, Rubb S, et al. Composite risk score for Kaposi's sarcoma based on a case-control and longitudinal study in the Multicenter AIDS Cohort Study (MACS) population. *Am J Epidemiol.* 1993;138(4):256–265.
[171] Beral V, Bull D, Darby S, et al. Risk of Kaposi's sarcoma and sexual practices associated with faecal contact in homosexual or bisexual men with AIDS. *Lancet.* 1992;339(8794):632–635.
[172] Sunil M, Reid E, Lechowicz MJ. Update on HHV-8-associated malignancies. *Curr Infect Dis Rep.* 2010;12(2):147–154.
[173] Sarid R, Flore O, Bohenzky RA, Chang Y, Moore PS. Transcription mapping of the Kaposi's sarcoma-associated herpesvirus (human herpesvirus 8) genome in a body cavity-based lymphoma cell line (BC-1). *J Virol.* 1998;72:1005–1012.
[174] Paulose-Murphy M, Ha NK, Xiang C, et al. Transcription program of human herpesvirus 8 (kaposi's sarcoma- associated herpesvirus). *J Virol.* 2001;75:4843–4853.
[175] Swanton C, Mann DJ, Fleckenstein B, Neipel F, Peters G, Jones N. Herpes viral cyclin/Cdk6 complexes evade inhibition by CDK inhibitor proteins. *Nature.* 1997;390:184–187.
[176] Luppi M, Barozzi P, Santagostino G, et al. Molecular evidence of organ-related transmission of Kaposi sarcoma-associated herpesvirus or human herpesvirus-8 in transplant patients. *Blood.* 2000;96:3279–3281.

[177] Friborg Jr J, Kong W, Hottiger MO, Nabel GJ. p53 inhibition by the LANA protein of KSHV protects against cell death. *Nature*. 1999;402:889–894. transparent image.
[178] Irmler M, Thome M, Hahne M, et al. Inhibition of death receptor signals by cellular FLIP. *Nature*. 1997;388:190–195.
[179] Djerbi M, Screpanti V, Catrina AI, et al. The inhibitor of death receptor signaling, FLICE-inhibitory protein defines a new class of tumor progression factors. *J Exp Med*. 1999;190:1025–1032.
[180] Ye FC, Zhou FC, Xie JP, et al. Kaposi's sarcoma-associated herpesvirus latent gene vFLIP inhibits viral lytic replication through NF-kappaB-mediated suppression of the AP-1 pathway: a novel mechanism of virus control of latency. *J Virol*. 2008;82:4235–4249.
[181] Sodhi A, Chaisuparat R, Hu J, et al. TSC2/mTOR pathway drives endothelial cell transformation induced by the Kaposi's sarcoma-associated herpesvirus G protein-coupled receptor. *Cancer Cell*. 2006;10:133–143.
[182] Bais C, Santomasso B, Coso O, et al. G-protein-coupled receptor of Kaposi's sarcoma-associated herpesvirus is a viral oncogene and angiogenesis activator. *Nature*. 1998;391:86–89.
[183] Weindel K, Marme D, Weich HA. AIDS-associated Kaposi's sarcoma cells in culture express vascular endothelial growth factor. *Biochem Biophys Res Commun*. 1992;183:1167–1174.
[184] Sarid R, Sato T, Bohenzky RA, Russo JJ, Chang Y. Kaposi's sarcoma-associated herpesvirus encodes a functional bcl-2 homologue. *Nat Med*. 1997;3:293–298.
[185] Friborg Jr J, Kong WP, Flowers CC, et al. Distinct biology of Kaposi's sarcoma-associated herpesvirus from primary lesions and body cavity lymphomas. *J Virol*. 1998;72:10073–10082.
[186] Foussat A, Wijdenes J, Bouchet L, et al. Human interleukin-6 is in vivo an autocrine growth factor for human herpesvirus-8-infected malignant B lymphocytes. *Eur Cytokine Netw*. 1999;10:501–508.
[187] Burger R, Neipel F, Fleckenstein B, et al. Human herpesvirus type 8 interleukin-6 homologue is functionally active on human myeloma cells. *Blood*. 1998;91(6):1858–1863.
[188] Brandt SJ, Bodine DM, Dunbar CE, Nienhuis AW. Dysregulated interleukin 6 expression produces a syndrome resembling Castleman's disease in mice. *J Clin Invest*. 1990;86(2):592–599.
[189] Samaniego F, Pati S, Karp JE, Prakash O, Bose D. Human herpesvirus 8K1-associated nuclear factor-kappa B-dependent promoter activity: role in Kaposi's sarcoma inflammation? *J Natl Cancer Inst Monogr*. 2001;(28):15–23.
[190] Zong JC, Ciufo DM, Alcendor DJ, et al. High-level variability in the ORF-K1 membrane protein gene at the left end of the Kaposi's sarcoma-associated herpesvirus genome defines four major virus subtypes and multiple variants or clades in different human populations. *J Virol*. 1999;73(5):4156–4170.
[191] Lurman A. Eine icterus epidemic. *Berl Klin Woschenschr*. 1885;22:20–23.
[192] Alter HJ, Blumberg BS. Further studies on a "new" human isoprecipitin system (Australia antigen). *Blood*. 1966;27(3):297–309.
[193] Prince AM. An antigen detected in the blood during the incubation period of serum hepatitis. *Proc Natl Acad Sci U S A*. 1968;60(3):814–821.
[194] Blumberg BS, Larouzé B, London WT, et al. The relation of infection with the hepatitis B agent to primary hepatic carcinoma. *Am J Pathol*. 1975;81(3):669–682.
[195] The ABCs of Hepatitis. Centers for Disease Control and Prevention. https://www.cdc.gov/hepatitis/resources/professionals/pdfs/abctable.pdf. Accessed August 21, 2016.

[196] Buynak EB, Roehm RR, Tytell AA, Bertland 2nd AU, Lampson GP, Hilleman MR. Vaccine against human hepatitis B. *JAMA*. 1976;235(26):2832–2834.
[197] Feinstone SM, Kapikian AZ, Purcell RH, Alter HJ, Holland PV. Transfusion-associated hepatitis not due to viral hepatitis type A or B. *N Engl J Med*. 1975;292(15):767–770.
[198] Alter HJ, Holland PV, Morrow AG, Purcell RH, Feinstone SM, Moritsugu Y. Clinical and serological analysis of transfusion-associated hepatitis. *Lancet*. 1975;2(7940): 838–841.
[199] Aach RD, Szmuness W, Mosley JW, et al. Serum alanine aminotransferase of donors in relation to the risk of non-A, non-B hepatitis in recipients: the transfusion-transmitted viruses study. *N Engl J Med*. 1981;304(17):989–994.
[200] Alter HJ, Purcell RH, Holland PV, Popper H. Transmissible agent in non-A, non-B hepatitis. *Lancet*. 1978;1(8062):459–463.
[201] Shimizu YK, Feinstone SM, Purcell RH, Alter HJ, London WT. Non-A, non-B hepatitis: ultrastructural evidence for two agents in experimentally infected chimpanzees. *Science*. 1979;205(4402):197–200.
[202] Feinstone SM, Mihalik KB, Kamimura T, Alter HJ, London WT, Purcell RH. Inactivation of hepatitis B virus and non-A, non-B hepatitis by chloroform. *Infect Immun*. 1983;41(2):816–821.
[203] Bradley DW, McCaustland KA, Cook EH, Schable CA, Ebert JW, Maynard JE. Posttransfusion non-A, non-B hepatitis in chimpanzees. Physicochemical evidence that the tubule-forming agent is a small, enveloped virus. *Gastroenterology*. 1985;88(3): 773–779.
[204] Houghton M. Discovery of the hepatitis C virus. *Liver Int*. 2009;29(suppl 1):82–88.
[205] Choo QL, Kuo G, Weiner AJ, Overby LR, Bradley DW, Houghton M. Isolation of a cDNA clone derived from a blood-borne non-A, non-B viral hepatitis genome. *Science*. 1989;244(4902):359–362.
[206] Kuo G, Choo QL, Alter HJ, et al. An assay for circulating antibodies to a major etiologic virus of human non-A, non-B hepatitis. *Science*. 1989;244(4902):362–364.
[207] Alter HJ, Purcell RH, Shih JW, et al. Detection of antibody to hepatitis C virus in prospectively followed transfusion recipients with acute and chronic non-A, non-B hepatitis. *N Engl J Med*. 1989;321(22):1494–1500.
[208] Colombo M, Kuo G, Choo QL, et al. Prevalence of antibodies to hepatitis C virus in Italian patients with hepatocellular carcinoma. *Lancet*. 1989;2(8670):1006–1008.
[209] Geller SA. Hepatitis B, and hepatitis C. *Clin Liver Dis*. 2002;6(2):317–334. v.
[210] Moriya K, Fujie H, Shintani Y, et al. The core protein of hepatitis C virus induces hepatocellular carcinoma in transgenic mice. *Nat Med*. 1998;4(9):1065–1067.
[211] Chen RF, Li ZH, Chen JS, Kong XH, Zou SQ. Human normal biliary epithelial cells transformation and tumor development induced by hepatitis C virus core protein. *Zhonghua Wai Ke Za Zhi*. 2005;43(3):153–156.
[212] Zemel R, Gerechet S, Greif H, et al. Cell transformation induced by hepatitis C virus NS3 serine protease. *J Viral Hepat*. 2001;8(2):96–102.
[213] Goossens N, Hoshida Y. Hepatitis C virus-induced hepatocellular carcinoma. *Clin Mol Hepatol*. 2015;21(2):105–114.

CHAPTER FIVE

The Interaction Between Human Papillomaviruses and the Stromal Microenvironment

B. Woodby, M. Scott, J. Bodily[1]
Louisiana State University Health Sciences Center, Shreveport, LA, United States
[1]Corresponding author: e-mail address: jbodil@lsuhsc.edu

Contents

Abstract

Human papillomaviruses (HPVs) are small, double-stranded DNA viruses that replicate in stratified squamous epithelia and cause a variety of malignancies. Current efforts in HPV biology are focused on understanding the virus–host interactions that enable HPV to persist for years or decades in the tissue. The importance of interactions between tumor

Progress in Molecular Biology and Translational Science, Volume 144
ISSN 1877-1173
http://dx.doi.org/10.1016/bs.pmbts.2016.09.003

cells and the stromal microenvironment has become increasingly apparent in recent years, but how stromal interactions impact the normal, benign life cycle of HPVs, or progression of lesions to cancer is less understood. Furthermore, how productively replicating HPV impacts cells in the stromal environment is also unclear. Here we bring together some of the relevant literature on keratinocyte–stromal interactions and their impacts on HPV biology, focusing on stromal fibroblasts, immune cells, and endothelial cells. We discuss how HPV oncogenes in infected cells manipulate other cells in their environment, and, conversely, how neighboring cells may impact the efficiency or course of HPV infection.

ABBREVIATIONS

(G)M-CSF (granulocyte) macrophage colony stimulating factor
CAF cancer-associated fibroblast
CAIX carbonic anhydrase IX
CDK cyclin-dependent kinase
CIN cervical intraepithelial neoplasia
CTLs cytotoxic t lymphocytes
DC dendritic cells
ECM extracellular matrix
EGF epidermal growth factor
EGFR epidermal growth factor receptor
EMT epithelial-to-mesenchymal transition
ERK extracellular signal-related kinase
ER-α estrogen receptor alpha
FGF fibroblast growth factor
FGFR fibroblast growth factor receptor
HB-EGF heparin-binding EGF
HDAC histone deacetylase
HGF hepatocyte growth factor
HIF hypoxia-inducible factor
HPV human papillomavirus
HRE hypoxia regulator element
IFN interferon
IFNAR interferon alpha/beta receptor
IGF Insulin like growth factor
IGFBP insulin-like growth factor binding protein
IKK IκB kinase
IL interleukin
IRF interferon response factor
ISG interferon-stimulated genes
ISRE interferon-sensitive response elements
KGF keratinocyte growth factor
KGFR keratinocyte growth factor receptor/
LAP latency-associated peptide
LC Langerhans cells

MAPK mitogen-activated protein kinase
MHC major histocompatibility complex
MMP matrix metalloproteinase
MT1-MMP membrane type 1 MMP
mTOR mammalian target of rapamycin
NFκB nuclear factor kappa B
NK natural killer cells
PDGF platelet-derived growth factor
PI3K phosphoinositol 3 kinase
Rb retonoblastoma
STAT signal transducer and activator of transcription
STING stimulator of interferon genes
TAM tumor-associated macrophage
TGFBRII TGFβ receptor
TGFβ transforming growth factor beta
Th1 (2) T helper 1 (2) cells
TLR Toll-like receptor
TNFα tumor necrosis factor alpha
Treg regulator T cells
TSP-1 thrombospondin-1
VEGF vascular endothelial growth factor
VEGFR VEGF receptor
VHL Von Hippel–Lindau

1. INTRODUCTION

Human papillomaviruses (HPVs[a]) are small, double-stranded DNA viruses that replicate as episomes in the keratinocytes of stratified squamous epithelia.[1] Genital HPVs are the most common sexually transmitted infections and are extremely widespread. Most cases are asymptomatic or consist of benign, self-limited lesions, but because of the sheer number of HPV infections, HPV-induced cancers are very prevalent worldwide.[1–3] Certain types of mucosal HPVs (denoted "high-risk" types) are responsible for essentially 100% of cervical cancers and significant proportions of anal, vaginal, vulvar, penile, and oropharyngeal cancers. About 500,000 cases of cervical cancer occur each year, especially in developing countries, and about half of these cases result in death.[4] Low-risk HPV types cause genital warts and other benign lesions.[5] Current prophylactic vaccines are extremely effective at preventing infection but are unable to clear

[a] For the convenience of the reader, abbreviations used in this chapter are listed in abbreviations list.

the virus once an infection is established.[6] Thus a critical next step in HPV biology is to understand the virus–host interactions that enable HPV to maintain itself in the tissue and how these mechanisms may be susceptible to intervention.

2. THE STROMAL MICROENVIRONMENT

Hyperproliferative lesions caused by productive HPV infections are not cancers, but HPV-infected cells display many of the characteristic hallmarks of cancer cells,[7] including immortalization,[8,9] resistance to apoptosis,[10] sustained proliferative signaling,[11,12] and changes in cellular metabolism.[13,14] However, cancers are not simply masses of proliferating cells. Rather, cancer acts like a dysregulated organ with a complex array of interactions between epithelial cells and fibroblasts, macrophages, endothelial cells, and immune cells in the stromal microenvironment (Fig. 1). The role of stromal cells and their products in cancer development is becoming more fully appreciated.[7,15–19]

Although HPVs infect keratinocytes exclusively, HPV regulates a wide array of growth factors, cytokines, and other paracrine mediators that have the potential to impact the behavior of cells in the stromal microenvironment,[20–22] including promotion of angiogenesis[23–25] and evasion of immune surveillance.[26] Paracrine factors produced by stromal cells may impact the growth and invasiveness of HPV-containing epithelia.[27] Much effort has been focused on how stromal interactions contribute to cancer development, but how stromal interactions impact the normal, benign life cycle of HPVs, or progression of benign lesions to cancer is less understood. Conversely, cell-intrinsic functions of HPV oncogenes are widely appreciated, but how productively replicating HPV impacts cells in the stromal environment is less clear.

The goal of this chapter is to bring together some of the relevant literature on keratinocyte–stromal interactions, especially pertaining to HPV biology, to create a more holistic picture of epithelial–stromal interactions in HPV infection. We will focus on how HPV oncogenes in infected cells manipulate other cells in their environment, and, conversely, how neighboring cells impact the efficiency or course of HPV infection. Since we cannot be comprehensive, we invite readers to refer back to primary and review literature cited throughout.

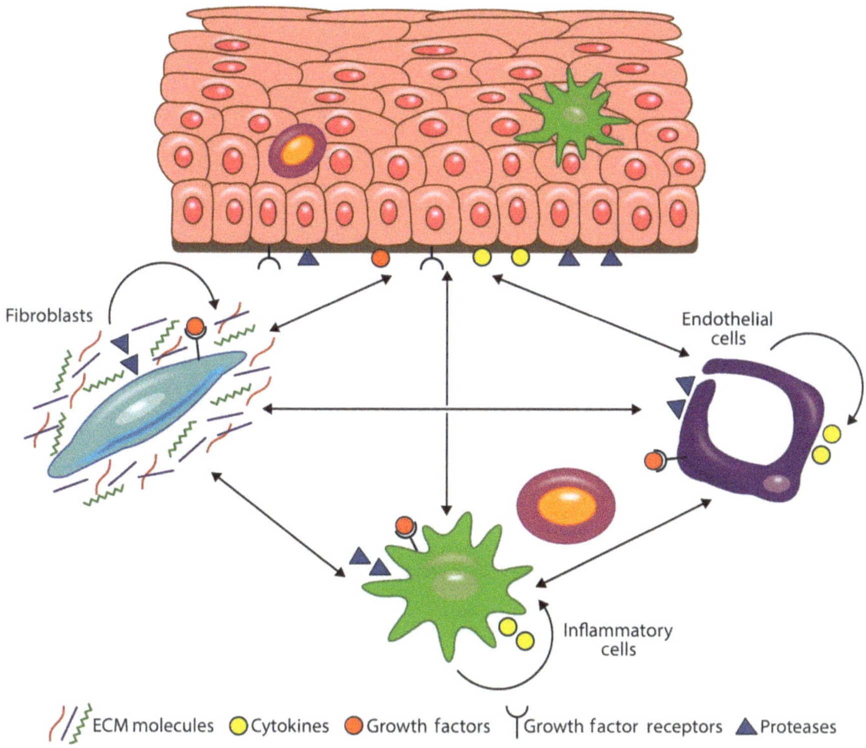

Fig. 1 Epithelial–stromal interactions. The epithelium consists primarily of keratinocytes with LC and T cells interspersed. In the stroma, fibroblasts contribute heavily to the maintenance of the extracellular matrix (ECM) and secretion of growth factors to support the epithelium. Also present are a variety of immune cells, including macrophages, lymphocytes, and others. The vasculature is composed primarily of endothelial cells, which proliferate and differentiate during angiogenesis. All of these cell types are in communication with each other through the exchange of growth factors, cytokines, and other molecules, some of which are sequestered and stored in the ECM.

3. THE HPV LIFE CYCLE

During the normal, productive life cycle, HPV gains access to the basal layer of the epithelium through a wound and infects keratinocytes of the epithelial basal layer[28–30] (Fig. 2). The basal layer contains the long-lived keratinocyte stem cells and is the only location in the normal epithelium where cell division is known to occur.[33] Following cell entry,[34,35] the virus undergoes genome replication to establish a stable pool of episomal viral genomes. Overall viral gene expression is suppressed. Following division

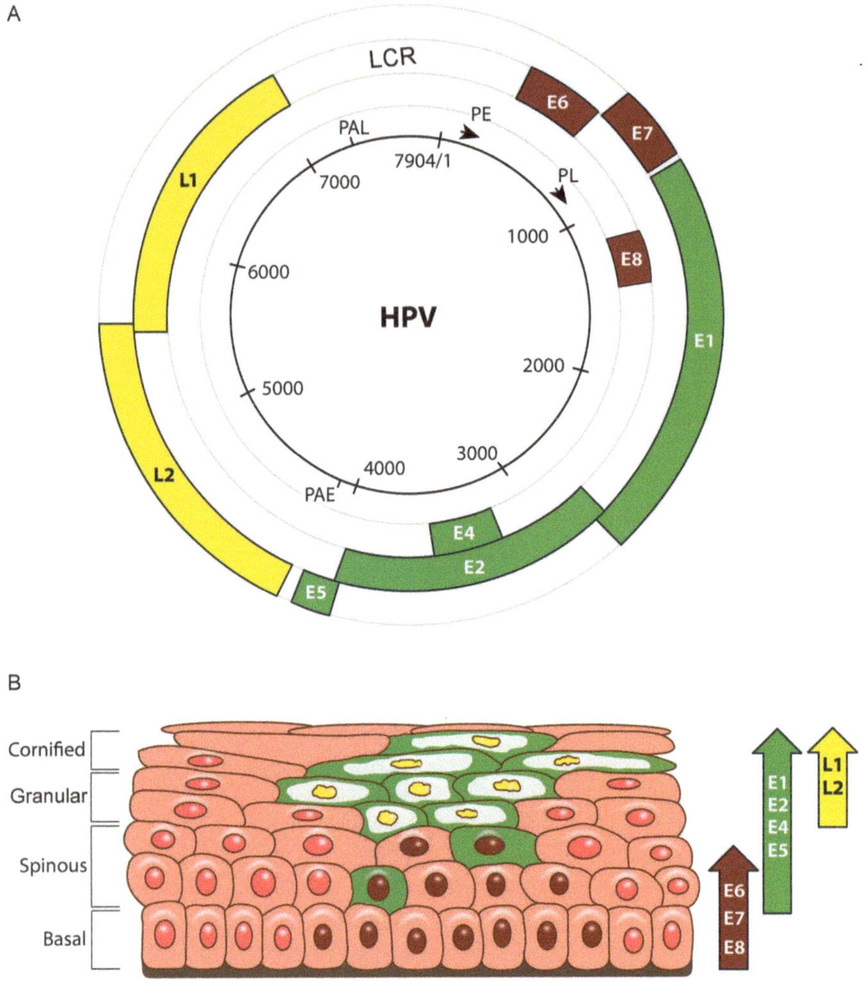

Fig. 2 The HPV life cycle. (A) Schematic of the HPV16 genome, showing the locations of the early and late promoters (PE and PL), the early and late polyadelylation signals (PAE and PAL), the long control region (LCR), the early genes (E6 and E7), intermediate genes (E1–E5 and E8), and late genes (L1 and L2). (B) Diagram of an HPV-infected epithelium. E6 and E7 are expressed at low levels in the basal layer. Expression of the intermediate genes is increased by late promoter activation in approximately the spinous layer. The capsid genes L1 and L2 are expressed in the granular layer, and virions are shed into the environment along with shedding corneocytes (for reviews, see Refs. 1,31,32).

of the basal cell, one of the daughter cells detaches from the basement membrane and begins the process of squamous differentiation.[33] In the course of differentiation, keratinocytes normally withdraw from the cell cycle; however, HPV oncogenes force the cell to reenter the cell cycle to make host

DNA synthesis machinery available to replicate the viral genome.[1] Cell cycle reentry contributes to the formation of a benign hyperproliferative lesion. At the same time, the virus responds to cellular differentiation signals to activate the viral late promoter, which drives expression of viral coat proteins L1 and L2. Virus particles are assembled as terminally differentiated keratinocytes die and the virus is released into the environment during desquamation.[1]

Benign, productive HPV infections typically last between 6 and 18 months in the face of constant immune surveillance and turnover of infected cells.[36–39] The ability of the virus to persist over this period is a feature of normal infection and results in part from how the HPV life cycle is organized, including the targeting of infection to long-lived basal cells,[26,31] maintenance of a stable pool of viral genomes through multiple mitoses,[40] restricting viral antigens to portions of the tissue with lower levels of immune surveillance,[41–43] and lack of cell lysis or inflammatory responses.[44]

Usually HPV infection is cleared through cell-mediated immunity, but in the minority of cases that go on to develop into cancers, infection persists for years or decades, providing ample opportunity for continued expression of viral oncogenes to facilitate accumulation of secondary cellular mutations.[1,31,39,45] For this reason, the single most important risk factor for developing HPV-induced cancers is long-term viral persistence.[38,39,46,47]

Cervical cancer development occurs through a clearly defined series of precursor lesions[6,48] (Fig. 3). The productive, differentiation-dependent virus life cycle occurs in benign, low-grade lesions called cervical intraepithelial neoplasia (CIN1)-1. Squamous intraepithelial lesion is an alternative term for CIN. CIN2 lesions are slightly more dysplastic and can either regress or progress to CIN3. CIN3 lesions are severely dysplastic, extending most of the way through the epithelium and are precursor lesions for cervical carcinoma in situ and eventually for invasive cervical cancer.[45] Aberrant integration of the viral genome into the host chromosomes is a frequent event in cervical cancer, but it is not strictly necessary for cancer development, at least for HPV16, despite widely held beliefs to the contrary.[49–52]

However, cancers are nonpermissive for productive HPV infection and are thus a dead-end for the virus.[48,53–55] The functions that the viral oncogenes evolved to perform in the normal, productive HPV life cycle are only partly understood. Many studies have uncovered molecular interactions responsible for the cellular effects of high-risk HPV oncoproteins (reviewed in Refs. 5,10,12,56). HPV oncogenes primarily activate, inhibit, or modify host gene expression patterns.[31] E6 and E7 are best known for promoting

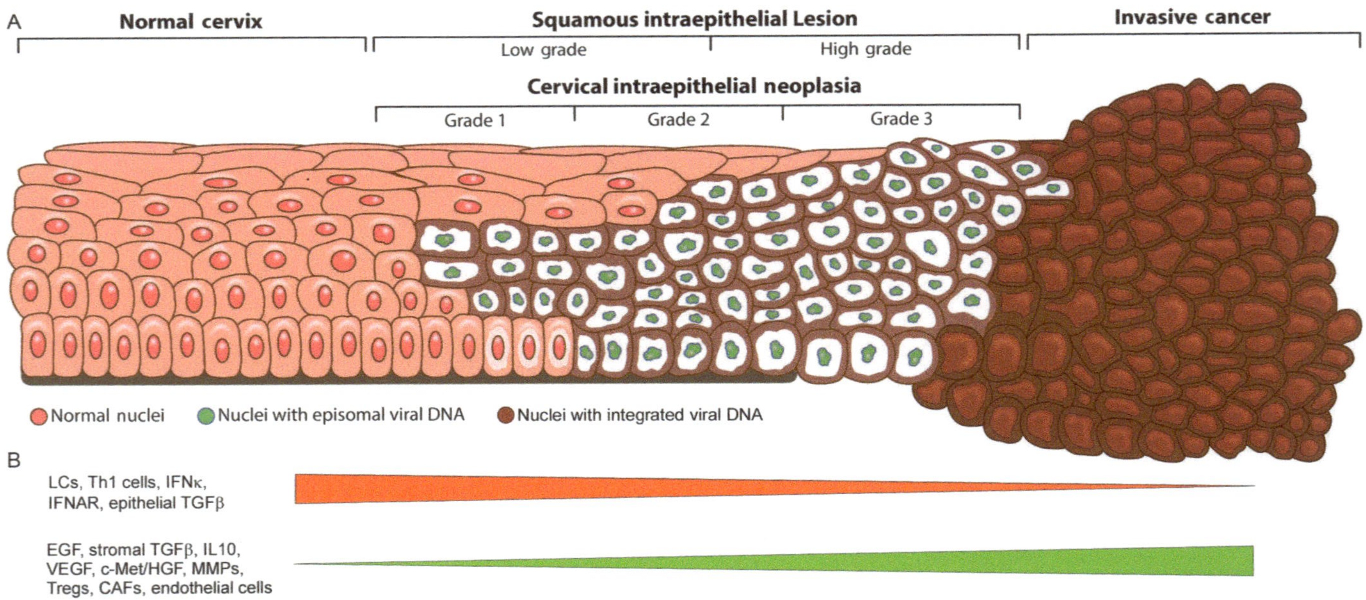

Fig. 3 Cervical cancer development. (A) The transition from normal stratified epithelium to invasive cancer is shown. The productive, benign HPV life cycle occurs in CIN1–CIN2 lesions, which show koilocytosis, or clearing around the nucleus. Transaction through CIN2 and CIN3 to cancer lesions is associated with increased nuclear to cytoplasmic ratios and loss of squamous differentiation. Integration of viral DNA into the host genome frequently but not always occurs in cancers. (B) Cell types and secreted factors whose presence in the tissue trend up or down during infection and cancer progression. See the abbreviations list.

cell cycle reentry through their ability to target p53 and the retinoblastoma (Rb)/E2F system, respectively.[1] High-risk HPVs also encode E5, a small, hydrophobic membrane protein, primarily found in the Golgi complex.[12,57] The best-understood molecular activity of E5 is to enhance signaling from the epidermal growth factor receptor (EGFR).[12] The viral E2 protein is a transcription factor that regulates viral gene expression, and recently has been found to regulate cellular promoters, as well,[58,59] although E2 does not act as an oncogene. The viral oncoproteins each have numerous other interactions and activities, many of which will be discussed in this chapter.[10,56]

Keratinocytes containing episomally replicating HPV genomes can be maintained in culture and can produce infectious virions if induced to differentiate.[60–62] One way to achieve keratinocyte differentiation is through organotypic, or "raft" cultures. Organotypic cultures are 3D cultures made by seeding uninfected or HPV-containing keratinocytes onto a collagen matrix containing fibroblasts. The construct is lifted to the air–liquid interface, where keratinocytes will stratify and differentiate until they form a full epithelial layer. Organotypic cultures recapitulate a fully differentiated epithelium that is similar to the morphology in vivo and allows the virus to undergo its complete differentiation-dependent life cycle.[61,62] The fact that a fibroblast-containing stroma is needed for raft culture development will be discussed later. Mice expressing HPV oncogenes in the epithelium have proven useful for studying HPV-induced carcinogenesis,[63–66] although there is no animal model of HPV infection.[67–69]

3.1 Two Caveats

Much of what we know about stratified epithelial–stromal interactions has been learned by studies of cutaneous skin, and thus will be included in this review. Mucosal sites infected by HPV, such as the cervix, are clearly different from cutaneous epidermis, particularly in hormone responsiveness and immunology.[70]

Much of the literature on stromal interactions in HPV biology relates to cancer, but less is known about the biology of cells in low-grade lesions that contain productively replicating episomal HPV. Given that only low-grade lesions produce new HPV progeny,[48,53,54] it is in these lesions rather than in cancers that the evolved functions of the HPV oncogenes operate. We will focus where possible on studies of productive HPV infection in CIN, with discussion of cancer primarily by way of contrast. One difference between benign HPV infection and HPV-associated cancers is that in productive low-grade lesions, HPV-infected cells are confined to the epithelial compartment, and thus communication between epithelium and stromal cells must depend on soluble factors that are capable of crossing the basement membrane. Exceptions to this rule are the skin-infiltrating immune cells, including primarily Langerhans cells (LCs) but also certain T cell populations. The epithelial basement membrane is semipermeable to ions, growth factors, and nutrients, and pores have been reported which could serve as conduits for both cells and soluble molecules.[71–74] However, it is not always clear the extent to which a given factor, whether produced by the stroma or epithelium, can penetrate through the intact basement membrane and impact cells in the opposite compartment. This fact should be kept in mind as we discuss the expression patterns of various factors.

4. GROWTH FACTORS

In our discussion of epithelial–stromal interactions, we first discuss several significant growth factors that regulate and are regulated by HPV.

4.1 Transforming Growth Factor-Beta

Throughout this chapter, transforming growth factor-beta (TGFβ) reappears constantly. In fact, microarray analysis shows that most of the genes regulated by E6/E7 in human cervical keratinocytes are targets of TGFβ.[20] TGFβ regulates many biological processes such as cell–cell communication, embryogenesis, wound healing, apoptosis, and epithelial-to-mesenchymal transition (EMT) (reviewed in Refs. 75–77). The TGFβ family consists of three related growth factors, of which TGFβ1 is the best understood. TGFβ is expressed in the basal and suprabasal layers in normal skin and in the stroma in cancers. TGFβ is secreted as an inactive complex noncovalently linked to a homodimer of its propeptide (called latency-associated peptide). Following secretion, the complex is sequestered on the extracellular matrix (ECM).[20,77–80] Latent TGFβ can be activated by variety of factors, including matrix metalloproteinases (MMPs), the anti-angiogenic factor thrombospondin-1 (TSP-1), and several integrins.[79] Because TGFβ is secreted as an inactive precursor and then activated from a stored reservoir in the ECM, the cells that secrete TGFβ can be distinct from those which produce the activator, thus increasing the flexibility of TGFβ signaling regulation.[81]

Following activation, TGFβ binds to TGFβ receptor II (TGFBR-II), which then recruits TGFBR-I (Fig. 4) (reviewed in Ref. 82). The specific ligand (TGFβ1, 2, or 3) can affect signaling strength.[75] TGFBRs are serine kinases that phosphorylate SMAD2 and SMAD3, which then associate with SMAD4 to translocate to the nucleus and bind to TGFβ-responsive promoters to regulate target gene transcription.[77] TGFβ can also signal through non-SMAD pathways, such as mitogen-activated protein kinase/extracellular signal-related kinase (MAPK/ERK).[77]

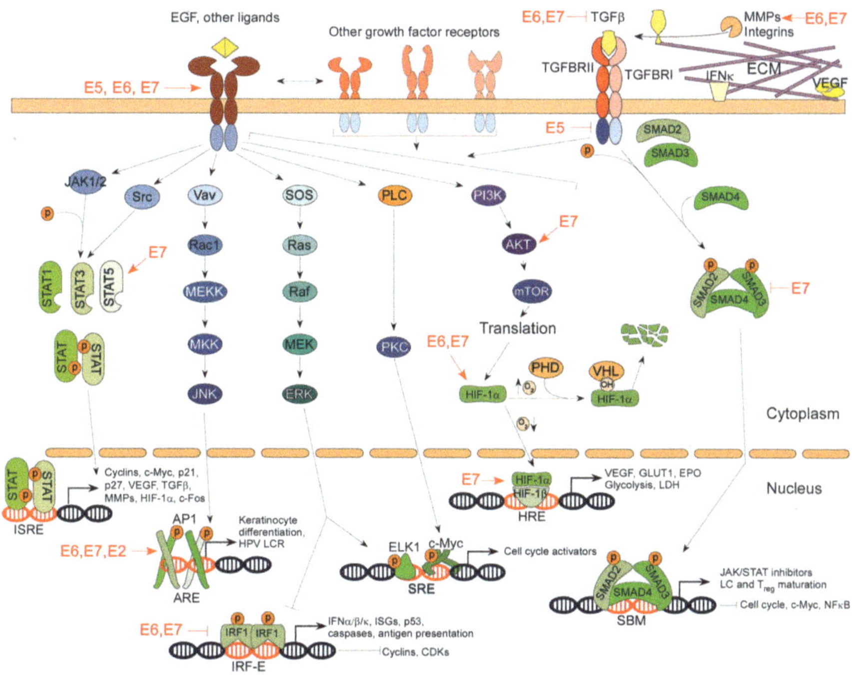

Fig. 4 Growth factor signaling pathways in HPV-containing keratinocytes. The EGF receptor is shown activating the STAT3/5, Jun kinase (JNK), MAPK, and phospholipase C (PLC) pathways, but other receptor tyrosine kinases such as c-Met and PDGF can activate similar signaling cascades (brackets). TGFBR-I/II signaling is also able to activate MAPKs (not shown) to induce ELK1 and other transcription factors. HIF-1α is regulated by both growth factor signaling and by the levels of O_2. Selected genes or processes regulated by each transcription factor are listed. Points of regulation by HPV oncogenes are indicated by *arrows* or *bars*. Color scheme in online version: ligands are shown in *yellows*, many of which are sequestered in the ECM. Receptors are *reds*, kinases are *blues*, transcription factors are *greens*, and other signaling factors are in other *colors*. ISRE, interferon-sensitive response element; IRF-E, IRF1 element; ARE, AP1 response element; SRE, serum response element; HRE, hypoxia response element; SBM, SMAD-binding motif; PHD, proline hydroxylase.

4.1.1 Functions as a Growth Inhibitor and Activator

TGFβ signaling depends strongly on cellular context. In normal keratinocytes, TGFβ acts as a growth suppressor and can promote differentiation.[75,83–89] However, during malignancy, TGFβ stimulates cell motility, invasion, and metastasis.[75,85,90–92] TGFβ activates genes involved in EMT and metastasis, increases deposition of ECM, and promotes expression of MMPs and angiogenesis markers.[75,93] TGFβ is one of the major tumor-derived factors that promote growth factor production by stromal fibroblasts (reviewed in Refs. 75,92,94). TGFβ is also very important in suppressing both innate and adaptive tumor immunity.[75,95] Both of these functions will be discussed in detail later.

4.1.2 Interaction With HPV

The interface of HPV with TGFβ is complex, in part because the various activities of TGFβ can have different effects on viral biology. High levels of TGFβ are found in cervical lesions.[80,93,96,97] On the other hand, both TGFβ1 and 2 can inhibit HPV early transcription,[98–100] reduce viral genome levels,[101,102] and prevent cell proliferation.[20,86,89] The mechanisms of inhibition remain unclear.

HPV has evolved several strategies to regulate TGFβ, presumably to minimize the negative effects on viral life cycle and maximize the positive. E7, with some participation by E6, can inhibit TGFβ signaling in cells containing HPV.[20,84,85,89,100,103–105] On the other hand, E6 and E7 are each reported to increase TGFβ1 promoter activity in cervical cancer cell lines.[106] E6 and E7 from low- and high-risk HPV types inhibit the promoter of the TGFBR-I gene, although the precise mechanism is unknown.[107–109] E7 can bind to SMAD2, SMAD3, and SMAD4, preventing DNA-binding and transcriptional activity.[104,110] TGFβ signaling is also regulated by E5, which downregulates TGFBR-II and phosphorylated SMAD levels in the nucleus after TGFβ1 treatment.[111]

Although HPV-containing cells are sensitive to TGFβ-induced growth inhibition in early passage, as HPV16-mediated cellular transformation progresses, decreased sensitivity to TGFβ is observed.[85,112,113] This switch in sensitivity is associated with decreased TGFBR-II expression and SMAD activity over time.[112,114] Resistance to TGFβ growth arrest appears to be primarily mediated through E7.[89,103] E7 alone can inhibit growth suppression in nonmalignant cells by blocking TGFβ expression and signaling.[104,105] Consequently, TGFβ treatment of HPV-containing cells can stimulate rather than arrest growth.[89,103,115] The specific activity of E7 that is responsible for resistance to TGFβ growth arrest is not clear, but other Rb-binding viral oncoproteins such as SV40 large T antigen and adenovirus E1A also confer resistance to TGFβ.[89]

We should point out that most of these studies of TGFβ and HPV have been done in the context of keratinocyte monoculture, and thus the effects of TGFβ on HPV in the presence of stromal cells is not known. In addition, most studies have focused on the effect of TGFβ on cell proliferation and tumorigenesis. As we will see, TGFβ has many other functions that could be relevant to HPV infections in vivo, but how regulation of TGFβ by HPV could impact those functions is not known.

4.2 Epidermal Growth Factor

4.2.1 EGFR Signaling

EGFR is a member of the ErbB receptor family and binds to several ligands of which epidermal growth factor (EGF) is best known. EGFR signaling is crucial for the development of several organs including skin (reviewed in Ref. 116). EGFR signaling is especially crucial to keratinocyte migration and proliferation during wound healing.[117] EGF-induced activation of EGFR in keratinocytes stimulates the MAPK signaling pathway, leading to growth stimulation (Fig. 4).[117,118] EGFR can also stimulate cell growth and invasion by inducing signal transducer and activator of transcription (STAT) signaling, primarily through STAT3 and STAT5 (reviewed in Ref. 119).

Like many growth factors, EGF contributes to the interaction between epithelium and stroma. For example, heparin-binding EGF (HB-EGF) is made by stromal fibroblasts in CIN and even more in cervical cancers[120] (Fig. 3). Secretion of HB-EGF by fibroblasts is stimulated by platelet-derived growth factor (PDGF) produced by the tumor cells, and HB-EGF in turn stimulates growth of the tumor.[120] EGFR also cross-talks with other growth factor signaling systems such as TGFβ.[121]

4.2.2 Relationship With HPV

The role of EGFR in keratinocyte growth and migration makes it a potential target of viral oncogenes. Additionally, the AP1 transcription factor is a downstream target of EGFR and is important for the regulation of the viral long control region (LCR), which contains the major HPV enhancer elements[122–126] (Fig. 2). EGFR signaling is needed for immortalization by either the HPV16 genome or E6/E7 alone.[127] The primary HPV protein that regulates EGFR signaling is E5. E5 prevents EGFR degradation through the endocytic pathway, promoting receptor recycling, and increasing EGFR levels at the cell surface.[128] E5 does not directly stimulate EGFR activity; rather, E5 potentiates activation by EGFR ligands.[12,128] The mechanisms of E5-dependent EGFR upregulation are controversial. Some researchers have found that E5 upregulates EGFR by inhibiting endosomal acidification,[128–130] while others observed that E5 has no such effect.[131,132] Whether E5 can bind to EGFR itself has also not been resolved.[12,133]

In any case, cells expressing HPV16 E5 show enhanced ERK1/2, Akt, and p38 MAPK activation and cell cycle progression in response to EGF treatment.[12,134–136] E5 acts synergistically with EGFR to induce anchorage independent growth, growth in low serum, DNA synthesis in keratinocytes, and tumors in transgenic mice.[12,137] Furthermore, HPV genomes that lack E5 are defective in productive replication and other events late in the viral life cycle.[138,139] E5 has also been reported to stimulate other ErbB family members, although as with EGFR, the role of these other receptors in HPV biology is poorly understood.[140]

Additionally, E6 increases EGFR mRNA levels.[118,141,142] E6- or E6/E7-expressing keratinocytes have higher total levels of EGFR phosphorylation than controls, increased mammalian target of rapamycin (mTOR) activation, and higher levels of migration and invasion in response to EGF[141,143] (Fig. 4).

4.3 Other Growth Factors

HPV also regulates other growth factor systems. c-Met is the receptor for hepatocyte growth factor (HGF) (reviewed in Refs. 15,16). c-Met/HGF is a classic example of an epithelial/stromal signaling system in which c-Met is typically found in the epithelium, while HGF is produced in the stroma.[15,16,144] c-Met/HGF signaling is especially important in situations involving cell movement, such as embryonic development, wound healing, and EMT.[15,145] c-Met levels are upregulated in CIN, with increased staining in cervical cancers.[144,146,147] The impact of HPV oncogenes or HPV infection on the c-Met/HGF axis is unknown, although E6 may activate c-Met through p53 inhibition.[148–151] Several studies have shown that HPV oncogenes can promote some aspects of EMT, such as reduced E-cadherin and increased vimentin, and changes in cell polarity.[43,152–161] It is not clear, however, what role c-Met might play in these changes, or what advantage HPV may gain from inducing EMT-like cell behavior.

HPV can regulate both overall levels and splicing of keratinocyte growth factor/fibroblast growth factor (KGF/FGF)-7 and FGF10 and their receptors, largely through E5.[162–166] In a fascinating series of studies of the role of the stroma in HPV biology, keratinocytes expressing E6/E7 were grown in organotypic culture with fibroblasts depleted of the tumor suppressor Rb. The resulting epithelia showed a striking increase in MMP expression and invasion into the stroma, which depended on increased KGF expression by Rb-deficient fibroblasts.[27,167] Although organotypic cultures containing Rb-depleted fibroblasts had normal gross morphology, differentiation markers were reduced and proliferation markers were increased,[167] showing that the proliferation–differentiation balance so central to the HPV life cycle can be directly regulated by stromal factors.

5. FIBROBLASTS

5.1 Fibroblasts as Epithelial Support Cells

Fibroblasts are mesenchymal cells in the dermis that engage in a wide variety of signaling, ECM maintenance, and other roles to promote tissue homeostasis (Fig. 1, reviewed in Ref. 168). Fibroblasts produce factors critical for epithelial differentiation, wound healing, angiogenesis, and regulation of inflammation. Despite their impact on these processes and their presence in the neighborhood of HPV-induced lesions, little is understood about the role of fibroblasts in regulating the HPV life cycle.

In vivo, normal keratinocyte differentiation requires growth factors expressed by fibroblasts,[169–171] and fibroblasts alone are sufficient to provide signals needed to support keratinocyte growth and differentiation in culture.[170–176] Organotypic keratinocyte cultures, which are the only method for producing HPV in vitro in a differentiation-dependent manner, require fibroblasts for proper stratification, differentiation, and basement membrane formation.[73,74,177] Stimulation of differentiation and growth of epithelia cannot be recapitulated with fibroblast-conditioned media, indicating that the supportive effects of fibroblasts requires a two way communication with keratinocytes.[171,172,178] The factors mediating this communication act locally, but can diffuse over some distance and do not require cell–cell contact, which is inhibited in vivo by the presence of the basement membrane.[73,177,179] Specific factors that contribute to fibroblast/keratinocyte communication have not been fully worked out, but include KGF,[180] HB-EGF,[120] and CXCL12.[181] In return, keratinocytes can trigger collagenase, KGF, granulocyte macrophage colony-stimulating factor

(GM-CSF), and interleukin (IL)6 mRNA expression in fibroblasts in cocultures.[171] Interestingly, endothelial cells can also function as feeders for keratinocytes,[73,171,179] suggesting that they may also provide some of the support functions that fibroblasts perform.

Fibroblasts can differentiate into myofibroblasts during wound healing, and they play a critical role in the activation of keratinocyte proliferation, migration, and differentiation (reviewed in Refs. 182,183). A wide range of growth factors are involved in wound healing,[145] but HGF originating in the stroma and signaling to epithelial c-Met is especially important.[149,184–186] Stromal cells such as macrophages, immune cells, and fibroblasts each play important roles in the wound healing process through autocrine and paracrine signaling loops, clearing microbes and debris, and secreting growth factors to promote epithelial cell growth.[182,183,187–189] It may be significant that HPV accesses basal keratinocytes through a wound in the epithelium,[28] so that initiation of infection occurs in a specialized growth factor environment. The consequences for HPV replication or gene expression are not known. It may also be significant that the cervix undergoes remarkable remodeling during pregnancy, childbirth, and healing, all orchestrated by hormones and associated with specific activities in the cervical stroma.[190] Again, the impact of the growth factor environment in the cervix on the HPV life cycle is not understood.

5.2 Cancer-Associated Fibroblasts

Stromal fibroblasts provide important signals in the development and progression of cancer. Cancer-associated fibroblasts (CAFs) are fibroblasts activated by paracrine mediators produced by cancer cells. CAFs are very similar to activated myofibroblasts present in wound healing environments.[191] In response to TGFβ, CXCL12, and other factors produced by tumor cells, CAFs can produce growth factors such as HGF and vascular endothelial growth factor (VEGF) that promote cancer growth and metastasis, angiogenesis, ECM remodeling, and immune evasion.[15–17,80,92,168,191–194] CAFs can also undergo mutations or epigenetic changes and coevolve with the tumor, increasing their tumorigenic effect.[17,92,195]

Although their impact on HPV replication is not known, myofibroblasts and markers of CAF differentiation are more prevalent in the stroma surrounding CIN1 to CIN3 than in normal cervix, and increase further when lesions progress to invasive carcinomas.[80,193,196,197] Conditioned media from cervical cancer-derived CAFs increase HPV16-immortalized keratinocyte invasion.[194] Coculture of cervical cancer cells with cervical CAFs increases TGFβ1 and TSP-1 secretion.[197] CAFs isolated from cervical cancers secrete more IL6, IL8, and VEGF and have more STAT3 activation than normal fibroblasts.[198] IL6/STAT3 is activated by E6 in cancer cells, and inhibiting the IL6/STAT3 pathway in stromal cells prevents tumor growth in xenografts.[198]

5.3 Estrogen in the Stroma

Estrogen signaling is another example of the role of stromal fibroblasts in HPV biology. The female reproductive tract is a hormone sensitive tissue, with immune and structural variations throughout the menstrual cycle (reviewed in Ref. 70). Estrogen has long been known as a cofactor in HPV-induced carcinogenesis and is an important regulator of the immunology and proliferation of the female reproductive tract.[65,70] Despite these observations, there is a dramatic loss of estrogen receptor alpha (ER)-α from keratinocytes during cancer progression, more than any other single gene; but there is a corresponding increase in ER-α in the fibroblasts adjacent to the lesion, which is necessary for cervical cancer development.[196,199] Thus the well-established role of estrogen in cervical cancer development is largely mediated through paracrine tumor–fibroblast interaction. The role in productive HPV infection is unknown.

6. IMMUNE INTERACTIONS

As a virus with persistent character, HPV has an evolutionary imperative to prevent the development of an effective immune response against itself. HPV regulates both soluble and cell-associated immune regulatory molecules. Importantly, some of the same soluble factors that promote immune/inflammatory processes also facilitate angiogenesis and cell proliferation. Thus there is a tension in the HPV life cycle to take advantage of the angiogenic and proliferative activities of host factors while at the same time subverting or suppressing immune processes.

6.1 Immunology of Stratified Epithelia

Keratinocytes, the most abundant cell type in stratified epithelia, are capable of actively participating in immune responses.[200] Keratinocytes constitutively express most Toll-like receptors (TLRs),[201] contain inflammasomes, and express a range of cytokines, including IL1α, IL1β, and IL18. They can be induced to express higher levels upon stimulation.[202–204] Interferon-gamma (IFNγ)-stimulated keratinocytes can express the major histocompatibility complex type II (MHC)-II antigen presentation machinery and ICAM1, and can thus efficiently present antigen to CD4+ T cells, stimulating T helper (Th)-1 and Th2 cytokine responses.[205–207] They can also present encoded or exogenous peptide to CD8+ T cells, promoting cytokine production and cell lysis.[206] All of these factors contribute to the interaction of normal or infected keratinocytes with immune and other cells in the stroma.

Cell-mediated immunity is critical for HPV clearance.[208] To evade clearance, HPV infection is locally immunosuppressive. HPV reduces both the number and function of immune cells in infected epithelium and in the associated stroma (discussed later). Despite being immunosuppressive, HPV infection is also mildly inflammatory, and inflammation increases during progression to high-grade lesions.[209] Biopsies of women with high-risk HPVs or mice expressing HPV oncogenes in the skin show greater inflammation within the epithelium compared to HPV-negative controls and moderate stromal inflammation as defined by immune cell infiltration.[210–213] The levels of a wide range of cytokines and other secreted factors are increased in the cervical secretions of women with HPV infection in the cervix, including IL8, VEGF, IL12, interferon (IFN)-α2, M-CSF, and G-CSF.[214] By contrast, IFNγ, a critical Th1 cytokine, is downregulated in infected women.[214] HPV can manipulate the quality of these infiltrates, preventing the development of effective immune responses.[215]

6.2 Innate Immune Responses Against HPV

The innate immune response provides a first round of protection against pathogens. This response is triggered by pathogen-associated molecules but is not antigen specific. Innate immune signaling facilitates the subsequent development of adaptive, or antigen specific, immunity.

6.2.1 TLRs and Nuclear Factor Kappa B

TLRs recognize pathogen-associated molecular patterns and drive innate immune responses, primarily through activating nuclear factor kappa B (NFκB) (reviewed in Refs. 216,217). Clearance of HPV16 infections is associated with increased TLR2, TLR3, TLR7, TLR8, and TLR9, suggesting evasion of TLRs may be necessary for persistent infections.[218] HPV16 E6 and E7 can inhibit transcription of TLR9, which recognizes unmethylated CpG DNA.[219]

In response to TLRs, cytokine receptors, and many other stimuli, NFκB is activated through signal-mediated degradation of the inhibitory IκB protein, which allows NFκB to translocate to the nucleus and increase expression of inflammatory cytokines, prosurvival factors, and IFNs (see Fig. 5) (reviewed in Refs. 222,223). The NFκB pathway is complex and regulates many downstream effects depending on the type of stimulating ligand and cell context. Relatively little is understood about how HPV interfaces with this pathway.[224] The effect of HPV products on NFκB is sometimes stimulatory, presumably to increase the levels of prosurvival targets of NFκB. E6, E5, and E2 can also upregulate aspects of NFκB signaling.[225–229] On the other hand, inhibition of NFκB increases the efficiency of transformation by complete HPV genomes or by E6/E7 alone.[230]

On the other hand, E7 represses NFκB by disrupting the IKK complex and by downregulating NEMO, thus preventing NFκB nuclear translocation.[231–234] Curiously, the efficiency and pattern of inhibition depends on the anatomical origin of the keratinocytes.[230] HPV can also block acetylation of the RelA subunit of NFκB, thus preventing activation of cytokines upon treatment with IFNγ and TNFα.[235] This effect is mediated through EGFR upregulation, presumably through E5.[235] How the virus achieves specificity within the complex NFκB pathway is not clear.

6.2.2 IFN Signaling

IFNs are antiviral cytokines critical to innate antiviral immunity (reviewed in Refs. 220,236). IFNs are grouped into the type I (IFNα, -β, -κ, -ε, and -ω), type II (IFNγ), or type III (IFNλ) families. The role of type III IFN in HPV biology is not known, so we will focus on types I and II. The production and functions IFNs are outlined in Fig. 5. Various stimuli, especially viral nucleic acids in the cytoplasm, are detected by cellular sensor proteins. The specific sensor triggered by HPV is not known, but TLR9 may recognize a CpG motif from the HPV16 E6 DNA sequence.[219] Sensors initiate a signaling cascade which culminates in the phosphorylation and activation of IFN response factors (IRFs). IRFs are transcription factors which drive IFNα/β expression in synergy with inflammatory signals delivered through the TLR/NFκB pathway. IFNs are secreted and bind to IFNα/β receptor (IFNAR) either on the same cell or neighboring cells, triggering phosphorylation of STAT proteins. STAT homo- or heterodimers associate with IRF9 and translocate to the nucleus where they bind to IFN-sensitive response elements (ISREs) to drive the expression of a wide range of genes, termed IFN-stimulated genes

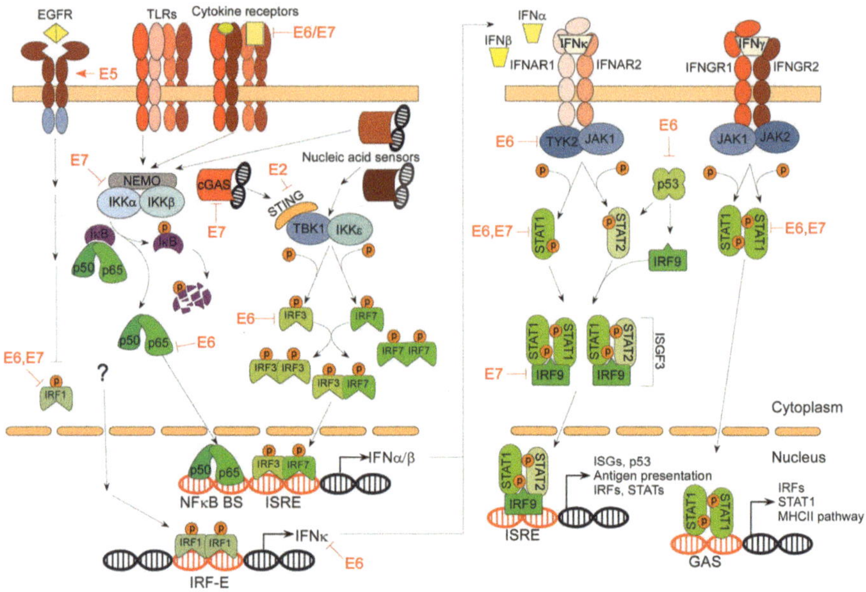

Fig. 5 Interferon signaling in HPV-containing keratinocytes. Pathogen detection proteins such as TLRs or cytoplasmic nucleic acid sensors detect the presence of viral infection and trigger the activation of IRF and NFκB signaling pathways. Type I IFNs are produced and bind in an autocrine or paracrine manner to IFNAR1/2, which activates the JAK/STAT1/2 pathway and induces ISGs. Type II IFN (IFNγ) binds to the IFN gamma receptor (IFNGR) to activate JAK/STAT1 pathway and induce a distinct set of genes. For reviews, see Refs. 220,221 EGFR inhibits IFN in part through suppressing IRF1. Unknown signals induce IFNκ expression in an IRF1-dependent manner. Targets of HPV oncogenes are shown. Color scheme in online version similar to Fig. 4. NFκB BS, NFκB-binding site; IRF-E, IRF1 element; GAS, gamma activation site.

(ISGs). ISGs have the combined effect of rendering cells resistant to viral infection. Because IFN can be either autocrine or paracrine, a virally infected cell can also induce a state of resistance in neighboring cells in the microenvironment. IFNγ is largely produced by immune cells, acts through a distinct signaling pathway involving STAT1, and controls a different set of genes (Fig. 5).

The molecular functions of the hundreds of individual ISGs are largely unknown, particularly with respect to DNA viruses, but IFN has many effects on the cell overall. Constitutive low levels of IFNs have physiological roles in maintaining basal cytokine signaling.[221] Because of its paracrine signaling abilities, IFN can influence the stroma, inhibiting fibroblast function and ECM production.[237] MHC class I molecules and members of the antigen processing and presentation pathway are all upregulated by

IFN,[205,238] facilitating surveillance and killing of infected cells by cytotoxic T lymphocytes (CTLs, discussed later).

Type I or type II IFN treatment of keratinocytes has been shown under various conditions to reversibly inhibit cell proliferation, promote differentiation, or cause cell death,[207,239–241] suggesting that IFN in intact epidermis may have additional antiproliferative roles besides that of antiviral factor. This notion is supported by the observation that IFN or ISG mutations appear frequently in cancers where loss of responsiveness to IFNs can permit immune escape.[221]

IFNs and growth factors tend to antagonize one another. Cancer cells and CAFs are relatively sensitive to virus infection, in part because the TGFβ1 and FGF produced by these cells can inhibit IFN production.[242] EGF and HGF can also suppress the antiviral activity of and IFNβ.[243] Conversely, IFNs can regulate signaling by growth factors. IFNα can downregulate c-Met,[239] and IFNγ slows or inhibits wound healing responses and angiogenesis.[244] Type I and type II IFNs suppress collagen deposition and TGFβ1 production at wound sites.[244,245]

6.2.2.1 IFN and HPV

IFN treatment can prevent infection of human keratinocytes with HPV particles, with IFNβ being more effective than IFNα.[246] IFN also inhibits HPV replication. Although transient increase in viral replication upon IFN treatment has been observed,[247] high and prolonged treatment with IFNβ promotes growth arrest of cells containing HPV31 episomes and reduces episome levels.[248] Although episomally replicating HPV is decreased by IFNβ treatment,[248] emergence of integrated viral genomes is associated with activation of ISGs by type I IFN.[102,249] Treatment of cells containing episomal HPV with IFNγ also results in rapid plasmid loss followed by the outgrowth of integrated clones.[101]

The molecular basis for these effects on HPV replication is not known. Type I or type II IFN can suppress E6/E7 transcripts,[250] although activation of E6/E7 transcripts has also been reported through IRF1 and IRF2 binding to the viral early promoter.[247] The only ISG that has been shown directly to affect HPV is IFIT1 (also called ISG56 or p56), which can inhibit DNA replication by interaction with viral helicase E1.[251] Whether IFIT1 can account for all of the effects of IFN on HPV is not clear.

One of the major pieces of evidence supporting a the critical role of IFN in suppressing HPV is the sheer number of mechanisms that the virus uses to interfere with IFN signaling. Suprabasal layers of normal epidermis express high levels of IFNβ and IFNARs, but HPV-induced lesions do not.[241,252] Intriguingly, although IFN levels in the epithelium are reduced in CIN and cancer, IFNβ and IFNγ mRNA levels increase in the cervical stroma, which correlates with increased stromal infiltration of monocytes and dendritic cells (DCs).[253]

Keratinocytes containing high-risk HPV episomes show reduced responsiveness to IFN signaling upon stimulation.[254,255] STAT1, which is central to the IFN pathway (Fig. 5), is reduced at both the protein and mRNA levels as compared to normal keratinocytes.[256] STAT1 normally increases upon differentiation but does so to a lesser degree in HPV-containing cells.[256] Restoration of STAT1 levels in HPV-containing cells significantly reduces genome replication upon differentiation and promotes viral genome integration.[256] Furthermore, expression of ISGs such as IFIT1, PKR, and MX1 are reduced in cell lines harboring HPV16, 18, and 31 genomes.[255]

Several HPV proteins mediate anti-IFN responses. E6 can reduce levels of cytoplasmic STAT1, and both E6 and E7 inhibit phosphorylation and nuclear translocation of STAT1 and ISG expression.[257] HPV18 E6 can interact with Tyk2 (Fig. 5) and interfere with STAT1 activation in response to IFNα.[258] In addition, HPV16 E6 interacts strongly with IRF3 and weakly with IRF1, preventing their functions.[259]

E6 is best known for promoting p53 degradation.[10] In addition to promoting cell cycle arrest, p53 also enhances the IFN system. The p53 promoter contains a functional ISRE, and both p53 and p53 targets are activated in response to IFN.[260,261] p53 by itself does not activate the IFN pathway, but primes the pathway for activation by increasing transcription of IRF9 through p53 response elements in the IRF9 promoter.[262] IRF5, ISG15, and TLR3 are also direct p53 targets.[263] p53 can promote STAT1 phosphorylation in a nontranscriptional manner, and p53 and STAT1 associate in a complex in cells.[264] IFNβ can counteract the effect of E6 on p53 levels and can interfere with E6-mediated transformation of fibroblasts.[261] Thus targeting of p53 by E6 could be an innate immune evasion mechanism.

E7 also suppresses IFN responses. E7 can inhibit the cytoplasmic DNA sensor cGAS.[265] E7 can also inhibit STAT1 phosphorylation in response to IFNγ and repress IFNγ induction of IRF1 protein levels.[266] Suppression of IRF1 by E7 can inhibit CTL-mediated keratinocyte lysis, and restoration of IRF1 expression can restore CTL-mediated killing.[266] In addition, E7 from both high and low-risk HPV types can physically interact with IRF1 and interfere with IRF1 transcriptional activity.[267] E7 can inhibit IFNα-inducible genes by binding to the IRF9 subunit of the ISGF3 complex and preventing translocation to the nucleus (Fig. 5); loss of this activity results in a loss of transformation capacity of HPV16.[268,269]

The impact of E5 on IFN signaling remains unclear. Overexpressed E5 can induce IFNβ by upregulating IRF1 expression.[270] On the other hand, E5 also promotes EGFR signaling, which inhibits IFN responses[271–274] (see later). Finally, E2 can transcriptionally suppresses stimulator of interferon genes (STING), which transduces in cytoplasmic DNA signals to the IRF pathway.[221,275]

6.2.2.2 IFNκ

Like other type I IFNs, IFNκ uses the IFNAR and can stimulate expression of ISGs through ISRE-mediated promoter upregulation[276]; but IFNκ also has unique properties and particular relationship with HPV. First, IFNκ is specific to keratinocytes and certain innate immune cells.[276] IFNκ is expressed by basal and parabasal keratinocytes, but it is downregulated in more differentiated layers of the epithelium.[41] Second, IFNκ is expressed to high levels in unstimulated, normal keratinocytes, while neither IFNβ nor IFNα is expressed in the absence of stimulation.[255,271,276] The fact that IFNκ is constitutively expressed positions it to serve an important surveillance role. Third, IFNκ is only weakly induced by stimuli that regulate other type I IFNs.[41,255,271] Other signals that may regulate IFNκ are unknown, except that it is readily induced by IFNγ[41] and upon EGFR inhibition through activation of IRF1.[271]

As a constitutive, keratinocyte-specific IFN, one would expect that IFNκ may be able to interfere with HPV. IFNκ expression inhibits growth of cells containing HPV31 episomes, reducing viral gene expression and copy number, although the molecular mechanisms remain unclear.[277] HPV, in turn, has multiple mechanisms to downregulate IFNκ levels. Although the presence of HPV indirectly triggers IFNκ expression in innate immune cells in the cervical stroma,[253] loss of IFNκ expression in the epithelium is an early event in HPV-induced carcinogenesis.[275,278] IFNκ mRNA and protein expression in the epithelium is diminished in CIN and absent in cervical cancer[253] and reduced in keratinocytes maintaining high-risk viral episomes.[255] Continuous expression of HPV16 E6 appears to be necessary to maintain IFNκ suppression through methylation of the IFNκ promoter.[255,278] In addition to E6, E2 expression also suppresses IFNκ mRNA at the transcriptional level, although the mechanisms remain unknown.[275]

6.3 HPV Effects on Immune Cells

Immune cells are present in the microenvironment of normal and HPV-infected epithelia (Fig. 1, reviewed in Ref. 279). The predominant lymphocytes in both the stromal and epithelial compartments of the normal uterine cervix are T cells, with an even distribution between CD4+ and CD8+.[280] T cells are more abundant in the ectocervix and vagina as compared to the endocervix and uterus, while natural killer (NK) cells and granulocytes are more common in the uterus.[70] LCs can be found in the suprabasal layers of the epithelium.[280] The configuration of cells and tissues in the cervix can change as a function of age, which may have an impact on the carcinogenic process.[281] Furthermore, the female reproductive tract is also hormone sensitive tissue, and the composition of immune cells changes through the menstrual cycle, with estrogen being somewhat immunosuppressive.[70]

The cancer stromal microenvironment is profoundly immunosuppressive (reviewed in Ref. 282), but to what extent suppressive interactions seen in cancers are also present in low grade, productive HPV infections is not known. The importance of the immune response in HPV infection is indicated by the increased risk of HPV infection and invasive cervical cancer in HIV-infected women and other immunosuppressed patients.[208,283–288] Women with persistent HPV infections have generally lower lymphocyte function, with lower LC function particularly.[289] This is especially true in older women when immune senescence takes a toll.[289]

Two cell types that are found in the epithelium under baseline conditions are LCs and resident memory T cells. Resident memory T cells (T_{RM}s) are memory cells positioned in the tissue, poised to reactivate in response to antigen.[290–292] Little is known about how these cells function in the context of HPV infection.

6.3.1 *Langerhans Cells*

LCs are professional antigen presenting DCs that are found primarily in the parabasal and spinous epithelial layers.[43] Upon activation, LCs mature, migrate to regional lymph nodes, and activate naive T cells.[293] Thus LCs are central to surveillance and initiation of immune responses in skin. LCs appear to be particular targets of HPV-induced immunosuppressive effects. A reduction of LCs has frequently been reported in HPV-containing CIN.[208,294–302] HPV may suppress LC numbers by inhibiting migration into the tissue, by preventing differentiation into mature LCs, or by promoting premature migration out of the epithelium.

6.3.1.1 Epithelial Recruitment

The precursor for LCs is controversial—they may be maintained by local proliferation under steady-state conditions,[293] but monocytes can also differentiate into LCs under inflammatory conditions.[303] Several factors including TGFβ, CCL2, CCL7, and CCL13 can induce directional migration of LCs into the tissue, but CCL20 (also called MIP-3α) seems to be most important chemoattractive factor produced by keratinocytes.[203,299,300,304–307] CCL20 is constitutively made by keratinocytes in the basal and spinous layers in normal cervical tissues, and so LCs are concentrated in those layers.[308,309] CCL20 can be upregulated by inflammatory cytokines to increase LC attraction to sites of inflammation.[306,308,310,311] CCL20 also has potent chemoattractive activity toward LCs and LC precursors in vitro and is necessary for chemotaxis toward normal keratinocyte supernatants.[306,312] Chemoattractive activity is more potent toward LC precursors than toward DC precursors,[306] indicating that DCs and LCs have different attraction profiles.

E6 and E7 are each able to reduce CCL20 expression and secretion,[300,301] and supernatants from HPV-containing cervical cancer cell lines or keratinocytes expressing E6 and E7 do not support efficient LC chemotaxis in vitro.[299,312] E6/E7 are also able to prevent LC infiltration into organotypic epithelial cultures of cervical cancer cells.[301] E6/E7 from low-risk HPV types also suppress CCL20, indicating that low-risk viruses may evade LC surveillance using mechanisms similar to high-risk types.[312]

6.3.1.2 Differentiation and Maintenance

Once in the epithelium, LC precursors differentiate into immature LCs. The central factor for LC differentiation and maintenance in the skin is TGFβ1. Loss of TGFβ1 expression by both LCs and keratinocytes completely prevents LC differentiation and results in the absence of LCs.[78,290,307,313–317] Maintenance of LCs in epidermis also requires TGFβ: loss of TGFβ signaling in LCs results in mass migration to regional lymph nodes.[290,315,316] Interestingly, autocrine TGFβ1 expression in LCs seems to be important in LC maintenance in the skin, but activation of TGFβ1 is achieved by KC-derived integrins.[290] Thus both cell types cooperate to maintain the pool of LCs needed for immune surveillance. T_{RM}s also require the integrins and TGFβ to develop and remain in the skin.[290–292] GM-CSF and IL34, which bind to the same receptor, seem to play a critical role in LC differentiation and maintenance combination with TGFβ.[303,314,318,319]

Finally, differentiation and maintenance of LCs in the epithelium requires adhesion between keratinocytes and LCs mediated by E-cadherin expressed on both cell types.[320,321]

Because TGFβ1, GM-CSF, and E-cadherin are critical for LC differentiation and maintenance, it is not surprising that HPV oncogenes have been found to suppress these three factors. Although whether regulation of TGFβ is important for the loss of LCs in HPV-containing lesions is not yet clear, E7 causes reduced production of TGFβ from infected cells.[20,84,85,89,100,103–105] HPV also inhibits GM-CSF production, preventing LC infiltration into the epithelium.[294,295,322] E-cadherin is reduced in both high and low-grade lesions in vivo as well as in vitro organotypic models.[155,294,295,297,298,323,324] High-risk E6 and E7 downregulate E-cadherin and do so through several mechanisms.[43,152,157,159,297,325,326] Downregulation of E-cadherin is also seen in lesions caused by low-risk HPV types.[298]

6.3.1.3 Maturation and Migration

Following antigen exposure, LCs mature and migrate out of the epithelium to the regional lymph nodes. IL1α and -β, TNFα, and IL10 promote outmigration.[203,208,327] These cytokines can be secreted by LCs, keratinocytes, or both[203,208] and act in part by downregulating E-cadherin.[327] Upregulation of LC activation markers like MHC-II, CD40, CD86 is coincident with E-cadherin downregulation.[327] In principle, the modest inflammation present in HPV-containing low-grade lesions[209,210,214] could act to clear out the LC population, but it is not yet known whether this is the case.

6.3.1.4 LC Function

In addition to their proper trafficking and maturation, LC function is a potential target for HPV oncogenes. When E7 is expressed in murine epidermis, LCs show a partially activated phenotype but are functionally impaired in both migration and antigen uptake, with reduced dendritic processes and fewer LCs per keratinocyte.[328] Extracted cells can be matured in culture and present antigen normally, indicating no cell-intrinsic defect.[328] Furthermore, LCs can be either immunogenic or tolerogenic depending on the context.[203] Under normal conditions, LCs are capable of activating regulatory T cells (Tregs), which can induce immune tolerance, but in the presence of a pathogen can activate T effector cells instead.[293,329] The difference between activating and tolerogenic effects may depend on the presence or absence of costimulation. For example, when HPV antigens derived from

HPV16 virus-like particles consisting of L1 and L2 are presented to T cells by LCs in the absence of costimulation, T cells fail to activate or to express MHC, CD86, or other markers.[313,330] However, addition of the IFN stimulant poly I:C can reverse the defect,[330] suggesting that reduction of IFN in infected tissues may have an impact on LC-mediated responses to HPV.

6.3.2 T Cells

T cells are another important population of cells in the microenvironment of HPV-infected epithelia. The majority of T cells in the cervical epithelium are CD8+, while the stroma has a more diverse population, with more NK cells and CD4+ and fewer CD8+ T cells.[331] It may be significant that the transformation zone between the columnar endocervical epithelium and the stratified ectocervix, which is the site of origin for the majority of cervical cancers,[332] has fewer T effector cells than normal ectocervix and transformation zone T cells make more immunosuppressive IL10 than other regions.[331]

Th1 cells are a subset of CD4+ T cells that secrete cytokines to promote antiviral immunity, especially the development of CD8+ CTLs. Regression of HPV-induced lesions and clearance of both high- and low-risk HPV infection is characterized by a Th1 response.[333–335] Lack of Th1 response is related to long-term viral persistence.[333,334,336] Stimulation of an effective cell-mediated immune response by therapeutic vaccination remains a major goal in HPV research,[337] but despite the fact that T cell responses against HPV early proteins are possible,[338] HPVs have developed numerous methods to circumvent effective T cell immunity.

6.3.2.1 HPV Interferes With Antigen Processing

In order for CTLs to kill an infected cell, viral antigens must be processed and presented to T cells via the major histocompatibility complex type I (MHC-I) pathway. Thus antigen processing and presentation are important targets for immune evasion by HPV, as for other viruses. Most of the factors in the antigen processing and presentation pathway are upregulated by IFN, and so HPV's ability to inhibit IFN responses (see earlier) may reduce the overall ability of the cell to present antigens.

High-risk (but not low risk) E7 proteins can repress MHC-I mRNA expression through recruitment of repressive HDAC complexes to the promoter.[339–341] HPV18 E7 can repress other components of the antigen processing pathway, such as TAP1,[340] but whether HPV16 E7 is able to do so is controversial.[254,341–343] E5 can bind to and sequester MHC-I complexes in the Golgi to reduce levels at the cell surface and inhibit T cell responses.[344–346] This effect is reversible with IFN treatment.[346] Interestingly, HPV16 E5 does not downregulate nonclassical MHC molecules (HLA-C/E),[346] which may prevent killing by NK cells, which recognize and eliminate cells lacking MHC expression.

6.3.2.2 T Cell Epitopes Are Poorly Immunogenic in the Context of Infection

The T cell response against HPV epitopes is reviewed in Ref. 208. The E6 and E2 proteins seem to be the primary T cell antigens and are most important for viral clearance in patients and animal models.[208,336,347–354] By contrast, E7, the viral protein most responsible for driving cellular proliferation, is poorly immunogenic.[348–350,353] Studies in E7 transgenic mice indicate that E7-specific CTLs ignore or become tolerant to keratinocytes persistently expressing E7, rendering CTLs ineffective in mediating antitumor immunity.[355,356] In CIN and cervical cancer patients, E7-specific T cells can be detected but they are ineffective in controlling neoplastic progression.[357–362] The reasons for these properties of E7 remain unknown.

6.3.2.3 HPV Skews the T Cell Response Away From Th1

HPV reduces the overall levels of T effector cells in CIN as compared to normal ectocervix.[331] HPV also inhibits the development of a Th1 response.[208,363,364] Instead, HPV promotes regulatory T cell (Treg) differentiation. Tregs recognize antigens as do other T cell types, but then suppress rather than activate immune responses (reviewed in Ref. 365). Migration of Tregs is promoted by the chemokines CCL17 and CCL22 produced by macrophages and LCs,[366] as well as by VEGF.[367] Tregs secrete immunosuppressive IL10 and TGFβ, factors that are present at high levels in benign HPV-containing lesions.[366] Depleting Tregs results in higher levels of proinflammatory cytokines.[366] CIN have higher levels of Tregs than normal tissues, and the presence of Tregs increases further in cancers.[366,368–371] Treg levels are correlated with a failure of lesions to regress.[335]

Mice-expressing E7 in their skins have abundant lymphocyte infiltrate consisting of activated CD8+ and CD4+ cells, but despite the presence of the viral antigen, E7-expressing skin is not rejected when transplanted.[372] Rather, lymphocytes in the transplanted skin actively suppress rejection, suggesting that Tregs may be recruited by E7,[372] but more work is necessary to clarify the mechanisms.

CD8+ T cells from cervical cancer express high levels of CD94/NKG2A, which is an NK inhibitory receptor and show minimal cytotoxicity as compared to normal CD8+.[373] Upregulation depends on TGFβ and IL15, which are highly expressed in cervical cancer stroma.[373] Finally, HPV-associated cancers express higher levels of ligands that suppress T cell responses, including PDL1.[374] The extent to which these ligands are present in lower grade, productive HPV infections is not clear.

6.3.3 Other Cell Types

Under steady-state conditions, macrophages are the most common immune cell in the skin.[279] Macrophages, like fibroblasts, can be activated in the presence of tumor cells to form tumor-associated macrophages (TAMs) (reviewed in Refs. 375,376). TAMs can promote tumor growth and angiogenesis by secreting growth factors such as EGF and VEGF,[376–378] and by releasing antiinflammatory cytokines, such as IL10 and TGFβ.[376] Whether HPV infection actually causes TAM development is not known, but macrophages with TAM markers infiltrate the epithelium and stroma of HPV-containing lesions, secrete proangiogenic IL8 and VEGF, and suppress T cell responses.[379–384] HPV can also interfere with the functions of DCs,[294,295,322] mast cells,[213] and NK cells.[346,385,386]

6.4 Soluble Immune Factors

6.4.1 Cytokines and Chemokines

Cytokines are cell–cell communication molecules especially important in coordinating immune function. Chemokines are cytokines which function specifically as chemotactic factors. Keratinocytes constitutively express low levels of a range of cytokines and chemokines but can be induced to express more upon stimulation.[203,204] Cytokine secretion by keratinocytes can vary depending on the anatomical source of the cells.[203,387]

HPV downregulates inflammatory cytokines, chemokines, and downstream pathogen signaling components in both stimulated and unstimulated cells.[254] TNFα and IL1β are classic inflammatory cytokines that that induce the NFκB pathway. IL1β appears to be a central inducer of other cytokines during HPV infections.[254] High grade lesions lack IL1β expression, and E6 is able to prevent IL1β induction.[388] HPV can also inhibit processing of IL1β, which is necessary for mature cytokine secretion.[254] E7 confers resistance to growth arrest by TNFα.[389,390]

On the other hand, HPV can also increase expression of anti-inflammatory cytokines such as TGFβ (see later) and IL10. IL10 mRNA levels are increased in CIN and expression increases with cancer progression.[97,391] Expression of IL10 in the stroma is also significantly higher in CIN2 and CIN3 than in normal cervix.[391,392] HPV can upregulate VEGF (see later) which may be antiinflammatory, resulting in reduced IL12, DC maturation, and NK T cells, and increased Tregs.[393]

Classical tumor suppressor genes inhibited by HPV are increasingly found to regulate immune signaling. For example, loss of p53 or PTEN in either tumor cells or stroma can cause chronic inflammation and persistent tissue damage.[394] The impact of tumor suppressor loss during HPV infection on immune or inflammatory processes is not well understood.

Chemokines are critical for movement of immune cells to the skin (reviewed in Ref. 305). Chemokines are diffusible molecules, but they can form a gradient by being immobilized on the ECM.[305] A variety of chemokines induce directional migration of LCs,[203] endothelial cells,[395] and T cells.[310] Most proinflammatory chemokines increase upon progression to cervical cancer, and some of these, including CXCL1, CXCL2, CXCL5, and CXCL6 are increased in CIN1/2 vs normal cervix, suggesting direct upregulation by HPV.[396] IL8, which acts on neutrophils and endothelial cells, is also upregulated.[208,396] By contrast, E7 suppresses expression of CXCL14 through hypermethylation of the CXCL14 promoter.[396] CXCL14 is expressed in normal suprabasal epithelial cells and stroma and inhibits angiogenesis by preventing endothelial cell chemotaxis.[395] CXCL14 can also promote chemotaxis of DCs.[395] Reexpressing CXCL14 in E6/E7-containing cells reduces cell motility and suppresses tumor growth by promoting infiltration of NK cells, CD4+, and CD8+ T cells to the tumor site.[396] As previously mentioned, the LC-attracting CCL20 is inhibited by HPV.[309,310]

In addition to their effects on the inflammatory and immune environment of a lesion, cytokines can act on HPV-containing cells directly: TNFα, IL1α, IL4, and TGFβ1 can inhibit HPV transcriptional activity in a dose-dependent manner.[99,397] The impact of this effect on HPV in vivo is not clear.

6.4.2 Immune Functions of TGFβ

TGFβ acts as a cytokine to regulate immune function during both innate and adaptive responses.[394]

6.4.2.1 Innate Immunity

TGFβ is antagonistic to type I and type II IFN responses. Epithelia (but not macrophages) treated with TGFβ are less responsive to IFNγ than untreated, perhaps due to upregulation of inhibitors of JAK/STAT signaling by TGFβ.[398,399] TGFβ2 causes decreased IFN secretion and less responsiveness to IFN agonists, which together dampen the antiviral functions of IFN.[398] Conversely, IFNγ inhibits TGFβ-induced SMAD phosphorylation and nuclear localization by upregulating the inhibitory SMAD7.[400] TGFβ inhibits NK cell proliferation and maturation.[401] TGFβ also attenuates IFNγ, TNFα, and GM-CSF production by NK cells and downregulates killer activating receptors.[70,95,401–403] In NK cells, inflammatory cytokines like IL12 and IL18 can inhibit TGFBR-II and SMAD expression.[402] TGFβ also promotes TAM differentiation.[75,95,375]

6.4.2.2 Cellular Immunity

TGFβ also has effects on cellular immune responses. TGFβ inhibits NK cells, promotes formation of tolerogenic DCs, inhibits antiviral T cell responses, and drives development of antitumor Tregs.[78,95] TGFβ is critical in the development, survival, and differentiation of some T cell lineages (reviewed in Refs. 78,81). TGFβ overall has a suppressive effect on T cells, since knockout of TGFBR-II in T cells results in enhanced lymphocyte proliferation and inflammation.[78,81] TGFβ suppresses Th1 differentiation and downregulates proliferation and cytolytic gene expression in CD8+ T cells.[78,81,95,404] Much of the effect of TGFβ occurs during T cell development in the thymus and lymph nodes,[81] but TGFβ can also have suppressive effects in the context of a tumor milleu.[95] For example, conditioned medium from cervical cancer cell lines decreases lymphocyte proliferation and causes apoptosis in CD4+ cells in a TGFβ-dependent manner in vitro.[405]

TGFβ is critical for the maturation of Tregs, which are found at high levels in HPV-induced lesions and cancers[95,366,368–371,392] and suggest a poor prognosis for patients.[95] Induced Tregs (iTreg) are generated in the periphery from naïve CD4+ cells,[78] and thus the TGFβ levels present in CIN and cervical cancers are likely to promote Treg differentiation. DCs are also an important source of TGFβ to promote iTregs.[78]

As discussed earlier, local TGFβ is critical for the differentiation and maintenance of LCs.[78,290,307,313–317] TGFβ also prevents DC migration, downregulates MHC-II, and promotes tolorogenic DC maturation which can facilitate tumor-specific Treg development.[95]

6.4.2.3 TGFβ in Cervical Lesions

As discussed earlier, HPV generally suppresses TGFβ signaling. However, the potent antiinflammatory activities of TGFβ would be predicted to benefit the virus. The expression of TGFβ in cervical lesions is not clear. Some studies have found that serum TGFβ levels are lower in patients with CIN and with cervical cancers as compared to controls,[406,407] while others have found levels to be higher.[369–371] Some have found that TGFβ1 levels increase CIN and cancers,[408–410] while others have found the opposite.[411–413] A solution to this paradox is that although HPV downregulates TGFβ in the infected keratinocytes, the virus upregulates it in the stroma.[80,93,96,97,411,412] Therefore the benefits for the virus of low TGFβ in the epithelium (e.g., loss of growth inhibition, reduced LC cells) could be combined with the benefits of increased TGFβ in the stroma (e.g., increased Tregs, more TAMs, suppressed IFN responses).

6.4.3 EGFR and the Immune Response

TGFβ is not the only growth factor that has direct effects on immune responses. EGFR antagonizes IFN signaling through the Ras/Raf/MEK pathway by inhibiting the activity of IRF1 (Fig. 4).[272–274,414,415] Inhibition of EGFR in keratinocytes grown in monolayer or in organotypic culture results in increased inflammatory and immune response genes, including IFN expression and NFκB signaling.[416,417] EGFR inhibition in patients causes increases in various chemokines and cytokines in the skin, including CCL2, CCL5, IL8, GM-CSF, and IFNκ.[271,418,419] EGFR can inhibit IRF1 activation, resulting in suppression of the IFN pathway and ISG synthesis, especially IFNκ.[271,420–422] Therapeutic inhibition of EGFR in humans consequently results in widespread inflammatory side effects, especially in the skin.[418,419,423] IFN inhibition can abolish the therapeutic effect of anti-EGFR therapies[424] (reviewed in Ref. 425), suggesting some antigrowth factor therapies may depend on IFN responses. EGFR suppresses MHC-I and MHC-II genes in cultured keratinocytes and in patients in response to IFNγ.[426] Thus upregulation of EGFR signaling by HPV oncogenes could be a critical mechanism of immune evasion, although further work will be needed to demonstrate this.

7. ANGIOGENESIS AND THE HYPOXIC RESPONSE

Hypoxia is a critical microenvironmental factor regulating epithelial–stromal interactions. Hypoxia refers specifically to low levels of oxygen in a tissue. In vertebrates, oxygen and other nutrients are delivered primarily by the circulation, which means that hypoxic tissues are also likely to have reduced levels of other nutrients, as well as buildup of waste compounds. The cellular hypoxic response serves to manage these stresses. Cellular proliferation results in hypoxia by increasing the number of cells competing for oxygen and nutrients and increasing the diffusion distance between cells and the vasculature. Since both high- and low-risk HPVs induce cellular proliferation, hypoxia is a factor that many HPV types will encounter in the course of their life cycles. Hypoxia can operate at a whole tissue level and not just on individual cells. Because it not a genetic effect, hypoxia can impact the behavior of tumor stroma beyond tumor cells themselves.[7,378]

The cellular response to hypoxia is largely regulated by the hypoxia-inducible factor (HIF) family of transcription factors, of which the best understood is HIF-1 (Fig. 4) (reviewed in Ref. 378). HIF-1 is a heterodimer of HIF-1α and HIF-1β. HIF-1β levels are more or less constant, but HIF-1α levels are regulated by oxygen. Following translation, proline residues in HIF-1α protein are hydroxylated in a reaction that requires molecular oxygen and α-ketoglutarate. Consequently, HIF-1α hydroxylation serves as a sensor for both oxygen and citric acid cycle metabolite levels. Hydroxylated prolines in HIF-1α form a recognition surface for the Von Hippel–Lindau (VHL) E3 ubiquitin ligase complex, which targets HIF-1α for proteasomal degradation.[378] Under hypoxic conditions, oxygen is unavailable to the proline hydroxylase enzymes and HIF-1α hydroxylation is reduced, preventing proteasomal degradation, and leading to increased HIF-1α levels. HIF-1α translocates to the nucleus and associates with HIF-1β to drive transcription of target genes containing a hypoxia regulatory element (HRE) in their promoters.[378] Translation of HIF-1α is also upregulated by PI3K/Akt/mTOR signaling which responds to increased levels of growth factors and nutrients—conditions in which cell growth is likely and thus increased angiogenesis may be necessary.[378,427,428]

Although the degree and pattern of hypoxic gene regulation varies between cell lines and cell types,[429] genes regulated by HIF-1 tend to regulate either metabolism or angiogenesis. Hypoxia can induce metabolic changes that affect stromal cells[7,378,430] but which are reviewed elsewhere.[378]

Angiogenesis is the production of new blood vessels through the proliferation, migration, and tube formation by endothelial cells.[18,393] In normal tissues, angiogenesis is quiescent, but angiogenesis is increased in conditions of cell proliferation, to meet the higher demand for oxygen, nutrients, and waste disposal.[393] Although physiological angiogenesis is needed throughout development and during wound healing, cancer cells can also acquire a proangiogenic phenotype as they encounter microenvironmental selection forces over time, including low oxygen (hypoxia), low pH, and competition for nutrients.[431] Failure to achieve an angiogenic phenotype (angiogenic switch) is thought to serve as a critical control to prevent cancer development.[18,432] Once a tumor has become malignant, angiogenesis is also important to provide an avenue for tumor metastasis.[393]

The level of angiogenesis is determined by the opposing actions of pro- and antiangiogenic molecules.[7,18,393] Among the most prominent proangiogenic factors is VEGF. There are several VEGF family members, but VEGFA is the most important for angiogenesis, and almost all tumors express it.[191,393] VEGF is produced by both normal and transformed epithelial cells in response to hypoxia, low pH, growth factors, and other stimuli (Fig. 4), but cancers can produce VEGF even in the absence of these conditions.[393,433] VEGF receptors (VEGFR1 and VEGFR2) are receptor tyrosine kinases expressed on endothelial cells; VEGFR signaling promotes proliferation, migration, and tube formation by endothelial cells during vascularization.[191] In addition to VEGF, PDGF, IL8, galectin 1, and FGF1 and FGF2 are potent angiogenic factors,[393,434,435] among many others. Not surprisingly, many proangiogenic genes are direct HIF-1 targets via HREs in their promoters.[436–440]

Of factors that inhibit angiogenesis, TSP-1 is particularly important, as are CXC chemokines and peptides derived from proteolytic degradation of collagen IV. These factors prevent angiogenesis by inhibiting endothelial cell migration and tube formation.[441,442] TSP-1 is also a HIF-1 target, resulting in negative feedback.[443,444] Additionally, some TLRs, antibacterial peptides, proinflammatory cytokines are HIF-1 targets and are upregulated by hypoxia.[438]

Stromal cells are important players in the coordination of angiogenesis. Stromal fibroblasts and macrophages in both wounds and tumors are a major source of VEGF and other angiogenesis regulators.[433,445,446] Tumor cells can promote VEGF expression in nontransformed cells in the microenvironment.[445] Conversely, stromal cells can regulate VEGF secretion by cancer cells,[435] and can also communicate directly with endothelial cells to promote the proper formation of vessels during angiogenesis[447] (Fig. 1). HIF-1 can promote expression and secretion of TGFβ, PDGF, CXCL2, and other factors that promote CAF differentiation and recruitment.[378] Hypoxic CAFs, in turn, produce CXCL12 which promotes tumor growth.[378]

7.1 Angiogenesis in HPV Lesions

In many cancers, angiogenesis happens in the later stages of tumor progression. In cervical cancer progression, however, angiogenesis is seen in early stages, consistent with the idea that HPV infection per se is angiogenic. Early CIN lesions have increased vascularity under the basement membrane as compared to normal cervical tissue.[380,448–455] Although vascular papillae extend into the epithelium in CIN, vessels themselves do not cross the basement membrane[448,449] and do not appear to be as disorganized as is often seen in tumor vasculature.[456,457] Whether changes in vascular function seen in tumors, such as reduced lymphocyte diapedesis or altered immune-associated adhesion molecules,[282] are also present in CIN, is not known. As cervical lesions progress to higher grades and cancers, the level of angiogenesis increases.[450–455,458–462] The particular stage at which the most increase in vascularity is seen depends on the study,[452,459–461,463,464] and correlations of microvascular density with cervical cancer survival are controversial.[369,393,452,453,462,464] Clearly, however, HPV infection has an angiogenic effect. Clinical studies show antiangiogenesis therapy can improve outcomes of cervical cancer patients (reviewed in Ref. 465). Because angiogenesis is seen in low-grade CIN, it is possible that disrupting angiogenesis may have an impact on low-grade lesions, as well.

Consistent with increased angiogenesis in HPV-containing lesions, HIF-1, angiogenic proteins, and other hypoxia-induced factors are found to be upregulated as compared to uninfected epithelium. Dysplastic cells of CIN1 and -2 lesions express HIF-1α and target genes like VEGF, glucose transporters, and erythropoietin.[409,451,458,466–469] HIF-1 mRNA and protein levels are increased along with cancer stage.[470,471] HIF-1α expression in cervical tumors increases with distance from vessels, suggesting that HIF-1α can respond to typical hypoxia-dependent upregulation in these lesions.[472] HIF-1 overexpression correlates with mortality, microvessel density, and radiation resistance in cervical cancer.[470,473] HIF-1α is also increased in HPV-induced oral squamous cell carcinomas as compared to HPV-negative lesions, and shows a significant correlation with E7.[474]

Normal human cervix expresses VEGF at low levels, but CIN1 lesions express more, with incremental increases as lesions progress, correlating with increased vascularization.[214,408,449,453,454,460,475,476] Serum VEGF levels also increase in CIN and cervical cancer patients.[407] Although VEGF is expressed in the keratinocytes of cervical lesions, stromal cells may also participate. Cervical cancer cell lines (including HeLa) and cervical CAFs upregulate VEGF and angiogenesis under hypoxic conditions in vitro, but the CAFs make more VEGF than the cancer cells in both normoxia and hypoxia.[477] IL8 levels are higher in dysplasias and cancers; in cancers TAMs appear to be the primary source.[208,379]

7.2 Regulation of Hypoxic Response and Angiogenesis by HPV

Because angiogenesis and other HIF-1-dependent processes are upregulated in HPV-containing lesions, we might expect that HPV oncogenes would promote the HIF-1 pathway in experimental models. Mice transgenic for the HPV16 early region show increased microvessel density in the immediate subepithelial region, with tufts of vessels extending up toward the epidermis, as is seen in cervical lesions in humans.[478,479] In vitro, cervical cancer cell lines have higher VEGF and IL8 mRNA levels than keratinocytes lacking HPV, and they secrete VEGF and promote endothelial cell proliferation.[480,481]

Both high- and low-risk episomal HPVs potentiate HIF-1α protein stabilization in keratinocytes during hypoxia, so that the levels of HIF-1α in the cells during hypoxia are higher than in controls.[25] Increased HIF-1α levels are reflected in increased levels of some HIF-1 target genes (e.g., VEGF) but not others (e.g., IL8).[23,25,482] E6 and E7 can each independently increase HIF-1α levels.[25,482] Some studies show stabilization specifically in hypoxia,[25] and others also see increased HIF-1α in normoxia, as well.[482]

In keratinocytes expressing HPV16 E6/E7, VEGF and IL8 mRNA and protein are increased and TSP1 is decreased.[23,24] Conditioned supernatants from E6/E7-containing keratinocytes can increase endothelial cell division and angiogenesis in vitro in a VEGF-dependent manner, but neither oncogene can do so alone.[23,483] However, both E6 and E7 do have independent effects on the HIF-1 pathway. E6 expression alone induces VEGF mRNA and protein levels and inhibits antiangiogenic factors.[24,481,484,485] This may be due in part to E6 counteracting the inhibitory effects of p53 on the HIF-1 pathway (see later),[484,486] but p53-independent mechanisms are also reported.[481,485] E7 expressed alone can also increase IL8 and VEGF production in keratinocytes.[483] E7 is able to prevent the association of HIF-1α with HDACs, and thus abrogate the negative effect of HDACs on HIF-1 activity.[486]

E6 and/or E7 may promote the PI3K/Akt/mTOR pathway, thus increasing HIF-1α translation[482] (Fig. 4). E5 can increase VEGF expression through EGFR–MEK–ERK and PI3K/Akt pathways in E5-expressing cervical cancer cells.[487] Cell lines containing episomal HPV promote angiogenesis more efficiently than those containing E6 and E7 alone, suggesting that E5 may be functionally significant in the regulation of angiogenesis by episomally replicating HPV[23]

HPV oncogenes are known to regulate several transcription factors that have an impact on HIF-1 activity.[10,56] p53, which is a target for the HPV E6 oncoprotein, antagonizes the HIF-1 pathway (reviewed in Ref. 436). p53 is stabilized by hypoxia and metabolic stress,[488–492] although the mechanisms and consequences are controversial.[489,491–498] p53 binds and destabilizes HIF-1α.[436,484,488–490,493,499–503] p53 also represses HIF-1-dependent transcription at some genes, including VEGF and metabolic genes such as carbonic anhydrase IX.[436,484,501,502,504–508] Repression may be through direct binding or through competition between p53 and HIF-1 for coactivators such as p300.[493,504] p53 can increase levels of TSP-1.[393,509–511] p53 increases expression of collagen prolyl hydroxylase expression and increases the antiangiogenic collagen fragment endostatin.[512] Thus p53 serves as an inhibitor of angiogenesis and metabolic changes during cancer progression.[436,509] Interestingly, there is selection pressure to inactivate p53 in tumor stroma, perhaps to promote angiogenesis.[436] Whether inactivation of p53 by E6 can fully account for the angiogenic activity of E6 is not known.

Finally, other transcription factors regulated by HPV can impact HIF-1 activity or independently respond to hypoxia, including E2F1, NFκB, AP1, and Myc.[228,438,513–522] IFNs can also decrease expression of angiogenic factors in part by inhibiting bFGF and VEGF production,[221,523] but the impact of HPV-induced suppression of IFN on angiogenesis specifically is not known.

8. ECM AND MMPs

Finally, many molecules that control intercellular communication are part of or sequestered in the ECM where they may remain latent until activated.[7] In this way, the cell that produces a molecule can be different from the one that produces the activator, allowing integration of signaling between cell types. Although this review has focused primarily on the cells that produce communication molecules, regulation of activation is also potentially regulated by HPV during infection.

One of the most important groups of activating molecules is the MMP family (reviewed in Refs. 524,525). MMPs are important factors in various aspects of biology from development, organogenesis, cell proliferation and differentiation, vascularization, as well as ECM remodeling and cell migration (reviewed in Ref. 526). MMPs create space for cells to migrate by degrading components of the ECM, and are thus widely understood to be critical factors in cancer metastasis and wound healing.[524,525] MMPs also regulate cell movement and interactions within the stroma (reviewed in Ref. 527). Most MMPs are secreted, but membrane type 1 MMP (MT1-MMP) is a transmembrane protein that is not secreted like soluble MMPs such as MMP2 or MMP9. MT1-MMP is well known for promoting tumor cell invasion via localization at the invasive front, along with MMP2 and MMP9, and can activate other MMPs (reviewed in Refs. 528,529). In addition to activation of MMPs and degradation of ECM components, MT1-MMP activity can also promote the transcriptional upregulation of VEGF and thus promotes angiogenesis.[530] Expression of some MMPs can be upregulated by TGFβ.[531,532]

PDGF, VEGF, and TGFβ are examples of factors cleaved and activated by MMPs.[191,525,533] Cleavage by MMPs can either activate or inactivate inflammatory cytokines.[525] MMP9 is also important for increasing angiogenesis by TAMs.[446] Interestingly, TGFβ2 is more sensitive than -β1 or -β3 to being activated by MMP9 and MMP2,[533] suggesting that activation may be one means by which the TGFβ isoforms may mediate different functions. IFNκ is also associated with heparin in the ECM, similar to IFN-β,[41,534] but whether IFNκ is regulated by MMPs is not known.

Whether the ability of MMPs to regulate ECM-bound growth factors or cytokines is important for the benign HPV life cycle in addition to promoting cancer invasiveness at later stages is currently unknown, but high-risk HPV oncogenes, particularly E7, have been reported to upregulate the expression of MMP1,[535] MMP2,[536,537] MMP7,[537] MMP9,[536,538,539] and MT1-MMP.[536,537] MMP upregulation is seen in premalignant HPV16- and HPV18-containing keratinocytes as well as in cancers.[535] The specific spectrum of MMPs regulated by HPV may depend on the specific virus type.[537] HPV-containing cells are thus more invasive than uninfected controls.[536,537]

9. UNRESOLVED QUESTIONS

As can be seen, HPVs regulate and are regulated by a huge array of factors that either arise from or have an impact on the stromal microenvironment. Although the ability of HPV to promote cell-autonomous proliferation is well appreciated, the regulation of stromal interactions is much less understood. Here are some outstanding questions:

To what extent are the protumorigenic activities of HPV oncoproteins in fact directly necessary for the viral life cycle, and which are side effects of immune suppression? Many of the factors discussed in this chapter have both cancer-promoting and immunological functions. Here are some examples touched on in this review:

- Although often thought of primarily as a tumor suppressor, p53 also supports antiviral immunity, with a direct role in priming the IFN response.[260–262,264,540] Similar immune roles for Rb have also been reported.[541]
- The many immune-related functions of TGFβ, especially for LC maturation, may be more critical to the evolved biology of HPV than avoiding its antiproliferative effects.[78,290,307,313–317]
- E-cadherin is thought to be downregulated in order to promote EMT, which is important for malignancy but does not benefit the virus. Instead, loss of E-cadherin may serve to prevent LC maintenance in the skin, thus facilitating viral immune evasion.[320,321]
- Activation of EGFR by E5 has usually been interpreted as a proproliferative function, but given that EGFR also inhibits the IFN response,[271,420–422] it is also likely to be an immune evasion mechanism.
- Upregulation of MMPs by HPV has been studied primarily focusing on their role in tumor invasion. However, productively replicating HPV has no selection pressure to promote invasion or metastasis, and thus MMP regulation activation of growth factors or cytokines stored in the ECM may be a more evolutionary reasonable explanation.[542,543]

Many of the effects of HPV oncogenes have been viewed through the lens of cancer with an emphasis on cell proliferation, using cervical cancer cell lines as a model. A more evolutionarily correct approach may be to view the effects of the HPV oncogenes through the lens of virology—i.e., what does the virus need to do to survive and reproduce—and use model systems with productively replicating virus genomes.

What is different about the natural history of high-risk HPVs that requires maintenance of potentially deleterious activities? By contrast with the high-risk HPVs, HPV11, a low-risk viral type, faces many of the same challenges as HPV16 in terms of infection, persistence, and replication, but tumor-promoting activities of its oncogenes have not been selected for by evolution.[5] Why?

What is the relative contribution of epithelium and stroma to TGFβ levels? Does differentially regulating of TGFβ in the different compartments enable the virus to take advantage of immunosuppressive effects while evading growth inhibition?

What is the contribution of growth factors such as TGFβ and EGFR to immune evasion by HPV?

What is the impact of HPV regulation of TGFβ on CAF differentiation in the stroma?

What is the role of factor sequestration in the ECM in regulating the HPV life cycle?

Are fibroblasts or endothelial cells from cervical lesions able to participate in innate immune signaling?

To what extent does the stroma contribute to anti-HPV innate immunity and how do innate immune pathways such as IFNs shape the subsequent development of an adaptive immune response?

Is it possible that the ability of IFN to inhibit HPV replication or gene expression may not be all bad for the virus? For example, IFN can inhibit adenovirus infection by repression of the E1A gene by increasing the association of Rb and p107 with the adenovirus genome, an effect that is reversed in the presence of E1A, an E7 homolog.[541] The consequence of repression by IFN is not abortive viral infection but rather persistent infection, which may promote long-term viral survival.[541] Whether IFN functions similarly to promote HPV persistence is not known, but given the importance of differentiation-dependent control of gene expression for HPV persistence and down-regulation of IFNκ upon differentiation, one may speculate that the virus may use IFNκ as a way to regulate its late genes. Further work will be needed to test this idea.

In summary, the activities and products of stromal cells can regulate the behavior of HPV during the benign life cycle and during carcinogenesis. Assembling a comprehensive picture of these noncell-autonomous interactions in HPV biology will occupy researchers for years to come.

REFERENCES

1. Moody CA, Laimins LA. Human papillomavirus oncoproteins: pathways to transformation. *Nat Rev Cancer*. 2010;10(8):550–560.
2. Parkin DM, Bray F. The burden of HPV-related cancers. *Vaccine*. 2006;24(suppl 3): S11–S25. chapter 2.
3. zur Hausen H. Papillomaviruses in the causation of human cancers—a brief historical account. *Virology*. 2009;384(2):260–265.
4. Torre LA, Bray F, Siegel RL, Ferlay J, Lortet-Tieulent J, Jemal A. Global cancer statistics, 2012. *CA Cancer J Clin*. 2015;65(2):87–108.
5. Klingelhutz AJ, Roman A. Cellular transformation by human papillomaviruses: lessons learned by comparing high- and low-risk viruses. *Virology*. 2012;424(2):77–98.
6. Markowitz LE, Dunne EF, Saraiya M, Lawson HW, Chesson H, Unger ER. Quadrivalent human papillomavirus vaccine: recommendations of the advisory committee on immunization practices (ACIP). *MMWR Recomm Rep*. 2007;56(RR-2):1–24.
7. Hanahan D, Weinberg RA. Hallmarks of cancer: the next generation. *Cell*. 2011;144(5):646–674.
8. Hawley-Nelson P, Vousden KH, Hubbert NL, Lowy DR, Schiller JT. HPV16 E6 and E7 proteins cooperate to immortalize human foreskin keratinocytes. *EMBO J*. 1989;8(12):3905–3910.
9. Bodily JM, Mehta KP, Cruz L, Meyers C, Laimins LA. The E7 open reading frame acts in cis and in trans to mediate differentiation-dependent activities in the human papillomavirus type 16 life cycle. *J Virol*. 2011;85(17):8852–8862.
10. Howie HL, Katzenellenbogen RA, Galloway DA. Papillomavirus E6 proteins. *Virology*. 2009;384(2):324–334.
11. Munger K, Werness BA, Dyson N, Phelps WC, Harlow E, Howley PM. Complex formation of human papillomavirus E7 proteins with the retinoblastoma tumor suppressor gene product. *EMBO J*. 1989;8(13):4099–4105.
12. Dimaio D, Petti LM. The E5 proteins. *Virology*. 2013;445(1–2):99–114.
13. Zwerschke W, Mazurek S, Massimi P, Banks L, Eigenbrodt E, Jansen-Durr P. Modulation of type M2 pyruvate kinase activity by the human papillomavirus type 16 E7 oncoprotein. *Proc Natl Acad Sci USA*. 1999;96(4):1291–1296.
14. Mazurek S, Zwerschke W, Jansen-Durr P, Eigenbrodt E. Effects of the human papilloma virus HPV-16 E7 oncoprotein on glycolysis and glutaminolysis: role of pyruvate kinase type M2 and the glycolytic-enzyme complex. *Biochem J*. 2001;356(pt 1):247–256.
15. Matsumoto K, Nakamura T. Hepatocyte growth factor and the Met system as a mediator of tumor-stromal interactions. *Int J Cancer*. 2006;119(3):477–483.
16. Steffan JJ, Coleman DT, Cardelli JA. The HGF-met signaling axis: emerging themes and targets of inhibition. *Curr Protein Pept Sci*. 2011;12(1):12–22.
17. Cirri P, Chiarugi P. Cancer-associated-fibroblasts and tumour cells: a diabolic liaison driving cancer progression. *Cancer Metastasis Rev*. 2012;31(1–2):195–208.
18. Folkman J, Kalluri R. Cancer without disease. *Nature*. 2004;427(6977):787.
19. Kankuri E, Cholujova D, Comajova M, Vaheri A, Bizik J. Induction of hepatocyte growth factor/scatter factor by fibroblast clustering directly promotes tumor cell invasiveness. *Cancer Res*. 2005;65(21):9914–9922.
20. Nees M, Geoghegan JM, Munson P, et al. Human papillomavirus type 16 E6 and E7 proteins inhibit differentiation-dependent expression of transforming growth factor-beta2 in cervical keratinocytes. *Cancer Res*. 2000;60(15):4289–4298.
21. Wan F, Miao X, Quraishi I, Kennedy V, Creek KE, Pirisi L. Gene expression changes during HPV-mediated carcinogenesis: a comparison between an in vitro cell model and cervical cancer. *Int J Cancer*. 2008;123(1):32–40.

22. Cortes-Malagon EM, Bonilla-Delgado J, Diaz-Chavez J, et al. Gene expression profile regulated by the HPV16 E7 oncoprotein and estradiol in cervical tissue. *Virology*. 2013;447(1–2):155–165.
23. Chen W, Li F, Mead L, et al. Human papillomavirus causes an angiogenic switch in keratinocytes which is sufficient to alter endothelial cell behavior. *Virology*. 2007;367(1):168–174.
24. Toussaint-Smith E, Donner DB, Roman A. Expression of human papillomavirus type 16 E6 and E7 oncoproteins in primary foreskin keratinocytes is sufficient to alter the expression of angiogenic factors. *Oncogene*. 2004;23(17):2988–2995.
25. Nakamura M, Bodily JM, Beglin M, Kyo S, Inoue M, Laimins LA. Hypoxia-specific stabilization of HIF-1alpha by human papillomaviruses. *Virology*. 2009;387(2): 442–448.
26. Frazer IH. Interaction of human papillomaviruses with the host immune system: a well evolved relationship. *Virology*. 2009;384(2):410–414.
27. Pickard A, Cichon AC, Barry A, et al. Inactivation of Rb in stromal fibroblasts promotes epithelial cell invasion. *EMBO J*. 2012;31(14):3092–3103.
28. Roberts JN, Buck CB, Thompson CD, et al. Genital transmission of HPV in a mouse model is potentiated by nonoxynol-9 and inhibited by carrageenan. *Nat Med*. 2007;13(7):857–861.
29. Culp TD, Budgeon LR, Christensen ND. Human papillomaviruses bind a basal extracellular matrix component secreted by keratinocytes which is distinct from a membrane-associated receptor. *Virology*. 2006;347(1):147–159.
30. Sapp M, Bienkowska-Haba M. Viral entry mechanisms: human papillomavirus and a long journey from extracellular matrix to the nucleus. *FEBS J*. 2009;276(24): 7206–7216.
31. Bodily J, Laimins LA. Persistence of human papillomavirus infection: keys to malignant progression. *Trends Microbiol*. 2011;19(1):33–39.
32. Doorbar J, Quint W, Banks L, et al. The biology and life-cycle of human papillomaviruses. *Vaccine*. 2012;30(suppl 5):F55–F70.
33. Fuchs E. Finding one's niche in the skin. *Cell Stem Cell*. 2009;4(6):499–502.
34. Sapp MJ. HPV virions hitchhike a ride on retromer complexes. *Proc Natl Acad Sci USA*. 2013;110(18):7116–7117.
35. DiGiuseppe S, Luszczek W, Keiffer TR, Bienkowska-Haba M, Guion LG, Sapp MJ. Incoming human papillomavirus type 16 genome resides in a vesicular compartment throughout mitosis. *Proc Natl Acad Sci USA*. 2016;113:6289–6294.
36. Trottier H, Mahmud S, Prado JC, et al. Type-specific duration of human papillomavirus infection: implications for human papillomavirus screening and vaccination. *J Infect Dis*. 2008;197(10):1436–1447.
37. Franco EL, Villa LL, Sobrinho JP, et al. Epidemiology of acquisition and clearance of cervical human papillomavirus infection in women from a high-risk area for cervical cancer. *J Infect Dis*. 1999;180(5):1415–1423.
38. Woodman CB, Collins S, Winter H, et al. Natural history of cervical human papillomavirus infection in young women: a longitudinal cohort study. *Lancet*. 2001;357(9271): 1831–1836.
39. Moscicki AB, Schiffman M, Burchell A, et al. Updating the natural history of human papillomavirus and anogenital cancers. *Vaccine*. 2012;30(suppl 5):F24–F33.
40. McBride AA, Oliveira JG, McPhillips MG. Partitioning viral genomes in mitosis: same idea, different targets. *Cell Cycle*. 2006;5(14):1499–1502.
41. Nardelli B, Zaritskaya L, Semenuk M, et al. Regulatory effect of IFN-kappa, a novel type I IFN, on cytokine production by cells of the innate immune system. *J Immunol*. 2002;169(9):4822–4830.

42. Gielen V, Schmitt D, Thivolet J. HLA class I antigen (heavy and light chain) expression by Langerhans cells and keratinocytes of the normal human epidermis: ultrastructural quantitation using immunogold labelling procedure. *Arch Dermatol Res.* 1988;280(3):131–136.
43. Caberg JH, Hubert PM, Begon DY, et al. Silencing of E7 oncogene restores functional E-cadherin expression in human papillomavirus 16-transformed keratinocytes. *Carcinogenesis*. 2008;29(7):1441–1447.
44. Stanley MA. Immune responses to human papilloma viruses. *Indian J Med Res.* 2009;130(3):266–276.
45. Moscicki AB, Schiffman M, Kjaer S, Villa LL. Updating the natural history of HPV and anogenital cancer. *Vaccine*. 2012;30(suppl 5):F24–F33. Chapter 5.
46. Schmeink CE, Melchers WJ, Siebers AG, Quint WG, Massuger LF, Bekkers RL. Human papillomavirus persistence in young unscreened women, a prospective cohort study. *PLoS One*. 2011;6(11):e27937.
47. Chen HC, Schiffman M, Lin CY, et al. Persistence of type-specific human papillomavirus infection and increased long-term risk of cervical cancer. *J Natl Cancer Inst.* 2011;103(18):1387–1396.
48. Park J, Sun D, Genest DR, Trivijitsilp P, Suh I, Crum CP. Coexistence of low and high grade squamous intraepithelial lesions of the cervix: morphologic progression or multiple papillomaviruses? *Gynecol Oncol.* 1998;70(3):386–391.
49. Woodman CB, Collins SI, Young LS. The natural history of cervical HPV infection: unresolved issues. *Nat Rev Cancer.* 2007;7(1):11–22.
50. Kristiansen E, Jenkins A, Holm R. Coexistence of episomal and integrated HPV16 DNA in squamous cell carcinoma of the cervix. *J Clin Pathol.* 1994;47(3):253–256.
51. Cheung JL, Cheung TH, Yu MY, Chan PK. Virological characteristics of cervical cancers carrying pure episomal form of HPV16 genome. *Gynecol Oncol.* 2013;131(2): 374–379.
52. Klaes R, Woerner SM, Ridder R, et al. Detection of high-risk cervical intraepithelial neoplasia and cervical cancer by amplification of transcripts derived from integrated papillomavirus oncogenes. *Cancer Res.* 1999;59(24):6132–6136.
53. Griffin H, Soneji Y, Van Baars R, et al. Stratification of HPV-induced cervical pathology using the virally encoded molecular marker E4 in combination with p16 or MCM. *Mod Pathol.* 2015;28(7):977–993.
54. Middleton K, Peh W, Southern S, et al. Organization of human papillomavirus productive cycle during neoplastic progression provides a basis for selection of diagnostic markers. *J Virol.* 2003;77(19):10186–10201.
55. Buckley CH, Butler EB, Fox H. Cervical intraepithelial neoplasia. *J Clin Pathol.* 1982;35(1):1–13.
56. Roman A, Munger K. The papillomavirus E7 proteins. *Virology*. 2013;445(1–2):138–168.
57. Maufort JP, Shai A, Pitot HC, Lambert PF. A role for HPV16 E5 in cervical carcinogenesis. *Cancer Res.* 2010;70(7):2924–2931.
58. Klymenko T, Hernandez-Lopez H, MacDonald AI, Bodily JM, Graham SV. Human papillomavirus E2 regulates SRSF3 (SRp20) to promote capsid protein expression in infected differentiated keratinocytes. *J Virol.* 2016;90(10):5047–5058.
59. Gauson EJ, Windle B, Donaldson MM, et al. Regulation of human genome expression and RNA splicing by human papillomavirus 16 E2 protein. *Virology*. 2014;468–470: 10–18.
60. Wilson R, Laimins LA. Differentiation of HPV-containing cells using organotypic "raft" culture or methylcellulose. *Methods Mol Med.* 2005;119:157–169.
61. Meyers C. Organotypic (raft) epithelial tissue culture system for the differentiation-dependent replication of papillomavirus. *Methods Cell Sci.* 1996;18:201–210.

62. Meyers C, Frattini MG, Hudson JB, Laimins LA. Biosynthesis of human papillomavirus from a continuous cell line upon epithelial differentiation. *Science*. 1992;257(5072): 971–973.
63. Balsitis SJ, Sage J, Duensing S, Munger K, Jacks T, Lambert PF. Recapitulation of the effects of the human papillomavirus type 16 E7 oncogene on mouse epithelium by somatic Rb deletion and detection of pRb-independent effects of E7 in vivo. *Mol Cell Biol*. 2003;23(24):9094–9103.
64. Herber R, Liem A, Pitot H, Lambert PF. Squamous epithelial hyperplasia and carcinoma in mice transgenic for the human papillomavirus type 16 E7 oncogene. *J Virol*. 1996;70(3):1873–1881.
65. Arbeit JM, Howley PM, Hanahan D. Chronic estrogen-induced cervical and vaginal squamous carcinogenesis in human papillomavirus type 16 transgenic mice. *Proc Natl Acad Sci USA*. 1996;93(7):2930–2935.
66. Arbeit JM, Munger K, Howley PM, Hanahan D. Progressive squamous epithelial neoplasia in K14-human papillomavirus type 16 transgenic mice. *J Virol*. 1994;68(7): 4358–4368.
67. Brandsma JL. The cottontail rabbit papillomavirus model of high-risk HPV-induced disease. *Methods Mol Med*. 2005;119:217–235.
68. Peh WL, Middleton K, Christensen N, et al. Life cycle heterogeneity in animal models of human papillomavirus-associated disease. *J Virol*. 2002;76(20):10401–10416.
69. Jain S, Moore RA, Anderson DM, Gough GW, Stanley MA. Cell-mediated immune responses to COPV early proteins. *Virology*. 2006;356(1–2):23–34.
70. Wira CR, Rodriguez-Garcia M, Patel MV. The role of sex hormones in immune protection of the female reproductive tract. *Nat Rev Immunol*. 2015;15(4):217–230.
71. Howat WJ, Holmes JA, Holgate ST, Lackie PM. Basement membrane pores in human bronchial epithelium: a conduit for infiltrating cells? *Am J Pathol*. 2001;158(2):673–680.
72. Alitalo K, Kuismanen E, Myllyla R, Kiistala U, Asko-Seljavaara S, Vaheri A. Extracellular matrix proteins of human epidermal keratinocytes and feeder 3T3 cells. *J Cell Biol*. 1982;94(3):497–505.
73. Smola H, Stark HJ, Thiekotter G, Mirancea N, Krieg T, Fusenig NE. Dynamics of basement membrane formation by keratinocyte-fibroblast interactions in organotypic skin culture. *Exp Cell Res*. 1998;239(2):399–410.
74. Marionnet C, Pierrard C, Vioux-Chagnoleau C, Sok J, Asselineau D, Bernerd F. Interactions between fibroblasts and keratinocytes in morphogenesis of dermal epidermal junction in a model of reconstructed skin. *J Invest Dermatol*. 2006;126(5): 971–979.
75. Pickup M, Novitskiy S, Moses HL. The roles of TGFbeta in the tumour microenvironment. *Nat Rev Cancer*. 2013;13(11):788–799.
76. Heldin CH, Landstrom M, Moustakas A. Mechanism of TGF-beta signaling to growth arrest, apoptosis, and epithelial-mesenchymal transition. *Curr Opin Cell Biol*. 2009;21(2):166–176.
77. Massague J. TGFbeta signalling in context. *Nat Rev Mol Cell Biol*. 2012;13(10): 616–630.
78. Travis MA, Sheppard D. TGF-beta activation and function in immunity. *Annu Rev Immunol*. 2014;32:51–82.
79. Todorovic V, Rifkin DB. LTBPs, more than just an escort service. *J Cell Biochem*. 2012;113(2):410–418.
80. Li Q, Huang W, Zhou X. Expression of CD34, alpha-smooth muscle actin and transforming growth factor-beta1 in squamous intraepithelial lesions and squamous cell carcinoma of the cervix. *J Int Med Res*. 2009;37(2):446–454.
81. Li MO, Flavell RA. TGF-beta: a master of all T cell trades. *Cell*. 2008;134(3):392–404.
82. Savage-Dunn C. TGF-beta signaling. *WormBook*. 2005;1–12.

83. Matsumoto K, Hashimoto K, Hashiro M, Yoshimasa H, Yoshikawa K. Modulation of growth and differentiation in normal human keratinocytes by transforming growth factor-beta. *J Cell Physiol.* 1990;145(1):95–101.
84. Ozbun MA, Meyers C. Transforming growth factor beta1 induces differentiation in human papillomavirus-positive keratinocytes. *J Virol.* 1996;70(8):5437–5446.
85. Creek KE, Geslani G, Batova A, Pirisi L. Progressive loss of sensitivity to growth control by retinoic acid and transforming growth factor-beta at late stages of human papillomavirus type 16-initiated transformation of human keratinocytes. *Adv Exp Med Biol.* 1995;375:117–135.
86. Shier MK, Neely EB, Ward MG, Meyers C, Howett MK. Transforming growth factor beta 1 (TGF beta 1) down-regulates expression and function of proliferation-inducing molecules in HPV-transformed cells. *Anticancer Res.* 1999;19(6B):4977–4982.
87. Shier MK, Neely EB, Ward MG, et al. Correlation of TGF beta 1 overexpression with down-regulation of proliferation-inducing molecules in HPV-11 transformed human tissue xenografts. *Anticancer Res.* 1999;19(6B):4969–4976.
88. Kim JW, Kim HS, Kim IK, et al. Transforming growth factor-beta 1 induces apoptosis through down-regulation of c-myc gene and overexpression of p27Kip1 protein in cervical carcinoma. *Gynecol Oncol.* 1998;69(3):230–236.
89. Pietenpol JA, Stein RW, Moran E, et al. TGF-beta 1 inhibition of c-myc transcription and growth in keratinocytes is abrogated by viral transforming proteins with pRB binding domains. *Cell.* 1990;61(5):777–785.
90. Derynck R, Akhurst RJ, Balmain A. TGF-beta signaling in tumor suppression and cancer progression. *Nat Genet.* 2001;29(2):117–129.
91. Drabsch Y, ten Dijke P. TGF-beta signaling in breast cancer cell invasion and bone metastasis. *J Mammary Gland Biol Neoplasia.* 2011;16(2):97–108.
92. Polanska UM, Orimo A. Carcinoma-associated fibroblasts: non-neoplastic tumour-promoting mesenchymal cells. *J Cell Physiol.* 2013;228(8):1651–1657.
93. Helleman J, Smid M, Jansen MP, van der Burg ME, Berns EM. Pathway analysis of gene lists associated with platinum-based chemotherapy resistance in ovarian cancer: the big picture. *Gynecol Oncol.* 2010;117(2):170–176.
94. Bhowmick NA, Moses HL. Tumor-stroma interactions. *Curr Opin Genet Dev.* 2005;15(1):97–101.
95. Flavell RA, Sanjabi S, Wrzesinski SH, Licona-Limon P. The polarization of immune cells in the tumour environment by TGFbeta. *Nat Rev Immunol.* 2010;10(8):554–567.
96. Alcocer-Gonzalez JM, Berumen J, Tamez-Guerra R, et al. In vivo expression of immunosuppressive cytokines in human papillomavirus-transformed cervical cancer cells. *Viral Immunol.* 2006;19(3):481–491.
97. Giannini SL, Al-Saleh W, Piron H, et al. Cytokine expression in squamous intra-epithelial lesions of the uterine cervix: implications for the generation of local immunosuppression. *Clin Exp Immunol.* 1998;113(2):183–189.
98. Baldwin A, Pirisi L, Creek KE. NFI-Ski interactions mediate transforming growth factor beta modulation of human papillomavirus type 16 early gene expression. *J Virol.* 2004;78(8):3953–3964.
99. Donalisio M, Cornaglia M, Landolfo S, Lembo D. TGF-beta1 and IL-4 downregulate human papillomavirus-16 oncogene expression but have differential effects on the malignant phenotype of cervical carcinoma cells. *Virus Res.* 2008;132(1–2):253–256.
100. Woodworth CD, Notario V, DiPaolo JA. Transforming growth factors beta 1 and 2 transcriptionally regulate human papillomavirus (HPV) type 16 early gene expression in HPV-immortalized human genital epithelial cells. *J Virol.* 1990;64(10):4767–4775.
101. Lace MJ, Anson JR, Haugen TH, Dierdorff JM, Turek LP. Interferon treatment of human keratinocytes harboring extrachromosomal, persistent HPV-16 plasmid genomes induces de novo viral integration. *Carcinogenesis.* 2015;36(1):151–159.

102. Pett MR, Herdman MT, Palmer RD, et al. Selection of cervical keratinocytes containing integrated HPV16 associates with episome loss and an endogenous antiviral response. *Proc Natl Acad Sci USA*. 2006;103(10):3822–3827.
103. Moses HL. TGF-beta regulation of epithelial cell proliferation. *Mol Reprod Dev*. 1992;32(2):179–184.
104. Habig M, Smola H, Dole VS, Derynck R, Pfister H, Smola-Hess S. E7 proteins from high- and low-risk human papillomaviruses bind to TGF-beta-regulated Smad proteins and inhibit their transcriptional activity. *Arch Virol*. 2006;151(10):1961–1972.
105. Murvai M, Borbely AA, Konya J, Gergely L, Veress G. Effect of human papillomavirus type 16 E6 and E7 oncogenes on the activity of the transforming growth factor-beta2 (TGF-beta2) promoter. *Arch Virol*. 2004;149(12):2379–2392.
106. Peralta-Zaragoza O, Bermudez-Morales V, Gutierrez-Xicotencatl L, Alcocer-Gonzalez J, Recillas-Targa F, Madrid-Marina V. E6 and E7 oncoproteins from human papillomavirus type 16 induce activation of human transforming growth factor beta1 promoter throughout Sp1 recognition sequence. *Viral Immunol*. 2006;19(3): 468–480.
107. Hypes MK, Pirisi L, Creek KE. Mechanisms of decreased expression of transforming growth factor-beta receptor type I at late stages of HPV16-mediated transformation. *Cancer Lett*. 2009;282(2):177–186.
108. Mi Y, Borger DR, Fernandes PR, Pirisi L, Creek KE. Loss of transforming growth factor-beta (TGF-beta) receptor type I mediates TGF-beta resistance in human papillomavirus type 16-transformed human keratinocytes at late stages of in vitro progression. *Virology*. 2000;270(2):408–416.
109. Popoca-Cuaya M, Diaz-Chavez J, Hernandez-Monge J, Alvarez-Rios E, Lambert PF, Gariglio P. The HPV16 E6 oncoprotein and UVB irradiation inhibit the tumor suppressor TGFbeta pathway in the epidermis of the K14E6 transgenic mouse. *Exp Dermatol*. 2015;24(6):430–435.
110. Lee DK, Kim BC, Kim IY, Cho EA, Satterwhite DJ, Kim SJ. The human papilloma virus E7 oncoprotein inhibits transforming growth factor-beta signaling by blocking binding of the Smad complex to its target sequence. *J Biol Chem*. 2002;277(41): 38557–38564.
111. French D, Belleudi F, Mauro MV, et al. Expression of HPV16 E5 down-modulates the TGFbeta signaling pathway. *Mol Cancer*. 2013;12:38.
112. Kowli S, Velidandla R, Creek KE, Pirisi L. TGF-beta regulation of gene expression at early and late stages of HPV16-mediated transformation of human keratinocytes. *Virology*. 2013;447(1–2):63–73.
113. De Geest K, Bergman CA, Turyk ME, Frank BS, Wilbanks GD. Differential response of cervical intraepithelial and cervical carcinoma cell lines to transforming growth factor-beta 1. *Gynecol Oncol*. 1994;55(3 pt 1):376–385.
114. Xu H, Zhang Y, Altomare D, et al. Six1 promotes epithelial-mesenchymal transition and malignant conversion in human papillomavirus type 16-immortalized human keratinocytes. *Carcinogenesis*. 2014;35(6):1379–1388.
115. Woodworth CD, Chung J, McMullin E, Plowman GD, Simpson S, Iglesias M. Transforming growth factor beta 1 supports autonomous growth of human papillomavirus-immortalized cervical keratinocytes under conditions promoting squamous differentiation. *Cell Growth Differ*. 1996;7(6):811–820.
116. Normanno N, De Luca A, Bianco C, et al. Epidermal growth factor receptor (EGFR) signaling in cancer. *Gene*. 2006;366(1):2–16.
117. Haase I, Evans R, Pofahl R, Watt FM. Regulation of keratinocyte shape, migration and wound epithelialization by IGF-1- and EGF-dependent signalling pathways. *J Cell Sci*. 2003;116(pt 15):3227–3238.

118. Akerman GS, Tolleson WH, Brown KL, et al. Human papillomavirus type 16 E6 and E7 cooperate to increase epidermal growth factor receptor (EGFR) mRNA levels, overcoming mechanisms by which excessive EGFR signaling shortens the life span of normal human keratinocytes. *Cancer Res*. 2001;61(9):3837–3843.
119. Quesnelle KM, Boehm AL, Grandis JR. STAT-mediated EGFR signaling in cancer. *J Cell Biochem*. 2007;102(2):311–319.
120. Murata T, Mizushima H, Chinen I, et al. HB-EGF and PDGF mediate reciprocal interactions of carcinoma cells with cancer-associated fibroblasts to support progression of uterine cervical cancers. *Cancer Res*. 2011;71(21):6633–6642.
121. Yamasaki K, Toriu N, Hanakawa Y, et al. Keratinocyte growth inhibition by high-dose epidermal growth factor is mediated by transforming growth factor beta autoinduction: a negative feedback mechanism for keratinocyte growth. *J Invest Dermatol*. 2003;120(6):1030–1037.
122. Chong T, Apt D, Gloss B, Isa M, Bernard HU. The enhancer of human papillomavirus type 16: binding sites for the ubiquitous transcription factors oct-1, NFA, TEF-2, NF1, and AP-1 participate in epithelial cell-specific transcription. *J Virol*. 1991;65(11): 5933–5943.
123. Kyo S, Klumpp DJ, Inoue M, Kanaya T, Laimins LA. Expression of AP1 during cellular differentiation determines human papillomavirus E6/E7 expression in stratified epithelial cells. *J Gen Virol*. 1997;78(pt 2):401–411.
124. Mack DH, Laimins LA. A keratinocyte-specific transcription factor, KRF-1, interacts with AP-1 to activate expression of human papillomavirus type 18 in squamous epithelial cells. *Proc Natl Acad Sci USA*. 1991;88(20):9102–9106.
125. O'Connor MJ, Tan SH, Tan CH, Bernard HU. YY1 represses human papillomavirus type 16 transcription by quenching AP-1 activity. *J Virol*. 1996;70(10):6529–6539.
126. Zhao W, Chow LT, Broker TR. Transcription activities of human papillomavirus type 11 E6 promoter-proximal elements in raft and submerged cultures of foreskin keratinocytes. *J Virol*. 1997;71(11):8832–8840.
127. Woodworth CD, Diefendorf LP, Jette DF, et al. Inhibition of the epidermal growth factor receptor by erlotinib prevents immortalization of human cervical cells by human papillomavirus type 16. *Virology*. 2011;421(1):19–27.
128. Straight SW, Hinkle PM, Jewers RJ, McCance DJ. The E5 oncoprotein of human papillomavirus type 16 transforms fibroblasts and effects the downregulation of the epidermal growth factor receptor in keratinocytes. *J Virol*. 1993;67(8):4521–4532.
129. Straight SW, Herman B, McCance DJ. The E5 oncoprotein of human papillomavirus type 16 inhibits the acidification of endosomes in human keratinocytes. *J Virol*. 1995;69(5):3185–3192.
130. Disbrow GL, Hanover JA, Schlegel R. Endoplasmic reticulum-localized human papillomavirus type 16 E5 protein alters endosomal pH but not trans-Golgi pH. *J Virol*. 2005;79(9):5839–5846.
131. Adam JL, Briggs MW, McCance DJ. A mutagenic analysis of the E5 protein of human papillomavirus type 16 reveals that E5 binding to the vacuolar H+-ATPase is not sufficient for biological activity, using mammalian and yeast expression systems. *Virology*. 2000;272(2):315–325.
132. Rodriguez MI, Finbow ME, Alonso A. Binding of human papillomavirus 16 E5 to the 16 kDa subunit c (proteolipid) of the vacuolar H+-ATPase can be dissociated from the E5-mediated epidermal growth factor receptor overactivation. *Oncogene*. 2000;19(33): 3727–3732.
133. Hwang ES, Nottoli T, Dimaio D. The HPV16 E5 protein: expression, detection, and stable complex formation with transmembrane proteins in COS cells. *Virology*. 1995;211(1):227–233.

134. Venuti A, Paolini F, Nasir L, et al. Papillomavirus E5: the smallest oncoprotein with many functions. *Mol Cancer*. 2011;10:140.
135. Crusius K, Auvinen E, Alonso A. Enhancement of EGF- and PMA-mediated MAP kinase activation in cells expressing the human papillomavirus type 16 E5 protein. *Oncogene*. 1997;15(12):1437–1444.
136. Pedroza-Saavedra A, Lam EW, Esquivel-Guadarrama F, Gutierrez-Xicotencatl L. The human papillomavirus type 16 E5 oncoprotein synergizes with EGF-receptor signaling to enhance cell cycle progression and the down-regulation of p27(Kip1). *Virology*. 2010;400(1):44–52.
137. Genther Williams SM, Disbrow GL, Schlegel R, Lee D, Threadgill DW, Lambert PF. Requirement of epidermal growth factor receptor for hyperplasia induced by E5, a high-risk human papillomavirus oncogene. *Cancer Res*. 2005;65(15):6534–6542.
138. Genther SM, Sterling S, Duensing S, Munger K, Sattler C, Lambert PF. Quantitative role of the human papillomavirus type 16 E5 gene during the productive stage of the viral life cycle. *J Virol*. 2003;77(5):2832–2842.
139. Fehrmann F, Klumpp DJ, Laimins LA. Human papillomavirus type 31 E5 protein supports cell cycle progression and activates late viral functions upon epithelial differentiation. *J Virol*. 2003;77(5):2819–2831.
140. Chen SL, Lin ST, Tsai TC, Hsiao WC, Tsao YP. ErbB4 (JM-b/CYT-1)-induced expression and phosphorylation of c-Jun is abrogated by human papillomavirus type 16 E5 protein. *Oncogene*. 2007;26(1):42–53.
141. da Rosa MR, Falcao AS, Fuzii HT, et al. EGFR signaling downstream of EGF regulates migration, invasion, and MMP secretion of immortalized cells derived from human ameloblastoma. *Tumour Biol*. 2014;35(11):11107–11120.
142. Sizemore N, Choo CK, Eckert RL, Rorke EA. Transcriptional regulation of the EGF receptor promoter by HPV16 and retinoic acid in human ectocervical epithelial cells. *Exp Cell Res*. 1998;244(1):349–356.
143. Spangle JM, Munger K. The HPV16 E6 oncoprotein causes prolonged receptor protein tyrosine kinase signaling and enhances internalization of phosphorylated receptor species. *PLoS Pathog*. 2013;9(3):e1003237.
144. Walker F, Kermorgant S, Darai E, et al. Hepatocyte growth factor and c-Met in cervical intraepithelial neoplasia: overexpression of proteins associated with oncogenic human papillomavirus and human immunodeficiency virus. *Clin Cancer Res*. 2003;9(1): 273–284.
145. Muller AK, Meyer M, Werner S. The roles of receptor tyrosine kinases and their ligands in the wound repair process. *Semin Cell Dev Biol*. 2012;23(9):963–970.
146. Shimabukuro K, Ichinose S, Koike R, et al. Hepatocyte growth factor/scatter factor is implicated in the mode of stromal invasion of uterine squamous cervical cancer. *Gynecol Oncol*. 2001;83(2):205–215.
147. Baykal C, Ayhan A, Al A, Yuce K. Overexpression of the c-Met/HGF receptor and its prognostic significance in uterine cervix carcinomas. *Gynecol Oncol*. 2003;88(2): 123–129.
148. Hwang CI, Matoso A, Corney DC, et al. Wild-type p53 controls cell motility and invasion by dual regulation of MET expression. *Proc Natl Acad Sci USA*. 2011;108(34): 14240–14245.
149. Gherardi E, Birchmeier W, Birchmeier C, Vande Woude G. Targeting MET in cancer: rationale and progress. *Nat Rev Cancer*. 2012;12(2):89–103.
150. Boccaccio C, Comoglio PM. Invasive growth: a MET-driven genetic programme for cancer and stem cells. *Nat Rev Cancer*. 2006;6(8):637–645.
151. Qian G, Wang D, Magliocca KR, et al. Human papillomavirus oncoprotein E6 upregulates c-Met through p53 downregulation. *Eur J Cancer*. 2016;65:21–32.

152. Hellner K, Mar J, Fang F, Quackenbush J, Munger K. HPV16 E7 oncogene expression in normal human epithelial cells causes molecular changes indicative of an epithelial to mesenchymal transition. *Virology*. 2009;391(1):57–63.
153. Jung YS, Kato I, Kim HR. A novel function of HPV16–E6/E7 in epithelial-mesenchymal transition. *Biochem Biophys Res Commun*. 2013;435:339–344.
154. Lee MY, Chou CY, Tang MJ, Shen MR. Epithelial-mesenchymal transition in cervical cancer: correlation with tumor progression, epidermal growth factor receptor overexpression, and snail up-regulation. *Clin Cancer Res*. 2008;14(15): 4743–4750.
155. Myong NH. Loss of E-cadherin and acquisition of vimentin in epithelial-mesenchymal transition are notable indicators of uterine cervix cancer progression. *Korean J Pathol*. 2012;46(4):341–348.
156. Geiger T, Sabanay H, Kravchenko-Balasha N, Geiger B, Levitzki A. Anomalous features of EMT during keratinocyte transformation. *PLoS One*. 2008;3(2):e1574.
157. Cheng YM, Chou CY, Hsu YC, Chen MJ, Wing LY. The role of human papillomavirus type 16 E6/E7 oncoproteins in cervical epithelial-mesenchymal transition and carcinogenesis. *Oncol Lett*. 2012;3(3):667–671.
158. Duffy CL, Phillips SL, Klingelhutz AJ. Microarray analysis identifies differentiation-associated genes regulated by human papillomavirus type 16 E6. *Virology*. 2003;314(1): 196–205.
159. D'Costa ZJ, Jolly C, Androphy EJ, Mercer A, Matthews CM, Hibma MH. Transcriptional repression of E-cadherin by human papillomavirus type 16 E6. *PLoS One*. 2012;7(11):e48954.
160. Latorre IJ, Roh MH, Frese KK, Weiss RS, Margolis B, Javier RT. Viral oncoprotein-induced mislocalization of select PDZ proteins disrupts tight junctions and causes polarity defects in epithelial cells. *J Cell Sci*. 2005;118(pt 18):4283–4293.
161. Pim D, Thomas M, Banks L. Chimaeric HPV E6 proteins allow dissection of the proteolytic pathways regulating different E6 cellular target proteins. *Oncogene*. 2002;21(53):8140–8148.
162. Savagner P, Valles AM, Jouanneau J, Yamada KM, Thiery JP. Alternative splicing in fibroblast growth factor receptor 2 is associated with induced epithelial-mesenchymal transition in rat bladder carcinoma cells. *Mol Biol Cell*. 1994;5(8):851–862.
163. Blunt AG, Lawshe A, Cunningham ML, Seto ML, Ornitz DM, MacArthur CA. Overlapping expression and redundant activation of mesenchymal fibroblast growth factor (FGF) receptors by alternatively spliced FGF-8 ligands. *J Biol Chem*. 1997;272(6): 3733–3738.
164. Purpura V, Belleudi F, Caputo S, Torrisi MR. HPV16 E5 and KGFR/FGFR2b interplay in differentiating epithelial cells. *Oncotarget*. 2013;4(2):192–205.
165. Belleudi F, Leone L, Purpura V, Cannella F, Scrofani C, Torrisi MR. HPV16 E5 affects the KGFR/FGFR2b-mediated epithelial growth through alteration of the receptor expression, signaling and endocytic traffic. *Oncogene*. 2011;30(50):4963–4976.
166. Cheng YM, Chou CY, Hsu YC, Chen MJ. Influence of HPV16 E6/7 on the expression of FGF2 and FGFR type B in cervical carcinogenesis. *Reprod Sci*. 2012;19(6): 580–586.
167. Pickard A, Cichon AC, Menges C, Patel D, McCance DJ. Regulation of epithelial differentiation and proliferation by the stroma: a role for the retinoblastoma protein. *J Invest Dermatol*. 2012;132(12):2691–2699.
168. Kalluri R, Zeisberg M. Fibroblasts in cancer. *Nat Rev Cancer*. 2006;6(5):392–401.
169. Schumacher M, Schuster C, Rogon ZM, et al. Efficient keratinocyte differentiation strictly depends on JNK-induced soluble factors in fibroblasts. *J Invest Dermatol*. 2013;134:1332–1341.

170. McBride AA, Dlugosz A, Baker CC. Production of infectious bovine papillomavirus from cloned viral DNA by using an organotypic raft/xenograft technique. *Proc Natl Acad Sci USA*. 2000;97(10):5534–5539.
171. Smola H, Thiekotter G, Fusenig NE. Mutual induction of growth factor gene expression by epidermal-dermal cell interaction. *J Cell Biol*. 1993;122(2):417–429.
172. Rheinwald JG, Green H. Serial cultivation of strains of human epidermal keratinocytes: the formation of keratinizing colonies from single cells. *Cell*. 1975;6(3):331–343.
173. Yaeger PC, Stiles CD, Rollins BJ. Human keratinocyte growth-promoting activity on the surface of fibroblasts. *J Cell Physiol*. 1991;149(1):110–116.
174. Limat A, Hunziker T, Boillat C, Bayreuther K, Noser F. Post-mitotic human dermal fibroblasts efficiently support the growth of human follicular keratinocytes. *J Invest Dermatol*. 1989;92(5):758–762.
175. Shipley GD, Pittelkow MR. Control of growth and differentiation in vitro of human keratinocytes cultured in serum-free medium. *Arch Dermatol*. 1987;123(11): 1541a–1544a.
176. Fu B, Quintero J, Baker CC. Keratinocyte growth conditions modulate telomerase expression, senescence, and immortalization by human papillomavirus type 16 E6 and E7 oncogenes. *Cancer Res*. 2003;63(22):7815–7824.
177. Prunieras M, Regnier M, Woodley D. Methods for cultivation of keratinocytes with an air-liquid interface. *J Invest Dermatol*. 1983;81(1 suppl):28s–33s.
178. El Ghalbzouri A, Ponec M. Diffusible factors released by fibroblasts support epidermal morphogenesis and deposition of basement membrane components. *Wound Repair Regen*. 2004;12(3):359–367.
179. el-Ghalbzouri A, Gibbs S, Lamme E, Van Blitterswijk CA, Ponec M. Effect of fibroblasts on epidermal regeneration. *Br J Dermatol*. 2002;147(2):230–243.
180. de Giorgi V, Sestini S, Massi D, Ghersetich I, Lotti T. Keratinocyte growth factor receptors. *Dermatol Clin*. 2007;25(4):477–485. vii.
181. Florin L, Maas-Szabowski N, Werner S, Szabowski A, Angel P. Increased keratinocyte proliferation by JUN-dependent expression of PTN and SDF-1 in fibroblasts. *J Cell Sci*. 2005;118(pt 9):1981–1989.
182. Guo S, Dipietro LA. Factors affecting wound healing. *J Dent Res*. 2010;89(3):219–229.
183. Raja SK, Garcia MS, Isseroff RR. Wound re-epithelialization: modulating keratinocyte migration in wound healing. *Front Biosci*. 2007;12:2849–2868.
184. Trusolino L, Bertotti A, Comoglio PM. MET signalling: principles and functions in development, organ regeneration and cancer. *Nat Rev Mol Cell Biol*. 2010;11(12): 834–848.
185. Shamis Y, Hewitt KJ, Carlson MW, et al. Fibroblasts derived from human embryonic stem cells direct development and repair of 3D human skin equivalents. *Stem Cell Res Ther*. 2011;2(1):10.
186. Chmielowiec J, Borowiak M, Morkel M, et al. c-Met is essential for wound healing in the skin. *J Cell Biol*. 2007;177(1):151–162.
187. Suter MM, Schulze K, Bergman W, Welle M, Roosje P, Muller EJ. The keratinocyte in epidermal renewal and defence. *Vet Dermatol*. 2009;20(5–6):515–532.
188. Lau K, Paus R, Tiede S, Day P, Bayat A. Exploring the role of stem cells in cutaneous wound healing. *Exp Dermatol*. 2009;18(11):921–933.
189. Wang Z, Wang Y, Farhangfar F, Zimmer M, Zhang Y. Enhanced keratinocyte proliferation and migration in co-culture with fibroblasts. *PLoS One*. 2012;7(7):e40951.
190. Timmons B, Akins M, Mahendroo M. Cervical remodeling during pregnancy and parturition. *Trends Endocrinol Metab*. 2010;21(6):353–361.
191. Lopes-Bastos BM, Jiang WG, Cai J. Tumour-endothelial cell communications: important and indispensable mediators of tumour angiogenesis. *Anticancer Res*. 2016;36(3):1119–1126.

192. Gray AL, Stephens CA, Bigelow RL, Coleman DT, Cardelli JA. The polyphenols (-)-epigallocatechin-3-gallate and luteolin synergistically inhibit TGF-beta-induced myofibroblast phenotypes through RhoA and ERK inhibition. *PLoS One*. 2014;9(10): e109208.
193. Barth PJ, Ramaswamy A, Moll R. CD34(+) fibrocytes in normal cervical stroma, cervical intraepithelial neoplasia III, and invasive squamous cell carcinoma of the cervix uteri. *Virchows Arch*. 2002;441(6):564–568.
194. Turner MA, Darragh T, Palefsky JM. Epithelial-stromal interactions modulating penetration of matrigel membranes by HPV 16-immortalized keratinocytes. *J Invest Dermatol*. 1997;109(5):619–625.
195. Drake LE, Macleod KF. Tumour suppressor gene function in carcinoma-associated fibroblasts: from tumour cells via EMT and back again? *J Pathol*. 2014;232(3): 283–288.
196. den Boon JA, Pyeon D, Wang SS, et al. Molecular transitions from papillomavirus infection to cervical precancer and cancer: role of stromal estrogen receptor signaling. *Proc Natl Acad Sci USA*. 2015;112(25):E3255–E3264.
197. Nagura M, Matsumura N, Baba T, et al. Invasion of uterine cervical squamous cell carcinoma cells is facilitated by locoregional interaction with cancer-associated fibroblasts via activating transforming growth factor-beta. *Gynecol Oncol*. 2015;136(1): 104–111.
198. Ren C, Cheng X, Lu B, Yang G. Activation of interleukin-6/signal transducer and activator of transcription 3 by human papillomavirus early proteins 6 induces fibroblast senescence to promote cervical tumourigenesis through autocrine and paracrine pathways in tumour microenvironment. *Eur J Cancer*. 2013;49(18):3889–3899.
199. Chung SH, Shin MK, Korach KS, Lambert PF. Requirement for stromal estrogen receptor alpha in cervical neoplasia. *Horm Cancer*. 2013;4(1):50–59.
200. Goldschmidt MH, Kennedy JS, Kennedy DR, et al. Severe papillomavirus infection progressing to metastatic squamous cell carcinoma in bone marrow-transplanted X-linked SCID dogs. *J Virol*. 2006;80(13):6621–6628.
201. Lebre MC, van der Aar AM, van Baarsen L, et al. Human keratinocytes express functional toll-like receptor 3, 4, 5, and 9. *J Invest Dermatol*. 2007;127(2):331–341.
202. Feldmeyer L, Keller M, Niklaus G, Hohl D, Werner S, Beer HD. The inflammasome mediates UVB-induced activation and secretion of interleukin-1beta by keratinocytes. *Curr Biol*. 2007;17(13):1140–1145.
203. Wang B, Amerio P, Sauder DN. Role of cytokines in epidermal Langerhans cell migration. *J Leukoc Biol*. 1999;66(1):33–39.
204. Williams IR, Kupper TS. Immunity at the surface: homeostatic mechanisms of the skin immune system. *Life Sci*. 1996;58(18):1485–1507.
205. Albanesi C, Cavani A, Girolomoni G. Interferon-gamma-stimulated human keratinocytes express the genes necessary for the production of peptide-loaded MHC class II molecules. *J Invest Dermatol*. 1998;110(2):138–142.
206. Black AP, Ardern-Jones MR, Kasprowicz V, et al. Human keratinocyte induction of rapid effector function in antigen-specific memory CD4+ and CD8+ T cells. *Eur J Immunol*. 2007;37(6):1485–1493.
207. Nickoloff BJ, Basham TY, Merigan TC, Morhenn VB. Antiproliferative effects of recombinant alpha- and gamma-interferons on cultured human keratinocytes. *Lab Invest*. 1984;51(6):697–701.
208. Scott M, Nakagawa M, Moscicki AB. Cell-mediated immune response to human papillomavirus infection. *Clin Diagn Lab Immunol*. 2001;8(2):209–220.
209. Castle PE, Hillier SL, Rabe LK, et al. An association of cervical inflammation with high-grade cervical neoplasia in women infected with oncogenic human papillomavirus (HPV). *Cancer Epidemiol Biomarkers Prev*. 2001;10(10):1021–1027.

210. Kovacic MB, Katki HA, Kreimer AR, Sherman ME. Epidemiologic analysis of histologic cervical inflammation: relationship to human papillomavirus infections. *Hum Pathol.* 2008;39(7):1088–1095.
211. Stone SC, Rossetti RA, Lima AM, Lepique AP. HPV associated tumor cells control tumor microenvironment and leukocytosis in experimental models. *Immun Inflamm Dis.* 2014;2(2):63–75.
212. Tran le S, Bergot AS, Mattarollo SR, Mittal D, Frazer IH. Human papillomavirus e7 oncoprotein transgenic skin develops an enhanced inflammatory response to 2,4-dinitrochlorobenzene by an arginase-1-dependent mechanism. *J Invest Dermatol.* 2014;134(9):2438–2446.
213. Bergot AS, Ford N, Leggatt GR, Wells JW, Frazer IH, Grimbaldeston MA. HPV16-E7 expression in squamous epithelium creates a local immune suppressive environment via CCL2- and CCL5-mediated recruitment of mast cells. *PLoS Pathog.* 2014;10(10):e1004466.
214. Zanotta N, Tornesello ML, Annunziata C, Stellato G, Buonaguro FM, Comar M. Candidate soluble immune mediators in young women with high-risk human papillomavirus infection: high expression of chemokines promoting angiogenesis and cell proliferation. *PLoS One.* 2016;11(3):e0151851.
215. Tummers B, Goedemans R, Jha V, et al. CD40-mediated amplification of local immunity by epithelial cells is impaired by HPV. *J Invest Dermatol.* 2014;134(12):2918–2927.
216. Meylan E, Tschopp J. Toll-like receptors and RNA helicases: two parallel ways to trigger antiviral responses. *Mol Cell.* 2006;22(5):561–569.
217. Vilaysane A, Muruve DA. The innate immune response to DNA. *Semin Immunol.* 2009;21(4):208–214.
218. Daud II, Scott ME, Ma Y, Shiboski S, Farhat S, Moscicki AB. Association between toll-like receptor expression and human papillomavirus type 16 persistence. *Int J Cancer.* 2011;128(4):879–886.
219. Hasan UA, Bates E, Takeshita F, et al. TLR9 expression and function is abolished by the cervical cancer-associated human papillomavirus type 16. *J Immunol.* 2007;178(5):3186–3197.
220. Schneider WM, Chevillotte MD, Rice CM. Interferon-stimulated genes: a complex web of host defenses. *Annu Rev Immunol.* 2014;32:513–545.
221. Cheon H, Borden EC, Stark GR. Interferons and their stimulated genes in the tumor microenvironment. *Semin Oncol.* 2014;41(2):156–173.
222. Gilmore TD. Introduction to NF-kappaB: players, pathways, perspectives. *Oncogene.* 2006;25(51):6680–6684.
223. Pahl HL. Activators and target genes of Rel/NF-kappaB transcription factors. *Oncogene.* 1999;18(49):6853–6866.
224. Da Costa RM, Bastos MM, Medeiros R, Oliveira PA. The NFkappaB signaling pathway in papillomavirus-induced lesions: friend or foe? *Anticancer Res.* 2016;36(5):2073–2083.
225. Xu M, Katzenellenbogen RA, Grandori C, Galloway DA. NFX1 plays a role in human papillomavirus type 16 E6 activation of NFkappaB activity. *J Virol.* 2010;84(21):11461–11469.
226. Yuan H, Fu F, Zhuo J, et al. Human papillomavirus type 16 E6 and E7 oncoproteins upregulate c-IAP2 gene expression and confer resistance to apoptosis. *Oncogene.* 2005;24(32):5069–5078.
227. An J, Mo D, Liu H, et al. Inactivation of the CYLD deubiquitinase by HPV E6 mediates hypoxia-induced NF-kappaB activation. *Cancer Cell.* 2008;14(5):394–407.
228. Kim SH, Oh JM, No JH, Bang YJ, Juhnn YS, Song YS. Involvement of NF-kappaB and AP-1 in COX-2 upregulation by human papillomavirus 16 E5 oncoprotein. *Carcinogenesis.* 2009;30(5):753–757.

229. Boulabiar M, Boubaker S, Favre M, Demeret C. Keratinocyte sensitization to tumour necrosis factor-induced nuclear factor kappa B activation by the E2 regulatory protein of human papillomaviruses. *J Gen Virol.* 2011;92(pt 10):2422–2427.
230. Vandermark ER, Deluca KA, Gardner CR, et al. Human papillomavirus type 16 E6 and E 7 proteins alter NF-kB in cultured cervical epithelial cells and inhibition of NF-kB promotes cell growth and immortalization. *Virology.* 2012;425(1):53–60.
231. Karim R, Tummers B, Meyers C, et al. Human papillomavirus (HPV) upregulates the cellular deubiquitinase UCHL1 to suppress the keratinocyte's innate immune response. *PLoS Pathog.* 2013;9(5):e1003384.
232. Thompson DA, Zacny V, Belinsky GS, et al. The HPV E7 oncoprotein inhibits tumor necrosis factor alpha-mediated apoptosis in normal human fibroblasts. *Oncogene.* 2001;20(28):3629–3640.
233. Li H, Zhan T, Li C, Liu M, Wang QK. Repression of MHC class I transcription by HPV16E7 through interaction with a putative RXRbeta motif and NF-kappaB cytoplasmic sequestration. *Biochem Biophys Res Commun.* 2009;388(2):383–388.
234. Spitkovsky D, Hehner SP, Hofmann TG, Moller A, Schmitz ML. The human papillomavirus oncoprotein E7 attenuates NF-kappa B activation by targeting the ikappa B kinase complex. *J Biol Chem.* 2002;277(28):25576–25582.
235. Tummers B, Goedemans R, Pelascini LP, et al. The interferon-related developmental regulator 1 is used by human papillomavirus to suppress NFkappaB activation. *Nat Commun.* 2015;6:6537.
236. Randall RE, Goodbourn S. Interferons and viruses: an interplay between induction, signalling, antiviral responses and virus countermeasures. *J Gen Virol.* 2008;89(pt 1):1–47.
237. Nedelec B, Dodd CM, Scott PG, Ghahary A, Tredget EE. Effect of interferon-alpha2b on guinea pig wound closure and the expression of cytoskeletal proteins in vivo. *Wound Repair Regen.* 1998;6(3):202–212.
238. White LC, Wright KL, Felix NJ, et al. Regulation of LMP2 and TAP1 genes by IRF-1 explains the paucity of CD8 + T cells in IRF-1-/- mice. *Immunity.* 1996;5(4):365–376.
239. Radaeva S, Jaruga B, Hong F, et al. Interferon-alpha activates multiple STAT signals and down-regulates c-Met in primary human hepatocytes. *Gastroenterology.* 2002;122(4): 1020–1034.
240. Yaar M, Karassik RL, Schnipper LE, Gilchrest BA. Effects of alpha and beta interferons on cultured human keratinocytes. *J Invest Dermatol.* 1985;85(1):70–74.
241. Bielenberg DR, McCarty MF, Bucana CD, et al. Expression of interferon-beta is associated with growth arrest of murine and human epidermal cells. *J Invest Dermatol.* 1999;112(5):802–809.
242. Ilkow CS, Marguerie M, Batenchuk C, et al. Reciprocal cellular cross-talk within the tumor microenvironment promotes oncolytic virus activity. *Nat Med.* 2015;21(5): 530–536.
243. Takahara T, Fukuyama Y, Saito S, Ogino T, Miyajima N, Kohase M. Il-1, EGF, and HGF suppress the antiviral activity of interferon in primary monkey hepatic parenchymal cells. *Jpn J Infect Dis.* 1999;52(2):45–48.
244. Ishida Y, Kondo T, Takayasu T, Iwakura Y, Mukaida N. The essential involvement of cross-talk between IFN-gamma and TGF-beta in the skin wound-healing process. *J Immunol.* 2004;172(3):1848–1855.
245. Tredget EE, Wang R, Shen Q, Scott PG, Ghahary A. Transforming growth factor-beta mRNA and protein in hypertrophic scar tissues and fibroblasts: antagonism by IFN-alpha and IFN-gamma in vitro and in vivo. *J Interferon Cytokine Res.* 2000;20(2):143–151.
246. Warren CJ, Griffin LM, Little AS, Huang IC, Farzan M, Pyeon D. The antiviral restriction factors IFITM1, 2 and 3 do not inhibit infection of human papillomavirus, cytomegalovirus and adenovirus. *PLoS One.* 2014;9(5):e96579.

247. Lace MJ, Anson JR, Klingelhutz AJ, et al. Interferon-beta treatment increases human papillomavirus early gene transcription and viral plasmid genome replication by activating interferon regulatory factor (IRF)-1. *Carcinogenesis*. 2009;30(8):1336–1344.
248. Chang YE, Pena L, Sen GC, Park JK, Laimins LA. Long-term effect of interferon on keratinocytes that maintain human papillomavirus type 31. *J Virol*. 2002;76(17): 8864–8874.
249. Herdman MT, Pett MR, Roberts I, et al. Interferon-beta treatment of cervical keratinocytes naturally infected with human papillomavirus 16 episomes promotes rapid reduction in episome numbers and emergence of latent integrants. *Carcinogenesis*. 2006;27(11):2341–2353.
250. Kim KY, Blatt L, Taylor MW. The effects of interferon on the expression of human papillomavirus oncogenes. *J Gen Virol*. 2000;81(pt 3):695–700.
251. Terenzi F, Saikia P, Sen GC. Interferon-inducible protein, P56, inhibits HPV DNA replication by binding to the viral protein E1. *EMBO J*. 2008;27(24):3311–3321.
252. Tirone NR, Peghini BC, Barcelos AC, Murta EF, Michelin MA. Local expression of interferon-alpha and interferon receptors in cervical intraepithelial neoplasia. *Cancer Immunol Immunother*. 2009;58(12):2003–2010.
253. DeCarlo CA, Severini A, Edler L, et al. IFN-kappa, a novel type I IFN, is undetectable in HPV-positive human cervical keratinocytes. *Lab Invest*. 2010;90(10):1482–1491.
254. Karim R, Meyers C, Backendorf C, et al. Human papillomavirus deregulates the response of a cellular network comprising of chemotactic and proinflammatory genes. *PLoS One*. 2011;6(3):e17848.
255. Reiser J, Hurst J, Voges M, et al. High-risk human papillomaviruses repress constitutive kappa interferon transcription via E6 to prevent pathogen recognition receptor and antiviral-gene expression. *J Virol*. 2011;85(21):11372–11380.
256. Hong S, Mehta KP, Laimins LA. Suppression of STAT-1 expression by human papillomaviruses is necessary for differentiation-dependent genome amplification and plasmid maintenance. *J Virol*. 2011;85(18):9486–9494.
257. Nees M, Geoghegan JM, Hyman T, Frank S, Miller L, Woodworth CD. Papillomavirus type 16 oncogenes downregulate expression of interferon-responsive genes and upregulate proliferation-associated and NF-kappaB-responsive genes in cervical keratinocytes. *J Virol*. 2001;75(9):4283–4296.
258. Li S, Labrecque S, Gauzzi MC, et al. The human papilloma virus (HPV)-18 E6 oncoprotein physically associates with Tyk2 and impairs Jak-STAT activation by interferon-alpha. *Oncogene*. 1999;18(42):5727–5737.
259. Ronco LV, Karpova AY, Vidal M, Howley PM. Human papillomavirus 16 E6 oncoprotein binds to interferon regulatory factor-3 and inhibits its transcriptional activity. *Genes Dev*. 1998;12(13):2061–2072.
260. Miciak J, Bunz F. Long story short: p53 mediates innate immunity. *Biochim Biophys Acta*. 2016;1865(2):220–227.
261. Takaoka A, Hayakawa S, Yanai H, et al. Integration of interferon-alpha/beta signalling to p53 responses in tumour suppression and antiviral defence. *Nature*. 2003;424(6948): 516–523.
262. Munoz-Fontela C, Macip S, Martinez-Sobrido L, et al. Transcriptional role of p53 in interferon-mediated antiviral immunity. *J Exp Med*. 2008;205(8):1929–1938.
263. Rivas C, Aaronson SA, Munoz-Fontela C. Dual role of p53 in innate antiviral immunity. *Viruses*. 2010;2(1):298–313.
264. Youlyouz-Marfak I, Gachard N, Le Clorennec C, et al. Identification of a novel p53-dependent activation pathway of STAT1 by antitumour genotoxic agents. *Cell Death Differ*. 2008;15(2):376–385.
265. Lau L, Gray EE, Brunette RL, Stetson DB. DNA tumor virus oncogenes antagonize the cGAS-STING DNA-sensing pathway. *Science*. 2015;350(6260):568–571.

266. Zhou F, Chen J, Zhao KN. Human papillomavirus 16-encoded E7 protein inhibits IFN-gamma-mediated MHC class I antigen presentation and CTL-induced lysis by blocking IRF-1 expression in mouse keratinocytes. *J Gen Virol.* 2013;94(pt 11): 2504–2514.
267. Park JS, Kim EJ, Kwon HJ, Hwang ES, Namkoong SE, Um SJ. Inactivation of interferon regulatory factor-1 tumor suppressor protein by HPV E7 oncoprotein. Implication for the E7-mediated immune evasion mechanism in cervical carcinogenesis. *J Biol Chem.* 2000;275(10):6764–6769.
268. Barnard P, McMillan NA. The human papillomavirus E7 oncoprotein abrogates signaling mediated by interferon-alpha. *Virology.* 1999;259(2):305–313.
269. Antonsson A, Payne E, Hengst K, McMillan NA. The human papillomavirus type 16 E7 protein binds human interferon regulatory factor-9 via a novel PEST domain required for transformation. *J Interferon Cytokine Res.* 2006;26(7):455–461.
270. Muto V, Stellacci E, Lamberti AG, et al. Human papillomavirus type 16 E5 protein induces expression of beta interferon through interferon regulatory factor 1 in human keratinocytes. *J Virol.* 2011;85(10):5070–5080.
271. Lulli D, Carbone ML, Pastore S. Epidermal growth factor receptor inhibitors trigger a type I interferon response in human skin. *Oncotarget.* 2016;7(30):47777–47793.
272. Battcock SM, Collier TW, Zu D, Hirasawa K. Negative regulation of the alpha interferon-induced antiviral response by the Ras/Raf/MEK pathway. *J Virol.* 2006;80(9):4422–4430.
273. Zhang Q, Gong R, Qu J, et al. Activation of the Ras/Raf/MEK pathway facilitates hepatitis C virus replication via attenuation of the interferon-JAK-STAT pathway. *J Virol.* 2012;86(3):1544–1554.
274. Komatsu Y, Christian SL, Ho N, Pongnopparat T, Licursi M, Hirasawa K. Oncogenic Ras inhibits IRF1 to promote viral oncolysis. *Oncogene.* 2015;34(30): 3985–3993.
275. Sunthamala N, Thierry F, Teissier S, et al. E2 proteins of high risk human papillomaviruses down-modulate STING and IFN-kappa transcription in keratinocytes. *PLoS One.* 2014;9(3):e91473.
276. LaFleur DW, Nardelli B, Tsareva T, et al. Interferon-kappa, a novel type I interferon expressed in human keratinocytes. *J Biol Chem.* 2001;276(43):39765–39771.
277. Habiger C, Jager G, Walter M, Iftner T, Stubenrauch F. Interferon kappa inhibits human papillomavirus 31 transcription by inducing Sp100 proteins. *J Virol.* 2015;90(2):694–704.
278. Rincon-Orozco B, Halec G, Rosenberger S, et al. Epigenetic silencing of interferon-kappa in human papillomavirus type 16-positive cells. *Cancer Res.* 2009;69(22): 8718–8725.
279. Tay SS, Roediger B, Tong PL, Tikoo S, Weninger W. The skin-resident immune network. *Curr Dermatol Rep.* 2014;3:13–22.
280. Poppe WA, Drijkoningen M, Ide PS, Lauweryns JM, Van Assche FA. Lymphocytes and dendritic cells in the normal uterine cervix. An immunohistochemical study. *Eur J Obstet Gynecol Reprod Biol.* 1998;81(2):277–282.
281. Castle PE, Jeronimo J, Schiffman M, et al. Age-related changes of the cervix influence human papillomavirus type distribution. *Cancer Res.* 2006;66(2):1218–1224.
282. Turley SJ, Cremasco V, Astarita JL. Immunological hallmarks of stromal cells in the tumour microenvironment. *Nat Rev Immunol.* 2015;15(11):669–682.
283. Palefsky JM, Minkoff H, Kalish LA, et al. Cervicovaginal human papillomavirus infection in human immunodeficiency virus-1 (HIV)-positive and high-risk HIV-negative women. *J Natl Cancer Inst.* 1999;91(3):226–236.
284. Serraino D, Carrieri P, Pradier C, et al. Risk of invasive cervical cancer among women with, or at risk for HIV infection. *Int J Cancer.* 1999;82(3):334–337.

285. Mbulaiteye SM, Katabira ET, Wabinga H, et al. Spectrum of cancers among HIV-infected persons in Africa: the Uganda AIDS-cancer registry match study. *Int J Cancer*. 2005;118(4):985–990.
286. Einstein MH, Kadish AS. Anogenital neoplasia in AIDS. *Curr Opin Oncol*. 2004;16(5): 455–462.
287. Clarke B, Chetty R. Postmodern cancer: the role of human immunodeficiency virus in uterine cervical cancer. *Mol Pathol*. 2002;55(1):19–24.
288. Ellerbrock TV, Chiasson MA, Bush TJ, et al. Incidence of cervical squamous intra-epithelial lesions in HIV-infected women. *JAMA*. 2000;283(8):1031–1037.
289. Garcia-Pineres AJ, Hildesheim A, Herrero R, et al. Persistent human papillomavirus infection is associated with a generalized decrease in immune responsiveness in older women. *Cancer Res*. 2006;66(22):11070–11076.
290. Mohammed J, Beura LK, Bobr A, et al. Stromal cells control the epithelial residence of DCs and memory T cells by regulated activation of TGF-beta. *Nat Immunol*. 2016;17(4):414–421.
291. Mackay LK, Rahimpour A, Ma JZ, et al. The developmental pathway for CD103(+) CD8+ tissue-resident memory T cells of skin. *Nat Immunol*. 2013;14(12):1294–1301.
292. Skon CN, Lee JY, Anderson KG, Masopust D, Hogquist KA, Jameson SC. Transcriptional downregulation of S1pr1 is required for the establishment of resident memory CD8+ T cells. *Nat Immunol*. 2013;14(12):1285–1293.
293. Sparber F. Langerhans cells: an update. *J Dtsch Dermatol Ges*. 2014;12(12):1107–1111.
294. Hubert P, Caberg JH, Gilles C, et al. E-cadherin-dependent adhesion of dendritic and Langerhans cells to keratinocytes is defective in cervical human papillomavirus-associated (pre)neoplastic lesions. *J Pathol*. 2005;206(3):346–355.
295. Delvenne P, Hubert P, Jacobs N, et al. The organotypic culture of HPV-transformed keratinocytes: an effective in vitro model for the development of new immunotherapeutic approaches for mucosal (pre)neoplastic lesions. *Vaccine*. 2001;19(17–19): 2557–2564.
296. Spinillo A, Tenti P, Zappatore R, De Seta F, Silini E, Guaschino S. Langerhans' cell counts and cervical intraepithelial neoplasia in women with human immunodeficiency virus infection. *Gynecol Oncol*. 1993;48(2):210–213.
297. Matthews K, Leong CM, Baxter L, et al. Depletion of Langerhans cells in human papillomavirus type 16-infected skin is associated with E6-mediated down regulation of E-cadherin. *J Virol*. 2003;77(15):8378–8385.
298. Leong CM, Doorbar J, Nindl I, Yoon HS, Hibma MH. Deregulation of E-cadherin by human papillomavirus is not confined to high-risk, cancer-causing types. *Br J Dermatol*. 2010;163(6):1253–1263.
299. Herman L, Hubert P, Caberg JH, et al. MIP3 alpha stimulates the migration of Langerhans cells in models of human papillomavirus (HPV)-associated (pre)neoplastic epithelium. *Cancer Immunol Immunother*. 2006;56(7):1087–1096.
300. Caberg JH, Hubert P, Herman L, et al. Increased migration of Langerhans cells in response to HPV16 E6 and E7 oncogene silencing: role of CCL20. *Cancer Immunol Immunother*. 2008;58(1):39–47.
301. Caberg JH, Hubert P, Herman L, et al. Increased migration of Langerhans cells in response to HPV16 E6 and E7 oncogene silencing: role of CCL20. *Cancer Immunol Immunother*. 2009;58(1):39–47.
302. Connor JP, Ferrer K, Kane JP, Goldberg JM. Evaluation of Langerhans' cells in the cervical epithelium of women with cervical intraepithelial neoplasia. *Gynecol Oncol*. 1999;75(1):130–135.
303. Ginhoux F, Tacke F, Angeli V, et al. Langerhans cells arise from monocytes in vivo. *Nat Immunol*. 2006;7(3):265–273.

304. Godefroy S, Guironnet G, Jacquet C, Schmitt D, Staquet MJ. A combination of MIP-3alpha and TGF-beta1 is required for the attraction of human Langerhans precursor cells through a dermal-epidermal barrier. *Eur J Cell Biol.* 2001;80(5):335–340.
305. Tan SY, Roediger B, Weninger W. The role of chemokines in cutaneous immunosurveillance. *Immunol Cell Biol.* 2015;93(4):337–346.
306. Dieu-Nosjean MC, Massacrier C, Homey B, et al. Macrophage inflammatory protein 3alpha is expressed at inflamed epithelial surfaces and is the most potent chemokine known in attracting Langerhans cell precursors. *J Exp Med.* 2000;192(5):705–718.
307. Thomas RM, Belsito DV, Huang C, et al. Appearance of Langerhans cells in the epidermis of Tgfb1(-/-) SCID mice: paracrine and autocrine effects of transforming growth factor-beta 1 and -beta 2(1). *J Invest Dermatol.* 2001;117(6):1574–1580.
308. Tohyama M, Shirakara Y, Yamasaki K, Sayama K, Hashimoto K. Differentiated keratinocytes are responsible for TNF-alpha regulated production of macrophage inflammatory protein 3alpha/CCL20, a potent chemokine for Langerhans cells. *J Dermatol Sci.* 2001;27(2):130–139.
309. Charbonnier AS, Kohrgruber N, Kriehuber E, Stingl G, Rot A, Maurer D. Macrophage inflammatory protein 3alpha is involved in the constitutive trafficking of epidermal Langerhans cells. *J Exp Med.* 1999;190(12):1755–1768.
310. Rollins BJ. Inflammatory chemokines in cancer growth and progression. *Eur J Cancer.* 2006;42(6):760–767.
311. Kao CY, Huang F, Chen Y, et al. Up-regulation of CC chemokine ligand 20 expression in human airway epithelium by IL-17 through a JAK-independent but MEK/NF-kappaB-dependent signaling pathway. *J Immunol.* 2005;175(10):6676–6685.
312. Guess JC, McCance DJ. Decreased migration of Langerhans precursor-like cells in response to human keratinocytes expressing human papillomavirus type 16 E6/E7 is related to reduced macrophage inflammatory protein-3{alpha} production. *J Virol.* 2005;79(23):14852–14862.
313. Fausch SC, Da Silva DM, Rudolf MP, Kast WM. Human papillomavirus virus-like particles do not activate Langerhans cells: a possible immune escape mechanism used by human papillomaviruses. *J Immunol.* 2002;169(6):3242–3249.
314. Larregina AT, Morelli AE, Spencer LA, et al. Dermal-resident CD14+ cells differentiate into Langerhans cells. *Nat Immunol.* 2001;2(12):1151–1158.
315. Bobr A, Igyarto BZ, Haley KM, Li MO, Flavell RA, Kaplan DH. Autocrine/paracrine TGF-beta1 inhibits Langerhans cell migration. *Proc Natl Acad Sci USA.* 2012;109(26):10492–10497.
316. Kel JM, Girard-Madoux MJ, Reizis B, Clausen BE. TGF-beta is required to maintain the pool of immature Langerhans cells in the epidermis. *J Immunol.* 2010;185(6):3248–3255.
317. Borkowski TA, Letterio JJ, Farr AG, Udey MC. A role for endogenous transforming growth factor beta 1 in Langerhans cell biology: the skin of transforming growth factor beta 1 null mice is devoid of epidermal Langerhans cells. *J Exp Med.* 1996;184(6):2417–2422.
318. Greter M, Lelios I, Pelczar P, et al. Stroma-derived interleukin-34 controls the development and maintenance of Langerhans cells and the maintenance of microglia. *Immunity.* 2012;37(6):1050–1060.
319. Wang Y, Szretter KJ, Vermi W, et al. IL-34 is a tissue-restricted ligand of CSF1R required for the development of Langerhans cells and microglia. *Nat Immunol.* 2012;13(8):753–760.
320. Tang A, Amagai M, Granger LG, Stanley JR, Udey MC. Adhesion of epidermal Langerhans cells to keratinocytes mediated by E-cadherin. *Nature.* 1993;361(6407):82–85.

321. Mayumi N, Watanabe E, Norose Y, et al. E-cadherin interactions are required for Langerhans cell differentiation. *Eur J Immunol.* 2013;43(1):270–280.
322. Hubert P, van den Brule F, Giannini SL, Franzen-Detrooz E, Boniver J, Delvenne P. Colonization of in vitro-formed cervical human papillomavirus-associated (pre)neoplastic lesions with dendritic cells: role of granulocyte/macrophage colony-stimulating factor. *Am J Pathol.* 1999;154(3):775–784.
323. Vessey CJ, Wilding J, Folarin N, et al. Altered expression and function of E-cadherin in cervical intraepithelial neoplasia and invasive squamous cell carcinoma. *J Pathol.* 1995;176(2):151–159.
324. Carico E, Fulciniti F, Giovagnoli MR, et al. Adhesion molecules and p16 expression in endocervical adenocarcinoma. *Virchows Arch.* 2009;455(3):245–251.
325. Al Moustafa AE, Foulkes WD, Benlimame N, et al. E6/E7 proteins of HPV type 16 and ErbB-2 cooperate to induce neoplastic transformation of primary normal oral epithelial cells. *Oncogene.* 2004;23(2):350–358.
326. Liao S, Deng D, Zhang W, et al. Human papillomavirus 16/18 E5 promotes cervical cancer cell proliferation, migration and invasion in vitro and accelerates tumor growth in vivo. *Oncol Rep.* 2013;29(1):95–102.
327. Jakob T, Udey MC. Regulation of E-cadherin-mediated adhesion in Langerhans cell-like dendritic cells by inflammatory mediators that mobilize Langerhans cells in vivo. *J Immunol.* 1998;160(8):4067–4073.
328. Abd Warif NM, Stoitzner P, Leggatt GR, Mattarollo SR, Frazer IH, Hibma MH. Langerhans cell homeostasis and activation is altered in hyperplastic human papillomavirus type 16 E7 expressing epidermis. *PLoS One.* 2015;10(5):e0127155.
329. Seneschal J, Clark RA, Gehad A, Baecher-Allan CM, Kupper TS. Human epidermal Langerhans cells maintain immune homeostasis in skin by activating skin resident regulatory T cells. *Immunity.* 2012;36(5):873–884.
330. Da Silva DM, Woodham AW, Rijkee LK, et al. Human papillomavirus-exposed Langerhans cells are activated by stabilized poly-I:C. *Papillomavirus Res.* 2015;1: 12–21.
331. Jacobs N, Renard I, Al-Saleh W, et al. Distinct T cell subsets and cytokine production in cultures derived from transformation zone and squamous intraepithelial lesion biopsies of the uterine cervix. *Am J Reprod Immunol.* 2003;49(1):6–13.
332. Herfs M, Yamamoto Y, Laury A, et al. A discrete population of squamocolumnar junction cells implicated in the pathogenesis of cervical cancer. *Proc Natl Acad Sci USA.* 2012;109(26):10516–10521.
333. Coleman N, Birley HD, Renton AM, et al. Immunological events in regressing genital warts. *Am J Clin Pathol.* 1994;102(6):768–774.
334. Tagami H, Oku T, Iwatsuki K. Primary tissue culture of spontaneously regressing flat warts. In vitro attack by mononuclear cells against wart-derived epidermal cells. *Cancer.* 1985;55(10):2437–2441.
335. Aiba S, Rokugo M, Tagami H. Immunohistologic analysis of the phenomenon of spontaneous regression of numerous flat warts. *Cancer.* 1986;58(6):1246–1251.
336. Woo YL, van den Hende M, Sterling JC, et al. A prospective study on the natural course of low-grade squamous intraepithelial lesions and the presence of HPV16 E2-, E6- and E7-specific T-cell responses. *Int J Cancer.* 2010;126(1):133–141.
337. Vici P, Pizzuti L, Mariani L, et al. Targeting immune response with therapeutic vaccines in premalignant lesions and cervical cancer: hope or reality from clinical studies. *Expert Rev Vaccines.* 2016;9:1–10.
338. Grabowska AK, Kaufmann AM, Riemer AB. Identification of promiscuous HPV16-derived T helper cell epitopes for therapeutic HPV vaccine design. *Int J Cancer.* 2015;136(1):212–224.

339. Li H, Ou X, Xiong J, Wang T. HPV16E7 mediates HADC chromatin repression and downregulation of MHC class I genes in HPV16 tumorigenic cells through interaction with an MHC class I promoter. *Biochem Biophys Res Commun*. 2006;349(4):1315–1321.
340. Georgopoulos NT, Proffitt JL, Blair GE. Transcriptional regulation of the major histocompatibility complex (MHC) class I heavy chain, TAP1 and LMP2 genes by the human papillomavirus (HPV) type 6b, 16 and 18 E7 oncoproteins. *Oncogene*. 2000;19(42):4930–4935.
341. Heller C, Weisser T, Mueller-Schickert A, et al. Identification of key amino acid residues that determine the ability of high risk HPV16-E7 to dysregulate major histocompatibility complex class I expression. *J Biol Chem*. 2011;286(13):10983–10997.
342. Deng XM, Li W, Zhang X, et al. RNA interference of human papillomavirus type 16 E7 increases HLA class I antigen expression in HaCaT-E7 cells. *Int J Gynecol Cancer*. 2011;21(1):28–34.
343. Li W, Deng XM, Wang CX, et al. Down-regulation of HLA class I antigen in human papillomavirus type 16 E7 expressing HaCaT cells: correlate with TAP-1 expression. *Int J Gynecol Cancer*. 2010;20(2):227–232.
344. Cortese MS, Ashrafi GH, Campo MS. All 4 di-leucine motifs in the first hydrophobic domain of the E5 oncoprotein of human papillomavirus type 16 are essential for surface MHC class I downregulation activity and E5 endomembrane localization. *Int J Cancer*. 2010;126(7):1675–1682.
345. Campo MS, Graham SV, Cortese MS, et al. HPV-16 E5 down-regulates expression of surface HLA class I and reduces recognition by CD8 T cells. *Virology*. 2010;407(1): 137–142.
346. Ashrafi GH, Haghshenas MR, Marchetti B, O'Brien PM, Campo MS. E5 protein of human papillomavirus type 16 selectively downregulates surface HLA class I. *Int J Cancer*. 2005;113(2):276–283.
347. Bagarazzi ML, Yan J, Morrow MP, et al. Immunotherapy against HPV16/18 generates potent TH1 and cytotoxic cellular immune responses. *Sci Transl Med*. 2012;4(155):155ra138.
348. Smyth LJ, Van Poelgeest MI, Davidson EJ, et al. Immunological responses in women with human papillomavirus type 16 (HPV-16)-associated anogenital intraepithelial neoplasia induced by heterologous prime-boost HPV-16 oncogene vaccination. *Clin Cancer Res*. 2004;10(9):2954–2961.
349. van Poelgeest MI, van Seters M, van Beurden M, et al. Detection of human papillomavirus (HPV) 16-specific CD4+ T-cell immunity in patients with persistent HPV16-induced vulvar intraepithelial neoplasia in relation to clinical impact of imiquimod treatment. *Clin Cancer Res*. 2005;11(14):5273–5280.
350. Welters MJ, de Jong A, van den Eeden SJ, et al. Frequent display of human papillomavirus type 16 E6-specific memory t-helper cells in the healthy population as witness of previous viral encounter. *Cancer Res*. 2003;63(3):636–641.
351. Kadish AS, Ho GY, Burk RD, et al. Lymphoproliferative responses to human papillomavirus (HPV) type 16 proteins E6 and E7: outcome of HPV infection and associated neoplasia. *J Natl Cancer Inst*. 1997;89(17):1285–1293.
352. Nakagawa M, Stites DP, Farhat S, et al. Cytotoxic T lymphocyte responses to E6 and E7 proteins of human papillomavirus type 16: relationship to cervical intraepithelial neoplasia. *J Infect Dis*. 1997;175(4):927–931.
353. Selvakumar R, Ahmed R, Wettstein FO. Tumor regression is associated with a specific immune response to the E2 protein of cottontail rabbit papillomavirus. *Virology*. 1995;208(1):298–302.
354. de Jong A, van der Burg SH, Kwappenberg KM, et al. Frequent detection of human papillomavirus 16 E2-specific T-helper immunity in healthy subjects. *Cancer Res*. 2002;62(2):472–479.

355. Melero I, Singhal MC, McGowan P, et al. Immunological ignorance of an E7-encoded cytolytic T-lymphocyte epitope in transgenic mice expressing the E7 and E6 oncogenes of human papillomavirus type 16. *J Virol.* 1997;71(5):3998–4004.
356. Borchers A, Braspenning J, Meijer J, Osen W, Gissmann L, Jochmus I. E7-specific cytotoxic T cell tolerance in HPV-transgenic mice. *Arch Virol.* 1999;144(8): 1539–1556.
357. Youde SJ, Dunbar PR, Evans EM, et al. Use of fluorogenic histocompatibility leukocyte antigen-A*0201/HPV 16 E7 peptide complexes to isolate rare human cytotoxic T-lymphocyte-recognizing endogenous human papillomavirus antigens. *Cancer Res.* 2000;60(2):365–371.
358. Evans EM, Man S, Evans AS, Borysiewicz LK. Infiltration of cervical cancer tissue with human papillomavirus-specific cytotoxic T-lymphocytes. *Cancer Res.* 1997;57(14): 2943–2950.
359. Bontkes HJ, de Gruijl TD, van den Muysenberg AJ, et al. Human papillomavirus type 16 E6/E7-specific cytotoxic T lymphocytes in women with cervical neoplasia. *Int J Cancer.* 2000;88(1):92–98.
360. Nakagawa M, Stites DP, Patel S, et al. Persistence of human papillomavirus type 16 infection is associated with lack of cytotoxic T lymphocyte response to the E6 antigens. *J Infect Dis.* 2000;182(2):595–598.
361. Ressing ME, van Driel WJ, Celis E, et al. Occasional memory cytotoxic T-cell responses of patients with human papillomavirus type 16-positive cervical lesions against a human leukocyte antigen-A *0201-restricted E7-encoded epitope. *Cancer Res.* 1996;56(3):582–588.
362. van der Burg SH, Ressing ME, Kwappenberg KM, et al. Natural T-helper immunity against human papillomavirus type 16 (HPV16) E7-derived peptide epitopes in patients with HPV16-positive cervical lesions: identification of 3 human leukocyte antigen class II-restricted epitopes. *Int J Cancer.* 2001;91(5):612–618.
363. Clerici M, Merola M, Ferrario E, et al. Cytokine production patterns in cervical intraepithelial neoplasia: association with human papillomavirus infection. *J Natl Cancer Inst.* 1997;89(3):245–250.
364. de Jong A, van Poelgeest MI, van der Hulst JM, et al. Human papillomavirus type 16-positive cervical cancer is associated with impaired CD4+ T-cell immunity against early antigens E2 and E6. *Cancer Res.* 2004;64(15):5449–5455.
365. Takeuchi Y, Nishikawa H. Roles of regulatory T cells in cancer immunity. *Int Immunol.* 2016;28(8):401–409.
366. Cao Y, Zhao J, Lei Z, et al. Local accumulation of FOXP3+ regulatory T cells: evidence for an immune evasion mechanism in patients with large condylomata acuminata. *J Immunol.* 2008;180(11):7681–7686.
367. Goel HL, Mercurio AM. VEGF targets the tumour cell. *Nat Rev Cancer.* 2013;13(12): 871–882.
368. Molling JW, de Gruijl TD, Glim J, et al. CD4(+)CD25hi regulatory T-cell frequency correlates with persistence of human papillomavirus type 16 and T helper cell responses in patients with cervical intraepithelial neoplasia. *Int J Cancer.* 2007;121(8):1749–1755.
369. Hou F, Li Z, Ma D, et al. Distribution of Th17 cells and Foxp3-expressing T cells in tumor-infiltrating lymphocytes in patients with uterine cervical cancer. *Clin Chim Acta.* 2012;413(23–24):1848–1854.
370. Ding AP, Zhang Y, Wei H, Luo QS, Zhang SL. Correlation study of HPV-16 existential status with Th17/Treg cytokines. *Zhonghua Yi Xue Za Zhi.* 2013;93(37): 2957–2960.
371. Chen Z, Ding J, Pang N, et al. The Th17/treg balance and the expression of related cytokines in Uygur cervical cancer patients. *Diagn Pathol.* 2013;8:61.

372. Choyce A, Yong M, Narayan S, et al. Expression of a single, viral oncoprotein in skin epithelium is sufficient to recruit lymphocytes. *PLoS One*. 2013;8(2):e57798.
373. Sheu BC, Chiou SH, Lin HH, et al. Up-regulation of inhibitory natural killer receptors CD94/NKG2A with suppressed intracellular perforin expression of tumor-infiltrating CD8+ T lymphocytes in human cervical carcinoma. *Cancer Res*. 2005;65(7):2921–2929.
374. Partlova S, Boucek J, Kloudova K, et al. Distinct patterns of intratumoral immune cell infiltrates in patients with HPV-associated compared to non-virally induced head and neck squamous cell carcinoma. *Oncoimmunology*. 2015;4(1):e965570.
375. De Vlaeminck Y, Gonzalez-Rascon A, Goyvaerts C, Breckpot K. Cancer-associated myeloid regulatory cells. *Front Immunol*. 2016;7:113.
376. Raggi C, Mousa HS, Correnti M, Sica A, Invernizzi P. Cancer stem cells and tumor-associated macrophages: a roadmap for multitargeting strategies. *Oncogene*. 2016;35(6):671–682.
377. Lin Y, Wei C, Liu Y, Qiu Y, Liu C, Guo F. Selective ablation of tumor-associated macrophages suppresses metastasis and angiogenesis. *Cancer Sci*. 2013;104(9):1217–1225.
378. LaGory EL, Giaccia AJ. The ever-expanding role of HIF in tumour and stromal biology. *Nat Cell Biol*. 2016;18(4):356–365.
379. Fujimoto J, Sakaguchi H, Aoki I, Tamaya T. Clinical implications of expression of interleukin 8 related to angiogenesis in uterine cervical cancers. *Cancer Res*. 2000;60(10):2632–2635.
380. Mazibrada J, Ritta M, Mondini M, et al. Interaction between inflammation and angiogenesis during different stages of cervical carcinogenesis. *Gynecol Oncol*. 2008;108(1):112–120.
381. Davidson B, Goldberg I, Gotlieb WH, et al. Macrophage infiltration and angiogenesis in cervical squamous cell carcinoma–clinicopathologic correlation. *Acta Obstet Gynecol Scand*. 1999;78(3):240–244.
382. Lepique AP, Daghastanli KR, Cuccovia IM, Villa LL. HPV16 tumor associated macrophages suppress antitumor T cell responses. *Clin Cancer Res*. 2009;15(13):4391–4400.
383. Heller DS, Hameed M, Cracchiolo B, et al. Presence and quantification of macrophages in squamous cell carcinoma of the cervix. *Int J Gynecol Cancer*. 2003;13(1):67–70.
384. Hammes LS, Tekmal RR, Naud P, et al. Macrophages, inflammation and risk of cervical intraepithelial neoplasia (CIN) progression–clinicopathological correlation. *Gynecol Oncol*. 2007;105(1):157–165.
385. Bonagura VR, Hatam LJ, Rosenthal DW, et al. Recurrent respiratory papillomatosis: a complex defect in immune responsiveness to human papillomavirus-6 and -11. *APMIS*. 2010;118(6–7):455–470.
386. Renoux VM, Bisig B, Langers I, et al. Human papillomavirus entry into NK cells requires CD16 expression and triggers cytotoxic activity and cytokine secretion. *Eur J Immunol*. 2011;41(11):3240–3252.
387. Fichorova RN, Anderson DJ. Differential expression of immunobiological mediators by immortalized human cervical and vaginal epithelial cells. *Biol Reprod*. 1999;60(2):508–514.
388. Niebler M, Qian X, Hofler D, et al. Post-translational control of IL-1beta via the human papillomavirus type 16 E6 oncoprotein: a novel mechanism of innate immune escape mediated by the E3-ubiquitin ligase E6-AP and p53. *PLoS Pathog*. 2013;9(8):e1003536.
389. Boccardo E, Noya F, Broker TR, Chow LT, Villa LL. HPV-18 confers resistance to TNF-alpha in organotypic cultures of human keratinocytes. *Virology*. 2004;328(2):233–243.

390. Basile JR, Zacny V, Munger K. The cytokines tumor necrosis factor-alpha (TNF-alpha) and TNF-related apoptosis-inducing ligand differentially modulate proliferation and apoptotic pathways in human keratinocytes expressing the human papillomavirus-16 E7 oncoprotein. *J Biol Chem*. 2001;276(25):22522–22528.
391. El-Sherif AM, Seth R, Tighe PJ, Jenkins D. Quantitative analysis of IL-10 and IFN-gamma mRNA levels in normal cervix and human papillomavirus type 16 associated cervical precancer. *J Pathol*. 2001;195(2):179–185.
392. Prata TT, Bonin CM, Ferreira AM, et al. Local immunosuppression induced by high viral load of human papillomavirus: characterization of cellular phenotypes producing interleukin-10 in cervical neoplastic lesions. *Immunology*. 2015;146(1):113–121.
393. Tomao F, Papa A, Rossi L, et al. Angiogenesis and antiangiogenic agents in cervical cancer. *Onco Targets Ther*. 2014;7:2237–2248.
394. Yang L, Karin M. Roles of tumor suppressors in regulating tumor-associated inflammation. *Cell Death Differ*. 2014;21(11):1677–1686.
395. Shellenberger TD, Wang M, Gujrati M, et al. BRAK/CXCL14 is a potent inhibitor of angiogenesis and a chemotactic factor for immature dendritic cells. *Cancer Res*. 2004;64(22):8262–8270.
396. Cicchini L, Westrich JA, Xu T, et al. Suppression of antitumor immune responses by human papillomavirus through epigenetic downregulation of CXCL14. *MBio*. 2016;7(3):e00270–e00316. http://dx.doi.org/10.1128/mBio.00270-16.
397. Kyo S, Inoue M, Hayasaka N, et al. Regulation of early gene expression of human papillomavirus type 16 by inflammatory cytokines. *Virology*. 1994;200(1):130–139.
398. Bedke N, Sammut D, Green B, et al. Transforming growth factor-beta promotes rhinovirus replication in bronchial epithelial cells by suppressing the innate immune response. *PLoS One*. 2012;7(9):e44580.
399. Reardon C, McKay DM. TGF-beta suppresses IFN-gamma-STAT1-dependent gene transcription by enhancing STAT1-PIAS1 interactions in epithelia but not monocytes/macrophages. *J Immunol*. 2007;178(7):4284–4295.
400. Ulloa L, Doody J, Massague J. Inhibition of transforming growth factor-beta/SMAD signalling by the interferon-gamma/STAT pathway. *Nature*. 1999;397(6721):710–713.
401. Bellone G, Aste-Amezaga M, Trinchieri G, Rodeck U. Regulation of NK cell functions by TGF-beta 1. *J Immunol*. 1995;155(3):1066–1073.
402. Yu J, Wei M, Becknell B, et al. Pro- and antiinflammatory cytokine signaling: reciprocal antagonism regulates interferon-gamma production by human natural killer cells. *Immunity*. 2006;24(5):575–590.
403. Eriksson M, Meadows SK, Wira CR, Sentman CL. Unique phenotype of human uterine NK cells and their regulation by endogenous TGF-beta. *J Leukoc Biol*. 2004;76(3):667–675.
404. Lin KY, Guarnieri FG, Staveley-O'Carroll KF, et al. Treatment of established tumors with a novel vaccine that enhances major histocompatibility class II presentation of tumor antigen. *Cancer Res*. 1996;56(1):21–26.
405. Lopez-Munoz H, Escobar-Sanchez ML, Lopez-Marure R, et al. Cervical cancer cells induce apoptosis in TCD4+ lymphocytes through the secretion of TGF-beta. *Arch Gynecol Obstet*. 2013;287(4):755–763.
406. Wu HS, Li YF, Chou CI, Yuan CC, Hung MW, Tsai LC. The concentration of serum transforming growth factor beta-1 (TGF-beta1) is decreased in cervical carcinoma patients. *Cancer Invest*. 2002;20(1):55–59.
407. Moon HS, Kim SC, Ahn JJ, Woo BH. Concentration of vascular endothelial growth factor (VEGF) and transforming growth factor-beta1 (TGF-beta1) in the serum of patients with cervical cancer: prediction of response. *Int J Gynecol Cancer*. 2000;10(2):151–156.

408. Soufla G, Sifakis S, Baritaki S, Zafiropoulos A, Koumantakis E, Spandidos DA. VEGF, FGF2, TGFB1 and TGFBR1 mRNA expression levels correlate with the malignant transformation of the uterine cervix. *Cancer Lett.* 2005;221(1):105–118.
409. Baritaki S, Sifakis S, Huerta-Yepez S, et al. Overexpression of VEGF and TGF-beta1 mRNA in Pap smears correlates with progression of cervical intraepithelial neoplasia to cancer: implication of YY1 in cervical tumorigenesis and HPV infection. *Int J Oncol.* 2007;31(1):69–79.
410. Fan DM, Tian XY, Wang RF, Yu JJ. The prognosis significance of TGF-beta1 and ER protein in cervical adenocarcinoma patients with stage Ib~IIa. *Tumour Biol.* 2014;35(11):11237–11242.
411. Comerci Jr JT, Runowicz CD, Flanders KC, et al. Altered expression of transforming growth factor-beta 1 in cervical neoplasia as an early biomarker in carcinogenesis of the uterine cervix. *Cancer.* 1996;77(6):1107–1114.
412. El-Sherif AM, Seth R, Tighe PJ, Jenkins D. Decreased synthesis and expression of TGF-beta1, beta2, and beta3 in epithelium of HPV 16-positive cervical precancer: a study by microdissection, quantitative RT-PCR, and immunocytochemistry. *J Pathol.* 2000;192(4):494–501.
413. Torng PL, Chan WY, Lin CT, Huang SC. Decreased expression of human papillomavirus E2 protein and transforming growth factor-beta1 in human cervical neoplasia as an early marker in carcinogenesis. *J Surg Oncol.* 2003;84(1):17–23.
414. Christian SL, Zu D, Licursi M, et al. Suppression of IFN-induced transcription underlies IFN defects generated by activated Ras/MEK in human cancer cells. *PLoS One.* 2012;7(9):e44267.
415. AbuSara N, Razavi S, Derwish L, Komatsu Y, Licursi M, Hirasawa K. Restoration of IRF1-dependent anticancer effects by MEK inhibition in human cancer cells. *Cancer Lett.* 2015;357(2):575–581.
416. Woodworth CD, Michael E, Marker D, Allen S, Smith L, Nees M. Inhibition of the epidermal growth factor receptor increases expression of genes that stimulate inflammation, apoptosis, and cell attachment. *Mol Cancer Ther.* 2005;4(4):650–658.
417. Blumenberg M. Profiling and metaanalysis of epidermal keratinocytes responses to epidermal growth factor. *BMC Genomics.* 2013;14:85.
418. Paul T, Schumann C, Rudiger S, et al. Cytokine regulation by epidermal growth factor receptor inhibitors and epidermal growth factor receptor inhibitor associated skin toxicity in cancer patients. *Eur J Cancer.* 2014;50(11):1855–1863.
419. Mascia F, Mariani V, Girolomoni G, Pastore S. Blockade of the EGF receptor induces a deranged chemokine expression in keratinocytes leading to enhanced skin inflammation. *Am J Pathol.* 2003;163(1):303–312.
420. Lupberger J, Duong FH, Fofana I, et al. Epidermal growth factor receptor signaling impairs the antiviral activity of interferon-alpha. *Hepatology.* 2013;58(4):1225–1235.
421. Kalinowski A, Ueki I, Min-Oo G, et al. EGFR activation suppresses respiratory virus-induced IRF1-dependent CXCL10 production. *Am J Physiol Lung Cell Mol Physiol.* 2014;307(2):L186–L196.
422. Zaheer RS, Koetzler R, Holden NS, Wiehler S, Proud D. Selective transcriptional down-regulation of human rhinovirus-induced production of CXCL10 from airway epithelial cells via the MEK1 pathway. *J Immunol.* 2009;182(8):4854–4864.
423. Campbell P, Morton PE, Takeichi T, et al. Epithelial inflammation resulting from an inherited loss-of-function mutation in EGFR. *J Invest Dermatol.* 2014;134(10):2570–2578.
424. Stagg J, Loi S, Divisekera U, et al. Anti-ErbB-2 mAb therapy requires type I and II interferons and synergizes with anti-PD-1 or anti-CD137 mAb therapy. *Proc Natl Acad Sci USA.* 2011;108(17):7142–7147.

425. Zitvogel L, Galluzzi L, Kepp O, Smyth MJ, Kroemer G. Type I interferons in anticancer immunity. *Nat Rev Immunol.* 2015;15(7):405–414.
426. Pollack BP, Sapkota B, Cartee TV. Epidermal growth factor receptor inhibition augments the expression of MHC class I and II genes. *Clin Cancer Res.* 2011;17(13):4400–4413.
427. Sullivan R, Graham CH. Hypoxia-driven selection of the metastatic phenotype. *Cancer Metastasis Rev.* 2007;26(2):319–331.
428. Mottet D, Dumont V, Deccache Y, et al. Regulation of hypoxia-inducible factor-1alpha protein level during hypoxic conditions by the phosphatidylinositol 3-kinase/Akt/glycogen synthase kinase 3beta pathway in HepG2 cells. *J Biol Chem.* 2003;278(33):31277–31285.
429. Tsuchihara K, Suzuki Y, Wakaguri H, et al. Massive transcriptional start site analysis of human genes in hypoxia cells. *Nucleic Acids Res.* 2009;37(7):2249–2263.
430. Sonveaux P, Vegran F, Schroeder T, et al. Targeting lactate-fueled respiration selectively kills hypoxic tumor cells in mice. *J Clin Invest.* 2008;118(12):3930–3942.
431. Gatenby RA, Gillies RJ. A microenvironmental model of carcinogenesis. *Nat Rev Cancer.* 2008;8(1):56–61.
432. Black WC, Welch HG. Advances in diagnostic imaging and overestimations of disease prevalence and the benefits of therapy. *N Engl J Med.* 1993;328(17):1237–1243.
433. Fukumura D, Duda DG, Munn LL, Jain RK. Tumor microvasculature and microenvironment: novel insights through intravital imaging in pre-clinical models. *Microcirculation.* 2010;17(3):206–225.
434. Matsuda Y, Yoshimura H, Suzuki T, Uchida E, Naito Z, Ishiwata T. Inhibition of fibroblast growth factor receptor 2 attenuates proliferation and invasion of pancreatic cancer. *Cancer Sci.* 2014;105(9):1212–1219.
435. Tang D, Gao J, Wang S, et al. Cancer-associated fibroblasts promote angiogenesis in gastric cancer through galectin-1 expression. *Tumour Biol.* 2016;37(2):1889–1899.
436. Amelio I, Melino G. The p53 family and the hypoxia-inducible factors (HIFs): determinants of cancer progression. *Trends Biochem Sci.* 2015;40(8):425–434.
437. Kim KS, Rajagopal V, Gonsalves C, Johnson C, Kalra VK. A novel role of hypoxia-inducible factor in cobalt chloride- and hypoxia-mediated expression of IL-8 chemokine in human endothelial cells. *J Immunol.* 2006;177(10):7211–7224.
438. Nizet V, Johnson RS. Interdependence of hypoxic and innate immune responses. *Nat Rev Immunol.* 2009;9(9):609–617.
439. Forsythe JA, Jiang BH, Iyer NV, et al. Activation of vascular endothelial growth factor gene transcription by hypoxia-inducible factor 1. *Mol Cell Biol.* 1996;16(9):4604–4613.
440. Semenza GL, Wang GL. A nuclear factor induced by hypoxia via de novo protein synthesis binds to the human erythropoietin gene enhancer at a site required for transcriptional activation. *Mol Cell Biol.* 1992;12(12):5447–5454.
441. Koskimaki JE, Karagiannis ED, Rosca EV, et al. Peptides derived from type IV collagen, CXC chemokines, and thrombospondin-1 domain-containing proteins inhibit neovascularization and suppress tumor growth in MDA-MB-231 breast cancer xenografts. *Neoplasia.* 2009;11(12):1285–1291.
442. Rosca EV, Lal B, Koskimaki JE, Popel AS, Laterra J. Collagen IV and CXC chemokine-derived antiangiogenic peptides suppress glioma xenograft growth. *Anticancer Drugs.* 2012;23(7):706–712.
443. Osada-Oka M, Ikeda T, Akiba S, Sato T. Hypoxia stimulates the autocrine regulation of migration of vascular smooth muscle cells via HIF-1alpha-dependent expression of thrombospondin-1. *J Cell Biochem.* 2008;104(5):1918–1926.
444. Distler JH, Jungel A, Pileckyte M, et al. Hypoxia-induced increase in the production of extracellular matrix proteins in systemic sclerosis. *Arthritis Rheum.* 2007;56(12):4203–4215.

445. Fukumura D, Xavier R, Sugiura T, et al. Tumor induction of VEGF promoter activity in stromal cells. *Cell.* 1998;94(6):715–725.
446. Huang S, Van Arsdall M, Tedjarati S, et al. Contributions of stromal metalloproteinase-9 to angiogenesis and growth of human ovarian carcinoma in mice. *J Natl Cancer Inst.* 2002;94(15):1134–1142.
447. Newman AC, Nakatsu MN, Chou W, Gershon PD, Hughes CC. The requirement for fibroblasts in angiogenesis: fibroblast-derived matrix proteins are essential for endothelial cell lumen formation. *Mol Biol Cell.* 2011;22(20):3791–3800.
448. Smith-McCune KK, Weidner N. Demonstration and characterization of the angiogenic properties of cervical dysplasia. *Cancer Res.* 1994;54(3):800–804.
449. Smith-McCune K, Zhu YH, Hanahan D, Arbeit J. Cross-species comparison of angiogenesis during the premalignant stages of squamous carcinogenesis in the human cervix and K14-HPV16 transgenic mice. *Cancer Res.* 1997;57(7):1294–1300.
450. Hanahan D, Folkman J. Patterns and emerging mechanisms of the angiogenic switch during tumorigenesis. *Cell.* 1996;86(3):353–364.
451. Obermair A, Bancher-Todesca D, Bilgi S, et al. Correlation of vascular endothelial growth factor expression and microvessel density in cervical intraepithelial neoplasia. *J Natl Cancer Inst.* 1997;89(16):1212–1217.
452. Dellas A, Moch H, Schultheiss E, et al. Angiogenesis in cervical neoplasia: microvessel quantitation in precancerous lesions and invasive carcinomas with clinicopathological correlations. *Gynecol Oncol.* 1997;67(1):27–33.
453. Dobbs SP, Hewett PW, Johnson IR, Carmichael J, Murray JC. Angiogenesis is associated with vascular endothelial growth factor expression in cervical intraepithelial neoplasia. *Br J Cancer.* 1997;76(11):1410–1415.
454. Lee JS, Kim HS, Park JT, Lee MC, Park CS. Expression of vascular endothelial growth factor in the progression of cervical neoplasia and its relation to angiogenesis and p53 status. *Anal Quant Cytol Histol.* 2003;25(6):303–311.
455. Nair P, Gangadevi T, Jayaprakash PG, Nair MB, Nair MK, Pillai MR. Increased angiogenesis in the uterine cervix associated with human papillomavirus infection. *Pathol Res Pract.* 1999;195(3):163–169.
456. Jain RK. Normalizing tumor microenvironment to treat cancer: bench to bedside to biomarkers. *J Clin Oncol.* 2013;31(17):2205–2218.
457. Chan DA, Giaccia AJ. Hypoxia, gene expression, and metastasis. *Cancer Metastasis Rev.* 2007;26(2):333–339.
458. Stepan D, Simionescu C, Stepan A, Muntean M, Voinea B. VEGF and CD105 immunoexpression in squamous cervical carcinomas and associated precancerous lesions. *Rom J Morphol Embryol.* 2012;53(3):585–589.
459. Abulafia O, Triest WE, Sherer DM. Angiogenesis in squamous cell carcinoma in situ and microinvasive carcinoma of the uterine cervix. *Obstet Gynecol.* 1996;88(6):927–932.
460. Guidi AJ, Abu-Jawdeh G, Berse B, et al. Vascular permeability factor (vascular endothelial growth factor) expression and angiogenesis in cervical neoplasia. *J Natl Cancer Inst.* 1995;87(16):1237–1245.
461. Sotiropoulou M, Diakomanolis E, Elsheikh A, Loutradis D, Markaki S, Michalas S. Angiogenic properties of carcinoma in situ and microinvasive carcinoma of the uterine cervix. *Eur J Gynaecol Oncol.* 2004;25(2):219–221.
462. Wiggins DL, Granai CO, Steinhoff MM, Calabresi P. Tumor angiogenesis as a prognostic factor in cervical carcinoma. *Gynecol Oncol.* 1995;56(3):353–356.
463. Triratanachat S, Niruthisard S, Trivijitsilp P, Tresukosol D, Jarurak N. Angiogenesis in cervical intraepithelial neoplasia and early-staged uterine cervical squamous cell carcinoma: clinical significance. *Int J Gynecol Cancer.* 2006;16(2):575–580.

464. Tjalma W, Sonnemans H, Weyler J, Van Marck E, Van Daele A, van Dam P. Angiogenesis in cervical intraepithelial neoplasia and the risk of recurrence. *Am J Obstet Gynecol.* 1999;181(3):554–559.
465. Krill LS, Tewari KS. Exploring the therapeutic rationale for angiogenesis blockade in cervical cancer. *Clin Ther.* 2015;37(1):9–19.
466. Lee WY, Huang SC, Hsu KF, Tzeng CC, Shen WL. Roles for hypoxia-regulated genes during cervical carcinogenesis: somatic evolution during the hypoxia-glycolysis-acidosis sequence. *Gynecol Oncol.* 2007;108(2):377–384.
467. Birner P, Schindl M, Obermair A, Plank C, Breitenecker G, Oberhuber G. Overexpression of hypoxia-inducible factor 1alpha is a marker for an unfavorable prognosis in early-stage invasive cervical cancer. *Cancer Res.* 2000;60(17):4693–4696.
468. Azuma Y, Chou SC, Lininger RA, Murphy BJ, Varia MA, Raleigh JA. Hypoxia and differentiation in squamous cell carcinomas of the uterine cervix: pimonidazole and involucrin. *Clin Cancer Res.* 2003;9(13):4944–4952.
469. Lu ZH, Wright JD, Belt B, Cardiff RD, Arbeit JM. Hypoxia-inducible factor-1 facilitates cervical cancer progression in human papillomavirus type 16 transgenic mice. *Am J Pathol.* 2007;171(2):667–681.
470. Fujimoto J, Sato E, Alam SM, et al. Plausible linkage of hypoxia-inducible factor (HIF) in uterine endometrial cancers. *Oncology.* 2006;71(1–2):95–101.
471. Luczak MW, Roszak A, Pawlik P, Kedzia H, Lianeri M, Jagodzinski PP. Increased expression of HIF-1A and its implication in the hypoxia pathway in primary advanced uterine cervical carcinoma. *Oncol Rep.* 2011;26(5):1259–1264.
472. Mayer A, Wree A, Hockel M, Leo C, Pilch H, Vaupel P. Lack of correlation between expression of HIF-1alpha protein and oxygenation status in identical tissue areas of squamous cell carcinomas of the uterine cervix. *Cancer Res.* 2004;64(16):5876–5881.
473. Semenza GL. Evaluation of HIF-1 inhibitors as anticancer agents. *Drug Discov Today.* 2007;12(19–20):853–859.
474. Rodolico V, Arancio W, Amato MC, et al. Hypoxia inducible factor-1 alpha expression is increased in infected positive HPV16 DNA oral squamous cell carcinoma and positively associated with HPV16 E7 oncoprotein. *Infect Agents Cancer.* 2011;6(1):18.
475. Lee JS, Kim HS, Jung JJ, Lee MC, Park CS. Expression of vascular endothelial growth factor in adenocarcinomas of the uterine cervix and its relation to angiogenesis and p53 and c-erbB-2 protein expression. *Gynecol Oncol.* 2002;85(3):469–475.
476. No JH, Jo H, Kim SH, et al. Expression of vascular endothelial growth factor and hypoxia inducible factor-1alpha in cervical neoplasia. *Ann NY Acad Sci.* 2009;1171:105–110.
477. Pilch H, Schlenger K, Steiner E, Brockerhoff P, Knapstein P, Vaupel P. Hypoxia-stimulated expression of angiogenic growth factors in cervical cancer cells and cervical cancer-derived fibroblasts. *Int J Gynecol Cancer.* 2001;11(2):137–142.
478. Garbow JR, Santeford AC, Anderson JR, Engelbach JA, Arbeit JM. Magnetic resonance imaging defines cervicovaginal anatomy, cancer, and VEGF trap antiangiogenic efficacy in estrogen-treated K14-HPV16 transgenic mice. *Cancer Res.* 2009;69(20):7945–7952.
479. Giraudo E, Inoue M, Hanahan D. An amino-bisphosphonate targets MMP-9-expressing macrophages and angiogenesis to impair cervical carcinogenesis. *J Clin Invest.* 2004;114(5):623–633.
480. Bequet-Romero M, Lopez-Ocejo O. Angiogenesis modulators expression in culture cell lines positives for HPV-16 oncoproteins. *Biochem Biophys Res Commun.* 2000;277(1):55–61.
481. Lopez-Ocejo O, Viloria-Petit A, Bequet-Romero M, Mukhopadhyay D, Rak J, Kerbel RS. Oncogenes and tumor angiogenesis: the HPV-16 E6 oncoprotein activates the vascular endothelial growth factor (VEGF) gene promoter in a p53 independent manner. *Oncogene.* 2000;19(40):4611–4620.

482. Tang X, Zhang Q, Nishitani J, Brown J, Shi S, Le AD. Overexpression of human papillomavirus type 16 oncoproteins enhances hypoxia-inducible factor 1 alpha protein accumulation and vascular endothelial growth factor expression in human cervical carcinoma cells. *Clin Cancer Res*. 2007;13(9):2568–2576.
483. Walker J, Smiley LC, Ingram D, Roman A. Expression of human papillomavirus type 16 E7 is sufficient to significantly increase expression of angiogenic factors but is not sufficient to induce endothelial cell migration. *Virology*. 2011;410(2):283–290.
484. Ravi R, Mookerjee B, Bhujwalla ZM, et al. Regulation of tumor angiogenesis by p53-induced degradation of hypoxia-inducible factor 1alpha. *Genes Dev*. 2000;14(1):34–44.
485. Clere N, Bermont L, Fauconnet S, et al. The human papillomavirus type 18 E6 oncoprotein induces vascular endothelial growth factor 121 (VEGF121) transcription from the promoter through a p53-independent mechanism. *Exp Cell Res*. 2007;313(15):3239–3250.
486. Bodily JM, Mehta KP, Laimins LA. Human papillomavirus E7 enhances hypoxia-inducible factor 1-mediated transcription by inhibiting binding of histone deacetylases. *Cancer Res*. 2011;71(3):1187–1195.
487. Kim SH, Juhnn YS, Kang S, et al. Human papillomavirus 16 E5 up-regulates the expression of vascular endothelial growth factor through the activation of epidermal growth factor receptor, MEK/ERK1,2 and PI3K/Akt. *Cell Mol Life Sci*. 2006;63(7–8):930–938.
488. Zhou J, Schmid T, Schnitzer S, Brune B. Tumor hypoxia and cancer progression. *Cancer Lett*. 2006;237(1):10–21.
489. Bardos JI, Ashcroft M. Negative and positive regulation of HIF-1: a complex network. *Biochim Biophys Acta*. 2005;1755(2):107–120.
490. An WG, Kanekal M, Simon MC, Maltepe E, Blagosklonny MV, Neckers LM. Stabilization of wild-type p53 by hypoxia-inducible factor 1alpha. *Nature*. 1998;392(6674): 405–408.
491. Alarcon R, Koumenis C, Geyer RK, Maki CG, Giaccia AJ. Hypoxia induces p53 accumulation through MDM2 down-regulation and inhibition of E6-mediated degradation. *Cancer Res*. 1999;59(24):6046–6051.
492. Vousden KH, Ryan KM. p53 and metabolism. *Nat Rev Cancer*. 2009;9(10):691–700.
493. Schmid T, Zhou J, Kohl R, Brune B. p300 relieves p53-evoked transcriptional repression of hypoxia-inducible factor-1 (HIF-1). *Biochem J*. 2004;380(pt 1):289–295.
494. Rempe DA, Lelli KM, Vangeison G, Johnson RS, Federoff HJ. In cultured astrocytes, p53 and MDM2 do not alter hypoxia-inducible factor-1alpha function regardless of the presence of DNA damage. *J Biol Chem*. 2007;282(22):16187–16201.
495. Pan Y, Oprysko PR, Asham AM, Koch CJ, Simon MC. p53 cannot be induced by hypoxia alone but responds to the hypoxic microenvironment. *Oncogene*. 2004;23(29):4975–4983.
496. Wenger RH, Camenisch G, Desbaillets I, Chilov D, Gassmann M. Up-regulation of hypoxia-inducible factor-1alpha is not sufficient for hypoxic/anoxic p53 induction. *Cancer Res*. 1998;58(24):5678–5680.
497. Hammond EM, Denko NC, Dorie MJ, Abraham RT, Giaccia AJ. Hypoxia links ATR and p53 through replication arrest. *Mol Cell Biol*. 2002;22(6):1834–1843.
498. Koumenis C, Alarcon R, Hammond E, et al. Regulation of p53 by hypoxia: dissociation of transcriptional repression and apoptosis from p53-dependent transactivation. *Mol Cell Biol*. 2001;21(4):1297–1310.
499. Hansson LO, Friedler A, Freund S, Rudiger S, Fersht AR. Two sequence motifs from HIF-1alpha bind to the DNA-binding site of p53. *Proc Natl Acad Sci USA*. 2002;99(16):10305–10309.
500. Blagosklonny MV, An WG, Romanova LY, Trepel J, Fojo T, Neckers L. p53 inhibits hypoxia-inducible factor-stimulated transcription. *J Biol Chem*. 1998;273(20): 11995–11998.

501. Yang J, Ahmed A, Poon E, et al. Small-molecule activation of p53 blocks hypoxia-inducible factor 1alpha and vascular endothelial growth factor expression in vivo and leads to tumor cell apoptosis in normoxia and hypoxia. *Mol Cell Biol.* 2009;29(8):2243–2253.
502. Carroll VA, Ashcroft M. Regulation of angiogenic factors by HDM2 in renal cell carcinoma. *Cancer Res.* 2008;68(2):545–552.
503. Kaluzova M, Kaluz S, Lerman MI, Stanbridge EJ. DNA damage is a prerequisite for p53-mediated proteasomal degradation of HIF-1alpha in hypoxic cells and down-regulation of the hypoxia marker carbonic anhydrase IX. *Mol Cell Biol.* 2004;24(13):5757–5766.
504. Blagosklonny MV, Giannakakou P, Romanova LY, Ryan KM, Vousden KH, Fojo T. Inhibition of HIF-1- and wild-type p53-stimulated transcription by codon Arg175 p53 mutants with selective loss of functions. *Carcinogenesis.* 2001;22(6):861–867.
505. Zhang L, Yu D, Hu M, et al. Wild-type p53 suppresses angiogenesis in human leiomyosarcoma and synovial sarcoma by transcriptional suppression of vascular endothelial growth factor expression. *Cancer Res.* 2000;60(13):3655 3661.
506. Schwartzenberg-Bar-Yoseph F, Armoni M, Karnieli E. The tumor suppressor p53 down-regulates glucose transporters GLUT1 and GLUT4 gene expression. *Cancer Res.* 2004;64(7):2627–2633.
507. Pal S, Datta K, Mukhopadhyay D. Central role of p53 on regulation of vascular permeability factor/vascular endothelial growth factor (VPF/VEGF) expression in mammary carcinoma. *Cancer Res.* 2001;61(18):6952–6957.
508. LaRusch GA, Jackson MW, Dunbar JD, Warren RS, Donner DB, Mayo LD. Nutlin3 blocks vascular endothelial growth factor induction by preventing the interaction between hypoxia inducible factor 1alpha and Hdm2. *Cancer Res.* 2007;67(2):450–454.
509. Teodoro JG, Evans SK, Green MR. Inhibition of tumor angiogenesis by p53: a new role for the guardian of the genome. *J Mol Med.* 2007;85(11):1175–1186.
510. Dameron KM, Volpert OV, Tainsky MA, Bouck N. Control of angiogenesis in fibroblasts by p53 regulation of thrombospondin-1. *Science.* 1994;265(5178):1582–1584.
511. Brat DJ, Kaur B, Van Meir EG. Genetic modulation of hypoxia induced gene expression and angiogenesis: relevance to brain tumors. *Front Biosci.* 2003;8:d100–d116.
512. Teodoro JG, Parker AE, Zhu X, Green MR. p53-mediated inhibition of angiogenesis through up-regulation of a collagen prolyl hydroxylase. *Science.* 2006;313(5789): 968–971.
513. Kenneth NS, Rocha S. Regulation of gene expression by hypoxia. *Biochem J.* 2008;414(1):19–29.
514. Pillai S, Kovacs M, Chellappan S. Regulation of vascular endothelial growth factor receptors by Rb and E2F1: role of acetylation. *Cancer Res.* 2010;70(12):4931–4940.
515. Lendahl U, Lee KL, Yang H, Poellinger L. Generating specificity and diversity in the transcriptional response to hypoxia. *Nat Rev Genet.* 2009;10(12):821–832.
516. Antinore MJ, Birrer MJ, Patel D, Nader L, McCance DJ. The human papillomavirus type 16 E7 gene product interacts with and trans-activates the AP1 family of transcription factors. *EMBO J.* 1996;15(8):1950–1960.
517. Hasan UA, Zannetti C, Parroche P, et al. The human papillomavirus type 16 E7 oncoprotein induces a transcriptional repressor complex on the Toll-like receptor 9 promoter. *J Exp Med.* 2013;210(7):1369–1387.
518. Subbaramaiah K, Dannenberg AJ. Cyclooxygenase-2 transcription is regulated by human papillomavirus 16 E6 and E7 oncoproteins: evidence of a corepressor/coactivator exchange. *Cancer Res.* 2007;67(8):3976–3985.
519. Kuner R, Vogt M, Sultmann H, et al. Identification of cellular targets for the human papillomavirus E6 and E7 oncogenes by RNA interference and transcriptome analyses. *J Mol Med.* 2007;85(11):1253–1262.

520. McMurray HR, McCance DJ. Human papillomavirus type 16 E6 activates TERT gene transcription through induction of c-Myc and release of USF-mediated repression. *J Virol.* 2003;77(18):9852–9861.
521. Wang YW, Chang HS, Lin CH, Yu WC. HPV-18 E7 conjugates to c-Myc and mediates its transcriptional activity. *Int J Biochem Cell Biol.* 2007;39(2):402–412.
522. Delcuratolo M, Fertey J, Schneider M, et al. Papillomavirus-associated tumor formation critically depends on c-Fos expression induced by viral protein E2 and bromodomain protein Brd4. *PLoS Pathog.* 2016;12(1):e1005366.
523. Gabellini C, Del Bufalo D, Zupi G. Involvement of RB gene family in tumor angiogenesis. *Oncogene.* 2006;25(38):5326–5332.
524. Sahebali S, Van den Eynden G, Murta EF, et al. Stromal issues in cervical cancer: a review of the role and function of basement membrane, stroma, immune response and angiogenesis in cervical cancer development. *Eur J Cancer Prev.* 2010;19(3): 204–215.
525. Page-McCaw A, Ewald AJ, Werb Z. Matrix metalloproteinases and the regulation of tissue remodelling. *Nat Rev Mol Cell Biol.* 2007;8(3):221–233.
526. Chang C, Werb Z. The many faces of metalloproteases: cell growth, invasion, angiogenesis and metastasis. *Trends Cell Biol.* 2001;11(11):S37–S43.
527. Kessenbrock K, Plaks V, Werb Z. Matrix metalloproteinases: regulators of the tumor microenvironment. *Cell.* 2010;141(1):52–67.
528. Itoh Y. MT1-MMP: a key regulator of cell migration in tissue. *IUBMB Life.* 2006;58(10):589–596.
529. Poincloux R, Lizarraga F, Chavrier P. Matrix invasion by tumour cells: a focus on MT1-MMP trafficking to invadopodia. *J Cell Sci.* 2009;122(pt 17):3015–3024.
530. Sounni NE, Devy L, Hajitou A, et al. MT1-MMP expression promotes tumor growth and angiogenesis through an up-regulation of vascular endothelial growth factor expression. *FASEB J.* 2002;16(6):555–564.
531. Stuelten CH, DaCosta Byfield S, Arany PR, Karpova TS, Stetler-Stevenson WG, Roberts AB. Breast cancer cells induce stromal fibroblasts to express MMP-9 via secretion of TNF-alpha and TGF-beta. *J Cell Sci.* 2005;118(pt 10):2143–2153.
532. Nuovo GJ. In situ detection of PCR-amplified metalloproteinase cDNAs, their inhibitors and human papillomavirus transcripts in cervical carcinoma cell lines. *Int J Cancer.* 1997;71(6):1056–1060.
533. Yu Q, Stamenkovic I. Cell surface-localized matrix metalloproteinase-9 proteolytically activates TGF-beta and promotes tumor invasion and angiogenesis. *Genes Dev.* 2000;14(2):163–176.
534. Buontempo PJ, Jubin RG, Buontempo CA, Wagner NE, Reyes GR, Baroudy BM. Antiviral activity of transiently expressed IFN-kappa is cell-associated. *J Interferon Cytokine Res.* 2006;26(1):40–52.
535. Smola-Hess S, Pahne J, Mauch C, Zigrino P, Smola H, Pfister HJ. Expression of membrane type 1 matrix metalloproteinase in papillomavirus-positive cells: role of the human papillomavirus (HPV) 16 and HPV8 E7 gene products. *J Gen Virol.* 2005;86(pt 5):1291–1296.
536. Zhu D, Ye M, Zhang W. E6/E7 oncoproteins of high risk HPV-16 upregulate MT1-MMP, MMP-2 and MMP-9 and promote the migration of cervical cancer cells. *Int J Clin Exp Pathol.* 2015;8(5):4981–4989.
537. Kaewprag J, Umnajvijit W, Ngamkham J, Ponglikitmongkol M. HPV16 oncoproteins promote cervical cancer invasiveness by upregulating specific matrix metalloproteinases. *PLoS One.* 2013;8(8):e71611.
538. Cardeal LB, Boccardo E, Termini L, et al. HPV16 oncoproteins induce MMPs/RECK-TIMP-2 imbalance in primary keratinocytes: possible implications in cervical carcinogenesis. *PLoS One.* 2012;7(3):e33585.

539. Barbaresi S, Cortese MS, Quinn J, Ashrafi GH, Graham SV, Campo MS. Effects of human papillomavirus type 16 E5 deletion mutants on epithelial morphology: functional characterization of each transmembrane domain. *J Gen Virol.* 2010;91(pt 2):521–530.
540. Moore PS, Chang Y. Why do viruses cause cancer? highlights of the first century of human tumour virology. *Nat Rev.* 2010;10(12):878–889.
541. Zheng Y, Stamminger T, Hearing P. E2F/Rb family proteins mediate interferon induced repression of adenovirus immediate early transcription to promote persistent viral infection. *PLoS Pathog.* 2016;12(1):e1005415.
542. Sahebali S, et al. Stromal issues in cervical cancer: a review of the role and function of basement membrane, stroma, immune response and angiogenesis in cervical cancer development. *Eur J Cancer Prev.* 2010;19(3):204–215.
543. Page-McCaw A, Ewald AJ, Werb Z. Matrix metalloproteinases and the regulation of tissue remodelling. *Nat Rev Mol Cell Biol.* 2007;8(3):221–233.

PART III

Pancreatic Cancer: Altered Signaling Networks and Emerging Treatment Strategies

CHAPTER SIX

Molecular Pathogenesis of Pancreatic Cancer

T.J. Grant, K. Hua, A. Singh[1]
Boston University School of Medicine, Boston, MA, United States
[1]Corresponding author: e-mail address: asingh3@bu.edu

Contents

Abstract

Pancreatic cancers arise predominantly from ductal epithelial cells of the exocrine pancreas and are of the ductal adenocarcinoma histological subtype (PDAC). PDAC is an aggressive disease associated with a poor clinical prognosis, weakly effective therapeutic options, and a lack of early detection methods. Furthermore, the genetic and phenotypic heterogeneity of PDAC complicates efforts to identify universally efficacious therapies. PDACs commonly harbor activating mutations in the *KRAS* oncogene, which is a potent driver of tumor initiation and maintenance. Inactivating mutations in tumor suppressor genes such as *CDKN2A*/p16, *TP53*, and *SMAD4* cooperate with *KRAS* mutations to cause aggressive PDAC tumor growth. PDAC can be classified into 3–4 molecular subtypes by global gene expression profiling. These subtypes can be distinguished by distinct molecular and phenotypic characteristics. This chapter will provide an overview of the current knowledge of PDAC pathogenesis at the genetic and molecular level as well as novel therapeutic opportunities to treat this highly aggressive disease.

Progress in Molecular Biology and Translational Science, Volume 144
ISSN 1877-1173
http://dx.doi.org/10.1016/bs.pmbts.2016.09.008

1. INTRODUCTION

The pancreas is a glandular organ with both endocrine and exocrine functions.[1,2] Its overall purpose is to maintain metabolic homeostasis by producing hormones that regulate blood glucose levels as well as enzymes that aid in digestion. The pancreas is derived from the embryonic foregut of the endodermal germ layer.[3] During embryonic development, two buds that ultimately give rise to the dorsal and ventral pancreas emerge from the foregut. As these buds expand, they are gradually repositioned over time until they come into contact and fuse together, forming the mature pancreas. Under the control of various developmental cues, pancreatic progenitor cells become acinar, endocrine, or ductal in function. Endocrine cells (α, β, and δ) secrete hormones-like insulin, glucagon, and somatostatin into the circulatory system to modulate blood glucose levels.[1] This homeostatic function ensures that the metabolic demands of various tissues and organs are met. Acinar cells in the ducts secrete enzymes like trypsinogen, chymotrypsinogen, lipase, and amylase into the pancreatic duct.[2] These enzymes subsequently enter into the small intestine, where they aid in the digestion of various dietary macromolecules such as proteins, carbohydrates, and lipids.

Pancreatic dysfunction can lead to a number of common diseases including diabetes, pancreatitis, and cancer.[4–6] Diabetes is the most prevalent of these diseases. However, cancer of the pancreas is by far the deadliest, and its etiology is often linked to other pancreatic disorders, including diabetes. Pancreatic cancer is the fourth leading cause of cancer-related deaths in the United States and is associated with a particularly poor prognosis.[7] Patients diagnosed with this disease exhibit a median overall survival of less than 6 months and a 5-year survival rate of roughly 8%. The poor prognosis associated with pancreatic cancer is attributed in part to poor methods of early detection.[8] Patients often remain asymptomatic until the disease has disseminated throughout the body. Additionally, the therapeutics used to treat pancreatic cancer are relatively ineffective, as they fail to extend patient survival more than several months.[9] Overcoming these challenges will be critical in the future treatment of the disease, as it is expected to become the second leading cause of cancer-related deaths in the United States by 2030.[10]

Pancreatic cancers can arise from either endocrine or exocrine cells. Thus, endocrine and exocrine tumors can be distinguished by histological appearance. Endocrine tumors are relatively uncommon and constitute less than 5% of all pancreatic cancers. They are associated with a median survival of 27 months and a 0.28-fold lower risk of mortality in comparison to the much more common pancreatic adenocarcinoma.[11] Endocrine tumors are commonly derived from pancreatic islet cells and often produce constitutively high levels of pancreatic hormones. They can be further categorized into insulinomas, glucagonomas, and gastrinomas depending on their cell of origin and the hormones that they secrete. Pancreatic endocrine tumors can be readily diagnosed due to their excessive hormone secretion, which leads to dramatic symptoms such as hypoglycemia or necrolytic migratory erythema (skin rash).[12] Pancreatic cancers derived from exocrine cells are much more common than endocrine tumors and can typically be classified into two histological subtypes. The pancreatic ductal adenocarcinoma (PDAC) subtype accounts for the majority of exocrine tumors and constitutes more than 90% of all pancreatic malignancies. PDACs are derived from epithelial cells that line the pancreatic duct and appear gland-like due to their origin.[11] These cancers frequently metastasize to the liver or lymph nodes.[13] Due to their lack of symptoms at the early stages of cancer development, PDACs are often diagnosed at a late stage, potentially after the cancer has already metastasized. As a result, anticancer therapeutics tend to be weakly effective due to the cancers having acquired strong cytoprotective mechanisms that promote drug resistance. Because of this aggressiveness and drug resistance, estimated median survival for PDAC can be as short as 4 months.[11]

PDACs are preceded by the development of hyperplastic lesions known as pancreatic intraepithelial neoplasias (PanINs) and intraductal papillary mucinous neoplasms (IPMNs) that are precancerous and exhibit a propensity to develop into cancer (Fig. 1). IPMNs look like papillae (finger-like

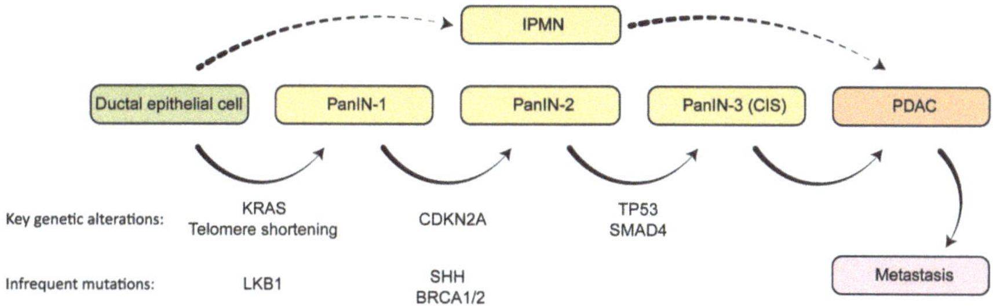

Fig. 1 Disease progression model of pancreatic cancer. Pancreatic cancer arises from two histological types of precursor lesions: PanINs and IPMNs. Through progressive stages of pathogenesis, molecular changes occur, leading to increasing degrees of nuclear and cytoskeletal abnormalities. Genetic alterations commonly observed in these lesions are indicated with respect to the stages in which they most often occur.

structures) that protrude into the pancreatic duct.[14] Mucinous tumors are the second most common histological subtype of pancreatic cancer, constituting less than 10% of cases. These tumors are usually much less invasive than adenocarcinomas at the time of diagnosis and have a 0.88-fold lower risk of mortality by comparison.[11] Mucinous tumors also originate from the pancreatic ductal epithelium, but secrete mucin, which can be seen in and around the cells, causing them to appear like they are "floating."[15,16] There are many other subtypes of pancreatic cancer, such as those that arise from acinar cells, which are undifferentiated and resemble liver cancers. However, these subtypes are rare and will not be discussed in this chapter.

This chapter will focus on the molecular etiology of PDAC, the most prevalent form of pancreatic malignancy. The core genetic alterations that contribute to PDAC pathogenesis will be discussed. Furthermore, the molecular subtypes of PDAC will be presented with a focus on their cellular origin and the genetic alterations associated with them. The diverse and deregulated signaling pathways that contribute to PDAC pathogenesis will also be described in detail. To conclude, the molecular characteristics of PDAC will be discussed in relation to the current therapeutic strategies employed to manage the disease in the clinic and future approaches that may further improve patient prognosis.

2. GENETIC ALTERATIONS IN PANCREATIC CANCER

Whole exome-sequencing studies have revealed that PDAC is a molecularly heterogeneous disease characterized by four common genetic alterations: oncogenic *KRAS* mutation and inactivation of the tumor suppressors *CDKN2A*, *TP53*, and *SMAD4* (Figs. 1 and 2A).[6] However, myriad additional genes are mutated in subsets of tumors, typically at a very low frequency ($\leq$10%), with many of these mutations not occurring in a recurrent manner. Further analysis of these infrequent alterations has revealed that they converge on a relatively small number of pathways and cellular processes including KRAS, transforming growth factor-β (TGF-β), WNT, NOTCH, and Hedgehog signaling as well as S-phase entry, axonal guidance, chromatin remodeling, DNA repair, and RNA processing.[17,18] Understanding how context-dependent interactions between these various genetically altered pathways contribute to PDAC progression is a key goal in the development of more selective and efficacious therapeutic modalities to treat the disease.

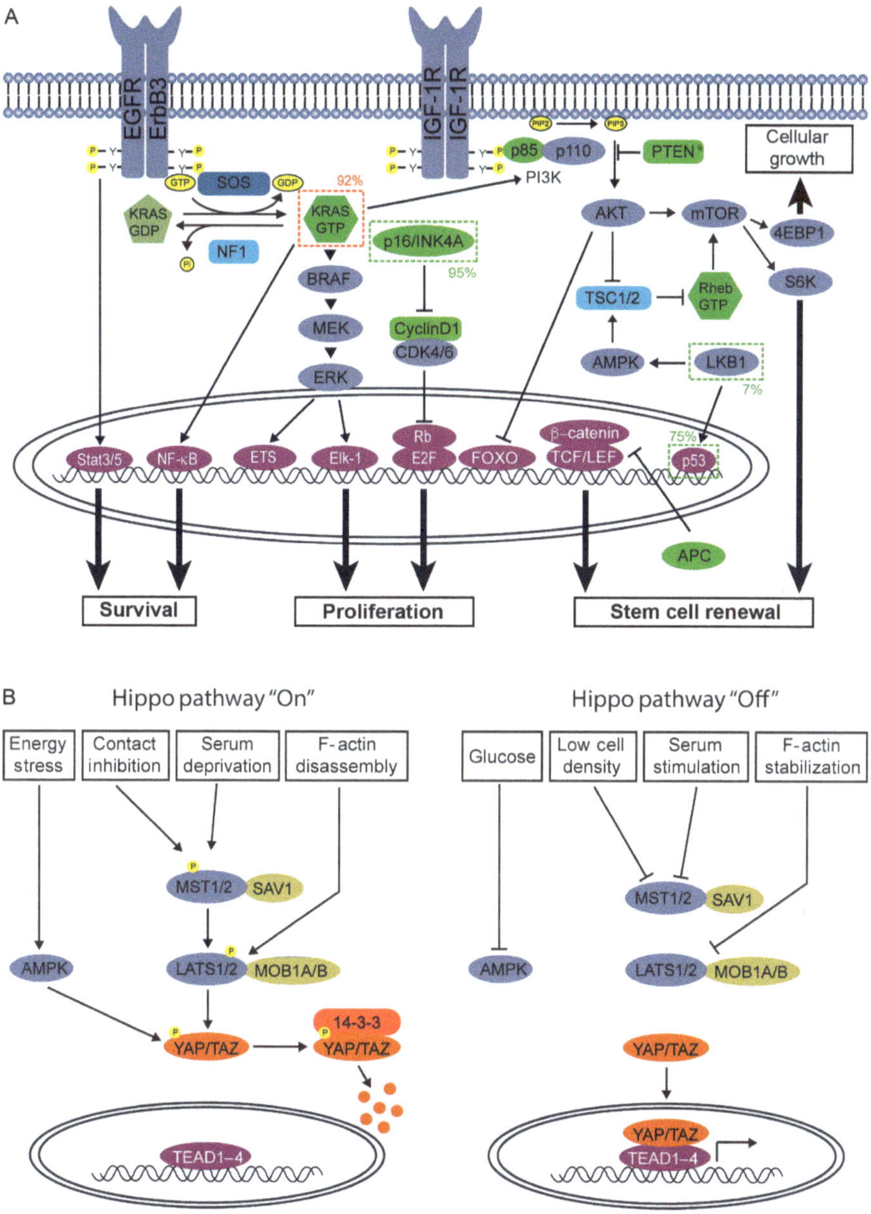

Fig. 2 Deregulated signaling networks in pancreatic cancer. (A) RTK and cell cycle-regulatory signaling networks frequently altered in pancreatic cancer. Oncogenes exhibiting gain-of-function mutations are indicated by a *dark dashed line*. Tumor suppressor genes altered in the disease are indicated by a *lighter dashed line*. The frequencies at which these genes are altered are also included. Oncogenic KRAS mutation cooperates with the loss of various tumor suppressor genes to promote cellular proliferation, growth, survival, and stem cell renewal. (B) Hippo signaling is frequently deregulated in pancreatic cancer. The ability of the pathway to restrict cell growth and induce apoptosis is mediated by a number of stimuli including cell density, glucose levels, serum levels, and cytoskeletal tension.

2.1 Oncogenic *KRAS* Mutations

Activating *KRAS* mutations are the defining genetic feature of PDAC progression and are found in approximately 92% of PDAC.[18] *KRAS*, a member of the *RAS* superfamily, encodes a small GTPase that regulates diverse cellular processes including cell proliferation, differentiation, survival, and migration (Fig. 2A).[19] Under normal physiological conditions, KRAS cycles through a GTP-bound active state and a GDP-bound inactive state. The transition between these two states is modulated by guanine nucleotide exchange factors (GEFs; e.g., Sos1), which catalyze the exchange of GDP for GTP, and GTPase-activating proteins (GAPs; e.g., NF1), which enhance the intrinsically weak ability of KRAS to hydrolyze GTP.[20,21] In quiescent cells, KRAS is predominantly GDP bound. However, upon growth factor stimulation, GEFs are able to bind to KRAS and promote its activation by catalyzing GDP–GTP exchange. In many solid cancers, KRAS or one of its closely related family members (HRAS and NRAS) undergo mutations that ultimately impair their ability to hydrolyze GTP. *HRAS* and *NRAS* mutations are typically not found in PDAC. Oncogenic RAS proteins are locked in an active state that results in constitutive stimulation of effector pathways capable of driving tumor development.[22] The major effector pathways downstream of active RAS are those mediated by RAF/MEK/ERK, PI3K, and RAL-GEF.

KRAS point mutations typically result in a single amino acid substitution at one of the three codons: G12, G13, or Q61.[21] All three substitutions occur in the catalytic domain of the GTPase. G12 and G13 mutations introduce a steric hindrance that blocks the formation of van der Waals interactions between KRAS and GAPs.[23] Q61 mutations disrupt the coordination of a water molecule required for GTP hydolysis.[24] Although all three mutations promote KRAS activation by impairing its GTPase activity, the frequencies at which they occur are vastly disparate. Substitutions at codon G12 (typically G12D or G12V) are the most prevalent (82%) followed by Q61 (14%) and G13 (<1%).[21] Interestingly, the three common KRAS mutant alleles exhibit contrasting functional properties. Tumors harboring KRAS Q61 substitutions display reduced MAPK activity and are associated with a better prognosis than substitutions at other codons.[25] This suggests that different *KRAS* mutations promote tumorigenesis in distinct ways.

Activating *KRAS* mutations are the earliest genetic alteration in the development of almost all PDACs (Fig. 1), as they are prevalent in greater than 90% of low-grade precursor PanIN lesions.[26] PDAC is preceded by the emergence of PanIN lesions, which are classified into three stages (I–III) based on the degree of architectural disorganization and nuclear abnormalities of the tissue.[6] In vivo studies of genetically engineered mouse models (GEMMs) have demonstrated that PanINs readily develop as a consequence of *KRAS*-activating mutations but do not ultimately give rise to PDAC. Rather, the subsequent and combined inactivation of multiple tumor suppressor genes (TSGs; i.e., *CDK2NA*, *TP53*, and *SMAD4*) is also required for malignant tumor progression.[6,27,28] These findings suggest that *KRAS* mutations alone are not sufficient to promote PDAC progression.

Although oncogenic KRAS promotes the initiation of PDAC development, its role in tumor maintenance is less clear. In a GEMM of PDAC, pancreas-specific induction of $KRAS^{G12D}$ expression coupled with loss of a single *TP53* allele results in the dedifferentiation of normal epithelial cells and the development of PanINs.[29] Extinction of oncogenic KRAS leads to the regression of these lesions, suggesting that it is required for PanIN progression. In a subsequent study, pancreas-specific induction of $KRAS^{G12D}$ expression coupled with the loss of one or both *TP53* alleles led to the development of PDAC.[30] Similar to the effects observed in PanINs, extinction of oncogenic KRAS in these PDAC lesions led to robust tumor regression. Taken together these studies suggest that oncogenic KRAS plays an essential role in both tumor initiation and maintenance.

In contrast to observations using GEMMs, which bear similarities to subsets of human PDAC, alternative studies using human PDAC-derived cancer cell lines suggest that KRAS is dispensable in certain contexts.[31,32] Approximately half of human-derived PDAC cell lines readily undergo apoptosis following RNAi-mediated depletion of the GTPase, indicating a state of KRAS oncogene "dependency." The discrepancy between these findings and those obtained in the GEMM studies may be attributed to a number of factors. First, pharmacological vulnerabilities in GEMMs do not often translate to clinical trials of human cancers, as has been observed with inhibitors of the MAPK and PI3K pathways.[20] This suggests that, although the same defined genetic mutations from human PDAC promote disease development in GEMMs, the resulting tumors likely display altered molecular characteristics compared to their human counterparts. Another potential explanation for the discrepancy between mouse and human cell line models is that some GEMM-derived tumors may be transiently dependent on KRAS during early stages of PDAC progression. In mouse xenografts, MAPK, PI3K, and RalGEF are all required for tumor initiation.[33] However, only the PI3K pathway is necessary for tumor maintenance, suggesting a reduced need for oncogenic KRAS.

Furthermore, activation of PI3K/AKT signaling by the tumor microenvironment is able to rescue the loss of mutant KRAS. This suggests that KRAS may become dispensable in the later stages of PDAC progression if the PI3K pathway is activated. Such a context may be difficult to model in GEMMs due to the rapid disease onset that results in a severe "dependency" on mutant KRAS for tumor maintenance.

More recent studies in GEMMs have demonstrated that the need for KRAS can be bypassed by enhanced mitochondrial activity or Hippo pathway deregulation.[34–36] In both cases, depletion of the KRAS protein resulted in robust tumor regression, consistent with previous in vivo work. However, subpopulations of cells were able to drive tumor relapse. In the case of Hippo pathway deregulation, surviving cells expressed lower KRAS protein levels and were unresponsive to KRAS depletion. Furthermore, the relapsed tumors exhibited a more mesenchymal morphology, reminiscent of the phenotype observed in PDAC cell lines that were insensitive to KRAS depletion.[31] Similarly, surviving cells that bypassed KRAS dependence by increased oxidative phosphorylation were also more stem-like in morphology. Taken together, these studies demonstrate that KRAS may be dispensable in subsets of tumors or subsets of cells within a tumor that can bypass the dependence on oncogenic KRAS signaling. As discussed later, this poses a new challenge to developing effective therapies to treat PDAC by targeting KRAS or its downstream effectors.

2.2 Tumor Suppressor Genes

TSGs restrict cell proliferation in the presence of oncogenic driver mutations by inducing cell cycle arrest, apoptosis, or senescence. Hence, the functional inactivation of TSGs is essential for tumorigenesis.[37] A number of TSGs are functionally lost in PDAC, the three most common of which are *CDKN2A*, *TP53*, and *SMAD4/DPC4*.[6] The inactivation of these genes occurs in a sequential manner following the appearance of oncogenic *KRAS* mutation in concordance with the multistage carcinogenesis model proposed by Vogelstein and colleagues (Fig. 1).[38] Together, TSG alterations constitute a major driving force of PDAC development.

CDKN2A encodes the p16/INK4A protein, a cyclin-dependent kinase inhibitor responsible for blocking entry into S-phase of the cell cycle.[39] It is the earliest and most frequently altered TSG observed in PDAC. Approximately 95% of tumors exhibit functional loss of this gene as a consequence of intragenic mutation coupled with loss of the second allele (40%), homozygous deletion (40%), or promoter hypermethylation (15%).[40,41] *CDKN2A* inactivation is typically detected in moderately advanced PanINs, prior to the development of PDAC.[42] *CDKN2A* loss is crucial in disease pathogenesis, as p16 induces senescence following the introduction of oncogenic *KRAS*.[43] For this reason, *CDKN2A* inactivation occurs immediately following the appearance of activating *KRAS* mutations and at a similar frequency to bypass the senescence response.

The *CDKN2A* locus also encodes the tumor suppressor p14ARF, which is expressed via a distinct first exon that introduces an alternate reading frame in downstream exons shared with *CDKN2A/p16*.[39] The p14ARF protein induces growth arrest or apoptosis by inhibiting MDM2-dependent p53 proteolysis. However, the frequent inactivation of p53 in PDAC, which often occurs concomitantly with p14 loss, suggests that this mechanism may not be relevant to disease progression. Furthermore, p14 inactivation only occurs as a consequence of *CDKN2A* deletion (40%).[40,41] Expression of this tumor suppressor is driven by an independent promoter that is unaffected by epigenetic changes influencing p16 transcription. Additionally, p14 appears to be spared by p16-inactivating mutations. These discrepancies indicate that loss of p16 is more critical in disease pathogenesis and that p14 loss is a byproduct. However, p14 may hinder the development of PDAC via p53-independent mechanisms, as it has been shown to repress ribosomal RNA processing, NF-κB transactivation, and c-Myc-induced hyperproliferation.[44–46] Furthermore, p14 can promote cell death by enhancing c-Myc-induced apoptosis as well proteasome-dependent degradation of the antiapoptotic transcriptional corepressor C-terminal binding proteins 1 and 2 (CtBP1/2).[46,47] Thus, p14 may hinder PDAC development via multiple mechanisms in specific contexts.

Another major TSG in PDAC is *TP53*, which encodes the transcription factor p53. Functional loss of this gene has been observed in up to 75% of tumors as a consequence of missense mutation and loss of heterozygosity (LOH).[40,48] Amino acid substitutions impair the ability of p53 to bind DNA, thus ablating its function as a transcription factor. As a consequence, mutant p53 is unable to induce the expression of genes that promote cell cycle arrest or apoptosis (e.g., *CDKN1A/p21*, *BAX*, *NOXA*, and *PUMA*) in response to cellular stress or DNA damage.[49] Inactivation of p53 is typically observed in advanced PanINs following the loss of *CDKN2A*.[43] At this stage of disease progression, the accumulation of DNA damage is believed to induce a selective pressure that necessitates the loss of p53 activity for the continued survival and proliferation of tumorigenic cells.

Emerging evidence suggests that *TP53* mutations in PDAC may also contribute to the highly metastatic nature of the disease, thus defining a gain-of-function role for mutant p53.[50] In one study, a specific tumor-associated mutant form of p53 ($p53^{R175H}$) was shown to bind the protein p73 and impair its ability to repress the expression of platelet-derived growth factor receptor β (PDGFRβ).[51] The resulting upregulation of PDGFR expression promotes an autocrine signaling loop that enhances the motility, invasiveness, and metastatic potential of tumor cells in mice. Additional studies have demonstrated that inactivation of p53 in other cancer types also promotes metastasis by rendering the transcription factor incapable of inducing the expression of genes that counteract cell migration, epithelial–mesenchymal transition (EMT), and stemness.[52,53] Thus, mutation of p53 may promote metastasis in PDAC via PDGFR-independent mechanisms. Collectively, these studies highlight the complexity of genetic alterations in PDAC, as p53 functions as both a tumor suppressor and an enhancer of metastasis.

The low-frequency inactivation (<10%) of several additional TSGs has been observed in PDAC.[54] The most notable of these genes, *STK11/LKB1*, encodes a serine/threonine (S/T) kinase that regulates cell polarity and metabolism.[55] Loss of LKB1 as a consequence of germline mutation is most frequently associated with Peutz–Jeghers syndrome (PJS), a disease characterized by the development of benign polyps in the gastrointestinal tract. Individuals with PJS have a 93% cumulative lifetime risk of developing other malignancies, the most frequent of which is PDAC (36% lifetime risk).[56] As both germline and somatic/sporadic *LKB1* mutations are found at low frequency in PDAC, LKB1 may play an important, context-specific role in disease pathogenesis.[57] One such context may be in the absence of p53 mutation, where *LKB1* haploinsufficiency was shown to cooperate with $KRAS^{G12D}$ in a GEMM to accelerate PDAC development by suppressing p21-dependent cell cycle arrest.[58] This limited but critical role highlights the potential importance of low-frequency genetic alterations in subsets of PDAC.

Several caretaker genes, which have tumor suppressor function, are also functionally lost in subsets of PDAC. Unlike *CDKN2A* and *TP53*, which are considered classical TSGs, caretaker genes do not directly regulate proliferation. Rather, their function is to maintain the integrity of the genome, preventing the accumulation of mutations that might otherwise promote tumor development. A number of caretaker genes are lost in PDAC including *BRCA1*, *BRCA2*, *hMLH1*, and *hMSH2*.[6,54] Similar to *LKB1*, inactivation of these TSGs can occur as a consequence of germline or somatic mutation and is more commonly associated with other malignancies. However, they play an implicit role in disease pathogenesis.

2.3 TGF-β/SMAD4 Alterations

Alterations in the TGF-β signaling pathway play context-dependent roles in PDAC pathogenesis, as the pathway can both induce apoptosis in some contexts and/or promote invasion and metastasis in others. The antiproliferative effect of TGF-β signaling depends on the activity of the transcriptional coactivator *SMAD4/DCP4*. *SMAD4* mutations and homozygous deletions lead to accelerated tumor development in a $KRAS^{G12D}$ GEMM of PDAC.[28] *SMAD4* reintroduction in *SMAD4*-deficient GEMM-derived tumor cells results in apoptosis upon treatment with TGF-β, highlighting its tumor suppressive properties. Approximately 90% of PDAC cases exhibit LOH for the *SMAD4* locus.[59] Furthermore, *SMAD4* biallelic inactivation has been observed in roughly 50% of tumors as a consequence of homozygous deletion or intragenic mutations.[60] These genetic alterations typically occur in advanced PanINs following the loss of *CDKN2A*, making *SMAD4* loss one of the final steps in tumor initiation.[43,61] Interestingly, *SMAD4* loss predicts poorer patient prognosis compared to tumors with intact, wild-type *SMAD4* expression.[62]

TGF-β is a known inducer of the developmental EMT program, discussed in detail later. Upon EMT induction, TGF-β can also promote the activation of an apoptotic program, which is referred to as "lethal EMT." It has recently been postulated that the tumor suppressor function of Smad4 in PDAC can be attributed to its involvement in the induction of lethal EMT.[63] In canonical TGF-β signaling, the TGF-β ligand binds to its receptor (TGFBR), resulting in activation of Smad2/3, which subsequently bind to Smad4. The resulting protein complex translocates to the nucleus and induces a Smad4-dependent transcriptional program that promotes EMT. As a consequence of this program, expression of the gastrointestinal-lineage master regulator Klf5 is repressed. Klf5 can cooperate with Sox4 to promote PDAC progression. However, loss of Klf5 expression dramatically alters the role of Sox4 to that of an apoptosis inducer. This mechanism highlights the complex role of Smad4 in context-dependent tumor suppression by the TGF-β pathway in the pathogenesis of PDAC.

Although TGF-β signaling clearly plays a role in tumor suppression, the pathway can also drive invasion and metastasis. At present, the underlying mechanism behind this alternative role is unclear. It has been suggested that TGF-β promotes invasion and metastasis via the induction of EMT.[64] However, more recent work in lung and pancreatic cancers has demonstrated that EMT is dispensable for metastasis.[65,66] These conflicting studies highlight the need to better understand how TGF-β signaling contributes to PDAC progression. EMT is clearly an important phenotypic outcome of TGF-β-mediated tumorigenic processes, as it can induce stemness and drug resistance.[67] However, other mechanisms downstream of TGF-β are likely to be involved in PDAC progression, including activation of the TGF-β-activated kinase, which activates antiapoptotic signaling mechanisms.[68] Low-frequency mutations have been observed in the *TGFBR1* and *TGFBR2* TGF-β receptor subunits.[69] However, like *SMAD4* inactivation, these alterations are most likely the result of a selective pressure to eliminate the tumor suppressor function of the pathway. Thus, the prometastatic functions of TGF-β remain intact in the vast majority of PDAC. Ultimately, these observations suggest that the TGF-β pathway is critical for PDAC pathogenesis, playing complex, and context-dependent roles in disease progression.

2.4 Telomere Abnormalities

Telomeres are specialized nucleoprotein structures that maintain the integrity of the genome by protecting the ends of linear chromosomes. These structures gradually shorten, or erode, over time with each successive round of cell division due to inefficient telomere-directed replication. When telomeres become critically short, sister chromatids can fuse together at their ends, forming a bridge.[70,71] This linkage poses a problem during anaphase, as the chromatids are unable to properly separate. Thus, when fused sister chromatids break during segregation in mitosis, the gain or loss of chromosomal fragments can occur.[72,73] This break-fusion-bridge process occurs in a cyclical manner, promoting genomic instability and a selective pressure to eliminate antiapoptotic pathways.

Telomere erosion has been observed in more than 90% of low-grade PanINs, suggesting that it is an early event in PDAC pathogenesis.[74] This genomic alteration would typically result in p53-induced senescence.[75] However, the frequent inactivation of p53 sustains continued tumor cell proliferation.[76] Robust telomere shortening precedes the loss of p53 and likely contributes to the selective pressure that results in its loss. Although telomeric loss can promote a genomically unstable state conducive to tumor formation, it may also be detrimental to disease progression if left unchecked, as the reactivation of telomerase has been observed in invasive PDAC.[77] This reactivation most likely occurs to prevent additional genomic alterations that would be catastrophic. Telomerase reactivation can promote tumor progression via induction of cellular immortalization, allowing for sustained telomere elongation during repeated rounds of DNA replication.

3. DEREGULATED EMT IN PANCREATIC CANCER

Epithelial cells are located at the surface of many tissues and organs that are derived from the endodermal and ectodermal embryonic germ layers. These cells form sheets that act as barriers against xenobiotic and pathogenic agents and serve specialized secretory functions in the intestine and the pancreas. Due to their location and function, epithelial cells exhibit a distinct apical–basolateral polarity created by macromolecular protein complexes at cell–cell contacts known as adherens and tight junctions. In contrast to epithelial cells, mesenchymal cells serve in anchoring or scaffolding roles and participate in early embryonic development, wound healing, and tissue repair. Mesenchymal cells lose apical–basolateral polarity as a consequence of the EMT transcriptional program. EMT is a developmental process in which the adherens and tight junctions of epithelial cells are degraded, resulting in a loss of cellular polarity and conversion to mesenchymal cells that are highly motile and invasive.[78] The key molecular changes associated with EMT are loss of epithelial protein marker expression, such as the adherens junction component E-cadherin, and gain of mesenchymal marker expression, including vimentin. During embryonic development, these cells can travel to distant sites and differentiate back into epithelial cells, known as mesenchymal-to-epithelial transition, enabling tissue morphogenesis, tissue repair, and wound healing. EMT is a transcriptional program regulated by specific extracellular factors and cytokines, including TGF-β, Wnt, and Notch, resulting in the activation of signaling pathways such as NF-κB. Many of these factors and pathways are all commonly dysregulated in cancer.[79] Of note, cells within tumors displaying EMT properties have been identified using immunohistological methods. Many parallels can be drawn between the processes of wound healing and tumorigenesis, since both involve the recruitment of mesenchymal stem cells and are associated with increased cellular invasiveness. It is believed that EMT is an important step in cancer invasion and metastasis, as cells exhibiting EMT-like characteristics are often observed at the invasive front of tumors.[80]

EMT-associated signaling networks promote activation of the transcriptional repressors Snail, Slug, Zeb1, Zeb2/SIP1, and Twist.[81] These factors bind to the promoter of E-cadherin and block transcription by promoting chromatin condensation via the activation of histone deacetylases and other corepressors. This results in the loss of adherens junctions as well as loss of cell polarity.[67] TGF-β is a cytokine that promotes EMT through its ability to drive Smad complex association with Zeb proteins, resulting in repression of E-cadherin expression.[82] TGF-β-mediated induction of EMT is accompanied by apoptosis and growth arrest, known as lethal EMT (see above). TGF-β can also induce transient activation of the RAS and PI3K/AKT pathways, which can help to block the apoptotic effects of the cytokine and produce a stable mesenchymal phenotype in cells.[83] Smad4, an important cofactor in TGF-β signal transduction, functions as a tumor suppressor in pancreatic cancer by promoting lethal EMT upon TGF-β stimulation.

In pancreatic cancer, as well as other malignancies, cancer stem cells (CSCs) are multipotent or pluripotent progenitor cells in the tumor. CSCs can be identified and distinguished by high expression of cell surface markers including CD133 and CD44, which are typically not expressed on bulk tumor cells. They can self-renew as well as divide asymmetrically to give rise to differentiated cells. These characteristics allow CSCs to initiate or regenerate a tumor. CSCs are thought to be derived from existing progenitor cells or dedifferentiated cells within a tumor.[67] EMT is known to promote CSC-like properties in pancreatic cancer by inducing a $CD44^{high}$ and $CD24^{low}$ cell surface marker expression profile characteristic of CSCs.[84] This observation suggests EMT plays a direct causal role in the emergence of CSCs. However, the mechanistic basis for the association between EMT and CSC induction remains to be fully elucidated.

Mesenchymal-like properties can render pancreatic cancers more resistant to anticancer therapeutics, especially cytotoxic agents that induce apoptotic cell death. When comparing mesenchymal-like cells in pancreatic cancer to epithelial-like cells, mesenchymal-like cells are more resistant to gemcitabine, 5-fluorouracil (5-FU), cisplatin, and epidermal growth factor receptor (EGFR) inhibitors as assessed by cellular growth and viability.[85] Furthermore, pancreatic cancer cells that are intrinsically resistant to gemcitabine express high levels of vimentin and low levels of E-cadherin, indicating that these drug-resistant cells are more mesenchymal in nature.[86]

As EMT has been associated with the emergence of CSCs, it has been noted that the use of cytotoxic agents such as gemcitabine can lead to an enrichment of $CD44^{high}$, $CD24^{low}$ CSC-like cells.[84] Finally, EMT is also associated with resistance to adjuvant or neoadjuvant radiotherapy, as resistant cells express high levels of vimentin and low levels of E-cadherin. These drug-resistant cells express high levels of the stem cell markers Oct4, CD24, and CD133, further indicating that EMT is associated with the acquisition of stem-like properties.[87] In the context of oncogenic KRAS signaling, studies indicate that loss of dependence on oncogenic KRAS for survival significantly correlates with mesenchymal-like phenotypic characteristics in PDAC cell lines.[31] These studies are supported by findings that primary PDAC tumors can be classified into distinct molecular subtypes based on the expression of EMT markers. These subtypes have been designated classical, exocrine-like, and quasimesenchymal. Of note, the cell lines with quasimesenchymal properties are weakly dependent on oncogenic KRAS to maintain viability.[32]

The clinical and pathophysiological significance of EMT in promoting metastasis and drug resistance remains controversial. A recent study has sought to tackle this question using EMT lineage tracing experiments in a spontaneous breast-to-lung metastasis model.[65] The study demonstrates that a small proportion of cells in primary tumors undergo EMT and that cells found in lung metastases are predominantly epithelial-like. Furthermore, blocking the EMT process via the overexpression of miR-200, a negative regulator of Zeb1, does not significantly impair the formation of distant lung metastases. These findings strongly suggest that EMT is not necessary for metastasis to occur efficiently. Using the same experimental conditions, the study also demonstrates that treatment of primary tumors with the chemotherapeutic agent cyclophosphamide results in the accumulation of mesenchymal-like cells which contribute more significantly to metastasis formation. Overexpression of miR-200 blocked this metastatic growth. Therefore, while EMT may not be required for metastasis under treatment naïve conditions, drug-resistant cells with EMT-like properties may emerge with an increased metastasis-forming ability. Such studies will be critical in addressing the key roles of EMT in driving cancer metastasis and drug resistance across a number of cancer types, including pancreatic cancer. Importantly, associations between the EMT program, the emergence of CSCs, cancer invasiveness, KRAS dependence, and drug resistance provide new therapeutic opportunities for pancreatic cancer treatment.

4. MOLECULAR SUBTYPE CLASSIFICATIONS OF PANCREATIC CANCER

PDACs harbor a number of recurrent genetic alterations, including activation of *KRAS* in addition to loss of *TP53*, *SMAD4*, and *CDKN2A*.[6] A multitude of other mutations occur to varying degrees in subsets of tumors, leading to dysregulation of cellular processes such as DNA damage repair, cell cycle regulation, TGF-β signaling, and chromatin modification.[18] However, some of the observed alterations may be "passenger" mutations that play minor roles in disease pathogenesis. Efforts to identify strategies to manage PDAC have been confounded by the molecular diversity of these aggressive tumors. Understanding the genetic variation in PDAC is now a major area of focus to develop more effective therapeutics. Although there are recurrent gene mutations, the transcriptional networks that are activated in tumors with similar genetic profiles can vary significantly as determined by global gene expression profiling. Using transcriptome data from RNA-seq analyses of primary PDACs, a recent study has generated a PDAC classification system of four major molecular subtypes. These are squamous (quasimesenchymal), pancreatic progenitor (classical), immunogenic, and aberrantly differentiated endocrine–exocrine (ADEX, exocrine-like).[18,32] Of note, these subtypes recapitulate the major subtypes identified by Collisson and colleagues, including those associated with EMT.[32]

The squamous PDAC subtype is characterized by high mesenchymal marker gene expression and has the worst prognosis in comparison to the other subtypes. This subtype primarily comprises PDACs that are histologically identified as adenosquamous carcinoma.[18,32] Squamous subtype tumors are enriched for *TP53* mutations. They display activation of the p53 family member TP63ΔN and its associated transcriptional network. TP63ΔN regulates squamous epithelial cellular differentiation in contrast to columnar differentiation that is characteristic of pancreatic ductal epithelial cells. Thus squamous transdifferentiation in this molecular subtype is driven by an EMT-like program. Hypermethylation of pancreatic endodermal cell-fate determination genes, including *PDX1*, *GATA6*, and *HNF1B*, further contributes to the dedifferentiated and mesenchymal nature of these tumors. A number of other phenotypic characteristics can distinguish squamous subtype PDAC tumors including increased prevalence *KDM6A* mutations, which affect chromatin remodeling, inflammation, the hypoxia response, metabolic reprogramming, TGF-β signaling, MYC activation, and autophagy. Finally, cell lines derived from squamous subtype tumors are more sensitive to the cytotoxic effects of gemcitabine.[32]

Pancreatic progenitor subtype tumors are typically more epithelial-like in nature, as determined by high expression levels of epithelial marker genes, including adhesion-associated genes such as *CDH1*/E-cadherin.[18] This subtype bears molecular similarities to KRAS-dependent PDAC cell lines.[31] Furthermore, high expression levels of genes that contribute to early pancreatic development are prevalent (*FOXA2/3*, *PDX2*, *MNX1*, and *GATA6*). These genes are important for terminal differentiation to pancreatic ductal epithelial cells. For example, PDX2 induces differentiation of ductal, exocrine, and endocrine cells of the pancreas. Cellular processes that are characteristic of the pancreatic progenitor subtype are fatty acid oxidation, steroid hormone biosynthesis, drug metabolism, and O-linked glycosylation of mucins. In contrast to the squamous subtype, pancreatic progenitor subtype tumor-derived cell lines are more sensitive to the EGFR inhibitor erlotinib.[32] Interestingly, development of pancreatic progenitor subtype tumors is linked to maturity onset diabetes of the young.[18]

The ADEX tumor subtype is characterized by the simultaneous expression of transcriptional programs observed in the endocrine and exocrine pancreas. Both programs are typically activated in the later stages of normal organ development and differentiation in a mutually exclusive manner. Genes upregulated in ADEX tumors play a role in acinar and endocrine differentiation as well as regeneration and pancreatitis. Furthermore, a number of these genes are associated with KRAS activation. ADEX tumors comprise a subclass of the pancreatic progenitor tumor subtype and are histologically associated with rare acinar cell carcinomas.[18,32]

Immunogenic subtype tumors exhibit many of the same molecular characteristics observed in the pancreatic progenitor subtype but can be distinguished by the upregulation of various immune-related transcriptional programs. These programs are associated with B- and T-cell receptor signaling, Toll-like receptor signaling, antigen presentation, and acquired immune suppression through immune checkpoint pathways such as CTLA4 and PD1. Additionally, immunogenic subtype tumors exhibit a notable increase in infiltrating B- and T-cells. Tumors of this subtype display histological characteristics observed in mucinous noncystic (colloid) adenocarcinomas and IPMN-derived carcinomas.[18] PDACs are generally nonresponsive to new classes of anti-PD1 immunomodulatory checkpoint inhibitors, such as pembrolizumab. However, tumors of the immunogenic PDAC subtype may be more responsive to these immune checkpoint blockers either alone or in combination with other chemotherapeutic agents.[88]

5. DEREGULATED SIGNALING NETWORKS IN PANCREATIC CANCER

5.1 The EGFR–KRAS Network

Receptor tyrosine kinases (RTKs) are cell surface receptors for many growth factor ligands, including epidermal growth factor.[89] RTK dysregulation plays a significant role in many cancers. Upon binding to a growth factor ligand, RTKs form homo- or heterodimers, bringing their intracellular kinase domains into close proximity. The intracellular receptor regions are trans-phosphorylated to create docking sites for SH2-domain containing adapter proteins and enzymes. This activates many downstream signaling cascades mediated by proteins including RAS and PI3K (Fig. 2A).[90] While EGFR mutations are rare in pancreatic cancer, inhibition of receptor kinase activity with erlotinib is moderately effective for treating a subset of PDACs.[89] Since *KRAS* is so frequently mutated in pancreatic cancer, the activation of RAS-mediated signaling pathways is thought to be a major driver of disease pathogenesis. Other RAS isoform genes such as *HRAS* and *NRAS* are mutated infrequently in pancreatic cancer, suggesting that the *KRAS* locus provides a unique advantage in the context of PDAC pathogenesis. Although not mutated in PDAC, HRAS and NRAS proteins may participate in EGFR/KRAS signaling networks to promote tumorigenesis via protection against DNA damage-induced stress pathways.[91] As discussed earlier, *KRAS* mutations occur early in PanIN lesions. Given the high prevalence of *KRAS* mutations, it is not surprising that the KRAS oncoprotein is a potent driver of PDAC development.[92]

RAS GAPs catalyze the hydrolysis of GTP to GDP, resulting in the inactivation of RAS.[20,21] Thus, GAPs, such as the *NF1* gene product, can serve as tumor suppressors, although they are not frequently mutated in PDAC. The GTP bound form of RAS undergoes a conformational change in which two "switch" regions (I and II) converge to form the effector-binding domain. This domain forms biochemical interactions with effectors to either promote their allosteric activation or enable recruitment to the membrane, allowing for initiation of downstream signal transduction cascades. One of the most well-studied classes of RAS effectors are RAF family S/T kinases, which initiate the MEK/ERK MAP kinase pathway through a cascade of sequential phosphorylation events.[19] This results in phosphorylation and nuclear localization of transcription factors, such as Elk-1, that drive cell proliferation, inflammatory signaling, differentiation, and cell survival. RAS can also activate the lipid kinase PI3K by associating with the p110 subunit of the PI3K complex. This complex is composed of a p110 catalytic subunit and a p85 regulatory subunit, which together regulate many key tumorigenic processes, including cell survival and proliferation. Activated RAS recruits PI3K to the plasma membrane and promotes its catalytic activity, which is to facilitate the conversion of the membrane phospholipid phosphatidylinositol-4,5-bisphosphate (PIP2) to phosphatidylinositol-3,4,5-triphosphate (PIP3). PIP3 serves as a binding site for proteins that contain a pleckstrin homology domain. To inactivate the PI3K pathway, the lipid phosphatase PTEN hydrolyzes PIP3 to PIP2. PTEN is often dysregulated in late stages of pancreatic cancer resulting in hyperactivation of the PI3K pathway and an acceleration of PDAC development.[93] A primary effector of PI3K activation is the AKT S/T kinase, which is activated by sequential phosphorylation mediated by PDK1 and mTORC2. AKT suppresses apoptosis by promoting MDM2-dependent p53 proteolysis. It can also activate mTORC1, which phosphorylates 4EBP1 and ribosomal S6 kinase to promote CAP-dependent mRNA translation (Fig. 2A). The activation of mTORC1 is a key feature of many PDACs that results in increased protein translation, stem cell renewal, proliferation, and inhibition of autophagy via ULK1 kinase activation.[94,95]

5.2 Hippo Signaling

The Hippo tumor suppressor pathway regulates organ size and maintains tissue homeostasis by restricting cell proliferation and, when necessary, inducing apoptosis.[96] The Hippo pathway is activated by a kinase cascade that was originally discovered via genetic screens aimed at identifying TSGs in the fruit fly *Drosophila melanogaster*. Mutations in various Hippo pathway component genes, such as *Hippo* (*Hpo*) itself, *Salvador* (*Sav*), and *Warts* (*Wts*) result in increased organ size as well as tissue overgrowth. The evolutionary conservation of Hippo signaling has been established in mammals, where pathway deregulation is frequently observed in many solid cancers, including PDAC.[97,98]

The induction of Hippo signaling occurs in response to a number of stimuli including contact inhibition, serum deprivation, energy stress, and actin depolymerization (Fig. 2B).[99] These signals ultimately result in the phosphorylation and activation of large tumor suppressor kinases 1/2 (LATS1/2; Wts orthologs) by the MST1/2 (Hpo orthologs) kinases. Once activated, LATS1/2 phosphorylate the transcriptional coactivator Yes-associated protein (YAP) and its vertebrate-specific paralog transcriptional coactivator TAZ, encoded by the *WWTR1* gene. Phosphorylation of YAP and TAZ promotes their association with 14-3-3 proteins, resulting in their cytoplasmic retention and subsequent degradation. In the absence of Hippo pathway activation, YAP and TAZ are able to translocate to the nucleus, where they interact with the TEAD family of transcription factors to induce the expression of genes associated with cell survival, proliferation, and migration. Thus, under normal physiological conditions in epithelial tissue such as the pancreatic duct, YAP/TAZ activity remains low.

Deregulation of Hippo signaling in PDAC is evidenced by increased YAP/TAZ protein levels and nuclear localization.[100–102] However, the underlying mechanisms accounting for increased YAP/TAZ activation remain an area of active investigation. Pan-cancer genomic analyses have revealed that mutations in the genes encoding Hippo pathway components occur at very low frequencies.[97,98] Given the oncogenic activity of YAP, this is a surprising finding.[103] The low frequency of Hippo pathway gene mutations in human cancer could indicate that that Hippo signaling is essential for both physiological and pathophysiological tissue homeostasis. This notion is supported by several studies in which the deletion of various Hippo pathway components in mice results in embryonic lethality.[104–108] Therefore, pathway output must be physiologically maintained within certain limits to ensure that cellular proliferation remains in check.

The mechanisms that control YAP/TAZ activation in PDAC have yet to be fully elucidated. One possible mechanism is loss of LATS1/2 expression as a consequence of promoter hypermethylation. Epigenetic regulation of the Hippo pathway, which has been documented in both colon and breast cancers,[109,110] functionally disconnects YAP/TAZ activity from upstream pathway induction. Another alteration that may account for the elevated YAP/TAZ levels in PDAC is gene amplification. *YAP* and *TAZ* gene copy number gains have been documented in several different cancer types.[97,111] In such contexts, increased YAP/TAZ protein levels result in constitutive activation even when the Hippo pathway is intact and active.

Recent studies of YAP function in GEMMs suggest that it plays a key role in PDAC initiation and maintenance. Ectopic expression of YAP in the developing mouse pancreas results in the appearance of metaplasia and impaired differentiation of both the endocrine and exocrine compartments.[112,113] However, increased organ size and tumor formation were not observed. Subsequent work in *Kras* and *Kras/Tp53*-mutant mice demonstrated that YAP is essential for the progression of neoplasia to PDAC.[102] Furthermore, YAP gene amplification is observed in some cases following loss of oncogenic KRAS expression in GEMMs and cancer cell lines, leading to increased cell survival and tumor recurrence.[35,36] In these studies, YAP activity can bypass the requirement for sustained KRAS signaling, conferring a loss of KRAS oncogene dependence. Furthermore, YAP can become activated as a consequence of MAPK signaling to promote survival and proliferation. Hence, the loss of KRAS expression in PDAC GEMMs initially results in robust tumor regression. However, a subpopulation of cells with amplification and overexpression of *YAP*, is able to repopulate the tumor, leading to disease relapse. Collectively, these studies suggest that Hippo pathway deregulation plays a critical role in PDAC progression as well as dependence on the KRAS–MAPK pathway. Thus, modulation of YAP and/or TAZ activity represents an attractive opportunity for therapeutic intervention.

5.3 Inflammation

Innate immune responses and inflammation have been associated with cancer etiology in many contexts, and pancreatic cancer is no exception. Recent studies using *KRAS*-driven GEMMs of PDAC have highlighted the role of acute pancreatitis-associated inflammation in driving the progression of PanIN lesions to full-blown PDAC.[114] Some studies estimate that up to 50% of the PDAC tumor cell mass can be composed of stromal and immune cells recruited to the tumor via paracrine cytokine signaling.[115] Furthermore, oncogenic KRAS signaling can promote the development of the inflammatory microenvironment found in PDAC tumors. It is hypothesized that inflammation promotes PDAC initiation and progression through several different mechanisms. Firstly, it promotes cell survival and proliferation through inflammatory mediators that activate transcription factors responsible for antiapoptotic signaling as well as cancer cell invasion and metastasis. Examples of such proinflammatory, prosurvival transcription factors include STAT3, the AP-1 complex (Jun/Fos), and NF-κB. KRAS induces IL-6 and IL-11 cytokine expression and secretion, leading to STAT3 transcriptional activation. KRAS also promotes NF-κB signaling via induction of the MAPK pathway, leading to increased TNFα and IL-1 levels, which drive a positive feedback loop for NF-κB activation. Reciprocally, NF-κB can enhance RAS activity through a RAS-NF-κB-cyclooxygenase-2 positive feedback loop.[68] Thus, inflammatory cytokines and RAS signaling are intimately linked. Secondly, proinflammatory cytokines, including TNFα, IL-6, and IL-1, can promote metastasis by inducing EMT as well as the acquisition of CSC-like traits.[115] These mechanisms, coupled with the ability of NF-κB to activate Notch and other oncogenic pathways, lead to accelerated PDAC development. Finally, inflammation-associated cytokines can impair immunosurveillance of tumor cells due to an increase in immune cell subsets that have immunosuppressive properties, including regulatory T-cells (T-regs) and myeloid-derived suppressor cells (MDSCs). These cell types can negatively regulate the numbers and functional activation of cytotoxic CD8+ T-cells, possibly via increased expression of granulocyte–macrophage colony-stimulating factor (GM-CSF).[116,117]

There are likely to be a number of additional and diverse mechanisms by which inflammation promotes cancer. Inflammation can cause cellular damage, for example, via the accumulation of reactive oxygen species that promote oxidative DNA damage.[115] This may lead to increased genetic evolution in PDAC development as tumor cells acquire somatic mutations that confer selective advantages to promote growth and survival. When cells acquire an activating oncogenic mutation (e.g., *KRAS*), a phenomenon known as oncogene-induced senescence is triggered as a consequence of *CDKN2A*/p16 induction. This senescence response can be suppressed by inflammatory signaling, representing a mechanism by which inflammation promotes cancer progression. Taken together, inflammation clearly plays a critical role in immunosuppression, KRAS pathway modulation, and cancer metastasis. Thus, targeting inflammatory pathways could represent an attractive avenue for therapeutic intervention.

5.4 Autophagy

Autophagy is a well-characterized metabolic, homeostatic process by which cellular constituents, such as proteins and organelles, are degraded and recycled to meet cellular demands under conditions of nutrient deprivation or stress. Such conditions occur during tumorigenesis, and indeed, constitutive activation of autophagy has been observed in many tumors including PDAC.[118,119] Three types of autophagy are typically described: macroautophagy, microautophagy, and chaperonin-mediated autophagy. Macroautophagy, henceforth referred to simply as autophagy, is the primary pathway, as well as the most significant in PDAC. During nutrient deprivation, misfolded and nonessential proteins, as well as organelles, are sequestered into a lipid bilayer known as a phagophore. This gives rise to a double-membrane structure known as the autophagosome, which subsequently fuses with a lysosome, forming an autophagolysosome or autolysosome. The presence of degradative enzymes combined with low pH within autolysosomes cause the breakdown of cellular macromolecules. Amino acids and other building blocks are then recycled, allowing for cellular homeostasis to be maintained. In normal pancreatic tissue, there is a basal level of autophagy that serves to maintain homeostasis.[118]

The role of autophagy in pancreatic cancer is complicated and has yet to be fully elucidated. PDACs typically exhibit high basal levels of autophagic activity, including increased number and size of both autophagosomes and autolysosomes, when compared to normal pancreatic tissue.[120] It is hypothesized that tumor cells gain a selective advantage when autophagy is activated, as it may allow them to cope with the stresses resulting from high rates of cell division as well as nutrient deprivation, which can occur from hypoxic conditions in poorly vascularized regions of the tumor.[121] Therefore, autophagy can drive the survival of PDAC cells under these conditions. Chloroquine, which inhibits autophagy by blocking the formation of autolysosomes, potently suppresses the growth of some PDAC cell lines. When autophagy is inhibited with chloroquine, or by genetic ablation of the key autophagy regulator Atg-7, *Kras*-induced progression of PanIN to PDAC is blocked as a consequence of cell death, growth arrest, or senescence.[122] However, when combined with *TP53* deletion, Atg-7 loss enhances, rather than suppresses, *Kras*-driven PDAC progression.[123] These findings highlight the highly context-dependent role of autophagy in PDAC pathogenesis, which depends on *TP53* status and possibly other TSG mutations. Lastly, autophagy could allow PDAC cells to cope with the deleterious effects of chemotherapeutics and radiation therapy. For example, treatment of the PDAC cell line PANC-1 with gemcitabine or 5-fluorouracil induces autophagy.[121] Interestingly, the combination of autophagy inhibitors, such as chloroquine, with chemotherapy or radiation therapy greatly enhances cytotoxicity in PDAC cell lines.

6. CURRENT AND FUTURE THERAPEUTIC STRATEGIES FOR PANCREATIC CANCER

PDACs are notoriously difficult to treat for a number of reasons.[9] Most patients with PDAC are often asymptomatic, and diagnoses are not usually made until after the tumors have become metastatic. Currently, there are few effective therapeutic options for PDAC patients. The only "curative" treatment is surgical resection, but its success rate in patients with operable tumors is low, with a 5-year survival rate of only 20%, a 60% rate of relapse within 6 months, and an overall relapse rate of more than 80%. Due to the typically late diagnosis, many patients are not candidates for surgical resection. In contrast, chemotherapy has marginal, but measurable, effects on overall survival in PDAC patients. The efficacy of chemoradiation as a PDAC therapeutic regimen remains unclear when compared to chemotherapy alone. With current PDAC therapeutic options, overall survival of 5 years or greater is estimated to be less than 5%, and these rates have not changed significantly in the past 30 years. Understanding PDAC etiology and pathogenesis at the detailed molecular level as a means to developing better therapeutics to treat the disease remains a pressing goal. PDAC is a complex disease with multiple stages that will respond to different sets of anticancer agents. The complexity of this disease is further highlighted by recent studies using a PDAC GEMM demonstrating that cells from PanIN lesions can metastasize. Pancreatic epithelial cells are able to disseminate from the pancreas at an early stage of the disease, when a primary lesion is not yet detectable. These cells are capable of seeding in the liver and potentially other distant sites. Therefore, therapeutics may be more successful if they are designed to treat pancreatic cancer as a systemic disease rather than localized one.[124] This could explain the low rate of success with local treatment, as cancer cells may be present at distal sites such as the liver or the lymph nodes. It is clear that new approaches to treat PDAC must be developed. A number of innovative avenues for therapeutic intervention are currently being evaluated in basic science and clinical studies.[9]

For PDAC patients with resectable lesions, the standard of care involves surgical resection and adjuvant chemotherapy with gemcitabine or 5-fluorouracil (5-FU). Both agents are pyrimidine analogs that block various stages of DNA replication, leading to cell cycle arrest and, in some cases, apoptosis. These conventional cytotoxic agents preferentially target rapidly dividing cells and can lead to the modest tumor regression or growth suppression. In metastatic disease, the standard of care is a drug regimen that consists of gemcitabine, FOLFIRINOX, or nab-paclitaxel. FOLFIRINOX is a combination chemotherapy made up of folinic acid (leucovorin), 5-FU, irinotecan, and oxaliplatin. These agents collectively interfere with DNA replication and transcription. Paclitaxel, on the other hand, binds to, and stabilizes, microtubules, preventing their disassembly and ultimately blocking progression through mitosis. Nab-paclitaxel is an albumin-bound form of the drug that has increased bioavailability. As these drugs indiscriminately target all rapidly dividing cells, many side effects and dose-limiting toxicities are associated with their utilization. For this reason, conventional cytotoxic chemotherapeutics provide only marginal increases in median patient survival on the order of weeks or months.[9] To improve the survival benefit conferred by these agents, many ongoing studies are investigating optimal dosing regimens and combinations for treating different stages of pancreatic cancer. Clinical trials involving chemoradiotherapy have returned inconsistent results, and the treatment remains controversial. This is possibly due to the lack of biomarkers to determine which patients will be responsive to radiation treatment. A newer strategy involves gemcitabine in combination with the EGFR tyrosine kinase inhibitor erlotinib, which results in a modest survival benefit of 12 days compared to gemcitabine alone. Thus, erlotinib is the only FDA-approved targeted therapy available for treatment of pancreatic cancer. Some patients respond much more favorably to erlotinib than others, suggesting that there could be biomarkers to identify PDAC patients who will likely benefit most from anti-EGFR therapies such as erlotinib. A recent study found that *TP53*-wild-type tumors may be more sensitive to EGFR inhibition. Thus, identifying the right patient population for a particular targeted therapy, in the interests of precision medicine, remains a key goal of PDAC therapeutics.[9]

Effective targeted therapeutics and precision medicine-based approaches for PDAC have yet to be identified. One approach under investigation is to take advantage of the tumor-specific environment. Due to the high density of fibrous connective tissue that is characteristic of pancreatic cancer, drug delivery to tumors is severely impaired. Hyaluronic acid is an extracellular matrix component found surrounding tumors that presents a physical barrier for drug delivery. The degradation of hyaluronic acid via hyaluronidase may enhance drug delivery. PEGPH20 is a PEGylated form of hyaluronidase that is currently being tested and shows prolonged survival of tumor-bearing mice when given in combination with gemcitabine.[125] DNA-damaging agents such as TH-302 (evofosfamide) can take advantage of the hypoxic environment in pancreatic tumors to increase their specificity. This drug is derived from nitrogen mustard that becomes activated under hypoxic conditions and releases its active form, dibromoisophosphoramide mustard (Br-IDM), a DNA alkylating agent. In combination with gemcitabine, TH-302 provides a 6-month survival benefit compared to 3.6 months with gemcitabine alone.[126,127] Drug modifications that provide improved delivery are also being evaluated, such as nab-paclitaxel or nanoliposomal formulations of irinotecan (MM-398). These modifications can increase the plasma half life of drugs and increase the availability of their active metabolites. MM-398 provides a 2-month survival advantage when given in combination with 5-FU and folinic acid.[128] To identify accurate biomarkers of response to particular agents, DNA sequencing of PDACs for sensitizing genetic alterations represents a major step in advancing precision medicines to treat the disease. Some examples currently under investigation include *SMAD4* for chemoradiotherapy, *STK11* for mTOR inhibitors, and the genes *PALB2*, *ATM*, and *BRCA2* for DNA-damaging agents.[9]

Another avenue for therapeutic intervention could be to exploit oncogene "addiction."[129] As mutant *KRAS* is a key linchpin in PDAC pathogenesis, it remains a major therapeutic target, albeit a stubborn one. Thus far, there has been little success in targeting the RAS-related proteins, due to their high affinity for GTP and the abundance of GTP in the cell, which prevent access to the protein active site. Nonetheless, many alternate methods of RAS inhibition are currently being investigated. The RAS protein must undergo several posttranslational modifications, including a farnesylation step, before the protein is functional. Blocking farnesylation of the protein is one potential method to inhibit the RAS pathway, but results have been disappointing thus far because KRAS can be alternatively isoprenylated with a geranylgeranyl group. After the RAS protein has been translated, it must be transported to the cell membrane, a process that requires the PDEδ protein. Thus, the PDEδ inhibitor Deltarasin has been developed, which can block RAS membrane translocation. This prevents the downstream activation of ERK, leading to suppression of KRAS-driven PDAC cell proliferation and viability. Although directly inhibiting KRAS has proven to be difficult, allosteric, covalent-modifying inhibitors that stabilize the GDP-bound form of the protein have been identified for the G12C isoform commonly found in nonsmall cell lung cancer (NSCLC). It is possible that a similar agent could be identified for the most common isoform found in pancreatic cancer, KRAS G12D. As direct inhibition of KRAS remains a challenge, downstream inhibition of the RAF-MEK-ERK cascade and/or PI3K/mTOR has become an active area of investigation.[9,19] Lastly, synthetic lethality is another approach that attempts to identify genetic interactors with mutant KRAS that cooperate to promote PDAC tumor cell survival.[129] Thus far, genome-wide KRAS synthetic lethality siRNA screens have failed to yield promising candidate therapeutic targets, perhaps due to the complexity and molecular heterogeneity of oncogenic *KRAS* mutant PDAC tumors.

Altered cellular metabolism is yet another area being studied to develop PDAC therapeutics. In the tumor microenvironment, dense desmoplastic regions surround the tumor, leading to hypoxia and decreased nutrient delivery to cells. Tumors with increased glycolysis gain a survival advantage in the hypoxic environment (known as the Warburg effect), and autophagy is induced in response to nutrient deprivation. As described earlier, chloroquine, or its derivative hydroxychloroquine, inhibits autophagy by blocking the formation of autolysosomes. While chloroquine alone is mildly effective in promoting PDAC cell death, studies indicate that it can sensitize tumors to MEK inhibitors, chemoradiotherapy, gemcitabine, and nab-paclitaxel.[130,131]

Finally, immunotherapy is an exciting field in cancer therapeutics that has demonstrated dramatic effects in other diseases, such as melanoma. This approach employs activation of the host immune system to combat tumors by promoting tumor cell clearance via cytotoxic T-cells. The immunosuppressive environment found in pancreatic cancer prevents immunosurveillance of tumors from occurring efficiently. Thus, supercharging the immune system to overcome this immunosuppression represents an innovative therapeutic strategy. One approach is to sensitize the immune system to pancreatic cancer cells through vaccination. GVAX pancreas is created from whole tumor cells, which are genetically engineered to express GM-CSF and then irradiated to prevent cell division. These engineered pancreatic cancer cells are capable of recruiting dendritic cells that can phagocytose the tumor cells, which are subsequently presented to T-cells to promote their activation and ability to recognize and clear tumor cells.[132] Along the same lines, T-cells can be modified to express chimeric antigen receptors that recognize tumor antigens. This approach has been successfully achieved in CD19-positive hematological malignancies but is still in early development for pancreatic cancer.[133] CD40, a cell surface protein on antigen presenting cells plays a key role in immune cell activation. Thus, CD40 agonists are being tested in combination with gemcitabine to promote accumulation of phagocytic macrophages in tumors.[134] PD-L1 and PD-L2 are ligands expressed by cancer cells capable of binding to the immune checkpoint receptor PD-1 on activated CD8+ T-cells, causing suppression of cytotoxic T-cell function. This impairs immunosurveillance of cancer cells. Monoclonal antibodies that target PD-L1 or PD-1, such as pembrolizumab, have been developed to promote CD8+ T-cell activation. These agents have yielded significant beneficial results in melanoma, NSCLC, and renal cell carcinoma. As is the case with these malignancies, PD-L1 expression in pancreatic tumors is also associated with a poor prognosis.[88] Monoclonal antibodies against CTLA4 (e.g., ipilimumab), another T-cell immune checkpoint receptor, have also been tested in pancreatic cancer.[135] Unfortunately, checkpoint inhibitors have had disappointing results as single therapies in PDAC thus far. However, there is hope that they will be effective in combination with other chemotherapeutics, such as gemcitabine and nab-paclitaxel. In summary, PDAC remains one of the most deadly of all human diseases due to a severe lack of effective therapeutics. Innovative therapeutic approaches for PDAC include optimizing the dosing regimens of current agents, targeting oncogene addiction, manipulating tumor metabolism, and harnessing the host immune system to fight this aggressive cancer type.

7. CONCLUSIONS

PDAC is a highly aggressive malignancy associated with very poor clinical prognosis. Although the core genetic alterations in PDAC are well documented, their contributions to PDAC pathogenesis remain to be fully determined at the molecular level. Next-generation sequencing has revealed the detailed complexity of the genomic landscape of PDAC, which is characterized by marked inter- and intratumor heterogeneity as well as a very high overall mutational burden. Gene mutations in PDAC have been shown to converge on a few critical signal transduction pathways and cellular processes including the KRAS–MAPK pathway, inflammation, and altered cellular metabolism. The complexity of PDAC pathogenesis is further illustrated by the classification of PDAC tumors into four major molecular subtypes that are distinguished by key phenotypic traits and pharmacological vulnerabilities. Detailed characterization of these subtypes could ultimately lead to the development of new precision medicines for treating PDAC. However, current therapeutic options for the disease remain limited. Understanding and attacking the complexity of PDAC pathogenesis will undoubtedly yield additional innovative therapeutic options for this aggressive and deadly disease.

REFERENCES

1. Roder PV, Wu B, Liu Y, Han W. Pancreatic regulation of glucose homeostasis. *Exp Mol Med*. 2016;48:e219.
2. Keller J, Layer P. Human pancreatic exocrine response to nutrients in health and disease. *Gut*. 2005;54(suppl 6):1–28.
3. Gittes GK. Developmental biology of the pancreas: a comprehensive review. *Dev Biol*. 2009;326(1):4–35.
4. Polonsky KS. The past 200 years in diabetes. *N Engl J Med*. 2012;367(14):1332–1340.

5. Braganza JM, Lee SH, McCloy RF, McMahon MJ. Chronic pancreatitis. *Lancet.* 2011;377(9772):1184–1197.
6. Hezel AF, Kimmelman AC, Stanger BZ, Bardeesy N, Depinho RA. Genetics and biology of pancreatic ductal adenocarcinoma. *Genes Dev.* 2006;20(10):1218–1249.
7. Siegel RL, Miller KD, Jemal A. Cancer statistics, 2016. *CA Cancer J Clin.* 2016;66(1):7–30.
8. Lennon AM, Wolfgang CL, Canto MI, et al. The early detection of pancreatic cancer: what will it take to diagnose and treat curable pancreatic neoplasia? *Cancer Res.* 2014;74(13):3381–3389.
9. Garrido-Laguna I, Hidalgo M. Pancreatic cancer: from state-of-the-art treatments to promising novel therapies. *Nat Rev Clin Oncol.* 2015;12(6):319–334.
10. Rahib L, Smith BD, Aizenberg R, Rosenzweig AB, Fleshman JM, Matrisian LM. Projecting cancer incidence and deaths to 2030: the unexpected burden of thyroid, liver, and pancreas cancers in the United States. *Cancer Res.* 2014;74(11):2913–2921.
11. Fesinmeyer MD, Austin MA, Li CI, De Roos AJ, Bowen DJ. Differences in survival by histologic type of pancreatic cancer. *Cancer Epidemiol Biomarkers Prev.* 2005;14(7): 1766–1773.
12. Phan GQ, Yeo CJ, Hruban RH, Lillemoe KD, Pitt HA, Cameron JL. Surgical experience with pancreatic and peripancreatic neuroendocrine tumors: review of 125 patients. *J Gastrointest Surg.* 1998;2(5):472–482.
13. Kern SE, Hruban RH, Hidalgo M, Yeo CJ. An introduction to pancreatic adenocarcinoma genetics, pathology and therapy. *Cancer Biol Ther.* 2002;1(6):607–613.
14. Hruban RH, Adsay NV, Albores-Saavedra J, et al. Pancreatic intraepithelial neoplasia: a new nomenclature and classification system for pancreatic duct lesions. *Am J Surg Pathol.* 2001;25(5):579–586.
15. Hruban RH, Takaori K, Klimstra DS, et al. An illustrated consensus on the classification of pancreatic intraepithelial neoplasia and intraductal papillary mucinous neoplasms. *Am J Surg Pathol.* 2004;28(8):977–987.
16. Sohn TA, Yeo CJ, Cameron JL, et al. Intraductal papillary mucinous neoplasms of the pancreas: an updated experience. *Ann Surg.* 2004;239(6):788–797. discussion 797–799.
17. Jones S, Zhang X, Parsons DW, et al. Core signaling pathways in human pancreatic cancers revealed by global genomic analyses. *Science.* 2008;321(5897):1801–1806.
18. Bailey P, Chang DK, Nones K, et al. Genomic analyses identify molecular subtypes of pancreatic cancer. *Nature.* 2016;531(7592):47–52.
19. Downward J. Targeting RAS, signalling pathways in cancer therapy. *Nat Rev Cancer.* 2003;3(1):11–22.
20. Stephen AG, Esposito D, Bagni RK, McCormick F. Dragging ras back in the ring. *Cancer Cell.* 2014;25(3):272–281.
21. Cox AD, Fesik SW, Kimmelman AC, Luo J, Der CJ. Drugging the undruggable RAS: mission possible? *Nat Rev Drug Discov.* 2014;13(11):828–851.
22. Pylayeva-Gupta Y, Grabocka E, Bar-Sagi D. RAS oncogenes: weaving a tumorigenic web. *Nat Rev Cancer.* 2011;11(11):761–774.
23. Scheffzek K, Ahmadian MR, Kabsch W, et al. The Ras-RasGAP complex: structural basis for GTPase activation and its loss in oncogenic Ras mutants. *Science.* 1997;277(5324):333–338.
24. Scheidig AJ, Burmester C, Goody RS. The pre-hydrolysis state of p21(ras) in complex with GTP: new insights into the role of water molecules in the GTP hydrolysis reaction of ras-like proteins. *Structure.* 1999;7(11):1311–1324.
25. Witkiewicz AK, McMillan EA, Balaji U, et al. Whole-exome sequencing of pancreatic cancer defines genetic diversity and therapeutic targets. *Nat Commun.* 2015;6: 6744.

26. Kanda M, Matthaei H, Wu J, et al. Presence of somatic mutations in most early-stage pancreatic intraepithelial neoplasia. *Gastroenterology*. 2012;142(4). 730–733.e9.
27. Aguirre AJ, Bardeesy N, Sinha M, et al. Activated Kras and Ink4a/Arf deficiency cooperate to produce metastatic pancreatic ductal adenocarcinoma. *Genes Dev*. 2003;17(24):3112–3126.
28. Bardeesy N, Cheng KH, Berger JH, et al. Smad4 is dispensable for normal pancreas development yet critical in progression and tumor biology of pancreas cancer. *Genes Dev*. 2006;20(22):3130–3146.
29. Collins MA, Bednar F, Zhang Y, et al. Oncogenic Kras is required for both the initiation and maintenance of pancreatic cancer in mice. *J Clin Invest*. 2012;122(2):639–653.
30. Ying H, Kimmelman AC, Lyssiotis CA, et al. Oncogenic Kras maintains pancreatic tumors through regulation of anabolic glucose metabolism. *Cell*. 2012;149(3):656–670.
31. Singh A, Greninger P, Rhodes D, et al. A gene expression signature associated with "K-Ras addiction" reveals regulators of EMT and tumor cell survival. *Cancer Cell*. 2009;15(6):489–500.
32. Collisson EA, Sadanandam A, Olson P, et al. Subtypes of pancreatic ductal adenocarcinoma and their differing responses to therapy. *Nat Med*. 2011;17(4):500–503.
33. Lim KH, Counter CM. Reduction in the requirement of oncogenic Ras signaling to activation of PI3K/AKT pathway during tumor maintenance. *Cancer Cell*. 2005;8(5):381–392.
34. Viale A, Pettazzoni P, Lyssiotis CA, et al. Oncogene ablation-resistant pancreatic cancer cells depend on mitochondrial function. *Nature*. 2014;514(7524):628–632.
35. Kapoor A, Yao W, Ying H, et al. Yap1 activation enables bypass of oncogenic Kras addiction in pancreatic cancer. *Cell*. 2014;158(1):185–197.
36. Shao DD, Xue W, Krall EB, et al. KRAS and YAP1 converge to regulate EMT and tumor survival. *Cell*. 2014;158(1):171–184.
37. Muller PA, Vousden KH. p53 mutations in cancer. *Nat Cell Biol*. 2013;15(1):2–8.
38. Vogelstein B, Fearon ER, Hamilton SR, et al. Genetic alterations during colorectal-tumor development. *N Engl J Med*. 1988;319(9):525–532.
39. Sherr CJ. The INK4a/ARF network in tumour suppression. *Nat Rev Mol Cell Biol*. 2001;2(10):731–737.
40. Rozenblum E, Schutte M, Goggins M, et al. Tumor-suppressive pathways in pancreatic carcinoma. *Cancer Res*. 1997;57(9):1731–1734.
41. Schutte M, Hruban RH, Geradts J, et al. Abrogation of the Rb/p16 tumor-suppressive pathway in virtually all pancreatic carcinomas. *Cancer Res*. 1997;57(15):3126–3130.
42. Wilentz RE, Geradts J, Maynard R, et al. Inactivation of the p16 (INK4A) tumor-suppressor gene in pancreatic duct lesions: loss of intranuclear expression. *Cancer Res*. 1998;58(20):4740–4744.
43. Maitra A, Adsay NV, Argani P, et al. Multicomponent analysis of the pancreatic adenocarcinoma progression model using a pancreatic intraepithelial neoplasia tissue microarray. *Mod Pathol*. 2003;16(9):902–912.
44. Sugimoto M, Kuo ML, Roussel MF, Sherr CJ. Nucleolar Arf tumor suppressor inhibits ribosomal RNA processing. *Mol Cell*. 2003;11(2):415–424.
45. Rocha S, Campbell KJ, Perkins ND. p53- and Mdm2-independent repression of NF-kappa B transactivation by the ARF tumor suppressor. *Mol Cell*. 2003;12(1):15–25.
46. Qi Y, Gregory MA, Li Z, Brousal JP, West K, Hann SR. p19ARF directly and differentially controls the functions of c-Myc independently of p53. *Nature*. 2004;431 (7009):712–717.
47. Paliwal S, Pande S, Kovi RC, Sharpless NE, Bardeesy N, Grossman SR. Targeting of C-terminal binding protein (CtBP) by ARF results in p53-independent apoptosis. *Mol Cell Biol*. 2006;26(6):2360–2372.

48. Barton CM, Staddon SL, Hughes CM, et al. Abnormalities of the p53 tumour suppressor gene in human pancreatic cancer. *Br J Cancer.* 1991;64(6):1076–1082.
49. Khoo KH, Verma CS, Lane DP. Drugging the p53 pathway: understanding the route to clinical efficacy. *Nat Rev Drug Discov.* 2014;13(3):217–236.
50. Morton JP, Timpson P, Karim SA, et al. Mutant p53 drives metastasis and overcomes growth arrest/senescence in pancreatic cancer. *Proc Natl Acad Sci U S A.* 2010;107(1): 246–251.
51. Weissmueller S, Manchado E, Saborowski M, et al. Mutant p53 drives pancreatic cancer metastasis through cell-autonomous PDGF receptor beta signaling. *Cell.* 2014;157(2):382–394.
52. Muller PA, Vousden KH, Norman JC. p53 and its mutants in tumor cell migration and invasion. *J Cell Biol.* 2011;192(2):209–218.
53. Powell E, Piwnica-Worms D, Piwnica-Worms H. Contribution of p53 to metastasis. *Cancer Discov.* 2014;4(4):405–414.
54. Maitra A, Hruban RH. Pancreatic cancer. *Annu Rev Pathol.* 2008;3:157–188.
55. Hezel AF, Bardeesy N. LKB1; linking cell structure and tumor suppression. *Oncogene.* 2008;27(55):6908–6919.
56. Giardiello FM, Brensinger JD, Tersmette AC, et al. Very high risk of cancer in familial Peutz–Jeghers syndrome. *Gastroenterology.* 2000;119(6):1447–1453.
57. Su GH, Hruban RH, Bansal RK, et al. Germline and somatic mutations of the STK11/ LKB1 Peutz–Jeghers gene in pancreatic and biliary cancers. *Am J Pathol.* 1999;154(6):1835–1840.
58. Morton JP, Jamieson NB, Karim SA, et al. LKB1 haploinsufficiency cooperates with Kras to promote pancreatic cancer through suppression of p21-dependent growth arrest. *Gastroenterology.* 2010;139(2). 586–597.e1–6.
59. Hahn SA, Seymour AB, Hoque AT, et al. Allelotype of pancreatic adenocarcinoma using xenograft enrichment. *Cancer Res.* 1995;55(20):4670–4675.
60. Hahn SA, Schutte M, Hoque AT, et al. DPC4, a candidate tumor suppressor gene at human chromosome 18q21.1. *Science.* 1996;271(5247):350–353.
61. Wilentz RE, Iacobuzio-Donahue CA, Argani P, et al. Loss of expression of Dpc4 in pancreatic intraepithelial neoplasia: evidence that DPC4 inactivation occurs late in neoplastic progression. *Cancer Res.* 2000;60(7):2002–2006.
62. Tascilar M, Skinner HG, Rosty C, et al. The SMAD4 protein and prognosis of pancreatic ductal adenocarcinoma. *Clin Cancer Res.* 2001;7(12):4115–4121.
63. David CJ, Huang YH, Chen M, et al. TGF-beta tumor suppression through a lethal EMT. *Cell.* 2016;164(5):1015–1030.
64. Padua D, Massague J. Roles of TGFbeta in metastasis. *Cell Res.* 2009;19(1):89–102.
65. Fischer KR, Durrans A, Lee S, et al. Epithelial-to-mesenchymal transition is not required for lung metastasis but contributes to chemoresistance. *Nature.* 2015;527 (7579):472–476.
66. Zheng X, Carstens JL, Kim J, et al. Epithelial-to-mesenchymal transition is dispensable for metastasis but induces chemoresistance in pancreatic cancer. *Nature.* 2015; 527(7579):525–530.
67. Singh A, Settleman J. EMT, cancer stem cells and drug resistance: an emerging axis of evil in the war on cancer. *Oncogene.* 2010;29(34):4741–4751.
68. Ling J, Kang Y, Zhao R, et al. KrasG12D-induced IKK2/beta/NF-kappaB activation by IL-1alpha and p62 feedforward loops is required for development of pancreatic ductal adenocarcinoma. *Cancer Cell.* 2012;21(1):105–120.
69. Goggins M, Shekher M, Turnacioglu K, Yeo CJ, Hruban RH, Kern SE. Genetic alterations of the transforming growth factor beta receptor genes in pancreatic and biliary adenocarcinomas. *Cancer Res.* 1998;58(23):5329–5332.

70. Blasco MA, Lee HW, Hande MP, et al. Telomere shortening and tumor formation by mouse cells lacking telomerase RNA. *Cell*. 1997;91(1):25–34.
71. Lee HW, Blasco MA, Gottlieb GJ, Horner 2nd JW, Greider CW, DePinho RA. Essential role of mouse telomerase in highly proliferative organs. *Nature*. 1998;392(6676): 569–574.
72. Artandi SE, Chang S, Lee SL, et al. Telomere dysfunction promotes non-reciprocal translocations and epithelial cancers in mice. *Nature*. 2000;406(6796):641–645.
73. O'Hagan RC, Chang S, Maser RS, et al. Telomere dysfunction provokes regional amplification and deletion in cancer genomes. *Cancer Cell*. 2002;2(2):149–155.
74. van Heek NT, Meeker AK, Kern SE, et al. Telomere shortening is nearly universal in pancreatic intraepithelial neoplasia. *Am J Pathol*. 2002;161(5):1541–1547.
75. Greenberg RA, Chin L, Femino A, et al. Short dysfunctional telomeres impair tumorigenesis in the INK4a(delta2/3) cancer-prone mouse. *Cell*. 1999;97(4): 515–525.
76. Chin L, Artandi SE, Shen Q, et al. p53 deficiency rescues the adverse effects of telomere loss and cooperates with telomere dysfunction to accelerate carcinogenesis. *Cell*. 1999;97(4):527–538.
77. Sato N, Mizumoto K, Nagai E, Tanaka M. Telomerase as a new target for pancreatic cancer treatment. *J Hepatobiliary Pancreat Surg*. 2002;9(3):322–327.
78. Thiery JP. Epithelial–mesenchymal transitions in development and pathologies. *Curr Opin Cell Biol*. 2003;15(6):740–746.
79. Wu Q, Miele L, Sarkar FH, Wang Z. The role of EMT in pancreatic cancer progression. *Pancreat Disord Ther*. 2012;2(3):e121.
80. Dvorak HF. Tumors: wounds that do not heal. Similarities between tumor stroma generation and wound healing. *N Engl J Med*. 1986;315(26):1650–1659.
81. Thiery JP. Epithelial–mesenchymal transitions in tumour progression. *Nat Rev Cancer*. 2002;2(6):442–454.
82. Peinado H, Olmeda D, Cano A. Snail, Zeb and bHLH factors in tumour progression: an alliance against the epithelial phenotype? *Nat Rev Cancer*. 2007;7(6): 415–428.
83. Gal A, Sjoblom T, Fedorova L, Imreh S, Beug H, Moustakas A. Sustained TGF beta exposure suppresses Smad and non-Smad signalling in mammary epithelial cells, leading to EMT and inhibition of growth arrest and apoptosis. *Oncogene*. 2008;27(9): 1218–1230.
84. Shah AN, Summy JM, Zhang J, Park SI, Parikh NU, Gallick GE. Development and characterization of gemcitabine-resistant pancreatic tumor cells. *Ann Surg Oncol*. 2007;14(12):3629–3637.
85. Arumugam T, Ramachandran V, Fournier KF, et al. Epithelial to mesenchymal transition contributes to drug resistance in pancreatic cancer. *Cancer Res*. 2009;69(14): 5820–5828.
86. Li Y, VandenBoom 2nd TG, Kong D, et al. Up-regulation of miR-200 and let-7 by natural agents leads to the reversal of epithelial-to-mesenchymal transition in gemcitabine-resistant pancreatic cancer cells. *Cancer Res*. 2009;69(16):6704–6712.
87. Du Z, Qin R, Wei C, et al. Pancreatic cancer cells resistant to chemoradiotherapy rich in "stem-cell-like" tumor cells. *Dig Dis Sci*. 2011;56(3):741–750.
88. Nomi T, Sho M, Akahori T, et al. Clinical significance and therapeutic potential of the programmed death-1 ligand/programmed death-1 pathway in human pancreatic cancer. *Clin Cancer Res*. 2007;13(7):2151–2157.
89. Troiani T, Martinelli E, Capasso A, et al. Targeting EGFR in pancreatic cancer treatment. *Curr Drug Targets*. 2012;13(6):802–810.
90. Pao W, Chmielecki J. Rational, biologically based treatment of EGFR-mutant non-small-cell lung cancer. *Nat Rev Cancer*. 2010;10(11):760–774.

91. Grabocka E, Pylayeva-Gupta Y, Jones MJ, et al. Wild-type H- and N-Ras promote mutant K-Ras-driven tumorigenesis by modulating the DNA damage response. *Cancer Cell.* 2014;25(2):243–256.
92. Eser S, Schnieke A, Schneider G, Saur D. Oncogenic KRAS signalling in pancreatic cancer. *Br J Cancer.* 2014;111(5):817–822.
93. Asano T, Yao Y, Zhu J, Li D, Abbruzzese JL, Reddy SA. The PI 3-kinase/Akt signaling pathway is activated due to aberrant Pten expression and targets transcription factors NF-kappaB and c-Myc in pancreatic cancer cells. *Oncogene.* 2004;23(53):8571–8580.
94. Baer R, Cintas C, Therville N, Guillermet-Guibert J. Implication of PI3K/Akt pathway in pancreatic cancer: when PI3K isoforms matter? *Adv Biol Regul.* 2015;59:19–35.
95. Soler A, Figueiredo AM, Castel P, et al. Therapeutic benefit of selective inhibition of p110alpha PI3-kinase in pancreatic neuroendocrine tumors. *Clin Cancer Res.* 2016, in press.
96. Zhao B, Li L, Lei Q, Guan KL. The Hippo-YAP pathway in organ size control and tumorigenesis: an updated version. *Genes Dev.* 2010;24(9):862–874.
97. Harvey KF, Zhang X, Thomas DM. The Hippo pathway and human cancer. *Nat Rev Cancer.* 2013;13(4):246–257.
98. Moroishi T, Hansen CG, Guan KL. The emerging roles of YAP and TAZ in cancer. *Nat Rev Cancer.* 2015;15(2):73–79.
99. Meng Z, Moroishi T, Guan KL. Mechanisms of Hippo pathway regulation. *Genes Dev.* 2016;30(1):1–17.
100. Morvaridi S, Dhall D, Greene MI, Pandol SJ, Wang Q. Role of YAP and TAZ in pancreatic ductal adenocarcinoma and in stellate cells associated with cancer and chronic pancreatitis. *Sci Rep.* 2015;5:16759.
101. Dong J, Feldmann G, Huang J, et al. Elucidation of a universal size-control mechanism in Drosophila and mammals. *Cell.* 2007;130(6):1120–1133.
102. Zhang W, Nandakumar N, Shi Y, et al. Downstream of mutant KRAS, the transcription regulator YAP is essential for neoplastic progression to pancreatic ductal adenocarcinoma. *Sci Signal.* 2014;7(324):ra42.
103. Overholtzer M, Zhang J, Smolen GA, et al. Transforming properties of YAP, a candidate oncogene on the chromosome 11q22 amplicon. *Proc Natl Acad Sci U S A.* 2006;103(33):12405–12410.
104. McPherson JP, Tamblyn L, Elia A, et al. Lats2/Kpm is required for embryonic development, proliferation control and genomic integrity. *EMBO J.* 2004;23(18): 3677–3688.
105. Morin-Kensicki EM, Boone BN, Howell M, et al. Defects in yolk sac vasculogenesis, chorioallantoic fusion, and embryonic axis elongation in mice with targeted disruption of Yap65. *Mol Cell Biol.* 2006;26(1):77–87.
106. Lee JH, Kim TS, Yang TH, et al. A crucial role of WW45 in developing epithelial tissues in the mouse. *EMBO J.* 2008;27(8):1231–1242.
107. Zhou D, Conrad C, Xia F, et al. Mst1 and Mst2 maintain hepatocyte quiescence and suppress hepatocellular carcinoma development through inactivation of the Yap1 oncogene. *Cancer Cell.* 2009;16(5):425–438.
108. Nishio M, Hamada K, Kawahara K, et al. Cancer susceptibility and embryonic lethality in Mob1a/1b double-mutant mice. *J Clin Invest.* 2012;122(12):4505–4518.
109. Wierzbicki PM, Adrych K, Kartanowicz D, et al. Underexpression of LATS1 TSG in colorectal cancer is associated with promoter hypermethylation. *World J Gastroenterol.* 2013;19(27):4363–4373.
110. Takahashi Y, Miyoshi Y, Takahata C, et al. Down-regulation of LATS1 and LATS2 mRNA expression by promoter hypermethylation and its association with biologically aggressive phenotype in human breast cancers. *Clin Cancer Res.* 2005;11(4):1380–1385.
111. Johnson R, Halder G. The two faces of Hippo: targeting the Hippo pathway for regenerative medicine and cancer treatment. *Nat Rev Drug Discov.* 2014;13(1):63–79.

112. Gao T, Zhou D, Yang C, et al. Hippo signaling regulates differentiation and maintenance in the exocrine pancreas. *Gastroenterology*. 2013;144(7). 1543–1553.e1.
113. George NM, Day CE, Boerner BP, Johnson RL, Sarvetnick NE. Hippo signaling regulates pancreas development through inactivation of Yap. *Mol Cell Biol*. 2012;32(24):5116–5128.
114. Guerra C, Schuhmacher AJ, Canamero M, et al. Chronic pancreatitis is essential for induction of pancreatic ductal adenocarcinoma by K-Ras oncogenes in adult mice. *Cancer Cell*. 2007;11(3):291–302.
115. Gukovsky I, Li N, Todoric J, Gukovskaya A, Karin M. Inflammation, autophagy, and obesity: common features in the pathogenesis of pancreatitis and pancreatic cancer. *Gastroenterology*. 2013;144(6). 1199–1209.e4.
116. Pylayeva-Gupta Y, Lee KE, Hajdu CH, Miller G, Bar-Sagi D. Oncogenic Kras-induced GM-CSF production promotes the development of pancreatic neoplasia. *Cancer Cell*. 2012;21(6):836–847.
117. Bayne LJ, Beatty GL, Jhala N, et al. Tumor-derived granulocyte–macrophage colony-stimulating factor regulates myeloid inflammation and T cell immunity in pancreatic cancer. *Cancer Cell*. 2012;21(6):822–835.
118. Iacobuzio-Donahue CA, Herman JM. Autophagy, p53, and pancreatic cancer. *N Engl J Med*. 2014;370(14):1352–1353.
119. Kimmelman AC. The dynamic nature of autophagy in cancer. *Genes Dev*. 2011;25(19):1999–2010.
120. Perera RM, Stoykova S, Nicolay BN, et al. Transcriptional control of autophagy-lysosome function drives pancreatic cancer metabolism. *Nature*. 2015;524(7565): 361–365.
121. Hashimoto D, Blauer M, Hirota M, Ikonen NH, Sand J, Laukkarinen J. Autophagy is needed for the growth of pancreatic adenocarcinoma and has a cytoprotective effect against anticancer drugs. *Eur J Cancer*. 2014;50(7):1382–1390.
122. Antonucci L, Fagman JB, Kim JY, et al. Basal autophagy maintains pancreatic acinar cell homeostasis and protein synthesis and prevents ER stress. *Proc Natl Acad Sci U S A*. 2015;112(45):E6166–E6174.
123. Rosenfeldt MT, O'Prey J, Morton JP, et al. p53 status determines the role of autophagy in pancreatic tumour development. *Nature*. 2013;504(7479):296–300.
124. Rhim AD, Mirek ET, Aiello NM, et al. EMT and dissemination precede pancreatic tumor formation. *Cell*. 2012;148(1–2):349–361.
125. Provenzano PP, Cuevas C, Chang AE, Goel VK, Von Hoff DD, Hingorani SR. Enzymatic targeting of the stroma ablates physical barriers to treatment of pancreatic ductal adenocarcinoma. *Cancer Cell*. 2012;21(3):418–429.
126. Sun JD, Liu Q, Wang J, et al. Selective tumor hypoxia targeting by hypoxia-activated prodrug TH-302 inhibits tumor growth in preclinical models of cancer. *Clin Cancer Res*. 2012;18(3):758–770.
127. Borad MJ, Reddy SG, Bahary N, et al. Randomized phase II trial of gemcitabine plus TH-302 versus gemcitabine in patients with advanced pancreatic cancer. *J Clin Oncol*. 2015;33(13):1475–1481.
128. Drummond DC, Noble CO, Guo Z, Hong K, Park JW, Kirpotin DB. Development of a highly active nanoliposomal irinotecan using a novel intraliposomal stabilization strategy. *Cancer Res*. 2006;66(6):3271–3277.
129. Singh A, Settleman J. Oncogenic K-ras "addiction" and synthetic lethality. *Cell Cycle*. 2009;8(17):2676–2677.
130. Wolpin BM, Rubinson DA, Wang X, et al. Phase II and pharmacodynamic study of autophagy inhibition using hydroxychloroquine in patients with metastatic pancreatic adenocarcinoma. *Oncologist*. 2014;19(6):637–638.

131. Chen Z, Cheng K, Walton Z, et al. A murine lung cancer co-clinical trial identifies genetic modifiers of therapeutic response. *Nature*. 2012;483(7391):613–617.
132. Laheru D, Lutz E, Burke J, et al. Allogeneic granulocyte macrophage colony-stimulating factor-secreting tumor immunotherapy alone or in sequence with cyclophosphamide for metastatic pancreatic cancer: a pilot study of safety, feasibility, and immune activation. *Clin Cancer Res*. 2008;14(5):1455–1463.
133. Beatty GL, Haas AR, Maus MV, et al. Mesothelin-specific chimeric antigen receptor mRNA-engineered T cells induce anti-tumor activity in solid malignancies. *Cancer Immunol Res*. 2014;2(2):112–120.
134. Beatty GL, Chiorean EG, Fishman MP, et al. CD40 agonists alter tumor stroma and show efficacy against pancreatic carcinoma in mice and humans. *Science*. 2011;331(6024):1612–1616.
135. Ozdemir BC, Pentcheva-Hoang T, Carstens JL, et al. Depletion of carcinoma-associated fibroblasts and fibrosis induces immunosuppression and accelerates pancreas cancer with reduced survival. *Cancer Cell*. 2014;25(6):719–734.

CHAPTER SEVEN

Current and Emerging Targeting Strategies for Treatment of Pancreatic Cancer

A.T. Baines*,†,1, P.M. Martin‡, C.J. Rorie‡
*Cancer Research Program, JLC-Biomedical/Biotechnology Research Institute, North Carolina Central University, Durham, NC, United States
†School of Medicine, UNC-Chapel Hill, Chapel Hill, NC, United States
‡North Carolina Agricultural and Technical State University, Greensboro, NC, United States
[1]Corresponding author: e-mail address: abaines@nccu.edu

Contents

Abstract

With a dismal 5-year survival rate of only 8%, pancreatic cancer still remains a very lethal disease. As with most cancers, pancreatic cancer is treated with different combinations of chemotherapeutic drugs which result in side effects and potential drug resistance leading in many cases to the unfortunate demise of the patient. Over recent years, a number of therapies have been developed against numerous molecular targets in cancers. Kinase inhibitors and monoclonal antibodies have been shown to target numerous kinases, growth factor receptors, and cell signaling pathways. This can lead to effects on tumor cell growth, angiogenesis, apoptosis, and the microenvironment. Most recent findings are very promising as they relate to the use of immunotherapy to treat certain cancers. Immune checkpoint inhibitors and cancer vaccines are currently being

Progress in Molecular Biology and Translational Science, Volume 144
ISSN 1877-1173
http://dx.doi.org/10.1016/bs.pmbts.2016.09.006

investigated. In this review, we will highlight some novel molecular targeted strategies that are being used or considered as potential therapeutics to treat patients with pancreatic cancer.

1. INTRODUCTION

Pancreatic cancer is one of the most lethal of all cancers with a 5-year survival rate of only 8%.[1] It is the seventh leading cause of cancer-related deaths worldwide with 280,000 new cases every year.[2] Unfortunately, there has been no significant improvement in survival over the past 40 years. In 2016, an estimated 53,070 people will be diagnosed with pancreatic cancer in the United States and approximately 42,680 will succumb to the disease.[3] Of all the cancers that can arise in the pancreas, pancreatic ductal adenocarcinoma (PDAC) is the most prevalent.[4] Most recently, PDAC changed from being the fourth to the third leading cause of cancer-related deaths in the United States, surpassing breast cancer.[4] In 2020, it is believed that it will rise to become the second leading cause of cancer-related death.[5] Pancreatic cancer has the highest mortality of all major cancers with 94% of the patients dying within 5 years of diagnosis and 74% dying within the first year of diagnosis. If metastatic disease is present, the average life expectancy after diagnosis ranges between 3 and 6 months. Surgery and chemotherapy are the main treatment strategies used to combat this disease, with only a small percentage (10–15%) of patients eligible for surgery. Usually, patients develop resistance to the chemotherapy and radiation treatments rendering it ineffective. Due to this dismal outcome, there is an urgent need to develop specific biomarkers to help diagnose PDAC at an early stage for surgery. Equally important, novel molecular drug targets need to be identified and validated in order to develop potential drugs that may be successful in treating pancreatic cancer. From a number of studies, some of the genetic mutations in oncogenes and tumor suppressor genes that have been commonly found in PDAC include *KRAS*, *p16/CDKN2A*, *TP53*, and *SMAD4*.[6] In addition, there are other genes and proteins of interest as it relates to development and tumor maintenance that could result in possible molecular targets in pancreatic cancer. In this chapter, we will discuss the traditional strategies that are currently being used to treat pancreatic cancer including surgery, chemotherapy, and radiotherapy. Along with the traditional methods of treatment, we will also discuss novel and emerging therapeutic options for combating this disease. They will include such topics as immunotherapy, epigenetic modulation, stromal disruption, and PARP inhibitors.

2. CURRENT THERAPEUTIC STRATEGIES

2.1 Surgery

Currently, surgical resection offers the only opportunity for a cure in patients with localized pancreatic cancer disease.[7] Even with this possibility of a surgical resection with negative tumor margins, there are still only about 15% of patients that fit in this group.[8] The success of the different PDAC surgeries (cephalic pancreaticoduodenectomy or Whipple procedure, distal pancreatectomy, or total pancreatectomy) depends on a number of factors. They include "tumor size and grade, lymph node involvement, CA 19-9 levels, and the resection margins."[9] Another major factor involves the quantity of surgeries conducted by well-experienced surgeons at specialized hospitals.[10] This factor alone can greatly decrease postoperative mortality and the risk to as low as 3%. However, even after the surgery, the 5-year survival rate rarely exceeds 20% with a less favorable prognosis.[7] To help increase the survival rate, adjuvant chemotherapy[11] or chemoradiotherapy[12,13] is usually administered postoperatively to the patient or neoadjuvant chemotherapy is provided presurgery against potential micrometastases.[14–18] Unfortunately for some patients, the cancer will be more advanced at the initial time of diagnosis. About 30% of patients will have borderline-resectable or unresectable locally advanced pancreatic cancer (LAPC).[1] They are "confined locoregionally with some degree of involvement of the nearby major vascular structures but without any evidence of distant metastases."[19] Depending on the relationship of the tumor with vascular structures, aggressive management with neoadjuvant chemotherapy can be administered to help shrink the tumor for some patients leading to a possible curative surgery.

As it relates to advances in surgical practice for pancreatic cancer, studies have demonstrated that minimally invasive laparoscopically approaches provide similar or more benefit to the patient than traditional open surgeries and have become a standard treatment.[20–23] The benefits can include a rapid recovery time and less time in the hospital and an opportunity to undergo other additional treatments sooner. Also, the use of robotic-assisted surgeries to perform the Whipple procedure has been studied and has shown promising results when compared to laparoscopy.[24] Further studies are needed to continue to evaluate the safety and efficiency of these surgical approaches. However, despite the advances in surgical practices in treating localized and some LAPCs, the majority of patients will present with metastatic pancreatic cancer. For these patients, the standard treatment option is chemotherapy.

2.2 Chemotherapy

One of the main reasons for the high mortality rate of patients with pancreatic cancer is due to the majority (50–60%) being initially diagnosed with metastatic disease.[1] The main chemotherapeutic drug that was commonly used starting in the 1990s and offered some benefit for patients was the nucleoside analogue gemcitabine (Gemzar).[25–30] In a very important clinical trial comparing gemcitabine alone with 5-fluorouracil (5-FU), a modest increase in survival was observed with gemcitabine (5.6 vs 4.4 months; $P=0.0025$).[25] The Food and Drug Administration (FDA) approved in 1997 the replacement of 5-FU with gemcitabine monotherapy as a standard treatment for metastatic pancreatic cancer despite having only a 1-month survival extension clinical benefit. Since then, gemcitabine has been the standard agent of choice for treatment of pancreatic cancer over the years.

To help improve overall survival, numerous attempts have been made to combine multiple agents with gemcitabine together and evaluate them. Unfortunately, there has been very little success observed combining gemcitabine with such drugs as platinum drugs, fluoropyrimidine, and topoisomerase inhibitors. The median overall survival with these combinations ranged from 6 to 9 months with a 10% increase in overall survival with the addition of the second drug. This led some French researchers to test a multidrug regiment against pancreatic cancer without the addition of gemcitabine. In the Actions Concertees dans les Cancer Colo-Rectaux et Digestifs (ACCORD) 11 trial, there was a 4.3-month improvement in median overall survival with patients on FOLFIRINOX than with gemcitabine (11.1 vs 6.8 months; $P<0.001$).[30] FOLFIRINOX is a four-drug regiment composed of 5-FU, leucovorin, irinotecan, and oxaliplatin. Currently, it is recommended as a first-line treatment option for metastatic pancreatic cancer despite having very serious side effects such as diarrhea, vomiting, and peripheral neuropathy.

A second drug that has been recommended by the FDA as a first-line treatment option for metastatic pancreatic cancer in combination with gemcitabine is Nab-paclitaxel (Abraxane), a 130-nM nanoparticle albumin-bound (nab) formulation of paclitaxel administered as a colloidal suspension. The origin of this drug is based on the development of an agent to overcome hypersensitivity effects suffered by patients due to the dissolving solvent of paclitaxel.[31] Since then, it has been approved by the FDA for metastatic breast and nonsmall cell lung cancers.[32,33] As it relates to mechanism of action, it is believed that the albumin in Nab-paclitaxel binds to a matrix glycoprotein called secreted protein acidic and rich in cysteine (SPARC) found in high levels in tumor stromal cells.[34] This allows for more of the drug to reach the actual tumor cells intratumorally. A second proposed mechanism is the possibility of Nab-paclitaxel being carried to the tumor interstitial space due to the albumin portion binding to an albumin receptor (gp60) on endothelial cells.[35] In the phase III Metastatic Pancreatic Adenocarcinoma Clinical Trial (MPACT), the median overall survival of Nab-paclitaxel plus gemcitabine compared to gemcitabine alone was 8.5 vs 6.7 months ($P<0.001$).[36] As compared to FOLFIRNOX, although the median overall survival was less, the toxicity safety profile is better and this combination provides another option for patients.

A third drug combination approved by the FDA as a biological agent for locally advanced or metastatic pancreatic cancer was erlotinib and gemcitabine. Clinical trials demonstrated that the combination gave an improvement in median overall survival of approximately 10 days (6.2 vs 5.9 months).[37] Although some patients do well with this regiment, the associated side effects and slight clinical benefit limit the use of this combination in patients. The two most preferred options for treating pancreatic cancer due to having greater clinical benefits include FOLFIRINOX and Nab-paclitaxel plus gemcitabine. Most recently, another drug was approved by the FDA for pancreatic cancer by the name Onivyde (irinotecan liposome injection). Studies show the benefit of using Onivyde in combination of 5-FU and leucovorin after gemcitabine-based therapy.[38] In a study between Onivyde, fluorouracil, and leucovorin and just fluorouracil and leucovorin, there was a significant increase in survival from 6.2 vs 4.2 months. However, despite having these newer drug combinations for treatment, the survival rate for most patients after treatment is unfortunately still very dismal and warrants the need for novel therapies.

2.3 Radiotherapy

Although radiation is a common treatment strategy for numerous cancers, there is much debate on the effectiveness of radiation in treating pancreatic cancers.[39–41] Gemcitabine or 5-FU-based chemoradiotherapy is usually used during postoperative treatment of localized disease. In one clinical trial by the Radiation Therapy Oncology Group (RTOG)-9704 trial, a statistical but nonsignificant improvement in median overall survival (20.5 vs 16.9 months) in the gemcitabine-containing arm was observed in resectable PDAC patients who received gemcitabine or 5-FU before and after 5-FU chemoradiation.[42] For locally advanced tumors of the pancreas, some studies demonstrate that resection rates are improved with radiation treatment, whereas other studies do not show any significant survival benefit. Based on results from the Eastern Cooperative Oncology Group (ECOG)-4201-trial[43] and the Taipei trial,[44] gemcitabine-based chemoradiation was approved as an acceptable treatment for LAPC. Studies are currently being evaluated to determine the effectiveness of chemotherapy followed by chemoradiation prior to surgery. Overall, there is still ongoing discussion on the effectiveness of radiotherapy being used as a treatment for pancreatic cancer.

3. NOVEL AND EMERGING THERAPEUTIC OPTIONS

3.1 Immunotherapy-Based Approaches

One of the recent and promising advances in cancer treatments receiving much attention is the area of immunotherapy. Immunotherapy involves prompting the immune system to become active in targeting and destroying any foreign substances in the human body. Despite past beliefs that pancreatic cancer was not immunogenic and would not be a candidate for immunotherapy, recent studies have demonstrated the contrary.[45–48] Studies have demonstrated that patients with pancreatic cancer produce immune B and T cells that are specific to a number of tumor cell antigens. They include antigens such as mucin I (MUC1),[49] survivin,[50] carcinoembryonic antigen (CEA),[51] mutated K-Ras,[52] and Wilms' tumor gene 1 (WT1).[53] Usually, cancers such as pancreatic cancer become quiescent and evade the defenses of the immune system by causing immune dysfunction.[54] Some examples include modulating "contact-dependent factors (e.g., immune system checkpoint ligands such as PD-L1), secretion of soluble immunosuppressive factors (TGF-β and VEGF), and interference with the MHC class I expression peptide presentation."[55] Immunotherapeutic treatments against cancer can be categorized as either passive or active immunotherapy. For the passive approach, the focus is on antibodies or effector cells generated in vitro. For the active approach, vaccines are developed to help stimulate an immune response against tumors in vivo.[55] Examples of both will be discussed later.

3.1.1 Passive Immunotherapy

As was mentioned earlier, antibodies are primary for passive immunotherapy and are important since they can bind to antigens found more abundant in cancer cells than in the normal cells (e.g., tumor-associated antigens). Also, they are involved in impeding molecules necessary for cancer growth, development, and progression.[55] One such antigen that has been shown to be expressed in normal mesothelial cells but overexpressed in 90–100% of pancreatic adenocarcinomas and be a favorable prognostic factor is the protein mesothelin.[56–58] A number of antibodies against mesothelin have been developed and are in different stages of testing. One example is the SS1P, a murine single-chain Fv, which has been fused to a 38-kDa portion of *Pseudomonas* exotoxin A (PE-A) and is human specific.[59] The binding of this antibody to mesothelin results in its internalization into the cell and inhibition of protein synthesis which can ultimately lead to apoptosis. SS1P has been demonstrated to be well tolerated in phase I clinical studies and has shown some promising results in mouse studies.[59–61] Other antibodies made against mesothelin include the monoclonal antibody MORAb-009[62] and the antihuman antibodies M912 and HN1.[63,64] Disease stabilization and a decrease in CA19-9 (carbohydrate antigen 19-9, used as an indicator for GI cancer progression) were observed in one pancreatic cancer patient in a clinical trial receiving MORAb-009 who progressed on gemcitabine.[55] A second antigen in pancreatic cancer that has been targeted by numerous antibodies includes the transmembrane glycoprotein receptor epidermal growth factor receptor (EGFR). It is overexpressed in 90% of pancreatic tumors, involved in many processes in cancer cells including growth and neovascularization,[65] and is negative prognosis factor.[66,67] Studies have demonstrated that blocking EGFR can disrupt cancer growth and make pancreatic cancer more susceptible to gemcitabine.[68,69] Two antibodies that have been developed against this receptor include cetuximab (Erbitux or IMC-C225) and matuzumab (EMD72000).[55] Both are monoclonal antibodies that have demonstrated promising laboratory results which led to various clinical trials. In one phase III study conducted by the Southwestern Oncology Group, cetuximab was administered in combination with gemcitabine for efficacy to patients with advanced pancreatic cancer. Unfortunately, there was found to be no clinical significant benefit with the addition of cetuximab to gemcitabine in this particular study.[70] In another clinical study, stable disease and a partial response were observed when matuzumab was combined with gemcitabine in advanced pancreatic cancer patients.[71] A third antigen involved in numerous processes of cancer growth transformation as it relates to angiogenesis and metastasis is the vascular endothelial growth factor (VEGF).[72] It is overexpressed in pancreatic cancer and has been shown to correlate with poor prognosis. One monoclonal antibody that has been made against VEGF is the recombinant humanized bevacizumab (Avastin). It has been tested in clinical trials in various combinations with gemcitabine and other drugs. In one pilot study, metastatic pancreatic cancer patients treated with bevacizumab in combination with gemcitabine showed significant improvement in response and survival.[73] Unfortunately, results from later studies including a phase III study showed no significant benefit with the addition of bevacizumab.[59] Other examples of antibodies currently under evaluation in animal models or clinical studies of pancreatic cancer have been made against include MUC1 (mucin-1, CD227), epidermal growth factor receptor 2 (HER2), and CEA.[55]

3.1.2 Active Immunotherapy

In contrast to passive immunotherapy where the focus is on monoclonal antibodies, vaccine therapy is the focus for active immunotherapy. For cancer vaccines, the goal is to express a tumor antigen to promote and stimulate tumor-specific immunity. Vaccines can be classified as whole cancer cell-based, antigen/peptide-specific, and dendritic cell-based.[55] For the whole cancer cell-based vaccines, irradiated cancer cells are inoculated back into the patient. Unfortunately due to the limited number of pancreatic cancer patients who are eligible for surgery (10–15% of patients) to provide cells, along with other factors such as time to culture and contamination concerns, availability of tumor cells is in short supply for most patients.[74] To address these important concerns, allogeneic tumor cell lines are considered as substitutes and have a number of advantages such as being good sources of tumor-associate antigens and being conducive to in vitro growth.[75] Two examples of allogeneic whole cell-based vaccines that have been studied in pancreatic cancer patients include the allogeneic granulocyte macrophage colony-stimulating factor (GM-CSF)-secreting vaccine (pancreatic GVAX, first to be developed against pancreatic cancer)[76] and the algenpantucel-L (hyperacute-PC vaccine).[77] In GVAX, two allogeneic human pancreatic cancer cell lines were engineered to express the immunomodulatory cytokine GM-CSF. GVAX has been studied in a variety of clinical studies to test for safety and anticancer activity.[76,77] In a recent clinical study

(ECLIPSE), GVAX was combined with the live-attenuated Listeria-based vaccine CRS-207 to enhance efficacy.[19] For this phase II clinical trial, participants included previously treated metastatic pancreatic cancer patients (NCT02004262). For algenpantucel-L immunotherapy, the vaccine consists of two irradiated human pancreatic cancer cell lines genetically modified to overexpress murine alpha (1,3)-galactosyltransferase. This results in the expression of alpha-galactosyl (Alpha-Gal) epitopes which are not found in human cells, resulting in an antitumor activity due to a strong T-cell response. In a recently completed phase III adjuvant trial, gemcitabine was compared with or without algenpantucel-L, followed by chemoradiation (NCT01072981). In another trial for borderline-resectable and unresectable LAPC, FOLFIRINOX is being evaluated with or without algenpantucel-L, followed by chemoradiation (NCT01836432). Despite all of the promising observations in animal models and clinical trials with antibodies and anticancer vaccines, there has not been observed any significant clinical benefit in overall patient survival with any immunotherapeutic strategy against pancreatic cancer in a randomized phase III clinical trial to date.

3.1.3 Immune Checkpoint Inhibitors

One of the most recent topics receiving much attention in the area of immunotherapy and promising new cancer treatments is the targeting of immunoinhibitory checkpoint pathways. They include such targets as cytotoxic T-lymphocyte-associated protein 4 (CTLA-4)/B7, programmed cell death-1 (PD-1)/programmed cell death ligand-1 (PD-L1). Unfortunately no significant antitumor activity was observed in a phase II trial when the checkpoint inhibitor ipilimumab was used as a single agent against advanced pancreatic cancer ([78]). Currently, there is ongoing phase I trial investigating the combination of ipilimumab with gemcitabine (NCT01473940). Despite all of these promising observations in animal models and clinical trials with immunotherapy, there has not been observed any clinical benefit with any immunotherapeutic strategy against pancreatic cancer in a randomized phase III clinical trial to date.

3.2 Chemoprevention and Neoadjuvant Strategies

Over the past four decades there has been no significant increase in the 5-year survival rate of pancreatic cancer patients, despite the increase of attention and research funding. With these disappointing developments, there has been renewed interest in developing novel adjuvant/neoadjuvant therapies and chemoprevention strategies for pancreatic cancer patients. Chemoprevention is defined as using agents that demonstrate the potential to delay and/or prevent the formation of cancer. Chemoprevention can be utilized to help prevent the formation of cancer in people more likely to be predisposed to cancer or by individuals seeking healthy lifestyle choices. Chemopreventive agents must be nontoxic to normal, noncancerous tissues so that these agents can be administered to noncancer patients as well as cancer patients who fall into high-risk categories. Additionally, it is advantageous for potential chemopreventive agents to have specific targets that are more abundantly expressed in cancer tissues than normal, noncancerous tissues.

3.2.1 Curcumin

In the past 10 years, several clinical trials have been established to investigate potential chemopreventive agents in the treatment of pancreatic cancer.[79,80] These trials have mainly focused on natural products, with most of the research focused on curcumin.[79,81] Curcumin is a bioactive component of turmeric and has been demonstrated to have several anticancer effects in several cancers including pancreatic cancer.[79] Curcumin has the capacity to prevent pancreatic cell proliferation through several different mechanisms. Specifically, curcumin is able to mediate the Notch and NFκB signaling pathways, respectively, and by associated cross talk between the well-studied signaling pathways.[82–84] Curcumin is able to significantly decrease the activation of NFκB and thus prevent pancreatic cancer cell proliferation while having little nonspecific targeting effects on normal pancreatic cells.[82] Curcumin has also been shown to increase the rate of apoptosis in certain pancreatic cell lines.[84] In addition to these targets, curcumin can also prevent ERK-mediated cell growth indirectly by reducing the expression of EGFR expression.[85] Recent studies have shown the relationship of curcumin with the ability to modulate microRNA (miRNA levels).[86] Curcumin was reported to increase the expression of miRNA-22 while decreasing the expression of miRNA-199a. These changes in miRNA expression by curcumin resulted in decreased cell growth.[86] The ability to modulate miRNA levels in cancers has been associated with potential therapeutic options.

In addition to demonstrating chemopreventive properties, curcumin has also shown very promising neoadjuvant properties. Curcumin has demonstrated the ability to sensitize pancreatic cancer cells to several common chemotherapeutic drugs such as celecoxib, gemcitabine, and paclitaxel.[82,83,85,87] Using curcumin as a neoadjuvant demonstrates promising results in xenograft models, specifically, 1 g/kg daily dose of curcumin increased the inhibitory aspects of gemcitabine in preventing cell proliferation and angiogenesis on MiaPaCa-2 (human pancreatic cancer cells). Together, curcumin and gemcitabine synergized to decrease survivin, BCl-2, and Bcl-xL.[83] Since curcumin has demonstrated reliable efficacy in numerous studies, additional studies have investigated its bioavailability. Normally, curcumin has a low bioavailability and does not induce negative side effects.[88–90] To increase its potential to serve as an adjuvant or preventive therapy for pancreatic cancer, three synthetic curcumin analogues have been synthesized. These analogues, FLLL11, FLLL12, and CDF have been tested in vitro and have been shown to be more effective in permeating pancreatic cancer cells, while also increasing the antiproliferative capacity of curcumin and inducing increased amounts of apoptosis.[91] These positive results initiated several groups to investigate using curcumin analogues in nanostructures, liposomes, and nanomicelles, to increase the permeability and target specificity of these analogues.[92,93] These studies were able to demonstrate even great efficacy and target acquisition of the curcumin analogues that correlated with increased apoptosis of pancreatic cancer cells in vitro and in vivo.[92,93]

The success on curcumin and curcumin analogues in cell culture studies and animal studies prompted the initiation of several clinical trials with curcumin in pancreatic cancer patients. One study in a phase II trial reported that 25 patients with advanced pancreatic cancer showed that there are no adverse side effects to curcumin, it is tolerated well, and demonstrates some biological effect on pancreatic cancer blood-related biomarkers, albeit curcumin was detected in low levels in blood plasma samples.[80]

3.2.2 Epiallocatechin-3-Gallate

Other natural products and bioactive compounds from natural products have been assayed for their potential antioncogenic properties against pancreatic cancer cells. Epiallocatechin-3-gallate (EGCG), polyphenols that comprise powerful antioxidants found in both black and green teas, is able to prevent in vitro growth of human pancreatic cancer cells.[94] Additionally, the target of the EGCG compound was the *KRAS* gene.[94] Animal model studies have demonstrated that EGCG can protect normal pancreas cells against oxidative stress induced by BOP (*N*-nitrosobis-2-(oxopropyl) amine).[95] Another separate study, showed the EGCG specifically from green tea, reduced the amount of hyperplastic pancreatic duct cells following BOP insults to hamsters.[96] These results prompted a series of correlative studies in humans relating the amount of tea consumption to pancreatic cancer risks. These studies provided a mixed bag of results with some studies demonstrating both a decrease in pancreatic cancer risk associated with tea consumption and studies that found no direct association between tea consumption and a decreased risk in pancreatic cancer.

Other natural products have also been investigated for preventative properties against pancreatic cancer such as β-carotene, vitamins D and E, as well as isothiocyanates. As with most natural products, these products are consumed in conjunction with a healthy diet, low in polyunsaturated fats correlate with lower risk of developing pancreatic cancer. However, these products in conjunction with a diet high in saturated fats, alcohol, and lowered physical activity are not able to decrease the risk of pancreatic cancer. Natural products have demonstrated strong adjuvant and preventative properties in relationship to pancreatic cancer and show promise in early stage clinical trials. However, the mainstream use of natural products as either a treatment or a preventative agent against pancreatic cancer formation and progression is still yet to be determined.

3.2.3 Nonsteroidal Antiinflammatory Drugs

Due to the recruitment of additional stromal cells associated with pancreatic cancer and the associated inflammation in the tumor microenvironment the use of antiinflammatory agents has been incorporated into therapeutic regiments for pancreatic cancer. Aspirin, ibuprofen, and other nonsteroidal antiinflammatory drugs (NSAIDs) agents have been tested to determine the efficacy in the treatment of pancreatic cancer as well as their regulation of the cellular and molecular changes associated with the formation and progression of pancreatic cancer. Initial data show that aspirin and other NSAIDs demonstrate significant promise as cancer chemoprevention agents, as an adjuvant therapy, as well as antiinflammatory properties in pancreatic cancer.[97] However, more recent comprehensive studies have demonstrated conflicting results.[98] Epidemiological-based studies have demonstrated no reduction in pancreatic cancer risk in people who regularly take aspirin or other NSAIDs.[98] Additionally, two other studies demonstrated a reduction in pancreatic cancer risk with the daily use of aspirin and/or NSAIDs.[98] An enhanced epidemiological study that summarized all of the recent publically available epidemiological evidence complicated the findings even further by reporting a null association between regular aspirin use and pancreatic risk.[99,100] Further analysis of these findings shows that people who take aspirin and not ibuprofen or other aspirin-like NSAIDs for longer than 5 years actually show a reduced risk of pancreatic cancer.[98,101]

In addition to these correlative studies, several mouse model studies have been conducted to test the role of aspirin in preventing pancreatic cancer. These studies demonstrated aspirin use is able to prevent the establishment of injected pancreatic cancer cells in these mice.[102] Further examination of these mice demonstrated an inhibition of inflammation associated with NFκB activation of proinflammatory signals.[102] Similar studies were conducted with aspirin in conjunction with low doses of nitric oxide. These studies demonstrated a protective response against the formation of orthotopic implanted pancreatic cancer cells.[103,104] NFκB is a major transcription factor related to promoting inflammation, especially inflammation associated with cancer formation and progression.[105] NFκB is constitutively activated in 70% of human pancreatic cancer tumor samples, and most human pancreatic cancer cell lines.[82,106] Since NFκB signaling is activated by Ras signaling it is not surprising that NFκB is constitutively active in pancreatic cancer cells due to the Kras oncogenic mutations that are frequent in pancreatic cancer.[107] Taking this knowledge into account experimental studies with kinase-inactive IκBα, the upstream kinase of NFκB, are able to slow pancreatic tumor growth in an orthotopic mouse model.[108]

Aspirin has also been shown to prevent other pancreatic cancer-associated signaling mechanisms, such as the mTOR signaling pathway and the Notch pathway.[109,110] In pancreatic cancer, the mTOR signaling pathway is dysregulated in most tumor samples and is associated with advanced tumors and poor prognosis.[109] Recent studies have also suggested that mTOR signaling may promote pancreatic stem cells to exhibit cancer stem cell properties.[109,111,112] Additionally, mTOR signaling has been shown to promote increased pancreatic cancer cell proliferation which correlates to mTOR being activated in tumor samples.[111,112] Rapamycin, a potent, specific inhibitor of mTOR has been used in the treatment of many cancers. In vitro studies have demonstrated that rapamycin can suppress PC-2 cell proliferation while increasing the expression levels of p53, BAX, and Beclin-1 proapoptotic genes.[113] A phase II clinical trial conducted with rapamycin in patients with advanced pancreatic cancer concluded that rapamycin as a single agent did not slow disease progression; however, in combination with other chemotherapeutic agents showed enough promise to warrant further studies.[114] Although initial studies have not shown rapamycin to be effective in treating pancreatic cancer, mTOR signaling still remains a viable option in chemotherapeutic cocktails for the treatment of pancreatic cancer.

Another signaling pathway that has been extensively studied in pancreatic cancer is the Notch signaling pathway.[115] Notch signaling and many of its effectors are expressed widely in pancreatic cancer. Although many studies have been conducted to better understand the manner in which Notch signaling may contribute to pancreatic cancer, several of these studies have results that contradict each other. Notch signaling and its effectors have been implicated in the coactivation of Kras-associated pancreatic cancer formation.[116] Whereas other studies demonstrate that activation of Notch signaling decreases formation of tumors in PDAC-bearing mice.[117,118] Notch signaling is also connected to NFκB signaling. Notch is thought to be downstream of NFκB and activated by NFκB-mediated phosphorylation which can lead to formation of pancreatic cancer development.[119] Although Notch is associated with pancreatic cancer formation few clinical and preclinical trials have tested anti-Notch therapeutic regimens. A preclinical study in the LSL-Kras(G12D/+); Pdx-1-Cre mouse model demonstrated a reduction of cancer cell formation as a result of necrosis after treatment with MRKK003, a γ-secretase inhibitor in combination with gemcitabine.[120] Since Notch signaling is involved in both procancer and anticancer regulation of pancreatic cancer cells developing Notch-based therapies may be challenging.

3.3 Hyaluronan and CD44

Hyaluronic acid (HA, hyaluronan) is a negatively charged glycosaminoglycan composed of repeating disaccharide units of *N*-acetylglucosamine and D-glucuronic acid. Excess hyaluronan is present in various tumor types including breast, prostate, and pancreatic cancers.[121–123] Accumulation of hyaluronan has been associated with increased aggressiveness, less treatable forms of the disease and poor prognoses.[124] Hyaluronan is continuously secreted and due to its chemical structure, hyaluronan absorbs interstitial fluid which increases the pressure in the microenvironment of the tumor. More importantly, this increased pressure in the tumor microenvironment promotes poor perfusion which leads hypoxia.[125] A hypoxic environment leads to increased production of hypoxia-inducible factor-1α (HIF-1α) which in turn increases the amount of hyaluronan production and accumulation, thus creating a positive feedback loop promoting increased interstitial pressure and hyaluronan accumulation.[126,127] Hyaluronan is found in extremely high levels in most human pancreatic cancer cases. Greater than 80% of pancreatic tumors contain high levels of hyaluronan.[122,123,125] Additionally, animal models of pancreatic cancer also demonstrate increased amounts on hyaluronan accumulation.[122,125] In vitro studies also demonstrate that hyaluronan increases the rate of tumor cell growth and increased migratory capacity. Pancreatic cancer cell models grown in culture in a matrix made with hyaluronan grow faster and migrate more readily than pancreatic cancer cells cultured in matrices lacking hyaluronan.[128,129] Hyaluronan removal from the interstitial fluid in cancer patients reduces interstitial fluid pressure, increases tumor perfusion, and decreases hyaluronan accumulation.[126,127] In animal model studies, reducing hyaluronan synthesis is able to prevent the growth of pancreatic cancer.[130,131] The hyaluronan synthesis inhibitor 4-methylumbelliferone and its derivatives suppress metastasis in various tumors.[132–134] Other methods of reducing hyaluronan or its synthesis have also shown to be able to reduce pancreatic cell growth, metastasis, and make pancreatic cancer more amenable to additional therapeutic options.[135,136]

Another clinical focus on targeting HA takes advantage of its putative cell receptor CD44, cluster of differentiation 44. CD44 is a ubiquitously expressed glycol protein on the surface of mammalian cells that participates in the regulation in a number of normal and diseased-related biological functions. Just as hyaluronan has been shown to be overexpressed in pancreatic cancer, so too has CD44.[137] Utilizing CD44 as a receptor-based target in treating aggressive cancers is in line with targeting receptors in other forms of less aggressive cancers. Recently, other receptors, transferrin, folate, epidermal growth factor along with CD44, have shown promise as selective targets for small-molecule therapy.[138–146] More importantly, recent approaches have focused on targeting CD44 using various alternative approaches such as antibodies, aptamers, as well as peptides.[147] CD44, a stem-like cell receptor, is known to be associated with promoting tumor progression in breast cancer cells.[146] CD44 is widely known to promote tumor cell migration as well as increased drug resistance in advanced breast cancers.[146] All of these characteristics are synonymous with aggressive forms of cancers, such as pancreatic cancer.

CD44 is a member of a family of glycoproteins that function in extracellular adhesion and signal transduction.[148] CD44 is found on chromosome 11 in humans and is highly conserved. The protein contains a cytoplasmic domain, transmembrane domain, as well as an extracellular domain. In addition to these domains, the protein also includes a variable region and an

N-terminal globular region that permits ligand binding, typically to HA.[149] Due to the variable region, the CD44 family is comprised of more than 20 different isoforms that are derived from posttranscriptional modification as well as alternative splicing.[148,150,151] The variability in CD44 expression is closely correlated to its role in tumor progression, metastasis, and drug resistance. CD44 variants 8–10 have been associated with increased invasive capacity in melanoma cancers.[152] Melanoma cells expressing CD44 variants 8–10 (CD44v8–10, variable exons 8–10) demonstrated a significant loss of endothelial junctions and increased invasive capacity. Melanoma cells that expressed the standard version of CD44 maintained endothelial junction integrity and were not invasive.[152] Another study, examining the expression level of CD44v9 isoform in human pancreatic cancer, demonstrated an increased expression of multidrug resistance protein 1 (MDR1).[153]

Since CD44 and more importantly its variants are associated with tumor progression and increased aggressiveness, efforts have been made to produce CD44-targeted therapies. The majority of CD44-based therapies have utilized monoclonal antibodies. CD44 antibody therapy has been shown to specifically target and inhibit CD44 receptors on cancer cells.[154] Anti-CD44 antibodies prevent CD44 from interacting with its receptors and the tumor matrix. HI44, an anti-CD44 monoclonal antibody is able to block the progression on acute myeloid leukemia (AML).[155] Other anti-CD44 monoclonal antibodies (H90 and A3D8) are able to block the progression of other forms of AML.[156] In relationship to pancreatic cancer, CD44v6 (variant isoform 6) is highly expressed and utilizing an anti-CD44v6 antibody is able to inhibit tumor growth and metastasis.[157,158] Moreover, this same anti-CD44v6 therapy has been tested in a Phase I clinical trial for head and neck squamous cell carcinoma, with promising results. Initial results demonstrate significant antitumor potential, accumulation of the antibody in tumors with no reported toxicity.[159,160] In addition to these anti-CD44 antibody studies, recent advances in nanomaterials have utilized RNA aptamers conjugated to liposomes to specifically target tumor cells.[161] Findings from these studies are promising albeit still in the experimental stage.

CD44 has become increasingly popular as an active target for cancer therapeutics. Although CD44 is ubiquitously expressed throughout the body, its expression is significantly higher in tumor cells, especially solid tumors like pancreatic and breast cancer.[146,162] Its binding partner HA can be used as a delivery vehicle since it is easily modified to accommodate functional groups necessary to attach potential therapeutic molecules.

Targeting the CD44/HA molecular relationship has revealed some very promising approaches. Many new trials are designed to target CD44/HA in various cancers and preliminary studies have demonstrated encouraging results.[163–166] Together CD44 and HA serve as convenient and promising therapeutic targets in aggressive cancers including pancreatic cancer.

3.4 Stromal Disruption

Most therapeutic options for pancreatic cancer are ultimately not successful. Pancreatic cancer is one of the most lethal cancers and one major reason for this is stromal displacement that occurs and the disease progresses. Advanced pancreatic cancer samples have a unique tissue formation with up to 90% of the tumor mass being comprised of stromal tissue.[167] It is this abnormal stromal structure and the concomitant progression of cancer cells that may lead to treatment resistance in pancreatic cancer. One major risk factor for pancreatic cancer is the formation of pancreatic inflammation that leads to chronic pancreatitis.[168] Patients diagnosed with chronic pancreatitis are 27 times more likely to develop pancreatic cancer.[168] Patients with topical pancreatitis are even more likely to develop pancreatic cancer, and they exhibit a 100-fold increase in risk.[169] More importantly persons with increased rates on pancreatic inflammation, specifically chronic and topical pancreatitis, are more likely to develop cancer much earlier than persons without these conditions.[169] Recently, a study demonstrating a link between pancreatic stromal cells undergoing epithelial–mesenchymal transition (EMT) to increased malignant formation of pancreatic cancer.[170] Dysregulation of stromal cells is also associated with hereditary forms of pancreatic cancer.[171] The majority of hereditary pancreatic cancers contain a specific ARG-HIS substitution at a specific residue in the cationic trypsinogen gene (PRSS1), but this mutation is not restricted to only pancreatic cells, but is also found in just about every cell in people with this mutation, it only leads to cancer of the pancreas.[172] Since these individuals only develop pancreatic cancer as a result of this mutation, it suggests additional factors specific to the pancreas that are promoting cancer formation. A recent hypothesis has suggested that stromal cells in the pancreas are participating in the transformation of the pancreatic cancer in these instances.

Since a correlation appears to exist between the formation of pancreatic cancer and disruption of the normal stromal interactions, several studies have been designed to investigate this potential. Coculture cell culture studies utilizing human pancreatic stellate cells (hPSCs) cultured with pancreatic tumor cells increased the rate of tumor cell proliferation, migration, and invasion.[173] Importantly, these tumor cells in coculture demonstrated increased resistance to common chemotherapeutic treatments.[173] Injecting hPSCs and pancreatic tumor cells into a mouse model result in more aggressive tumor formation, increased tumor volume, and increased metastasis.[172,173] All of these studies surmise that stromal cells participate in promoting pancreatic cancer formation and increased malignancy. However, these studies do not address how the tumor microenvironment regulates increased pancreatic stromal cell formation. Additionally, these above-referenced studies do not address the mechanisms that permit the tumor microenvironment to coadapt the pancreatic stroma into supporting the formation of pancreatic cancer.

Pancreatic stromal cells are generated from four major sources and several growth factor signaling pathways participate in regulating the generation of these stromal cells. Stromal cells are generated from preexisting stromal cells, EMT transition, as well as derivation from cancer stem cells and differentiation from precursor cells. Studies have shown that fibroblast may be induced to produce myofibroblast properties.[174,175] This allows pancreatic tumor cells to activate local tumor stromal cells as well as stromal cells from adjacent regions to participate in tumor progression.[176] Pancreatic cells have also been shown to recruit mesenchymal stem cells (MSCs).[177] These connective tissue progenitors are heterogeneous and can be found in adipose tissue in the pancreas.[177,178] Studies have shown that pancreatic tumor cells are able to secrete chemoattractant factors to recruit nearby MSCs and potential change these stem cells into tumor stem cells.[167,177] Although studies have been conducted in vivo with other cancers to demonstrate the ability of cancer cells to recruit MSCs to tumors and become active components of the tumor microenvironment, few have demonstrated this with pancreatic cancer cells and work is still needed to confirm this possibility.[170,179,180] EMT can also be utilized to transition pancreatic stromal cells into cancer cells. Specifically, myofibroblasts which are abundant in the stromal environment have recently been shown to be derived from epithelial cells.[181,182] This would permit an EMT to take place and thus resulting in greater numbers of pancreatic cells.[183] Associated with an EMT leading to cancer, it has also been suggested that pancreatic tumor cells also may promote endothelial cells located in the tumor microenvironment to form tumors.[184,185] Although there is no direct evidence of this in pancreatic cells, it may be possible and could create another mechanism to target in the treatment of pancreatic tumors.

In addition to the various forms of cells that may be used as precursor cells there are multiple growth factor signaling pathways that are important in promoting pancreatic cancer development and progression. Several common growth factor signaling pathways have been implicated in the formation of pancreatic cancer from stromal precursors, such as the transforming growth factor-β (TGF-β) and platelet-derived growth factor (PDGF). Specifically, pancreatic cancer cells are capable of expressing TGF-β1 which results in increased fibroblast proliferation and influence the production of VEGF.[186–188] Increasing the amount of VEGF and fibroblast support the theory that pancreatic cancer cells can drive tumor growth by recruiting both fibroblast and endothelial cells to the tumor and potentiating their transition into tumor stromal cells. Since TGF-β is a potent stimulator of pancreatic cell growth and tumor microenvironment changes, several therapeutic options are in various stages of clinical trials.[189–191] However, these treatments have demonstrated various levels of success. PDGF family members are ubiquitously expressed in tumors and are directly correlated with the desmoplasia associated with pancreatic tumors.[192,193] Pancreatic cancer cells exposed to PDGF, initiate PDGF secretion which then recruits fibroblast to the tumor microenvironment and increases the rate of tumor growth and migration.[192,193] This response to PDGF correlates well with the observation of fibroblast cells being recruited to the adjacent stroma of pancreatic cancer cells to increase the tumor burden. Initial studies have shown that PDGF signaling may be a possible therapeutic target. Introducing prebound PDGF receptor, bound to IgG, was able to bind to PDGF receptors present on normal stromal cells and prevent PDGF activation in the tumor microenvironment.[192] Together, these new growth factor-based therapies have shown early promise in disrupting major signaling mechanisms associated with pancreatic cancer cell formation and progression; specifically, the promotion of pancreatic cancer formation by stromal cells.

3.5 Epigenetics

Genes that play a role in driving the functions of a cell including growth and division, metabolic regulation, and signal transduction mechanisms can be regulated epigenetically and ultimately lead to the dysregulation of these processes which can have deleterious effects on normal cell function and behavior.[194] Epigenetic changes can be inherited and refers to changes in the regulation of the genes responsible for the indicated behaviors of a cell due to chemical modifications of the DNA, RNA, or proteins without direct changes to the DNA sequence of those genes. Some epigenetic modifications are considered transient or nonpermanent, and could be reversed depending on the modification and the known mechanism used to reverse the epigenetic modification and restoring normal functions of some genes such as tumor suppressor genes. The field of epigenomics and the investigation of the consequence of epigenetic changes in disorders and diseases such as cancer have provided valuable insight into how cancer, for example, is initiated and progresses.[195,196] There have been a number of epigenetic alterations that have been observed and reported to play a role in pancreatic cancer.[196] The epigenetic modifications in PDACs may provide drug targets that may help in the development of better therapeutic options for PDAC patients.

3.5.1 DNA Methylation

DNA hypermethylation is a type of epigenetic modification that can have a negative or repressive effect on the expression of crucial genes that regulate various signaling pathways in the cell. The DNA is methylated at cytosine nucleotide residues that precede guanine residues in CpG islands in the promoter regions of tumor suppressor genes which lead to the recruitment of methyl-CpG-binding domain proteins or MBDs resulting in chromatin compaction.[197] Chromatin compaction causes gene silencing of the methylated tumor suppressor genes due to the interference of the methyl groups located in the promoter region of the genes to transcription factors and RNA polymerases. DNA methyltransferases (DNMTs) are a group of enzymes that assist in the maintenance of the DNA-methylated status of genes.[198] DNA methyltransferase DNMT1 reportedly maintains the DNA methylation pattern of genes during DNA replication of the parent DNA strand to the newly synthesized daughter DNA strand and are thought to primarily play a role in the silencing of tumor suppressor genes leading to cancer including pancreatic cancer.[199] DNMT3a and DNMT3b are other DNMTs that are reportedly responsible for the methylation of genes de novo and play a greater role in embryonic development.

Aberrant methylation patterns of tumor suppressor genes have been reported in pancreatic cancers. For example, the p16 tumor suppressor gene is hypermethylated and inactivated in over 95% of pancreatic cancers.[200,201] The inactivation of the p16 gene in PDAC leads to the phosphorylation of the Retinoblastoma (pRb) tumor suppressor protein and an increase in cell growth and division or cell proliferation.[200,202,203] Methylation analysis reveals the hypermethylation of another candidate PDAC tumor suppressor gene, the Kruppel-like factor 11 or KLF11 gene that inhibits cell growth and causes apoptosis.[204,205] A number of other genes (e.g., UCHL1, NPTX2, SARP2, CLDN5, reprimo, LHX1, WNT7A, FOXE1, TJP2, CDH3, and ST14) have also been reported after using high-throughput microarray analysis to be hypermethylated.[206] A study using primary pancreatic tumors expanded in immunocompromised mice to establish PDAC cell lines. The PDAC cell lines were then exposed to both DNMT1 inhibitors zebularine and decitabine resulting in a decrease in DNMT1 protein levels, a reduction in the cancer stem cell population, and an increase in epithelial differentiation. Furthermore, the FDA has approved the DNMT1 inhibitor 5-azacytidine for the treatment of myelodysplastic syndrome and therefore, DNMT1 may be used as a target for PDAC treatment and therapy.[207]

3.5.2 Histone Modification

Histones are proteins that play a role in the organization of eukaryotic DNA into nucleosomes, which are the basic structures that form chromosomes.[208,209] The organization of DNA into chromosomes plays a crucial role in the protection of DNA from damage and allows for the proper progression of cells through the process of mitosis, and for the regulation of gene expression. Approximately 160 nucleotide base pairs of DNA wraps twice around a core of histone proteins made up of eight polypeptides or two molecules each of four different histone proteins to form a nucleosome linked and held together through linker DNA and the H1 histone protein to form euchromatin. The four histone proteins that make up the core of the nucleosomes are the H2A, H2B, H3, and H4 proteins.[210,211] The nucleosomes and histone H1 assist in the further organization of DNA into 30 nm fibers that are then compacted into higher-ordered heterochromatin structures. Less active or silenced genes are typically found in the heterochromatin regions of DNA, while the more actively transcribed genes are located in the euchromatic regions of DNA. Chromatin is remodeled through the modification of histones in the nucleosomes to allow for the regulation and transcription of genes found in both the euchromatic and some heterochromatic regions of the chromatin.[210,212] The dysregulation of chromatin remodeling results in the silencing of tumor suppressor genes and in the overexpression of oncogenes resulting in cancer, including PDAC.[210,212] Some of the common chromatin remodeling proteins including p300, HDACs, and Brg1 have been reported to be mutated or aberrantly expressed in pancreatic cancer.[198] The most widely investigated and best understood histone that is modified in the regulation of genes is the H3 histone protein.

Histone H3 is modified primarily by two mechanisms: acetylation and methylation.[213]

3.5.3 Histone Acetylation/Deacetylation

The acetylation and deacetylation of lysine residues in both histones and nonhistone proteins play a role in controlling the expression of genes.[214,215] Acetylation typically results in the activation of genes, while deacetylation results in the silencing of genes. Histone acetyltransferases or HATs transfer acetyl groups from the donor acetyl-coA to lysine groups of genes, while histone deacetyltransferases or HDACs reverse this reaction by removing acetyl groups from lysine residues of genes.[216] The acetylation and deacetylation of histones promote chromatin remodeling, whereas the acetylated state of genes results in less condensed chromatin to allow for the binding of transcription factors and RNA polymerases, while deacetylation results in the condensation of chromatin. There has been an effort to find HAT inhibitors to use as therapies against HATs such as the CREB-Binding Protein or CBP/p300; however, most efforts remain in the preclinical stage except curcumin which is being in clinical trials a potential anticancer drug and could possibly be used in the future as a chemoprevention or treatment for PDAC.[195,206]

In PDAC, the transcription factors nuclear factor of activated t-cells (NFAT) and GLI3 have been implicated in the recruitment of the HAT p300 enzyme to induce the expression of the CMYC oncogene resulting in hyperacetylation and an increase in cell growth.[214] The NFAT is a family of proteins that have been reported to be transcriptionally active in early pancreatic cancer precursor lesions and highly expressed in advanced pancreatic cancer. GLI3 and p300 transactivate the vacuole membrane protein 1 (VMP1) resulting in a cancer phenotype autophagy. Therefore, targeting acetylation could provide a way to target PDAC for treatment.[214]

HDAC activity is highest in pancreatic cancer tissues as compared to normal and chronically inflamed pancreas tissue. PDAC has been reported to overexpress HDACs 1, 2, 3, and 7 resulting in the silencing of tumor suppressor genes such as the p53 gene.[217,218] HDAC1 is recruited to repress the E-cadherin gene by EMT-inducing genes and subsequently result in EMT.[219] HDAC1 has been reported to repress the TGFβRII gene, and the inhibition of HDAC1 resulted in an increase in apoptosis if a PDAC derived cell line. HDAC2 is activated by CMYC and has been shown to be overexpressed in poorly differentiated PDAC.[217,219] In another study, it was determined that in over 80% of PDAC tissue samples the HDAC7 gene was overexpressed. HDAC inhibitors tend to function broadly as some are currently being used in preclinical and clinical trials including abexinostat, pracinostat, resminostat, givinostat, panobinostat, and CUDC-101.[220–223] Therefore, the HDAC family of proteins may be a good target to treat PDAC.

3.5.4 Histone Methylation

The methylation and demethylation of histones result in the modification of chromatin to promote or silence gene expression. Methylation of histones may occur in various residues of DNA where gene expression is controlled including the gene promoter and enhancer regions. Methylation may include between one and three methyl groups add on both lysine (K) and arginine (R) residues found in the histones.[210,224] Commonly methylated residues in histones resulting in gene silencing include the lysine (K) residues H3-K9, H3-K27, and H4-K20, while the methylated residues in histones resulting in gene activation include H3-K4, H3-K36, and H3-K79.[210,224] The enzyme *S*-adenosylmethionine or SAM-dependent methyltransferases have been implicated in the methylation of both lysine and arginine residues.[225] The protein arginine methyltransferases or PRMTs are a family of enzymes classified as either types I, II, or III, and is in the early stages of being developed as drugs targets used to inhibit arginine methylation for treatment.[226]

There are over 50 lysine methyltransferases (KMTs) with most of them containing a SET domain which characterizes how the histone residues are methylated.[225,226] The enzyme enhancer of zest homolog 2 (EZH2) protein is part of the Polycomb PRC2 complex and is responsible for trimethylation of H3-K27 histone residues.[227,228] In PDAC, the loss of EZH2 has been associated with poor prognosis. Nuclear accumulation of EZH2 has been reported in poorly differentiated PDAC and is associated with an increase in cell growth.[229,230] The p16 tumor suppressor gene has been shown to be silenced in PDAC due to the interaction of the Polycomb complex and the p16 promoter, which in turn results in the recruitment of methyltransferases to further silence the p16 tumor suppressor gene due to methylation.[203,231] Methyltransferase inhibitors adenosine dialdehyde (AdOx) and 3-deazaneplanocin (DZNep) inhibit EZH2 and result in an induction of cell death. Gemcitabine and DZNep have revealed a synergistic antiproliferative effect in both PDAC tumors and PDAC cell lines. SAM competitive compound inhibitors GSK126, EPZ-6438, UNC-1999, and CPI-169 have been developed and have been reported as being highly selective as EZH2 inhibitors and are potential therapeutic options for PDAC treatment.[198,226,228,232]

3.6 Noncoding RNAs

It is estimated that only about 35% of all genes are transcribed and eventually translated into proteins, while the other 65% are transcribed into noncoding RNAs.[233,234] Noncoding ribonucleic acids/RNAs or ncRNAs are not translated into proteins; however, they play vital roles in various cellular processes including cell growth and division, cellular and tissue differentiation, and cellular responses to stress signals resulting in cell cycle arrests and cell death including necrosis and apoptosis. The more commonly described ncRNAs that are biologically meaningful include the ribosomal RNAs (rRNAs) and transfer RNAs (tRNAs) that play major and important roles in the translation of messenger RNA (mRNA) into protein. Other commonly described ncRNAs include small nuclear or snRNAs that are involved in splicing of primary transcripts into mRNAs, and small nucleolar or snoRNAs that are involved in rRNA processing. Historically, small RNA molecules were ignored as nucleic acid remnants that had no biological relevance, but that sentiment has changed since the completion of the Human Genome project.[235,236]

The sequencing of the human genome along with the technological advances that resulted from the Human Genome Project has led to the discovery of other biologically relevant ncRNAs, and has greatly expanded the knowledge and understanding of these other ncRNAs and has implicated them in various molecular signaling pathways including their involvement in diseases as causative agents and as potential targets for the treatment of those diseases, including pancreatic cancer. Some of these ncRNAs are transcribed from genes, while some are transcribed from regions found within the introns and exons of other genes.[236] Some of the ncRNAs are made as the result of splicing, while others come from the degradation or cleavage of other RNAs. The linear ncRNAs are classified into two major subcategories, the long noncoding RNAs and the short or small noncoding RNAs.[235–237] The long noncoding RNAs are typically linear; however, there are also circular or circRNAs that form circle-like RNA noncoding structures.[238] The small noncoding RNAs are subclassified into the following categories: sdRNAs, piRNAs, and miRNAs.[235,236,239] Both categories of ncRNAs have genes that have tumor suppressive and/or oncogenic characteristics. The role that ncRNAs play in pancreatic cancer are gaining interests as biomarkers that could be used to diagnose and offer prognosis to patients, but more importantly could be used as potential targets to treat PDAC.

3.6.1 Long Noncoding RNA and PDAC

Long noncoding RNAs or lncRNAs are a heterogeneous group of linear transcripts that are typically longer than 200 nucleotides and are possibly precursors due to splicing to another type of circular ncRNAs called circRNAs.[237,240] The majority of noncoding RNAs fall into the category of lncRNAs which include rRNAs, but most of their functions have yet to be elucidated or characterized extensively. There are approximately 13,000 genes that encode for lncRNAs, but a small number of those genes and the resulting lncRNAs have been studied as it relates to their molecular consequences in the cell. LncRNAs are very diverse in their structure and how they function in the cells; however, lncRNAs are structured similarly to mRNAs in that they have a 5′ cap, a 3′ polyadentylated tail, and undergo splicing. LncRNAs are found in both of the major subcompartments of the cell, the nucleus and cytoplasm, but it has yet to be established what signal controls its localization in the cell at a given time or under a given condition. LncRNAs are classified into five categories based on their location in the genome: sense, antisense, bidirectional, intronic, and intergenic.[237,240,241]

There are a number of lncRNAs that have been reported as being upregulated in PDAC suggesting a role in the promotion and progression of pancreatic cancer. One of the first lncRNAs identified to play a role in human disease is the maternally imprinted gene H19 located on chromosome 11p15.5. The H19 gene is usually expressed only during embryonic development, but has been shown to be expressed again in various types of cancers including pancreatic cancer tissue samples and pancreatic cancer cell lines.[242] H19 overexpression has also been correlated to tumor invasion and metastasis through its inhibitory effects on the miRNA let-7's suppression of HMGA2-mediated EMT. Studies have shown that pancreatic cancer tissues and cells that overexpress the H19 gene can be targeted using a DNA-based therapy where the tissues or cancer cells are injected with the BC-819 plasmid containing the diphtheria toxin A chain (DTA) gene. The BC-819 plasmid uses the H19 gene regulatory system to express the DTA gene which is cytotoxic to the pancreatic cancer cells. This DNA-based therapy along with the sequential use of gemcitabine was significantly more effective than using either treatment alone.[241–243]

There are more lncRNAs that are upregulated or overexpressed in pancreatic cancers that have been targeted for downregulation in order to elicit an arrest in the pancreatic cancer growth or in the shrinkage of the pancreatic cancer as a result of cell death. The HOX transcript antisense intergenic RNA (HOTAIR) and the HOXA transcript at the distal tip (HOTTIP) are bot HOX gene-associated lncRNAs that have been targeted in knockdown studies resulting in the cell cycle arrest, apoptosis, and the inhibition of invasion and metastasis of pancreatic cancer cell lines.[244] The metastasis-associated lung adenocarcinoma transcript-1 (MALAT-1) has been shown to be overexpressed in pancreatic cancer tissue samples and in pancreatic cancer cell lines. MALAT-1 overexpression is associated with a poor prognosis and shorter survival due to advanced clinical stages and metastasis. The knockdown of MALAT-1 in pancreatic cancer cells resulted in the reduction of cell proliferation, migration, invasion, and metastasis which suggest that this gene could be a good candidate for drug targeting in the treatment of pancreatic cancer.[243,245]

3.6.2 MicroRNAs and PDAC

MicroRNAs or miRNAs are the most extensively studied category of noncoding RNAs that reveal to play a major role in the regulation of gene expression by directly blocking mRNA translation or by causing the degradation of mRNA along with the protein RISC. MicroRNAs are typically 22 nucleotides in length and can be found in viruses, bacteria, plants, and animals. It is believed that the human genome contains over 1000 different miRNAs. miRNAs are produced from their own genes or within the introns of other protein-coding or nonprotein-coding genes, with the majority of the miRNAs being classified as intergenic.

MicroRNAs have been implicated in a number of diseases and disorders including heart disease, kidney disease, neurological disorders, and cancer.[243,246–251] A number of miRNAs have been reported as playing a role in the initiation and progression of pancreatic cancer, and therefore could provide opportunities in the treatment of pancreatic cancer.[243,252–254] MicroRNAs may fall into either an oncogenic category as oncomiRs or into a tumor suppressor category as TSmiRs, providing the traditional approaches that one would take in treating most types of cancers with established oncogenic expressions or the downregulation or mutation of tumor suppressor genes.[247,248,255–259]

The upregulation or overexpression of multiple oncomiRs has been reported as being associated with PDAC including miR-21, miR-221, miR-155, miR-10b, and miR-208.[252,253,260–263] MiR-21 is one of the most extensively studied microRNAs, has been associated with increased proliferation and invasion, and has been implicated in the development of chemoresistance to gemcitabine.[264,265] The overexpression of miR-21 has been associated with the mTOR pathway which promotes pancreatic cancer cell survival, proliferation, and progression. It has been demonstrated using a xenograft model that targeting miR-21 resulted in the slowing of pancreatic cancer cell proliferation, in the increase of apoptosis, and in the inhibition of tumor growth.[218,266] The overexpression of the remaining oncomiRs all similarly result in the increase in cell proliferation, increase in migration and invasion, and the increase in metastasis; and they all present valuable drug targets to reverse or inhibit the mechanisms of the oncomiRs in the progression of pancreatic cancer.[248]

TSmiRs are microRNAs with tumor suppressor capabilities that prevent cancer progression by blocking oncogenic activity and inducing cell cycle arrest and apoptosis.[267–269] The dysregulation or downregulation of multiple TSmiRs has been implicated in pancreatic cancer including the miR-200 family of microRNAs, miR-34a, miR-124, and miR-203.[269,270] The miR-200 family of microRNAs play a pivotal role in the regulation of EMT resulting in the inhibition of invasion and metastasis; however, the miR-200 family of miRNAs have been shown to be downregulated in a number of different cancers including pancreatic cancer.[271,272] Another TSmiR that has been reported as being downregulated in PDAC is the miR-34a. MiR-34 restoration resulted in the reduction of pancreatic stem cell growth and survival, and activation of miR-34a resulted in the inhibition of cell proliferation, the inhibition of cell cycle progression, and the inhibition of invasion. In vivo and in vitro studies have revealed that the delivery of miR-34a resulted in the inhibition of cell proliferation and in the increase in apoptosis of pancreatic cancer cells in xenograft models.[273–276]

3.6.3 Other Potential PDAC ncRNA Targets

There are a number of other subtypes of ncRNAs that require further studies and investigation to determine their roles in their promotion in the progression of pancreatic cancer. The circular ncRNAs or circRNAs are a novel type of ncRNA distinguishable from linear ncRNAs in that the circRNAs form a continuous loop structure.[238,277] CircRNAs have been reported to play roles in alternative splicing, in the regulation of parental genes, and as competitors to other endogenous RNA regulators, or as miRNA sponges. It has been reported that circRNA global expression is dysregulated in PDAC requiring further studies to elucidate the specific changes and how they relate to PDAC initiation and progression.[238]

PiRNAs are the largest class of small noncoding RNAs in animal cells that form complexes with piwi proteins that have been linked to epigenetic and posttranscriptional gene silencing.[239] Next-generation sequencing revealed that a piwi-interacting RNA or piRNA, piR-017061, was downregulated in pancreatic cancer; however, the role that piR-017061 plays in the cell has yet to be elucidated.[239] The same researchers revealed through next-generation sequencing the downregulation of 14 small nucleolar RNAs (sno-derived RNAs), and the upregulation of four sno-derived RNAs (sdRNAs) with the most significant change in the fivefold downregulation of sno-HBII-296B. This recent report is the first to implicate the possible dysregulation of piRNAs, sno-RNAs, and sdRNAs in pancreatic cancer.[239] Therefore, these sncRNAs could prove to play a vital role in the near future in the diagnosis, prognosis, and treatment of PDAC.

3.7 PARP1 Inhibitors

DNA is subjected to genotoxic stress that causes various types of damage to DNA thousands of times a day. Some of the types of DNA damage that are caused by stress include nucleotide dimerization; nucleotide modifications due to oxidation, alkylation, hydrolysis, and bulky adduct formations; and both single- and double-stranded DNA breaks.[278,279] Unrepaired DNA damage can cause improper DNA folding which may lead to the impediment of DNA replication and ultimately result in cell death under normal circumstances. Some of the DNA damage is repaired through various pathways that are initiated and controlled by tumor suppressor genes including the BRCA1 and BRCA2 genes.[279–282] BRCA1 is located at chromosome 17q and BRCA2 is located on chromosome 13q, and they both have been associated with a number of cancers including breast, ovarian, and a subset of pancreatic cancer.[283] BRCA1 and BRAC2 are tumor suppressor genes that play a role in promoting the repair of double-stranded DNA breaks through nonhomologous end joining (NHEJ) and homologous recombination. When the DNA of normal cells undergoes double-strand DNA (dsDNA) breaks, the BRCA genes promote the homologous recombination repair mechanism of these breaks. However, when BRCA genes are mutated or have a compromised function, poly ADP-ribose polymerase 1 or PARP1 which normally functions in the repair of single-strand DNA (ssDNA) breaks takes over the role of BRCA. The repair of DNA strand breaks in cancerous cells leads to more genetic instability in the cancer cells which can lead to the reduced effectiveness in chemotherapy treatments for breast, ovarian, and pancreatic cancers.[279,284]

It has been reported that in Ashkenazi Jewish families, there is a high risk of pancreatic cancer associated with BRCA mutations. Specifically, BRCA mutations in the Ashkenazi Jewish families have been associated with a two- to fourfold increase of pancreatic cancer risk, and thus may account for up to 17% of pancreatic cancer caused by inherited germline mutations.[283,285] Research has revealed that targeting PARP1 in some patients with BRCA mutations help to elicit stronger and more effective responses to chemotherapy treatments, and in 2014 Olaparib, a PARP inhibitor, was approved as treatment for advanced ovarian cancer in patients with BRCA mutations providing researchers with more clinical options in the treatment of cancers with DNA repair abrogations.[286–288]

In vitro studies comparing the response of BRCA2-deficient and BRCA2-proficient pancreatic cancer cell lines to a panel of DNA cross-linking agents (DCL) and PARP inhibitors revealed that the BRCA2-deficient cells were more sensitive to all of the DCL and PARP inhibitors tested.[289] It was also revealed that the BRCA2-deficient cells had an increase in sensitivity to the cisplatin and PARP inhibitor BMN673 as compared to other DCLs and PARPi veliparib. Knockdown of BRCA2 in proficient cell lines showed similar increases in sensitivities of the BRCA2 knockdown cells to cisplatin and PARP inhibitor BMN673, without sensitivity increases to veliparib. Xenographic murine models with biallelic BRCA2 mutations also revealed an increase in sensitivity to cisplatin and BMN673, but not veliparib thereby suggesting that the combination of DCL and PARP inhibitor used may require a personalized therapeutic approach.[289]

There are a number of clinical trial studies that are using novel PARP inhibitors in the treatment of pancreatic cancer with BRCA mutations and the results are limited, but early reports are encouraging and indicate that BRCA-associated pancreatic cancer has a partial response to the use of PARP inhibitors in treatment.[290] Some other reports revealed that PARP inhibitors alone resulted initially in either a partial response of the cancer or a disease-stabilizing effect, but in some cases, the disease progressed again after 5 months. While these results are limited in that only a small percentage of pancreatic cancer patients present with the BRCA germline mutations thereby limiting the number of patients using PARP inhibitors in treatment, the preliminary data in patients indicate that it is important to continue to investigate this method as an option for treatment.[290–292]

4. CONCLUDING REMARKS

Despite the advances made over the past 40 years in the diagnosing and treatment of cancers, patients with pancreatic cancer have not significantly benefited as it relates to overall survival. Unfortunately, there is still no cure for this lethal disease and most patients succumb to the disease from 6 months to a little over a year after diagnosis. This warrants the urgent need to identify and validate novel drug targets and treatment strategies to help significantly improve overall patient survival. Early detection using specific biomarkers is essential to identifying more patients with early stage and localized cancers that can be eligible for surgical intervention. This population could potentially benefit from chemopreventive compounds in helping the patient stay healthy and prevent the cancer from becoming metastatic. For those who do have surgically, inoperable cancers, research may be able to help to make those cancers more susceptible to stromal disruptions and immunotherapeutic strategies. For example, studies by Zheng *et al.* demonstrated how radiotherapy and vaccination overcame immune checkpoint blockade resistance in pancreatic cancer.[293] Additionally, the more we can learn about the genetics and DNA repair mechanisms of pancreatic cancers, the more we can potentially select for certain populations of pancreatic cancer patients who may be susceptible to certain treatments. For example, studies suggest that patients with BRCA mutations could potentially be ideal for therapy with PARP inhibitors and be the first example of personalized medicine seen in pancreatic cancer. The hope is to either cure pancreatic cancer or make it a more chronic and manageable disease. This chapter highlighted some of the amazing work that is being done to help ensure that the next 40 years will be significantly better for those diagnosed with pancreatic cancer.

REFERENCES

1. Siegel RL, Miller KD, Jemal A. Cancer statistics, 2015. *CA Cancer J Clin.* 2015;65(1):5–29.
2. Pancreatic Cancer Action Network. *World Pancreatic Cancer Day.* 2016, Oakhanger, Hampshire, 2016.
3. American Cancer Society. *Cancer Facts & Figures 2016.* Atlanta, GA: American Cancer Society; 2016.
4. Pancreatic Cancer Action Network. *Pancreatic Cancer Facts 2016*, Oakhanger, Hampshire, 2016.
5. Rahib L, et al. Projecting cancer incidence and deaths to 2030: the unexpected burden of thyroid, liver, and pancreas cancers in the United States. *Cancer Res.* 2014;74(11):2913–2921.

6. Yachida S, et al. Distant metastasis occurs late during the genetic evolution of pancreatic cancer. *Nature.* 2010;467(7319):1114–1117.
7. Wagner M, et al. Curative resection is the single most important factor determining outcome in patients with pancreatic adenocarcinoma. *Br J Surg.* 2004;91(5):586–594.
8. Ni X, Yang J, Li M. Imaging-guided curative surgical resection of pancreatic cancer in a xenograft mouse model. *Cancer Lett.* 2012;324(2):179–185.
9. Hidalgo M. Pancreatic cancer. *N Engl J Med.* 2010;362(17):1605–1617.
10. Schmidt CM, et al. Effect of hospital volume, surgeon experience, and surgeon volume on patient outcomes after pancreaticoduodenectomy: a single-institution experience. *Arch Surg.* 2010;145(7):634–640.
11. Neoptolemos JP, et al. Adjuvant chemotherapy with fluorouracil plus folinic acid vs gemcitabine following pancreatic cancer resection: a randomized controlled trial. *JAMA.* 2010;304(10):1073–1081.
12. Neoptolemos JP, et al. A randomized trial of chemoradiotherapy and chemotherapy after resection of pancreatic cancer. *N Engl J Med.* 2004;350(12):1200–1210.
13. Regine WF, et al. Fluorouracil-based chemoradiation with either gemcitabine or fluorouracil chemotherapy after resection of pancreatic adenocarcinoma: 5-year analysis of the U.S. Intergroup/RTOG 9704 phase III trial. *Ann Surg Oncol.* 2011;18(5):1319–1326.
14. Evans DB, et al. Preoperative chemoradiation and pancreaticoduodenectomy for adenocarcinoma of the pancreas. *Arch Surg.* 1992;127(11):1335–1339.
15. Pisters PW, et al. Rapid-fractionation preoperative chemoradiation, pancreaticoduodenectomy, and intraoperative radiation therapy for resectable pancreatic adenocarcinoma. *J Clin Oncol.* 1998;16(12):3843–3850.
16. Pisters PW, et al. Preoperative paclitaxel and concurrent rapid-fractionation radiation for resectable pancreatic adenocarcinoma: toxicities, histologic response rates, and event-free outcome. *J Clin Oncol.* 2002;20(10):2537–2544.
17. Evans DB, et al. Preoperative gemcitabine-based chemoradiation for patients with resectable adenocarcinoma of the pancreatic head. *J Clin Oncol.* 2008;26(21): 3496–3502.
18. Varadhachary GR, et al. Preoperative gemcitabine and cisplatin followed by gemcitabine-based chemoradiation for resectable adenocarcinoma of the pancreatic head. *J Clin Oncol.* 2008;26(21):3487–3495.
19. Goel G, Sun W. Novel approaches in the management of pancreatic ductal adenocarcinoma: potential promises for the future. *J Hematol Oncol.* 2015;8:44.
20. Kooby DA. Laparoscopic pancreatic resection for cancer. *Expert Rev Anticancer Ther.* 2008;8(10):1597–1609.
21. Kooby DA, et al. Left-sided pancreatectomy: a multicenter comparison of laparoscopic and open approaches. *Ann Surg.* 2008;248(3):438–446.
22. Eom BW, et al. Clinical outcomes compared between laparoscopic and open distal pancreatectomy. *Surg Endosc.* 2008;22(5):1334–1338.
23. Kim SC. Surgical management of pancreatic cancer. *Korean J Gastroenterol.* 2008;51(2):89–100.
24. Zureikat AH, et al. Robotic-assisted major pancreatic resection and reconstruction. *Arch Surg.* 2011;146(3):256–261.
25. Burris 3rd HA, et al. Improvements in survival and clinical benefit with gemcitabine as first-line therapy for patients with advanced pancreas cancer: a randomized trial. *J Clin Oncol.* 1997;15(6):2403–2413.
26. Hertel LW, et al. Evaluation of the antitumor activity of gemcitabine (2′,2′-difluoro-2′-deoxycytidine). *Cancer Res.* 1990;50(14):4417–4422.
27. Cullinan SA, et al. A comparison of three chemotherapeutic regimens in the treatment of advanced pancreatic and gastric carcinoma. Fluorouracil vs fluorouracil and

doxorubicin vs fluorouracil, doxorubicin, and mitomycin. *JAMA.* 1985;253(14):2061–2067.
28. DeCaprio JA, et al. Fluorouracil and high-dose leucovorin in previously untreated patients with advanced adenocarcinoma of the pancreas: results of a phase II trial. *J Clin Oncol.* 1991;9(12):2128–2133.
29. Hansen R, et al. Continuous 5-fluorouracil (5FU) infusion in carcinoma of the pancreas: a phase II study. *Am J Med Sci.* 1988;295(2):91–93.
30. Conroy T, et al. FOLFIRINOX versus gemcitabine for metastatic pancreatic cancer. *N Engl J Med.* 2011;364(19):1817–1825.
31. Peddi PF, et al. Nab-paclitaxel monotherapy in refractory pancreatic adenocarcinoma. *J Gastrointest Oncol.* 2013;4(4):370–373.
32. Gradishar WJ, et al. Phase III trial of nanoparticle albumin-bound paclitaxel compared with polyethylated castor oil-based paclitaxel in women with breast cancer. *J Clin Oncol.* 2005;23(31):7794–7803.
33. Socinski MA, et al. Weekly nab-paclitaxel in combination with carboplatin versus solvent-based paclitaxel plus carboplatin as first-line therapy in patients with advanced non-small-cell lung cancer: final results of a phase III trial. *J Clin Oncol.* 2012;30(17):2055–2062.
34. Edmonds C, Cengel KA. Tumor-stroma interactions in pancreatic cancer: will this SPARC prove a raging fire? *Cancer Biol Ther.* 2008;7(11):1816–1817.
35. Cucinotto I, et al. Nanoparticle albumin bound Paclitaxel in the treatment of human cancer: nanodelivery reaches prime-time? *J Drug Deliv.* 2013;2013:905091.
36. Von Hoff DD, Ervin T, Arena FP, et al. Randomized phase III study of weekly nab-paclitaxel plus gemcitabine vs. gemcitabine alone in patients with metastatic adenocarcinoma of the pancreas (MPACT). In: *ASCO Meeting*; 2013.
37. Moore MJ, et al. Erlotinib plus gemcitabine compared with gemcitabine alone in patients with advanced pancreatic cancer: a phase III trial of the National Cancer Institute of Canada Clinical Trials Group. *J Clin Oncol.* 2007;25(15):1960–1966.
38. Zhang H. Onivyde for the therapy of multiple solid tumors. *Onco Targets Ther.* 2016;9:3001–3007.
39. Gillen S, et al. Preoperative/neoadjuvant therapy in pancreatic cancer: a systematic review and meta-analysis of response and resection percentages. *PLoS Med.* 2010;7(4):e1000267.
40. Kim EJ, et al. A multi-institutional phase 2 study of neoadjuvant gemcitabine and oxaliplatin with radiation therapy in patients with pancreatic cancer. *Cancer.* 2013;119(15):2692–2700.
41. Kharofa J, et al. Neoadjuvant chemoradiation with IMRT in resectable and borderline resectable pancreatic cancer. *Radiother Oncol.* 2014;113(1):41–46.
42. Regine WF, et al. Fluorouracil vs gemcitabine chemotherapy before and after fluorouracil-based chemoradiation following resection of pancreatic adenocarcinoma: a randomized controlled trial. *JAMA.* 2008;299(9):1019–1026.
43. Loehrer Sr PJ, et al. Gemcitabine alone versus gemcitabine plus radiotherapy in patients with locally advanced pancreatic cancer: an Eastern Cooperative Oncology Group trial. *J Clin Oncol.* 2011;29(31):4105–4112.
44. Li CP, et al. Concurrent chemoradiotherapy treatment of locally advanced pancreatic cancer: gemcitabine versus 5-fluorouracil, a randomized controlled study. *Int J Radiat Oncol Biol Phys.* 2003;57(1):98–104.
45. Yokokawa J, et al. Identification of novel human CTL epitopes and their agonist epitopes of mesothelin. *Clin Cancer Res.* 2005;11(17):6342–6351.
46. Andersen MH, et al. Identification of a cytotoxic T lymphocyte response to the apoptosis inhibitor protein survivin in cancer patients. *Cancer Res.* 2001;61(3):869–872.

47. Johnston FM, et al. Circulating mesothelin protein and cellular antimesothelin immunity in patients with pancreatic cancer. *Clin Cancer Res.* 2009;15(21):6511–6518.
48. Kotera Y, et al. Humoral immunity against a tandem repeat epitope of human mucin MUC-1 in sera from breast, pancreatic, and colon cancer patients. *Cancer Res.* 1994;54(11):2856–2860.
49. Ueda M, et al. MUC1 overexpression is the most reliable marker of invasive carcinoma in intraductal papillary-mucinous tumor (IPMT). *Hepatogastroenterology.* 2005;52(62):398–403.
50. Wobser M, et al. Complete remission of liver metastasis of pancreatic cancer under vaccination with a HLA-A2 restricted peptide derived from the universal tumor antigen survivin. *Cancer Immunol Immunother.* 2006;55(10):1294–1298.
51. Yamaguchi K, Enjoji M, Tsuneyoshi M. Pancreatoduodenal carcinoma: a clinicopathologic study of 304 patients and immunohistochemical observation for CEA and CA19-9. *J Surg Oncol.* 1991;47(3):148–154.
52. Gjertsen MK, et al. Vaccination with mutant ras peptides and induction of T-cell responsiveness in pancreatic carcinoma patients carrying the corresponding RAS mutation. *Lancet.* 1995;346(8987):1399–1400.
53. Oji Y, et al. Overexpression of the Wilms' tumor gene WT1 in primary astrocytic tumors. *Cancer Sci.* 2004;95(10):822–827.
54. Zou W. Immunosuppressive networks in the tumour environment and their therapeutic relevance. *Nat Rev Cancer.* 2005;5(4):263–274.
55. Amedei A, Niccolai E, Prisco D. Pancreatic cancer: role of the immune system in cancer progression and vaccine-based immunotherapy. *Hum Vaccin Immunother.* 2014;10(11):3354–3368.
56. Hellstrom I, et al. Anti-mesothelin antibodies and circulating mesothelin relate to the clinical state in ovarian cancer patients. *Cancer Epidemiol Biomarkers Prev.* 2008;17(6):1520–1526.
57. Ho M, et al. Humoral immune response to mesothelin in mesothelioma and ovarian cancer patients. *Clin Cancer Res.* 2005;11(10):3814–3820.
58. Abizaid A, et al. The sirolimus-eluting Cypher Select coronary stent for the treatment of bare-metal and drug-eluting stent restenosis: insights from the e-SELECT (Multicenter Post-Market Surveillance) registry. *JACC Cardiovasc Interv.* 2012;5(1):64–71.
59. Hassan R, et al. Phase I study of SS1P, a recombinant anti-mesothelin immunotoxin given as a bolus I.V. infusion to patients with mesothelin-expressing mesothelioma, ovarian, and pancreatic cancers. *Clin Cancer Res.* 2007;13(17):5144–5149.
60. Filpula D, Zhao H. Releasable PEGylation of proteins with customized linkers. *Adv Drug Deliv Rev.* 2008;60(1):29–49.
61. Kreitman RJ, et al. Phase I trial of continuous infusion anti-mesothelin recombinant immunotoxin SS1P. *Clin Cancer Res.* 2009;15(16):5274–5279.
62. Hassan R, et al. Phase I clinical trial of the chimeric anti-mesothelin monoclonal antibody MORAb-009 in patients with mesothelin-expressing cancers. *Clin Cancer Res.* 2010;16(24):6132–6138.
63. Feng Y, et al. A novel human monoclonal antibody that binds with high affinity to mesothelin-expressing cells and kills them by antibody-dependent cell-mediated cytotoxicity. *Mol Cancer Ther.* 2009;8(5):1113–1118.
64. Ho M, et al. A novel high-affinity human monoclonal antibody to mesothelin. *Int J Cancer.* 2011;128(9):2020–2030.
65. Lemoine NR, et al. The epidermal growth factor receptor in human pancreatic cancer. *J Pathol.* 1992;166(1):7–12.

66. Yamanaka Y, et al. Coexpression of epidermal growth factor receptor and ligands in human pancreatic cancer is associated with enhanced tumor aggressiveness. *Anticancer Res.* 1993;13(3):565–569.
67. Bruns CJ, et al. Blockade of the epidermal growth factor receptor signaling by a novel tyrosine kinase inhibitor leads to apoptosis of endothelial cells and therapy of human pancreatic carcinoma. *Cancer Res.* 2000;60(11):2926–2935.
68. Ng SS, et al. Effects of the epidermal growth factor receptor inhibitor OSI-774, Tarceva, on downstream signaling pathways and apoptosis in human pancreatic adenocarcinoma. *Mol Cancer Ther.* 2002;1(10):777–783.
69. Baselga J, Arteaga CL. Critical update and emerging trends in epidermal growth factor receptor targeting in cancer. *J Clin Oncol.* 2005;23(11):2445–2459.
70. Philip PA, et al. Phase III study comparing gemcitabine plus cetuximab versus gemcitabine in patients with advanced pancreatic adenocarcinoma: Southwest Oncology Group-directed intergroup trial S0205. *J Clin Oncol.* 2010;28(22):3605–3610.
71. Graeven U, et al. Phase I study of the humanised anti-EGFR monoclonal antibody matuzumab (EMD 72000) combined with gemcitabine in advanced pancreatic cancer. *Br J Cancer.* 2006;94(9):1293–1299.
72. Korc M. Pathways for aberrant angiogenesis in pancreatic cancer. *Mol Cancer.* 2003;2:8.
73. Kindler HL, et al. Phase II trial of bevacizumab plus gemcitabine in patients with advanced pancreatic cancer. *J Clin Oncol.* 2005;23(31):8033–8040.
74. Koido S, et al. Dendritic/pancreatic carcinoma fusions for clinical use: comparative functional analysis of healthy- versus patient-derived fusions. *Clin Immunol.* 2010;135(3):384–400.
75. Koido S, et al. Dendritic cells fused with allogeneic colorectal cancer cell line present multiple colorectal cancer-specific antigens and induce antitumor immunity against autologous tumor cells. *Clin Cancer Res.* 2005;11(21):7891–7900.
76. Jaffee EM, et al. Novel allogeneic granulocyte-macrophage colony-stimulating factor-secreting tumor vaccine for pancreatic cancer: a phase I trial of safety and immune activation. *J Clin Oncol.* 2001;19(1):145–156.
77. Laheru D, et al. Allogeneic granulocyte macrophage colony-stimulating factor-secreting tumor immunotherapy alone or in sequence with cyclophosphamide for metastatic pancreatic cancer: a pilot study of safety, feasibility, and immune activation. *Clin Cancer Res.* 2008;14(5):1455–1463.
78. Royal RE, et al. Phase 2 trial of single agent Ipilimumab (anti-CTLA-4) for locally advanced or metastatic pancreatic adenocarcinoma. *J Immunother.* 2010;33(8):828–833.
79. Kanai M. Therapeutic applications of curcumin for patients with pancreatic cancer. *World J Gastroenterol.* 2014;20(28):9384–9391.
80. Dhillon N, et al. Phase II trial of curcumin in patients with advanced pancreatic cancer. *Clin Cancer Res.* 2008;14(14):4491–4499.
81. Goel A, Kunnumakkara AB, Aggarwal BB. Curcumin as "Curecumin": from kitchen to clinic. *Biochem Pharmacol.* 2008;75(4):787–809.
82. Wang W, et al. The nuclear factor-kappa B RelA transcription factor is constitutively activated in human pancreatic adenocarcinoma cells. *Clin Cancer Res.* 1999;5(1):119–127.
83. Kunnumakkara AB, et al. Curcumin potentiates antitumor activity of gemcitabine in an orthotopic model of pancreatic cancer through suppression of proliferation, angiogenesis, and inhibition of nuclear factor-kappaB-regulated gene products. *Cancer Res.* 2007;67(8):3853–3861.
84. Wang Z, et al. Notch-1 down-regulation by curcumin is associated with the inhibition of cell growth and the induction of apoptosis in pancreatic cancer cells. *Cancer.* 2006;106(11):2503–2513.

85. Lev-Ari S, et al. Curcumin augments gemcitabine cytotoxic effect on pancreatic adenocarcinoma cell lines. *Cancer Invest.* 2007;25(6):411–418.
86. Sun M, et al. Curcumin (diferuloylmethane) alters the expression profiles of microRNAs in human pancreatic cancer cells. *Mol Cancer Ther.* 2008;7(3):464–473.
87. Lev-Ari S, et al. Curcumin synergistically potentiates the growth inhibitory and proapoptotic effects of celecoxib in pancreatic adenocarcinoma cells. *Biomed Pharmacother.* 2005;59(suppl 2):S276–S280.
88. Soni KB, Kuttan R. Effect of oral curcumin administration on serum peroxides and cholesterol levels in human volunteers. *Indian J Physiol Pharmacol.* 1992;36(4):273–275.
89. Sharma RA, et al. Phase I clinical trial of oral curcumin: biomarkers of systemic activity and compliance. *Clin Cancer Res.* 2004;10(20):6847–6854.
90. Cheng AL, et al. Phase I clinical trial of curcumin, a chemopreventive agent, in patients with high-risk or pre-malignant lesions. *Anticancer Res.* 2001;21(4B):2895–2900.
91. Friedman L, et al. Curcumin analogues exhibit enhanced growth suppressive activity in human pancreatic cancer cells. *Anticancer Drugs.* 2009;20(6):444–449.
92. Kesharwani P, et al. Hyaluronic acid engineered nanomicelles loaded with 3,4-difluorobenzylidene curcumin for targeted killing of CD44+ stem-like pancreatic cancer cells. *Biomacromolecules.* 2015;16(9):3042–3053.
93. Li L, Braiteh FS, Kurzrock R. Liposome-encapsulated curcumin: in vitro and in vivo effects on proliferation, apoptosis, signaling, and angiogenesis. *Cancer.* 2005;104(6): 1322–1331.
94. Lyn-Cook BD, et al. The effects of phytoestrogens on human pancreatic tumor cells in vitro. *Cancer Lett.* 1999;142(1):111–119.
95. Takabayashi F, et al. Effect of green tea catechins on the amount of 8-hydroxydeoxyguanosine (8-OHdG) in pancreatic and hepatic DNA after a single administration of N-nitrosobis(2-oxopropyl)amine (BOP). *Pancreas.* 1997;15(2): 109–112.
96. Majima T, et al. Inhibitory effects of beta-carotene, palm carotene, and green tea polyphenols on pancreatic carcinogenesis initiated by N-nitorsobis(2-oxopropyl)amine in Syrian golden hamsters. *Pancreas.* 1998;16(1):13–18.
97. Ulrich CM, Bigler J, Potter JD. Non-steroidal anti-inflammatory drugs for cancer prevention: promise, perils and pharmacogenetics. *Nat Rev Cancer.* 2006;6(2):130–140.
98. Tan XL, et al. Aspirin, nonsteroidal anti-inflammatory drugs, acetaminophen, and pancreatic cancer risk: a clinic-based case-control study. *Cancer Prev Res (Phila).* 2011;4(11):1835–1841.
99. Larsson SC, et al. Aspirin and nonsteroidal anti-inflammatory drug use and risk of pancreatic cancer: a meta-analysis. *Cancer Epidemiol Biomarkers Prev.* 2006;15(12):2561–2564.
100. Capurso G, et al. Meta-analysis: the use of non-steroidal anti-inflammatory drugs and pancreatic cancer risk for different exposure categories. *Aliment Pharmacol Ther.* 2007;26(8):1089–1099.
101. Rothwell PM, et al. Effect of daily aspirin on long-term risk of death due to cancer: analysis of individual patient data from randomised trials. *Lancet.* 2011;377(9759):31–41.
102. Sclabas GM, et al. Nuclear factor kappa B activation is a potential target for preventing pancreatic carcinoma by aspirin. *Cancer.* 2005;103(12):2485–2490.
103. Fendrich V, et al. The angiotensin-I-converting enzyme inhibitor enalapril and aspirin delay progression of pancreatic intraepithelial neoplasia and cancer formation in a genetically engineered mouse model of pancreatic cancer. *Gut.* 2010;59(5):630–637.
104. Rao CV, et al. Inhibition of pancreatic intraepithelial neoplasia progression to carcinoma by nitric oxide-releasing aspirin in p48(Cre/+)-LSL-Kras(G12D/+) mice. *Neoplasia.* 2012;14(9):778–787.

105. Bollrath J, Greten FR. IKK/NF-kappaB and STAT3 pathways: central signalling hubs in inflammation-mediated tumour promotion and metastasis. *EMBO Rep.* 2009;10(12):1314–1319.
106. Niu J, et al. Identification of an autoregulatory feedback pathway involving interleukin-1alpha in induction of constitutive NF-kappaB activation in pancreatic cancer cells. *J Biol Chem.* 2004;279(16):16452–16462.
107. di Magliano MP, Logsdon CD. Roles for KRAS in pancreatic tumor development and progression. *Gastroenterology.* 2013;144(6):1220–1229.
108. Fujioka S, et al. Inhibition of constitutive NF-kappa B activity by I kappa B alpha M suppresses tumorigenesis. *Oncogene.* 2003;22(9):1365–1370.
109. Matsubara S, et al. mTOR plays critical roles in pancreatic cancer stem cells through specific and stemness-related functions. *Sci Rep.* 2013;3:3230.
110. Deng XS, et al. Metformin targets Stat3 to inhibit cell growth and induce apoptosis in triple-negative breast cancers. *Cell Cycle.* 2012;11(2):367–376.
111. Asano T, et al. The rapamycin analog CCI-779 is a potent inhibitor of pancreatic cancer cell proliferation. *Biochem Biophys Res Commun.* 2005;331(1):295–302.
112. Pham NA, et al. Immunohistochemical analysis of changes in signaling pathway activation downstream of growth factor receptors in pancreatic duct cell carcinogenesis. *BMC Cancer.* 2008;8:43.
113. Dai Z-J, et al. Antitumor effects of rapamycin in pancreatic cancer cells by inducing apoptosis and autophagy. *Int J Mol Sci.* 2013;14(1):273.
114. Andreu-Perez P, et al. Methylthioadenosine (MTA) inhibits melanoma cell proliferation and in vivo tumor growth. *BMC Cancer.* 2010;10:265.
115. Avila JL, Kissil JL. Notch signaling in pancreatic cancer: oncogene or tumor suppressor? *Trends Mol Med.* 2013;19(5):320–327.
116. De La OJ, et al. Notch and Kras reprogram pancreatic acinar cells to ductal intraepithelial neoplasia. *Proc Natl Acad Sci USA.* 2008;105(48):18907–18912.
117. Hanlon L, et al. Notch1 functions as a tumor suppressor in a model of K-ras-induced pancreatic ductal adenocarcinoma. *Cancer Res.* 2010;70(11):4280–4286.
118. Mazur PK, et al. Notch2 is required for progression of pancreatic intraepithelial neoplasia and development of pancreatic ductal adenocarcinoma. *Proc Natl Acad Sci USA.* 2010;107(30):13438–13443.
119. Maniati E, et al. Crosstalk between the canonical NF-kappaB and Notch signaling pathways inhibits Ppargamma expression and promotes pancreatic cancer progression in mice. *J Clin Invest.* 2011;121(12):4685–4699.
120. Cook N, et al. Gamma secretase inhibition promotes hypoxic necrosis in mouse pancreatic ductal adenocarcinoma. *J Exp Med.* 2012;209(3):437–444.
121. Cheng XB, et al. Prognostic impact of hyaluronan and its regulators in pancreatic ductal adenocarcinoma. *PLoS One.* 2013;8(11):e80765.
122. Jacobetz MA, et al. Hyaluronan impairs vascular function and drug delivery in a mouse model of pancreatic cancer. *Gut.* 2013;62(1):112–120.
123. Kultti A, et al. Therapeutic targeting of hyaluronan in the tumor stroma. *Cancers (Basel).* 2012;4(3):873–903.
124. Tammi RH, et al. Hyaluronan in human tumors: pathobiological and prognostic messages from cell-associated and stromal hyaluronan. *Semin Cancer Biol.* 2008;18(4):288–295.
125. Provenzano PP, et al. Enzymatic targeting of the stroma ablates physical barriers to treatment of pancreatic ductal adenocarcinoma. *Cancer Cell.* 2012;21(3):418–429.
126. Gao F, et al. Hypoxia-induced alterations in hyaluronan and hyaluronidase. *Adv Exp Med Biol.* 2005;566:249–256.
127. Semenza GL. Targeting HIF-1 for cancer therapy. *Nat Rev Cancer.* 2003;3(10):721–732.

128. Torgenson MJ, et al. Natural history of pancreatic cancer recurrence following "curative" resection in athymic mice. *J Surg Res.* 2008;149(1):57–61.
129. Scaife CL, et al. Synthetic extracellular matrix enhances tumor growth and metastasis in an orthotopic mouse model of pancreatic adenocarcinoma. *J Gastrointest Surg.* 2008;12(6):1074–1080.
130. Simpson MA, Wilson CM, McCarthy JB. Inhibition of prostate tumor cell hyaluronan synthesis impairs subcutaneous growth and vascularization in immunocompromised mice. *Am J Pathol.* 2002;161(3):849–857.
131. Udabage L, et al. Antisense-mediated suppression of hyaluronan synthase 2 inhibits the tumorigenesis and progression of breast cancer. *Cancer Res.* 2005;65(14):6139–6150.
132. Hajime M, et al. Inhibitory effect of 4-methylesculetin on hyaluronan synthesis slows the development of human pancreatic cancer in vitro and in nude mice. *Int J Cancer.* 2007;120(12):2704–2709.
133. Urakawa H, et al. Inhibition of hyaluronan synthesis in breast cancer cells by 4-methylumbelliferone suppresses tumorigenicity in vitro and metastatic lesions of bone in vivo. *Int J Cancer.* 2012;130(2):454–466.
134. Yoshihara S, et al. A hyaluronan synthase suppressor, 4-methylumbelliferone, inhibits liver metastasis of melanoma cells. *FEBS Lett.* 2005;579(12):2722–2726.
135. Olive KP, et al. Inhibition of Hedgehog signaling enhances delivery of chemotherapy in a mouse model of pancreatic cancer. *Science.* 2009;324(5933):1457–1461.
136. Twarock S, et al. Inhibition of oesophageal squamous cell carcinoma progression by in vivo targeting of hyaluronan synthesis. *Mol Cancer.* 2011;10:30.
137. Misra S, et al. Hyaluronan-CD44 interactions as potential targets for cancer therapy. *FEBS J.* 2011;278(9):1429–1443.
138. Niitsu Y, et al. Transferrin receptors in human cancerous tissues. *Tohoku J Exp Med.* 1987;153(3):239–243.
139. Qian ZM, et al. Targeted drug delivery via the transferrin receptor-mediated endocytosis pathway. *Pharmacol Rev.* 2002;54(4):561–587.
140. Hilgenbrink AR, Low PS. Folate receptor-mediated drug targeting: from therapeutics to diagnostics. *J Pharm Sci.* 2005;94(10):2135–2146.
141. Leamon CP, Reddy JA. Folate-targeted chemotherapy. *Adv Drug Deliv Rev.* 2004;56(8):1127–1141.
142. Selvaggi G, et al. Epidermal growth factor receptor overexpression correlates with a poor prognosis in completely resected non-small-cell lung cancer. *Ann Oncol.* 2004;15(1):28–32.
143. Schrevel M, et al. Molecular mechanisms of epidermal growth factor receptor overexpression in patients with cervical cancer. *Mod Pathol.* 2011;24(5):720–728.
144. Resnick NM, et al. Expression of the cell adhesion molecule CD44 in human lung tumors and cell lines. *Mol Diagn.* 1998;3(2):93–103.
145. Shipitsin M, et al. Molecular definition of breast tumor heterogeneity. *Cancer Cell.* 2007;11(3):259–273.
146. Hiscox S, et al. Overexpression of CD44 accompanies acquired tamoxifen resistance in MCF7 cells and augments their sensitivity to the stromal factors, heregulin and hyaluronan. *BMC Cancer.* 2012;12:458.
147. Wang L, et al. CD44 antibody-targeted liposomal nanoparticles for molecular imaging and therapy of hepatocellular carcinoma. *Biomaterials.* 2012;33(20):5107–5114.
148. Ponta H, Sherman L, Herrlich PA. CD44: from adhesion molecules to signalling regulators. *Nat Rev Mol Cell Biol.* 2003;4(1):33–45.
149. Prochazka L, Tesarik R, Turanek J. Regulation of alternative splicing of CD44 in cancer. *Cell Signal.* 2014;26(10):2234–2239.
150. Zoller M. CD44: can a cancer-initiating cell profit from an abundantly expressed molecule? *Nat Rev Cancer.* 2011;11(4):254–267.

151. Louderbough JM, Schroeder JA. Understanding the dual nature of CD44 in breast cancer progression. *Mol Cancer Res.* 2011;9(12):1573–1586.
152. Zhang P, et al. CD44 variant, but not standard CD44 isoforms, mediate disassembly of endothelial VE-cadherin junction on metastatic melanoma cells. *FEBS Lett.* 2014;588(24):4573–4582.
153. Kiuchi S, et al. Pancreatic cancer cells express CD44 variant 9 and multidrug resistance protein 1 during mitosis. *Exp Mol Pathol.* 2015;98(1):41–46.
154. Mattheolabakis G, et al. Hyaluronic acid targeting of CD44 for cancer therapy: from receptor biology to nanomedicine. *J Drug Target.* 2015;23(7–8):605–618.
155. Song G, et al. HI44a, an anti-CD44 monoclonal antibody, induces differentiation and apoptosis of human acute myeloid leukemia cells. *Leuk Res.* 2004;28(10):1089–1096.
156. Gadhoum Z, et al. The effect of anti-CD44 monoclonal antibodies on differentiation and proliferation of human acute myeloid leukemia cells. *Leuk Lymphoma.* 2004;45(8):1501–1510.
157. Matzke-Ogi A, et al. Inhibition of tumor growth and metastasis in pancreatic cancer models by interference with CD44v6 signaling. *Gastroenterology.* 2016;150(2). 513. e10–525.e10.
158. Wang Z, et al. CD44v6-competent tumor exosomes promote motility, invasion and cancer-initiating cell marker expression. *Oncotarget.* 2016;7(34):55409–55436.
159. Stroomer JW, et al. Safety and biodistribution of 99mTechnetium-labeled anti-CD44v6 monoclonal antibody BIWA 1 in head and neck cancer patients. *Clin Cancer Res.* 2000;6(8):3046–3055.
160. Colnot DR, et al. Safety, biodistribution, pharmacokinetics, and immunogenicity of 99mTc-labeled humanized monoclonal antibody BIWA 4 (bivatuzumab) in patients with squamous cell carcinoma of the head and neck. *Cancer Immunol Immunother.* 2003;52(9):576–582.
161. Alshaer W, et al. Functionalizing liposomes with anti-CD44 aptamer for selective targeting of cancer cells. *Bioconjug Chem.* 2015;26(7):1307–1313.
162. Wei HJ, et al. Expression of CD44, CD24 and ESA in pancreatic adenocarcinoma cell lines varies with local microenvironment. *Hepatobiliary Pancreat Dis Int.* 2011;10(4): 428–434.
163. Bassi PF, et al. Paclitaxel-hyaluronic acid for intravesical therapy of bacillus Calmette-Guerin refractory carcinoma in situ of the bladder: results of a phase I study. *J Urol.* 2011;185(2):445–449.
164. Colnot DR, et al. Phase I therapy study of 186Re-labeled chimeric monoclonal antibody U36 in patients with squamous cell carcinoma of the head and neck. *J Nucl Med.* 2000;41(12):1999–2010.
165. Postema EJ, et al. Dosimetric analysis of radioimmunotherapy with 186Re-labeled bivatuzumab in patients with head and neck cancer. *J Nucl Med.* 2003;44(10): 1690–1699.
166. Borjesson PK, et al. Phase I therapy study with (186)Re-labeled humanized monoclonal antibody BIWA 4 (bivatuzumab) in patients with head and neck squamous cell carcinoma. *Clin Cancer Res.* 2003;9(10 pt 2):3961S–3972S.
167. Kong X, et al. Targeted destruction of the orchestration of the pancreatic stroma and tumor cells in pancreatic cancer cases: molecular basis for therapeutic implications. *Cytokine Growth Factor Rev.* 2012;23(6):343–356.
168. Whitcomb DC. Inflammation and cancer V. Chronic pancreatitis and pancreatic cancer. *Am J Physiol Gastrointest Liver Physiol.* 2004;287(2):G315–G319.
169. Whitcomb DC, Pogue-Geile K. Pancreatitis as a risk for pancreatic cancer. *Gastroenterol Clin North Am.* 2002;31(2):663–678.
170. Rhim AD, et al. EMT and dissemination precede pancreatic tumor formation. *Cell.* 2012;148(1–2):349–361.

171. Haddad A, et al. Hereditary pancreatic and hepatobiliary cancers. *Int J Surg Oncol.* 2011;2011:154673.
172. Whitcomb DC, et al. Hereditary pancreatitis is caused by a mutation in the cationic trypsinogen gene. *Nat Genet.* 1996;14(2):141–145.
173. Hwang RF, et al. Cancer-associated stromal fibroblasts promote pancreatic tumor progression. *Cancer Res.* 2008;68(3):918–926.
174. Ronnov-Jessen L, et al. The origin of the myofibroblasts in breast cancer. Recapitulation of tumor environment in culture unravels diversity and implicates converted fibroblasts and recruited smooth muscle cells. *J Clin Invest.* 1995;95(2):859–873.
175. Tomasek JJ, et al. Myofibroblasts and mechano-regulation of connective tissue remodelling. *Nat Rev Mol Cell Biol.* 2002;3(5):349–363.
176. Udagawa T, et al. Analysis of tumor-associated stromal cells using SCID GFP transgenic mice: contribution of local and bone marrow-derived host cells. *FASEB J.* 2006;20(1):95–102.
177. Spaeth EL, et al. Mesenchymal stem cell transition to tumor-associated fibroblasts contributes to fibrovascular network expansion and tumor progression. *PLoS One.* 2009;4(4):e4992.
178. Studeny M, et al. Mesenchymal stem cells: potential precursors for tumor stroma and targeted-delivery vehicles for anticancer agents. *J Natl Cancer Inst.* 2004;96(21): 1593–1603.
179. Nakamizo A, et al. Human bone marrow-derived mesenchymal stem cells in the treatment of gliomas. *Cancer Res.* 2005;65(8):3307–3318.
180. Hung SC, et al. Mesenchymal stem cell targeting of microscopic tumors and tumor stroma development monitored by noninvasive in vivo positron emission tomography imaging. *Clin Cancer Res.* 2005;11(21):7749–7756.
181. Kim KK, et al. Alveolar epithelial cell mesenchymal transition develops in vivo during pulmonary fibrosis and is regulated by the extracellular matrix. *Proc Natl Acad Sci USA.* 2006;103(35):13180–13185.
182. Gudjonsson T, et al. Myoepithelial cells: their origin and function in breast morphogenesis and neoplasia. *J Mammary Gland Biol Neoplasia.* 2005;10(3):261–272.
183. Selman M, Pardo A. Role of epithelial cells in idiopathic pulmonary fibrosis: from innocent targets to serial killers. *Proc Am Thorac Soc.* 2006;3(4):364–372.
184. Wang R, et al. Glioblastoma stem-like cells give rise to tumour endothelium. *Nature.* 2010;468(7325):829–833.
185. Ricci-Vitiani L, et al. Tumour vascularization via endothelial differentiation of glioblastoma stem-like cells. *Nature.* 2010;468(7325):824–828.
186. Vogelmann R, et al. Effects of fibrogenic mediators on the development of pancreatic fibrosis in a TGF-beta1 transgenic mouse model. *Am J Physiol Gastrointest Liver Physiol.* 2001;280(1):G164–G172.
187. Lohr M, et al. Transforming growth factor-beta1 induces desmoplasia in an experimental model of human pancreatic carcinoma. *Cancer Res.* 2001;61(2):550–555.
188. Jones S, et al. Core signaling pathways in human pancreatic cancers revealed by global genomic analyses. *Science.* 2008;321(5897):1801–1806.
189. Melisi D, et al. LY2109761, a novel transforming growth factor beta receptor type I and type II dual inhibitor, as a therapeutic approach to suppressing pancreatic cancer metastasis. *Mol Cancer Ther.* 2008;7(4):829–840.
190. Medicherla S, et al. Antitumor activity of TGF-beta inhibitor is dependent on the microenvironment. *Anticancer Res.* 2007;27(6B):4149–4157.
191. Schlingensiepen KH, et al. Targeted tumor therapy with the TGF-beta 2 antisense compound AP 12009. *Cytokine Growth Factor Rev.* 2006;17(1–2):129–139.
192. Blaine SA, et al. Epidermal growth factor receptor regulates pancreatic fibrosis. *Am J Physiol Gastrointest Liver Physiol.* 2009;297(3):G434–G441.

193. Dong J, et al. VEGF-null cells require PDGFR alpha signaling-mediated stromal fibroblast recruitment for tumorigenesis. *EMBO J.* 2004;23(14):2800–2810.
194. Reik W. Genomic imprinting and genetic disorders in man. *Trends Genet.* 1989;5(10):331–336.
195. McCleary-Wheeler AL, et al. Insights into the epigenetic mechanisms controlling pancreatic carcinogenesis. *Cancer Lett.* 2013;328(2):212–221.
196. Lomberk GA, Iovanna J, Urrutia R. The promise of epigenomic therapeutics in pancreatic cancer. *Epigenomics.* 2016;8(6):831–842.
197. Reik W, Allen ND. Genomic imprinting. Imprinting with and without methylation. *Curr Biol.* 1994;4(2):145–147.
198. Ghoshal K, Bai S. DNA methyltransferases as targets for cancer therapy. *Drugs Today (Barc).* 2007;43(6):395–422.
199. Gao J, et al. Aberrant DNA methyltransferase expression in pancreatic ductal adenocarcinoma development and progression. *J Exp Clin Cancer Res.* 2013;32:86.
200. Shapiro GI, et al. p16INK4A participates in a G1 arrest checkpoint in response to DNA damage. *Mol Cell Biol.* 1998;18(1):378–387.
201. Delpu Y, et al. Genetic and epigenetic alterations in pancreatic carcinogenesis. *Curr Genomics.* 2011;12(1):15–24.
202. Sherr CJ, McCormick F. The RB and p53 pathways in cancer. *Cancer Cell.* 2002;2(2):103–112.
203. Schutte M, et al. Abrogation of the Rb/p16 tumor-suppressive pathway in virtually all pancreatic carcinomas. *Cancer Res.* 1997;57(15):3126–3130.
204. Ellenrieder V, et al. KLF11 mediates a critical mechanism in TGF-beta signaling that is inactivated by Erk-MAPK in pancreatic cancer cells. *Gastroenterology.* 2004;127(2):607–620.
205. Buck A, et al. The tumor suppressor KLF11 mediates a novel mechanism in transforming growth factor beta-induced growth inhibition that is inactivated in pancreatic cancer. *Mol Cancer Res.* 2006;4(11):861–872.
206. Hinton J, et al. Potential epigenetic biomarkers for the diagnosis and prognosis of pancreatic ductal adenocarcinomas. *Expert Rev Mol Diagn.* 2013;13(5):431–443.
207. Zagorac S, et al. DNMT1 inhibition reprograms pancreatic cancer stem cells via upregulation of the miR-17-92 cluster. *Cancer Res.* 2016;76(15):4546–4558.
208. Heslop-Harrison JS, Schwarzacher T. Nucleosomes and centromeric DNA packaging. *Proc Natl Acad Sci USA.* 2013;110(50):19974–19975.
209. Cutter AR, Hayes JJ. A brief review of nucleosome structure. *FEBS Lett.* 2015;589(20 pt A):2914–2922.
210. Becker PB, Workman JL. Nucleosome remodeling and epigenetics. *Cold Spring Harb Perspect Biol.* 2013;5(9):1–19.
211. Harshman SW, et al. H1 histones: current perspectives and challenges. *Nucleic Acids Res.* 2013;41(21):9593–9609.
212. Annunziato AT. Assembling chromatin: the long and winding road. *Biochim Biophys Acta.* 2013;1819(3–4):196–210.
213. Kitamoto S, et al. DNA methylation and histone H3-K9 modifications contribute to MUC17 expression. *Glycobiology.* 2011;21(2):247–256.
214. Schneider G, et al. Acetylation as a transcriptional control mechanism—HDACs and HATs in pancreatic ductal adenocarcinoma. *J Gastrointest Cancer.* 2011;42(2):85–92.
215. Juliano CN, et al. H4K12 and H3K18 acetylation associates with poor prognosis in pancreatic cancer. *Appl Immunohistochem Mol Morphol.* 2016;24(5):337–344.
216. Horikoshi M. Histone acetylation: from code to web and router via intrinsically disordered regions. *Curr Pharm Des.* 2013;19(28):5019–5042.
217. Schuler S, et al. HDAC2 attenuates TRAIL-induced apoptosis of pancreatic cancer cells. *Mol Cancer.* 2010;9:80.

218. Arlt A, Muerkoster SS, Schafer H. Targeting apoptosis pathways in pancreatic cancer. *Cancer Lett*. 2013;332(2):346–358.
219. Aghdassi A, et al. Recruitment of histone deacetylases HDAC1 and HDAC2 by the transcriptional repressor ZEB1 downregulates E-cadherin expression in pancreatic cancer. *Gut*. 2012;61(3):439–448.
220. Peulen O, et al. The anti-tumor effect of HDAC inhibition in a human pancreas cancer model is significantly improved by the simultaneous inhibition of cyclooxygenase 2. *PLoS One*. 2013;8(9):e75102.
221. Mazur PK, et al. Combined inhibition of BET family proteins and histone deacetylases as a potential epigenetics-based therapy for pancreatic ductal adenocarcinoma. *Nat Med*. 2015;21(10):1163–1171.
222. Li Y, Seto E. HDACs and HDAC inhibitors in cancer development and therapy. *Cold Spring Harb Perspect Med*. 2016;6(10).
223. Clawson GA. Histone deacetylase inhibitors as cancer therapeutics. *Ann Transl Med*. 2016;4(15):287.
224. Morgan HD, et al. Epigenetic reprogramming in mammals. *Hum Mol Genet*. 2005;14(Spec. No. 1):R47–R58.
225. Martin C, Zhang Y. The diverse functions of histone lysine methylation. *Nat Rev Mol Cell Biol*. 2005;6(11):838–849.
226. Tian X, et al. Histone lysine-specific methyltransferases and demethylases in carcinogenesis: new targets for cancer therapy and prevention. *Curr Cancer Drug Targets*. 2013;13(5):558–579.
227. Qazi AM, et al. Laser capture microdissection of pancreatic ductal adenocarcinoma cells to analyze EzH2 by Western Blot analysis. *Methods Mol Biol*. 2011;755:245–256.
228. Verma SK, et al. Identification of potent, selective, cell-active inhibitors of the histone lysine methyltransferase EZH2. *ACS Med Chem Lett*. 2012;3(12): 1091–1096.
229. Avan A, et al. Molecular mechanisms involved in the synergistic interaction of the EZH2 inhibitor 3-deazaneplanocin A with gemcitabine in pancreatic cancer cells. *Mol Cancer Ther*. 2012;11(8):1735–1746.
230. Maftouh M, et al. A polymorphism in the promoter is associated with EZH2 expression but not with outcome in advanced pancreatic cancer patients. *Pharmacogenomics*. 2014;15(5):609–618.
231. Wang Z, et al. Activated K-Ras and INK4a/Arf deficiency promote aggressiveness of pancreatic cancer by induction of EMT consistent with cancer stem cell phenotype. *J Cell Physiol*. 2013;228(3):556–562.
232. Mody H, et al. Inhibition of S-adenosylmethionine-dependent methyltransferase attenuates TGF-beta1-induced EMT and metastasis in pancreatic cancer: putative roles of miR-663a and miR-4787-5p. *Mol Cancer Res*. September 2016.
233. de Almeida RA, et al. Non-coding RNAs and disease: the classical ncRNAs make a comeback. *Biochem Soc Trans*. 2016;44(4):1073–1078.
234. Chen X, Fan S, Song E. Noncoding RNAs: new players in cancers. *Adv Exp Med Biol*. 2016;927:1–47.
235. Veneziano D, et al. Noncoding RNA: current deep sequencing data analysis approaches and challenges. *Hum Mutat*. September 2016.
236. Hombach S, Kretz M. Non-coding RNAs: classification, biology and functioning. *Adv Exp Med Biol*. 2016;937:3–17.
237. Schmitt AM, Chang HY. Long noncoding RNAs in cancer pathways. *Cancer Cell*. 2016;29(4):452–463.
238. Qu S, et al. Circular RNA: a new star of noncoding RNAs. *Cancer Lett*. 2015;365(2):141–148.

239. Muller S, et al. Next-generation sequencing reveals novel differentially regulated mRNAs, lncRNAs, miRNAs, sdRNAs and a piRNA in pancreatic cancer. *Mol Cancer.* 2015;14:94.
240. Beckedorff FC, et al. Long non-coding RNAs and their implications in cancer epigenetics. *Biosci Rep.* 2013;33(4):667–675.
241. Lavorgna G, et al. Long non-coding RNAs as novel therapeutic targets in cancer. *Pharmacol Res.* 2016;110:131–138.
242. Ma C, et al. H19 promotes pancreatic cancer metastasis by derepressing let-7's suppression on its target HMGA2-mediated EMT. *Tumour Biol.* 2014;35(9):9163–9169.
243. Taucher V, Mangge H, Haybaeck J. Non-coding RNAs in pancreatic cancer: challenges and opportunities for clinical application. *Cell Oncol (Dordr).* 2016;39(4): 295–318.
244. Li Z, et al. The long non-coding RNA HOTTIP promotes progression and gemcitabine resistance by regulating HOXA13 in pancreatic cancer. *J Transl Med.* 2015;13:84.
245. Liu JH, et al. Expression and prognostic significance of lncRNA MALAT1 in pancreatic cancer tissues. *Asian Pac J Cancer Prev.* 2014;15(7):2971–2977.
246. D'Ippolito E, Iorio MV. MicroRNAs and triple negative breast cancer. *Int J Mol Sci.* 2013;14(11):22202–22220.
247. John B, et al. Human microRNA targets. *PLoS Biol.* 2004;2(11):e363.
248. Hammond SM. RNAi, microRNAs, and human disease. *Cancer Chemother Pharmacol.* 2006;58(suppl 1):s63–s68.
249. Godfrey SS, et al. Microarray analysis reveal differential expression of microRNAs in triple negative breast cancer and normal breast cells. *Int J Sci Res Publ (IJSRP).* 2015;5(4):1–6.
250. Godfrey SS, Moses MM, Rorie CJ. Cancer associated microRNAs are differentially expressed in triple negative breast cancer and normal breast cells. *Int J Sci Res (IJSR).* 2015;4(9):73–78.
251. Stallings VL, Gurley KM, Rorie CJ. Triple negative breast cancer cell lines with TP53 mutations are able to undergo cell death. *Int J Sci Res.* 2015;4(9):1024–1028.
252. Mardin WA, Mees ST. MicroRNAs: novel diagnostic and therapeutic tools for pancreatic ductal adenocarcinoma? *Ann Surg Oncol.* 2009;16(11):3183–3189.
253. Hwang JH, et al. Identification of microRNA-21 as a biomarker for chemoresistance and clinical outcome following adjuvant therapy in resectable pancreatic cancer. *PLoS One.* 2010;5(5):e10630.
254. Diab M, et al. The role of microRNAs in the diagnosis and treatment of pancreatic adenocarcinoma. *J Clin Med.* 2016;5(6):1–13.
255. Myhre S, et al. Influence of DNA copy number and mRNA levels on the expression of breast cancer related proteins. *Mol Oncol.* 2013;7(3):704–718.
256. Lee Y, et al. MicroRNA maturation: stepwise processing and subcellular localization. *EMBO J.* 2002;21(17):4663–4670.
257. Hammond SM. MicroRNAs as oncogenes. *Curr Opin Genet Dev.* 2006;16(1):4–9.
258. Melo SA, Esteller M. Dysregulation of microRNAs in cancer: playing with fire. *FEBS Lett.* 2011;585(13):2087–2099.
259. Jansson MD, Lund AH. MicroRNA and cancer. *Mol Oncol.* 2012;6(6):590–610.
260. Steele CW, et al. Clinical potential of microRNAs in pancreatic ductal adenocarcinoma. *Pancreas.* 2011;40(8):1165–1171.
261. Setoyama T, et al. microRNA-10b: a new marker or the marker of pancreatic ductal adenocarcinoma? *Clin Cancer Res.* 2011;17(17):5527–5529.
262. Bhatti I, et al. Knockdown of microRNA-21 inhibits proliferation and increases cell death by targeting programmed cell death 4 (PDCD4) in pancreatic ductal adenocarcinoma. *J Gastrointest Surg.* 2011;15(1):199–208.

263. Funel N. The role of miR-21 and miR-211 on MMP9 regulation in pancreatic ductal adenocarcinoma: cooperation in invasiveness behaviors? *Epigenomics*. 2015;7(3): 333–335.
264. Ammerpohl O, et al. Complementary effects of HDAC inhibitor 4-PB on gap junction communication and cellular export mechanisms support restoration of chemosensitivity of PDAC cells. *Br J Cancer*. 2007;96(1):73–81.
265. Song WF, et al. MiR-21 upregulation induced by promoter zone histone acetylation is associated with chemoresistance to gemcitabine and enhanced malignancy of pancreatic cancer cells. *Asian Pac J Cancer Prev*. 2013;14(12):7529–7536.
266. Xiao Q, et al. Cancer associated fibroblasts in pancreatic cancer are reprogrammed by tumor-induced alterations in genomic DNA methylation. *Cancer Res*. 2016;76(18): 5395–5404.
267. Mees ST, et al. Analyzing miRNAs in ductal adenocarcinomas of the pancreas. *J Surg Res*. 2011;169(2):241–246.
268. Otsuka K, Ochiya T. Genetic networks lead and follow tumor development: microRNA regulation of cell cycle and apoptosis in the p53 pathways. *Biomed Res Int*. 2014;2014:749724.
269. Frampton AE, et al. MicroRNAs cooperatively inhibit a network of tumor suppressor genes to promote pancreatic tumor growth and progression. *Gastroenterology*. 2014;146(1). 268.e18–277.e18.
270. Brunetti O, et al. MicroRNA in pancreatic adenocarcinoma: predictive/prognostic biomarkers or therapeutic targets? *Oncotarget*. 2015;6(27):23323–23341.
271. Bryant JL, et al. A microRNA gene expression signature predicts response to erlotinib in epithelial cancer cell lines and targets EMT. *Br J Cancer*. 2012;106(1):148–156.
272. Singh SK, et al. Antithetical NFATc1-Sox2 and p53-miR200 signaling networks govern pancreatic cancer cell plasticity. *EMBO J*. 2015;34(4):517–530.
273. Misso G, et al. Mir-34: a new weapon against cancer? *Mol Ther Nucleic Acids*. 2014;3:e194.
274. Garajova I, et al. Molecular mechanisms underlying the role of microRNAs in the chemoresistance of pancreatic cancer. *Biomed Res Int*. 2014;2014:678401.
275. Frampton AE, et al. microRNAs with prognostic significance in pancreatic ductal adenocarcinoma: a meta-analysis. *Eur J Cancer*. 2015;51(11):1389–1404.
276. Li CH, et al. EZH2 coupled with HOTAIR to silence microRNA-34a by the induction of heterochromatin formation in human pancreatic ductal adenocarcinoma. *Int J Cancer*. September 2016.
277. Salzman J. Circular RNA expression: its potential regulation and function. *Trends Genet* 2016;32(5):309–316.
278. O'Connor KW, et al. PARI overexpression promotes genomic instability and pancreatic tumorigenesis. *Cancer Res*. 2013;73(8):2529–2539.
279. Mateo J, et al. DNA repair in prostate cancer: biology and clinical implications. *Eur Urol*. 2016;1–9.
280. Rozenblum E, et al. Tumor-suppressive pathways in pancreatic carcinoma. *Cancer Res*. 1997;57(9):1731–1734.
281. Golan T, et al. Overall survival and clinical characteristics of pancreatic cancer in BRCA mutation carriers. *Br J Cancer*. 2014;111(6):1132–1138.
282. de Mestier L, et al. Pancreatic ductal adenocarcinoma in BRCA2 mutation carriers. *Endocr Relat Cancer*. 2016;23(10):T57–T67.
283. Lucas AL, et al. BRCA1 and BRCA2 germline mutations are frequently demonstrated in both high-risk pancreatic cancer screening and pancreatic cancer cohorts. *Cancer*. 2014;120(13):1960–1967.
284. Maginn EN, et al. Opportunities for translation: targeting DNA repair pathways in pancreatic cancer. *Biochim Biophys Acta*. 2014;1846(1):45–54.

285. Lucas AL, et al. High prevalence of BRCA1 and BRCA2 germline mutations with loss of heterozygosity in a series of resected pancreatic adenocarcinoma and other neoplastic lesions. *Clin Cancer Res.* 2013;19(13):3396–3403.
286. Choy E, et al. Phase II study of olaparib in patients with refractory Ewing sarcoma following failure of standard chemotherapy. *BMC Cancer.* 2014;14:813.
287. Liu JF, et al. Combination cediranib and olaparib versus olaparib alone for women with recurrent platinum-sensitive ovarian cancer: a randomised phase 2 study. *Lancet Oncol.* 2014;15(11):1207–1214.
288. Orta ML, et al. The PARP inhibitor Olaparib disrupts base excision repair of 5-aza-2′-deoxycytidine lesions. *Nucleic Acids Res.* 2014;42(14):9108–9120.
289. Andrei AZ, et al. Increased in vitro and in vivo sensitivity of BRCA2-associated pancreatic cancer to the poly(ADP-ribose) polymerase-1/2 inhibitor BMN 673. *Cancer Lett.* 2015;364(1):8–16.
290. Wang YQ, et al. An update on poly(ADP-ribose)polymerase-1 (PARP-1) inhibitors: opportunities and challenges in cancer therapy. *J Med Chem.* July 2016.
291. Awasthi N, Schwarz MA, Schwarz RE. Combination effects of bortezomib with gemcitabine and EMAP II in experimental pancreatic cancer. *Cancer Biol Ther.* 2010;10(1):99–107.
292. Sonnenblick A, et al. An update on PARP inhibitors—moving to the adjuvant setting. *Nat Rev Clin Oncol.* 2015;12(1):27–41.
293. Zheng W, et al. Combination of radiotherapy and vaccination overcome checkpoint blockade resistance. *Oncotarget.* 2016;7(28):43039–43051.

PART IV

Soft Tissue Sarcomas: Genomic and Epigenomic Alterations

CHAPTER EIGHT

Molecular Changes Associated With Tumor Initiation and Progression of Soft Tissue Sarcomas: Targeting the Genome and Epigenome

P.W. Halcrow*, M. Dancer*, M. Panteah*, C. Walden*, J.E. Ohm[†,1]
*University of North Dakota School of Medicine and Health Sciences, Grand Forks, ND, United States
[†]Roswell Park Cancer Institute, Buffalo, NY, United States
[1]Corresponding author: e-mail address: joyce.ohm@roswellpark.org

Contents

Abstract

Soft tissue sarcomas are rare, but generally aggressive tumors which disproportionately affect children and young adults. They represent less than 10% of all cancers, but are one of the most frequently diagnosed cancers in pediatric patients. These cancers have a high rate of morbidity and mortality, and their overall incidence has been increasing at an estimated rate of 26% over the last 2 decades. The cause of this increased incidence is unknown but various environmental factors have been implicated. Establishing

Progress in Molecular Biology and Translational Science, Volume 144
ISSN 1877-1173
http://dx.doi.org/10.1016/bs.pmbts.2016.10.001

standard therapeutic strategies is challenging for soft tissue sarcomas as more than 50 different histological subtypes exist, each with their own molecular alterations and clinical characteristics, and this combination of tumor heterogeneity and a limited number of clinical cases make detailed omics level molecular studies particularly challenging. This chapter will focus on the unique genetic and epigenetic changes which characterize these cancers, with an emphasis on translocation-associated sarcomas involving primary gene fusions with the RNA chaperone protein EWSR1. We will highlight current therapeutic approaches and discuss opportunities for targeted molecular therapeutics.

1. INTRODUCTION

Soft tissue sarcomas are malignant tumors that are thought to originate from mesenchymal lineage cells in the soft tissues of the body, including muscle, tendons, fat, lymph vessels, blood vessels, nerves, and tissue around joints. While these tumors can be found anywhere throughout the body, the most common sites of primary soft tissue sarcomas are the arms, legs, chest, or abdomen. The most common presentation of a soft tissue sarcoma is the appearance of a noticeable lump or area of swelling within the soft tissue. In many cases, the tumor is not diagnosed until it begins to press on local nerves or other organs causing pain or interfering with normal organ function. As a result, many soft tissue sarcomas are fairly large at the time of diagnosis. Prior radiation therapy, certain environmental exposures such as exposure to thorium dioxide, vinyl chloride, or arsenic, and certain preexisting diseases and inherited conditions including Retinoblastoma, Neurofibromatosis type 1 (NF1), Tuberous sclerosis, Familial adenomatous polyposis (FAP), Li–Fraumeni syndrome, Werner syndrome (adult progeria), Nevoid basal cell carcinoma syndrome (Gorlin syndrome), and acquired immune deficiency syndrome (AIDS) are thought to increase the risk of developing these cancers.[1]

Soft tissue sarcoma is not a single disease, but it is instead a heterogeneous group of cancers lumped together due to their mesenchymal lineage. While there are many overlapping features of these tumors in terms of diagnosis, treatment, and prognosis; each cancer subtype is characterized by unique molecular changes in the tumor which contribute to tumor initiation and progression and each tumor subtype has unique features with regard to their epidemiology, pathophysiology, and prognosis. When biopsied, many of these tumors often present as "small round blue cell tumors" and diagnosis can be challenging. Differential diagnoses include Ewing sarcoma (EWS) family tumors, rhabdomyosarcoma, neuroblastoma, lymphoma, synovial sarcoma, ectomesenchymoma, and Wilms' tumor in children; and small cell carcinoma, carcinoid tumor, neuroendocrine carcinomas, merkel cell carcinoma, and small cell mesothelioma in adults. The stage of the sarcoma is determined by a combination of grade (how aggressive the cells appear), tumor size, location, and metastatic spread, and it is used by oncologists to develop a treatment plan. Characteristic molecular alterations are often used in confirming the diagnosis, and in these cancers they include both genetic and epigenetic changes which will be discussed throughout this chapter.

Soft tissue sarcomas are generally diagnosed by performing a biopsy of a suspicious lesion followed by immunohistochemistry, cytogenetics, fluorescence in situ hybridization (FISH), and/or flow cytometry. Due to the diversity of these tumors, they can be challenging to diagnose. Treatment and prognosis depend on: tumor size, grade, stage, tumor location, the overall age and health of the patient, and whether the tumor is a primary tumor or a recurrent and/or metastatic lesion. Like most solid tumors, the standard of care for localized soft tissue sarcoma cases is wide-margin surgical resection. For some soft tissue sarcomas, removal of the tumor in surgery may be the only treatment needed. In other cases, patients may also undergo chemotherapy and radiation as part of front-line standard-of-care therapy.

One common feature of many soft tissue sarcomas is the presence of characteristic driver mutations involving proteins such as EWS RNA-binding protein 1 (EWSR1). The endogenous EWSR1 protein, whose function is not completely understood, has two domains that are thought to contribute to its function: The first domain, the transcriptional activation domain, allows the EWS protein to activate transcription, and the other region is an RNA-binding domain. EWSR1 is required in some model organisms for cell survival in the central nervous system.[2] EWSR1 plays a role in transcription initiation: EWSR1 is able to associate with the basal transcription factor TFIID (a multiprotein complex composed of the TATA-binding protein (TBP) and TBP-associated factors (TAFIIs)) and the RNA polymerase II complex.[3] It also has been shown to associate with heterogeneous RNA-binding proteins (hnRNPs), such as RBM38 and RBM39 (RNA-binding motif proteins 38 and 39, 20q13 and 20q11, respectively), as well as EP300 and CREBBP. EWSR1 has also been shown to help mediate the action of other transcription factors and has been shown to function as a coactivator of CREBBP-dependent transcription factors where EWSR1-EP300/CREBBP mediates both FOS activation and HNF4 activation.[4,5] CREBBP is a transcription coactivator which enables the interaction between various transcription factors and RNA Pol II, which in turn bring enzymes to the promoter and remodels the chromatin favoring an open status. Endogenous EWSR1 is thought to have the ability to remodel chromatin via this interaction and it is expected that EWSR1 translocation fusion proteins will affect this normal function.[6] EWSR1 is phosphorylated by PKC (protein kinase C), and this posttranslational modification inhibits RNA binding of EWSR1; also, CALM (calmodulin) binding to EWSR1 is known to inhibit PKC phosphorylation.[7] Downstream targeting of EWSR1 is also known to activate other transcription factors such as POU4F1 (BRN3A)[8] and POU5F1 (OCT4),[9] important proteins with known roles in development and tumorigenesis, and EWSR1 interacts with POU4F2 (4q31), a regulator of neuronal cell differentiation. Loss of EWSR1 results in premature cellular senescence,[10] and this observation may be related to the observation that EWSR1 is required for proper localization of aurora B during mitosis and maintains mitotic spindle integrity.[10]

Additionally, EWSR1 is thought to play a significant role in RNA splicing via interactions with multiple known splicing factors and regulatory proteins. EWSR1 interacts with BARD1 (BRCA1-associated RING domain 1),[11] CCNL1 (cyclin L1, 3q25) and TFIP11 (tuftelin-interacting protein 11, 22q12), the latter being functionally related to the spliceosome and involved in pre-mRNA splicing.[12] EWSR1 has been shown to function as a docking molecule by recruiting serine–arginine (SR) splicing factors such as SRSF10 (serine/arginine-rich splicing factor 10, 1p36, or TASR, which represses pre-mRNA splicing) to RNA Pol II. This suggests a role for EWSR1 in coupling gene transcription to RNA splicing by binding to hyperphosphorylated RNA Pol II through its N-terminal domain, and SR splicing factors through its C-terminal domain.[13] YBX1 (Y box-binding protein 1, 1p34), a multifunctional protein that shuttles between the cytoplasm (where it binds to mRNA and regulates mRNA translation) and the nucleus (where it regulates transcription of diverse target genes), interacts with the C-terminal domain of EWSR1. This results in the interaction of YBX1 and RNA Pol II which helps to facilitate pre-mRNA splicing.[14] SMN1 (survival of motor neuron 1, telomeric, 5q13) plays a major role in the pre-mRNA splicing machinery (role in spliceosomal snRNP assembly in the cytoplasm and in pre-mRNA splicing in the nucleus),[15] and SMN1, through its Tudor region, binds the RG1 region of EWSR1.[16] RNU1–2 (RNA, U1 small nuclear 2, 1p36, also called U1C), another splicing protein, also interacts with EWSR1.[17] Taken together, these data suggest that endogenous EWSR1 plays several key roles in transcription, splicing, chromatin remodeling, and also has been shown to interact with multiple proteins that have been implicated in tumorigenesis. Additional work is needed to fully appreciate how EWSR1 translocation fusion proteins disrupt these normal processes and contribute to tumorigenesis.

2. CLINICAL PARAMETERS OF EWSR1-TRANSLOCATION-ASSOCIATED SOFT TISSUE SARCOMA SUBTYPES

The most common subtype of soft tissue sarcoma in pediatric patients is rhabdomyosarcoma, which is a cancer that initiates in the muscles. It is usually diagnosed in children under the age of 10 and generally present with lumps that form near the surface of the body resulting in early diagnosis. Adults are more likely to be diagnosed with GIST (gastrointestinal stromal tumors), which occur most frequently in the soft tissue of the gastrointestinal tract, primarily in the stomach or small intestines.[18] GIST may be either benign or malignant and are often not diagnosed until they are very large due to their relative lack of symptoms. Other common sarcoma subtypes include Kaposi sarcoma and uterine sarcoma, but this chapter will focus on six soft tissue sarcoma subtypes that have each been described to frequently present with EWSR1 translocations. In each of these cancers, the c-terminus of EWSR1 is fused with the n-terminus of a known transcription factor. These subtypes are highlighted in Table 1 with a discussion of clinical parameters of these tumors to follow.

2.1 Desmoplastic Small Round Cell Tumor (DSRCT)

DSRCT is a soft tissue sarcoma with a median age of onset of 14 years.[19–22] To date, more than 200 DSRCT cases have been described in the medical literature. DSRCT exhibits a male predominance of 90%, and 85% of patients are Caucasian. The 5-year survival rate is estimated to be ~15%.[19,23] One of the primary considerations driving this poor prognosis is the fact that greater than 40% of patients diagnosed with DSRCT already have distant metastasis present at the time of diagnosis and are therefore diagnosed with late-stage disease. This tumor tends to grow along a serosal

Table 1 Soft Tissue Sarcoma Subtypes With Known EWSR1 Translocations

EWSR1 Translocation-Associated Tumor	Peak Onset (Decade)	Most Common Primary Site of Origin	5-Year Survival Rate	Gender (M:F)	Incidence Rate Among Soft Tissue Sarcomas
Desmoplastic Small Round Cell Tumor	Second–third	Peritoneal Lining	15%	4:1	Rare
Ewing's Sarcoma	First–third	Bone	70–75%	1:1	5%
Clear Cell Sarcoma of the Soft Tissue	Third–fourth	Aponeuroses and tendons of extremities	50%	1:1	Rare
Extraskeletal Myxoid Chondrosarcoma	Sixth	Proximal extremities and limb girdles	90%	2:1	3%
Angiomatoid Fibrous Histiocytoma	Second–third	Deep dermis of the extremities	Generally Nonfatal	1:1	0.3%
Myxoid Liposarcoma	Third–fifth	Soft tissues of the extremities	60%	1:1	10–35%

lining in the later stages of the disease, and the most common primary sites for DSRCT are the peritoneum or soft tissue of the abdomen and pelvis, but other primary sites have been described, including the ovary/fallopian tube, cerebellum, pleura, ethmoid sinuses, scalp, hand, posterior cranial fossa, pancreas, ovary, paratesticular, kidney, and cerebral ventricle.[23–26] Symptoms related to extraabdominal tumors vary, and include scoliosis, chronic sinusitis, and pain.[20,21,27–33] At the histological level, this tumor presents as variably sized clusters of small round or spindled cells lying within a desmoplastic, collagenous stroma.[23] DSRCT is more commonly diagnosed in males, where the male to female ratio is 4:1,[34] and there is a slightly increased incidence in African-Americans.[27] More than 40% of patients have distant metastases at the time of diagnosis, mostly located in the liver, lungs, and lymph nodes. The most commonly described genetic alteration found in DSRCT is a translocation which results in the c-terminus of the EWSR1 fused with the n-terminus of the Wilm's tumor 1 (WT1) gene. The resulting translocation (EWSR1–WT1) is thought to be a driver mutation in this soft tissue sarcoma subtype.[35]

DSRCTs often reach considerable size in the abdominal cavity prior to diagnosis: a mean size of 10 cm has been described.[35] Macroscopically, they are solid, firm, multilobulated gray-white masses within which localized cystic areas can be found. The correct diagnosis of DSRCT on fine needle aspiration specimens is challenging and requires expertise in the utilization of ancillary techniques such as immunocytochemistry and flow-cytometric immunophenotyping. When available, the use of RT-PCR detection of the EWS–WT1 transcript is a way of increasing diagnostic accuracy using less invasive techniques. Histologic examination shows small cells that can be round, ovoid, or spindled usually grouped in clumps, cords, nests, or sheets. These cells contain hyperchromatic nuclei with condensed chromatin and eosinophilic cytoplasm, and have frequently observable mitotic figures. The extensive collagenous stroma is referred to a desmoplasia and is a distinctive feature of this tumor on pathology.

DSRCT is a very aggressive neoplasm. Treatment options include surgery, radiotherapy, chemotherapy with or without stem cell transplantation, and scientists are beginning to test a variety of molecularly targeted therapies in clinical trials for this tumor. Unfortunately, a true gold standard therapeutic regimen has not been described since these tumors are rare and no single chemotherapy regimen has been shown to be superior. Chemotherapeutic agents utilized include cyclophosphamide, doxorubicin, vincristine, ifosfamide, and etoposide (a combination known as P6 protocol) as well as cisplatin, carboplatin, topotecan, temozolamide, vinorelbine, and irinotecan. DSRCTs are generally chemosensitive, but these tumors frequently reoccur, necessitating multimodality therapy with radiotherapy, surgery, and/or high-dose chemotherapy (HDCT) with stem cell rescue. Surgery is usually extensive and due to the invasive nature of this tumor, a complete resection with negative margins is often not possible without irreparably damaging adjacent essential organ structures. Debulking surgery has been described as an attempt to eliminate 90% of the tumor bulk.[21] Studies have shown that aggressive and complete surgical resection is a major determinant in patient survival; however, since DSRCT most commonly presents as a metacentric abdominal mass, complete upfront resection is not often possible. In addition to surgery and radiation therapy, local control options for DSRCT (particularly metastatic disease) include radiofrequency ablation, gammaknife, cryoablation, embolization, and chemoembolization. These are usually performed in academic centers after careful consideration of individual cases. HDCT with autologous stem cell rescue has also been attempted[36]; however, no significant impact in long-term survival has been achieved after transplant.

Although DSRCTs are generally sensitive to chemotherapy, the response is not enough to achieve cure since patients almost invariably relapse. This could potentially be a reflection of the heterogeneity of the cells within the tumor; where a distinct population of cells, often referred to as "cancer stem cells" are less sensitive to chemotherapy and radiotherapy, possess the ability to self-renew, and retain the capacity to regenerate the tumor bulk after it has been eradicated.[37,38] This represents an attractive hypothesis in terms of developing novel treatment strategies since it could explain tumor behavior and lead to the identification of new molecular targets for more effective therapies. Unlike cancer stem cells (CSCs) described in other small round blue cell tumors like EWS (described later),[39] such a stem cell has not yet been clearly identified in DSRCTs.

In a series of 66 patients reported by MSKCC, individuals who received multimodal treatment consisting of debulking surgery, whole abdominopelvic radiation[40] and the P6 chemotherapy regimen had a 3-year survival of 55% (29/66), compared to 27% in those patients who did not receive all three therapies. In this series, patients who underwent tumor resection had a 58% 3-year survival rate whereas no survivors were found in the nonsurgical group. This may represent a difference in tumor extension at the time of diagnosis; however, a therapeutic advantage of surgical resection has been suggested by others. A retrospective series of 11 pediatric patients (10 males, median age 14 years) treated at St. Jude Children's Hospital with multimodal therapy included two long-term survivors (2 and 8 years from diagnosis), which presented with localized extra abdominal disease. It is still unclear whether extra abdominal disease is less aggressive or if a more effective debulking surgery had an impact in this positive outcome.

2.2 Ewing's Sarcoma (EWS)

EWS is an aggressive malignant small round cell tumor that primarily occurs in the bone, soft tissue, and the parenchyma. EWS is the second most common tumor of bone in childhood and adolescence, behind osteosarcoma. This tumor was first described by James Ewing in 1921 as "diffuse endothelioma of bone,"[41] which most commonly metastasizes to the lung or distant bony sites. For EWS, specifically on radiographs, this tumor often can be visualized as a destructive growth arising in the middle of the bone. Advanced imaging techniques including magnetic resonance imaging (MRI) are generally used to help establish the diagnosis, especially for cases where the suspected lesion is thought to reside outside of the bone. EWS has an early median onset of 15 years old and an overall estimated 5-year survival rate of 70–75%.[42–45] Prognosis has dramatically improved, from less than 10% of long-term survivors, but unfortunately, for the approximately 25% of EWS patients which present with distant metastasis at diagnosis, 5-year event-free, and overall survival rates remain poor at around 20%.[46] This may be attributed to the fact that patients who relapse following an initial clinical remission have very limited options for successful second line therapy.[47,48] The most commonly described genetic alteration in EWS is a translocation in which the EWSR1 gene is fused with the Friend leukemia integration 1 (FLI1) transcription factor (EWSR1-FLI1). This gene fusion is found in >90% of cases of EWS and research has demonstrated that overexpression of this fusion in preclinical models largely recapitulates the clinical and pathological features of the human disease.

Like DSRCT, treatment involves three potential types of therapy: chemotherapy, radiation therapy, and surgery. EWS is generally more responsive to chemotherapy than DSRCT and many adult-type sarcomas, and as a result, physicians will generally prescribe chemotherapy as part of front-line therapy for all individuals diagnosed with EWS. For patients with a single localized tumor site, an initial round of cytoreductive chemotherapy may be recommended to shrink the tumor and target micrometastases in order to prevent further spread. After completing this initial course of chemotherapy, patients with tumors that are amenable to surgical resection will undergo surgery with curative intent. For patients with tumors that cannot be removed surgically, targeted radiation therapy is used to kill the localized tumor. It is also very common for patients to undergo another round of consolidation chemotherapy after surgery or radiation with the goal of eradicating any residual tumor cells and reduce both local recurrence rates and metastasis.[49–51] The role of neoadjuvant chemotherapy for patients with EWS remains an area of active investigation. Establishing appropriate doses to achieve optimal cytoreduction and to prevent amputation are major goals of these studies and are designed to assess the effectiveness of various chemotherapy regimens.[52] In certain cases, both surgery and radiation are used. This stratified, combined approach has been credited with increasing overall survival to 70% in localized disease, although outcome for metastatic disease remains poor.

2.3 Clear Cell Sarcoma of Soft Tissue (CCSST)

CCSST is a rare malignant small round blue cell tumor which is thought to primarily originate from the tendons and aponeurosis of the distal extremities such as the hands and feet.[53] The histological clear cell appearance of this tumor is due to the accumulation of glycogen in the cells.[54] In 1983, this soft tissue sarcoma subtype was originally classified as a "malignant melanoma of the soft parts" due to its histological similarities to malignant melanoma, and many of these tumors and cell lines derived from these cancers express melanin at relatively high levels.[53] CCSST has a median onset of 30 years old, and approximately 30% of patients present with preexisting metastatic lesions at diagnosis. Local recurrence occurs in as many as 84% of patients and late metastasis occurs in up to 63% of patients. The primary genetic lesion associated with CCSST is a translocation in which the c-terminus of the EWSR1 is fused with a cyclic AMP-dependent transcription factor called ATF1.[55–57] The resulting fusion protein (EWSR1-ATF1) is characteristic of this sarcoma subtype and the presence of this translocation combined with this characteristic histology is considered diagnostic for CCSST. A minority of patients may present with an alternate EWSR1/CREB1 translocation.[58]

For CCSST, after discovery of the tumor, one or more biopsies will be performed in order to confirm a diagnosis. However, clear cell sarcoma's close similarity to malignant melanoma can still make it difficult to correctly diagnose without resorting to additional genetic tests that can identify mutations indicative of specific diseases. Clear cell sarcoma tumors can be characterized further using an injected MRI contrast agent (which can sometimes be taken orally for gastrointestinal tumors). Chest X-ray or CT scans are used to check for metastases in the lungs, a common site of spread. A PET scan uses intravenous administration of a mildly radioactive sugar to detect metastases almost anywhere in the body. Presently, radiation and chemotherapy have not been shown to be effective primary treatment modalities for CCSST. Surgical removal is the mainstay of treatment, and wide local excision, aggressive resection, or amputation often combined with adjuvant radiotherapy is the current standards of care.

The prognosis of clear cell sarcoma is guarded, mainly because the disease is difficult to catch early and is prone to relapse and spread long after initial diagnosis. The 5-, 10-, and 20-year disease-specific survival rates have been reported as ~50%, 33%, and 10%, respectively.[59] The largest determinant in patient prognosis is size of the tumor before surgery: tumors less than 5 cm in diameter are associated with much better long-term survival than larger tumors. Due to the rarity of clear cell sarcoma, it is challenging to conduct statistically significant clinical trials to prove the benefit of existing or novel drugs. While the disease is currently difficult to treat, experimental therapies have shown promise in case studies and are under active investigation in clinical trials of soft tissue sarcomas.

2.4 Extraskeletal Myxoid Chondrosarcoma (EMCS)

EMCS is an intermediate-grade tumor that represents less than 3% of all soft tissue sarcomas. Myxoid chondrosarcoma has a median age at diagnosis of 50 years and is more commonly diagnosed in males than females at a ratio of approximately 2:1 (males:females). EMCS is found to arise in both the bone and soft tissues, but this subtype is primarily found in sub-fascia soft tissues of the proximal extremities and limb girdles. The thigh and popliteal fossa are the most common primary location with less common areas including the distal extremities, trunk, paraspinal region, and head and neck region. EMCS presents clinically as a slow-growing soft tissue mass associated with pain and tenderness. As with all of these sarcoma subtypes, wide local resection is currently the standard of care and is thought to be the best treatment option for EMCS. Existing radiation and chemotherapy regimens have shown to only be minimally effective. Patients with EMCS have a prolonged clinical course with 5-, 10-, and 15-year survival rates of 90%, 70%, and 60%, respectively, and local recurrences occur in ~48% of patients.[60] Metastasis occurs in approximately 50% of cases with the most common site for distant metastasis being the lung, a common site of metastasis for all sarcomas. The most frequent genetic alteration associated with EMCS is a translocation involving the EWSR1 fused with nuclear receptor subfamily 4, group A (NR4A3).[61]

2.5 Angiomatoid Fibrous Histiocytoma (AFH)

AFH is a rare intermediate-grade tumor that comprises of approximately 0.3% of all soft tissue tumors.[62] AFH equally affects males and females and a patient population with wide age-range including reported cases in both infants and in patients in their eighth and ninth decades. The majority of cases present in children and young adults with a median age of diagnosis of 20 years of age. The most common primary locations for AFH include the deep dermis and superficial subcutaneous tissue of the extremities, followed by the trunk and head and neck region. This sarcoma subtype presents clinically as a slow-growing mass that is usually without pain or tenderness, and as a result, this cancer is sometimes misdiagnosed as a hematoma or hemangioma. In rare instances, AFH can present with systemic symptoms including anemia, fever, and weight loss. Wide local excision with clear margins is considered curative in the majority of patients, as the vast majority is diagnosed with local AFH and who do not present with metastatic lesions at diagnosis. AFH recurs in less than 15% of cases and metastatic spread to regional lymph nodes occurs in less than 1% of cases. AFH is associated with the same translocation events as CCSST with both EWSR1-CREB1 and EWSR1-ATF1 fusions reportedly occurring in both tumor types. As a result, the pathology of these tumors cannot be solely driven by the EWSR1 translocation and additional genetic, epigenetic, cell of origin, or microenvironment questions must be addressed.

2.6 Myxoid Liposarcoma (MLS)

MLS is a malignant tumor that is thought to be of adipocyte origin. MLS is the second most commonly diagnosed liposarcoma, accounting for 30–35% of all liposarcomas and approximately 5–10% of soft tissue sarcomas.[63] MLS affects males and females equally and usually occurs between the third and fifth decades. Even though peak incidence is in adults, it is also known to affect the pediatric population and is the leading subtype of liposarcoma to be diagnosed in children. Clinically, this sarcoma subtype most commonly presents as a slow-growing mass in the deep soft tissues of the extremities with 66–75% of cases occurring subfacially in the musculature of the thigh. The remaining tumors generally arise in the retroperitoneum. Local recurrence is common in MLS and approximately 1/3 of these tumors will metastasize to other soft tissue locations, bone, and/or the lungs. Wide local excision is considered to be the best treatment available, but in cases with aggressive lesions, adjuvant therapy may be used in addition to surgery. MLS is associated with the fusion of DDIT3(CHOP) with FUS in 90% of cases, but can also be associated with a DDIT3(CHOP) fusion with EWSR1.[64–66]

3. DEVELOPING NOVEL MOLECULAR THERAPEUTIC STRATEGIES FOR THE TREATMENT OF EWSR1 TRANSLOCATION SOFT TISSUE SARCOMA SUBTYPES

In addition to surgery, radiation, and chemotherapy, a wide range of clinical trials are ongoing to treat these soft tissue sarcoma subtypes.

A summary of these ongoing clinical trials is included in Table 2, but these strategies are frequently designed to use the following strategies to target tumor cells that are resistant to front-line therapies.

3.1 Developing Molecular Strategies to Overcome Resistance to Chemotherapy

There are several common mechanisms by which tumor cells are able to acquire resistance to a broad range of anticancer drugs. These mechanisms have been reviewed in detail elsewhere,[67–70] but some of the best understood of these mechanisms is the abnormal expression of one or more energy-dependent transporters that detect and eject anticancer drugs from cells, insensitivity of cells to drug-induced apoptosis, or induction of drug-detoxifying mechanisms. Activation of these mechanisms may be linked to either host factors that are specific to the individual's normal cells and/or organs or to specific genetic or epigenetic alterations which are unique to the cancer cells themselves.

Much work has been done over the last several decades in understanding the general mechanisms by which tumor cells become resistant to chemotherapy. In some cases, there is the acquisition of resistance to a specific drug, perhaps by loss of a transporter protein or cell surface receptor. In these cases, combination chemotherapy using multiple different agents is often effective. There are, however, many cases where cancer cells have acquired multiple simultaneous mechanisms of resistance or they have altered one or more mechanisms that inhibit the effectiveness of many different classes of drugs. This phenomenon is called multidrug resistance and can result from changes that limit accumulation of drugs within cells by limiting uptake, enhancing efflux, or affecting general properties of the cell membrane.[71] Most often, these changes block apoptosis, which should be induced by many chemotherapy drugs, activate generalized responses which result in detoxification or altered DNA repair, or disrupt the normal cell cycle checkpoints.

Another common mechanism by which cancer cells become resistant to chemotherapy is the upregulation of a variety of different transporter proteins including multi-drug resistance gene 1 (MDR1). MDR1 is an ATP-binding cassette (ABC) transporter protein which is a member of a super family of channel proteins and multidrug resistance-related proteins (MRPs) whose normal functions include the efflux of ions, nutrients, lipids, amino acids, peptides, proteins, and drugs. A lot of research has been done in recent years on MDR1 and the three-dimensional structures of bacterial and human ABC transporters demonstrate that these proteins are

Table 2 Ongoing Clinical Trials for Soft-Tissue Sarcomas

Clinical Trial Number	Study	Condition	Intervention	Phase	Special Considerations
Ewing Sarcoma					
NCT00987636	Study in Localized and Disseminated Ewing Sarcoma	Ewing's Sarcoma	Drug: Zoledronic acid; Drug: Busulfan; Drug: Treosulfan	3	
NCT02063022	Efficacy of Dose Intensification in Patients With Nonmetastatic Ewing Sarcoma	Ewing's Sarcoma	Drug: Standard Treatment (protocol ISG SSG III); Drug: Intensified Chemotherapy	3	
NCT02581384	Sterotactic Body Radiotherapy for Pulmonary Metastases in Ewing Sarcoma, Rhabdomyosarcoma, and Wilms' Tumor	Ewing's Sarcoma, Rhabdomyosarcoma, Wilms' Tumor	Radiation: Stereotactic Body Radiotherapy	1/2	
NCT02243605	Cabozantinib-s-malate in Treating Patients With Relapsed Osteosarcoma or Ewing Sarcoma	Metastatic Ewing Sarcoma/Peripheral Primitive Neuroectodermal Tumor; Metastatic Osteosarcoma; Recurrent Ewing Sarcoma/Peripheral Primitive Neuroectodermal Tumor; Recurrent Osteosarcoma	Drug: Cabozantinib-S-malate; Other: Laboratory Biomarker Analysis	2	
NCT01969942	A Phase I Study to Examine the Toxicity of Allogenic Stem Cell Transplantation for Relapsed or Therapy Refractory Ewing Sarcoma	Ewing Sarcoma	Biological: Allogeneic stem cell transplantation; Drug: Clyclophosphamide; Drug: Fludarabina; Drug: Busulfan; Drug: Melphalan	1	

NCT02511132	Randomized Phase IIb Trial of Vigil™ Versus Gemcitabine +Docetaxel for Ewing's Sarcoma	Ewing's Sarcoma	Biological: Vigil; Drug: Gemcitabine and docetaxel	2
NCT00743496	A Phase I Trial of the Humanized Anti-GD2 Antibody in Children and Adolescents With Neuroblastoma, Osteosarcoma, Ewing Sarcoma and Melanoma	Neuroblastoma; Melanoma; Osteosarcoma; Ewing Sarcoma	Biological: Anti-GD2 antibody	1
NCT01876303	Genetic Biomarkers in Saliva Samples From Patients With Ewing Sarcoma	Ewing Sarcoma/Peripheral Primitive Neuroectodermal Tumor (PNET)	Other: Laboratory biomarker analysis; Other: Questionnaire administration	
NCT00899990	Collecting and Storing Biological Samples From Patients With Ewing Sarcoma	Askin Tumor; Localized Ewing Sarcoma/Peripheral Primitive Neuroectodermal Tumor; Metastatic Ewing Sarcoma/ Peripheral Primitive Neuroectodermal Tumor; Recurrent Ewing Sarcoma/ Peripheral Primitive Neuroectodermal Tumor	Other: Laboratory biomarker analysis	

Continued

Table 2 Ongoing Clinical Trials for Soft-Tissue Sarcomas—cont'd

Clinical Trial Number	Study	Condition	Intervention	Phase	Special Considerations
NCT02736565	Pbi-shRNA™ EWS/FLI1 Type 1 LPX in Subjects With Advanced Ewing's Sarcoma	Ewing's Sarcoma	Biological: pbi-shRNA™ EWS/ FLI1 Type 1 LPX	1	
NCT01858168	Phase I Study of Olaprib and Temozolomide for Ewing's Sarcoma	Ewing Sarcoma	Drug: Olaparib; Drug: Temozolomide	1	
NCT01864109	Irinotecan and Temozolomide in Combination With Existing High-Dose Alkylator-Based Chemotherapy for Treatment of Patients With Newly Diagnosed Ewing Sarcoma	Newly Diagnosed Ewing Sarcoma	Drug: Cyclophosphamide; Device: Doxorubicin; Drug: Vincristine; Device: Ifosfamide; Drug: Etoposide; Procedure: Surgery; Radiation: Radiation Therapy[a]; Drug: Temozolomide; Drug: Irinotecan	2	
NCT02727387	Protocol for the Treatment of Metastatic Ewing Sarcoma	Ewing's Sarcoma (ES)	Drug: TEMIRI; Drug: ADM; Drug: IFO; Drug: CYC; Drug: ETO; Drug: BUMEL; Drug: VIN	2	

NCT02306161	Combination Chemotherapy With or Without Ganitumab in Treating Patients With Newly Diagnosed Metastatic Ewing Sarcoma	Metastatic Ewing Sarcoma; Metastatic Malignant Neoplasm in the Bone; Metastatic Malignant Neoplasm in the Bone Marrow; Metastatic Malignant Neoplasm in the Lung; Metastatic Peripheral Primitive Neuroectodermal Tumor of Bone; Peripheral Primitive Neuroectodermal Tumor of Soft Tissues	Drug: Cyclophosphamide; Drug: Doxorubicin Hydrochloride; Drug: Etoposide; Radiation: External Beam Radiation Therapy; Biological: Ganitumab; Drug: Ifosfamide; Other: Laboratory Biomarker Analysis; Other: Pharmacological Study; Radiation: Stereotactic Radiosurgery; Procedure: Therapeutic Surgical Procedure; Drug: Vincristine Sulfate	2
NCT02044120	ESP1/SARC025 Global Collaboration: A Phase I Study of a Combination of the PARP Inhibitor, Niraparib and Temozolomide or Irinotecan in Patients With Previously Treated, Incurable Ewing Sarcoma	Ewing Sarcoma	Drug: Niraparib; Drug: Temozolomide; Drug: Irinotecan	1
NCT02657005	TK216 in Patients With Relapsed or Refractory Ewing Sarcoma	Sarcoma, Ewing	Drug: TK216	1

Continued

Table 2 Ongoing Clinical Trials for Soft-Tissue Sarcomas—cont'd

Clinical Trial Number	Study	Condition	Intervention	Phase	Special Considerations
NCT02472392	Study to Examine Toxicity of Allogeneic Stem Cell Transplantation for Relapsed or Therapy Refractory Ewings Sarcoma	Ewings Sarcoma	Drug: Rabbit antithymocyte globulin; Drug: Busulfan; Drug: Melphalan; Drug: Fludarabine; Drug: Cyclophosphamide	1	
NCT02100891	Phase 2 STIR Trial: Haploidentical Transplant and Donor Natural Killer Cells for Solid Tumors	Ewing Sarcoma; Neuroblastoma; Rhabdomyosarcoma; Osteosarcoma; CNS Tumors	Procedure: Allogeneic HCT; Drug: Donor NK Cell Infusion	2	
NCT02304458	Nivolumab With or Without Ipilimumab in Treating Younger Patients With Recurrent or Refractory Solid Tumors or Sarcomas	Childhood Solid Neoplasm; Metastatic Melanoma; Recurrent Ewing Sarcoma/Peripheral Primitive Neuroectodermal Tumor; Recurrent Hodgkin Lymphoma; Recurrent Melanoma; Recurrent Neuroblastoma; Recurrent Non-Hodgkin Lymphoma; Recurrent Osteosarcoma; Recurrent Rhabdomyosarcoma; Refractory Hodgkin Lymphoma; Refractory Non-Hodgkin Lymphoma; Stage IIIA Skin Melanoma; Stage IIIB Skin Melanoma; Stage IIIC Skin Melanoma; Stage IV Skin Melanoma	Biological: Ipilimumab; Other: Laboratory Biomarker Analysis; Biological: Nivolumab; Other: Pharmacological Study	1/2	

NCT01962103	To Find a Safe Dose and Show Early Clinical Activity of Weekly Nab-paclitaxel in Pediatric Patients With Recurrent/Refractory Solid Tumors	Neuroblastoma; Rhabdomyosarcoma; Ewing's Sarcoma; Ewing's Tumor; Sarcoma, Ewing's; Sarcomas, Epitheliod; Sarcoma, Soft Tissue; Sarcoma, Spindle Cell; Melanoma; Malignant Melanoma; Clinical Oncology; Oncology, Medical; Pediatrics, Osteosarcoma; Osteogenic Sarcoma; Osteosarcoma Tumor; Sarcoma, Osteogenic; Tumors; Cancer; Neoplasia; Neoplasm; Histiocytoma; Fibrosarcoma; Dermatofibrosarcoma	Drug: Nab-paclitaxel	1/2

Continued

Table 2 Ongoing Clinical Trials for Soft-Tissue Sarcomas—cont'd

Clinical Trial Number	Study	Condition	Intervention	Phase	Special Considerations
NCT02116777	Talazoparib and Temozolomide in Treating Younger Patients With Refractory or Recurrent Malignancies	Adult Solid Neoplasm; Childhood Solid Neoplasm; Recurrent Adult Acute Lymphoblastic Leukemia; Recurrent Childhood Acute Lymphoblastic Leukemia; Recurrent Childhood Central Nervous System Neoplasm; Recurrent Ewing Sarcoma/ Peripheral Primitive Neuroectodermal Tumor; Refractory Central Nervous System Neoplasm	Other: Laboratory Biomarker Analysis; Other: Pharmacological Study; Drug: Talazoparib; Drug: Temozolomide	1/2	
NCT02389244	A Phase II Study Evaluating Efficacy and Safety of Regorafenib in Patients With Metastatic Bone Sarcomas	Ewing Sarcomas; Chondrosarcomas; Osteosarcomas; Chondroma	Drug: Regorafenib; Drug: Placebo	2	
NCT02409576	Pilot Study of Expanded, Activated Haploidentical Natural Killer Cell Infusions for Sarcomas	Ewing Sarcoma; Osteosarcoma; Rhabdomyosarcoma	Biological: Expanded, Activated NK cells	1/2	
NCT02520128	A Study of IMRT in Primary Bone and Soft Tissue Sarcoma	Soft Tissue Sarcoma; Ewing Sarcoma; Bone Sarcoma; Chordoma	Radiation: Intensity modulated radiotherapy (IMRT)		

NCT01050296	Molecular Analysis of Solid Tumors	Pediatric Solid Tumors	
NCT00931931	Advanced MR Imaging in Sarcoma Patients	Sarcoma	
NCT01336803	Differentiation of Bone Sarcomas and Osteomyelitis With Ferumoxytol-Enhanced MRI	Bone Cancer; Chondrosarcoma; Ewing's Sarcoma; Osteosarcoma; Rhabdomyosarcoma; Bone Necrosis; Bone Sarcoma; Osteomyelitis	Drug: Feraheme; Procedure: MR Scan
NCT00840047	Methionine PET/CT Studies In Patients With Cancer	Brain Tumors and/or Solid Tumors Including: Brain Stem Glioma; High Grade CNS Tumors; Ependymoma; Medulloblastoma; Craniopharyngioma; Low Grade CNS Tumors; Hodgkin Lymphoma; Non-Hodgkin Lymphoma; Ewing Sarcoma; Osteosarcoma; Rhabdomyosarcoma; Neuroblastoma; Other	Drug: Methionine

Continued

Table 2 Ongoing Clinical Trials for Soft-Tissue Sarcomas—cont'd

Clinical Trial Number	Study	Condition	Intervention	Phase	Special Considerations
NCT02390843	Simvastatin With Topotecan and Cyclophosphamide in Relapsed and/or Refractory Pediatric Solid and CNS Tumors	Retinoblastoma; Clear Cell Sarcoma; Renal Cell Carcinoma; Rhabdoid Tumor; Wilms' Tumor; Hepatoblastoma; Neuroblastoma; Germ Cell Tumors; Ewings Sarcoma; Nonrhabdomyosarcoma Soft Tissue Sarcoma; Osteosarcoma; Rhabdomyosarcoma	Drug: Simvastatin; Drug: Cyclophosphamide; Drug: Topotecan; Drug: Myeloid growth factor	1	
NCT02624388	Study of Genistein in Pediatric Oncology Patients (UVA-Gen001)	Lymphoma; Childhood Lymphoma; Solid Tumor; Childhood Solid Tumor; Neuroblastoma; Ewing Sarcoma; Hodgkin Lymphoma; Non-Hodgkin Lymphoma; Rhabdomyosarcoma; Soft Tissue Sarcoma; Medulloblastoma; Germ Cell Tumor; Wilms' Tumor; Brain Neoplasms; Medulloblastoma, Childhood; Neuroectodermal Tumors, Primitive	Drug: Genistein; Drug: Placebo	2	

NCT02162732	Molecular-Guided Therapy for Childhood Cancer	Neuroblastoma; Medulloblastoma; Glioma; Ependymoma; Choroid Plexus Neoplasms; Craniopharyngioma; Dysembryoplastic Neuroepithelial Tumor; Meningioma; Primitive Neuroectodermal Tumors (PNETs); Germ Cell Tumors; Rhabdomyosarcoma; Nonrhabdomyosarcoma; Ewings Sarcoma; Osteosarcoma; Wilms' Tumor; Renal Cell Carcinoma; Malignant Rhabdoid Tumor; Clear Cell Sarcoma; Liver Tumors	Device: Guided Therapy	
NCT02508038	TCRαβ+/CD19+ Depleted Haploidentical HSCT +Zoledronate	Acute Myeloid Leukemia; Acute Lymphoblastic Leukemia; Hodgkin Lymphoma; Non-Hodgkin Lymphoma; Myelodysplastic Syndrome; Myeloproliferative Syndrome; Rhabdomyosarcoma; Ewing Sarcoma; Primitive Neuroectodermal Tumor; Osteosarcoma; Neuroblastoma	Procedure: TCRαβ+/CD19+ depleted Haploidentical HSCT; Drug: Zoledronate	1

Continued

Table 2 Ongoing Clinical Trials for Soft-Tissue Sarcomas—cont'd

Clinical Trial Number	Study	Condition	Intervention	Phase	Special Considerations
NCT02013336	Phase 1 Study of MM-398 Plus Cyclophosphamide in Pediatric Solid Tumors	Recurrent or Refractory Solid Tumors; Ewing Sarcoma; Rhabdomyosarcoma; Neuroblastoma; Osteosarcoma	Drug: MM-398 (Irinotecan Sucrosofate Liposome Injection) plus cyclophosphamide	1	
NCT01625351	A Study of CD45RA+ Depleted Haploidentical Stem Cell Transplantation in Children With Relapsed or Refractory Solid Tumors and Lymphomas	Ewing Sarcoma; Gastrointestinal Tumor; Germ Cell Tumor; Hepatic Tumor; Lymphoma; Wilms' Tumor; Rhabdoid Tumor; Clear Cell Carcinoma; Renal Cell Carcinoma; Melanoma; Neuroblastoma; Rhabdomyosarcoma; Nonrhabdomyosarcoma	Drug: Alemtuzumab; Drug: Fludarabine; Drug: Sirolimus; Drug: Busulfan; Drug: Melphalan; Biological: Stem cells	1	
NCT02689336	Erlotinib in Combination With Temozolomide in Treating Relapsed/Recurrent/Refractory Pediatric Solid Tumors	Glioma; Rhabdomyosarcoma; Osteosarcoma; Medulloblastoma; Neuroectodermal Tumor; Ependymoma; Ewing's Sarcoma; Wilms' Tumor	Biological: Erlotinib; Drug: Temozolomide	2	

NCT02644460	Abemaciclib in Children With DIPG or Recurrent/Refractory Solid Tumors	Diffuse Intrinsic Pontine Glioma; Brain Tumor, Recurrent; Solid Tumor, Recurrent; Neuroblastoma, Recurrent, Refractory; Ewing Sarcoma, Recurrent, Refractory; Rhabdomyosarcoma, Recurrent, Refractory; Osteosarcoma, Recurrent, Refractory; Rhabdoid Tumor, Recurrent, Refractory	Drug: Abemaciclib	1
NCT02048371	SARC024: A Blanket Protocol to Study Oral Regorafenib in Patients With Refractory Liposarcoma, Osteogenic Sarcoma, and Ewing Sarcomas	Liposarcoma; Osteogenic Sarcoma; Ewing/Ewing-like Sarcoma	Drug: Regorafenib; Drug: Placebo	2
NCT02415816	Diffusion-Weighted Magnetic Resonance in Imaging Younger Patients With Newly Diagnosed Bone or Soft Tissue Sarcomas	Sarcoma, Bone; Sarcoma, Soft Tissue; Osteosarcoma; Sarcoma, Ewing; Rhabdomyosarcoma; Nonrhabdomyosarcoma Soft tissue sarcoma	Procedure: Diffusion-Weighted Magnetic Resonance Imaging	

Continued

Table 2 Ongoing Clinical Trials for Soft-Tissue Sarcomas—cont'd

Clinical Trial Number	Study	Condition	Intervention	Phase	Special Considerations
NCT00898755	Collecting and Storing Tissue From Young Patients With Cancer	Cancer	Genetic: DNA analysis; Genetic: Reverse transcriptase-polymerase chain reaction; Other: Biologic sample preservation procedure; Other: Flow cytometry; Other: Laboratory biomarker analysis		
NCT02557854	HIFU Hyperthermia With Liposomal Doxorubicin (DOXIL) for Relapsed or Refractory Pediatric and Young Adult Solid Tumors	Rhabdomyosarcoma; Neuroblastoma; Sarcoma; Sarcoma, Ewing; Osteosarcoma	Drug: Doxorubicin HCl liposomal injection; Device: Philips Sonalleve MR-HIFU Hyperthermia	1	
NCT01661400	Antiangiogenic Therapy Post Transplant (ASCR) for Pediatric Solid Tumors	Glioma; Neuroectodermal Tumors, Primitive; Wilms' Tumor; Rhabdomyosarcoma; Sarcoma, Ewing; Osteosarcoma; Retinoblastoma	Drug: Metronomic Cyclophosphamide; Drug: Thalidomide	0	

NCT02159443	Pretreatment Antitherapeutic Antibodies (PATA) in Patients Treated With hu14.18K322A Antibody	Ewing Family of Tumors; Melanoma; Neuroblastoma; Osteosarcoma		
NCT00445965	Iodine I 131 Monoclonal Antibody 3F8 in Treating Patients With Central Nervous System Cancer or Leptomeningeal Cancer	Brain and Central Nervous System Tumors; Intraocular Melanoma; Lung Cancer; Melanoma (Skin); Metastatic Cancer; Neuroblastoma; Ovarian Cancer; Retinoblastoma; Sarcoma; Small Intestine Cancer	Genetic: DNA analysis; Other: Immunologic technique; Other: Pharmacological study; Radiation: Iodine I 131 monoclonal antibody 3F8; Radiation: 131I-3F8	2
NCT00089245	Radiolabeled Monoclonal Antibody Therapy in Treating Patients With Refractory, Recurrent, or Advanced CNS or Leptomeningeal Cancer	Brain and Central Nervous System Tumors; Neuroblastoma; Sarcoma	Radiation: Iodine I 131 monoclonal antibody 8H9	1
NCT02574728	Sirolimus in Combination With Metronomic Chemotherapy in Children With Recurrent and/or Refractory Solid and CNS Tumors	Cancer	Drug: Sirolimus; Drug: Celecoxib; Drug: Etoposide; Drug: Cyclophosphamide	2

Continued

Table 2 Ongoing Clinical Trials for Soft-Tissue Sarcomas—cont'd

Clinical Trial Number	Study	Condition	Intervention	Phase	Special Considerations
NCT01804634	A Phase II Trial of Reduced Intensity Conditioning and Haploidentical BMT for High-Risk Solid Tumors	Refractory and/or Relapsed Metastatic Solid Tumors	Drug: Cyclophosphamide; Drug: Fludarabine; Radiation: Low dose total body irradiation; Drug: Melphalan; Drug: Sirolimus	2	
NCT01047878	Genomic Analysis of Pediatric Bone Tumors	Bone Cancer			
NCT02441062	Impact of Ga-68 DOTATOC PET-CT Imaging in Management of Neuroendocrine Tumors	Neuroendocrine Tumors; Carcinoid Tumors; Neuroblastoma; Medulloblastoma	Drug: 68Ga-DOTATOC PET/CT	2	
NCT02392793	Talazoparib Plus Irinotecan With or Without Temozolomide in Children With Refractory or Recurrent Solid Malignancies	Childhood Solid Tumors	Drug: Talazoparib; Drug: Irinotecan; Drug: Temozolomide	1	
NCT02595255	AMH as a Predictor of Infertility Risk in Children With Cancer (CHANCE)	Fertility Preservation; Lymphoma; Pediatrics Cancer; Gonadotropin-releasing Hormone Agonist	Other: No intervention		

NCT01956669	Pazopanib Pediatric Phase II Trial Children's Oncology Group (COG) in Solid Tumors	Solid Tumors	Drug: Pazopanib GW786034	2
NCT02454972	Clinical Trial of Lurbinectedin (PM01183) in Selected Advanced Solid Tumors	Advanced Solid Tumors	Drug: Lurbinectedin (PM01183)	2
NCT00876031	Trial for Localized High-Risk Rhabdomyosarcoma and Rhabdomyosarcoma-Like Soft Tissue Sarcoma	Soft Tissue Sarcoma	Drug: Trofosfamide, idarubicin, etoposide	3
NCT01505569	Auto Transplant for High Risk or Relapsed Solid or CNS Tumors	Ewing's Family Tumors; Renal Tumors; Hepatoblastoma; Rhabdomyosarcoma; Soft Tissue Sarcoma; Primary Malignant Brain Neoplasms; Retinoblastoma; Medulloblastoma; Supra-tentorial Primative Neuro-Ectodermal Tumor (PNET); Atypical Teratoid/Rhabdoid Tumor (AT/RT); CNS Tumors; Germ Cell Tumors	Drug: Ifosfamide; Drug: Etoposide; Drug: Mesna; Biological: G-CSF; Drug: Busulfan; Drug: Melphalan; Drug: Thiotepa; Biological: Autologous stem cell infusion; Radiation: Radiation; Drug: Carboplatin; Drug: Paclitaxel; Procedure: Leukapheresis	

Continued

Table 2 Ongoing Clinical Trials for Soft-Tissue Sarcomas—cont'd

Clinical Trial Number	Study	Condition	Intervention	Phase	Special Considerations
Desmoplastic Small Round Cell Tumor					
NCT01189643	Irinotecan, Temozolomide and Bevacizumab in Combination With Existing High-Dose Alkylator-Based Chemotherapy for Treatment of Newly Diagnosed Patients With Desmoplastic Small Round Cell Tumor	Sarcoma; Desmoplastic Small Round Cell Tumor (DSRCT)	Drug: Irinotecan, temozolomide, and bevacizumab incorporated into an existing schedule of high-dose alkylator		
NCT01099644	Intraperitoneal Radioimmunotherapy With 131I-8H9 for Patients With Desmoplastic Small Round Cell Tumors and Other Solid Tumors Involving the Peritoneum	Peritoneal Cancer	Biological: 131 I-8H9	1	
NCT00919269	Collecting and Storing Tissue, Blood, and Bone Marrow Samples From Patients With Rhabdomyosarcoma or Other Soft Tissue Sarcoma	Adult Rhabdomyosarcoma; Childhood Desmoplastic Small Round Cell Tumor; Chordoma; Desmoid Tumor; Metastatic Childhood Soft Tissue Sarcoma; Nonmetastatic Childhood Soft Tissue Sarcoma; Previously Treated Childhood Rhabdomyosarcoma; Previously Untreated Childhood Rhabdomyosarcoma; Recurrent	Other: Biologic sample preservation procedure; Other: Laboratory biomarker analysis		

		Adult Soft Tissue Sarcoma; Recurrent Childhood Rhabdomyosarcoma; Recurrent Childhood Soft Tissue Sarcoma; Stage I Adult Soft Tissue Sarcoma; Stage II Adult Soft Tissue Sarcoma; Stage III Adult Soft Tissue Sarcoma; Stage IV Adult Soft Tissue Sarcoma			
NCT01532687	Gemcitabine Hydrochloride With or Without Pazopanib Hydrochloride in Treating Patients With Refractory Soft Tissue Sarcoma	Adult Alveolar Soft Part Sarcoma; Adult Angiosarcoma; Adult Desmoplastic Small Round Cell Tumor; Adult Epithelioid Hemangioendothelioma; Adult Epithelioid Sarcoma; Adult Extraskeletal Myxoid Chondrosarcoma; Adult Extraskeletal Osteosarcoma; Adult Fibrosarcoma; Adult Leiomyosarcoma; Adult Liposarcoma; Adult Malignant Mesenchymoma; Adult Malignant Peripheral Nerve Sheath Tumor; Adult Rhabdomyosarcoma; Adult Synovial Sarcoma; Adult Undifferentiated Pleomorphic Sarcoma; Malignant Adult	Drug: Gemcitabine Hydrochloride; Other: Laboratory Biomarker Analysis; Drug: Pazopanib Hydrochloride; Other: Placebo	2	

Continued

Table 2 Ongoing Clinical Trials for Soft-Tissue Sarcomas—cont'd

Clinical Trial Number	Study	Condition	Intervention	Phase	Special Considerations
		Hemangiopericytoma; Recurrent Adult Soft Tissue Sarcoma; Stage III Adult Soft Tissue Sarcoma; Stage IV Adult Soft Tissue Sarcoma			
NCT00502177	Quality of Life Questionnaire for Pediatric Patients Undergoing Peritoneal Perfusion	Peritoneal Neoplasms	Behavioral: Questionnaire		
Clear Cell Sarcoma					
NCT01524926	CREATE: Cross-Tumoral Phase 2 With Crizotinib	Locally Advanced and/or Metastatic Anaplastic Large Cell Lymphoma; Locally Advanced and/or Metastatic Inflammatory Myofibroblastic Tumor; Locally Advanced and/or Metastatic Papillary Renal Cell Carcinoma Type 1; Locally Advanced and/or Metastatic Alveolar Soft Part Sarcoma; Locally Advanced and/or Metastatic Clear Cell Sarcoma; Locally Advanced and/or Metastatic Alveolar Rhabdomyosarcoma	Drug: Crizotinib (PF-02341066)	2	

NCT02180867	Radiation Therapy With or Without Combination Chemotherapy or Pazopanib Hydrochloride Before Surgery in Treating Patients With Newly Diagnosed Nonrhabdomyosarcoma Soft Tissue Sarcomas That can be Removed by Surgery	Adult Fibrosarcoma; Alveolar Soft Part Sarcoma; Angiomatoid Fibrous Histiocytoma; Atypical Fibroxanthoma; Clear Cell Sarcoma of Soft Tissue; Epithelioid Malignant Peripheral Nerve Sheath Tumor; Epithelioid Sarcoma; Extraskeletal Myxoid Chondrosarcoma; Extraskeletal Osteosarcoma; Fibrohistiocytic Neoplasm; Glomus Tumor of the Skin; Inflammatory Myofibroblastic Tumor; Intimal Sarcoma; Leiomyosarcoma; Liposarcoma; Low Grade Fibromyxoid Sarcoma; Low Grade Myofibroblastic Sarcoma; Malignant Cutaneous Granular Cell Tumor; Malignant Peripheral Nerve Sheath Tumor; Malignant Triton Tumor;	Drug: Doxorubicin Hydrochloride; Drug: Ifosfamide; Other: Laboratory Biomarker Analysis; Drug: Pazopanib Hydrochloride; Other: Pharmacological Study; Radiation: Radiation Therapy; Procedure: Therapeutic Conventional Surgery	2/3	

Continued

Table 2 Ongoing Clinical Trials for Soft-Tissue Sarcomas—cont'd

Clinical Trial Number	Study	Condition	Intervention	Phase	Special Considerations
		Mesenchymal Chondrosarcoma; Myxofibrosarcoma; Myxoid Chondrosarcoma; Myxoinflammatory Fibroblastic Sarcoma; Nerve Sheath Neoplasm; PEComa; Pericytic Neoplasm; Plexiform Fibrohistiocytic Tumor; Sclerosing Epithelioid Fibrosarcoma; Stage IB Soft Tissue Sarcoma; Stage IIB Soft Tissue Sarcoma; Stage III Soft Tissue Sarcoma; Stage IV Soft Tissue Sarcoma; Synovial Sarcoma; Undifferentiated (Embryonal) Sarcoma; Undifferentiated High Grade Pleomorphic Sarcoma of Bone			

NCT00898365	Study of Kidney Tumors in Younger Patients	Clear Cell Sarcoma of the Kidney; Congenital Mesoblastic Nephroma; Diffuse Hyperplastic Perilobar Nephroblastomatosis; Rhabdoid Tumor of the Kidney; Stage I Renal Cell Cancer; Stage I Wilms' Tumor; Stage II Renal Cell Cancer; Stage II Wilms' Tumor; Stage III Renal Cell Cancer; Stage III Wilms' Tumor; Stage IV Renal Cell Cancer; Stage IV Wilms' Tumor; Stage V Wilms' Tumor	Other: Laboratory biomarker analysis; Other: Cytology specimen collection procedure	
NCT01164228	Sunitinib Malate With or Without Gemcitabine Hydrochloride in Treating Patients With Advanced Kidney Cancer That Cannot be Removed by Surgery	Kidney Cancer	Drug: Gemcitabine hydrochloride; Drug: Sunitinib malate	2

Continued

Table 2 Ongoing Clinical Trials for Soft-Tissue Sarcomas—cont'd

Clinical Trial Number	Study	Condition	Intervention	Phase	Special Considerations
NCT02475213	Safety Study of MGA271 in Combination With Pembrolizumab in Refractory Cancer	Melanoma; Head and Neck Cancer; Non Small Cell Lung Cancer; Mesothelioma; Urethelial Carcinoma; Clear Cell Renal Cell Carcinoma; Ovarian Cancer; Thyroid Cancer; Triple Negative Breast Cancer; Pancreatic Cancer; Colon Cancer; Soft Tissue Sarcoma; Prostate Cancer	Biological: MGA271; Biological: Pembrolizumab	1	
NCT02381314	Safety Study of MGA271 in Combination With Ipilimumab in Refractory Cancer	Melanoma; Head and Neck Cancer; Nonsmall Cell Lung Cancer; Mesothelioma; Urothelial Carcinoma; Clear Cell Renal Cell Carcinoma; Ovarian Cancer; Thyroid Cancer; Triple Negative Breast Cancer; Pancreatic Cancer; Colon Cancer; Soft Tissue Sarcoma; Prostate Cancer	Biological: MGA271 plus ipilimumab	1	

NCT02628535	Safety Study of MGD009 in B7-H3-expressing Tumors	Mesothelioma; Bladder Cancer; Melanoma; Squamous Cell Carcinoma of the Head and Neck; Non Small Cell Lung Cancer; Clear Cell Renal Cell Carcinoma; Ovarian Cancer; Thyroid Cancer; Breast Cancer; Pancreatic Cancer; Prostate Cancer; Colon Cancer; Soft Tissue Sarcoma	Biological: MGD009	1
NCT02020707	Paclitaxel Albumin-Stabilized Nanoparticle Formulation and Bevacizumab in Treating Patients With Stage IV Melanoma That Cannot be Removed by Surgery or Gynecological Cancers	Cervical Adenosarcoma; Cervical Adenosquamous Carcinoma; Cervical Carcinosarcoma; Cervical Squamous Cell Carcinoma; Endometrial Clear Cell Adenocarcinoma; Endometrial Endometrioid Adenocarcinoma; Endometrial Mixed Adenocarcinoma; Endometrial Mucinous Adenocarcinoma; Endometrial Squamous Cell Carcinoma; Endometrial Transitional Cell Carcinoma; Endometrial Undifferentiated Carcinoma; Fallopian Tube Adenocarcinoma; Fallopian Tube Clear Cell	Biological: Bevacizumab; Other: Laboratory Biomarker Analysis; Drug: Paclitaxel Albumin-Stabilized Nanoparticle Formulation; Other: Pharmacological Study	1

Continued

Table 2 Ongoing Clinical Trials for Soft-Tissue Sarcomas—cont'd

Clinical Trial Number	Study	Condition	Intervention	Phase	Special Considerations
		Adenocarcinoma; Fallopian Tube Mucinous Adenocarcinoma; Fallopian Tube Serous Adenocarcinoma; Fallopian Tube Transitional Cell Carcinoma; Malignant Ovarian Epithelial Tumor; Malignant Peritoneal Neoplasm; Ovarian Carcinosarcoma; Ovarian Clear Cell Adenocarcinoma; Ovarian Endometrioid Adenocarcinoma; Ovarian Mucinous Adenocarcinoma; Ovarian Serous Adenocarcinoma; Ovarian Transitional Cell Carcinoma; Primary Peritoneal Serous Adenocarcinoma; Recurrent Fallopian Tube Carcinoma; Recurrent Melanoma; Recurrent Ovarian Carcinoma; Recurrent Primary Peritoneal Carcinoma; Stage IV Skin Melanoma; Undifferentiated Fallopian Tube Carcinoma; Undifferentiated Ovarian Carcinoma; Uterine Corpus Carcinosarcoma			

NCT00989651	Carboplatin, Paclitaxel, Bevacizumab, and Veliparib in Treating Patients With Newly Diagnosed Stage II–IV Ovarian Epithelial, Fallopian Tube, or Primary Peritoneal Cancer	Fallopian Tube Carcinosarcoma; Fallopian Tube Clear Cell Adenocarcinoma; Fallopian Tube Endometrioid Adenocarcinoma; Fallopian Tube Mucinous Adenocarcinoma; Fallopian Tube Serous Neoplasm; Fallopian Tube Transitional Cell Carcinoma; Ovarian Brenner Tumor; Ovarian Carcinosarcoma; Ovarian Clear Cell Adenocarcinoma; Ovarian Endometrioid Adenocarcinoma; Ovarian Mucinous Adenocarcinoma; Ovarian Seromucinous Tumor; Ovarian Serous Adenocarcinoma; Ovarian Transitional Cell Carcinoma; Primary Peritoneal Serous Adenocarcinoma; Stage IIA Fallopian Tube Cancer; Stage IIA Ovarian Cancer; Stage IIB Fallopian Tube Cancer; Stage IIB Ovarian Cancer; Stage IIC Fallopian Tube Cancer;	Biological: Bevacizumab; Drug: Carboplatin; Drug: Cisplatin; Other: Laboratory Biomarker Analysis; Drug: Paclitaxel; Drug: Veliparib	1

Continued

Table 2 Ongoing Clinical Trials for Soft-Tissue Sarcomas—cont'd

Clinical Trial Number	Study	Condition	Intervention	Phase	Special Considerations
		Stage IIC Ovarian Cancer; Stage IIIA Fallopian Tube Cancer; Stage IIIA Ovarian Cancer; Stage IIIA Primary Peritoneal Cancer; Stage IIIB Fallopian Tube Cancer; Stage IIIB Ovarian Cancer; Stage IIIB Primary Peritoneal Cancer; Stage IIIC Fallopian Tube Cancer; Stage IIIC Ovarian Cancer; Stage IIIC Primary Peritoneal Cancer; Stage IV Fallopian Tube Cancer; Stage IV Ovarian Cancer; Stage IV Primary Peritoneal Cancer; Undifferentiated Fallopian Tube Carcinoma; Undifferentiated Ovarian Carcinoma			
NCT02713386	Ruxolitinib Phosphate, Paclitaxel, and Carboplatin in Treating Patients With Stage III–IV Epithelial Ovarian, Fallopian Tube, or Primary Peritoneal Cancer	Fallopian Tube Carcinosarcoma; Fallopian Tube Clear Cell Adenocarcinoma; Fallopian Tube Endometrioid Adenocarcinoma; Fallopian Tube Serous Neoplasm; High Grade Ovarian Serous	Drug: Carboplatin; Other: Laboratory Biomarker Analysis; Drug: Paclitaxel; Drug: Ruxolitinib Phosphate; Procedure: Therapeutic Conventional Surgery	1/2	

		Adenocarcinoma; Ovarian Carcinosarcoma; Ovarian Clear Cell Adenocarcinoma; Ovarian Endometrioid Adenocarcinoma; Primary Peritoneal Serous Adenocarcinoma; Stage IIIA Fallopian Tube Cancer; Stage IIIA Ovarian Cancer; Stage IIIA Primary Peritoneal Cancer; Stage IIIB Fallopian Tube Cancer; Stage IIIB Ovarian Cancer; Stage IIIB Primary Peritoneal Cancer; Stage IIIC Fallopian Tube Cancer; Stage IIIC Ovarian Cancer; Stage IIIC Primary Peritoneal Cancer; Stage IV Fallopian Tube Cancer; Stage IV Ovarian Cancer; Stage IV Primary Peritoneal Cancer		
NCT02101775	Gemcitabine Hydrochloride With or Without WEE1 Inhibitor MK-1775 in Treating Patients With Recurrent Ovarian, Primary Peritoneal, or Fallopian Tube Cancer	Ovarian Brenner Tumor; Ovarian Carcinosarcoma; Ovarian Clear Cell Cystadenocarcinoma; Ovarian Endometrioid Adenocarcinoma; Ovarian Mucinous Cystadenocarcinoma; Ovarian	Drug: Gemcitabine Hydrochloride; Other: Laboratory Biomarker Analysis; Other: Pharmacological Study; Other: Placebo; Drug: WEE1 Inhibitor AZD1775	2

Continued

Table 2 Ongoing Clinical Trials for Soft-Tissue Sarcomas—cont'd

Clinical Trial Number	Study	Condition	Intervention	Phase	Special Considerations
		Seromucinous Carcinoma; Ovarian Serous Cystadenocarcinoma; Ovarian Serous Surface Papillary Adenocarcinoma; Recurrent Fallopian Tube Carcinoma; Recurrent Ovarian Carcinoma; Recurrent Primary Peritoneal Carcinoma; Undifferentiated Ovarian Carcinoma			
Extraskeletal Myxoid Chondrosarcoma					
NCT02066285	Trial of Pazopanib in Patients With Solitary Fibrous Tumor and Extraskeletal Myxoid Chondrosarcoma	Solitary Fibrous Tumor; Extraskeletal Myxoid Chondrosarcoma	Drug: Pazopanib	2	
NCT02601937	A Phase 1 Study of the EZH2 Inhibitor Tazemetostat in Pediatric Subjects With Relapsed or Refractory INI1-Negative Tumors or Synovial Sarcoma	Rhabdoid Tumors; INI1-negative Tumors; Synovial Sarcoma; Malignant Rhabdoid Tumor of Ovary	Drug: Tazemetostat	1	

NCT02601950	A Phase II, Multicenter Study of the EZH2 Inhibitor Tazemetostat in Adult Subjects With INI1-Negative Tumors or Relapsed/ Refractory Synovial Sarcoma	Malignant Rhabdoid Tumors (MRT); Rhabdoid Tumors of the Kidney (RTK); Atypical Teratoid Rhabdoid Tumors (ATRT); Selected Tumors With Rhabdoid Features; Synovial Sarcoma; INI1-negative Tumors; Malignant Rhabdoid Tumor of Ovary; Renal Medullary Carcinoma; Epithelioid Sarcoma	Drug: Tazemetostat	2
NCT02821507	Sirolimus and Cyclophosphamide in Metastatic or Unresectable Myxoid Liposarcoma and Chondrosarcoma	Conventional Chondrosarcoma; Myxoid Liposarcoma; Mesenchymal Chondrosarcoma; Dedifferentiated Chondrosarcoma	Drug: Sirolimus and cyclophosphamide	2
Myxoid Liposarcoma				
NCT02106312	Dose Reduction of Preoperative Radiotherapy in Myxoid Liposarcomas	Myxoid Liposarcoma of Soft Tissue	Radiation: Dose reduction of preoperative radiotherapy in MLS	2
NCT01957709	Recombinant Interferon Gamma in Treating Patients With Soft Tissue Sarcoma	Myxoid Liposarcoma; Round Cell Liposarcoma; Synovial Sarcoma	Other: Laboratory Biomarker Analysis; Biological: Recombinant Interferon Gamma	

Continued

Table 2 Ongoing Clinical Trials for Soft-Tissue Sarcomas—cont'd

Clinical Trial Number	Study	Condition	Intervention	Phase	Special Considerations
NCT01710176	Localized High-Risk Soft Tissue Sarcomas Of The Extremities And Trunk Wall In Adults: An Integrating Approach Comprising Standard Vs Histotype-Tailored Neoadjuvant Chemotherapy	Localized High-risk Soft Tissue Sarcomas of the Extremities and Trunk Wall in Adults	Drug: Epirubicin 60 $mg/m^2/day$ (days 1 and 2) and ifosfamide 3 $g/m^2/day$ (days 1, 2, and 3); Drug: Gemcitabine 900 mg/m^2 (days 1 and 8) and docetaxel 75 mg/m^2 (day 8); Drug: Trabectedin 1.3 mg/m^2; Drug: High-dose ifosfamide 14 g/m^2, given in 14 days; Drug: Etoposide 150 $mg/m^2/day$ (days 1, 2, and 3) and ifosfamide 3 $g/m^2/day$ (days 1, 2, and 3); Drug: Gemcitabine 1800 mg/m^2 (day 1) and dacarbazine 500 mg/m^2 (day 1)	3	
NCT02275286	Phase I–II Trial, Multicenter, Open, Exploring Trabectedin Plus Radiotherapy in Soft Tissue Sarcoma Patients	Liposarcoma, Myxoid; Sarcoma, Soft Tissue	Drug: Trabectedin; Radiation: Radiotherapy	1/2	
NCT02609984	Trial of CMB305 and Atezolizumab in Patients With Sarcoma	Sarcoma; Myxoid/Round Cell Liposarcoma; Synovial Sarcoma; Metastatic Sarcoma; Recurrent Adult Soft Tissue Sarcoma; Locally Advanced Sarcoma; Liposarcoma	Biological: CMB305; Biological: Atezolizumab	2	
Angiomatoid fibrous histocytoma					

[a]No new studies found.

energy-dependent molecular machines whose main function is to scan the inner membrane of cells for lipids/drugs and then flip those drugs to the outer cell membrane (reviewed in Ref. 72). In many human cancers, scientists have measured the level of expression of MDR1 and shown that this protein is an important independent prognostic factor that is strongly associated with the response of cancer cells to combination chemotherapy. In cases where there is a high level of MDR1 expression, drug efflux out of the cell is significant increased, the result can be both an intrinsic or acquired resistance to a variety of chemotherapy drugs. In some cases, MDR1 expression is also associated with poor clinical outcome and lower complete remission (CR) rates.[72] Recent clinical trials in a variety of different cancer types have shown promise for a prolonged remission and improved overall survival when the MDR1 protein is inhibited in combination with chemotherapy.

Host factors which may be linked to drug resistance include things like poor absorption into the tissues, rapid metabolism, or excretion of a drug resulting in low serum concentrations, inaccessibility of the tumor to drug due to the location of the tumors and/or the characteristics of the surrounding tissues, or poor tolerance to side effects of any given drug.[70,73] This limits the ability of physicians to prescribe a drug in sufficient amounts to effectively target the cancer cells. The inability to deliver sufficient dosages of a drug to the site of a tumor is particularly troublesome as it may result in a high level of side effects while having little to no effect on the tumor itself. This is common with bulky tumors, with biological agents of high molecular weight and low tissue penetration, or with tumors with unusual features controlling the blood supply to the tumor.

Just as individuals are unique based on their genetic and epigenetic makeup, cancer cells each have unique characteristics that govern their behavior and response to therapy. Genetic mutations are common in all types of cancer cells, and the term epigenetic refers to a variety of different regulatory mechanisms which control gene expression in a cell. Epigenetic changes in cancer cells are able to alter the ability of a cell to respond properly to chemotherapy, and these changes make excellent potential targets for anticancer therapy and are described in more detail later. The genetic and epigenetic characteristics of a tumor are based on both the tissue of origin of the cancer, in this case presumably mesenchymal lineage cells, but also on the pattern of altered gene expression within the various cells within a tumor. This includes both the activation of tumor growth promoting genes called oncogenes and the abnormal suppression of genes called tumor suppressor genes. These are proteins which help to regulate cell growth and division in normal cells, but are lost in cancer cells. Further complicating this is the fact that even within a tumor, there is generally a very high degree of heterogeneity or variability from cell to cell. This translates into a differential response to drugs even within a single tumor and is a key component of the "cancer stem cell hypothesis" as it pertains to tumor recurrence and metastasis.

Epigenetics is the study of heritable changes in gene expression that occur independently of alterations in the primary DNA sequence. Increasingly, epigenetics has been recognized as a key component to the onset and progression of many devastating human diseases including cancer. Gene expression changes occurring during normal development are largely epigenetic. Assuming that the genetic code is faithfully replicated and remains uniform from cell to cell within an individual, epigenetic factors are left to orchestrate the activation and inactivation of genes required for the development of trillions of cells from a fertilized egg. Epigenetic remodeling also plays a key role in both the conversion of adult stem/progenitor cells into terminally differentiated cells and the reprogramming of differentiated cells for tissue repair and regeneration. Epigenetic gene silencing, and associated promoter CpG island DNA hypermethylation, is an alternative mechanism to mutations by which tumor suppressor genes may be inactivated within a cancer cell.[74–80] In addition to this targeted hypermethylation, cancer cells also demonstrate global hypomethylation,[81] as well as significant changes in chromatin structure[82] and the expression of miRNA.[83] These epigenetic changes are prevalent in all types of cancer, and their appearance may precede genetic changes in premalignant cells and foster the accumulation of additional genetic and epigenetic hits.[84] These epigenetically modified genes constitute important categories of tumor suppressor genes including cell cycle regulators, prodifferentiation factors, and antiapoptotic genes,[76] and many of these genes are known to play a role in normal development.[85–87] The dynamic, coordinated regulation of these genes during development suggests that a normal process gone awry might be responsible for instructively silencing these genes in cancer. There is much we can learn from the research done in cancer, and this knowledge, particularly with regards to mechanisms of gene silencing and methodology can be applied to other diseases with suspected links to abnormal epigenomic remodeling.

Abnormal epigenetic changes in cancer also have the distinct advantage of being potentially reversible, and as a result, epigenetic therapies are being actively investigated in both preclinical cell- and animal-based models, as well as in the clinic through clinical trials. Some of the earliest successes associated with epigenetic therapies have been in hematopoietic cancers, and hypomethylating agents which are capable of removing the abnormal DNA methylation described earlier have been shown to be effective in the treatment of relapsed AML after allo-SCT.[88–90] Building upon these successes epigenetic therapies are currently being tested in a variety of solid tumors as well, including soft tissue sarcomas. One of the best benefits of treatment with the DNA-demethylating agent azacytidine when given in low doses is that it is associated with only marginal nonhematopoietic toxicity, and as a result, these types of epigenetic demethylating agents may be considered as possible interventions as either single agents or in combination with other epigenetic therapies to target residual disease and prevent recurrence in cancer patients.[91] The two best-studied hypomethylating agents are the structurally similar nucleoside analogs decitabine and azacitidine, while a variety of histone deacetylase inhibitors are currently being tested in preclinical and clinical models, including Vorinostat (Zolinza), Trichostatin A, Panobinostat, and sodium butyrate.

3.2 Targeting the Cancer Stem Cell

Many groups have suggested that stem cells or early progenitor cells may be the precursors from which cancer cells are derived. All of these studies cite the fact that tumors harbor stem/precursor cell population markers, have heterogeneous (nonuniform, mixed) cell populations among which only a small percentage of cells have tumor regeneration capacity, and are driven by the upregulation of many pathways important to maintenance of normal stem/precursor cells including the Wnt, Hedgehog, and Notch pathways.[92] This cancer stem cell (CSC) hypothesis may be partially responsible for why cures for many types of cancer have remained so elusive, and it is thought that acquiring stem cell like properties may help cancer cells to become resistant to chemotherapy and ultimately metastasize.[92] Drugs, like chemotherapy, which target the rapidly proliferating bulk tumor population may leave the relatively quiescent (resting or even hiding) CSCs untouched and allow for tumor regrowth at a distant point in the future. While front-line therapy may create a "complete response" to the eye, it does not touch the root or "stem cell" of the cancer which invariably will regrow without further targeted treatment. Both normal and malignant stem cells are also known to be comparatively resistant to chemotherapy, and therefore a strong stem cell signature may contribute significantly to the mechanisms associated with chemotherapeutic resistance described earlier.

In 2008, a seminal paper was published by the laboratory of Robert Weinberg in Boston which showed that poorly differentiated aggressive human tumors like hormone receptor negative breast cancers have an embryonic stem cell-like gene expression signature.[93] This research was focused on the hypothesis that cancer cells possess a variety of key traits that are similar to those found in normal stem cells. They found that histologically poorly differentiated tumors show preferential overexpression of genes normally enriched in ES cells, combined with preferential repression of a group of stem cell-related polycomb-regulated genes. In addition, activation targets of Nanog, Oct4, Sox2, and c-Myc are more frequently overexpressed in poorly differentiated tumors. Targeted therapies which are focused on killing CSCs rely on some of the unique characteristics that these tumor cells have in common with normal stem cells. One of the advantages of these types of therapies is that these pathways are generally not activated in the vast majority of cells in the human body, and as a result, side effects are generally manageable.

Self-renewal is a characteristic process of stem cells which is also shared with cancer cells. Normal stem cells must regulate the balance between self-renewal and differentiation. CSCs are generally impaired in their ability to differentiate, and instead just self-renew without any regulatory checks and balances. The balance between self-renewal and differentiation of stem cells is known to be regulated by many signaling pathways, some of which have also been shown to be associated with tumor initiation and progression including the Notch, Shh, BMI1, and Wnt signaling pathways.[94]

One characteristic of CSCs that differentiates them from other normal cells in the tumor is that CSCs have high levels of ABC transporter proteins such as MDR1, described earlier. This high expression of ABC transporter proteins is also associated with several additional stem cell traits including the active exclusion of the fluorescent dyes Hoechst 33342 and Rhodamine 123 from the cells. These cells are often referred to as side population cells based on their detection via a method called flow cytometry. Normal and malignant stem cells also have an active DNA repair capacity and a resistance to apoptosis.

CSCs may be eliminated by selectively targeted therapies against various self-renewal signaling pathways including Notch, Shh, BMI1, and Wnt signaling pathways.[95] LGK974 and DKN-01 are selective Wnt pathway inhibitor being tested in clinical trials for advanced human cancers, while drugs such as RO4929097 and MK0752 are inhibitors of Notch. Both the Wnt pathway and the Notch pathway have been hypothesized to play a role in tumor initiation and progression of many cancer types, including sarcoma.[96,97]

Smoothened (Smo) is a GPCR-like (G protein-coupled receptor) molecule that positively regulates signal transduction of a signaling pathway called Sonic Hedgehog (Hh). This pathway plays a critical role in stem cell biology and function and the development of many organs and tissues, and genetic activation of the Hh pathway is linked to tumorigenesis of many types of cancers, including sarcoma.[98] Because aggressive, soft tissue sarcomas are thought to be enriched for CSCs or tumor-initiating cells (TICs), it has been hypothesized that they would also have upregulated multiple stem cell pathways such as those that have been investigated in other solid tumors with a strong stem cell signature such as triple negative breast cancer (TNBC). For example, researchers found that vascular endothelial growth factor (VEGF) receptors expressed on TNBC cells mediate cell signaling through the VEGF receptor neuropilin-2 (NRP2), and that this signaling is required for tumor initiation. This occurs via a complex signaling mechanism by which NRP2 signaling promotes tumor initiation via stimulation of the $\alpha6\beta1$ integrin, activation of Ras/MEK signaling by focal adhesion kinase-(FAK)-mediated activation, and finally increased expression of the Hedgehog effector GLI1.[99] GLI1 then induces BMI-1, a key stem cell factor, which in turn enhances NRP2 expression and the function of $\alpha6\beta1$.[99] This creates a positive feedback loop with many key players that can be targeted in the clinic including NRP2, FAK, and GLI1. Additional examples of Hh pathway inhibitors currently being evaluated for other cancer types include LDE225, a selective, orally bioavailable Smo antagonist from a novel structural class and has been shown to potentially inhibit Hh- and Smo-dependent proliferation in vivo and Saridegib (IPI-926), an experimental drug candidate undergoing clinical trials for the treatment of various types of cancer which inhibits Smo.[100]

Overall, these stem cell-related signaling pathways represent logical therapeutic targets for patients with soft tissue sarcomas, but in all cases, these approaches need clinical validation of efficacy before they can be used as part of standard therapy for advanced sarcomas.

3.3 Identification and Direct Targeting of Driver Mutations and Their Known Pathway Components

There are two related hypotheses in the cancer research world that are at present the focus of intense research, and when considered together, these concepts are leading to impressive advances in the world of experimental cancer therapeutics. The first is the concept of "driver" vs "passenger" mutations. The second is the concept of oncogenic pathway addiction. Both are leading to rapid advances in the development of novel, targeted biological therapies that are utilized to target tumors in an individual way and with minimal side effects.

Recent advances in next-generation sequencing have allowed multiple labs around the world to begin to sequence the complete genome of individual tumors to gain an understanding of the genomic mutational burden in these cells and to identify novel pathways that may be driving tumor initiation and progression.[101] During the course of these studies, research has shown that individual tumors generally have dozens of individual mutations in their genomes, and yet functional studies suggest that only a limited number of these mutations are actually driving tumor growth, while others are simply passenger mutations that are present, but are not necessarily essential for the malignant characteristics of the tumor. In fact, tumor cells may become so dependent on these driver mutations to facilitate tumor growth that they become addicted to the dysregulation of these pathways. This form of dependence on a limited number of genetic or epigenetic changes in a cell leads to oncogenic pathway addiction, and if the correct pathways can be identified and targeted, long-term remissions and/or cures may be found for many challenging tumor types.[102]

Several clear examples of these concepts have now emerged for challenging tumor types. Now, a classic example is the identification of frequent mutations within the kinase domain of the epidermal growth factor receptor (EGFR) which make certain nonsmall cell lung cancers (NSCLC) heavily dependent on EGF signaling, but also make these tumors extremely sensitive to small molecular receptor tyrosine kinase inhibitors of EGFR.[103,104] This research has made dramatic changes in patient management for NSCLC in the last decade and has produced durable responses in patient who previously lacked good options for therapy.

A variety of novel targeted biological therapies have been suggested as potential strategies for soft tissue sarcomas based on their known, unique molecular, and biochemical properties of these cells. These are often small molecule inhibitors that are expected to have minimal side effects and may be prescribed either alone or in combination with chemotherapeutic agents or each other. Aberrant regulation of the cell cycle is a hallmark of cancer, and multiple mechanisms contribute to the deregulation of the G1-to-S checkpoint.[105,106] These mechanisms include amplification or mutation of the CDK4 and CDK6 genes, amplification of the genes encoding D-type cyclins, and deletion or epigenetic silencing by proximal promoter DNA methylation of the CDKN2A/B genes, which encodes for the INK4 inhibitors p16ink4a and p15ink4b.[107–110] Additionally, the aberrant expression of a variety of oncogenes, growth factors, or their respective growth factor receptors can activate downstream signaling molecules that drive the expression of cyclin D1. The disruption of normal cell cycle regulation is crucial for various oncogenic transformation processes, and data suggest that many cancer cells are dependent on cellular alterations which lead to high CDK4/6 activity. This data are strengthened by multiple studies involving the use of genetically modified mouse strains which further argue for a prominent role of CDK4 in tumor initiation. The loss of CDK4 has also been implicated in the inability of KRasG12-induced lung tumors and c-Myc-induced skin tumors to develop.[111] Collectively, these studies, combined with research showing that most normal tissues can development in the absence of CDK4/6 activity, suggests that this pathway represents a promising therapeutic target for the treatment of a variety of different cancers, including soft tissue sarcomas.[112–115] Several highly selective inhibitors of CDK4 and CDK6 are currently being tested in phase II/III clinical trials against a variety of chemotherapy resistant cancers.[116–118] Palbociclib is the first highly selective inhibitor of CDK4/6 to be tested and approved in humans for use in combination with letrozole for the treatment of postmenopausal women with estrogen receptor (ER)-positive human epidermal growth factor receptor 2 (HER2)-negative advanced breast cancer as an initial endocrine-based therapy for metastatic disease.[118] Palbociclib has been tested in rabdomyosarcoma[119] and liposarcoma harboring elevated CDK4 expression.[120] Recently, palbociclib has entered a phase II trial in patients with advanced CDK4-amplified or well-differentiated liposarcoma.[121]

For EWS the overall survival rate has improved with new therapies, but in advanced stages the outcome is still dismal and fatal. Both genetic and epigenetic molecular alterations have been implicated in EWS and other childhood cancers. While we currently lack safe, effective, thoroughly tested gene therapy approaches which may someday be used to correct for the underlying genetic driver mutations; epigenetic changes are by their nature potentially reversible via therapeutic intervention, and epigenetic strategies may be particularly effective in these types of low mutational load tumors. Studies have found that the polycomb group (PcG) protein EZH2 is upregulated as a downstream effect of EWS-FLI1 fusion in EWS.[122] PcG proteins are transcriptional repressors which are known to play a key role in modifying chromatin and influencing gene expression.[123] Using novel RNA interference strategies in preclinical EWS models, researchers have shown that the inhibition of EZH2 is sufficient to suppressed tumor development and metastasis in vivo. Furthermore, the EWSR1-FLI1 fusion protein has been shown to deregulate several additional key regulators of the epigenome in EWS including a class of bromodomain containing proteins which each include an approximately 110 amino acid protein domain that has been shown to recognize acetylated lysine residues.[124] These residues are frequently located on the N-terminal tails of histones and help control gene transcription. JQ1 is an inhibitor of this class of bromodomain containing proteins which causes the detachment of these proteins from chromatin and the inhibition of gene transcription of a variety of important genes that have been implicated in tumorigenesis including BCL2, MYC, and CDK6. A recent study showing that the use of the bromodomain inhibitor JQ1 significantly blocked in vitro proliferation of different ES cell lines has been published,[124] and interestingly, in this study, researchers observed a strong downregulation of the EWSR1-FLI1 fusion protein itself. The feedback mechanism for this downregulation is unknown, but JQ1 treatment has also been shown to block key stem cell signaling pathways known to be critical for embryonic stem cell signaling and increased apoptosis of cancer cell lines when treated in vitro.

4. DISCUSSION

The development of treatments against sarcomas has been a long process given the relative rarity of these tumor subtypes. It is difficult to recruit sufficient numbers of patients of any given soft tissue sarcoma subtype for large-scale clinical trials of new drugs, and the lack of sufficient numbers of primary patient samples also makes genetic and epigenetic studies of the molecular alterations driving tumor initial and progression challenging. However, continued research in this area is expected to increase our knowledge of these cancers with the ultimate goal of developing novel, targeted molecular therapeutics to address the lack of effective therapies for patients who are not successfully treated by surgical resection. Focusing our efforts on key regulators of malignancy within sarcomas will not only expedite the development of treatment but should also yield effective therapies for a multitude of different sarcoma subtypes, particularly in cases where the molecular alterations may converge on common cellular processes such as those seen in diverse subtypes which all have EWSR1 translocations such as those described in this chapter.

REFERENCES

[1] Farid M, Ngeow J. Sarcomas associated with genetic cancer predisposition syndromes: a review. *Oncologist.* 2016;21(8):1002–1013.

[2] Azuma M, Embree LJ, Sabaawy H, Hickstein DD. Ewing sarcoma protein ewsr1 maintains mitotic integrity and proneural cell survival in the zebrafish embryo. *PLoS One.* 2007;2(10):e979.

[3] Bertolotti A, Melot T, Acker J, Vigneron M, Delattre O, Tora L. EWS, but not EWS-FLI-1, is associated with both TFIID and RNA polymerase II: interactions between two members of the TET family, EWS and hTAFII68, and subunits of TFIID and RNA polymerase II complexes. *Mol Cell Biol.* 1998;18(3):1489–1497.

[4] Rossow KL, Janknecht R. The Ewing's sarcoma gene product functions as a transcriptional activator. *Cancer Res.* 2001;61(6):2690–2695.

[5] Araya N, Hirota K, Shimamoto Y, et al. Cooperative interaction of EWS with CREB-binding protein selectively activates hepatocyte nuclear factor 4-mediated transcription. *J Biol Chem.* 2003;278(7):5427–5432.

[6] Pattenden SG, Simon JM, Wali A, et al. High-throughput small molecule screen identifies inhibitors of aberrant chromatin accessibility. *Proc Natl Acad Sci U S A.* 2016;113(11):3018–3023.

[7] Deloulme JC, Prichard L, Delattre O, Storm DR. The prooncoprotein EWS binds calmodulin and is phosphorylated by protein kinase C through an IQ domain. *J Biol Chem.* 1997;272(43):27369–27377.

[8] Gascoyne DM, Thomas GR, Latchman DS. The effects of Brn-3a on neuronal differentiation and apoptosis are differentially modulated by EWS and its oncogenic derivative EWS/Fli-1. *Oncogene.* 2004;23(21):3830–3840.

[9] Lee J, Rhee BK, Bae GY, Han YM, Kim J. Stimulation of Oct-4 activity by Ewing's sarcoma protein. *Stem Cells.* 2005;23(6):738–751.

[10] Li H, Watford W, Li C, et al. Ewing sarcoma gene EWS is essential for meiosis and B lymphocyte development. *J Clin Invest.* 2007;117(5):1314–1323.

[11] Spahn L, Petermann R, Siligan C, Schmid JA, Aryee DN, Kovar H. Interaction of the EWS NH2 terminus with BARD1 links the Ewing's sarcoma gene to a common tumor suppressor pathway. *Cancer Res.* 2002;62(16):4583–4587.

[12] Tannukit S, Wen X, Wang H, Paine ML. TFIP11, CCNL1 and EWSR1 protein–protein interactions, and their nuclear localization. *Int J Mol Sci.* 2008;9(8):1504–1514.

[13] Yang L, Chansky HA, Hickstein DD. EWS. Fli-1 fusion protein interacts with hyperphosphorylated RNA polymerase II and interferes with serine–arginine protein-mediated RNA splicing. *J Biol Chem.* 2000;275(48):37612–37618.

[14] Chansky HA, Hu M, Hickstein DD, Yang L. Oncogenic TLS/ERG and EWS/Fli-1 fusion proteins inhibit RNA splicing mediated by YB-1 protein. *Cancer Res.* 2001;61(9):3586–3590.

[15] Pellizzoni L, Kataoka N, Charroux B, Dreyfuss G. A novel function for SMN, the spinal muscular atrophy disease gene product, in pre-mRNA splicing. *Cell.* 1998;95(5):615–624.

[16] Young PJ, Francis JW, Lince D, Coon K, Androphy EJ, Lorson CL. The Ewing's sarcoma protein interacts with the Tudor domain of the survival motor neuron protein. *Brain Res Mol Brain Res.* 2003;119(1):37–49.

[17] Knoop LL, Baker SJ. The splicing factor U1C represses EWS/FLI-mediated transactivation. *J Biol Chem.* 2000;275(32):24865–24871.

[18] Bhatt NR, Collins D, Crotty P, Ridgway PF. Prognosis and management of adult wild type gastrointestinal stromal tumours (GISTs): a pooled analysis and review of literature. *Surg Oncol.* 2016;25(3):152–157.

[19] Saab R, Khoury JD, Krasin M, Davidoff AM, Navid F. Desmoplastic small round cell tumor in childhood: the St. Jude Children's Research Hospital experience. *Pediatr Blood Cancer.* 2007;49(3):274–279.

[20] Lee YS, Hsiao CH. Desmoplastic small round cell tumor: a clinicopathologic, immunohistochemical and molecular study of four patients. *J Formos Med Assoc.* 2007; 106(10):854–860.

[21] Lal DR, Su WT, Wolden SL, Loh KC, Modak S, La Quaglia MP. Results of multimodal treatment for desmoplastic small round cell tumors. *J Pediatr Surg.* 2005;40(1):251–255.

[22] Speleman F, Delattre O, Peter M, Hauben E, Van Roy N, Van Marck E. Malignant melanoma of the soft parts (clear-cell sarcoma): confirmation of EWS and ATF-1 gene fusion caused by a t(12;22) translocation. *Mod Pathol.* 1997;10(5):496–499.

[23] Lae ME, Roche PC, Jin L, Lloyd RV, Nascimento AG. Desmoplastic small round cell tumor: a clinicopathologic, immunohistochemical, and molecular study of 32 tumors. *Am J Surg Pathol.* 2002;26(7):823–835.

[24] Bisogno G, Ferrari A, Rosolen A, et al. Sequential intensified chemotherapy with stem cell rescue for children and adolescents with desmoplastic small round-cell tumor. *Bone Marrow Transplant.* 2010;45(5):907–911.

[25] Bisogno G, Roganovich J, Sotti G, et al. Desmoplastic small round cell tumour in children and adolescents. *Med Pediatr Oncol.* 2000;34(5):338–342.

[26] Cummings OW, Ulbright TM, Young RH, Dei Tos AP, Fletcher CD, Hull MT. Desmoplastic small round cell tumors of the paratesticular region. A report of six cases. *Am J Surg Pathol.* 1997;21(2):219–225.

[27] Stuart-Buttle CE, Smart CJ, Pritchard S, Martin D, Welch IM. Desmoplastic small round cell tumour: a review of literature and treatment options. *Surg Oncol.* 2008;17(2):107–112.

[28] Eaton SH, Cendron MA. Primary desmoplastic small round cell tumor of the kidney in a 7-year-old girl. *J Pediatr Urol.* 2006;2(1):52–54.

[29] Karavitakis EM, Moschovi M, Stefanaki K, et al. Desmoplastic small round cell tumor of the pleura. *Pediatr Blood Cancer.* 2007;49(3):335–338.

[30] Wang LL, Perlman EJ, Vujanic GM, et al. Desmoplastic small round cell tumor of the kidney in childhood. *Am J Surg Pathol.* 2007;31(4):576–584.

[31] Ryan A, Razak A, Graham J, et al. Desmoplastic small round-cell tumor of the pancreas. *J Clin Oncol.* 2007;25(11):1440–1442.

[32] Doros L, Kaste SC, Rodriguez-Galindo C. Sister Mary Joseph's nodule as presenting sign of a desmoplastic small round cell tumor. *Pediatr Blood Cancer.* 2008;50(2):388–390.

[33] Albano EA, Kanter J. Images in clinical medicine. Sister Mary Joseph's nodule. *N Engl J Med.* 2005;352(18):1913.

[34] Philippe-Chomette P, Kabbara N, Andre N, et al. Desmoplastic small round cell tumors with EWS-WT1 fusion transcript in children and young adults. *Pediatr Blood Cancer.* 2012;58(6):891–897.

[35] Thway K, Noujaim J, Zaidi S, et al. Desmoplastic small round cell tumor: pathology, genetics, and potential therapeutic strategies. *Int J Surg Pathol.* September 12, 2016; http://dx.doi.org/10.1177/1066896916668637.

[36] Forlenza CJ, Kushner BH, Kernan N, et al. Myeloablative chemotherapy with autologous stem cell transplant for desmoplastic small round cell tumor. *Sarcoma.* 2015;2015:269197.

[37] Visvader JE, Lindeman GJ. Cancer stem cells in solid tumours: accumulating evidence and unresolved questions. *Nat Rev Cancer.* 2008;8(10):755–768.

[38] Rosen JM, Jordan CT. The increasing complexity of the cancer stem cell paradigm. *Science.* 2009;324(5935):1670–1673.

[39] Suva ML, Riggi N, Stehle JC, et al. Identification of cancer stem cells in Ewing's sarcoma. *Cancer Res.* 2009;69(5):1776–1781.

[40] Goodman KA, Wolden SL, La Quaglia MP, Kushner BH. Whole abdominopelvic radiotherapy for desmoplastic small round-cell tumor. *Int J Radiat Oncol Biol Phys.* 2002;54(1):170–176.
[41] Burchill SA. Ewing's sarcoma: diagnostic, prognostic, and therapeutic implications of molecular abnormalities. *J Clin Pathol.* 2003;56(2):96–102.
[42] Bacci G, Longhi A, Ferrari S, Mercuri M, Versari M, Bertoni F. Prognostic factors in non-metastatic Ewing's sarcoma tumor of bone: an analysis of 579 patients treated at a single institution with adjuvant or neoadjuvant chemotherapy between 1972 and 1998. *Acta Oncol.* 2006;45(4):469–475.
[43] Huang M, Lucas K. Current therapeutic approaches in metastatic and recurrent ewing sarcoma. *Sarcoma.* 2011;2011:863210.
[44] Esiashvili N, Goodman M, Marcus Jr RB. Changes in incidence and survival of Ewing sarcoma patients over the past 3 decades: surveillance epidemiology and end results data. *J Pediatr Hematol Oncol.* 2008;30(6):425–430.
[45] Gupta AA, Pappo A, Saunders N, et al. Clinical outcome of children and adults with localized Ewing sarcoma: impact of chemotherapy dose and timing of local therapy. *Cancer.* 2010;116(13):3189–3194.
[46] Grier HE, Krailo MD, Tarbell NJ, et al. Addition of ifosfamide and etoposide to standard chemotherapy for Ewing's sarcoma and primitive neuroectodermal tumor of bone. *N Engl J Med.* 2003;348(8):694–701.
[47] Arnaldez FI, Helman LJ. New strategies in Ewing sarcoma: lost in translation? *Clin Cancer Res.* 2014;20(12):3050–3056.
[48] Kelleher FC, Thomas DM. Molecular pathogenesis and targeted therapeutics in Ewing sarcoma/primitive neuroectodermal tumours. *Clin Sarcoma Res.* 2012;2(1):6.
[49] Hustu HO, Holton C, James Jr D, Pinkel D. Treatment of Ewing's sarcoma with concurrent radiotherapy and chemotherapy. *J Pediatr.* 1968;73(2):249–251.
[50] Pinkel D. Cyclophosphamide in children with cancer. *Cancer.* 1962;15:42–49.
[51] Rosen G, Wollner N, Tan C, et al. Proceedings: disease-free survival in children with Ewing's sarcoma treated with radiation therapy and adjuvant four-drug sequential chemotherapy. *Cancer.* 1974;33(2):384–393.
[52] Wunder JS, Paulian G, Huvos AG, Heller G, Meyers PA, Healey JH. The histological response to chemotherapy as a predictor of the oncological outcome of operative treatment of Ewing sarcoma. *J Bone Joint Surg Am.* 1998;80(7):1020–1033.
[53] Chung EB, Enzinger FM. Malignant melanoma of soft parts. A reassessment of clear cell sarcoma. *Am J Surg Pathol.* 1983;7(5):405–413.
[54] Enzinger FM. Clear-cell sarcoma of tendons and aponeuroses. An analysis of 21 cases. *Cancer.* 1965;18:1163–1174.
[55] Fletcher JA. Translocation (12;22)(q13–14;q12) is a nonrandom aberration in soft-tissue clear-cell sarcoma. *Genes Chromosom Cancer.* 1992;5(2):184.
[56] Punnett HH, Halligan GE, Zaeri N, Karmazin N. Translocation 10;17 in clear cell sarcoma of the kidney. A first report. *Cancer Genet Cytogenet.* 1989;41(1):123–128.
[57] Reeves BR, Fletcher CD, Gusterson BA. Translocation t(12;22)(q13;q13) is a nonrandom rearrangement in clear cell sarcoma. *Cancer Genet Cytogenet.* 1992;64(2): 101–103.
[58] Antonescu CR, Nafa K, Segal NH, Dal Cin P, Ladanyi M. EWS-CREB1: a recurrent variant fusion in clear cell sarcoma—association with gastrointestinal location and absence of melanocytic differentiation. *Clin Cancer Res.* 2006;12(18): 5356–5362.
[59] Sara AS, Evans HL, Benjamin RS. Malignant melanoma of soft parts (clear cell sarcoma). A study of 17 cases, with emphasis on prognostic factors. *Cancer.* 1990;65(2): 367–374.

[60] Ogura K, Fujiwara T, Beppu Y, et al. Extraskeletal myxoid chondrosarcoma: a review of 23 patients treated at a single referral center with long-term follow-up. *Arch Orthop Trauma Surg*. 2012;132(10):1379–1386.

[61] Nishio J, Iwasaki H, Nabeshima K, Naito M. Cytogenetics and molecular genetics of myxoid soft-tissue sarcomas. *Genet Res Int*. 2011;2011:497148.

[62] Thway K, Fisher C. Angiomatoid fibrous histiocytoma: the current status of pathology and genetics. *Arch Pathol Lab Med*. 2015;139(5):674–682.

[63] Crago AM, Dickson MA. Liposarcoma: multimodality management and future targeted therapies. *Surg Oncol Clin N Am*. 2016;25(4):761–773.

[64] Turc-Carel C, Limon J, Dal Cin P, Rao U, Karakousis C, Sandberg AA. Cytogenetic studies of adipose tissue tumors. II. Recurrent reciprocal translocation t(12;16)(q13; p11) in myxoid liposarcomas. *Cancer Genet Cytogenet*. 1986;23(4):291–299.

[65] Smith S, Reeves BR, Wong L. Translocation t(12;16) in a case of myxoid liposarcoma. *Cancer Genet Cytogenet*. 1987;26(1):185–186.

[66] Bode-Lesniewska B, Frigerio S, Exner U, Abdou MT, Moch H, Zimmermann DR. Relevance of translocation type in myxoid liposarcoma and identification of a novel EWSR1-DDIT3 fusion. *Genes Chromosom Cancer*. 2007;46(11):961–971.

[67] Pan ST, Li ZL, He ZX, Qiu JX, Zhou SF. Molecular mechanisms for tumour resistance to chemotherapy. *Clin Exp Pharmacol Physiol*. 2016;43(8):723–737.

[68] Mitra A, Mishra L, Li S. EMT, CTCs and CSCs in tumor relapse and drug-resistance. *Oncotarget*. 2015;6(13):10697–10711.

[69] Kathawala RJ, Gupta P, Ashby Jr CR, Chen ZS. The modulation of ABC transporter-mediated multidrug resistance in cancer: a review of the past decade. *Drug Resist Updat*. 2015;18:1–17.

[70] Tomasetti C. Drug resistance. *Adv Exp Med Biol*. 2014;844:303–316.

[71] Wijdeven RH, Pang B, Assaraf YG, Neefjes J. Old drugs, novel ways out: drug resistance toward cytotoxic chemotherapeutics. *Drug Resist Updat*. 2016;28:65–81.

[72] Mahadevan D, Shirahatti N. Strategies for targeting the multidrug resistance-1 (MDR1)/P-gp transporter in human malignancies. *Curr Cancer Drug Targets*. 2005;5(6):445–455.

[73] Alfarouk KO, Stock CM, Taylor S, et al. Resistance to cancer chemotherapy: failure in drug response from ADME to P-gp. *Cancer Cell Int*. 2015;15:71.

[74] Jones PA, Laird PW. Cancer epigenetics comes of age. *Nat Genet*. 1999;21:163–167.

[75] Feinberg AP, Tycko B. The history of cancer epigenetics. *Nat Rev Cancer*. 2004;4(2):143–153.

[76] Herman JG, Baylin SB. Gene silencing in cancer in association with promoter hypermethylation. *N Engl J Med*. 2003;349(21):2042–2054.

[77] Jones PA, Baylin SB. The fundamental role of epigenetic events in cancer. *Nat Rev Genet*. 2002;3(6):415–428.

[78] Hahn WC, Counter CM, Lundberg AS, Beijersbergen RL, Brooks MW, Weinberg RA. Creation of human tumour cells with defined genetic elements. *Nature*. 1999;400(6743):464–468.

[79] Aaltonen LA, Peltomaki P, Leach FS, et al. Clues to the pathogenesis of familial colorectal cancer. *Science*. 1993;260(5109):812–816.

[80] Kinzler KW, Vogelstein B. Cancer-susceptibility genes. Gatekeepers and caretakers. *Nature*. 1997;386:761–763.

[81] Feinberg AP, Vogelstein B. Hypomethylation distinguishes genes of some human cancers from their normal counterparts. *Nature*. 1983;301(5895):89–92.

[82] Baylin SB, Jones PA. A decade of exploring the cancer epigenome—biological and translational implications. *Nat Rev Cancer*. 2011;11(10):726–734.

[83] Kasinski AL, Slack FJ. Epigenetics and genetics. MicroRNAs en route to the clinic: progress in validating and targeting microRNAs for cancer therapy. *Nat Rev Cancer*. 2011;11(12):849–864.

[84] Jair KW, Bachman KE, Suzuki H, et al. De novo CpG island methylation in human cancer cells. *Cancer Res.* 2006;66(2):682–692.
[85] Gregorieff A, Clevers H. Wnt signaling in the intestinal epithelium: from endoderm to cancer. *Genes Dev.* 2005;19(8):877–890.
[86] Furukawa Y. Cell cycle control genes and hematopoietic cell differentiation. *Leuk Lymphoma.* 2002;43(2):225–231.
[87] Burch JB. Regulation of GATA gene expression during vertebrate development. *Semin Cell Dev Biol.* 2005;16(1):71–81.
[88] Pusic I, Choi J, Fiala MA, et al. Maintenance therapy with decitabine after allogeneic stem cell transplantation for acute myelogenous leukemia and myelodysplastic syndrome. *Biol Blood Marrow Transplant.* 2015;21(10):1761–1769.
[89] Kim DY, Lee JH, Park YH, et al. Feasibility of hypomethylating agents followed by allogeneic hematopoietic cell transplantation in patients with myelodysplastic syndrome. *Bone Marrow Transplant.* 2012;47(3):374–379.
[90] de Lima M, Ravandi F, Shahjahan M, et al. Long-term follow-up of a phase I study of high-dose decitabine, busulfan, and cyclophosphamide plus allogeneic transplantation for the treatment of patients with leukemias. *Cancer.* 2003;97(5):1242–1247.
[91] Burke MJ, Lamba JK, Pounds S, et al. A therapeutic trial of decitabine and vorinostat in combination with chemotherapy for relapsed/refractory acute lymphoblastic leukemia. *Am J Hematol.* 2014;89(9):889–895.
[92] Borah A, Raveendran S, Rochani A, Maekawa T, Kumar DS. Targeting self-renewal pathways in cancer stem cells: clinical implications for cancer therapy. *Oncogenesis.* 2015;4:e177.
[93] Ben-Porath I, Thomson MW, Carey VJ, et al. An embryonic stem cell-like gene expression signature in poorly differentiated aggressive human tumors. *Nat Genet.* 2008;40(5):499–507.
[94] Jordan CT. Cancer stem cell biology: from leukemia to solid tumors. *Curr Opin Cell Biol.* 2004;16(6):708–712.
[95] Massard C, Deutsch E, Soria JC. Tumour stem cell-targeted treatment: elimination or differentiation. *Ann Oncol.* 2006;17(11):1620–1624.
[96] Matushansky I, Maki RG, Cordon-Cardo C. A context dependent role for Wnt signaling in tumorigenesis and stem cells. *Cell Cycle.* 2008;7(6):720–724.
[97] Rota R, Ciarapica R, Miele L, Locatelli F. Notch signaling in pediatric soft tissue sarcomas. *BMC Med.* 2012;10:141.
[98] Kelleher FC, Cain JE, Healy JM, Watkins DN, Thomas DM. Prevailing importance of the hedgehog signaling pathway and the potential for treatment advancement in sarcoma. *Pharmacol Ther.* 2012;136(2):153–168.
[99] Goel HL, Pursell B, Chang C, et al. GLI1 regulates a novel neuropilin-2/alpha6beta1 integrin based autocrine pathway that contributes to breast cancer initiation. *EMBO Mol Med.* 2013;5(4):488–508.
[100] Tremblay MR, Lescarbeau A, Grogan MJ, et al. Discovery of a potent and orally active hedgehog pathway antagonist (IPI-926). *J Med Chem.* 2009;52(14):4400–4418.
[101] Vogelstein B, Papadopoulos N, Velculescu VE, Zhou S, Diaz Jr LA, Kinzler KW. Cancer genome landscapes. *Science.* 2013;339(6127):1546–1558.
[102] Baylin SB, Ohm JE. Epigenetic gene silencing in cancer—a mechanism for early oncogenic pathway addiction? *Nat Rev Cancer.* 2006;6(2):107–116.
[103] Paez JG, Janne PA, Lee JC, et al. EGFR mutations in lung cancer: correlation with clinical response to gefitinib therapy. *Science.* 2004;304(5676):1497–1500.
[104] Lynch TJ, Bell DW, Sordella R, et al. Activating mutations in the epidermal growth factor receptor underlying responsiveness of non-small-cell lung cancer to gefitinib. *N Engl J Med.* 2004;350(21):2129–2139.
[105] Hanahan D, Weinberg RA. The hallmarks of cancer. *Cell.* 2000;100(1):57–70.

[106] Hanahan D, Weinberg RA. Hallmarks of cancer: the next generation. *Cell.* 2011;144(5):646–674.
[107] Gil J, Peters G. Regulation of the INK4b-ARF-INK4a tumour suppressor locus: all for one or one for all. *Nat Rev Mol Cell Biol.* 2006;7(9):667–677.
[108] Malumbres M, Barbacid M. Cell cycle kinases in cancer. *Curr Opin Genet Dev.* 2007;17(1):60–65.
[109] Malumbres M, Barbacid M. Is Cyclin D1-CDK4 kinase a bona fide cancer target? *Cancer Cell.* 2006;9(1):2–4.
[110] Malumbres M, Sotillo R, Santamaria D, et al. Mammalian cells cycle without the D-type cyclin-dependent kinases Cdk4 and Cdk6. *Cell.* 2004;118(4):493–504.
[111] Miliani de Marval PL, Macias E, Rounbehler R, et al. Lack of cyclin-dependent kinase 4 inhibits c-myc tumorigenic activities in epithelial tissues. *Mol Cell Biol.* 2004;24(17):7538–7547.
[112] Gallorini M, Cataldi A, di Giacomo V. Cyclin-dependent kinase modulators and cancer therapy. *BioDrugs.* 2012;26(6):377–391.
[113] Dickson MA. Molecular pathways: CDK4 inhibitors for cancer therapy. *Clin Cancer Res.* 2014;20(13):3379–3383.
[114] Carnero A. Targeting the cell cycle for cancer therapy. *Br J Cancer.* 2002;87(2): 129–133.
[115] Shapiro GI. Cyclin-dependent kinase pathways as targets for cancer treatment. *J Clin Oncol.* 2006;24(11):1770–1783.
[116] Logan JE, Mostofizadeh N, Desai AJ, et al. PD-0332991, a potent and selective inhibitor of cyclin-dependent kinase 4/6, demonstrates inhibition of proliferation in renal cell carcinoma at nanomolar concentrations and molecular markers predict for sensitivity. *Anticancer Res.* 2013;33(8):2997–3004.
[117] Sausville EA. Cyclin-dependent kinase modulators studied at the NCI: pre-clinical and clinical studies. *Curr Med Chem Anticancer Agents.* 2003;3(1):47–56.
[118] Finn RS, Dering J, Conklin D, et al. PD 0332991, a selective cyclin D kinase 4/6 inhibitor, preferentially inhibits proliferation of luminal estrogen receptor-positive human breast cancer cell lines in vitro. *Breast Cancer Res.* 2009;11(5):R77.
[119] Saab R, Bills JL, Miceli AP, et al. Pharmacologic inhibition of cyclin-dependent kinase 4/6 activity arrests proliferation in myoblasts and rhabdomyosarcoma-derived cells. *Mol Cancer Ther.* 2006;5(5):1299–1308.
[120] Zhang YX, Sicinska E, Czaplinski JT, et al. Antiproliferative effects of CDK4/6 inhibition in CDK4-amplified human liposarcoma in vitro and in vivo. *Mol Cancer Ther.* 2014;13(9):2184–2193.
[121] Dickson MA, Tap WD, Keohan ML, et al. Phase II trial of the CDK4 inhibitor PD0332991 in patients with advanced CDK4-amplified well-differentiated or dedifferentiated liposarcoma. *J Clin Oncol.* 2013;31(16):2024–2028.
[122] Richter GH, Plehm S, Fasan A, et al. EZH2 is a mediator of EWS/FLI1 driven tumor growth and metastasis blocking endothelial and neuro-ectodermal differentiation. *Proc Natl Acad Sci U S A.* 2009;106(13):5324–5329.
[123] Laugesen A, Hojfeldt JW, Helin K. Role of the polycomb repressive complex 2 (PRC2) in transcriptional regulation and cancer. *Cold Spring Harb Perspect Med.* September 1, 2016;6(9). http://dx.doi.org/10.1101/cshperspect.a026575. pii: a026575.
[124] Jacques C, Lamoureux F, Baud'huin M, et al. Targeting the epigenetic readers in Ewing sarcoma inhibits the oncogenic transcription factor EWS/Fli1. *Oncotarget.* 2016;7(17):24125–24140.

PART V

Leukemias: Molecular Changes and Treatment Strategies

CHAPTER NINE

Molecular Changes During Acute Myeloid Leukemia (AML) Evolution and Identification of Novel Treatment Strategies Through Molecular Stratification

E. Karjalainen, G.A. Repasky[1]
Institute for Molecular Medicine Finland FIMM, Nordic EMBL Partnership for Molecular Medicine, University of Helsinki, Helsinki, Finland
[1]Corresponding author: e-mail address: gretchen.repasky@fimm.fi

Contents

Progress in Molecular Biology and Translational Science, Volume 144
ISSN 1877-1173
http://dx.doi.org/10.1016/bs.pmbts.2016.09.005

Abstract

Acute myeloid leukemia (AML) is a hematopoietic malignancy characterized by impaired differentiation and uncontrollable proliferation of myeloid progenitor cells. Due to high relapse rates, overall survival for this rapidly progressing disease is poor. The significant challenge in AML treatment is disease heterogeneity stemming from variability in maturation state of leukemic cells of origin, genetic aberrations among patients, and existence of multiple disease clones within a single patient. Disease heterogeneity and the lack of biomarkers for drug sensitivity lie at the root of treatment failure as well as selective efficacy of AML chemotherapies and the emergence of drug resistance. Furthermore, standard-of-care treatment is aggressive, presenting significant tolerability concerns to the commonly advanced-age AML patient. In this review, we examine the concept and potential of molecular stratification, particularly with biologically relevant drug responses, in identifying low-toxicity precision therapeutic combinations and clinically relevant biomarkers for AML patient care as a way to overcome these challenges in AML treatment.

1. ACUTE MYELOID LEUKEMIA

Acute myeloid leukemia (AML) is an aggressive and rapidly progressing hematopoietic malignancy that originates from the myeloid progenitor cells in the bone marrow. Normally these progenitor cells differentiate into myeloid lineage blood cells, mainly granulocytes and monocytes. Due to genetic aberrations in AML, the normal hematopoietic differentiation process is impaired and cells proliferate uncontrollably. These abnormal immature cells, blasts, rapidly accumulate in the bone marrow and blood stream, eventually resulting in disruption of normal hematopoiesis, bone marrow failure, fatal infections, bleeding, and organ infiltration.[1]

AML is the most common adult leukemia, occurring in approximately 4 out of 100,000 people.[2] The 5-year survival rate is 40–50% for patients under age 60 and less than 10% for the elderly and high-risk patients with secondary AML arising from preceding myeloproliferative disorders or due to cytotoxic chemotherapy.[1] Secondary, AML has especially poor prognosis associated with multidrug resistance.[1]

AML incidence is clearly tied to age, with the average AML patient being older than 60 years.[2] Age relatedness has been associated with accumulation of somatic mutations and epigenetic drift in the hematopoietic stem cell/progenitor cell (HSPC) compartment.[3,4] The standard care for AML consists of first line chemotherapy, mainly cytarabine (araC) followed by anthracycline to achieve remission, defined as a significant reduction in the leukemic burden.[5] Thereafter, second line treatment options include bone marrow transplantation or consolidation therapy, often consisting of additional courses of intensive chemotherapy.[5] Although the majority of AML patients under 60 years old are initially responsive to the standard care conventional cytotoxic chemotherapy and achieve disease remission, long-term disease-free survival remains poor, and the majority relapse with the emergence of chemoresistance.[1] Furthermore, the older and high-risk AML patients are most often refractory to standard treatment or ineligible to receive conventional chemotherapy due to its harsh toxic nature.[1] Thus, older AML patients are left without effective treatment options resulting in intolerably high mortality rates among older patients.[1] Undeniably, the need for novel treatment options, especially low-toxicity ones, is dire.

The greatest challenge in AML care is disease heterogeneity. According to textbooks, AML is justly described as an assortment of several discernible disease subtypes. The classical AML subtyping scheme, the French–American–British system, is based on differentiation state-related blast morphology and cytochemistry and recognizes eight subtypes.[1] Later, when the genetic landscape of AML had been extensively charted, the World Health Organization (WHO) guidelines recognized the high prognostic value of several molecular markers such as recurrent genetic aberrations (see Table 1) and implemented them in the subtyping system.[1] This system created a minimum of 17 recognizable subclasses of AML.

Even though the current subclassification system has contributed greatly to defining patient prognostics it has very limited clinical usage and patient benefit, since in most cases the standard therapy is still similar for all subtypes. However, efforts for stratifying AML patients into clinically relevant groups with specific effective treatment strategies are ongoing, and with no doubt, the breakthroughs in understanding the epigenetic-, clonal-, and cell maturation architecture of AML will contribute significantly to AML care in the future. Indeed, to appreciate AML heterogeneity fully, one needs to consider both inter- and intrapatient differences on the genetic, epigenetic, clonal, and maturational level that lie at the root of treatment failure, relapse, and the emergence of drug resistance.

Table 1 Most Frequent Genetic Aberrations in AML

Genetic Alteration	Prevalence %	Functional Class/Mutation Type
Point Mutations		
NPM1	27	Multifunctional nucleolar/GOF, LOF[6]
FLT3	28	Activated signaling/GOF[7]
DNMT3A	26	Chromatin modifying/DN[8]
IDH1 or *IDH2*	20	Chromatin modifying/GOF[9]
NRAS or *KRAS*	12	Activated signaling/GOF[10]
RUNX1	10	Myeloid transcription factor/LOF[11]
TET2	8	Chromatin modifying/LOF[12]
TP53	6–8	Tumor suppressor/LOF[13]
CEBPA	6	Myeloid transcription factor/LOF[14]
Oncogenic Gene Fusions		
t(15;17); (*PML-RARA*)	9	Activated signaling fusion
t(8;21); (*AML1-ETO*)	4–10	Myeloid transcription factor fusion
inv(16); (*CBFB-MYH11*)	5	Myeloid transcription factor fusion
Chromosomal Losses/Deletions		
Loss of 5 or del(5q)	7–8	–
Loss of 7 or del(7q)	10	–

Prevalence refers to frequency of occurrence as percentage of AML cases. Prevalence percentage and functional classification are based on publications by Papaemmanuil *et al.* (2016) and Cancer Genome Atlas Research Network, 2013.[15,16] Abbreviations: *LOF*, loss of function; *GOF*, gain of function; *DN*, dominant negative.

2. CHALLENGES IN AML TREATMENT STEM FROM DISEASE HETEROGENEITY

2.1 The Genomic Landscape in AML

Compared to most other adult cancers, AML carries a low mutational load.[15] Based on the Cancer Genome Atlas Research Network study of 200 de novo AML patients, on average, only 13 somatic mutations were found in gene coding sequences per AML sample.[15] The number of mutations varies between individual patients and most likely contains passenger mutations as well as leukemia driver genes. Clearly, the mutational load itself does not explain the vast disease heterogeneity in AML; what matters is the wealth of different combinations of mutations that translate into different characteristics of leukemic cell function and thus, treatment responsiveness and resistance.

Roughly half of all AML patients harbor an abnormal karyotype, with large-scale chromosomal rearrangements.[15] Many of these aberrations are recurrent, well-characterized events, such as the translocation *t*(15;17) leading to expression of the oncogenic fusion protein promyelocytic leukemia–retinoic acid receptor alpha (*PML-RARα*) that occurs in 10% of AML cases and defines a distinct AML subtype called acute promyelocytic leukemia (APL) with specific clinical features and distinct treatment options.[1] Also, many of the genetic aberrations represent balanced translocations or inversions that lead to distinct gene fusions and as follows, the expression of oncogenic fusion proteins.[1] Based on their cytogenetic profile, AML patients can be divided into risk categories that are valuable in predicting the response to standard AML therapy.[1]

In addition to large chromosomal level aberrations, several recurring smaller-scale mutations with leukemic potential have been defined, especially in the AML cases with normal karyotype. Since several of these mutations have significant prognostic value, they have been incorporated into the AML subtype classification system and the risk stratification scheme, for example, the three most recurrent mutations in AML: Fms-like tyrosine kinase 3 (*FLT3*), nucleophosmin 1 (*NPM1*), and DNA (cytosine-5-)-methyltransferase 3 alpha (*DNMT3A*) that each occur in approximately 30% of AML cases (Table 1). Overall, the genetic aberrations observed in AML patients can be divided into functional categories, as introduced by the Cancer Genome Atlas Research Network,[15] having a putative role in AML leukemogenesis: transcription factor fusions (18%), NPM1 encoding genes (27%), tumor suppressor genes (18%), DNA-methylation-related genes (44%), activated signaling genes (59%), chromatin-modifying genes (30%), myeloid transcription factor genes (22%), spliceosome complex genes (14%), and cohesin-complex genes (13%). The most frequent genetic aberrations in AML are listed in Table 1 and some described in further detail in Section 3.3.

AML genomes most often harbor more than one driver mutation.[15] Since the prognostic effect of a driver mutation can be altered by the presence or absence of another mutation,[16] the interplay of mutations is extremely important in determining the course of AML. Despite the vast variety of AML-associated mutations, there seems to exist a set of rules, rather than a random assortment, governing the mixing-and-matching of the mutations in patients.[15,17,18] For example NPM1, tumor suppressor protein 53 (*TP53*), or Runt-related transcription factor 1 (*RUNX1*, encoding a protein typically referred to as AML1 protein or core-binding factor alpha 2) mutations seem to be mutually exclusive, and *NPM1*, *FLT3*, and *DNMT3A* mutations cooccur more frequently than expected.[3,18] Also, preferences for comutations have been observed, not only in gene level but in mutational hotspots within genes as well.[16] For example, the *NPM1* mutation was shown to occur preferentially with mutations encoding $NRAS^{G12/13}$, but not $NRAS^{Q6116}$. The rules of mixing-and-matching might be related to the sequence of mutation acquisition. Specifically, the first lesion enhances or decreases the probability of the second mutation, or the combination of lesions confers synergistic, strong effects on driving leukemogenesis.[17] Understanding the functional or evolutionary link between frequently cooccurring mutations might prove very useful for charting new targetable vulnerabilities in AML cells.

Distinct comutational patterns have emerged as important predictors of prognosis and clinical outcome and will have a profound effect on the genetics-based classification of AML in the future.[16] The recent study by Papaemmanuil *et al.* (2016) of >1500 AML patients provides the most extensive basis for genetic classification and prognosis of AML as of today, identifying 11 nonoverlapping subgroups of AML based on comutational patterning.[16] The largest subgroup was *NPM1*-mutated AML (27%) with the majority also carrying mutations in DNA-methylation or hydroxymethylation genes. Mutations encoding regulators of RNA-splicing, transcription, and chromatin modification define the second largest subgroup (18%), and the third largest subgroup (13%) was identified by the presence of *TP53* mutations and/or chromosomal aneuploidy. The clinical implications of this genomic classification scheme and molecular subgroups for AML as well as how the genetic landscape in AML will contribute to clinical treatment must be further validated in prospective clinical trials.

In addition to the aforementioned recurrent mutations in AML, there is an even wider spectrum of nonrecurring rare mutations, and their roles in disease development and progress are poorly defined. It has been suggested that many of these mutations occur already in hematopoietic stem cells before the actual initiation of leukemia and thus might be passenger mutations not crucial for disease development; however, they might present unlikely but eligible roads to AML leukemogenesis.[4]

2.2 Epigenetic Dysregulation Appears Early in AML Pathogenesis and Contributes to Disease Plasticity

A strong connection between aging and changes in DNA methylation, resulting in a diverging epigenomic landscape, or "epigenetic drift" has been observed.[19] Moreover, from the initiating steps through disease evolution, epigenetic dysfunction is a common occurrence in AML and is demonstrated by aberrant methylation as well as mutations affecting chromatin modification and organization.[15,20] For example, in a study of over 300 AML patients a set of genes was identified to be aberrantly methylated in a substantial portion of patients.[20] Moreover, specific methylation patterns have been shown to be strongly associated with distinct genetic aberrations.[15,20] In addition, genome-wide DNA-methylation profiling identified novel subtypes of AML with distinct methylation patterns with prognostic significance.[20]

Epigenetic allele burden varies during the course of AML evolution and is linked to survival.[21] In particular, leukemic cells show plasticity in epigenetic patterning where epigenetic diversification may occur over time and through exposure to external stimuli.[21] It is possible that genetic and epigenetic heterogeneity can occur independently with distinct kinetics affecting the disease progression in different ways.[21]

2.3 Clonal Evolution and Aberrant Cell Differentiation Create Subclonal Heterogeneity in AML

2.3.1 *Clonal Evolution in AML*

AML is a prime example of clonal evolution in cancer, an iterative process of clonal cell expansion which includes genetic and epigenetic diversification and clonal selection, requiring dynamic adaptability, and resulting in a complex clonal architecture.[3] Clonal evolution of AML is thought to begin with one founding cell clone with high leukemic potential that can give rise to distinct subclones harboring additional mutations.[3] In fact, more than half of the AML cases have at least one, and up to five, subclones in addition to the founding clone.[15] AML subclones have been proven to differ in their functional properties, and the diversity and dominance of different clonal cell populations can vary during the progression of AML, suggesting that the development of improved treatment strategies requires a better understanding of these clonal cell populations.[3,22,23]

As illustrated by Ding *et al.*, there is evidence for both linear and clonal, branching evolution in AML.[22] Specifically, whole-genome sequencing of eight AML samples was performed before and after relapse to characterize the changes in clonal subpopulation structure. In linear evolution, the leukemia-founding cell population acquired new mutations to enhance fitness in the face of chemotherapy. In branching evolution, a distinct subclone survived therapy either by harboring or acquiring mutations, conferring a survival advantage. However, the emergence of an entirely mutationally distinct AML clone during chemotherapy is extremely rare.[4,15,23,24]

A factor that clearly affects the evolutionary course of AML is chemotherapy itself. Although eradicating leukemic cells, DNA damaging, cytotoxic chemotherapy can act as a mutagen and possibly contribute to disease relapse by changing the spectrum of somatic single nucleotide variants, thereby generating new mutations that boost clonal evolution (see Section 4.1.2).[22] Competition exists among cooccurring subpopulations. Chemotherapy can add selective pressure, favoring the survival of subpopulations with certain characteristics, or reset the relationships between different clonal populations allowing minor subclones to outcompete populated ones.[3] Although it is clear that the functionally heterogeneous subclonal structure is a risk factor for relapse, the effects of small molecules, i.e., drug compounds, on clonal relationships in AML remains to be determined. A better understanding of both the development of clonal populations as well as their interplay might enable the design of targeted treatment strategies, perhaps in combination, that modulate these relationships for the benefit of the patient. In the future, treatment decisions most likely will be based on detailed information of the clonal architecture already in diagnostic as well as later disease stages.

2.3.2 Impaired Cell Differentiation in AML

AML is characterized by impaired cell differentiation; however, the specific stage at which differentiation is halted varies among patient cases. It is clear that to understand AML one has to understand the process of normal myeloid hematopoiesis (Fig. 1) and how the process can go awry. Normal myeloid hematopoiesis starts with rarely dividing, pluripotent hematopoietic stem cells (HSC) residing in the bone marrow. HSCs are an extremely rare cell population that can give rise to either new HSCs or immature, but committed, multipotent progenitor (MPP) cells. According to the classical dichotomous model of hematopoiesis, the MPPs can commit to either the lymphoid or myeloid lineage, although revised models of hematopoiesis propose more complex lineage relationships, reviewed by Nimmo *et al.*[25] Cells committed to the myeloid lineage are called common myeloid

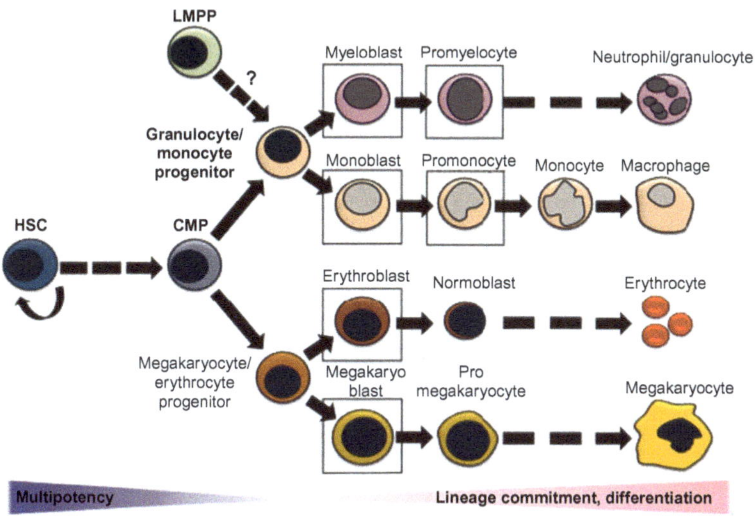

Fig. 1 A simplified schematic representation of myeloid hematopoiesis. *Boxed differentiation stages* represent the variability in stages of AML differentiation blockade. Abbreviations: *HSC*, hematopoietic stem cell; *CMP*, common myeloid progenitor; *LMPP*, lymphomyeloid-restricted progenitors.

precursor (CMP) cells that can further commit to erythrocytic, megakaryocytic, monocytic, or granulocytic differentiation. The CMPs are highly proliferative cells, regulated by different external and internal signals, through successive stages of differentiation until they become a specific type of mature, nondividing blood cells.[26] A schematic representation of the myeloid hematopoiesis is shown in Fig. 1.

The specific hematopoietic transcription factors have a crucial role in hematopoietic lineage commitment, and the network of protein interactions and patterns of gene expression regulating these factors is complex. The orderly expression cascade of transcription factors induces lineage commitment through autoregulation, inhibition of alternative pathways, and activation of lineage-specific genes resulting in a largely irreversible progression through distinct differentiation stages.[26] For example, transcription factor AML1 is involved in the differentiation of almost all hematopoietic lineages, whereas factors such as PU.1, GATA-binding protein 1 (GATA1), and CCAAT/enhancer-binding protein-α (C/EBPα) have lineage-specific expression patterns.[26] Disrupted myeloid differentiation can be commonly traced to these hematopoietic transcription factors, frequently mutated or deregulated in AML, and each can be associated with a differentiation block at a specific cell differentiation stage.[26] Still, the specific cause for impaired cell differentiation cannot be easily determined in all cases, owing, for example, to the complexity of epigenetic dysregulation in AML (see Section 2.2).

The transition from one differentiation stage to another is accompanied by tremendous change, ranging from epigenetic and transcriptomic to the functional level.[26] Such change is easily demonstrated by the lineage- and differentiation stage-specific functionality, signaling capacities, and drug responsiveness of hematopoietic cells.[27] The leukemic population often is a mix of cells at various differentiation stages; the leukemic population can retain a hierarchical structure somewhat similar to normal hematopoiesis (Fig. 1).[28] Thus, the leukemic cell population can be vastly functionally heterogeneous. In addition, specific subclones might attain differences in their ability to differentiate into more mature cells creating even more complexity in the leukemic population structure.[23]

The functional consequence of a mutation is dependent on the differentiation stage of the cell. This relationship is clearly observed in AML, since the characteristics of leukemic cells are defined by the interplay of the transcriptional environment imposed by differentiating cells, as well as the acquired leukemogenic mutations. For example, inducing the *RUNX1-ETO* (Runt-related transcription factor (RUNX1/AML1)-Eight-Twenty One oncoprotein (ETO)/MTG8) translocation, a well-known genetic aberrancy in AML, at specific stages of blood cell development failed or succeeded in inducing leukemogenesis, depending on the transcriptional network within target cells, in other words, depending on the cell differentiation stage.[29] Furthermore, the stage of cell differentiation at which the leukemic driver mutation is acquired and the cell differentiation stage at which the selective advantage of the mutation is apparent, might not always be one and the same.[3] To simplify, a mutation acquired at an early differentiation stage might only have functional consequences at later maturational stages. This observation implies that selective forces could drive cells with the same mutational background but different maturational states to diverse evolutionary pathways.[3]

Genetic variability, epigenetic plasticity, clonal complexity, and differentiation stage variance mold into vast functional heterogeneity that current chemotherapies fail to confront and erase. Still, perhaps not all of the leukemic cell populations observed in a patient possess the potential to initiate leukemia anew. Thus, specifically targeting the leukemia initiating cells could lessen the chances of relapse. In cancer, relapse is often attributed to the existence of cancer stem cells (CSC), a subpopulation of tumor cells capable of limitless self-renewal.[3] The existence of leukemic stem cells (LSCs) in AML is a controversial subject. While the LSC model is supported by experiments in T-acute lymphoblastic leukemia (T-ALL), the evidence for LSC-mediated relapse in AML is still ambiguous.[30] Also, current evidence suggests that a latent preleukemic state, in which the healthy hematopoietic stem or progenitor cells acquire somatic mutations but can still maintain differentiation, precedes a full-blown leukemia.[30] This preleukemic cell population might also represent the source of relapse. Be it the dominant leukemic population, the rare LSCs or the preleukemic clones that initiate leukemia anew after chemotherapy, there is lot to gain in solving the identities and means of chemoresistance in these relapse-initiating cells in AML.

2.4 Disease Heterogeneity Underlines the Need for Precision Medicine and Novel Targeted Chemotherapy in AML Based on Stratified Patient Responses

To summarize the previous sections, the failure of standard-of-care AML therapy to provide cures, for the reasons outlined earlier, along with the lack of low-toxicity treatment options for elderly AML patients begs the development of novel treatment strategies. One solution to lessen the toxicity of treatment is to use mechanism based, targeted inhibitors to attack cancer cell-specific vulnerabilities, subjects of the following sections. Furthermore, to tackle the challenge of disease heterogeneity, a combination of these targeted treatments to achieve enhanced treatment efficacy, and lower the likelihood of toxicity may provide logical avenues to improved treatments. Clearly, patients with the same clinical disease respond differently to the same treatments, and we are beginning to better understand the reasons why. Precision medicine aims for improved, more personalized treatments for each patient as a result of a better understanding of the molecular nature of their disease. To this end, we need systematic, integrative approaches to find powerful ways to stratify AML cases by their specific molecular vulnerabilities, and define the most efficient combinations for targeting these vulnerabilities in the hopes these efforts will result in clinically beneficial biomarker-guided treatments.

Although we have examples of how powerful genetic stratification can be for guiding therapies, such as imatinib (Gleevec/Glivac) for (Breakpoint Cluster Region-Abelson murine Leukemia viral oncogene homolog 1) *BCR-ABL1* positive chronic myeloid leukemia (CML), or all *trans*-retinoic acid (ATRA)/ATO therapy for *PML-RARA* positive AML (see Sections 3.1 and 3.3, respectively), in most cases we still lack models and approaches that would allow us to find predictive genetic biomarkers from a complex genetic background. Also, all of the evidence about AML heterogeneity urges us to search for biomarkers not just from genetic information, but also from transcriptomic and epigenetic levels from which predictive biomarkers are even harder to identify with our current capabilities.

Stratifying AML patients according to biologically relevant drug responses and integrating drug–response profiles with molecular profiles, such as exome sequencing and RNA-sequencing, represents a powerful way of exploring new avenues for biomarker-guided therapies. As a specific example approach, the drug sensitivity and resistance testing (DSRT) platform was developed as a research tool to support tailoring individualized treatment options for chemorefractory AML patients.[31] The DSRT platform enables ex vivo drug response testing for primary AML patient samples over a wide array (>500) of different anticancer agents in a 3-day assay. The drug responses are scored using a systematic and quantitative drug sensitivity scoring (DSS; see Section 4.2) model, which allows the responses of individual drugs to be easily compared within a single patient or across multiple patients. Comparing DSS-scored drug response profiles of a set of AML patients allows for functional drug response-based stratification of AML cases. This stratification scheme integrated with patient-wise molecular data and the exploration of drug–target interactions can provide invaluable insights into the basis of drug sensitivity and drug resistance and help discover biomarkers to guide AML therapy. The power of such an approach was demonstrated by the discovery of axitinib, a designated vascular endothelial growth factor receptor (VEGFR) tyrosine kinase inhibitor (TKI) approved originally for the treatment of renal cell carcinoma, as a selective and effective inhibitor of BCR–ABL1-driven CML harboring a specific gatekeeper mutation (T315I) in ABL1.[32]

For targeted and personalized cancer medicine to continue to develop, both research and implementation feasibility efforts are essential. A large part of the potential lies in determining if targeted and personalized cancer treatments will produce the benefits expected.[33] New efforts to create large-scale datasets and databases connecting multiple centers and efforts to look at efficacy across populations are developing and will be important as the clinical utility of drug combination treatment is expanded. For example, in the Targeted Agent and Profiling Utilization Registry (TAPUR) project at 30 sites in the USA, aims to assess, in large scale, if the use of a particular targeted therapy produces an improved outcome. If 25% of patients respond, the drug will be considered useful for the indication. Ideally, these data will then be available for the larger research community to more deeply examine the molecular mechanisms. Similarly, additional projects such as the Genomics, Evidence, Neoplasia, Information, Exchange (GENIE) project with the goal to gather evidence of clinical utility of next-generation sequencing data for personalized medicine for those who fit a certain molecular profile, and especially the Med-C project which aims to determine mutations linked to targeted treatments and patient outcome will be instrumental in building globally accessible knowledge of effective treatments and potential biomarkers of disease.[33] Recently, a pilot clinical trial was established by the University of Chicago in collaboration with the National Cancer Institute (clinical trial identifier: NCT0255171) aiming to study the feasibility of choosing treatment based on a high-throughput ex vivo drug sensitivity assay in combination with mutation analysis for patients with relapsed/refractory acute leukemia. These kinds of efforts are invaluable in aiding implementation of, as much as research in, personalized medicine for cancer.

3. TARGETED CHEMOTHERAPY AND MECHANISM BASED, TARGETED DRUG DISCOVERY IN AML

3.1 Targeted, Mechanism-Based Drug Discovery

As disease targets, most commonly protein enzymes, are now discovered and characterized at a high rate, molecular oncology and targeted, mechanism-based drug discovery seek to study tumors at the molecular level and find small molecule agents that will inhibit the function or expression of these disease targets, thereby lessening disease progression, combating drug resistance, and extending prognosis and patient survival.[34] With the uptick of rational drug discovery in the last 15 years and a better understanding of oncogene addiction, the focus has been on constitutively activated signaling proteins as targets of designed therapeutics. For example, the imatinib (Gleevac/Glivac) targeting the BCR–ABL1 fusion protein, the driver oncoprotein in CML, has transformed CML from a highly lethal into a highly curable disease.[35] Both the success of imatinib as well as the emergence of imatinib resistance have inspired the development of more potent second- and third-generation ABL1-targeting TKIs improving CML care even further (see Section 5.1).[35] In some cases, like that of BCR-ABL1-driven CML, molecular oncology efforts have been highly effective, but in other cases there is either a lack of effect in patients or a rapid development of drug resistance. However, it is clear that targeted, rational drug discovery is here to stay as we understand more and more about the molecular drivers and landscape of cancer including mechanisms of drug resistance.

3.2 Therapeutic Index

The primary goal of targeted drug discovery is to identify agents that are both effective and have few toxic side effects, as indicated by a high therapeutic index (also known as therapeutic window), a ratio of toxic dose over therapeutic dose (LD50/ED50).[36] A key to efficacious therapy, therapeutic index provides an indication of disease cell sensitivity compared to toxic effects on healthy cells. However, selectivity toward malignant cells does not always mean that a drug is not systemically toxic, since much depends on the pharmacokinetics, as well. Therapeutic index relies on the affinity of a drug for its efficacy target (most often the primary and intended protein target; target that, when affected by drug, gives rise to the primary biological response), as well as off-target, binding, and general toxicity.[36] Efforts to functionally annotate bioactivities resulting from drug–target and drug-off–target binding with a common ontology have been lacking previously and are essential to building a consensus on cellular effects of small molecules (Drug Target Commons, personal communications, https://drugtargetcommons.fimm.fi).

3.3 Targeted Chemotherapies in AML

Defining recurrent mutations and characterizing the genomic landscape in AML has opened doors for designing targeted chemotherapies to be added to the arsenal of available AML treatments. These recurrent mutations include the fusion protein PML-RARα, N-Ras, signaling pathways driven by mutated FLT3, as well as frequently affected epigenetic regulation and cell cycle abnormalities in AML (Table 1). The success with targeting specific disease proteins and pathways in AML subtypes has boosted the search and promise of finding novel targeted therapy strategies for AML treatment. In the following sections, we examine the utility of and advances with targeting these most common AML disease-related oncoproteins and signaling pathways.

3.3.1 ATRA Leads the Way in AML-Targeted Therapy

Since AML is characterized by impaired cell maturation, driving cells toward terminal differentiation is one viable therapeutic path and has already proven to be an effective treatment strategy, highlighted by the success of APL, a distinct AML subtype, that was transformed from a highly fatal into a highly curable disease (90% remission rate and 80% cure rate) by affecting differentiation.[37] With APL, a differentiation block can be inferred to a specific genetic translocation that results in the production of the fusion protein PML-RARα and leads to repression of expression of genes essential for myelocytic differentiation.[1,38] This differentiation block was demonstrated in both a mouse model system of APL and a human in vitro system of purified hematopoietic progenitor stem cells in which retroviral vector-mediated expression of PML-RARα, via a nuclear receptor corepressor (N-CoR)-dependent mechanism, resulted in a block in terminal myeloid cell differentiation and differentiation arrest to the promyelocytic stage.[38,39]

PML-RARα-mediated repression of gene expression and cell differentiation can be overcome by ATRA. Specifically, ATRA binds to the retinoic acid-binding domain on the retinoic acid receptor, including that of RARα of the PML-RARα fusion protein. This binding leads to a change of corepressors to coactivators on the gene activation elements of target genes whose expression promotes the differentiation toward mature granulocytes.[40] The discovery of arsenic trioxide (As_2O_3, ATO) as an enhancer of the ubiquitination and as follows, protein degradation of PML-RARα, opened a possibility for combinatorial treatment with ATRA and has further improved the outcome of both ATRA-responsive as well as ATRA-nonresponsive patients.[41] To this day, ATRA/ATO therapy for PML-RARα-positive AML is the only example of successful targeted treatment able to confer long-lasting cures in AML, although several new targeted approaches are actively being investigated and show promising results.

3.3.2 FLT3 Inhibitors as Promising Candidates for AML-Targeted Therapy

As stated in Section 2.1, mutations in the *FLT3* gene are one of the most common genetic aberrancies in AML and confer a particularly poor prognosis.[15,18] The FLT3 receptor protein belongs to the family of class III receptor-tyrosine kinases (RTK) and functions in growth and differentiation of hematopoietic precursor cells. Binding of FLT3 ligand to the FLT3-receptor elicits receptor dimerization on the plasma membrane, leading to transautophosphorylation, and activation of downstream signaling through the Ras/MEK/ERK, PI3K/Akt/mTOR, and STAT-5 pathways, stimulating proliferation and promoting survival of hematopoietic cells.[42–44] Mutations in *FLT3* can be divided into two categories: (1) internal tandem duplications (ITD) in the juxtamembrane domain of the receptor and (2) point mutations in the receptor activation loop of the tyrosine kinase domain (TKD). The ITD mutations are of variable length and occur in approximately 23% of de novo AML patients, whereas TKD mutations, such as D835Y mutation, are found in 7–10% of AML patients.[7,43] These mutant forms of FLT3 are often expressed at higher levels and demonstrate ligand independent, constitutive autophosphorylation, making mutant FLT3 an attractive target for therapeutic inhibition.[7]

RTK inhibitors most often function through competitive inhibition of ATP binding to the target kinase active site. The first RTK inhibitors tested for FLT3 receptor inhibition, sunitinib, sorafenib, midostaurin, and lestaurinib were initially developed against other RTKs often abnormally activated in cancers, e.g., mast/stem cell growth factor receptor (SCFR or c-Kit), platelet-derived growth factor receptor, VEGFR, and Janus kinase 2 (JAK2), but have demonstrated activity against FLT3 kinase activity as well.[7] Although these first-generation FLT3 inhibitors failed to show sustainable responses as monotherapies in phase 1/2 clinical trials for relapsed/refractory AML, they clearly suggested benefits of FLT3 receptor inhibition for patients carrying *FLT3* mutations,[7] prompting the development of the more specific and potent second-generation FLT3 inhibitors.

The farthest advanced second-generation FLT3 inhibitor, quizartinib, is involved in Phases 1–3 clinical trials, both as monotherapy and in combination with standard AML chemotherapy, as well as in maintenance therapy for patients who have gone through hematopoietic transplant.[7] Quizartinib was identified as a potent and selective FLT3 inhibitor and demonstrated to have a long half life in vivo and capacity for sustained FLT3 inhibition, as well as greater monotherapeutic efficacy and lower toxicity compared to first-generation FLT3 inhibitors.[7] These benefits have been apparent also with elderly patients with relapsed/refractory disease, adding to the promise of the drug.[7] Of note, other second-generation FLT3 inhibitors, such as PLX3397 and crenolanib, are also actively investigated in Phases 2 and 3 settings. Despite these advances with second- and third-generation FLT3 inhibitors, selective FLT3 inhibition can be easily circumvented. In fact, the only FLT3 inhibitor to date that has shown significant improvement in AML patient survival in a Phase 3 clinical trial is midostaurin (interim report of ClinicalTrials.gov Identifier NCT00651261), and these effects are thought to be a result of blockage of critical salvage pathway signaling providing midostaurin with breakthrough therapy designation by the FDA.[45]

Acquired resistance to RTK inhibitors is a significant challenge in targeted treatment. This challenge has been marked as well with FLT3 inhibition, as demonstrated by the appearance of secondary TKD mutations during FLT3-targeted therapy.[7] Interestingly, different FLT3 inhibitors seem to give rise to distinct, mostly nonoverlapping spectra of resistance-conferring mutations that retain sensitivity toward other FLT3 inhibitors, thus opening a possibility of using two FLT3 inhibitors in combination to prevent the emergence of FLT3 resistance.[46] In addition, upregulated PI3K/AKT, STAT, and MEK/ERK signaling as a cell intrinsic or stromal-dependent compensatory mechanism has also been associated with resistance to FLT3 inhibitors.[7] For example, Parmar and colleagues used a coculture model to demonstrate that the stroma can abrogate the antiproliferative effect of FLT3 inhibition in primary FLT3–ITD+/CD34+ progenitor blasts.[47] Also, it is important to note that after chemotherapy-induced loss of blood cell generation, systemic homeostasis mechanisms (see Section 4.1) stimulate hematopoiesis in the bone marrow through elevated proliferative signaling which might also affect the success of FLT3 inhibition. Taken together, FLT3 inhibitors have fallen prey to the same problems of drug resistance observed with other RTK inhibitors, necessitating development of effective combination therapies, in particular powered by stratification of patient profiles (see Section 2.4).

3.3.3 Epigenetic Modifiers: Reshaping Epigenetic Dysregulation in AML

The links among epiallele burden, AML outcome, and the wealth of AML driver mutations associated with epigenetic regulation of transcription are a testimony to the importance of exploring epigenetic modifiers as a novel AML treatment strategy. For example, mutations in *DNMT3A*, Tet Methylcytosine Dioxygenase 2 (*TET2*), and isocitrate dehydrogenase (*IDH*) 1/2 are frequent events in AML, associated with epigenetic dysregulation (see Section 2.2), and are of major interest to AML molecular oncology research.[9] As such, several epigenetic regulators have been selected as rational targets for AML treatment, such as inhibitors of DNA methyltransferases, histone deacetylases (HDACs), and histone-binding proteins and specific inhibitors of mutant IDH enzymes.

6–19% of AML patients harbor an *IDH1/2* mutation that confers a change in IDH catalytic activity through an amino acid mutation occurring primarily in residue R132 in IDH1 and R140/R172 in IDH2.[9] Whereas wild-type IDH1/2 are key metabolic enzymes converting isocitrate to α-ketoglutarate (α-KG), the mutant enzymes gain the ability to convert α-KG into an oncometabolite, 2-hydroxyglutarate (2-HG).[9] 2-HG disturbs the α-KG-dependent activities of several histone and DNA methylases, in a competitive manner, resulting in epigenetic dysregulation and, as follows, impaired hematopoietic cell differentiation.[9] Preclinical testing has demonstrated that inhibition of mutant IDH can reverse epigenetic dysregulation and induce cellular differentiation.[48] Prompted by the compelling evidence, the first mutant IDH2-selective inhibitor enasidenib (AG-221) entered Phase 1 clinical trials during 2013, and was followed by ivosidenib (AG-120), targeting mutant IDH1, in 2014. Since then, IDH-selective inhibitors have been involved in several clinical trials with evidence of efficacy and low toxicity.[48]

The hypomethylating agents (HMAs), azacitidine and decitabine are the most studied DNA methyltransferase inhibitors in AML care and have demonstrated benefits in response rate as well as overall survival compared to conventional care regimens, especially in elderly patients.[49] Of particular note to the theme of this chapter, several novel targeted therapies for AML are being tested in combination with a hypomethylating compound (see Section 5.2).[48] Histone deacetylase inhibitors (HDACIs) represent another class of compounds actively investigated in AML. By inhibiting posttranslational deacetylation of DNA-bound histones, HDACIs induce a conformational change in DNA, rendering it more accessible to transcription factor binding. Since, in AML, HDAC expression is frequently deregulated and there is evidence for subtype-specific HDAC-binding patterns, there are ample reasons for use of HDACIs in AML treatment. Also, HDACIs are known to disrupt the DNA damage response through multiple mechanisms, presenting a targetable vulnerability.

Although several HDACIs, such as valproic acid, pracinostat, and panobinostat, have been explored in clinical trials, single-agent responses to HDACIs have proven only moderate and as such, driven forth testing of HDACIs in combination with other anticancer agents.[48] Other interesting AML-relevant epigenetic regulators include the bromodomain and extraterminal (BET) proteins that control gene expression by binding to acetylated lysines in histones.[48] Numerous investigational BET inhibitors have been in or recently entered Phases 1–3 clinical testing. For example, the BET inhibitor birabresib (MK-8628/OTX015) has progressed into Phase 1 clinical trials for relapsed/refractory AML, and preclinical evidence suggests synergism between HDACI panobinostat and FLT3 inhibitors (see Section 5.2).[48] Thus, consideration of drug combinations with HDACIs might hold promise in development of new approaches to AML chemotherapeutic treatment, especially after patient stratification.

3.3.4 The Usual Suspect: Targeting the Canonical RAS Pathway in AML

Mutations in Ras protooncogenes are one of the most frequent genetic aberrancies seen across all cancer types, and as such, Ras proteins and the canonical Ras/Raf/MEK/ERK signaling pathway are among the most studied subjects in the history of cancer research. Ras family proteins belong to the superfamily of small GTPases and act as membrane-bound signal transducers. The three most clinically relevant Ras isoforms are H-Ras, N-Ras, and K-Ras, from which N-Ras has been most frequently associated with hematological malignancies including approximately 10% of AML cases.[10] Several upstream activators of Ras are frequently mutated in AML, for example, the aforementioned FLT3 as well as c-Kit RTK, and the Ras pathway has been implicated in regulating CSC fates.[10] All in all, constitutive phosphorylation of ERK has been observed in more than 50% of AML cases, giving weight to the importance of Ras-mediated signaling in AML, although much is still unclear about the genetic basis and degree of Ras-dependency in AML.[10]

Direct Ras targeting has proven challenging due to the small size of Ras proteins and their high affinity for GTP and GDP. Attempts to inhibit Ras processing have also failed to show ample clinical value.[10] Thus, much effort has gone into developing inhibitors for other Ras/Raf/MEK/ERK signaling components, making it the most targeted pathway in cancer. Despite the wealth of targeted compounds already available, the overall success with targeting the Ras pathway in AML has been moderate, at most.[10] In our experience with ex vivo drug sensitivity testing of primary AML samples, inhibition of Raf, MEK, and ERK mostly results in cytostatic responses and only rarely exhibits cytotoxicity (unpublished data).

Only a few Ras pathway inhibitors have been involved in clinical testing for AML. For example, in a Phase 2 study of advanced relapsed/refractory AML, selumetinib, a non-ATP-competitive inhibitor of the mitogen-activated protein kinase kinase (MEK) 1/2 demonstrated only modest single-agent antileukemic activity, and none of the *NRAS* mutant cases responded to MEK1/2 inhibition.[50] In addition, patients carrying the FLT3–ITD mutation were nonresponsive to MEK inhibition, although Ras pathway activation has been previously associated with such mutations.[50] A recent study where the potent MEK1/2 inhibitor trametinib was tested as a single-agent treatment with relapsed/refractory AML patients reported a slightly higher response rate with patients harboring Ras mutations; however, the responses failed to translate into a survival benefit.[51]

Attesting to the moderate success of Ras pathway inhibitors as monotherapies, most are involved in clinical trials as combinations. Nevertheless, the Ras pathway is seen as a valuable target for AML treatment. Although Ras may be considered one of the most studied "kings of oncoproteins," the complex protein interactions and feedback mechanisms involved in the regulation of the Ras signaling pathway in AML have to be elucidated before we can reap the full benefits of targeting the Ras signaling pathway.

3.3.5 Rationale for DNA Damage Repair Inhibitors and Cell Cycle Inhibitors in AML-Targeted Therapy

It is known that several recurrent chromosomal translocations typical for AML (*AML-ETO1*, *PML-RARA*, *MLL*-fusions), as well as specific point mutations (e.g., in *FLT3*, *NPM1*), are linked to defects in DNA damage repair (DDR) and replication stress and contribute to DNA damage accumulation, increased mutational load and as such, AML disease progression.[52] Increased DNA damage accumulation activates cell cycle control checkpoints and DNA repair mechanisms that drive leukemic cells toward senescence or apoptosis unless the leukemic cells are able to escape by inactivating checkpoints and/or hyperactivating DDR.[52] These escape mechanisms lead to genomic instability, a mutator phenotype, and confer resistance to genotoxic treatments.[52] The mutator phenotype is associated with a complex karyotype AML that represents one of the poorest prognostic subgroups and is mostly observed in secondary and therapy-resistant AML patients.[52,53]

Although the mutator phenotype with dysfunctional DDR can contribute to disease evolution and chemoresistance, it also bestows an actionable therapeutic opportunity: targeting of the DNA repair machinery. The logic behind the use of DNA repair inhibitors is based on the simultaneous genetic and chemical inhibition of two different pathways of DNA damage response in the cancer cells which represents a road to synthetic lethality that for cancer cells is hard to escape, but would spare normal cells.[52] The synthetic lethality stemming from inhibiting two pathways of DDR was first observed with breast cancer (BRCA) 1/2-deficient BRCAs treated with poly(ADP-ribose) polymerase (PARP) inhibitors.[54] Esposito *et al.* demonstrated similar synthetic lethal effects using PARP inhibitors with AML carrying *RUNX1-ETO* and *PML-RARA* translocations that confer a DDR defective phenotype.[54] With a promise for synthetic lethal effects, PARP inhibitors are in clinical trials for AML treatment.

Aberrant proliferative signaling is essential for cancer cells; however, blocking just one signaling pathway can often be quickly overcome by other parallel signaling routes (see Section 4.1). Ultimately, all of the pathways leading to uncontrolled proliferation will converge in the signals that activate the procession of several unidirectional sequential events leading from G1 to M.[55] Thus, in theory, blocking cell cycle progression should simultaneously annul the effect of all hyperactive proliferative signaling and stop aberrant proliferation. As hyperactivated proliferative signaling through constitutively activated kinases is a common occurrence in AML and the role of cyclins and cyclin-dependent kinases (CDK) has proven important in proliferative and cell differentiation aspects of leukemogenesis, targeting cell cycle components seems like a rational therapeutic strategy.[55]

An example of a cell cycle inhibitor, volasertib, polo-like kinase inhibitor, was granted breakthrough therapy status by FDA for the use with low-dose cytarabine in patients with high-risk AML ineligible for standard therapy.[48] Volasertib blocks spindle formation and induces cell cycle arrest at M phase, and combining volasertib with a low dose of cytarabine resulted in higher response rates and small increases in median event-free survival.[48] As it was shown that *MLL-AF9* (mixed-lineage leukemia (MLL)-myeloid/lymphoid or mixed-lineage leukemia translocated to 3 (MLLT3/AF-9)) translocation driven AML and *FLT3–ITD*-positive AML seem to be associated with oncogenic activation of CDK6,[55] another designated breakthrough therapy compound, CDK4/6 inhibitor palbociclib, is currently being tested in the treatment of MLL-driven AML. Other emerging targets include the Wee1 kinase that catalyzes inhibitory phosphorylation of CDC2/cyclin B kinase resulting in G2 cell cycle arrest. Wee1 was identified as a critical mediator of cell fate and a novel therapeutic target in AML through integrated genomic analysis.[48] Some targets might provide dual benefits through mechanisms of hampering both cell cycle and DDR, hand-in-hand.[55] Thus, the rationale for combining such inhibitors in the treatment of cancer, takes advantage of the convergence of cellular pathways (see Section 4).

3.3.6 Targeting the Regulators of Apoptosis: BH3 Mimetics and Mdm2 Antagonists

As cancer cells are often masters of evading apoptosis, reactivating apoptosis pathways is seen as viable tactic for battling cancer. The mitochondrial apoptosis pathway is regulated by an intricate balance of proapoptotic effectors such as Bak and Bax (Bcl-2 homologous antagonist/killer; apoptosis regulator Bax) and antiapoptotic proteins such as Bcl-2 and Bcl-XL (apoptosis regulator Bcl-2; Bcl-2-like protein 1).[56] The antiapoptotic proteins inhibit apoptosis by sequestering the proapoptotic effectors, thereby preventing the permeabilization of the mitochondrial membrane and the release of cytochrome *c*, the activator of the apoptotic caspase cascade.[56] In AML, defects in apoptosis regulation represent an important chemoresistance mechanism and are associated with a poor survival. Also, previous studies suggest that IDH-mutated leukemic cells are substantially dependent on Bcl-2-mediated survival[56] and recently it was found that ex vivo BCL-2 inhibitor sensitivity is associated with mutations in *WT1* (Wilms tumor 1) and *IDH1/IDH2* and overexpression of specific *HOXA* and *HOXB* (homeobox A and B) gene transcripts, opening a possibility for establishing biomarkers for Bcl-2 sensitive AML patients and giving weight to the feasibility of Bcl-2 targeting in AML.[57]

Several apoptotic regulators interact functionally through their Bcl2 homology domain 3 (BH3) and thus, the BH3 domain presents a targetable vulnerability.[57] Indeed, "BH3 mimetics" or small molecular inhibitors that bind to the BH3 domain of antiapoptotic proteins have been developed for releasing the apoptosis blockade in cancer. The first-generation BH3 mimetics, such as navitoclax, that target a wide range of BH3-containing apoptotic regulators were found to exhibit severe on-target toxicity, especially through inhibition of Bcl-XL.[56] Venetoclax (ABT-199) represents the second generation of BH3 mimetics and was designed to be specific for Bcl-2, but lack affinity for Bcl-XL.[56] Venetoclax has been tested as single-agent treatment in clinical trial with relapsed/refractory AML and demonstrated antileukemic activity, especially among patients with IDH mutations.[56] Venetoclax is currently being explored further in multiple clinical trials.

Another interesting mechanism for inducing cell cycle arrest and apoptosis in leukemic cells involves the p53 tumor suppressor and its negative regulator, E3 ubiquitin-protein ligase Mdm2 (*MDM2*). The tumor suppressor p53 is a major regulator and convergence point for cellular stress signaling, including oncogenic signaling, preventing pathogenic cellular transformation by inducing cell cycle arrest, senescence, and/or apoptosis.[58] Inactivation of p53 is considered to be one of the most common oncogenic transforming events. Intriguingly, only a small fraction of AML patients carry an inactivating p53 mutation, and the loss of p53 function can most often be attributed to defects in the p53 regulatory network, such as overexpression of *MDM2*.[58] Mdm2 binds directly to p53, inhibiting the transcription of p53-regulated genes, and promoting its nuclear export and ubiquitination, leading to its degradation. The recurrent Mdm2-mediated p53 dysfunction in leukemias has marked Mdm2 as a valuable disease target and led to the development of the first inhibitor of p53/Mdm2 interaction, the nutlin.[58] From there on, several nutlin derivatives have been tested in clinical trials, for example, idasanutlin, that has already reached Phase 3 clinical trials (ClinicalTrials.gov Identifier: NCT02545283). Importantly considering the topic of this chapter, idasanutlin in combination with venetoclax, two targeted chemotherapies, is currently in clinical testing for relapsed/refractory AML patients (ClinicalTrials.gov Identifier: NCT02670044).

3.3.7 Targeting MLL-Rearranged AML Through DOT1L and the HOX Pathway

Translocations at locus 11q23 lead to rearrangements of the mixed-lineage leukemia gene (*MLL1*), also known as lysine-specific methyltransferase 2A (*KMT2A*).[59] This translocation event is fairly common in AML, occurring in approximately 5–10% of AML cases and conferring a poor prognosis.[48] Normally, the MLL protein functions as histone H3K4 methyltransferase and is part of a protein complex involved in transcription regulation, specifically the regulation of *HOX* cluster genes in the HSPC compartment.[59] As the dysregulation of *HOX* genes has previously been implicated in driving tumorigenesis, it is not surprising that the *MLL1*-rearrangement contributes to leukemogenesis through direct regulation of the homeobox genes.

The mechanism of transcriptional dysregulation in *MLL*-rearranged AML has been related to elevated levels of methylation in H3K79 (histone 3, lysine 79), promoting active transcription of the MLL fusion protein target genes.[59] DOT1L (disruptor of telomeric silencing 1-like; lysine N-methyltransferase 4/KMT4) is the only known mammalian enzyme for H3K79 methylation and was found to directly interact with several *MLL* translocation partners.[59] In brief, the MLL segment of the fusion protein is directed to its target genes, such as HOX genes, and the MLL translocation partner recruits DOT1L, which promotes gene transcription through lysine methylation. The discovery of this mechanism provided the rationale for targeting DOT1L in *MLL*-rearranged AML. As such, the small molecular inhibitor of DOT1L, EPZ-5676, has been tested in Phase 1 clinical trials for MLL-rearranged relapsed/refractory AML patients.[48] Also, other targeted anticancer agents, such as CDK and BET inhibitors are being tested as disruptors of DOT1L-mediated leukemogenesis.[48]

3.3.8 The Future of Targeted Therapies in AML Lies in Biomarker-Guided Drug Combinations

Although several new, targeted agents are emerging as promising candidates for AML therapy, the reality remains that *all* targeted treatments, except for ATRA/ATO-targeted therapy, for AML have failed to show sufficient efficacy as monotherapies to be approved for clinical use and, as a consequence, are either discarded or explored in combinatorial therapies. Table 2 summarizes the currently approved anticancer agents and combination treatment approaches for AML. Given the formidable disease complexity in AML, the need for combinatorial therapies is not surprising. Nevertheless, if enhanced efficacy and selectivity can be achieved by using drug combinations it will not be enough to overcome the brick wall of AML heterogeneity, meaning no drug combination will prove to be the "silver bullet" that can be used to treat each and every subtype of AML. What is urgently needed for a drug or a drug combination to succeed is a response-predicting molecular biomarker signature to guide the selection of treatment-responsive patients. In the next section, we explore further the biological basis for combination therapies that was highlighted in several of the examples earlier.

4. COMBINATORIAL CONTROL AND DISCOVERY OF RELEVANT DRUG COMBINATIONS

Two major obstacles in cancer chemotherapy are (1) partial, poorly sustained therapeutic responses due to cancer heterogeneity and plasticity and (2) the emergence of drug resistance.[34] Drug combinations or multicomponent drug treatments are seen as a powerful way to enhance treatment responses without increasing toxicity or to overcome the emergence of drug resistance. Thus, systematic approaches to predict and experimentally determine effective drug combinations are essential in order to identify and advance translatable treatment options.[60] To do so, approaches must integrate clinical, genomic, transcriptomic, and proteomic signaling pathway analyses.

Combination therapy is defined as disease treatment with two or more drugs to achieve efficacy with lower doses or lower toxicity drugs, chemosensitize cells so that an additional compound can be more potent, gain additive or synergistic effects, or combat expected acquired resistance or minimize the possibility for development of drug resistance. Successful combinations reach one or more of these goals with low toxicity.[36,61]

Table 2 Currently Approved Anticancer Agents and Commonly Applied Drug Combinations in the Treatment of AML

Nonproprietary Name	Proprietary Name	Mechanism/Target	Class
All *trans*-retinoic acid (ATRA)	Tretinoin	Retinoic acid receptor agonist	Targeted agent
Arsenic trioxide (ATO)	Trisenox	Thioredoxin reductase inhibitor	Conventional chemotherapy
Cladribine	Leustatin	Antimetabolite, purine analog	Conventional chemotherapy
Cyclophosphamide	Clafen/ Cytoxan/ Neosar	DNA-alkylating agent	Conventional chemotherapy
Cytarabine	Cytosar-U/ Tarabine PFS	Antimetabolite, modified cytidine analog	Conventional chemotherapy
Daunorubicin hydrochloride	Cerubidine/ rubidomycin	Topoisomerase II inhibitor	Conventional chemotherapy
Doxorubicin hydrochloride	Adriamycin	Topoisomerase II inhibitor	Conventional chemotherapy
Fludarabine	Fludara	Antimetabolite, purine analog	Conventional chemotherapy
Idarubicin hydrochloride	Idamycin	Topoisomerase II inhibitor	Conventional chemotherapy
Mitoxantrone hydrochloride	Mitozantrone	Topoisomerase II inhibitor	Conventional chemotherapy
Thioguanine	Tabloid	Antimetabolite, modified guanine analog	Conventional chemotherapy
Vincristine sulfate	Vincasar PFS/ Oncovin	Mitotic inhibitor, microtubule depolymerizer	Conventional chemotherapy

Combination Treatment	Clinical Application[a]
Cytarabine + Daunorubicin or Idarubicin	Induction treatment for patients <60 years[b]
Cytarabine + Daunorubicin + Cladribine	Induction treatment for patients <60 years
Cytarabine + Daunorubicin + Etoposide	Induction treatment for pediatric AML

Continued

Table 2 Currently Approved Anticancer Agents and Commonly Applied Drug Combinations in the Treatment of AML—cont'd

Combination Treatment	Clinical Application
ATRA + ATO	Induction/consolidation treatment for APL
ATRA + Cytarabine + Daunorubicin	Induction treatment for APL
Mitomycin + Etoposide + Cytarabine	Treatment for relapsed/refractory patients
Etoposide + Mitoxantrone + Cytarabine	Treatment for relapsed/refractory patients
Cladribine + Cytarabine + Mitoxantrone + Filgrastim	Treatment for relapsed/refractory patients
Fludarabine + Cytarabine + Idarubicin + Filgrastim	Treatment for relapsed/refractory patients

[a]Clinical applications vary highly according to patient risk status and overall fitness and, to some extent, by country. Actual treatment regimens vary with regard to dosing and duration of treatment period.
[b]Classical first line therapy for AML consists of cytarabine as a 7-day continuous infusion with three doses of anthracycline (daunorubicin or idarubicin), commonly referred to as the "7 + 3" regimen.

Contemporary chemotherapeutic treatments have for many years employed several drugs in combination for treatment of most cancers, and combinations of targeted signaling inhibitors with conventional cytotoxic chemotherapy has proven useful, to a limited extent (~20%).[62] As an example, childhood ALL is one of the cancers that has shown dramatic improvement in prognosis during the latter half of the 20th century, with an increase in survival from 10% to 90%. This improvement was due not to development of new therapies but to combinations of existing chemotherapies.[63] In contrast, characterization of the molecular nature of the disease can provide deeper understanding to the rational design of combinations of appropriately targeted signaling inhibitors.[62] Appropriate targeting and discovery of unique treatment regimens will require a stratification of patient responses (see Section 3.3) as well as an understanding of the combinatorial control of biological processes (see Section 4.1) that goes awry in disease.

4.1 Combinatorial Control of Biological Processes

Biological processes are most often controlled through a combination of regulatory factors, events, or activities that follow the principles of redundancy, homeostasis, and multifunctionality.[36] Such combinatorial control of biological processes lends challenges to the efficacy of targeted chemotherapies, even those that are the most selective and potent, and must be considered in the search for effective treatment strategies.

4.1.1 Redundancy

Redundancy, often involved in signal transduction pathways, refers to the existence of compensatory pathways that achieve the same or a similar biological response. Compensatory routes can provide avenues for development of drug resistance and partial response to single-agent treatment. For example, cell growth and proliferation are controlled by numerous growth factor receptor signaling pathways. Inhibition of a single pathway typically fails to halt all proliferative activity.[36,62] Furthermore, inhibition of one signaling molecule or pathway can trigger a compensatory response in another. For example, inhibition of MEK1/2 with the kinase inhibitor trametinib resulted in an increase in phosphorylated fibroblast growth factor receptor substrate 2 (FRS2), a marker of fibroblast growth factor receptor 1 (FGFR1) activation and thus compensatory signaling involving FGFR1 and concomitant drug resistance in a K-Ras mutant lung cancer cell model and patient cells.[64] Genetic or pharmacological approaches to inhibit MEK1/2 and FGFR1 in combination resulted in a decrease in FRS2 phosphorylation and cell death in activated K-Ras-mediated pancreatic and lung cancer models.[64] Thus, compensatory mechanisms of combinatorial control can provide clues to combating adaptive drug resistance.

It is clear that compensatory signaling events increase the challenge of developing effective single target-based treatments, but blocking compensatory, bypass signaling pathways with combinatorial drug treatment could be effective in preventing redundancy-based drug resistance, as in the example earlier. Specifically, the goal then is twofold: (1) determining the redundant, compensatory pathways and (2) finding the right combination to block the potential escape routes without lowering the therapeutic index and increasing the chances for drug resistance. Searches for global changes in phosphoproteomic and gene expression profiles can help to identify compensatory signaling pathways on a large scale.[65] However, using these approaches to determine effective, translatable drug combinations are challenging due to the crosstalk, network-like structure of most signaling events. Therefore, both experimental and computationally predicted approaches are essential to identify useful drug combinations.

4.1.2 Homeostasis

Homeostasis refers to the physiological tendency toward the maintenance of a stable environment controlled internally and coordinately to buffer external changes. For example, metabolic and signaling pathways are commonly regulated by numerous levels of homeostatic control, such as feedback activation and inhibition.[36] An example of feedback control of hematopoietic cell proliferation and homeostasis involves a loop of three components: Trib2, a Tribbles family pseudokinase; C/EBPα, CCAAT enhancer-binding protein alpha; and the E2F1 transcription factor. The altered state of homeostasis established during leukemic cell proliferation and maintenance of leukemogenesis is partially driven by this three component loop. Under normal, noncancerous conditions, Trib2 regulates cell proliferation, survival, and metabolism, but a subset of AML shows elevated Trib2 and loss of C/EBPα-p42,[66] likely because Trib2 inhibits C/EBPα p42. The consequent loss of C/EBPα-p42 is required for Trib2-mediated leukemogenesis.[66] Furthermore, loss of C/EBPα-p42 results in loss of transcription factor E2F1-mediated regulation, and E2F1 appears to regulate Trib2, most likely through a C/EBPα-p42-dependent manner in normal granulocyte/macrophage progenitor cells.[67] Thus, the feedback loop where E2F1-regulated C/EBPα p42 normally attenuates Trib2-mediated control of proliferation, survival, and metabolism is clearly at play in normal hematopoietic cell proliferation and homeostasis, and an altered state is established in leukemic cell proliferation and maintenance of leukemogenesis when Trib2 mediates inactivation of C/EBPα p42.[67]

Both cancer and the chemotherapy designed to combat cancer cause abundant physiological and molecular imbalances. For example, chemotherapy damages mitochondria, which leads to an imbalance of water, ions, minerals, and salts and therefore affects osmotic homeostasis. Resistance to chemotherapy can result from the ability of cancer cells to reestablish homeostasis, compensate to restore the original state, or adapt a new steady state, and thus, homeostatic control is a challenge to developing effective target-based chemotherapies. For example, as described in Section 3.3, the initial antiproliferative effect of FLT3 inhibitors could stimulate hematopoiesis in the bone marrow through elevated proliferative signaling which could, in turn, affect the success of FLT3 inhibition.

4.1.3 Multifunctionality

Multifunctionality refers to the possibility of a single protein to control multiple pathways or cellular processes through one or more of its domain structures and activities.[36] Multifunctionality plays a strong role in combinatorial control of biological processes. In fact, many key factors in pathophysiological signaling control a plethora of pathways and processes. As described in Section 3.3, FLT3 receptor activation stimulates leukemic cell survival and proliferation and controls apoptosis via the activation of Ras/MEK/ERK1, Stat5, and PI3K/Akt downstream signaling pathways, illustrating the classical multifunctionality of RTKs.[42,43] Multifunctionality poses a challenge to developing effective target-based drugs because even the most potent and selective inhibitors, if targeting a multifunctional modality, could be problematic by inadvertently disrupting potentially essential processes and increasing undesirable side effects.

Since biological processes are controlled in combination, perhaps the best approach for tackling the aberrancies in cancer is through development of novel combinations of targeted chemotherapeutics based on stratified patient responses (see Section 3.3). Such a coordinated effort to inhibit aberrant pathway activation, and therefore cellular processes contributing to disease, aims to achieve efficacy while permitting some target activity to offset toxicity due to multifunctional activities. Furthermore, mutant disease targets, as is often the case in drug-resistant disease, narrow the spectrum of combinatorial control challenges, and focus the search for effective chemotherapies.

4.2 Discovery of Relevant Drug Combinations and Evaluation of Synergy and Efficacy

While examining single agents for inhibitory effect is reasonably straightforward with IC50 or area under the curve (AUC) types of measurements, accurately assessing synergy of small molecule combinations is more complex, and numerous models and approaches have been developed and proposed.[36,68–71] On simple consideration, two small molecules that show no interaction in an assay can be thought of as additivity ($1+1=2$). Similarly, a positive interaction or synergy can be considered as $1+1>2$, and a negative interaction or antagonism as $1+1<2$. However, a cellular situation is much more complex, and whether a measurement of synergy, or rather, observable increased efficacy is needed to identify novel combinations for potential further study remains to be seen. In the following sections, we examine the use of reference models, experimental approaches, and the integration of in silico prediction and mathematical approaches with experimental tactics to evaluating synergy and efficacy, and in Section 5, we explore combination therapy with targeted therapeutics as a way forward with AML.

4.2.1 *Measuring the Effect of Drug Combinations With Reference Models*

Methods such as Loewe additivity and Bliss independence are classic toxicological models used in biochemical analyses of two or more agonists of a simple enzymatically stimulated reaction. Reviewed in Tang *et al.*, Greco *et al.*, and Fitzgerald *et al.*, these models provide a numerical measure of additivity, synergy, or antagonism.[36,72,73] The basic assumption in Loewe additivity is that two inhibitors act on one target through the same or a similar mechanism.[70,74] The assumption is that no drug agent can interact with itself and will produce an additive effect if combined with itself.[75] In contrast, the Bliss independence model assumes that two inhibitors act on one target through different, independent mechanisms.[76,77] An attempt to develop consensus terminology for two agent combination effects was broached in 1992 and the "Saariselkä agreement" produced.[69] Specifically, combinations providing stronger responses than predicted result in "synergism," either Loewe or Bliss synergism, if both agents are effective individually according to the respective prediction equations. Furthermore, combination effects that are less than predicted are considered antagonism, either Loewe or Bliss antagonism. Finally, if the combination effect is equal to the prediction of Loewe additivity or Bliss independence the inhibitors are said to be additive. Another way of looking at the models is by comparison of the predicted value based on Loewe's additivity or Bliss independence to the experimental result: no interaction of the inhibitors indicates that the inhibitors match the prediction and the effect is the sum of the two individual inhibitors. If the experimental data show values greater than the predicted, there is a positive interaction or synergism. If the experimental data show less than the prediction, then there is antagonism.[36]

During the last tens of years, a heavy debate on the definitions and applications of drug interaction models has kept the field a lively one. Numerous models in addition to those described briefly above have been proposed, discussed, and debated. Calls for a consensus on "what is an additive effect" and "what is synergy" have even been launched.[68] In addition, a variety of newer methods have been developed. For example, the combination index is a computational method with algorithmic quantitative output potentially widely applicable. As a result of the divergent models, many researchers use their own interpretations of synergism, based on their own needs.

This chapter does not intend to review all models of synergism or to propose a preference. For the readers' interest, these models are reviewed for similarities, differences, strengths, and weaknesses in Greco *et al.* and Tang *et al.*[72,73]

4.2.2 Measuring the Effect of Drug Combinations Experimentally

Most complex cellular processes that are at play in disease states may arguably not be well modeled by the pharmacological predictions of Loewe additivity and Bliss independence. For example, compounds tend to affect multiple pathways in a cellular environment that is not implemented in Loewe- and Bliss-based predictions. The success of drug combinations in cancer treatment has commonly relied on clinical evaluation. For example, in a study of the TKI erlotinib in combination with conventional chemotherapeutic gemcitabine for treatment of advanced pancreatic cancer, the primary endpoint was overall survival,[78] which lacks informative value of the biological mechanisms behind treatment success or failure in vivo. To truly elucidate, the biological endpoints of the drug combination in vivo, it is important to consider relevant biological parameters in addition to standard clinical parameters and establish collaborations between clinicians involved in clinical testing and the biology/pharmacology research. Also, clinical testing is time consuming and costly and, especially in regard to drug combinations, carries a risk of adverse effects due to toxicity. Thus, experimental testing is essential from the stage of identification of novel low-toxic combinations, e.g., through high-throughput drug-screening campaigns, all the way to the clinical translation.

As an example of a high-throughput combinatorial drug-screening effort, Axelrod *et al.* aimed to identify pathways that generate resistance to cytotoxic agents and show synergistic cytotoxicity upon inhibition.[60] Specifically, synergistic cytotoxicity was examined in 14 human tumor derived, epithelial cell lines of three different cancers using 420 drug combinations. Percent synergy was defined by Bliss independence predictions for additivity compared to the actual experimental readout for cytotoxicity. In practice, synergism was defined as approximately 30% above additivity. "Hits" of the screen were determined to show 30–35% synergy. Follow-up demonstrated compensatory crosstalk as well as nodes of convergence on a single signaling molecule. Although no cancer lineage specificity or underlying genetic sensitivity useful for prediction of synergy was observed, this study turned up several synergistic drug combinations, both expected and novel.

As another example, the DSRT platform, introduced in Section 3.3, permits systematic and standardized screening of >500 approved and investigational oncology compounds in dose–response matrix layout in combination with selected targeted therapeutics.[31] This powerful platform, used with patient samples, and primary and established cell cultures provide a valuable approach to identifying new combinations that are active, especially on patient cells. The DSRT platform has been used in large-scale efforts, such as to identify synergistic drug combinations for the treatment of highly lethal and treatment resistant Ras-driven pancreatic ductal adenocarcinoma. The library of >500 drugs was screened in combination with potent and selective inhibitors of Ras/Raf/MEK/ERK and PI3K/AKT/mTOR pathways. Cytotoxicity and cell viability were measured after 3-day combination treatment of 20 cell lines, most derived from patient tumor–xenograft models. Through this screening effort, several potentially relevant drug combinations could be identified and are being explored further (Wennerberg K, Gautam P, personal communication). These studies provide a systematic basis for high-throughput combinatorial drug screening that can be applied to multiple tumor types including patient-derived cells to search for drug combinations and provide valuable information on stratification of samples based on drug responses and sensitivities.

In practical terms, efficacy of a single agent in a bioassay is commonly explored in a number of ways, including (1) concentration of the inhibitor needed to achieve 50% inhibition (IC50), (2) dissociation constant of the target–inhibitor complex (K_i), and (3) area under the dose–response curve (AUC) types of measurements, such as DSS. For example, two drugs given in combination may result in a lower IC50 than either drug alone in a bioassay with cell viability or toxicity as an endpoint (Fig. 2). The assessment of drug combination efficacy most often boils down to comparing the efficacy values of the single agent with that of the combination resulting in (1) no change, (2) additive effect, (3) greater than additive effect = synergistic, and (4) less than additive effect = antagonistic. While IC50 and K_i values are important measures for efficacy they are hard to scale when comparing relative efficacies between different drugs. In particular, K_i is a biochemical measurement and cannot reflect drug sensitivity in a cell-based context. Thus, in large drug-screening efforts it is essential to utilize a parameter that easily allows comparisons.

DSS is a systematic, quantitative output for multiparametric scoring of drug sensitivity of a single agent based on AUC calculations but further normalized to favor compounds that show potency in a wide therapeutic

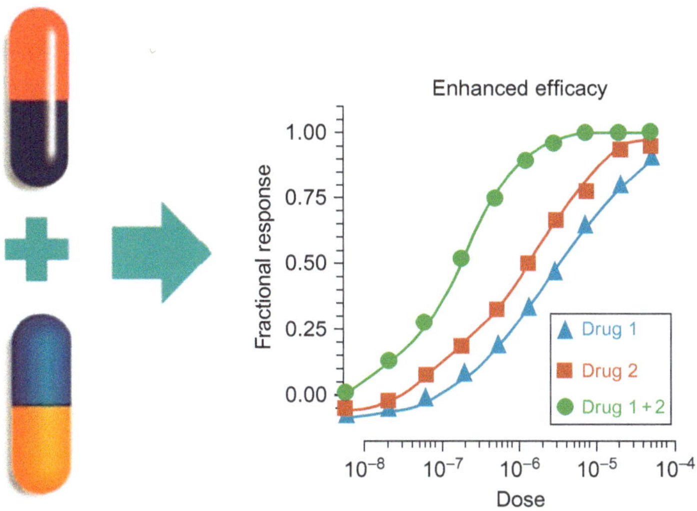

Fig. 2 Potential synergy of drugs in combination. Dose–response curves plotted with mock data demonstrate a synergistic effect of drug 1 and drug 2 in combination. IC50 is reduced dramatically in cases of synergy.

window.[71] One key advantage of DSS, over potency measures such as IC50 (Fig. 2), is that DSS can detect changes in efficacy or maximum response. Furthermore, in addition to providing a single score by which to rank single-agent responses, DSS also provides an easy way to compare responses by studying change in DSS (ΔDSS) that can be used to assess (a) drug combinations, when comparing DSS values between single agents vs combination and (b) drug selectivity, when comparing drug–responses in healthy cells vs malignant cells. The assessment of selectivity becomes increasingly important, since simply comparing efficacies may not be the best way to find clinically relevant drug combinations as we try to demonstrate in the following section.

4.2.3 Drug Efficacy vs Therapeutic Synergy

As summarized in a recent review, it is important to distinguish between drug efficacy and therapeutic synergy. Therapeutic synergy was first and simply defined as the occasion that a combination of drugs produces an improvement in therapy compared to the best single agent alone.[79] Kashif *et al.* revisited the concept of therapeutic synergy by comparing isogenic cancer cell lines, in which one clinically relevant allele had been silenced, and examining in vitro cell culture drug resistance.[70] Through their analyses, they showed that therapeutic synergy is a result of an increase in therapeutic index (see Section 3.2), a ratio of therapeutic dose over toxic dose compared to single agents alone. In short, the most viable drug combination candidates may be those that both show high synergy as well as an increased therapeutic index, considering that a compound may in fact lessen adverse effects.[36] This notion leads us to the concept of combinatorial selectivity.

4.2.4 Combinatorial Selectivity

Combinatorial selectivity refers to achieving treatment efficacy by targeting several disease processes with combinatorial drug treatment and achieving selectivity by leaving each target sufficiently active in normal tissues to minimize toxicity.[36] Ideally, the target processes should be disease-specific, thereby lessening the risk of toxicity toward healthy cells and thus increasing therapeutic index.[36] The element of selectivity can be implemented in in silico models, for example, by predicting disease specificity (see section below) or in experimental approaches by including drug responses from healthy/normal cells/tissues as controls (subsequent section).

4.2.5 Experimental Approaches for Assessing Drug Combinations

The most common experimental approaches for assessing drug combinations rely on cell viability and/or cytotoxicity as the end-point measurement. The response to conventional cytotoxic chemotherapy observed in ex vivo drug sensitivity testing (section above) may not correlate with the response seen in vivo, arguing that the mechanism of action of many types of chemotherapy in vivo is more complex than simple inhibition of cell growth or cell killing and may be mediated through interactions with the immune system and surrounding tumor environment. Thus, it is important to also evaluate other therapeutically relevant drug responses, such as immunomodulatory effects or induced cell differentiation, especially when considering AML.

The study of Zhou *et al.* provides a great example of systematic exploration of drug-induced cell differentiation.[80] In this study, the metastatic BRCA cell line MCF-7 was used in a large-scale drug screen to identify compounds that drive cells from a proliferative to a nonproliferative, differentiated state. The transcriptomes of breast tumor-derived MCF-7 cells were measured at days 1, 3, and 5 posttreatment with identified differentiation-inducing compounds to characterize high-dimensional gene expression trajectories along which the cells move to a differentiated state. The major finding was that drug-induced transcriptome changes diverged after day 1 but converged remarkably at day 5, although the chemical structures and target profiles of the compounds overlapped minimally. To simplify, perturbing MCF-7 cells with nonsimilar compounds led to a similar transcriptomic state but through different transcriptomic trajectories. This study highlights how different drugs may trigger cells to exit a proliferative state and enter a "differentiation state" via different mechanisms, and partial responses may indicate cell population heterogeneity as well as different gene expression trajectories.[80]

In summary, one of the challenges in the search for drug combinations for novel treatment strategies is the assessment of drug interactions and clear identification of relevant compound pairs that include the element of selectivity. Also, considering only cell viability or cell growth will likely result in therapeutically relevant compound pairs being overlooked. In this light, it is not surprising that classical models of assessing synergism and drug interactions fall short in identifying therapeutically relevant combinations that will provide an increased therapeutic window compared to the single agents alone.

4.2.6 Predicting Drug Combinations In Silico

Since there is a wealth of possible combinations of different chemicals, targets and doses, and any screening campaign quickly becomes overwhelming and practically impossible to test experimentally, in silico models are urgently needed to prioritize experimental approaches.[81] Complex genotype/molecular signature-to-phenotype relationships pose challenges for creating in silico models fine-tuned enough to predict efficacy of combinatorial inhibition. Polypharmacological compounds, with their ability to cause an explosion of different primary and off-target effects, pose another challenge.[81] Also, most in silico models utilize large databases as their primary source of knowledge leading to a biased approach where the canonical disease pathways, "the usual suspects," are overrepresented and the global kinetics of disease processes most often are not available. Thus, pathway crosstalk, novel cancer dependencies, and context dependency are often overlooked.[81]

The in silico models that are built upon experimental data on a specific disease model system provide an alternative approach to computational prediction of drug effects. As an example, the NCI-DREAM (Dialogue for Reverse Engineering Assessments and Methods) consortium posed a challenge to computationally predict and rank 91 drug combinations on a scale of most synergistic to most antagonistic based on single-agent drug effects on gene expression in OCI-LY3, human diffuse large B-cell lymphoma cell line.[82] This crowd-sourcing approach encouraged all interested scientists to participate and as an end result, 31 predictive models were admitted and assessed. Many of the generated models predicted synergy significantly more reliably than random predictions. It was also evident that the context dependency of synergy does not allow results to be generalized across cancer types.[82] Still, this study demonstrated the power and utility of crowd-sourcing and community efforts to advance in silico modeling which will no doubt prove very important in the future.

As another example, published earlier this year, Nguyen and colleagues developed a method to predict efficacy of drug combinations by utilizing an *Schizosaccharomyces pombe* gene network for drug resistance.[83] For example, looking at the gene network for doxorubicin resistance, they identified six other drugs acting in the same network. Previously unlinked with doxorubicin, cisplatin was determined to slow growth in doxorubicin-resistant mutants. Thus, drug combination candidates were identified by extent of overlap of fission yeast gene networks along with synthetic lethal sensitivities.[83]

5. COMBINATION CHEMOTHERAPY AS A WAY TO OVERCOME CHALLENGES IN AML TREATMENT

5.1 Examples of Efforts Toward Identification of Drug Combinations

Even though the idea of using drugs in combination for cancer treatment has been around for a long time, there is still a long road to travel before combination strategies will be easily translated to clinical benefits. Combination chemotherapies show promise in both blood- and epithelial-derived cancers, but in most cases the best combinations to-date are those that combine a targeted therapy with traditional cytotoxic chemotherapy (Table 2). There are few clinical studies combining two or more targeted chemotherapies, at least in AML, although the urgent need for low-toxicity treatment options would require such studies. At the moment, the most compelling evidence of combining targeted treatments exists from preclinical studies. In the following section, we briefly describe advances made in CML as well as with solid tumors, as examples of success stories and provide examples of targetable combinatorial opportunities in AML.

5.1.1 Chronic Myeloid Leukemia

CML is a clonal disease of hematopoietic stem cells displaying enhanced production of myeloid precursor cells and is most commonly a result of a reciprocal chromosomal translocation of *t*(9;22), the Philadelphia chromosome.[84] Although initially successful, drug resistance often emerges with first- and second-generation ABL1 kinase inhibitors and leads to the persistence of BCR-ABL1-driven disease with CML progenitor cells.[84] Thus, in addition to examining third-generation ABL1 kinase inhibitors, it is also relevant to look downstream of ABL1, as the primary disease driver and consider targeted therapy combinations. Doing so, Pellicano *et al.*, determined that inhibition of ERK pathway signaling using the ATP noncompetitive inhibitor PD184352 increased the apoptotic effect of the farnesyltransferase inhibitor BMS-214662-induced apoptosis in CD34+ CML progenitor cells.[84] The drug combination appears to have an enhanced effect on all assays examined, so the extent to which promising results such as these are relevant for clinical use remains to be determined.

Combination targeted therapy has been taken as an approach in CML, and although numerous examples can be found, results are slim. For example, combination of the third-generation ABL1 inhibitor nilotinib with the JAK1/2 inhibitor ruxolitinib is in clinical testing in CML and Ph+ALL patients (ClinicalTrials.gov Identifiers: NCT02253277 and NCT01702064). Nilotinib is also under Phase 1/2 consideration in combination with MEK inhibitor MEK-162 (ClinicalTrials.gov Identifier: NCT02225574). With an angle of drug repurposing, a Phase 1 safety study of zileuton, approved for asthma treatment, with imatinib in an approach to target LSCs, which escape imatinib, was terminated due to lack of enrollment, but in 2016 entered safety testing in combination with dasatinib (ClinicalTrials.gov Identifier: NCT01130688 and NCT02047149). As a last example, ABL1 inhibitor dasatinib in combination with the smoothened (SMO) inhibitor BMS-833923 recently completed clinical safety testing and determination of recommended dose for Phase 2 testing (ClinicalTrials.gov Identifier: NCT01218477), and although some participants showed a major cytogenetic response, participant numbers were small, study results are not readily available in the scientific literature, and Bristol–Myers Squibb terminated its SMO inhibitor program, completing the clinical study early. Thus, although some success has been demonstrated, CML patients, especially those with acquired drug resistance mutations are in need of new therapies.

5.1.2 Carcinomas

Successful examples of drug combination therapies or the promise of such from cell-based preclinical experimental models of epithelial solid tumors in pancreatic, ovarian, liver, lung, and gastric carcinomas and melanoma are growing. For example, in patients with local, but advanced, inoperable pancreatic cancer patients, low-dose nucleoside analog gemcitabine in combination with the EGFR-targeted inhibitor erlotinib has been approved as a first line chemotherapy.[78,85,86] Recently, the FDA approved a targeted therapy combination for the treatment of renal cancer: an mTOR inhibitor, everolimus, combined with a VEGFR TKI lenvatinib for second line treatment.[87] Additional examples from ovarian cancer, liver cancer, and lung cancer are well documented.[61,88,89]

B-Raf inhibitors have been particularly successful in targeted chemotherapy development and in combination with MEK and EGFR inhibition, in melanoma and colon cancer, respectively, resulting in longer progression free and overall survival when compared to B-Raf inhibition alone.[90,91] In preclinical testing in activated K-Ras-driven nonsmall cell lung and pancreatic carcinoma cell models, navitoclax, targeting Bcl-2/Bcl-XL, in combination with inhibition of MEK enhanced efficacy compared to MEK inhibition alone, and a Phase 1/2 study has been initiated to examine safety and efficacy in patients (ClinicalTrials.gov Identifiers: NCT02079740 and NCT01989585).[92] Additional examples can be observed with Met inhibitors in combination with other growth factor receptor inhibitors.[93] Genotype-directed high-throughput drug screening for effective combinations of melanoma has demonstrated that prospects for unique treatment regimens can be experimentally determined.[94]

Combination approaches have also been useful for combating multidrug resistance. For example, gastric cancer patients commonly suffer from multidrug resistance resulting in a poor prognosis. Zhang *et al.* demonstrated that treatment of a gastric cancer cell line with the antimutagenic flavonoid naringenin and the ABT-737 Bcl-2 inhibitor in combination resulted in a decrease in cell growth and colony formation as well as initiation of the early molecular stages of apoptotic cell death, decreased Akt activation, and increased p53 expression.[95] Through these examples, it is clear that the concept of targeted chemotherapies in combination may provide new treatment options, and researchers can hopefully parley the successes in carcinomas to look for advances through combination therapy for AML, especially through enhanced patient stratification for efficacy of targeted therapy.

5.2 Targeted Chemotherapeutic Combinations in AML

5.2.1 HMAs in Combinations for AML

As of today, several anticancer agents have been tested in combination with HMAs, most commonly azacitidine and decitabine, to find novel therapeutic options for AML. Azacitidine and decitabine have demonstrated both a low-toxicity profile and suitability for treating AML patients unfit for conventional chemotherapy, as well as reasonable efficacy as a single-agent treatments, making these two HMAs attractive partners for combination therapies.[96] To start, dual epigenetic targeting by combining HDACIs with HMAs was previously seen as a promising combination to explore, considering the mechanistic convergence of these types of inhibitors. Unfortunately, several clinical trials have failed to show the benefits of this "dual epigenetic therapy" in AML, although sequential administration rather than concurrent administration might be the key for successful combination therapy with HMAs and HDACIs.[96] HMAs are also involved in several clinical trials with other targeted anticancer agents. For example, Phase 2 clinical trials have already seen combinations of azacitidine with CD33-targeted chemotherapeutics (GO) and decitabine with the proteasome inhibitor bortezomib.[96] The latter combination was prompted by the report of bortezomib-mediated upregulation of certain microRNAs that downregulate transcriptional activation of several genes relevant to myeloid leukemogenesis.[96] Decitabine has also been tested in combination with the TKI sorafenib and protein kinase inhibitor midostaurin in clinical trials, especially with FLT3-mutated AML.[96] Lastly, sequentially combining azacitidine with lenalidomide, a targeted chemotherapy targeting the E3 ubiquitin ligase, in higher risk AML patients with del(5q) demonstrated feasibility and potential efficacy.[96]

HMAs have also been explored in combination with conventional chemotherapeutics. Azacitidine combined with daunorubicin and a cytosine arabinoside showed no benefit in overall survival but induced increased toxicity with elderly AML patients.[97] Decitabine is currently in Phase 3 trials with sapacitabine, a novel cytotoxic nucleoside analog that showed promising results when used in a sequential manner with decitabine in a lead-in phase of a Phase 3 clinical trial.[56]

5.2.2 Synergism Between FLT3 and HDACIs Stems From Dual Inhibition of FLT3

Preclinical evidence points to possible synergism between HDAC and FLT3 inhibitors.[98–100] HDACs have been implicated in the regulation and activation of nonhistone proteins, for example, heat shock proteins (hsp) that function as protein chaperones. It is known that acetylation of hsp90 inactivates the chaperone function, leading to erroneous folding, and proteosomal degradation of its client proteins, including FLT3.[98] In addition, the HDACI panobinostat has been shown to contribute to FLT3 degradation by upregulating ubiquitin conjugase UBCH8.[100] Thus, HDAC inhibition might carry a dual benefit by reversing epigenetic dysregulation and promoting FLT3 degradation.[98]

5.2.3 Overriding Stromal-Mediated TKI-Resistance by Targeting the AKT and JAK/STAT Axis

The bone marrow stromal environment plays an important role in acquired drug resistance that is frequently associated with targeted chemotherapy, especially TKIs.[101] The stromal influence is exemplified by clinical trial data showing that while malignant cells in the peripheral blood respond strongly to treatment, the response is attenuated in the bone marrow.[101] As a specific example, it was shown that cell-to-cell stromal contact and soluble factors, such as cytokines and increased levels of FLT3 ligand contribute to the lack of cytotoxic responses in early leukemic FLT3–ITD-mutated stem/progenitor cells during TKI treatment.[47,102] Thus, stromal-mediated drug resistance mechanisms represent an important therapeutic target and a logical direction when considering drug combinations.

To this end, Weisberg *et al.* used a cell-based model mimicking stromal protection in a high-throughput chemical screen, setting out to identify kinase inhibitors with the potential to override microenvironment-mediated drug resistance in mutant FLT3-positive AML.[101] Stromal protection was mimicked by using cell medium conditioned by the HS-5 stromal cell line, and the screen itself measured viability of MOLM-13 AML cells after a 2-day incubation with fixed concentrations of single-agent TKIs or combinations of TKIs with FLT3 inhibitors. With this screen, Weisberg *et al.* identified synergistic FLT3 inhibitor/TKI combinations, e.g., a combination of the multikinase targeting drug dasatinib with midostaurin, that overcame stromal-mediated drug resistance and were shown to be selective toward malignant cells over stromal cells. In addition, the results hinted at the importance of JAK/STAT- and AKT-mediated signaling in stromal–leukemic cell interactions. With further investigation, several selective AKT inhibitors were also identified as synergistic partners with midostaurin in FLT3 positive AML.[103]

In addition to stromal-mediated drug resistance to FLT3–ITD inhibitors, wild-type FLT3 hyperactivation through elevated levels of FLT3 ligand also plays an important role for attenuating the effects of FLT3–ITD inhibition and, as such, contributes to drug resistance.[104] AML cells frequently coexpress FLT3 and its ligand, establishing an autocrine or paracrine signaling loop resulting in constitutive FLT3 signaling and AKT activation.[104] A dual inhibitor of FLT3 and AKT, A674563, was recently found to overcome AKT-mediated FLT3 drug resistance.[105] Since there are hints to the importance of AKT signaling in both stromal- and paracrine signaling-mediated drug resistance in *FLT3*-mutated AML, combining AKT and FLT3 inhibitors may provide a powerful treatment strategy for AML in the future.

5.3 Immunotherapeutic Targeted Treatments in AML

5.3.1 Gemtuzumab Ozogamicin as the First CD33-Targeted Therapy for AML

Targeting only cancerous cells, selectivity toward cancer, is a key goal in combinatorial selectivity, but is hard to achieve: cytotoxic chemotherapies knockout all proliferatively active cells. The key for cancer selectivity is defining cancer-specific markers, be it cell-intrinsic processes or external properties such as cell surface markers. The idea of using leukemia-associated antigens (LAA) to increase treatment selectivity has been around for a long time and finally antibody-based immunotherapies are coming strong into the AML treatment front.

The LAA with most clinical experience in AML is CD33 (sialic acid-binding lectin CD33, SIGLEC-3) that is predominantly expressed in myeloid cells and, as such, is an ideal marker for targeting myeloid-derived malignancies-like AML and CML. During the disease course, the malignant blasts can lose the expression of normal myeloid antigens, but it has been approximated that around 90% of AML patients have CD33+ disease, or at least harbor a CD33 positive blast subpopulation.[106] Also, the amount of CD33 expressed on leukemic blasts varies significantly between individual AML patients with the variation being at least somewhat accountable to specific molecular abnormalities. For example, high levels of CD33 are associated with *NPM1* as well as *FLT3/ITD* mutations, while CD33 expression is generally low with core-binding factor translocations.[106,107] Nevertheless, CD33 is a good surface marker for targeting AML blasts as was demonstrated with the first antibody-drug conjugate (ADC) developed for AML treatment, gemtuzumab ozogamicin (GO).

GO is a humanized recombinant monoclonal antibody recognizing CD33 antigen and conjugated to a cytotoxic antibiotic calicheamicin.[108] The mode of action relies on the specific binding of GO to CD33 positive cells and internalization of the ADC. The encouraging results from three Phase 2 clinical trials, 30% complete response rate observed with relapsed AML patients, led to FDA approval of GO in year 2000. Unfortunately, following trials failed to prove an overall efficacy benefit for GO combined with standard therapy when compared to standard therapy alone, and indicated increased fatal toxicity associated with GO treatment. Subsequently, FDA approval for GO was withdrawn in 2010. Nevertheless, GO has been actively involved in clinical trials and after the dose regimen was revised, the use of GO in AML induction therapy was not found excessively toxic and demonstrated somewhat prolonged disease-free survival, at least with patients who have favorable risk cytogenetics.[108]

Thus, it seems that GO is here to stay and perhaps in the future will be added to the available treatment options for AML to the benefit of the right patients. Still, the challenges with GO treatment, (a) dependence on abundant CD33 expression on malignant cells which varies within and between patients, (b) slow internalization of the ADC, and (c) drug transporter activity in AML cells, will most likely require GO to be used in combination with other drugs.[108]

In addition to GO, other LAAs are of interest as potential targeted therapies for use in AML. Specifically, other anti-CD33 ADCs are currently under investigation and some have reached clinical testing. For example, vadastuximab talirine (SGN-CD33A), is currently in a Phase 3 clinical trial to test combination of SGN-CD33A with HMAs in older AML patients (ClinicalTrials.gov Identifier: NCT02785900). Additional targets, such as CD123/IL-3 receptor alpha subunit, are also being explored for use in ADCs. The fact that elevated expression of CD123 on leukemic cells is associated with poor prognosis indicated that CD123 may be a good target for intervention and ADCs have been developed and show promise as targeted therapies.[109–111] A Phase 1 safety study trial will soon begin for SGN-CD123A in relapsed or refractory AML (ClinicalTrials.gov Identifier: NCT02848248). Thus, ADCs appear to hold promise in targeted therapies for AML.

All in all, several ADCs are providing hope for targeted treatment in relapsed/refractory AML with inotuzumab ozogamicin possibly able to induce complete remission and under investigation in combinations (ClinicalTrials.gov Identifiers: NCT01925131, NCT01363297, and NCT01564784) and vadastuximab talirine passing safety testing in a Phase 1 trial and beginning combination testing (ClinicalTrials.gov Identifiers: NCT01902329 and NCT02326584).[112]

5.3.2 Bispecific Antibody AMG 330 Is Being Tested in the Treatment of AML

Bispecific T cell engaging antibodies are fusion proteins consisting of two adjoined single-chain antibody variable fragments, one with specificity toward CD3, belonging to the T cell receptor (TCR) complex, and the other with specificity toward a selected LAA.[113] In short, bispecific antibody molecules can engage both a leukemic cell and a CD3 positive T cell, bringing these two cells together and eliciting T cell activation. In the case of cytotoxic T cells, bispecific antibody-mediated activation leads to T cell cytotoxicity toward the leukemic cell by inducing the formation of a transient cytolytic synapse between the cells, secretion of lysolytic granzyme- and perforin-containing vesicles by T cell that induce the lysis of the leukemic target cell.[113] Normally, only T cells that harbor TCR receptors able to recognize a specific tumor antigen can elicit a cytotoxic response toward the tumor. Bispecific antibody molecules directly engage any cytotoxic T cell thus bypassing the need for TCR antigen specificity,[113] a significant advance and highly modulatory technology.

Inspired by the success of blinatumomab, a CD19/CD3-specific bispecific antibody developed for the treatment of ALL, a bispecific antibody with CD33- and CD3-binding modalities was developed for targeting CD33 positive AML. Preclinical ex vivo studies with AML patient samples and cell lines clearly demonstrated that AMG 330 can effectively recruit T cells and direct T cell cytotoxicity toward AML blasts in low concentrations and without the need for T cell costimulation.[106] Also, AMG 330 was found to prolong the survival in a xenograft mice model transplanted with AML cell line MOLM-13 cells.[114] AMG 330 has already advanced to Phase 1 clinical trials in patients with relapsed/refractory AML but it has become evident that the successful application of bispecific antibody, such as AMG 330 depends on defining the factors that might contribute to clinical response or resistance.

As expected, among the most important determinants of AMG 330 activity are treatment exposure time, AMG 330 dose, and the effector–target ratio (E:T), the ratio between cytotoxic T cells vs malignant blasts.[106] As the normal bone marrow function is endangered in AML, the effector–target ratio will most likely vary from AML patient to patient many fold. For example, a study by Harrington *et al.* of 40 AML patient bone marrow samples found autologous T cell to AML cell ratio to range from 1:2 to as low as 1:>2700.[107] Results from a large multicenter Phase 2 trial of blinatumomab monotherapy suggested that the tumor burden present at the time of therapy may be one critical determinant for the clinical activity of bispecific antibody constructs[115] and this will most probably hold true for AMG 330, as well.

Based on preclinical testing, it is evident that factors beyond dosage and E:T ratio also contribute to the efficacy of bispecific antibody treatment. As mentioned earlier, the amount of CD33 expressed on leukemic blasts varies significantly between individual AML patients and might represent a factor contributing to bispecific antibody responsiveness. The number of surface-expressed CD33 was found to correlate with EC50 values for T cell cytotoxicity when explored by cell lines overexpressing CD33 in various levels.[106] However, when tested in patient samples this dependency was evident only with high AMG 330 doses and E:T ratios,[107] suggesting that CD33 expression levels might not be a critically limiting factor for AMG 330 cytotoxicity. Also, CD33 levels might be chemically enhanced with epigenetic modifiers.[106]

The first and most advanced CD33/CD3-targeting bispecific antibody is AMG 330, but others are also in development, e.g., CD3xCD123-targeting bispecific antibody called DART (dual-affinity retargeting/MGD006), after demonstrating dose-dependent AML cell killing in preclinical analyses and has now entered safety testing in Phase 1 clinical trial in relapsed and refractory AML (ClinicalTrials.gov Identifier: NCT02152956).[116,117]

Normally, T cell activation is modulated by the target cells or other immune cells by the display of several inhibitory and stimulatory signal molecules, for example, activating ligands CD80 and CD86 and inhibitory ligands PD-L1 and 2 (Programmed death ligands 1 and 2; CD274 and CD273).[115] These regulatory signals represent immune checkpoints that in cancer are often dysregulated, favoring cancer immune evasion. As such, it is reasonable to assume that these stimulatory and inhibitory signals might also affect bispecific antibody-mediated T cell activation. It is known that bispecific antibodies are able to elicit potent target cell cytolysis in the absence of stimulatory signaling through T cell coreceptors which in normal situation is needed for sufficient T cell activation.[115]

Although stimulatory signaling might not be a strict requirement for bispecific antibody-mediated T cell activation, still the role of stimulatory and inhibitory signaling might be the factor that tips the scale in favor of successful or unsuccessful clinical response. Bearing this in mind, a study using engineered AML cell lines by Laszlo *et al.* found that overexpression of PD-L1 and PD-L2 reduced the T cell cytotoxic activity of AMG 330, whereas expression of the activating ligands CD80 and CD86 augmented the cytotoxic activity of AMG 330 toward AML cells.[115] Furthermore, in primary AML samples, T cell stimulation through T cell coactivating receptor CD28 increased the activity of AMG 330.[115] Taken together, the data suggest that modulation of bispecific antibody efficacy is possible through T cell activators and inhibitors, signifying a possible role as response-predictive biomarkers as well as modifiable targets to enhance bispecific antibody efficacy. Increasing evidence indicates that such ligands can be induced by various stimuli such as cytokines, HDACIs, DNA methyltransferase inhibitors, and conventional chemotherapeutics.[115] All in all, the drug/antibody-mediated modulation of immune environment to enhance the efficacy of immune therapies will be very important in the future and will, without doubt, add to the successful treatment of AML.

5.3.3 Use of Immune-Based Approaches in Combination Therapy for Novel AML Treatment Strategies

Immune-based approaches to cancer therapy have become one of the most active areas of investigation, and the search for modalities that work in combination with cancer chemotherapy holds high potential.[63] Cytotoxic chemotherapy indirectly affects the immune system in ways that could be exploited with high potential for successful combination treatment regimens. Reviewed in Bracci *et al.*, these mechanisms include directly or indirectly stimulating immune effectors, attenuating immunosuppression, influencing immune cell populations by affecting hematopoietic homeostasis, enhancing tumor cell visibility to the immune system, and indirectly stimulating tumor cell death through activation of signaling pathways that stimulate phagocytosis and immune cell stimulation.[118]

In addition, numerous targeted therapies have also been shown to elicit immunomodulatory effects, such as attenuating immunosuppression, stimulating immune cell maturation and priming, and enabling immune recognition.[118] For example, bortezomib, a proteasome inhibitor approved for treatment of multiple myeloma, has been shown to sensitize cells to NK-mediated cell lysis[119] and the imatinib TKI amplifies antitumor T cell responses by activating CD8+ T cells and inducing regulatory T cells by inhibiting tumor cell expression of indoleamine 2,3-dioxygenase (*IDO*), an immunosuppressive enzyme.[120] Thus, through these multiple mechanisms, logical and optimal combinations of immune and nonimmune-based therapies may provide strategies for better cancer treatment.

The combination of targeted chemotherapeutics with immune-boosting checkpoint inhibitors such as ipilimumab, that relieve the normal T cell repression, revived the search for effective combinations of existing drugs and has been a major source of new melanoma clinical trials examining effective drug combinations since 2010.[63] The future of anticancer immunotherapy may lie in combination with nanomedicine and the development of nanovehicles of cancer and immune system-targeted chemotherapeutics.[121,122]

6. SUMMARY

In this chapter, we aimed to familiarize the reader with the molecular evolution of AML and the challenges presented by the disease heterogeneity. To summarize, the genomic landscape of AML is very complex and the comutational patterning plays a great role in AML disease pathogenesis. Also, epigenetic dysregulation appears early in leukemogenesis and has emerged as a defining characteristic of AML. Impaired cell differentiation and clonal evolutionary mechanisms create complex subpopulation structures with vast functional heterogeneity. The inter- and intrapatient disease heterogeneity in turn translates into emergence of drug resistance and as follows, high relapse rates and mortality. An additional challenge with development of new, targeted cancer therapies is the high cost of resulting treatment options. Thus, efforts of repurposing drugs with lox toxicity, particularly in combination, may hold some promise to also make cancer therapy more accessible for more patients. Considering the various points of challenge raised in the chapter, the need for efficacious, high precision combination chemotherapeutics for AML is dire and future possibilities lie in targeting cell differentiation pathways gone awry, as well as releasing epigenetic repression.

Systems medicine, as a combination of high-throughput screening, computational analyses and prediction models, and functional and clinical studies allows researchers to more fully examine the scope of molecular evolution of disease and to bring that knowledge to individualized medicine practice.[123] Integrated computational and systems biomedicine approaches hold the key to identifying disease pathway regulators and developing translatable drug combinations for unique treatment regimens. Since compounds bind their efficacy targets as well as other epistatic proteins, off-targets, and proteins involved in compound transport or modification, target deconvolution, and better knowledge of drug–target interactions are essential especially to determine resistance mutations and use combination drug therapies to overcome resistance mechanisms. Experimental methods as well as prediction approaches, such as variomics to systematically search for resistance mutations, and knowledge-based resources of complete and well annotated and curated databases of drug–target interactions will all be required.[124]

With new and rapid advances in the understanding of tumor immunology, combination of either conventional, commonly cytotoxic, or targeted chemotherapy with therapies that modulate adaptive and innate immune responses holds tremendous promise in clinical testing.[63,118] Contemporary chemotherapeutic treatments already employ two or more drugs, both conventional and targeted chemotherapy, in combination for most cancers.

While the concept of combining two more drugs may seem simple at first consideration, it truly is a complex challenge affected by the biological consequences, drug interactions, and even the timing of the individual drugs in combination therapy (e.g., simultaneous or sequential, treatment), especially concerning combination treatments with epigenetic modifiers or immunologics, when epigenetic priming and immunological sensitizing are key factors. With a promise of large scale, disease focused, drug combination screening campaigns, the future of AML molecular oncology lies clearly in developing a deeper understanding of genetic, epigenetic, differentiation, and immunological disease mechanisms and is focused on identifying high precision, high therapeutic index combination chemotherapies.

ACKNOWLEDGMENTS

We would like to thank Krister Wennerberg and Maija Wolf for critical reading of the manuscript and many fruitful and insightful discussions and Kimmo Porkka for advice on clinical treatment regimens for AML treatment. E.K. is supported through the Doctoral Programme in Biomedicine at the University of Helsinki.

REFERENCES

1. Kumar CC. Genetic abnormalities and challenges in the treatment of acute myeloid leukemia. *Genes Cancer*. 2011;2(2):95–107.
2. Howlader N, Noone AM, Krapcho M, et al. *SEER Cancer Statistics Review, 1975–2013*. Bethesda, MD: National Cancer Institute; April 2016 based on November 2015 SEER data submission, posted to the SEER website*http://seer.cancer.gov/csr/1975_2013/*.
3. Grove CS, Vassiliou GS. Acute myeloid leukaemia: a paradigm for the clonal evolution of cancer? *Dis Model Mech*. 2014;7(8):941–951.
4. Welch JS, Ley TJ, Link DC, et al. The origin and evolution of mutations in acute myeloid leukemia. *Cell*. 2012;150(2):264–278.

5. O'Donnell MR, Appelbaum FR, Baer MR, et al. Acute myeloid leukemia clinical practice guidelines in oncology. *J Natl Compr Canc Netw*. 2006;4(1):16–36.
6. Grisendi S, Mecucci C, Falini B, Pandolfi PP. Nucleophosmin and cancer. *Nat Rev Cancer*. 2006;6(7):493–505.
7. Wander SA, Levis MJ, Fathi AT. The evolving role of FLT3 inhibitors in acute myeloid leukemia: quizartinib and beyond. *Ther Adv Hematol*. 2014;5(3):65–77.
8. Russler-Germain DA, Spencer DH, Young MA, et al. The R882H DNMT3A mutation associated with AML dominantly inhibits wild-type DNMT3A by blocking its ability to form active tetramers. *Cancer Cell*. 2014;25(4):442–454.
9. Im AP, Sehgal AR, Carroll MP, et al. DNMT3A and IDH mutations in acute myeloid leukemia and other myeloid malignancies: associations with prognosis and potential treatment strategies. *Leukemia*. 2014;28(9):1774–1783.
10. Zebisch A, Czernilofsky AP, Keri G, Smigelskaite J, Sill H, Troppmair J. Signaling through RAS-RAF-MEK-ERK: from basics to bedside. *Curr Med Chem*. 2007;14(5):601–623.
11. Goyama S, Schibler J, Cunningham L, et al. Transcription factor RUNX1 promotes survival of acute myeloid leukemia cells. *J Clin Invest*. 2013;123(9):3876–3888.
12. Langemeijer SM, Kuiper RP, Berends M, et al. Acquired mutations in TET2 are common in myelodysplastic syndromes. *Nat Genet*. 2009;41(7):838–842.
13. Kadia TM, Jain P, Ravandi F, et al. TP53 mutations in newly diagnosed acute myeloid leukemia: clinicomolecular characteristics, response to therapy, and outcomes. *Cancer* 2016 [Epub ahead of print]. http://dx.doi.org/10.1002/cncr.30203.
14. Lin LI, Chen CY, Lin DT, et al. Characterization of CEBPA mutations in acute myeloid leukemia: most patients with CEBPA mutations have biallelic mutations and show a distinct immunophenotype of the leukemic cells. *Clin Cancer Res*. 2005;11(4):1372–1379.
15. Cancer Genome Atlas Research Network. Genomic and epigenomic landscapes of adult de novo acute myeloid leukemia. *N Engl J Med*. 2013;368(22):2059–2074.
16. Papaemmanuil E, Gerstung M, Bullinger L, et al. Genomic classification and prognosis in acute myeloid leukemia. *N Engl J Med*. 2016;374(23):2209–2221.
17. Shih AH, Jiang Y, Meydan C, et al. Mutational cooperativity linked to combinatorial epigenetic gain of function in acute myeloid leukemia. *Cancer Cell*. 2015;27(4):502–515.
18. Metzeler KH, Herold T, Rothenberg-Thurley M, et al. Spectrum and prognostic relevance of driver gene mutations in acute myeloid leukemia. *Blood*. 2016;128:686–698.
19. Hannum G, Guinney J, Zhao L, et al. Genome-wide methylation profiles reveal quantitative views of human aging rates. *Mol Cell*. 2013;49(2):359–367.
20. Figueroa ME, Lugthart S, Li Y, et al. DNA methylation signatures identify biologically distinct subtypes in acute myeloid leukemia. *Cancer Cell*. 2010;17(1):13–27.
21. Li S, Garrett-Bakelman FE, Chung SS, et al. Distinct evolution and dynamics of epigenetic and genetic heterogeneity in acute myeloid leukemia. *Nat Med*. 2016;22(7):792–799.
22. Ding L, Ley TJ, Larson DE, et al. Clonal evolution in relapsed acute myeloid leukaemia revealed by whole-genome sequencing. *Nature*. 2012;481(7382):506–510.
23. Klco JM, Spencer DH, Miller CA, et al. Functional heterogeneity of genetically defined subclones in acute myeloid leukemia. *Cancer Cell*. 2014;25(3):379–392.
24. Parkin B, Ouillette P, Li Y, et al. Clonal evolution and devolution after chemotherapy in adult acute myelogenous leukemia. *Blood*. 2013;121(2):369–377.
25. Nimmo RA, May GE, Enver T. Primed and ready: understanding lineage commitment through single cell analysis. *Trends Cell Biol*. 2015;25(8):459–467.
26. Tenen DG. Disruption of differentiation in human cancer: AML shows the way. *Nat Rev Cancer*. 2003;3(2):89–101.

27. Bendall SC, Simonds EF, Qiu P, et al. Single-cell mass cytometry of differential immune and drug responses across a human hematopoietic continuum. *Science.* 2011;332(6030):687–696.
28. Behbehani GK, Samusik N, Bjornson ZB, Fantl WJ, Medeiros BC, Nolan GP. Mass cytometric functional profiling of acute myeloid leukemia defines cell-cycle and immunophenotypic properties that correlate with known responses to therapy. *Cancer Discov.* 2015;5(9):988–1003.
29. Regha K, Assi SA, Tsoulaki O, Gilmour J, Lacaud G, Bonifer C. Developmental-stage-dependent transcriptional response to leukaemic oncogene expression. *Nat Commun.* 2015;6:7203.
30. Shlush LI, Mitchell A. AML evolution from preleukemia to leukemia and relapse. *Best Pract Res Clin Haematol.* 2015;28(2–3):81–89.
31. Pemovska T, Kontro M, Yadav B, et al. Individualized systems medicine strategy to tailor treatments for patients with chemorefractory acute myeloid leukemia. *Cancer Discov.* 2013;3(12):1416–1429.
32. Pemovska T, Johnson E, Kontro M, et al. Axitinib effectively inhibits BCR-ABL1 (T315I) with a distinct binding conformation. *Nature.* 2015;519(7541):102–105.
33. Chakradhar S. Group mentality: determining if targeted treatments really work for cancer. *Nat Med.* 2016;22(3):222–224.
34. Ellis LM, Hicklin DJ. Resistance to targeted therapies: refining anticancer therapy in the era of molecular oncology. *Clin Cancer Res.* 2009;15(24):7471–7478.
35. Granatowicz A, Piatek CI, Moschiano E, El-Hemaidi I, Armitage JD, Akhtari M. An overview and update of chronic myeloid leukemia for primary care physicians. *Korean J Fam Med.* 2015;36(5):197–202.
36. Fitzgerald JB, Schoeberl B, Nielsen UB, Sorger PK. Systems biology and combination therapy in the quest for clinical efficacy. *Nat Chem Biol.* 2006;2(9):458–466.
37. Testa U, Lo-Coco F. Targeting of leukemia-initiating cells in acute promyelocytic leukemia. *Stem Cell Investig.* 2015;2:8. http://dx.doi.org/10.3978/j.issn.2306-9759.2015.04.03.
38. Minucci S, Monestiroli S, Giavara S, et al. PML-RAR induces promyelocytic leukemias with high efficiency following retroviral gene transfer into purified murine hematopoietic progenitors. *Blood.* 2002;100(8):2989–2995.
39. Grignani F, Valtieri M, Gabbianelli M, et al. PML/RAR alpha fusion protein expression in normal human hematopoietic progenitors dictates myeloid commitment and the promyelocytic phenotype. *Blood.* 2000;96(4):1531–1537.
40. Nitto T, Sawaki K. Molecular mechanisms of the antileukemia activities of retinoid and arsenic. *J Pharmacol Sci.* 2014;126(3):179–185.
41. Tomita A, Kiyoi H, Naoe T. Mechanisms of action and resistance to all-trans retinoic acid (ATRA) and arsenic trioxide (As2O3) in acute promyelocytic leukemia. *Int J Hematol.* 2013;97(6):717–725.
42. Chen W, Drakos E, Grammatikakis I, et al. mTOR signaling is activated by FLT3 kinase and promotes survival of FLT3-mutated acute myeloid leukemia cells. *Mol Cancer.* 2010;9:292. http://dx.doi.org/10.1186/1476-4598-9-292.
43. Nogami A, Oshikawa G, Okada K, et al. FLT3-ITD confers resistance to the PI3K/akt pathway inhibitors by protecting the mTOR/4EBP1/mcl-1 pathway through STAT5 activation in acute myeloid leukemia. *Oncotarget.* 2015;6(11):9189–9205.
44. Kothe S, Muller JP, Bohmer SA, et al. Features of ras activation by a mislocalized oncogenic tyrosine kinase: FLT3 ITD signals through K-ras at the plasma membrane of acute myeloid leukemia cells. *J Cell Sci.* 2013;126(Pt 20):4746–4755.
45. Mazzarella L. Orlando magic: report from the 57th meeting of the American Society of Haematology, 5–7 December 2015, Orlando, USA. *ecancermedicalscience.* 2016;10:612.

46. von Bubnoff N, Engh RA, Aberg E, Sanger J, Peschel C, Duyster J. FMS-like tyrosine kinase 3-internal tandem duplication tyrosine kinase inhibitors display a non-overlapping profile of resistance mutations in vitro. *Cancer Res*. 2009;69(7):3032–3041.
47. Parmar A, Marz S, Rushton S, et al. Stromal niche cells protect early leukemic FLT3-ITD+ progenitor cells against first-generation FLT3 tyrosine kinase inhibitors. *Cancer Res*. 2011;71(13):4696–4706.
48. Shafer D, Grant S. Update on rational targeted therapy in AML. *Blood Rev*. 2016;30(4):275–283.
49. Yun S, Vincelette ND, Abraham I, Robertson KD, Fernandez-Zapico ME, Patnaik MM. Targeting epigenetic pathways in acute myeloid leukemia and myelodysplastic syndrome: a systematic review of hypomethylating agents trials. *Clin Epigenetics*. 2016;8:68. http://dx.doi.org/10.1186/s13148-016-0233-2.
50. Jain N, Curran E, Iyengar NM, et al. Phase II study of the oral MEK inhibitor selumetinib in advanced acute myelogenous leukemia: a University of Chicago phase II consortium trial. *Clin Cancer Res*. 2014;20(2):490–498.
51. Borthakur G, Popplewell L, Boyiadzis M, et al. Activity of the oral mitogen-activated protein kinase kinase inhibitor trametinib in RAS-mutant relapsed or refractory myeloid malignancies. *Cancer*. 2016;122(12):1871–1879.
52. Esposito MT, So CW. DNA damage accumulation and repair defects in acute myeloid leukemia: implications for pathogenesis, disease progression, and chemotherapy resistance. *Chromosoma*. 2014;123(6):545–561.
53. Sheikhha MH, Tobal K, Liu Yin JA. High level of microsatellite instability but not hypermethylation of mismatch repair genes in therapy-related and secondary acute myeloid leukaemia and myelodysplastic syndrome. *Br J Haematol*. 2002;117(2):359–365.
54. Esposito MT, Zhao L, Fung TK, et al. Synthetic lethal targeting of oncogenic transcription factors in acute leukemia by PARP inhibitors. *Nat Med*. 2015;21(12):1481–1490.
55. Aleem E, Arceci RJ. Targeting cell cycle regulators in hematologic malignancies. *Front Cell Dev Biol*. 2015;3:16.
56. Kadia TM, Ravandi F, Cortes J, Kantarjian H. New drugs in acute myeloid leukemia. *Ann Oncol*. 2016;27(5):770–778.
57. Kontro M, Kumar A, Majumder MM, et al. HOX gene expression predicts response to BCL-2 inhibition in acute myeloid leukemia. *Leukemia*. 2016;1–9.
58. Kojima K, Ishizawa J, Andreeff M. Review: pharmacological activation of wild-type p53 in the therapy of leukemia. *Exp Hematol*. 2016;44:791–798.
59. Chen CW, Armstrong SA. Targeting DOT1L and HOX gene expression in MLL-rearranged leukemia and beyond. *Exp Hematol*. 2015;43(8):673–684.
60. Axelrod M, Gordon VL, Conaway M, et al. Combinatorial drug screening identifies compensatory pathway interactions and adaptive resistance mechanisms. *Oncotarget*. 2013;4(4):622–635.
61. Bukowska B, Gajek A, Marczak A. Two drugs are better than one. a short history of combined therapy of ovarian cancer. *Contemp Oncol (Pozn)*. 2015;19(5):350–353.
62. Tortora G, Bianco R, Daniele G. Strategies for multiple signalling inhibition. *J Chemother*. 2004;16(Suppl 4):41–43.
63. Ledford H. Cocktails for cancer with a measure of immunotherapy. *Nature*. 2016;532(7598):162–164.
64. Manchado E, Weissmueller S, Morris 4th JP, et al. A combinatorial strategy for treating KRAS-mutant lung cancer. *Nature*. 2016;534(7609):647–651.
65. Gioeli D, Wunderlich W, Sebolt-Leopold J, et al. Compensatory pathways induced by MEK inhibition are effective drug targets for combination therapy against castration-resistant prostate cancer. *Mol Cancer Ther*. 2011;10(9):1581–1590.

66. Keeshan K, He Y, Wouters BJ, et al. Tribbles homolog 2 inactivates C/EBPalpha and causes acute myelogenous leukemia. *Cancer Cell.* 2006;10(5):401–411.
67. Rishi L, Hannon M, Salome M, et al. Regulation of Trib2 by an E2F1-C/EBPalpha feedback loop in AML cell proliferation. *Blood.* 2014;123(15):2389–2400.
68. Chou TC. Preclinical versus clinical drug combination studies. *Leuk Lymphoma.* 2008;49(11):2059–2080.
69. Greco W, Unkelbach H, Poch G, Suhnel J, Kundi M, Bodeker W. Consensus on concepts and terminology for interaction assesssment: the saanselka agreement. *Arch Complex Environ Studies.* 1992;4:65–69.
70. Kashif M, Andersson C, Aberg M, et al. A pragmatic definition of therapeutic synergy suitable for clinically relevant in vitro multicompound analyses. *Mol Cancer Ther.* 2014;13(7):1964–1976.
71. Yadav B, Pemovska T, Szwajda A, et al. Quantitative scoring of differential drug sensitivity for individually optimized anticancer therapies. *Sci Rep.* 2014;4:5193.
72. Greco WR, Bravo G, Parsons JC. The search for synergy: a critical review from a response surface perspective. *Pharmacol Rev.* 1995;47(2):331–385.
73. Tang J, Wennerberg K, Aittokallio T. What is synergy? The Saariselka agreement revisited. *Front Pharmacol.* 2015;6:181.
74. Loewe S. The problem of synergism and antagonism of combined drugs. *Arzneimittelforschung.* 1953;3(6):285–290.
75. Loewe S, Muischnek N. Effect of combinations: mathematical basis of the problem. *Arch Exp Pathol Pharmakol.* 1926;114:313–326.
76. BLISS CI. The calculation of microbial assays. *Bacteriol Rev.* 1956;20(4):243–258.
77. Bliss C. The toxicity of poisons applied jointly. *Ann Appl Biol.* 1939;26(3):585.
78. Moore MJ, Goldstein D, Hamm J, et al. Erlotinib plus gemcitabine compared with gemcitabine alone in patients with advanced pancreatic cancer: a phase III trial of the National Cancer Institute of Canada Clinical Trials Group. *J Clin Oncol.* 2007;25(15):1960–1966.
79. GOLDIN A, HUMPHREYS SR, MANTEL N, VENDITTI JM. Combined treatment of advanced leukemia (L1210) in mice with amethopterin and 6-mercaptopurine. *J Natl Cancer Inst.* 1956;17(5):631–638.
80. Zhou JX, Isik Z, Xiao C, et al. Systematic drug perturbations on cancer cells reveal diverse exit paths from proliferative state. *Oncotarget.* 2016;7(7):7415–7425.
81. Tang J, Karhinen L, Xu T, et al. Target inhibition networks: predicting selective combinations of druggable targets to block cancer survival pathways. *PLoS Comput Biol.* 2013;9(9):e1003226.
82. Bansal M, Yang J, Karan C, et al. A community computational challenge to predict the activity of pairs of compounds. *Nat Biotechnol.* 2014;32(12):1213–1222.
83. Nguyen TT, Chua JK, Seah KS, et al. Predicting chemotherapeutic drug combinations through gene network profiling. *Sci Rep.* 2016;6:18658.
84. Pellicano F, Simara P, Sinclair A, et al. The MEK inhibitor PD184352 enhances BMS-214662-induced apoptosis in CD34+ CML stem/progenitor cells. *Leukemia.* 2011;25(7):1159–1167.
85. Yang ZY, Yuan JQ, Di MY, et al. Gemcitabine plus erlotinib for advanced pancreatic cancer: a systematic review with meta-analysis. *PLoS One.* 2013;8(3): e57528.
86. Czarnecka AM, Korzen P, Nowak-Dement A, Kukwa W, Korniluk J, Szczylik C. Prolonged complete response following gemcitabine-erlotinib combined therapy in advanced pancreatic cancer. *Oncol Lett.* 2016;11(2):1101–1104.
87. No author. FDA approves drug combo for kidney cancer. *Cancer Discov.* 2016;6(7):687–688.

88. Peng YC, Lu SD, Zhong JH, et al. Combination of 5-fluorouracil and 2-morphilino-8-phenyl-4H-chromen-4-one may inhibit liver cancer stem cell activity. *Tumour Biol.* 2016;37:10943–10958.
89. de Oliveira VA, da Motta LL, De Bastiani MA, et al. In vitro evaluation of antitumoral efficacy of catalase in combination with traditional chemotherapeutic drugs against human lung adenocarcinoma cells. *Tumour Biol.* 2016;37:10775–10784.
90. Grob JJ, Amonkar MM, Karaszewska B, et al. Comparison of dabrafenib and trametinib combination therapy with vemurafenib monotherapy on health-related quality of life in patients with unresectable or metastatic cutaneous BRAF Val600-mutation-positive melanoma (COMBI-v): results of a phase 3, open-label, randomised trial. *Lancet Oncol.* 2015;16(13):1389–1398.
91. Al-Marrawi MY, Saroya BS, Brennan MC, Yang Z, Dykes TM, El-Deiry WS. Off-label use of cetuximab plus sorafenib and panitumumab plus regorafenib to personalize therapy for a patient with V600E BRAF-mutant metastatic colon cancer. *Cancer Biol Ther.* 2013;14(8):703–710.
92. Tan N, Wong M, Nannini MA, et al. Bcl-2/bcl-xL inhibition increases the efficacy of MEK inhibition alone and in combination with PI3 kinase inhibition in lung and pancreatic tumor models. *Mol Cancer Ther.* 2013;12(6):853–864.
93. Liu L, Shi H, Liu Y, et al. Synergistic effects of foretinib with HER-targeted agents in MET and HER1- or HER2-coactivated tumor cells. *Mol Cancer Ther.* 2011;10(3):518–530.
94. Held MA, Langdon CG, Platt JT, et al. Genotype-selective combination therapies for melanoma identified by high-throughput drug screening. *Cancer Discov.* 2013;3(1):52–67.
95. Zhang H, Zhong X, Zhang X, Shang D, Zhou YI, Zhang C. Enhanced anticancer effect of ABT-737 in combination with naringenin on gastric cancer cells. *Exp Ther Med.* 2016;11(2):669–673.
96. Bose P, Grant S. Rational combinations of targeted agents in AML. *J Clin Med.* 2015;4(4):634–664.
97. Diesch J, Zwick A, Garz AK, Palau A, Buschbeck M, Gotze KS. A clinical-molecular update on azanucleoside-based therapy for the treatment of hematologic cancers. *Clin Epigenetics.* 2016;8:71. http://dx.doi.org/10.1186/s13148-016-0237-y.
98. Bali P, George P, Cohen P, et al. Superior activity of the combination of histone deacetylase inhibitor LAQ824 and the FLT-3 kinase inhibitor PKC412 against human acute myelogenous leukemia cells with mutant FLT-3. *Clin Cancer Res.* 2004;10(15):4991–4997.
99. Lin WH, Yeh TK, Jiaang WT, et al. Evaluation of the antitumor effects of BPR1J-340, a potent and selective FLT3 inhibitor, alone or in combination with an HDAC inhibitor, vorinostat, in AML cancer. *PLoS One.* 2014;9(1):e83160.
100. Pietschmann K, Bolck HA, Buchwald M, et al. Breakdown of the FLT3-ITD/STAT5 axis and synergistic apoptosis induction by the histone deacetylase inhibitor panobinostat and FLT3-specific inhibitors. *Mol Cancer Ther.* 2012;11(11): 2373–2383.
101. Weisberg E, Liu Q, Nelson E, et al. Using combination therapy to override stromal-mediated chemoresistance in mutant FLT3-positive AML: synergism between FLT3 inhibitors, dasatinib/multi-targeted inhibitors and JAK inhibitors. *Leukemia.* 2012;26(10):2233–2244.
102. Yang X, Sexauer A, Levis M. Bone marrow stroma-mediated resistance to FLT3 inhibitors in FLT3-ITD AML is mediated by persistent activation of extracellular regulated kinase. *Br J Haematol.* 2014;164(1):61–72.
103. Weisberg E, Liu Q, Zhang X, et al. Selective akt inhibitors synergize with tyrosine kinase inhibitors and effectively override stroma-associated cytoprotection of mutant FLT3-positive AML cells. *PLoS One.* 2013;8(2):e56473.

104. Smith AM, Dun MD, Lee EM, et al. Activation of protein phosphatase 2A in FLT3+ acute myeloid leukemia cells enhances the cytotoxicity of FLT3 tyrosine kinase inhibitors. *Oncotarget*. 2016;7(30).
105. Wang A, Wu H, Chen C, et al. Dual inhibition of AKT/FLT3-ITD by A674563 overcomes FLT3 ligand-induced drug resistance in FLT3-ITD positive AML. *Oncotarget*. 2016;7:29131–29142.
106. Laszlo GS, Gudgeon CJ, Harrington KH, et al. Cellular determinants for preclinical activity of a novel CD33/CD3 bispecific T-cell engager (BiTE) antibody, AMG 330, against human AML. *Blood*. 2014;123(4):554–561.
107. Harrington KH, Gudgeon CJ, Laszlo GS, et al. The broad anti-AML activity of the CD33/CD3 BiTE antibody construct, AMG 330, is impacted by disease stage and risk. *PLoS One*. 2015;10(8):e0135945.
108. Grosso DA, Hess RC, Weiss MA. Immunotherapy in acute myeloid leukemia. *Cancer*. 2015;121(16):2689–2704.
109. Vergez F, Green AS, Tamburini J, et al. High levels of CD34+CD38low/-CD123+ blasts are predictive of an adverse outcome in acute myeloid leukemia: a Groupe Ouest-Est des Leucemies Aigues et Maladies du Sang (GOELAMS) study. *Haematologica*. 2011;96(12):1792–1798.
110. Sutherland MSK, Yu C, Walter RB, et al. SGN-CD123A, a pyrrolobenzodiazepine dimer linked anti-CD123 antibody drug conjugate, demonstrates effective anti-leukemic activity in multiple preclinical models of AML. *Blood*. 2015;126(23):330.
111. Testa U, Pelosi E, Frankel A. CD 123 is a membrane biomarker and a therapeutic target in hematologic malignancies. *Biomark Res*. 2014;2(1):4. http://dx.doi.org/10.1186/2050-7771-2-4.
112. ADCs show promise in leukemias. *Cancer Discov*. 2016;6(9):939. doi: 10.1158/2159-8290.CD-NB2016-088. Epub 2016 Jul 7.
113. Zugmaier G, Klinger M, Schmidt M, Subklewe M. Clinical overview of anti-CD19 BiTE(®) and ex vivo data from anti-CD33 BiTE(®) as examples for retargeting T cells in hematologic malignancies. *Mol Immunol*. 2015;67(2 Pt A):58–66.
114. Friedrich M, Henn A, Raum T, et al. Preclinical characterization of AMG 330, a CD3/CD33-bispecific T-cell-engaging antibody with potential for treatment of acute myelogenous leukemia. *Mol Cancer Ther*. 2014;13(6):1549–1557.
115. Laszlo GS, Gudgeon CJ, Harrington KH, Walter RB. T-cell ligands modulate the cytolytic activity of the CD33/CD3 BiTE antibody construct, AMG 330. *Blood Cancer J*. 2015;5:e340.
116. Al-Hussaini M, Rettig MP, Ritchey JK, et al. Targeting CD123 in acute myeloid leukemia using a T-cell-directed dual-affinity retargeting platform. *Blood*. 2016;127(1):122–131.
117. Chichili GR, Huang L, Li H, et al. A CD3xCD123 bispecific DART for redirecting host T cells to myelogenous leukemia: preclinical activity and safety in nonhuman primates. *Sci Transl Med*. 2015;7(289):289ra82.
118. Bracci L, Schiavoni G, Sistigu A, Belardelli F. Immune-based mechanisms of cytotoxic chemotherapy: implications for the design of novel and rationale-based combined treatments against cancer. *Cell Death Differ*. 2014;21(1):15–25.
119. Seeger JM, Schmidt P, Brinkmann K, et al. The proteasome inhibitor bortezomib sensitizes melanoma cells toward adoptive CTL attack. *Cancer Res*. 2010;70(5):1825–1834.
120. Balachandran VP, Cavnar MJ, Zeng S, et al. Imatinib potentiates antitumor T cell responses in gastrointestinal stromal tumor through the inhibition of ido. *Nat Med*. 2011;17(9):1094–1100.
121. Da Silva CG, Rueda F, Lowik CW, Ossendorp F, Cruz LJ. Combinatorial prospects of nano-targeted chemoimmunotherapy. *Biomaterials*. 2016;83:308–320.

122. Jiang X, Bugno J, Hu C, et al. Eradication of acute myeloid leukemia with FLT3 ligand-targeted miR-150 nanoparticles. *Cancer Res*. 2016;76(15):4470–4480.
123. Benson M. Clinical implications of omics and systems medicine: focus on predictive and individualized treatment. *J Intern Med*. 2016;279(3):229–240.
124. Schirle M, Jenkins JL. Identifying compound efficacy targets in phenotypic drug discovery. *Drug Discov Today*. 2016;21(1):82–89.

CHAPTER TEN

Myeloproliferative Neoplasms: Molecular Drivers and Therapeutics

G.W. Reuther[1]
H. Lee Moffitt Cancer Center, Tampa, FL, United States
University of South Florida, Tampa, FL, United States
[1]Corresponding author: e-mail address: gary.reuther@moffitt.org

Contents

Abstract

Activating mutations in genes that drive neoplastic cell growth are numerous and widespread in cancer, and specific genetic alterations are associated with certain types of cancer. For example, classic myeloproliferative neoplasms (MPNs) are hematopoietic stem cell disorders that affect cells of the myeloid lineage, including erythrocytes, platelets, and granulocytes. An activating mutation in the JAK2 tyrosine kinase is prevalent in these diseases. In MPN patients that lack such a mutation, other genetic changes that lead to activation of the JAK2 signaling pathway are present, indicating deregulation of JAK2 signaling plays an etiological driving role in MPNs, a concept supported by significant evidence from in vivo experimental MPN systems. Thus, small

Progress in Molecular Biology and Translational Science, Volume 144
ISSN 1877-1173
http://dx.doi.org/10.1016/bs.pmbts.2016.09.004

molecules that inhibit JAK2 activity are ideal drugs to impede the progression of disease in MPN patients. However, even though JAK inhibitors provide significant symptomatic relief, they have failed as a remission-inducing therapy. Nonetheless, the progress made understanding the molecular etiology of MPNs since 2005 is significant and has provided insight for the development and testing of novel molecular targeted therapeutic approaches. The current understanding of driver mutations in MPNs and an overview of current and potential therapeutic strategies for MPN patients will be discussed.

1. MYELOPROLIFERATIVE NEOPLASMS

Myeloproliferative neoplasms (MPNs) are blood disorders that affect the myeloid lineage of the hematopoietic system.[1,2] MPNs were previously termed myeloproliferative disorders until 2011.[3] The World Health Organization classifies chronic myeloid leukemia (CML), polycythemia vera (PV), essential thrombocythemia (ET), and primary myelofibrosis (PMF) as classic MPNs. Other MPNs include chronic neutrophilic leukemia, chronic eosinophilic leukemia-not otherwise specified, and MPN unclassifiable.[4] The classic MPNs are further classified as Philadelphia chromosome-positive (or *BCR-ABL*-positive) (CML) and Philadelphia chromosome-negative (*BCR-ABL*-negative) (PV, ET, and PMF). The focus of this chapter will be the Philadelphia chromosome-negative (*BCR-ABL*-negative) classic MPNs, PV, ET, and PMF, which are the three diseases generally referred to by the simplified term, MPNs.[2]

In 1951, Dr. Dameshek described[5] a group of bone marrow disorders that, while characterized as clinically distinct diseases, had a number of similar features including expansion of mature myeloid cells. While PV classically was considered a red blood cell disease, PV, ET, and PMF all could exhibit erythrocytosis, thrombocytosis, and leukocytosis, as well as splenomegaly. Dr. Dameshek suggested that "It is possible that these various conditions—myeloproliferative disorders—are all somewhat variable manifestations of proliferative activity of the bone marrow cells, perhaps due to a hitherto undiscovered stimulus."[5] Over the next half century, it was determined that CML was caused by a chromosomal translocation that generated the Philadelphia chromosome, encoding a constitutively activated tyrosine kinase, BCR-ABL, and that this activity provided a therapeutic target for CML.[6] In 2005, a mutation in the Janus Kinase-2 (JAK2) tyrosine kinase was prevalently identified in patients with PV, ET, and PMF.[7–11] While different than BCR-ABL in that this mutation is a single nucleotide and not a chromosomal translocation, it too imparts a deregulated activity on the tyrosine kinase. While Dr. Dameshek did not know about BCR-ABL or JAK2, his suggestion of a common cause of what would become MPNs was indeed prescient.[5]

The major observations of Dr. William Dameshek included myeloproliferation of three different cell lineages in PV, ET, and PMF.[5] These include myeloid, erythroid, and megakaryocytic cells. Thus, individual MPNs may exhibit signs of elevated red blood cells, granulocytes, and platelets. PV is an MPN that manifests itself in increased red blood cell and hemoglobin levels. However, elevated platelets and granulocytes are also observed. Other characteristics of PV pathology include decrease erythropoietin levels as well as the formation of erythropoietin-independent erythroid colonies in culture. ET is characterized by elevated platelet levels, due to the concurrent observation of increased megakaryocyte proliferation. PMF patients exhibit not only elevated megakaryocytes and platelets but also have fibrosis of the bone marrow due to reticulin and collagen fibers. Bone marrow failure can lead to anemia. Splenomegaly and hepatomegaly, due to extramedullary hematopoiesis, not only are common in PMF patients but also observed in PV and ET patients. Both PV and ET can progress to a myelofibrotic state.[2,12]

In addition to these clinical and pathological features of MPNs, more recent genetic characterization has determined common genetic features of MPNs. MPNs are clonal diseases with common etiologically relevant mutations. Not only have these mutations enhanced understanding of the mechanisms by which MPNs may develop, they have also provided insights into therapeutic intervention as well as genetic markers that contribute to diagnosis.[12,13] The most prominent mutations in MPNs are in the genes that encode for JAK2, MPL, and CALR. Of these, the *JAK2-V617F* mutation was the first identified, and, is present in 95% of PV, and about 55% and 60% of ET and PMF patients, respectively. *MPL* mutations are present in about 3% ET patients and about 7% of PMF patients.[12] Interestingly, *CALR* mutations are present in the majority of ET and PMF patients who lack *JAK2* and *MPL* mutations, accounting for about 25% of ET and about 30% of PMF patients having a *CALR* mutation.[12–15] The mutual exclusivity of these mutations indicates a likely overlapping cellular mechanism of MPN formation. Details regarding these mutations and the molecular basis for their roles in MPNs will be presented later in this chapter.

Epidemiological estimates for MPN were determined by analyzing data from two large health plans covering the years 2008–2010.[16] The prevalence of PV in the United States during this time was about 44–57 per 100,000 people and that of ET was about 38–57 per 100,000.[16] Based on this data, the projected number of people in the United States afflicted with PV and ET at the end of 2010 would be at least 148,000 and 133,000, respectively. PMF is the least common of these three MPNs, with a prevalence of about 2 per 100,000 people in the United States.[16] However, as PV and ET can precede myelofibrosis, the prevalence of myelofibrosis was about 5 per 100,000, or about 15,000 people in the United States at the end of 2010.[16]

MPNs are generally diagnosed over the age of 60. The median life expectancies for such patients are 5.9 years for PMF, 13.7 years for PV, and 19.8 years for ET.[13] Of note, the median life expectancy for ET is still inferior to that of age- and sex-matched controls.[13] Each MPN is classified into different risk subgroups.[2] These risk groups are based on a variety of factors including age and a history of thrombosis. Importantly, as details of the molecular genetics of MPNs have developed over the last decade, factors such as genotype are now being assessed as to their prognostic value.[13] The high cellular burden in MPN patients can lead to thrombotic and hemorrhagic complications. Bone marrow failure and transformation to an incurable acute myeloid leukemia (AML) lead to poor outcomes in MPNs, especially for patients with myelofibrosis. The cumulative rate of PV transformation to leukemia 20 years after diagnosis, with death as a competing risk, is estimated to be about 7–8%.[17] For ET this rate to leukemic transformation is about half of that of PV.[13] For PMF on the other hand, this rate is reported to be about 14%.[13] Thus, the increased risk of MPN patients for developing AML is a major concern.

2. ETIOLOGIC DRIVER AND NONDRIVER MUTATIONS IN MPNs

2.1 JAK2 Mutations

The expression of BCR-ABL from the Philadelphia chromosome is the etiological basis of CML. The constitutively activated tyrosine kinase, which is present in almost all cases of CML, provides deregulated cell signaling that leads to CML formation.[6] The 2005 discovery of an activating mutation in the JAK2 tyrosine kinase (JAK2-V617F) in MPNs[7–11] drew parallels with

BCR-ABL in CML, which is also a myeloproliferative neoplasm, as it was eventually determined this mutationally activated JAK2 tyrosine kinase is expressed in about 95% of PV patients.[12] This suggested mutation of JAK2 as an immediate suspect as a contributor to MPN formation. This mutation of *JAK2* is also present in about 55–60% of patients with ET and PMF,[12] suggesting a single mutation can contribute to clinically distinct MPNs. In addition to JAK2-V617F, mutations within exon 12 of *JAK2* are also present in *JAK2-V617F*-negative PV patients.[18] The frequencies of *JAK2-V617F* (95%) and exon 12 mutations in *JAK2* (3%)[12] suggest that mutationally active JAK2 is the primary driving mutation in PV.

2.2 MPL Mutations

The myeloproliferative leukemia virus oncogene, *MPL*, is mutated in about 3% and 7% of patients with ET and PMF, respectively.[12] These include mutations altering tryptophan-515 of MPL.[19] The major signaling pathway downstream of the MPL protein, which is the thrombopoietin receptor (TpoR), is initiated by the JAK2 tyrosine kinase.[20] Mutations of MPL in MPNs have been shown to induce constitutive signaling through the JAK2 tyrosine kinase pathway.[19] Thus, in the absence of direct mutational activation of JAK2, constitutive JAK2 signaling can be induced by mutational activation of upstream activators of JAK2. This further demonstrates the importance of the JAK2 pathway in MPNs.

2.3 CALR Mutations

The identification of *JAK2* mutations and *MPL* mutations in MPN suggested that JAK2 is a major driver of MPNs. However, about one-third of ET and PMF mutations lack JAK2 or MPL mutations. Whole-exome sequencing studies of samples from such patients identified a mutation in the gene that encodes calreticulin (CALR).[14,15] Approximately 25% of ET and 30% of PMF patients harbor a *CALR* mutation, which represents the majority of *JAK2-V617F*-negative and *MPL* mutation-negative ET and PMF patients.[12–15] At the time of the identification of *CALR* mutations in MPN, the functions of CALR did not suggest an apparent connection with the JAK2 signaling pathway. However, subsequent studies of mutated CALR have indeed provided a connection between the two proteins and provide further demonstration of the importance of JAK2 in MPNs. These details will be discussed later in this chapter.

2.4 Other Mutations

As already discussed, mutation of JAK2 is observed in nearly all patients with PV. Interestingly, one of the three major MPN driving mutations (in *JAK2*, *MPL*, and *CALR*) is present in approximately 90% of ET and PMF patients.[12] These mutations are considered MPN driver mutations because of their prevalence and demonstration that mouse models expressing these mutant proteins induce an MPN-like phenotype. Recent next-generation sequencing studies have been performed on so-called triple-negative ET and PMF samples. These studies have demonstrated that a fraction of "triple-negative" ET and PMF patients harbor other gain of function MPL and JAK2 mutations.[21,22] Thus, "triple-negative" ET and PMF patients likely harbor other genetic variations that impart activation of the JAK2 signaling pathway.

Other potential driver mutations are likely present in patients that lack mutations in *JAK2*, *MPL*, and *CALR*. For example, a small number of MPN patients have a mutation in the gene that encodes the LNK protein.[23] LNK is a negative regulator of JAK/STAT signaling, and LNK mutations identified in MPNs are inactivating. Inactivation of LNK by mutation leads to activation of JAK2 and, thus, highlights the potential existence of other driver mutations, albeit rare, that lead to alternative mechanisms by which JAK2 can be activated in MPNs. Interestingly, LNK mutations may be more prominent in MPN patients following leukemic transformation than before this change in disease state.[24]

While mutations of *JAK2*, *MPL*, and *CALR* are major driving factors of MPNs, other recurring mutations are found in MPNs. These mutations often alter proteins that function in epigenetic control of gene expression. For example, mutations in *TET2*, *DNMT3A*, *AXSL1*, *EZH2*, *IDH1*, and *IDH2* are present in MPNs.[12] In addition, also mutated in MPNs are a number of splicesome regulators as well as genes whose products participate in the transcriptional regulation of gene expression. Mutations in these genes are not considered to be drivers of MPNs (they cooccur with JAK2-activating mutations and are found in other myeloid cancers) and likely contribute to MPN formation by cooperating with driver mutations, dictating various aspects of disease such as MPN phenotype and progression. For example, it was shown that loss of *EZH2* drives myelofibrosis formation in JAK2-V617F transgenic mice, suggesting mutation of this methyltransferase can influence MPN phenotype.[25] Importantly, such mutations will likely provide prognostic value to clinicians (see next section). Details of these mutations will not be described in this chapter, and readers are referred to literature reviews for additional information.[26–28]

2.5 Prognostic Value of MPN Mutations

The genetics of MPNs provides clear diagnostic and prognostic value, particularly with respect to the driver mutations in *JAK2*, *MPL*, and *CALR*, as well as insight into potential therapeutic targets, the details of which will be discussed later in this chapter. With respect to disease prognosis, ET has a median survival about 20 years, while the median survival of PV patients is about 14 years, and this survival advantage was not affected by the genotype of the *JAK2*, *MPL*, or *CALR* driver mutations.[13] However, the survival rate of PMF patients is dependent on driver mutation status. While the overall median survival of JAK2-mutated PMF patients is about 6 years, PMF patients that harbor a *CALR* mutation have a median survival rate of almost 16 years. In addition, *MPL*-mutation-positive patients have a median survival of 10 years, while PMF patients that are considered "triple-negative" have the worst survival at a little over 2 years.[13] Another important outcome of this same study was the identification of a similar relative risk to leukemic blast transformation, where *CALR*-mutated patients had the lowest risk (6.5%) and triple-negative patients the highest (25%).[13] Thus, driver mutation status in PMF is prognostically informative.

While the role of, nondriving, genetic variants in MPNs is less understood, such variants may also provide prognostic value to clinicians. For example, mutations in *ASXL1* appear to provide prognostic value in myelofibrosis, as *ASXL1* mutations predict for poor survival.[29] In fact, in 2014 it was reported that an *ASXL1* mutation in the absence of a *CALR* mutation is the most significant risk factor for survival in myelofibrosis.[30] Also, mutations in the genes for the splicesome factor SRSF2 and for isocitrate dehydrogenase 1 (IDH1) may be predictive of leukemic transformation,[29] potentially providing an extremely important prognostic tool to MPN clinicians. In addition, loss of function of the p53 tumor suppressor is common following MPN transformation to AML.[31] However, the absence of *p53* mutations before leukemic transformation may limit prognostic value of this mutation. Thus, specific combinations of mutations likely dictate disease phenotype and patient outcome, and this perhaps may be significantly defined by nondriver mutational genotype.

3. MOLECULAR BIOLOGY OF MPN DRIVER MUTATIONS

3.1 JAK2: Mutational Activation

3.1.1 JAK Tyrosine Kinases and Cytokine Receptor Signaling

The JAK family of nonreceptor tyrosine kinase contains four members: JAK1, JAK2, JAK3, and Tyk2.[32] A distinguishing feature of these kinases is the presence of two kinase domains at the carboxy terminus. The distal kinase domain is an active tyrosine kinase domain, while the upstream kinase, which was originally described as a "pseudokinase" domain, acts as an autoinhibitory domain for the active tyrosine kinase domain.[33] While originally members of this family may have been considered "just another kinase," these kinases are termed Janus kinases because the presence of these two apparent kinase domains is reminiscent of the two-faced depiction of the Roman God Janus.

As nonreceptor tyrosine kinases, members of the JAK family of tyrosine kinases play roles downstream of various transmembrane receptors. The primary signaling mechanism controlled by members of the JAK family of tyrosine kinases is via activation of a family of proteins called Signal Transducers and Activators of Transcription, or STATs.[32] There are seven STAT family members and this along with the different JAK family members create a plethora of potential signaling combinations. This is often dictated by the specific receptor that activates JAK/STAT signaling.[32]

JAKs constitutively bind to receptors, such as cytokine receptors, utilizing a receptor-binding domain (FERM domain) at their amino terminal and a specific sequence on the cytoplasmic portion of receptors called Box 1 and Box 2 motifs. Like receptor tyrosine kinases, JAKs become activated by ligand-induced dimerization of the receptors.[32] This ligand binding and receptor dimerization are believed to bring receptor-bound JAKs into close proximity and induce a conformational change in the receptor bound JAKs. Together, this allows the kinases to transphosphorylate each other, which ultimately leads to their fully activated state.

In addition to JAK transphosphorylation, another important phosphorylation step in signaling by JAKs is the phosphorylation of STAT proteins.[32] This can be preceded by JAK phosphorylation of the receptor cytoplasmic domain at specific tyrosine sites. The phosphotyrosine-binding SH2 domain of STATs can then recognize and bind to these sites. This places STATs in proximity of the activated JAK proteins, where STATs are then phosphorylated on a specific tyrosine residue. This leads to dimerization, which can be homo- or heterodimerization, of STATs and subsequent translocation to the nucleus. There, STATs function as transcription factors and contribute to the regulation of gene expression by binding specific sequences in target genes.[32] While STAT phosphorylation by receptor-mediated JAK activation is the dogmatic description of STAT activation, important activity and function of nonphosphorylated STATs continue to be delineated.[34]

Notably, JAKs play roles in the signaling of cytokine receptors that regulate blood cell development and immune responses. Individual JAK signaling is not restricted to specific receptors, as multiple JAKs can be activated by the same receptor. For example, while JAK2 is thought to be the only JAK activated by the erythropoietin receptor (EpoR), IL-12 receptor activation leads to both Tyk2 and JAK2 activation. Likewise, multiple STATs may be activated by a single receptor.[32] There are extensive details understood about JAK/STAT signaling and the receptors that mediate this signaling. As this is not the focus of this chapter, readers are referred here to literature that reviews the functional details of these important proteins in cell signaling mechanisms and cell biology.[32,35]

3.1.2 JAKs in Cancer

Given the regulated enzymatic nature of a kinase to induce a cascade of signaling events in the cell, it is not surprising that deregulation of the control of kinase activity, especially tyrosine kinases, plays a major role in the development of cancer. JAK activity is deregulated in a number of different cancer types, and by a variety of mechanisms.[36] For example, chromosomal translocations that involve genes that encode members of the JAK family are found in cancer. The TEL-JAK2 chimeric fusion protein is the result of a chromosomal translocation that contributes to some cases of leukemia. This chimeric protein results in a loss of the normal regulatory mechanisms of JAK2, leading to constitutive tyrosine kinase activation and downstream signaling. The loss of the control of this signaling plays an important role in the progression of the disease. Another mechanism by which JAK proteins are deregulated in cancer is by mutation or inactivation of proteins that are involved in the negative regulation of JAK activity. Such proteins include SOCS and PIAS proteins, as well as protein tyrosine phosphatases such as the SHPs. Inactivation of such negative regulators can lead to deregulated control of JAK/STAT pathway activation, resulting in enhanced signaling. Such proteins are found mutated in a variety of hematological malignancies including various childhood leukemia (SHP) as well as lymphomas (SOCS, PIAS), among others. Finally, JAK proteins can be activated by point mutation exemplified by the presence of JAK1 and JAK3 mutations in B-cell and T-cell leukemia and JAK2 in myeloid neoplasms including myeloid leukemia and myelodysplastic syndrome.[36]

3.1.3 Signaling Pathways Activated by JAK2

As JAK2 activation drives MPN formation, it is important to understand some of the details of signaling by this tyrosine kinase. As briefly described earlier, JAK2 activation is prototypically described in the context of activation of homodimeric cytokine receptors such as the receptors for erythropoietin (EpoR) and thrombopoietin (TpoR). As JAK2 is constitutively bound to the cytoplasmic region of these receptors, the key initiating step of JAK2 signaling is the binding of ligand to the receptor. The subsequent conformational change of the receptor and associated JAK2 facilitates phosphorylation of the tyrosine kinase as well as the receptor and other downstream targets. The sequences surrounding the specific phosphorylated tyrosines in the receptor dictate what SH2 domains bind to it. Thus, one receptor that is bound to JAK2 may signal through different STATs than another receptor, depending on the sequence near the phosphorylated tyrosine.[32] Generically speaking, activation of STATs is one of three major signaling pathways often described for JAK2 signaling, the other two being the RAS/ERK pathway and the PI3K/Akt pathway.[37]

JAK2 can activate multiple STATs in response to signaling, including STAT5, STAT3, and STAT1, and this is dependent on the cytokine receptor that activates JAK2.[32] For example, in response to activation of the EpoR by erythropoietin, STAT5 is the major activated STAT family member canonically described. Activation of STATs by JAK2 leads to transcriptional regulation of STAT target genes. For example, activation of STAT5 leads to enhanced transcription of a variety of target genes that contain STAT5-binding sites in their transcriptional regulatory regions (e.g., promoters). Together, these genes have a wide effect on a variety of cellular mechanisms that regulate cell proliferation, cell survival, and cell differentiation.

Activation of JAK2 also leads to activation of the ERK pathway. ERK is a serine threonine kinase that can phosphorylate a large number of proteins in both the cytoplasm as well as the nucleus. Specific details of ERK targets are beyond the scope of this discussion but suffice it to say that in the cytoplasm, ERK contributes to the regulation of metabolism, apoptosis, receptor signaling, ion transport, protein translation, as well as feedback regulation of RAS signaling, among others. ERK can translocate to the nucleus where it has an additional plethora of targets including regulators of cell cycle and a large variety of transcription factors. In and of itself, ERK is central to a variety if signaling pathways and a key regulator of just about every facet of cell biology. Thus, deregulation of this pathway is thought to be a major contributor to cancer via loss of growth control, enhanced survival, and just about every other hallmark of cancer.[38]

JAK2 activation feeds positive signaling into the PI3K/Akt pathway.[39] Akt is widely known as a cell survival kinase as it is well documented to inhibit apoptosis by phosphorylating the proapoptotic protein BAD. This phosphorylation inactivates BAD and prevents it from inducing apoptotic. In addition, AKT phosphorylates and inactivates the transcription factor FOXO3A, which enhances expression of genes whose products induce apoptosis. Activation of PI3K/Akt is common in cancer as this pathway potently suppresses apoptosis, thus rendering cancer cells more susceptible to surviving under a variety of stresses, including loss of contact inhibition required for cancer cell metastasis as well as in response to chemotherapy, among others. Akt activation however has numerous other effects in the cell, including activation of signaling that controls protein translation via mTORC1, activation of cell cycle machinery, regulation of the p53 tumor suppressor protein, regulation of metabolism, and activation of the NFkB pathway, among others.

3.1.4 The Effect of the JAK2-V617F Mutation

The V617F mutation of JAK2 is present in the JAK2 pseudokinase domain. Structural studies and simulations have demonstrated that this mutation induces an altered interaction of the pseudokinase domain and the tyrosine kinase domain.[40] This change effectively diminishes the ability of the pseudokinase domain to inhibit the tyrosine kinase domain. The effect of this is a less stringently regulated kinase that is now much more sensitive to hyperactivation. This is somewhat different than many other potent-activating tyrosine kinase mutations observed in cancer (e.g., BCR-ABL, EGFR), as JAK2-V617F itself is not a particularly strong oncoprotein in cell-based assays. In fact, this mutant JAK2 still requires the presence of cytokine receptors to act as scaffolds upon which it can be activated.[41,42] It has been demonstrated that homodimeric receptors expression is required for JAK2-V617F to transform hematopoietic cells in culture. Overexpression of components of heterodimeric receptors can provide a similar functional role for JAK2-V617F activation and transforming ability.[43] Thus, the simple expression of JAK2-V617F may not be sufficient to significantly induce activation of cell signaling and the context in which it is expressed (i.e., what cytokine receptors are present) may affect the activity of this mutant JAK2.

Like JAK2, JAK2-V617F is constitutively bound to cytokine receptors. As described earlier, it is believed cytokine binding to its receptor leads to JAK2 dimerization and activation. In this process there is a conformational change of the receptor and its bound JAK2 molecules,[44] the latter of resulting in release of the negative regulation of the pseudokinase domain on the kinase domain. This conformational change of JAK2 is effectively provided by the V617F mutation.[40] Since the conformational change of JAK2-V617F is less dependent on cytokine binding to the receptor, this single point mutation provides cytokine-independent JAK2 activation. This is demonstrated in a variety of systems, most notably by transformation of cytokine-dependent cell lines[42] as well as erythropoietin-independent growth of hematopoietic progenitor cells of MPN patients, most notably, PV patients.[45] This growth has actually been used as a pathologic/diagnostic test for many decades, long before the identification of the V617F mutation of JAK2.

Activation of JAK2 via JAK2 mutation leads to enhanced signaling of downstream JAK2 effector pathways.[1] Activation of STATs, ERK, and Akt is observed in models of JAK2-driven neoplastic transformation, MPN mouse models, and cells from MPN patients. There is a variety of evidence that suggests each of these pathways may play a role downstream of JAK2 in MPNs. For some pathways (STAT5) the evidence is strong. For example, STAT5-deficient mice have been used to show the requirement of this JAK2 effector for JAK2-V617F-induced MPN phenotypes in mice.[46,47] Small molecule inhibitors of Akt[48–50] and MEK,[51,52] the direct activator of ERK, have also suggested roles for these effectors in JAK2-driven MPNs. Importantly, no matter what the genetic driver or accompanying mutations, a JAK/STAT gene signature is observed in primary cells of MPN patients.[53] This further demonstrates the importance of STAT activation in MPNs. Additional description of signaling downstream of JAK2-V617F will be presented later as the basis of anti-MPN therapeutic studies.

3.1.5 JAK2-V617F in Mouse Models of MPN

Important and compelling experimental data demonstrating the significance of JAK2-V617F in the formation of the MPN phenotype in humans has come from mouse model studies. The first mouse model study was presented in one of the initial reports describing the identification of JAK2-V617F in MPN patients.[9] In this, JAK2-V617F was expressed in bone marrow of a donor mouse, and these cells were then transplanted into a recipient mouse. The recipient mouse developed erythrocytosis, a hallmark of human PV. Additional reports reporting the results of bone marrow transplant models of JAK2-V617F-mediated MPNs have been performed, and each highlights the significance of the *JAK2-V617F* mutation in the pathogenesis of MPNs.[9,54–57] These models demonstrated that mice receiving bone marrow-expressing JAK2-V617F from a retroviral promoter developed erythrocytosis, leukocytosis, thrombocytosis, myelofibrosis, and splenomegaly. The recapitulation of a multitude of pathologic parameters associated with human MPN in a mouse model is significant in the context of elucidating etiological factors for MPN formation. Not surprisingly, expression of JAK2-V617F in these models resulted in activation of JAK2 and JAK2 signaling. An interesting observation in this MPN model system is the differences obtained with different mouse strains.[55] This suggests that the degree of the pathological phenotypes that are observed in human MPN may be affected by the specific genetics of the patients.

In addition, JAK2-V617F-expressing transgenic mice have been generated and display phenotypes resembling human MPN.[58–60] These models, like the retroviral promoter used in the bone marrow transplant models, used heterologous promoter/expression systems, which led to variability in JAK2-V617F expression. Interestingly, this variability suggested that the expression level of JAK2-V617F protein could influence the MPN phenotype. For example, higher expression led to a more PV-like phenotype,[59] which is interesting given the fact that homozygosity for *JAK2-V617F*, due to mitotic recombination resulting in uniparental disomy, is observed in PV and not other MPNs.[61]

In order to circumvent the use of heterologous promoter systems to express JAK2-V617F in a mouse model system, several knock-in models of JAK2-V617F-driven MPN have been developed.[62–64] These models further demonstrated that JAK2-V617F expression led to MPN-like phenotypes in mice. As the JAK2-V617F expression was present in hematopoietic stem cells in knock-in and transgenic models, researchers demonstrated that MPN phenotypes were transplantable and that hematopoietic stem cells and not progenitor cells were responsible as MPN-initiating cells. Expression of this mutant JAK2 may provide a competitive advantage to these cells. This could affect the disease burden observed in human MPN.

One of the most important uses of animal models in cancer is a xenograft system in immune compromised mice such as the nonobese diabetic/severe combined immunodeficient (NOD/SCID) mouse. The advantage of such a model is the ability to study primary cancer cells from patients in an in vivo system, where disease progression and therapeutic studies can be performed.[65] MPN researchers have not had widespread success using this system. While primary human cells from PV and ET patients were demonstrated to poorly engraft in NOD/SCID mice,[66,67] cells from myelofibrosis patients were capable of engraftment.[68] Interestingly, these bone marrow repopulating cells were from the peripheral blood of the patient. This engraftment however did not lead to normal hematopoietic repopulation, as more cells of the myeloid lineage were observed. Such a model may be useful to study the transformation of MPN progenitor cells into AML.[68]

In summary, JAK2-V617F expression in a variety of mouse model systems clearly indicates that this JAK2 mutant plays a significant role in the formation and progression of MPN. Similarly, expression of other MPN driver mutations that activate JAK2 (e.g., *JAK2* exon 12 mutation) in similar mouse models also leads to MPN phenotypes.[18,19] These model systems have wide ranging value, including studying etiological assessment of MPN drivers (alone and in combination with other genetic mutations/modifiers or loss of function of JAK2 effectors), assessing details of clonal hematopoiesis induced by JAK2-V617F expression, as well as in vivo therapeutic systems to assess novel MPN therapeutics. For example, experiments using STAT loss-of-function mice indicate a profound requirement for STAT5,[46,47] while surprisingly the loss of STAT3 actually enhanced the observed MPN phenotype.[69]

3.2 MPL: Mutational Activation

3.2.1 Thrombopoietin Cytokine Receptor Signaling

The oncogenic activity of the *v-MPL* oncogene was initially shown to immortalize a range of different cell types of the hematopoietic system.[70] Human c-*MPL* was identified as the gene that encodes TpoR, providing further insight into its ability to alter signaling pathways in blood cells.[71] MPL, as TpoR is commonly referred to, is a transmembrane receptor that functions as a receptor for thrombopoietin, or megakaryocyte growth and development factor. As its name implies, thrombopoietin functions to stimulate growth of megakaryocytes, which go on to form platelets.[71] Genetic removal of MPL from mice leads to a thrombocytopenia, establishing MPL as an important component of platelet formation.[72]

MPL functions as a homodimeric receptor for thrombopoietin. Binding of thrombopoietin to MPL activates cell-signaling pathways, including

JAK/STAT, ERK, and Akt, as well as the adapter molecule SHC.[73] Most notably, JAK2 plays a major role in eliciting signaling initiated by MPL. JAK2 activation by MPL is conceptually similar to that described for EpoR. The homodimerization of JAK2-bound receptor leads to conformational changes in the receptor and JAK2, facilitating transphosphorylation and full activation of JAK2 tyrosine kinase activity.[73]

3.2.2 Mutational Activation of MPL in MPNs

MPL mutations are found in familial thrombocytosis, or familial ET. This mutation changes amino acid 505 from a serine to an asparagine within the transmembrane domain of the receptor, resulting in activation of the receptor in the absence of thrombopoietin.[74] Thus, the identification of somatic *MPL* mutations in ET and PMF is not surprising. The most common somatic mutation of MPL observed in MPNs is at amino acid 515, changing a tryptophan to a leucine.[19] While near the familial ET mutation at amino acid 505, this mutation is present in the juxtamembrane region of MPL, which is just inside the plasma membrane at the beginning of the cytoplasmic portion of the receptor. This mutation disrupts a helix that is important in keeping the receptor in an inactive conformation.[75] Thus, this mutation leads to a receptor that is constitutively active and leads to activation of associated JAKs, including JAK2.[19] Somatic mutation of MPL may enhance dimerization of receptors as well as conformational changes in the receptor and/or the associated JAKs.[75] Thus, such a mutation of the receptor circumvents the need for ligand, thrombopoietin, as the binding of ligand is what is normally required to induce these changes for activation of associated JAKs and subsequent signal transduction from the receptor.

Expression of the MPN-associated MPL-W515L protein, but not wild-type MPL, leads to cytokine-independent transformation of cytokine-dependent cells.[19] This expression leads to pronounced activation of JAK2 and downstream JAK2 effector pathways, including STAT5 and STAT3 activation, as well as activation of the PI3K/Akt and ERK pathways, compared to expression of wild-type MPL. Hematopoietic cell transformation to cytokine independence by MPL-W515L is sensitive to JAK inhibition, demonstrating the requisite nature of this pathway for MPL-W515L-mediated neoplastic signaling and cell growth. Furthermore, in bone marrow transplant assays, MPL-W515L induces a strong MPN phenotype, including thrombocytosis, splenomegaly, and fibrosis of the bone marrow.[19] This phenotype can be ameliorated by treatment of animals with a JAK2 inhibitor.[76] Once again, in vivo studies demonstrate that an oncogenic driving mutation found in MPNs can recapitulate many aspects of the disease in an in vivo animal model. Furthermore, MPL-W515L studies have clearly demonstrated activation of JAK2 by, and a dependency on JAK2 for, neoplastic signaling by this mutant receptor. This further solidified the importance of JAK2 in MPNs and explains why activating MPL and JAK2 mutations in MPNs are generally mutually exclusive.

3.3 CALR: Altered Function by Mutation

3.3.1 The CALR–MPN Connection

The presence of activating *JAK2* mutations in essentially all cases of PV and about half of ET and PMF patients, along with the approximate 3–7% of ET and PMF cases having an activating mutation in *MPL* that enhances JAK2 signaling, clearly suggested that aberrant regulation of JAK2 was a driving force behind human MPN. As described earlier, animal models and small molecule JAK2 inhibitor studies demonstrated the significance of deregulation of JAK2 in the etiology of MPN. However, with this said, about 30–40% of ET and PMF patients do not contain mutated JAK2 or MPL. This suggested there were likely other mutations that led to activated JAK2 signaling, in non-JAK2 and non-MPL-mutated MPNs, that had yet to be identified. Mutations of other cytokine receptors or other kinases were likely culprits. In 2013, mutations in the *CALR* gene, which encodes for the protein calreticulin, were identified as recurring somatic mutation in MPNs, specifically ET and PMF.[14,15] These mutations were found in the subset of *JAK2/MPL*-mutation-negative patients. Unlike well-known signaling proteins such as kinases, adapter proteins, and cell surface receptors, calreticulin is a protein with a wide variety of functions, none of which include phosphorylating or dephosphorylating other proteins, or participating in well-known neoplastic signaling cascades.[77] The identification of mutated *CALR* in MPNs[14,15] thrust the spotlight on the function of CALR and how its mutation could impinge on the MPN phenotype. Before describing the CALR mutations found in MPNs in more detail, it will be beneficiary to describe the function of CALR.

3.3.2 Cellular Functions of CALR

While a rather unknown molecular player in cancer, CALR was actually first identified in the early 1970s. This work identified CALR as a protein binder of calcium in the sarcoplasmic reticulum of muscle cells.[78,79] In fact, CALR is a major calcium-binding protein found in a variety of locations of the cell, including intracellular, extracellular, and on the cell surface. Major functions of CALR include regulating calcium homeostasis, protein folding, adhesion, as well as immune cell function, among others.[77] CALR protein has a number of specific domains important to its cellular functions.[77,80] The N-domain of CALR is a region of the protein required for intramolecular interactions as well as interactions with other proteins. The P-domain is the central portion of the protein. This domain provides a high-affinity-binding site for calcium, and thus is a critical domain of CALR. Together the N-domain and the P-domain contribute to the molecular chaperone function of CALR. Calcium binding is also a function of the C-terminus of CALR. The C-terminus of CALR is rich in acidic amino acids that impart a high-capacity calcium-binding activity for CALR. Importantly, this binding is low affinity and thus plays an important role in the calcium homeostasis function of CALR. Notably, CALR contains an endoplasmic retention sequence (KDEL motif) at its very C-terminus. This KDEL motif plays the critical role of assuring CALR is properly localized in the endoplasmic reticulum.[77,80]

CALR regulates calcium homeostasis in the endoplasmic reticulum. In fact, it is reported that CALR binds over 50% of the Ca^{2+} in the lumen of the endoplasmic reticulum, which is the major location of calcium in the cell. Thus, CALR plays a role in the storage of calcium that can be released during calcium signaling.[77,80] As such, overexpression of CALR elevates the level of calcium in the endoplasmic reticulum,[81–83] while cells lacking CALR have diminished calcium storage in the endoplasmic reticulum.[80] Altering the levels of CALR, up or down, in cells affects calcium homeostasis with little affect on protein folding,[84] another major function of CALR. This suggests calcium homeostasis is a more significant function for CALR than its role in protein folding. In fact, mice deficient in CALR die due to impaired cardiac development due to the loss of the calcium homeostasis function of CALR.[85]

Another important function of CALR is as a molecular chaperone in the endoplasmic reticulum.[77,86] Proteins synthesized in the endoplasmic reticulum go through processing that leads to their proper delivery to subcellular compartments. During this process, molecular chaperones bind to newly synthesized proteins to ensure the proper folding of the protein and subsequent processing and localization.[87] Improperly folded or misfolded proteins can be targeted for proteolysis to control their levels. Deregulation of this system (e.g., an increase in the levels of misfolded proteins) can lead to endoplasmic reticulum stress, which can play a role in a variety of diseases, including cancer.[88] CALR is one of many molecular chaperones that participate in this complex, yet vital, regulation of newly synthesized proteins. CALR participates with calnexin in a process called the calreticulin/calnexin cycle, which regulates the processing of newly synthesized glycosylated proteins.[77] This is a complex cycle of glucosylation and deglucosylation that ensues to ensure newly synthesized glycoproteins fold properly.[77] Interestingly, calnexin is highly homologous in amino acid sequence and functional domains,[77] but calnexin mutations have yet to be identified in MPNs.

While CALR plays a critical role in calcium homeostasis and functions as a molecular chaperone in the endoplasmic reticulum, CALR also directly functions extracellularly. Outside the cell CALR functions in complexes that regulate adhesion as well as interactions with immune cells. CALR can regulate expression of adhesion proteins, although this is actually from within the endoplasmic reticulum. It also plays a role outside of the cell, where it is involved in the regulation of the loss of focal adhesion, and thus may contribute to migratory properties of cells. In addition, CALR plays a role in a cell's interaction with the immune system.[89,90] For example, the presence of CALR on the surface of apoptotic cells is required for engulfment by phagocytic cells. In addition, CALR plays roles in T-cell function, tumor necrosis factor signaling, and cancer immunity. Readers are referred to recent reviews on CALR function, as, not surprisingly, a multifunctional protein that regulates calcium homeostasis, protein folding, adhesion, and immune responses among other activities cannot be adequately described here.[77,89–92]

3.3.3 Roles of CALR in Cancer

As a multifunctional protein that regulates calcium homeostasis and adhesion, CALR has, not surprisingly, been suggested to play roles in malignant cell growth in cancer.[91,92] CALR has been reported to be overexpressed in a wide variety of cancers, including breast, stomach, and pancreatic cancer, among others.[91,92] Overexpression of CALR in cancer is not limited to solid tumors, as it is also reported to be elevated in hematopoietic cancers such as AML.[91] One of the better understood roles for CALR in cancer involves its ability to regulate adhesion. In fact, an increased degree of invasion and metastasis is associated with overexpression of CALR in various cancers.[91,92] Mechanistically, it has been suggested that CALR can regulate adhesion of neoplastic cells by regulating the expression of E-cadherin via enhanced activity of the Slug transcription factor,[93] which functions as a negative regulator of E-cadherin.[94] Decreased E-cadherin is associated with enhanced invasive properties of cancer cells.[38,95] In addition, CALR can regulate focal adhesion formation, which could alter migratory and invasive properties of cancer cells.[96,97] There is also evidence that CALR may regulate motility and inhibit anoikis (apoptosis upon loss of adhesion to extracellular matrix) by regulating STAT3 and Akt activity in cancer cells.[98] In addition to possibly contributing to neoplastic cell growth in cancer via pro-metastatic activities, the functions of CALR in phagocytosis of apoptotic cells by immune cells and in inflammatory responses of the immune system are mechanisms by which this protein may play roles in immunological responses to cancer cells, especially in the context of chemotherapeutic treatment.[89,90] Upregulation of VEGF[99] and enhancing cell proliferation[91,92,100] may also contribute to cancer cell growth. However, as can be expected with a protein like CALR, which has a large variety of functional activities ascribed to it, the contribution of CALR in cancer may be cancer cell-type dependent, as CALR has also demonstrated antiproliferative effects in cancer cells.[92,101,102]

3.3.4 CALR Mutations in MPNs

Mutations in the gene that encodes CALR were identified in MPN patients, more specifically, ET and PMF patients that lacked activating mutations in *JAK2* or *MPL*.[14,15] In fact, *JAK2*, *MPL*, and *CALR* mutations are generally mutually exclusive, as *CALR* is mutated in the majority of *JAK2/MPL* mutation-negative patients, which corresponds to about a quarter of all ET and PMF patients.[12] Considering the known role of JAK2 activation as a driver for MPN formation, whether this is in the context of *JAK2-V617F* mutation in PV, ET, and PMF, *JAK2* exon 12 mutations in PV, or *MPL* mutations in ET and PMF, it was a reasonable assumption that additional drivers of MPNs would have a connection to JAK2 activation. Because CALR did not have an obvious, or rather was not clearly connected to JAK2 activation, the identification of the high frequency recurring somatic *CALR* mutations was a bit puzzling.

The initial identification of frequently recurring *CALR* mutations in MPNs in late 2013 provided additional insight into the genetics of MPNs.[14,15] Mutated CALR was initially thought to be a driver of ET and PMF because of its mutual exclusivity with *JAK2* and *MPL* mutations, its high prevalence in patients who do not have these MPN driver mutations, and its expression in different stages of myeloid cell development, including hematopoietic stem cells in these patients.[14,15] With a dearth of knowledge that would clearly indicated how CALR might activate JAK2, it was important that this connection, if real, be determined, which would solidify CALR is a bona fide driving mutation in MPNs.

CALR is mutated in MPNs by DNA insertion as well as deletion.[14,15] Interestingly, and significantly, these apparently disparate mutational mechanisms result in very similar structural changes to CALR. Dozens of distinct insertion or deletion mutations were identified the in exon 9 of *CALR*, which encodes the carboxy terminus of CALR. The two most common mutations identified included a 52-bp deletion mutation and a 5-bp insertion mutation. Fascinatingly, each of these mutations actually leads to very similar structural changes to the amino acid sequence of CALR.[14,15] These mutations cause a +1 change in the *CALR* reading frame, resulting in a unique amino acid sequence at the carboxy terminus of mutant CALR protein. The carboxy terminus of nonmutated CALR protein, as described earlier, is extremely rich in negatively charged amino acids that are important for calcium-binding function of CALR. Also, the KDEL endoplasmic retention motif is located at the very carboxy terminus of CALR.[80] The exon 9 mutations found in MPN thus disrupt an important region of CALR that is requisite for its function. As calcium homeostasis is a major function of CALR that depends on the ability of CALR to localize to the endoplasmic reticulum, disruption of CALR's calcium-binding capacity and localization to the endoplasmic reticulum would be expected to have a drastic affect on its function in the cell. In fact, not only does mutant CALR lack the KDEL endoplasmic retention sequence, but its unique carboxy terminus consists of essentially all positively charged amino acids, the complete opposite of the negatively charged rich tail of wild-type CALR.[14,15]

The initial reports that identified *CALR* mutations in MPNs suggested that mutant CALR is still present in the endoplasmic reticulum, even though it lacks the KDEL endoplasmic retention sequence.[15] However, it was suggested that this localization was less prominent than nonmutated CALR, which would be consistent with a loss of the KDEL motif.[14] In an effort to identify a role for mutant CALR in JAK2 activation, mutant CALR was expressed in hematopoietic cells that require interleukin-3 (IL-3), and subsequent JAK2 activation, for growth. Expression of mutant CALR led to IL-3 hypersensitivity, showing enhanced STAT5 phosphorylation.[14] Cells expressing exogenous mutant CALR showed proliferative capacity in the complete absence of cytokine, while cells expressed exogenous nonmutated CALR could not. Finally, these effects were reversed by the use of a small molecule JAK2 inhibitor.[14] Taken together, the altered functionality of mutant CALR found in MPNs appeared to, at least in part, impinge on JAK2/STAT5 signaling. However, the mechanism by which this occurred was not determined in these studies.

3.3.5 Mechanism of Gain of Function of Mutant CALR in MPNs

Once again animal models have proven the go to system to assess the potential role an MPN oncogenic driver may have in formation and progression of MPN phenotypes. In a bone marrow transplant assay, mutant CALR was shown to induce an MPN phenotype in mice.[103,104] This phenotype was highlighted by thrombocytosis and megakaryocytic hyperplasia and thus similar to ET in human MPN patients. In one report, myelofibrosis was also observed in mice expressing mutant CalR.[104] Mutant CALR expression appeared to affect the number of hematopoietic stem cells[104] but not the numbers of myeloid progenitor cells in this model.[103] However, mutant CALR did increase the number of megakaryocyte progenitors and megakaryocytes,[103,104] further demonstrating a driving force behind thrombocytosis in MPN. The MPN phenotype observed with expression of mutant CALR was not observed with expression of wild-type CALR, suggesting the ability of mutated CALR was indeed a function of the mutated carboxy terminus.[103,104] These studies solidified *CALR* mutation as a MPN driving oncogene, joining mutational activation of JAK2 and MPL as the three major driving genetic events in PV, ET, and PMF.

In addition to the animal model studies of MPN-associated mutant CALR, insight into the potential mechanism by which mutant CALR might contribute to ET and PMF has been provided. The ability of mutant CALR to activate JAK2 signaling is likely requisite in the mechanism by which it can function as a molecular driver of MPN. In this regard, numerous groups have shown that mutant CALR can enhance signaling from MPL, the receptor for thrombopoietin.[103–109] Expression of mutant CALR was shown to transform cytokine-dependent cells in vitro.[103–109] This transformation was dependent on the expression of MPL.[103–109] Expression of other type cytokine receptors, such as EpoR or GCSFR, could not support the transforming properties of mutant CALR.[103–108] Protein interaction studies have determined that the MPN mutations of CALR lead to an interaction of mutant CALR with MPL.[103–105] Because mutant CALR interacts with MPL preferentially over nonmutated CALR, it was suspected the novel carboxy terminus played a role in this interaction. The loss of the KDEL endoplasmic retention sequence is one major alteration of mutant

CALR, but this was shown to not be sufficient to mimic the frameshift-mutated CALR.[103] In fact, it is the alteration of the carboxy terminus to a highly positively charged state that bestows its functional alteration as a MPL-interacting and activating protein.[103,107] CALR is understood to have both intracellular and extracellular roles, and frameshift mutant CALR found in MPNs has altered subcellular localization. This may be in part because of the altered repertoire of proteins mutant CALR interacts with, such as MPL, compared to wild-type CALR. To date, the mechanistic consensus drawn from numerous reports is that the interaction of mutant CALR with MPL requires the positively charged novel carboxy terminus to enhance its interaction with MPL via the N-domain of CALR.[103–109] Interestingly, MPL, like other cytokine receptors, is a glycosylated protein, and the glycan-binding function of the N-domain of CALR likely plays a role in the interaction of CALR with MPL.[105,107] Alteration of MPL glycosylation processing leading to a decrease of MPL localized to the cell surface may play a role in the mechanism by which mutant CALR activates MPL signaling.[106,107] Thus, mutation of CALR in MPNs may actually lead to interaction of CALR and the extracellular region of MPL, inducing activation of MPL homodimeric receptor complexes associated with JAK2, and subsequent activation of JAK2 signaling. In a sense, mutant CALR functions like a ligand for MPL, in that it interacts with the extracellular region of MPL and activates receptor signaling. Of course, this "ligand" appears to be constitutively bound to MPL from the early stages of MPL processing in the endoplasmic reticulum and Golgi vesicles. Furthermore, mutant CALR could not function in a paracrine manner to activate MPL.[105] Importantly, while MPL is required for mutant CALR-induced signaling and transformation and cannot be replaced by other cytokine receptors such as EpoR or GCSFR, a chimeric protein encompassed of the extracellular and transmembrane domain of MPL, and the intracellular domain of EpoR, could be activated by mutant CALR and support mutant CALR-mediated cell transformation.[107] The opposite chimeric protein could not, suggesting it is a unique property of the MPL extracellular domain that is required for mutant CALR to activate MPL signaling.[107]

Alteration of the function, subcellular localization, protein-binding partners, etc., of mutated CALR could result in multiple mechanistic aspects that contribute to mutant CALR function in MPNs. With that, while CALR plays a role in phagocytic responses in the immune system, it does not appear that the ability of immune cells to phagocytose cells expressing mutant CALR is affected.[110] Thus, the mechanism by which CALR contributes to MPN formation may not involve alteration of this immunological response. Myeloperoxidase deficiency, which is observed in a small fraction of MPN patients, was shown to be associated with homozygous mutation of *CALR* in myelofibrosis patients.[111] Loss of CALR function led to proteasomal degradation of myeloperoxidase, a client protein of CALR chaperone activity. Thus, while mutant CALR may drive MPN formation via MPL activation, loss of CALR chaperone activity by homozygous *CALR* mutation may contribute to additional clinical features observed in MPN patients.[111] Also, CALR expression regulates calcium-dependent apoptosis.[112,113] A decrease in CALR sensitizes cells to apoptosis because of the requirement of ER calcium stores for apoptosis.[112] Altered CALR function may alter regulation of apoptotic pathways in *CALR* mutation-positive MPNs. To date, the effect of altered calcium homeostasis in *CALR* mutation-positive MPNs is poorly understood.

4. MPN THERAPEUTICS

4.1 Standard of Care of MPNs

The development of the standard of care of MPNs has a long history, although many recent significant clinical studies have contributed to its development. While a brief description is provided here, readers are referred to a 2016 review that provides details for current standard MPN therapies.[2] PV and ET are chronic neoplastic diseases that have near-normal life expectancies. This, as a generalization, dictates a nonaggressive frontline therapy for these diseases.[2] PV and ET patients are at increased risk of complications such as thrombosis and bleeding, and treatments for these neoplasms are generally focused on decreasing the risk of such complications and alleviating symptomology.[2] Lower hematocrit levels provide a clear survival advantage for PV patients, and thus phlebotomy is performed in these patients to control hematocrit.[114] Low-dose aspirin and cytoreductive therapy, generally with hydroxyurea treatment, are used to decrease thrombotic complications and other symptoms.[2] The specific frontline therapy recommended for an individual PV or ET patient depends on a variety of factors that effectively classify patients into risk groups.[2] For example, these groups range from the very low risk group, consisting of patients less than 60 years old with no history of thrombotic complications and no *JAK2* mutation, to a high risk group, consisting of patients who are older than 60 and have a *JAK2* mutation and history of thrombosis.

As described earlier, PMF is a much more serious MPN, with patients having a significantly reduced life expectancy. Unfortunately no treatment has proven to significantly increase survival for PMF patients,[2] although long-term analysis of patients treated with the JAK2 inhibitor ruxolitinib (described in the next section) is promising in this regard.[115] The lone exception to this is an allogeneic stem cell transplant. If even an option, many cannot benefit from it because of the associated high risk.[12] Again PMF patients are treated based on risk stratification. Patients are treated witha range of drugs, again primarily to reduce symptomology.[2] As examples, these include hydroxyurea in cases of symptomatic splenomegaly,[116] prednisone with or without lenalidomide for anemia,[117] and splenectomy for drug-insensitive splenomegaly.[118] Notably, the JAK2 inhibitor ruxolitinib has recently been approved for use in higher risk myelofibrosis patients.[119]

4.2 Molecular Targeted and Investigational Therapies for MPN

Standard of care for MPNs has centered on controlling symptomology and reducing risk factors for survival. However, the identification of the JAK2-V617F mutation in MPNs in 2005 ushered in a new era of investigational therapies. Most notably is the development of JAK2 tyrosine kinase inhibitors. Mutational activation of the JAK2 tyrosine kinase in chronic myeloid neoplasms is reminiscent of the mutational activation of the ABL tyrosine kinase in CML, as CML, in fact, is an MPN. This similarity along with the success of the tyrosine kinase inhibitor imatinib, and subsequent ABL kinase inhibitors, for the targeted therapy for CML, suggested JAK2 inhibitors might be effective therapies for MPNs. Thus, JAK2 inhibitors were rapidly developed and assessed for potential therapeutic efficacy against MPNs.[119] In addition to JAK2 inhibitors, additional targeted therapies have been tested for MPNs, including numerous combination targeted therapeutic approaches.

4.2.1 JAK2 Inhibitors

4.2.1.1 Ruxolitinib

Ruxolitinib (INCB018424) is a potent dual JAK1 and JAK2 inhibitor that exhibits low single digit nanomolar biochemical IC_{50}s for both kinases. The selectivity within the JAK family members is represented by a sixfold selectivity over Tyk2 and approximately 130-fold selectivity over JAK3.[45] Ruxolitinib demonstrates dose-dependent inhibition of JAK2/STAT signaling and inhibition of cell growth that is dependent on JAK2 activation (e.g., JAK2-V617F). Ruxolitinib has little effect on cells that are not dependent on JAK2, suggesting specificity of the kinase may be suitable for its use to target JAK2-driven neoplasms such as MPNs. It effectively inhibits the neoplastic growth of erythropoietin-independent erythroid colonies derived from myeloid progenitor cells from MPN patients. Ruxolitinib has demonstrated efficacy in numerous animal models of MPNs, where it leads to improvement of parameters of disease burden. However, JAK2 inhibition is not curative in these models.

Ruxolitinib has been tested extensively in clinical trials. Most notably the controlled myelofibrosis study with oral JAK inhibitor treatment (COMFORT) trials, COMFORT-I[120] and COMFORT-II,[121] were two randomized Phase III trials conducted to test ruxolitinib treatment vs placebo (COMFORT-I) or best available therapy (COMFORT-II). In short, significant responses in splenomegaly reduction and improvement of quality of life were observed. Anemia and thrombocytopenia were the most common adverse effects, and likely represent on-target JAK2 inhibition as JAK2 is utilized for signaling by thrombopoietin receptor (MPL) and the EpoR, thus regulating platelet and red blood cell production, respectively. Positive effects of ruxolitinib were observed independently of the presence of JAK2-V617F, possibly demonstrating the conserved importance of aberrant JAK2 activation in MPN patients.[119–121] Because of the successful improvement of the quality of life observed in the COMFORT-I and COMFORT-II trials, ruxolitinib was approved for use in myelofibrosis patients in the United States in 2011, and subsequently in Canada and Europe.[119] Follow-up studies suggested a survival improvement in ruxolitinib-treated patients in both trials.[115,122,123] While clearly significant for patients and clinicians, the reason for this survival improvement is not yet clear, as it may be at least partly a reflection of the improved quality of life (e.g., decreased satiety leading to improvement in nutrition and overall health).[119] A decrease, but not elimination, in allele burden was observed in most patients from COMFORT-II,[123,124] suggesting alternative strategies need to be tested if ruxolitinib, or perhaps other JAK2 inhibitors, is going to be part of a treatment regimen that offers hope for a potential cure for patients. Ruxolitinib is not specific for mutated JAK2 and likely has on-target and off-target effects that contribute to both positive (e.g., decrease in proinflammatory cytokine signaling) as well as adverse effects, suggesting improvements in not only efficacy but also specificity may lead to a drug that will truly target malignant cells in myelofibrosis and other MPNs.[125]

With its success in clinical trials for myelofibrosis, the RESPONSE (randomized study of efficacy and safety in PV with JAK inhibitor INCB018424 vs best supportive care) trial assessed the efficacy of ruxolitinib vs best available therapy in PV.[126] PV patients in this trial were resistant or intolerant to hydroxyurea. This trial demonstrated long-term safety of ruxolitinib treatment for hydroxyurea patients. Data analysis after all patients passed 80 weeks of treatment demonstrated an approximately 90% probability of ruxolitinib-treated patients maintaining primary responses, including at least 35% reduction in spleen size and phlebotomy-independent control of hematocrit.[127] Complete hematological response was observed in a significant fraction of patients, and most of the patients that were randomized to the best available therapy arm of this trial crossed over to ruxolitinib.[127] Adverse effects were relatively rare and mild, and thromboembolic events were reduced in ruxolitinib-treated patients.[127] Ruxolitinib was approved for use in hydroxyurea-resistant or intolerant PV patients in the United States (2014),[128] and more recently in Europe (2015), and Canada (2016).

4.2.1.2 Fedratinib

Fedratinib (TG101348, SAR302503) is a JAK2 inhibitor with a similar reported IC_{50} (~3 nM)[129] toward JAK2 kinase activity as ruxolitinib. However, unlike ruxolitinib, fedratinib demonstrates selectivity (~35-fold) over JAK1 and thus provides a more specific JAK2 inhibitor in this biochemical assessment.[129] It is also reported to exhibit about 135-fold and 334-fold lower an IC_{50} forJAK2 than Tyk2 and JAK3, respectively.[129] Fedratinib blocks JAK2/STAT signaling and induces growth inhibition and apoptosis in JAK2-driven cells.[129,130] In addition, it inhibits neoplastic growth of myeloid progenitor cells from MPN patients and demonstrates efficacy in MPN animal models.[62,129–131] However, it is unable to eliminate the MPN-initiating stem cell pool in an MPN mouse model.[62]

Fedratinib clinical trial studies for myelofibrosis demonstrated reduction in splenomegaly and other symptomology, including decreased circulation cytokines.[132,133] No significant reduction in JAK2-V617F allele burden was observed.[132,133] Adverse effects included anemia and diarrhea.[132,133] Interestingly, fedratinib was also reported to demonstrate clinical benefits to patients who had previously failed ruxolitinib treatment.[134] This suggests that patients who do not benefit from one JAK2 inhibitor may benefit from an alternative JAK2 inhibitor therapy. While fedratinib looked like a promising candidate as a new therapy for MPNs, its clinical development was halted after Wernicke's encephalopathy, a neurological side effect brought upon by thiamine deficiency,[135] was identified in patients.[132] Interestingly, fedratinib, and not other JAK2 inhibitors tested, was shown to bind to thiamine receptors and inhibit thiamine uptake, thus highlighting a possible off-target mechanism that could contribute to Wernicke's encephalopathy observed in fedratinib-treated patients.[136]

4.2.1.3 Momelotinib

Momelotinib (CYT387) is a dual JAK1 and JAK2 inhibitor with biochemical IC_{50}s of approximately 10 nM for each.[137] Like other JAK2 inhibitors, momelotinib inhibits JAK2 signaling in cells as well as JAK2-driven cell proliferation.[137] In addition, this inhibitor blocks neoplastic colony growth of myeloid progenitor cells from MPN patients.[137] Clinical studies in advanced myelofibrosis patients demonstrated momelotinib reduces splenomegaly and constitutional symptoms. Notably, momelotinib actually improved anemia in myelofibrosis patients. Complications included thrombocytopenia as well as a treatment-induced neuropathy which was observed in about one in four patients.[138] While momelotinib is a JAK1 and JAK2 inhibitor, it also inhibits the IKBK-epsilon and TBK1 kinases. It is possible the inhibition of these or other kinases may play a role in the improvement of anemia and/or perhaps the induction of peripheral neuropathy in momelotinib-treated patients.

4.2.1.4 Pacritinib

Pacritinib (SB1518) is a JAK2 inhibitor that also targets Fms-like tyrosine kinase-3 (FLT3). The IC_{50} for these kinases is about 20 nM. Unlike a number of other JAK2 inhibitors, pacritinib exhibits extreme selectivity (about 60-fold) of JAK2 over JAK1, while also inhibiting JAK3 with about 25-fold less potency than JAK2.[139] A phase 2 trial of pacritinib in myelofibrosis patients again demonstrated reduction in splenomegaly and improved constitutional symptoms. Importantly, anemia and thrombocytopenia were observed but were not significant enough to alter drug treatment, and, in fact, pacritinib did not significantly affect preexisting anemia or thrombocytopenia, suggesting pacritinib may be a better JAK2 inhibitor option for patients with these cytopenias.[140]

4.2.1.5 Others

There are numerous other JAK2 inhibitors that have been evaluated for their efficacy against JAK2 and in MPN model systems, including BMS-911543, an inhibitor that is selective for JAK2 over JAK1.[141] Interestingly, this drug did not result in inhibition of disease phenotype, including no effect on hematocrit or allele burden, in an MPN animal model, suggesting JAK2 inhibition alone may be insufficient for treating MPNs.[142] Other JAK inhibitors being assessed in MPN model systems or other diseases include AZD1480 and tofacitinib.[143]

4.2.1.5.1 JAK2 Kinase Inhibitor Resistance/Persistence The lack of significant efficacy of JAK2 inhibitors to decrease neoplastic allele burden in animal models and humans suggests that neoplastic MPN cells are inherently resistant to JAK2 inhibitors. This inherent resistance is rapid, in that there is not a remission period followed by a relapse of disease. As such, it was not surprising that patients that are undergoing chronic ruxolitinib treatment do not develop mutants of JAK2 that induce drug resistance,[144] which is commonly observed in other kinase inhibitor-based anticancer therapies. Thus, there appears to be no selective pressure for developing kinase inhibitor resistance mutations. Chronic exposure of MPN model cell lines to type I JAK2 inhibitors leads to cells that are cross-resistant to other JAK2 inhibitors.[144] This resistance is referred to as persistence, because while these cells persistently grow in the presence of JAK2 inhibitor, this phenotype is reversible, as removal of drug can resensitize cells to the effects of JAK2 inhibition.[144] The development of JAK2 inhibitor persistent cells does not select for drug-resistant mutations within *JAK2*, as such mutations are not found in these cells.[144] This mimics the lack of drug resistance-inducing *JAK2* mutations in patients undergoing chronic JAK2 inhibitor treatment. The mechanism for this JAK2 inhibitor persistent phenotype has been attributed to the recruitment of other JAK family members, namely, JAK1 and Tyk2, into heterodimeric complexes with JAK2, where these members contribute to JAK2 pathway activation.[144] Importantly, JAK2 is still required for this persistent state[144] and for the maintenance of the MPN phenotype in mice undergoing ruxolitinib treatment.[145] These data suggest not only that inefficient JAK2 inhibition by current JAK2 inhibitors may provide an escape of JAK2-driven neoplastic cells in MPNs but that the development of alternative approaches to inhibit JAK2 and JAK2 signaling still hold promise as anti-MPN therapies. Such therapies are described in the next section and include combination therapies of JAK2 inhibition with small molecules that target important proteins in signaling pathways downstream of JAK2 or other pathways, as well as alternative approaches to target JAK2 more directly.

4.2.2 *Other Investigational, Small Molecule, and Combination Therapies*

While JAK2 inhibitors have provided hope and symptomatic relief for MPN patients, these kinase inhibitors are unable to induce clinical remission. JAK2 inhibitors that function by a different mechanism may prove to be more effective. Many other small molecule inhibitors of signaling proteins have been tested for their efficacy in MPN model systems. Most notably, these include inhibitors of signaling pathways downstream of JAK2 (e.g., PI3K, Akt, mTOR, PIM, BCL2, etc.) as well as other inhibitors that provide less specificity with respect to target activity (e.g., heat shock protein (HSP) inhibitors and histone deacetylase inhibitors (HDAC)), but clearly offer a rationale for their potential use and mechanism of action in the setting of JAK2-driven neoplasms. As MPNs driven by aberrant JAK2 activity, combination therapies with JAK2 inhibitors are widely being studied, as it is believed that targeting parallel signaling pathways, or multiple points in critical signaling pathways, could enhance therapeutic efficacy and perhaps prevent the onset of drug resistance.

4.2.2.1 Type II JAK2 Kinase Inhibition

The kinase inhibitors discussed so far are referred to as type I kinase inhibitors because these small molecules bind to their target kinase domains when these domains are in an active confirmation and subsequently inhibit the activated kinase. JAK2 inhibitor resistance is observed in cells chronically exposed to type I kinase inhibitors as well as in MPN patients who are nonresponsive to ruxolitinib. Interestingly, JAK2-driven cells that are resistant to one type I JAK2 inhibitor are cross-resistant to other type I JAK2 inhibitors.[144] Unlike type I kinase inhibitors, type II kinase inhibitors bind to their target when the kinase is in an inactive confirmation, thus preventing kinase activation. CHZ868 is a type II JAK2 kinase inhibitor that has been assessed as an alternative approach to JAK2 inhibitor development for MPNs.[146] Cells that are resistant to type I kinase inhibitors remain sensitive to CHZ868, as this inhibitor reversed the JAK2-inhibitor state of drug-resistant cells.[146] CHZ868 demonstrated impressive efficacy in MPN animal models, including reduction of disease allele burden.[146] The latter is significant as the reductions in allele burden are much more significant than previous studies with type I JAK2 kinase inhibitors. The clinical assessment of type II JAK2 kinase inhibitors in MPNs will be very important to the field of anti-MPN therapeutics.

4.2.2.2 PI3K/Akt/mTOR Inhibitors

PI3K is a lipid kinase that plays a key role in cell surface receptor signaling to activate various signaling pathways, most notably activation of the Akt serine threonine kinase. Akt affects a variety of cellular activities, including inhibiting apoptosis by phosphorylating and inactivating the proapoptotic protein BAD.[39] Another activity of the Akt serine threonine kinase is to activate the mTOR kinase within the mTORC1 protein complex. mTORC1 signals to regulate a wide variety of cellular activities, including regulation of protein synthesis, ribosome biogenesis, metabolism, and autophagy. Inhibitors of these signaling proteins have been shown to be effective in JAK2-driven MPN model systems.[48–50,147,148] For example, the dual PI3K/mTOR inhibitor BEZ235 suppresses JAK2-driven MPN model cell growth, induces apoptosis, and synergizes with JAK2 inhibition in these cells.[49,148] CD34+ cells from MPN patients are more sensitive to BEZ235 than similar cells from healthy individuals. Importantly, BEZ235 also induces apoptosis in MPN model cells that are resistant to JAK2 inhibitors,[49] and combination of BEZ235 and JAK2 inhibition was more effective at reducing disease phenotypes in in vivo models of MPN.[148] Similar effects have been observed with the mTOR inhibitors RAD001 (everolimus) and PP242, which also suppressed the neoplastic erythroid colony growth of primary cells from PV patients.[147] Finally, the Akt inhibitor MK-2206 inhibits MPN model cell growth alone and synergistically with ruxolitinib, and enhances suppression of disease burden by ruxolitinib in a mouse model of MPN.[50] The mTOR inhibitor everolimus has shown clinical responses in clinical trials for myelofibrosis.[149]

4.2.2.3 PIM Inhibitors

The PIM (proviral insertion in murine lymphomas) family of serine threonine kinases are unique kinases in that they are regulated by transcription and protein stability and not phosphorylation by upstream signaling. PIMs can function as hematopoietic oncogenes in conjunction with Myc activation.[39] Members of the PIM family are regulated by JAK2/STAT5 signaling and thus are upregulated when this pathway is activated.[39] PIM activation leads to a number of effects in cells through the phosphorylation and inactivation of the proapoptotic BAD protein, the phosphorylation and inactivation of 4EBP1, leading to enhanced initiation of translation, and the regulation of Myc, among others.[39] Because PIMs are activated by JAK2/STAT5 signaling and possess properties similar to activation of Akt,[39] PIM kinase inhibitors have been assessed in MPN model systems. PIM inhibition is an attractive approach for human therapeutics, as inactivation of all three PIM genes in the mouse demonstrated these genes are dispensable for viability,[150] although these animals do have detectable hematopoietic abnormalities.[150,151] In fact, PIM1 has been shown to play a role in regulating the function and number of hematopoietic stem cells,[152] thus further providing support for targeting PIMs in MPNs. Interestingly, PIM inhibition with the PIM inhibitor AZD1208 has little effect alone on MPN model cells in vitro but synergizes with ruxolitinib to induce apoptosis in these cells.[153] AZD1208 alone does, however, inhibit the neoplastic growth of erythroid colonies from MPN patient hematopoietic progenitor cells but not from healthy control cells, and it synergizes with ruxolitinib to this end.[153] Importantly, PIM inhibitors can resensitize JAK2 inhibitor resistant cells to undergo cell death and growth inhibition in the presence of both drugs.[153,154] The effect of PIM inhibition in MPN cells may be via the regulation of Myc activity.[154] Combination of PIM inhibition and ruxolitinib enhanced the effects in an MPN mouse model compared to monotherapy with each drug.[155] The combination of ruxolitinib, the PIM inhibitor PIM447, and the CDK4/6 inhibitor ribociclib (LEE011) has shown promising results in preclinical studies.[156] This combination therapy is currently being tested in a clinical trial for myelofibrosis (ClinicalTrials.gov, NCT02370706).

4.2.2.4 HSP Inhibitors

One approach to improve MPN therapies has been to target JAK2 to reduce the levels of the protein, as opposed to targeting JAK2 kinase activity directly. JAK2 is a client protein for the chaperone activity of heat shock protein 90 (HSP90),[157,158] which facilitates the stabilization of JAK2 protein. HSP inhibitors block the chaperone/protein stabilizing activity of these proteins.[159] Thus, it was considered that inhibition of HSP90 might decrease JAK2 protein levels and thus be effective in an anti-MPN therapy. Treatment of MPN cells with the HSP90 inhibitors PU-H71[158] or AUY922[160] in fact decreases JAK2 protein and makes cells more susceptible to JAK2 inhibition. PU-H71 also reduces disease burden, including allele burden, in in vivo models of MPN[158] and also enhances the effect of ruxolitinib therapy in vivo.[145] Studies of JAK2 inhibitor resistance suggested that cells still require JAK2 protein, even in the face of high levels of JAK2 inhibitors.[144] This suggested perhaps JAK2 activity was not completely inhibited in cells that persistently grow in high levels of JAK2 inhibitors.[144] In support of this, such persistent cells are sensitive to treatment with

PU-H71[144] or AUY922,[160] which lead to decreases in JAK2 protein and signaling in these cells. Thus, the incomplete inhibition of JAK2 kinase activity by small molecule JAK2 inhibitors in persistent cells may be responsible for the failure of JAK2 inhibitors to eliminate neoplastic cells from MPN animal models and human MPNs. The concept that JAK2 kinase activity may not be completely inhibited by JAK2 inhibitors led researchers to determine that genetic *JAK2* deletion in an MPN mouse model was more effective than JAK2 inhibitor treatment at ameliorating the MPN phenotype, including significantly reducing allele burden.[145] This provides significant further support for the use of agents that target JAK2 protein stability as part of an anti-MPN therapy. PU-H71 is currently being assessed in clinical trials for cancer, including MPN patients (ClinicalTrials.gov, NCT01393509).

4.2.2.5 HDAC Inhibitors

Another approach to decrease JAK2 protein has been the use of HDAC inhibition,[157] as the chaperone activity of HSP90 is inactivated by acetylation.[161,162] Treatment of MPN cells with the pan-HDAC inhibitor panobinostat led to degradation of JAK2, inhibition of JAK2-dependent signaling, and induction of apoptosis.[157] Importantly, HDAC inhibition synergized with JAK2 inhibition to induce apoptosis of MPN model cells, and CD34+ cells from myelofibrosis patients were more sensitive to panobinostat treatment than similar cells from healthy individuals.[157] Combination of ruxolitinib and panobinostat improved efficacy in an in vivo model of JAK2-driven disease, compared to monotherapeutic treatment of the individual drugs.[163] HDAC inhibitors have been and continue to be tested in numerous clinical trials (ClinicalTrials.gov). However, results to date of clinical studies using various HDAC inhibitors have not provided much optimism for these classes of drugs.[164–167] These results and the pleiotropic effects of pan-HDAC inhibitors suggest HDAC-specific inhibitors might provide more optimistic effects in MPNs.

4.2.2.6 Hedgehog Inhibitors

The hedgehog pathway signals to regulate a variety of processes, including aspects of embryogenesis as well as normal and malignant cell growth. Hedgehog is the ligand for Patched, a transmembrane protein that inhibits the G-protein coupled receptor, Smoothened. This pathway is activated in a variety of cancer types and may play a significant role in hematopoietic cancers including MPNs.[168,169] Hedgehog inhibitors have been tested in various cancer models, including as anti-MPN therapies. For example, the Smoothened inhibitor Sonidegib (LDE225) and ruxolitinib provide a more efficacious inhibition together than each as a monotherapy, in an MPN mouse model.[170] Notably, a decrease in allele burden and enhanced inhibition of bone marrow fibrosis were observed in the combination therapy.[170] Initial clinical assessment of this combination therapy provided positive results in myelofibrosis patients.[171] PF-04449913 is a smoothened inhibitor that has demonstrated activity in a small number of myelofibrosis patients.[172] Sonidegib and PF-04449913 are being testing in clinical trials for MPN patients, including combination studies with JAK2 inhibition in myelofibrosis (ClinicalTrials.gov, NCT02226172, NCT01787552, NCT02129101).

4.2.2.7 Telomerase Inhibition: Imetelstat

Telomerase activity is deregulated in cancer, preventing the normal shortening of telomeres, the ends of chromosomes, that is seen in mammalian cells over time. This shortening of telomeres plays a role in cellular senescence and aging, and overcoming this contributes to cancer cells overcoming replicative senescence and obtaining a limitless replicative potential, which is a hallmark of cancer.[38] Thus, telomerase inhibitors could potentially reverse this state and convey a cancer cell back into a replicative senescent state. Imetelstat is a telomerase inhibitor that is being tested in a variety of cancers. Recent work with imetelstat in MPNs studies includes clinical assessment in ET[173] and myelofibrosis.[174] The results of these early studies were quite interesting as all patients ($n=18$) in the ET study had a hematological response, with 89% of them having a complete hematological response. Molecular responses were also observed, including reduction in JAK2-V617F, mutant CALR, or mutant MPL allele burden.[173] The myelofibrosis study reported 21% (7 of 33) patients obtained a complete ($n=4$) or partial ($n=3$) remission, with median duration of responses being 18 months for complete responders and 10 months for partial responders.[174] Importantly, bone marrow fibrosis was reversed in all of the complete responders.[174] A major concern with these studies was the observed myelosuppression,[173,174] and this and other data suggest that the molecular mechanism by which imetelstat functions to elicit clinical responses in MPN patients remains unknown.[175] Thus, further studies are required to understand the potential of, and optimize the use of, imetelstat as an anti-MPN therapy.

4.2.2.8 Interferon-Alpha

Interferon-alpha has long been used as a myelosuppressive agent for MPN patients, exhibiting durable hematological responses, albeit with significant toxicity concerns.[176] A phase 2 clinical trial of interferon-alpha suggested the toxicity of this agent was too great to consider it as an option to replace hydroxyurea as a myelosuppressive agent.[177] More recent studies have suggested this agent may be capable of eliminating the malignant population of cells from patients, an activity never seen with any other MPN therapy.[178] The mechanism of action of interferon-alpha in MPN patients is unclear and may be broad in nature, but studies in animal models of MPN suggest this may involve selected depletion of neoplastic hematopoietic stem cells over normal stem cells.[179] The associated toxicity of interferon-alpha treatment in patients is still a concern. Nonetheless, pegylated-interferon-alpha (Pegasys®), which has less toxicity than nonpegylated interferon-alpha, is actively being investigated in numerous clinical trials (ClinicalTrials.gov). There is anecdotal evidence that interferon-alpha in combination with ruxolitinib may be promising,[180] and a clinical trial has been designed to assess this combination in PMF (ClinicalTrials.gov, NCT02742324).

4.2.2.9 MPL Antagonists

Recent discoveries have demonstrated an expanding significance for the role of the TpoR, MPL, in MPN formation. These include the identification of direct aberrant activation of MPL via somatic mutation of *MPL* itself as well as the indirect activation of MPL by CALR mutants in MPNs, as described earlier in this chapter. While MPL signaling is important in the development of megakaryocytes,[181] which go on to produce platelets, it is also critical in hematopoietic stem cells and pluripotent progenitor cells.[72,182] Interestingly, it has been shown that endogenous MPL expression is required for JAK2-V617F to induce disease in an MPN mouse model.[183] Furthermore, transplantation of bone marrow from mice engineered to express JAK2-V617F restricted to the megakaryocyte lineage only leads to myeloproliferation in recipient mice, as well as expansion of pan-progenitor cells and hematopoietic stem cells.[184] Thus, megakaryocyte expansion could contribute to myeloproliferation in MPNs, further demonstrating a potentially critical role for MPL signaling in the development of these neoplasms. Thus, specifically targeting MPL signaling may provide an effective anti-MPN therapy.[184] In this regard, a cyclic peptide antagonist of MPL exerts greater efficacy at targeting stem and progenitor cells from myelofibrosis patients than similar cells from healthy controls.[185] This may be due to the ability of this MPL antagonist to selectively induce apoptosis in CD34+ cells from myelofibrosis patients compared to CD34+ cells from healthy individuals.[185] A therapy that targets malignant stem cells will likely be critical to alter the course of MPNs and provide hope for a remission-inducing therapy for MPN patients.

4.2.2.10 Aurora Kinase Inhibitors

A recent study has indicated that Aurora kinase A (AURKA) may be a therapeutic target in MPNs.[186] AURKA plays a role in the regulation of chromosomal segregation during mitosis.[187] Activation of this kinase was demonstrated in cells that express oncogenic driver mutations of MPNs including primary cells from MPN patients. A selective AURKA inhibitor completely blocked the MPN phenotype of a bone marrow transplant MPL model of MPN.[186] Significantly, this included inhibition of bone marrow fibrosis. Likewise, AURKA inhibition blocked the MPN phenotype induced by the transgenic expression of JAK2-V617F.[186] The targeting of AURKA by genetic means confirmed inhibition of this kinase could provide a novel target for an anti-MPN therapy.[186] This work further suggests an important role of megakaryocytes in the fibrotic phenotype of myelofibrosis.

4.2.2.11 Immunomodulatory Drugs

Immunomodulatory agents such as thalidomide,[188] lenalidomide,[117,189] and pomalidomide[190–193] have been assessed for anti-MPN activity. In short, data from these studies suggest Immunomodulatory Drugs (IMIDs) are active agents in myelofibrosis and durable responses are possible, but overall the data do not clearly suggest IMIDs will provide an effective anti-MPN therapy. A clinical study of lenalidomide with ruxolitinib in myelofibrosis patients was terminated, and the authors of the study suggested alternate dosing schemes should be considered.[194] Nonetheless, the numerous clinical trials (ClinicalTrials.gov) ongoing for testing IMIDs, alone and in combination with targeted therapies such as ruxolitinib, in MPNs suggest the enthusiasm for incorporating an IMID into an effective anti-MPN therapy remains high.

4.2.2.12 Statins

Recently, it was reported that a PV patient exhibited a significant hematological response while on simvastatin and alendronate over 56 months.[195] This response also included a molecular response, as the allele burden decreased over threefold.[195] It has been speculated that statins may be a potential agent in anti-MPN therapy as the antiinflammatory nature of statins could combat the inflammatory response seen in MPNs.[196,197] It is possible that such antiinflammatory activity of statin and the bisphosphonate, alendronate, contributed to the response observed in this patient,[195] as TNF-alpha levels are known to decrease upon treatment with statins,[198] and TNF-alpha has been shown to facilitate clonal expansion of MPN progenitor cells.[199] Also, statin treatment may alter cholesterol-rich lipid rafts and lipid raft-associated JAK2-V617F signaling.[200] Statin treatment induces apoptosis in JAK2-V617F-driven MPN model cells as well as inhibits neoplastic erythroid colony growth of myeloid progenitor cells from MPN patients, but not from healthy individuals.[200] The combination of statins and JAK2 inhibition has yet to be tested in patients.

4.2.2.13 Immunotherapy

Harnessing a patient's own immune system to fight cancer has become an extremely exciting, technically advanced, and optimistic strategy to develop anticancer therapies.[201] Chimeric antigen receptor (CAR) T-cell and antiimmune checkpoint therapies are examples of such approaches. CAR T-cell therapy involves expressing a CAR in a patient's own T-cells such that they can then specifically target neoplastic cells. Antiimmune checkpoint therapies overcome the activation of signals that inhibit immune responses to cancer cells. CAR T-cell therapy can be envisioned for MPNs by developing CARs that, for example, recognize epitopes within the novel carboxy terminus of mutant CALR that may be expressed on the cell surface of neoplastic cells of *CALR*–mutant-positive MPNs. Similarly, vaccine therapeutic approaches could take advantage of novel antigens in MPNs. An immune checkpoint therapeutic approach is being studied in a clinical trial for myelofibrosis (ClinicalTrials.gov, NCT02421354). Immune therapeutics are challenging and costly, and their general and long-term success in cancer remains to be determined, although it is an exciting field that is rapidly advancing. Thus, understanding the potential such therapies may have for MPN patients may take considerable patience.

4.2.2.14 Epigenetic Modulators

Although not the focus of this chapter, epigenetic alterations play an important role in MPNs, as the genes for many epigenetic regulators (e.g., TET2, IDH1/2, DNMT3A, EZH2, ASXL1) are mutated in MPNs.[27,28] Small molecules that modulate the activity of epigenetic regulators have been tested in MPN model systems. These include DNA methyl transferase inhibitors and BET inhibitors. DNA methyltransferase inhibitors are being assessed clinically in combination with ruxolitinib in MPN patients (ClinicalTrials.gov, NCT01787487, NCT02076191). BET inhibitors have been shown to enhance the effects of JAK2 inhibition on MPN cells[202] and are currently being tested in clinical trials for myeloid malignancies, including MPNs (ClinicalTrials.gov, NCT02158858, NCT02431260).

5. CONCLUDING REMARKS

The identification of aberrant activation of JAK2 signaling in MPNs provides a molecular basis supporting William Dameshek's suggestion, from over half a century ago, that such neoplasms had a similar underlying cause. It also ushered in a new era of potential experimental therapeutics and hope for MPN patients. Considering the clinical success of kinase inhibitor treatment of CML over the past couple of decades, and the similarities between CML and MPNs, it has been disappointing that JAK2 inhibitors have not been more successful in MPN patients. However, modern genetic technology allows for further elucidation of patient MPN genotypes, which includes mutations that not only likely affect disease phenotype but also and perhaps more importantly may provide clinicians with significantly enhanced prognostic ability and guidance in the personalization of patient care for current and future treatment options. One aspect of MPNs that has not been discussed in this chapter is the significant role of inflammation and the bone marrow microenvironment in MPNs. Many targeted therapies tested in MPN patients have improved constitutional symptoms, which was the basis for the approval of ruxolitinib for use in certain MPN patients. This is due to suppression of the inflammatory cytokine storm observed in MPN patients. However, the role of inflammation also likely plays a role in the expansion of the malignant clone and contributes to bone marrow fibrosis. Interestingly, JAK/STAT signaling in nonmalignant cells also contributes to the MPN phenotype and its inhibition to therapeutic response.[203] Thus, anti-MPN therapies may need to target both neoplastic and nonmalignant cells to thwart MPN phenotypes. Ultimately, a successful remission-inducing targeted therapy for MPNs will involve one that can eliminate the malignant stem cell in patients. While progress has been made on this front, there is still a long way to go, and while the large number of experimental therapies may provide hope, it also exemplifies the bottleneck that was encountered following the inability of JAK2 inhibitors to alter the course of disease for MPN patients. Nonetheless, the extensive list of experimental therapies also reflects the significant number of clinical and basic researchers who are dedicating their careers to the plight of MPN patients, which should also provide extreme optimism to current and future MPN patients.

REFERENCES

1. Levine RL, Pardanani A, Tefferi A, Gilliland DG. Role of JAK2 in the pathogenesis and therapy of myeloproliferative disorders. *Nat Rev Cancer*. 2007;7(9):673–683.
2. Tefferi A. Myeloproliferative neoplasms: a decade of discoveries and treatment advances. *Am J Hematol*. 2016;91(1):50–58.
3. Vardiman JW, Thiele J, Arber DA, et al. The 2008 revision of the World Health Organization (WHO) classification of myeloid neoplasms and acute leukemia: rationale and important changes. *Blood*. 2009;114(5):937–951.
4. Arber DA, Orazi A, Hasserjian R, et al. The 2016 revision to the World Health Organization classification of myeloid neoplasms and acute leukemia. *Blood*. 2016; 127(20):2391–2405.
5. Dameshek W. Some speculations on the myeloproliferative syndromes. *Blood*. 1951; 6(4):372–375.
6. Druker BJ. Translation of the Philadelphia chromosome into therapy for CML. *Blood*. 2008;112(13):4808–4817.
7. Levine RL, Wadleigh M, Cools J, et al. Activating mutation in the tyrosine kinase JAK2 in polycythemia vera, essential thrombocythemia, and myeloid metaplasia with myelofibrosis. *Cancer Cell*. 2005;7(4):387–397.
8. Kralovics R, Passamonti F, Buser AS, et al. A gain-of-function mutation of JAK2 in myeloproliferative disorders. *N Engl J Med*. 2005;352(17):1779–1790.
9. James C, Ugo V, Le Couedic JP, et al. A unique clonal JAK2 mutation leading to constitutive signalling causes polycythaemia vera. *Nature*. 2005;434(7037):1144–1148.
10. Baxter EJ, Scott LM, Campbell PJ, et al. Acquired mutation of the tyrosine kinase JAK2 in human myeloproliferative disorders. *Lancet*. 2005;365(9464):1054–1061.
11. Zhao R, Xing S, Li Z, et al. Identification of an acquired JAK2 mutation in polycythemia vera. *J Biol Chem*. 2005;280(24):22788–22792.
12. Tefferi A, Pardanani A. Myeloproliferative neoplasms: a contemporary review. *JAMA Oncol*. 2015;1(1):97–105.
13. Tefferi A, Guglielmelli P, Larson DR, et al. Long-term survival and blast transformation in molecularly annotated essential thrombocythemia, polycythemia vera, and myelofibrosis. *Blood*. 2014;124(16):2507–2513. quiz 2615.
14. Klampfl T, Gisslinger H, Harutyunyan AS, et al. Somatic mutations of calreticulin in myeloproliferative neoplasms. *N Engl J Med*. 2013;369(25):2379–2390.
15. Nangalia J, Massie CE, Baxter EJ, et al. Somatic CALR mutations in myeloproliferative neoplasms with nonmutated JAK2. *N Engl J Med*. 2013;369(25):2391–2405.
16. Mehta J, Wang H, Iqbal SU, Mesa R. Epidemiology of myeloproliferative neoplasms in the United States. *Leuk Lymphoma*. 2014;55(3):595–600.
17. Tefferi A, Rumi E, Finazzi G, et al. Survival and prognosis among 1545 patients with contemporary polycythemia vera: an international study. *Leukemia*. 2013;27(9): 1874–1881.
18. Scott LM, Tong W, Levine RL, et al. JAK2 exon 12 mutations in polycythemia vera and idiopathic erythrocytosis. *N Engl J Med*. 2007;356(5):459–468.
19. Pikman Y, Lee BH, Mercher T, et al. MPLW515L is a novel somatic activating mutation in myelofibrosis with myeloid metaplasia. *PLoS Med*. 2006;3(7):e270.

20. Wendling F, Vainchenker W. Thrombopoietin and its receptor. *Eur Cytokine Netw.* 1998;9(3):221–231.
21. Milosevic Feenstra JD, Nivarthi H, Gisslinger H, et al. Whole-exome sequencing identifies novel MPL and JAK2 mutations in triple-negative myeloproliferative neoplasms. *Blood.* 2016;127(3):325–332.
22. Cabagnols X, Favale F, Pasquier F, et al. Presence of atypical thrombopoietin receptor (MPL) mutations in triple-negative essential thrombocythemia patients. *Blood.* 2016;127(3):333–342.
23. Oh ST, Simonds EF, Jones C, et al. Novel mutations in the inhibitory adaptor protein LNK drive JAK-STAT signaling in patients with myeloproliferative neoplasms. *Blood.* 2010;116(6):988–992.
24. Pardanani A, Lasho T, Finke C, Oh ST, Gotlib J, Tefferi A. LNK mutation studies in blast-phase myeloproliferative neoplasms, and in chronic-phase disease with TET2, IDH, JAK2 or MPL mutations. *Leukemia.* 2010;24(10):1713–1718.
25. Sashida G, Wang C, Tomioka T, et al. The loss of Ezh2 drives the pathogenesis of myelofibrosis and sensitizes tumor-initiating cells to bromodomain inhibition. *J Exp Med.* 2016;213(8):1459–1477.
26. Them NC, Kralovics R. Genetic basis of MPN: beyond JAK2-V617F. *Curr Hematol Malig Rep.* 2013;8(4):299–306.
27. Shih AH, Abdel-Wahab O, Patel JP, Levine RL. The role of mutations in epigenetic regulators in myeloid malignancies. *Nat Rev Cancer.* 2012;12(9):599–612.
28. Reuther GW. Recurring mutations in myeloproliferative neoplasms alter epigenetic regulation of gene expression. *Am J Cancer Res.* 2011;1(6):752–762.
29. Vannucchi AM, Lasho TL, Guglielmelli P, et al. Mutations and prognosis in primary myelofibrosis. *Leukemia.* 2013;27(9):1861–1869.
30. Tefferi A, Guglielmelli P, Lasho TL, et al. CALR and ASXL1 mutations-based molecular prognostication in primary myelofibrosis: an international study of 570 patients. *Leukemia.* 2014;28(7):1494–1500.
31. Rampal R, Ahn J, Abdel-Wahab O, et al. Genomic and functional analysis of leukemic transformation of myeloproliferative neoplasms. *Proc Natl Acad Sci USA.* 2014;111(50): E5401–E5410.
32. Baker SJ, Rane SG, Reddy EP. Hematopoietic cytokine receptor signaling. *Oncogene.* 2007;26(47):6724–6737.
33. Saharinen P, Silvennoinen O. The pseudokinase domain is required for suppression of basal activity of Jak2 and Jak3 tyrosine kinases and for cytokine-inducible activation of signal transduction. *J Biol Chem.* 2002;277(49):47954–47963.
34. Park HJ, Li J, Hannah R, et al. Cytokine-induced megakaryocytic differentiation is regulated by genome-wide loss of a uSTAT transcriptional program. *EMBO J.* 2016; 35(6):580–594.
35. Liongue C, Sertori R, Ward AC. Evolution of cytokine receptor signaling. *J Immunol.* 2016;197(1):11–18.
36. Springuel L, Renauld JC, Knoops L. JAK kinase targeting in hematologic malignancies: a sinuous pathway from identification of genetic alterations towards clinical indications. *Haematologica.* 2015;100(10):1240–1253.
37. Levine RL, Wernig G. Role of JAK-STAT signaling in the pathogenesis of myeloproliferative disorders. *Hematology Am Soc Hematol Educ Program.* 2006;2006(1):233–239. http://dx.doi.org/10.1182/asheducation-2006.1.233.
38. Hanahan D, Weinberg RA. Hallmarks of cancer: the next generation. *Cell.* 2011; 144(5):646–674.
39. Amaravadi R, Thompson CB. The survival kinases Akt and Pim as potential pharmacological targets. *J Clin Invest.* 2005;115(10):2618–2624.

40. Shan Y, Gnanasambandan K, Ungureanu D, et al. Molecular basis for pseudokinase-dependent autoinhibition of JAK2 tyrosine kinase. *Nat Struct Mol Biol.* 2014;21(7):579–584.
41. Lu X, Huang LJ, Lodish HF. Dimerization by a cytokine receptor is necessary for constitutive activation of JAK2V617F. *J Biol Chem.* 2008;283(9):5258–5266.
42. Lu X, Levine R, Tong W, et al. Expression of a homodimeric type I cytokine receptor is required for JAK2V617F-mediated transformation. *Proc Natl Acad Sci USA.* 2005;102(52):18962–18967.
43. Pradhan A, Lambert QT, Reuther GW. Transformation of hematopoietic cells and activation of JAK2-V617F by IL-27R, a component of a heterodimeric type I cytokine receptor. *Proc Natl Acad Sci USA.* 2007;104(47):18502–18507.
44. Lu X, Gross AW, Lodish HF. Active conformation of the erythropoietin receptor: random and cysteine-scanning mutagenesis of the extracellular juxtamembrane and transmembrane domains. *J Biol Chem.* 2006;281(11):7002–7011.
45. Quintas-Cardama A, Vaddi K, Liu P, et al. Preclinical characterization of the selective JAK1/2 inhibitor INCB018424: therapeutic implications for the treatment of myeloproliferative neoplasms. *Blood.* 2010;115(15):3109–3117.
46. Yan D, Hutchison RE, Mohi G. Critical requirement for Stat5 in a mouse model of polycythemia vera. *Blood.* 2012;119(15):3539–3549.
47. Walz C, Ahmed W, Lazarides K, et al. Essential role for Stat5a/b in myeloproliferative neoplasms induced by BCR-ABL1 and JAK2(V617F) in mice. *Blood.* 2012;119(15):3550–3560.
48. Bartalucci N, Guglielmelli P, Vannucchi AM. Rationale for targeting the PI3K/Akt/mTOR pathway in myeloproliferative neoplasms. *Clin Lymphoma Myeloma Leuk.* 2013;13(Suppl 2):S307–S309.
49. Fiskus W, Verstovsek S, Manshouri T, et al. Dual PI3K/AKT/mTOR inhibitor BEZ235 synergistically enhances the activity of JAK2 inhibitor against cultured and primary human myeloproliferative neoplasm cells. *Mol Cancer Ther.* 2013;12(5):577–588.
50. Khan I, Huang Z, Wen Q, et al. AKT is a therapeutic target in myeloproliferative neoplasms. *Leukemia.* 2013;27(9):1882–1890.
51. Winter PS, Sarosiek KA, Lin KH, et al. RAS signaling promotes resistance to JAK inhibitors by suppressing BAD-mediated apoptosis. *Sci Signal.* 2014;7(357):ra122.
52. Oku S, Takenaka K, Kuriyama T, et al. JAK2 V617F uses distinct signalling pathways to induce cell proliferation and neutrophil activation. *Br J Haematol.* 2010;150(3):334–344.
53. Rampal R, Al-Shahrour F, Abdel-Wahab O, et al. Integrated genomic analysis illustrates the central role of JAK-STAT pathway activation in myeloproliferative neoplasm pathogenesis. *Blood.* 2014;123(22):e123–e133.
54. Zaleskas VM, Krause DS, Lazarides K, et al. Molecular pathogenesis and therapy of polycythemia induced in mice by JAK2 V617F. *PLoS One.* 2006;1:e18.
55. Wernig G, Mercher T, Okabe R, Levine RL, Lee BH, Gilliland DG. Expression of Jak2V617F causes a polycythemia vera-like disease with associated myelofibrosis in a murine bone marrow transplant model. *Blood.* 2006;107(11):4274–4281.
56. Lacout C, Pisani DF, Tulliez M, Gachelin FM, Vainchenker W, Villeval JL. JAK2V617F expression in murine hematopoietic cells leads to MPD mimicking human PV with secondary myelofibrosis. *Blood.* 2006;108(5):1652–1660.
57. Bumm TG, Elsea C, Corbin AS, et al. Characterization of murine JAK2V617F-positive myeloproliferative disease. *Cancer Res.* 2006;66(23):11156–11165.
58. Xing S, Wanting TH, Zhao W, et al. Transgenic expression of JAK2V617F causes myeloproliferative disorders in mice. *Blood.* 2008;111(10):5109–5117.
59. Tiedt R, Hao-Shen H, Sobas MA, et al. Ratio of mutant JAK2-V617F to wild-type Jak2 determines the MPD phenotypes in transgenic mice. *Blood.* 2008;111(8):3931–3940.
60. Shide K, Shimoda HK, Kumano T, et al. Development of ET, primary myelofibrosis and PV in mice expressing JAK2 V617F. *Leukemia.* 2008;22(1):87–95.

61. Wang L, Swierczek SI, Lanikova L, et al. The relationship of JAK2(V617F) and acquired UPD at chromosome 9p in polycythemia vera. *Leukemia*. 2014;28(4): 938–941.
62. Mullally A, Lane SW, Ball B, et al. Physiological Jak2V617F expression causes a lethal myeloproliferative neoplasm with differential effects on hematopoietic stem and progenitor cells. *Cancer Cell*. 2010;17(6):584–596.
63. Marty C, Lacout C, Martin A, et al. Myeloproliferative neoplasm induced by constitutive expression of JAK2V617F in knock-in mice. *Blood*. 2010;116(5):783–787.
64. Akada H, Yan D, Zou H, Fiering S, Hutchison RE, Mohi MG. Conditional expression of heterozygous or homozygous Jak2V617F from its endogenous promoter induces a polycythemia vera-like disease. *Blood*. 2010;115(17):3589–3597.
65. Siolas D, Hannon GJ. Patient-derived tumor xenografts: transforming clinical samples into mouse models. *Cancer Res*. 2013;73(17):5315–5319.
66. James C, Mazurier F, Dupont S, et al. The hematopoietic stem cell compartment of JAK2V617F-positive myeloproliferative disorders is a reflection of disease heterogeneity. *Blood*. 2008;112(6):2429–2438.
67. Ishii T, Zhao Y, Sozer S, et al. Behavior of CD34+ cells isolated from patients with polycythemia vera in NOD/SCID mice. *Exp Hematol*. 2007;35(11):1633–1640.
68. Xu M, Bruno E, Chao J, et al. The constitutive mobilization of bone marrow-repopulating cells into the peripheral blood in idiopathic myelofibrosis. *Blood*. 2005;105(4):1699–1705.
69. Grisouard J, Shimizu T, Duek A, et al. Deletion of Stat3 in hematopoietic cells enhances thrombocytosis and shortens survival in a JAK2-V617F mouse model of MPN. *Blood*. 2015;125(13):2131–2140.
70. Souyri M, Vigon I, Penciolelli JF, Heard JM, Tambourin P, Wendling F. A putative truncated cytokine receptor gene transduced by the myeloproliferative leukemia virus immortalizes hematopoietic progenitors. *Cell*. 1990;63(6):1137–1147.
71. de Sauvage FJ, Hass PE, Spencer SD, et al. Stimulation of megakaryocytopoiesis and thrombopoiesis by the c-Mpl ligand. *Nature*. 1994;369(6481):533–538.
72. Alexander WS, Roberts AW, Nicola NA, Li R, Metcalf D. Deficiencies in progenitor cells of multiple hematopoietic lineages and defective megakaryocytopoiesis in mice lacking the thrombopoietic receptor c-Mpl. *Blood*. 1996;87(6):2162–2170.
73. Hitchcock IS, Kaushansky K. Thrombopoietin from beginning to end. *Br J Haematol*. 2014;165(2):259–268.
74. Ding J, Komatsu H, Wakita A, et al. Familial essential thrombocythemia associated with a dominant-positive activating mutation of the c-MPL gene, which encodes for the receptor for thrombopoietin. *Blood*. 2004;103(11):4198–4200.
75. Defour JP, Itaya M, Gryshkova V, et al. Tryptophan at the transmembrane-cytosolic junction modulates thrombopoietin receptor dimerization and activation. *Proc Natl Acad Sci USA*. 2013;110(7):2540–2545.
76. Koppikar P, Abdel-Wahab O, Hedvat C, et al. Efficacy of the JAK2 inhibitor INCB16562 in a murine model of MPLW515L-induced thrombocytosis and myelofibrosis. *Blood*. 2010;115(14):2919–2927.
77. Michalak M, Groenendyk J, Szabo E, Gold LI, Opas M. Calreticulin, a multi-process calcium-buffering chaperone of the endoplasmic reticulum. *Biochem J*. 2009;417(3): 651–666.
78. Johnson S, Michalak M, Opas M, Eggleton P. The ins and outs of calreticulin: from the ER lumen to the extracellular space. *Trends Cell Biol*. 2001;11(3):122–129.
79. Ostwald TJ, MacLennan DH. Isolation of a high affinity calcium-binding protein from sarcoplasmic reticulum. *J Biol Chem*. 1974;249(3):974–979.
80. Nakamura K, Zuppini A, Arnaudeau S, et al. Functional specialization of calreticulin domains. *J Cell Biol*. 2001;154(5):961–972.

81. Arnaudeau S, Frieden M, Nakamura K, Castelbou C, Michalak M, Demaurex N. Calreticulin differentially modulates calcium uptake and release in the endoplasmic reticulum and mitochondria. *J Biol Chem.* 2002;277(48):46696–46705.
82. Bastianutto C, Clementi E, Codazzi F, et al. Overexpression of calreticulin increases the Ca2+ capacity of rapidly exchanging Ca2+ stores and reveals aspects of their lumenal microenvironment and function. *J Cell Biol.* 1995;130(4):847–855.
83. Mery L, Mesaeli N, Michalak M, Opas M, Lew DP, Krause KH. Overexpression of calreticulin increases intracellular Ca2+ storage and decreases store-operated Ca2+ influx. *J Biol Chem.* 1996;271(16):9332–9339.
84. Molinari M, Eriksson KK, Calanca V, et al. Contrasting functions of calreticulin and calnexin in glycoprotein folding and ER quality control. *Mol Cell.* 2004;13(1): 125–135.
85. Mesaeli N, Nakamura K, Zvaritch E, et al. Calreticulin is essential for cardiac development. *J Cell Biol.* 1999;144(5):857–868.
86. Jiang Y, Dey S, Matsunami H. Calreticulin: roles in cell-surface protein expression. *Membranes (Basel).* 2014;4(3):630–641.
87. Diaz-Villanueva JF, Diaz-Molina R, Garcia-Gonzalez V. Protein folding and mechanisms of proteostasis. *Int J Mol Sci.* 2015;16(8):17193–17230.
88. Chevet E, Hetz C, Samali A. Endoplasmic reticulum stress-activated cell reprogramming in oncogenesis. *Cancer Discov.* 2015;5(6):586–597.
89. Raghavan M, Wijeyesakere SJ, Peters LR, Del Cid N. Calreticulin in the immune system: ins and outs. *Trends Immunol.* 2013;34(1):13–21.
90. de Bruyn M, Wiersma VR, Helfrich W, Eggleton P, Bremer E. The ever-expanding-immunomodulatory role of calreticulin in cancer immunity. *Front Oncol.* 2015;5:35.
91. Zamanian M, Veerakumarasivam A, Abdullah S, Rosli R. Calreticulin and cancer. *Pathol Oncol Res.* 2013;19(2):149–154.
92. Lu YC, Weng WC, Lee H. Functional roles of calreticulin in cancer biology. *Biomed Res Int.* 2015;2015:526524.
93. Hayashida Y, Urata Y, Muroi E, et al. Calreticulin represses E-cadherin gene expression in Madin-Darby canine kidney cells via Slug. *J Biol Chem.* 2006;281(43): 32469–32484.
94. Phillips S, Kuperwasser C. SLUG: critical regulator of epithelial cell identity in breast development and cancer. *Cell Adh Migr.* 2014;8(6):578–587.
95. Talmadge JE, Fidler IJ. AACR centennial series: the biology of cancer metastasis: historical perspective. *Cancer Res.* 2010;70(14):5649–5669.
96. Villagomez M, Szabo E, Podcheko A, Feng T, Papp S, Opas M. Calreticulin and focal-contact-dependent adhesion. *Biochem Cell Biol.* 2009;87(4):545–556.
97. Lu YC, Chen CN, Wang B, et al. Changes in tumor growth and metastatic capacities of J82 human bladder cancer cells suppressed by down-regulation of calreticulin expression. *Am J Pathol.* 2011;179(3):1425–1433.
98. Du XL, Yang H, Liu SG, et al. Calreticulin promotes cell motility and enhances resistance to anoikis through STAT3-CTTN-Akt pathway in esophageal squamous cell carcinoma. *Oncogene.* 2009;28(42):3714–3722.
99. Chen CN, Chang CC, Su TE, et al. Identification of calreticulin as a prognosis marker and angiogenic regulator in human gastric cancer. *Ann Surg Oncol.* 2009;16(2): 524–533.
100. Sheng W, Chen C, Dong M, et al. Overexpression of calreticulin contributes to the development and progression of pancreatic cancer. *J Cell Physiol.* 2014;229(7):887–897.
101. Weng WC, Lin KH, Wu PY, et al. Calreticulin regulates VEGF-A in neuroblastoma cells. *Mol Neurobiol.* 2015;52(1):758–770.
102. Alur M, Nguyen MM, Eggener SE, et al. Suppressive roles of calreticulin in prostate cancer growth and metastasis. *Am J Pathol.* 2009;175(2):882–890.

103. Elf S, Abdelfattah NS, Chen E, et al. Mutant calreticulin requires both its mutant C-terminus and the thrombopoietin receptor for oncogenic transformation. *Cancer Discov.* 2016;6(4):368–381.
104. Marty C, Pecquet C, Nivarthi H, et al. Calreticulin mutants in mice induce an MPL-dependent thrombocytosis with frequent progression to myelofibrosis. *Blood.* 2016;127(10):1317–1324.
105. Araki M, Yang Y, Masubuchi N, et al. Activation of the thrombopoietin receptor by mutant calreticulin in CALR-mutant myeloproliferative neoplasms. *Blood.* 2016;127(10):1307–1316.
106. Balligand T, Achouri Y, Pecquet C, et al. Pathologic activation of thrombopoietin receptor and JAK2-STAT5 pathway by frameshift mutants of mouse calreticulin. *Leukemia.* 2016;30(8):1775–1778.
107. Chachoua I, Pecquet C, El-Khoury M, et al. Thrombopoietin receptor activation by myeloproliferative neoplasm associated calreticulin mutants. *Blood.* 2016;127(10): 1325–1335.
108. Han L, Schubert C, Kohler J, et al. Calreticulin-mutant proteins induce megakaryocytic signaling to transform hematopoietic cells and undergo accelerated degradation and Golgi-mediated secretion. *J Hematol Oncol.* 2016;9(1):45.
109. Nivarthi H, Chen D, Cleary C, et al. Thrombopoietin receptor is required for the oncogenic function of CALR mutants. *Leukemia.* 2016;30(8):1759–1763.
110. Daitoku S, Takenaka K, Yamauchi T, et al. Calreticulin mutation does not contribute to disease progression in essential thrombocythemia by inhibiting phagocytosis. *Exp Hematol.* 2016;44(9):817.e3–825.e3.
111. Theocharides AP, Lundberg P, Lakkaraju AK, et al. Homozygous calreticulin mutations in patients with myelofibrosis lead to acquired myeloperoxidase deficiency. *Blood.* 2016;127(25):3253–3259.
112. Nakamura K, Bossy-Wetzel E, Burns K, et al. Changes in endoplasmic reticulum luminal environment affect cell sensitivity to apoptosis. *J Cell Biol.* 2000;150(4):731–740.
113. Lim S, Chang W, Lee BK, et al. Enhanced calreticulin expression promotes calcium-dependent apoptosis in postnatal cardiomyocytes. *Mol Cells.* 2008;25(3):390–396.
114. Marchioli R, Finazzi G, Specchia G, et al. Cardiovascular events and intensity of treatment in polycythemia vera. *N Engl J Med.* 2013;368(1):22–33.
115. Vannucchi AM, Kantarjian HM, Kiladjian JJ, et al. A pooled analysis of overall survival in COMFORT-I and COMFORT-II, 2 randomized phase III trials of ruxolitinib for the treatment of myelofibrosis. *Haematologica.* 2015;100(9):1139–1145.
116. Martinez-Trillos A, Gaya A, Maffioli M, et al. Efficacy and tolerability of hydroxyurea in the treatment of the hyperproliferative manifestations of myelofibrosis: results in 40 patients. *Ann Hematol.* 2010;89(12):1233–1237.
117. Quintas-Cardama A, Kantarjian HM, Manshouri T, et al. Lenalidomide plus prednisone results in durable clinical, histopathologic, and molecular responses in patients with myelofibrosis. *J Clin Oncol.* 2009;27(28):4760–4766.
118. Tefferi A. Primary myelofibrosis: 2014 update on diagnosis, risk-stratification, and management. *Am J Hematol.* 2014;89(9):915–925.
119. Ferreira BV, Harrison C. How many JAK inhibitors in myelofibrosis? *Best Pract Res Clin Haematol.* 2014;27(2):187–195.
120. Verstovsek S, Mesa RA, Gotlib J, et al. A double-blind, placebo-controlled trial of ruxolitinib for myelofibrosis. *N Engl J Med.* 2012;366(9):799–807.
121. Harrison C, Kiladjian JJ, Al-Ali HK, et al. JAK inhibition with ruxolitinib versus best available therapy for myelofibrosis. *N Engl J Med.* 2012;366(9):787–798.
122. Verstovsek S, Mesa RA, Gotlib J, et al. Efficacy, safety, and survival with ruxolitinib in patients with myelofibrosis: results of a median 3-year follow-up of COMFORT-I. *Haematologica.* 2015;100(4):479–488.

123. Harrison CN, Vannucchi AM, Kiladjian JJ, et al. Long-term findings from COMFORT-II, a phase 3 study of ruxolitinib vs best available therapy for myelofibrosis. *Leukemia.* 2016;30(8):1701–1707.
124. Cervantes F, Vannucchi AM, Kiladjian JJ, et al. Three-year efficacy, safety, and survival findings from COMFORT-II, a phase 3 study comparing ruxolitinib with best available therapy for myelofibrosis. *Blood.* 2013;122(25):4047–4053.
125. Tefferi A. JAK inhibitors for myeloproliferative neoplasms: clarifying facts from myths. *Blood.* 2012;119(12):2721–2730.
126. Vannucchi AM, Kiladjian JJ, Griesshammer M, et al. Ruxolitinib versus standard therapy for the treatment of polycythemia vera. *N Engl J Med.* 2015;372(5):426–435.
127. Verstovsek S, Vannucchi AM, Griesshammer M, et al. Ruxolitinib versus best available therapy in patients with polycythemia vera: 80-week follow-up from the RESPONSE trial. *Haematologica.* 2016;101(7):821–829.
128. Raedler LA. Jakafi (Ruxolitinib): first FDA-approved medication for the treatment of patients with polycythemia vera. *Am Health Drug Benefits.* 2015;8(Spec Feature):75–79.
129. Wernig G, Kharas MG, Okabe R, et al. Efficacy of TG101348, a selective JAK2 inhibitor, in treatment of a murine model of JAK2V617F-induced polycythemia vera. *Cancer Cell.* 2008;13(4):311–320.
130. Lasho TL, Tefferi A, Hood JD, Verstovsek S, Gilliland DG, Pardanani A. TG101348, a JAK2-selective antagonist, inhibits primary hematopoietic cells derived from myeloproliferative disorder patients with JAK2V617F, MPLW515K or JAK2 exon 12 mutations as well as mutation negative patients. *Leukemia.* 2008;22(9):1790–1792.
131. Geron I, Abrahamsson AE, Barroga CF, et al. Selective inhibition of JAK2-driven erythroid differentiation of polycythemia vera progenitors. *Cancer Cell.* 2008;13(4): 321–330.
132. Pardanani A, Tefferi A, Jamieson C, et al. A phase 2 randomized dose-ranging study of the JAK2-selective inhibitor fedratinib (SAR302503) in patients with myelofibrosis. *Blood Cancer J.* 2015;5:e335.
133. Harrison CN, Cortes JE, Cervantes F, et al. Results of a randomized, double-blind, placebo-controlled phase III study (JAKARTA) of the JAK2-selective inhibitor fedratinib (SAR302503) in patients with myelofibrosis (MF). *Blood.* 2013;122(21):393.
134. Schaap NP, Zweegman S, Jourdan E, et al. Efficacy and safety of fedratinib (SAR302503/TG101348) in patients with intermediate- or high-risk myelofibrosis (MF), post-polycythemia vera (PV) MF, or post-essential thrombocythemia (ET) MF previously treated with ruxolitinib: interim results from a phase II study (JAKARTA-2). *Blood.* 2013;122(21):661.
135. Day GS, del Campo CM. Wernicke encephalopathy: a medical emergency. *CMAJ.* 2014;186(8):E295.
136. Zhang Q, Zhang Y, Diamond S, et al. The Janus kinase 2 inhibitor fedratinib inhibits thiamine uptake: a putative mechanism for the onset of Wernicke's encephalopathy. *Drug Metab Dispos.* 2014;42(10):1656–1662.
137. Pardanani A, Lasho T, Smith G, Burns CJ, Fantino E, Tefferi A. CYT387, a selective JAK1/JAK2 inhibitor: in vitro assessment of kinase selectivity and preclinical studies using cell lines and primary cells from polycythemia vera patients. *Leukemia.* 2009;23(8):1441–1445.
138. Pardanani A, Laborde RR, Lasho TL, et al. Safety and efficacy of CYT387, a JAK1 and JAK2 inhibitor, in myelofibrosis. *Leukemia.* 2013;27(6):1322–1327.
139. William AD, Lee AC, Blanchard S, et al. Discovery of the macrocycle 11-(2-pyrrolidin-1-yl-ethoxy)-14,19-dioxa-5,7,26-triaza-tetracyclo[19.3.1.1(2,6). 1(8,12)] heptacosa-1(25),2(26),3,5,8,10,12(27),16,21,23-decaene (SB1518), a potent Janus kinase 2/fms-like tyrosine kinase-3 (JAK2/FLT3) inhibitor for the treatment of myelofibrosis and lymphoma. *J Med Chem.* 2011;54(13):4638–4658.

140. Komrokji RS, Seymour JF, Roberts AW, et al. Results of a phase 2 study of pacritinib (SB1518), a JAK2/JAK2(V617F) inhibitor, in patients with myelofibrosis. *Blood.* 2015;125(17):2649–2655.
141. Purandare AV, McDevitt TM, Wan H, et al. Characterization of BMS-911543, a functionally selective small-molecule inhibitor of JAK2. *Leukemia.* 2012;26(2):280–288.
142. Pomicter AD, Eiring AM, Senina AV, et al. Limited efficacy of BMS-911543 in a murine model of Janus kinase 2 V617F myeloproliferative neoplasm. *Exp Hematol.* 2015;43(7):537–545. e531-511.
143. Meyer SC, Levine RL. Molecular pathways: molecular basis for sensitivity and resistance to JAK kinase inhibitors. *Clin Cancer Res.* 2014;20(8):2051–2059.
144. Koppikar P, Bhagwat N, Kilpivaara O, et al. Heterodimeric JAK-STAT activation as a mechanism of persistence to JAK2 inhibitor therapy. *Nature.* 2012;489(7414):155–159.
145. Bhagwat N, Koppikar P, Keller M, et al. Improved targeting of JAK2 leads to increased therapeutic efficacy in myeloproliferative neoplasms. *Blood.* 2014;123(13): 2075–2083.
146. Meyer SC, Keller MD, Chiu S, et al. CHZ868, a type II JAK2 inhibitor, reverses type I JAK inhibitor persistence and demonstrates efficacy in myeloproliferative neoplasms. *Cancer Cell.* 2015;28(1):15–28.
147. Bogani C, Bartalucci N, Martinelli S, et al. mTOR inhibitors alone and in combination with JAK2 inhibitors effectively inhibit cells of myeloproliferative neoplasms. *PLoS One.* 2013;8(1):e54826.
148. Bartalucci N, Tozzi L, Bogani C, et al. Co-targeting the PI3K/mTOR and JAK2 signalling pathways produces synergistic activity against myeloproliferative neoplasms. *J Cell Mol Med.* 2013;17(11):1385–1396.
149. Guglielmelli P, Barosi G, Rambaldi A, et al. Safety and efficacy of everolimus, a mTOR inhibitor, as single agent in a phase 1/2 study in patients with myelofibrosis. *Blood.* 2011;118(8):2069–2076.
150. Mikkers H, Nawijn M, Allen J, et al. Mice deficient for all PIM kinases display reduced body size and impaired responses to hematopoietic growth factors. *Mol Cell Biol.* 2004;24(13):6104–6115.
151. An N, Kraft AS, Kang Y. Abnormal hematopoietic phenotypes in Pim kinase triple knockout mice. *J Hematol Oncol.* 2013;6:12.
152. An N, Lin YW, Mahajan S, et al. Pim1 serine/threonine kinase regulates the number and functions of murine hematopoietic stem cells. *Stem Cells.* 2013;31(6):1202–1212.
153. Mazzacurati L, Lambert QT, Pradhan A, Griner LN, Huszar D, Reuther GW. The PIM inhibitor AZD1208 synergizes with ruxolitinib to induce apoptosis of ruxolitinib sensitive and resistant JAK2-V617F-driven cells and inhibit colony formation of primary MPN cells. *Oncotarget.* 2015;6(37):40141–40157.
154. Huang SM, Wang A, Greco R, et al. Combination of PIM and JAK2 inhibitors synergistically suppresses MPN cell proliferation and overcomes drug resistance. *Oncotarget.* 2014;5(10):3362–3374.
155. Saci A, Pinzon-Ortiz M, Wang D, et al. The combination of JAK inhibitor, Ruxolitinib, and PIM inhibitor, LGH447, in preclinical models of myeloproliferative neoplasia. *Blood.* 2013;122(21):4100.
156. Pinzon-Ortiz M, Rong X, Saci A, et al. Abstract 3684: the combination of JAK inhibitor, ruxolitinib, pan-PIM inhibitor, LGH447, and CDK4/6 inhibitor, LEE011, in a preclinical mouse model of myeloproliferative neoplasia. *Cancer Res.* 2014;74-(19 Suppl):3684.
157. Wang Y, Fiskus W, Chong DG, et al. Cotreatment with panobinostat and JAK2 inhibitor TG101209 attenuates JAK2V617F levels and signaling and exerts synergistic cytotoxic effects against human myeloproliferative neoplastic cells. *Blood.* 2009; 114(24):5024–5033.

158. Marubayashi S, Koppikar P, Taldone T, et al. HSP90 is a therapeutic target in JAK2-dependent myeloproliferative neoplasms in mice and humans. *J Clin Invest.* 2010;120(10):3578–3593.
159. Solarova Z, Mojzis J, Solar P. Hsp90 inhibitor as a sensitizer of cancer cells to different therapies (review). *Int J Oncol.* 2015;46(3):907–926.
160. Fiskus W, Verstovsek S, Manshouri T, et al. Heat shock protein 90 inhibitor is synergistic with JAK2 inhibitor and overcomes resistance to JAK2-TKI in human myeloproliferative neoplasm cells. *Clin Cancer Res.* 2011;17(23):7347–7358.
161. Scroggins BT, Robzyk K, Wang D, et al. An acetylation site in the middle domain of Hsp90 regulates chaperone function. *Mol Cell.* 2007;25(1):151–159.
162. Li J, Buchner J. Structure, function and regulation of the hsp90 machinery. *Biomed J.* 2013;36(3):106–117.
163. Evrot E, Ebel N, Romanet V, et al. JAK1/2 and Pan-deacetylase inhibitor combination therapy yields improved efficacy in preclinical mouse models of JAK2V617F-driven disease. *Clin Cancer Res.* 2013;19(22):6230–6241.
164. Rambaldi A, Dellacasa CM, Finazzi G, et al. A pilot study of the histone-deacetylase inhibitor givinostat in patients with JAK2V617F positive chronic myeloproliferative neoplasms. *Br J Haematol.* 2010;150(4):446–455.
165. Mascarenhas J, Lu M, Li T, et al. A phase I study of panobinostat (LBH589) in patients with primary myelofibrosis (PMF) and post-polycythaemia vera/essential thrombocythaemia myelofibrosis (post-PV/ET MF). *Br J Haematol.* 2013;161(1): 68–75.
166. DeAngelo DJ, Mesa RA, Fiskus W, et al. Phase II trial of panobinostat, an oral pan-deacetylase inhibitor in patients with primary myelofibrosis, post-essential thrombocythaemia, and post-polycythaemia vera myelofibrosis. *Br J Haematol.* 2013;162(3):326–335.
167. Andersen CL, Mortensen NB, Klausen TW, Vestergaard H, Bjerrum OW, Hasselbalch HC. A phase II study of vorinostat (MK-0683) in patients with primary myelofibrosis and post-polycythemia vera myelofibrosis. *Haematologica.* 2014;99(1): e5–e7.
168. Tibes R, Mesa RA. Targeting hedgehog signaling in myelofibrosis and other hematologic malignancies. *J Hematol Oncol.* 2014;7:18.
169. Klein C, Zwick A, Kissel S, et al. Ptch2 loss drives myeloproliferation and myeloproliferative neoplasm progression. *J Exp Med.* 2016;213(2):273–290.
170. Keller MD, Rampal RK, Shank K, et al. Improved efficacy of combination of JAK2 and hedgehog inhibitors in myelofibrosis. *Blood.* 2013;122(21):666.
171. Gupta V, Harrison CN, Hasselbalch H, et al. Phase 1b/2 study of the efficacy and safety of Sonidegib (LDE225) in combination with ruxolitinib (INC424) in patients with myelofibrosis. *Blood.* 2015;126(23):825.
172. Martinelli G, Oehler VG, Papayannidis C, et al. Treatment with PF-04449913, an oral smoothened antagonist, in patients with myeloid malignancies: a phase 1 safety and pharmacokinetics study. *Lancet Haematol.* 2015;2(8):e339–e346.
173. Baerlocher GM, Oppliger Leibundgut E, Ottmann OG, et al. Telomerase inhibitor imetelstat in patients with essential thrombocythemia. *N Engl J Med.* 2015;373(10): 920–928.
174. Tefferi A, Lasho TL, Begna KH, et al. A pilot study of the telomerase inhibitor imetelstat for myelofibrosis. *N Engl J Med.* 2015;373(10):908–919.
175. Armanios M, Greider CW. Treating myeloproliferation—on target or off? *N Engl J Med.* 2015;373(10):965–966.
176. Gilbert HS. Long term treatment of myeloproliferative disease with interferon-alpha-2b: feasibility and efficacy. *Cancer.* 1998;83(6):1205–1213.

177. Tefferi A, Elliot MA, Yoon SY, et al. Clinical and bone marrow effects of interferon alfa therapy in myelofibrosis with myeloid metaplasia. *Blood*. 2001;97(6):1896.
178. Kiladjian JJ, Giraudier S, Cassinat B. Interferon-alpha for the therapy of myeloproliferative neoplasms: targeting the malignant clone. *Leukemia*. 2016;30(4): 776–781.
179. Mullally A, Bruedigam C, Poveromo L, et al. Depletion of Jak2V617F myeloproliferative neoplasm-propagating stem cells by interferon-alpha in a murine model of polycythemia vera. *Blood*. 2013;121(18):3692–3702.
180. Bjorn ME, de Stricker K, Kjaer L, Ellemann K, Hasselbalch HC. Combination therapy with interferon and JAK1-2 inhibitor is feasible: proof of concept with rapid reduction in JAK2V617F-allele burden in polycythemia vera. *Leuk Res Rep*. 2014;3(2):73–75.
181. Geddis AE, Linden HM, Kaushansky K. Thrombopoietin: a pan-hematopoietic cytokine. *Cytokine Growth Factor Rev*. 2002;13(1):61–73.
182. Solar GP, Kerr WG, Zeigler FC, et al. Role of c-mpl in early hematopoiesis. *Blood*. 1998;92(1):4–10.
183. Sangkhae V, Etheridge SL, Kaushansky K, Hitchcock IS. The thrombopoietin receptor, MPL, is critical for development of a JAK2V617F-induced myeloproliferative neoplasm. *Blood*. 2014;124(26):3956–3963.
184. Zhan H, Ma Y, Lin CH, Kaushansky K. JAK2V617F-mutant megakaryocytes contribute to hematopoietic stem/progenitor cell expansion in a model of murine myeloproliferation. *Leukemia*. 2016. http://dx.doi.org/10.1038/leu.2016.114.
185. Wang X, Haylock D, Hu CS, et al. A thrombopoietin receptor antagonist is capable of depleting myelofibrosis hematopoietic stem and progenitor cells. *Blood*. 2016;127(26): 3398–3409.
186. Wen QJ, Yang Q, Goldenson B, et al. Targeting megakaryocytic-induced fibrosis in myeloproliferative neoplasms by AURKA inhibition. *Nat Med*. 2015;21(12): 1473–1480.
187. Goldenson B, Crispino JD. The aurora kinases in cell cycle and leukemia. *Oncogene*. 2015;34(5):537–545.
188. Thapaliya P, Tefferi A, Pardanani A, et al. International working group for myelofibrosis research and treatment response assessment and long-term follow-up of 50 myelofibrosis patients treated with thalidomide-prednisone based regimens. *Am J Hematol*. 2011;86(1):96–98.
189. Chihara D, Masarova L, Newberry KJ, et al. Long-term results of a phase II trial of lenalidomide plus prednisone therapy for patients with myelofibrosis. *Leuk Res*. 2016;48:1–5.
190. Begna KH, Pardanani A, Mesa R, et al. Long-term outcome of pomalidomide therapy in myelofibrosis. *Am J Hematol*. 2012;87(1):66–68.
191. Daver N, Shastri A, Kadia T, et al. Phase II study of pomalidomide in combination with prednisone in patients with myelofibrosis and significant anemia. *Leuk Res*. 2014;38(9):1126–1129.
192. Gowin KL, Mesa RA. Profile of pomalidomide and its potential in the treatment of myelofibrosis. *Ther Clin Risk Manag*. 2015;11:549–556.
193. Passamonti F, Barbui T, Barosi G, et al. Phase 3 study of pomalidomide in myeloproliferative neoplasm (MPN)-associated myelofibrosis with RBC-transfusion-dependence. *Blood*. 2013;122(21):394.
194. Daver N, Cortes J, Newberry K, et al. Ruxolitinib in combination with lenalidomide as therapy for patients with myelofibrosis. *Haematologica*. 2015;100(8):1058–1063.
195. Sorensen AL, Kallenbach K, Hasselbalch HC. A remarkable hematological and molecular response pattern in a patient with polycythemia vera during combination therapy with simvastatin and alendronate. *Leuk Res Rep*. 2016;6:20–23.

196. Hasselbalch HC. Perspectives on the impact of JAK-inhibitor therapy upon inflammation-mediated comorbidities in myelofibrosis and related neoplasms. *Expert Rev Hematol.* 2014;7(2):203–216.
197. Bjorn ME, Hasselbalch HC. The role of reactive oxygen species in myelofibrosis and related neoplasms. *Mediators Inflamm.* 2015;2015:648090.
198. Khattri S, Zandman-Goddard G. Statins and autoimmunity. *Immunol Res.* 2013;56(2–3):348–357.
199. Fleischman AG, Aichberger KJ, Luty SB, et al. TNFalpha facilitates clonal expansion of JAK2V617F positive cells in myeloproliferative neoplasms. *Blood.* 2011;118(24): 6392–6398.
200. Griner LN, McGraw KL, Johnson JO, List AF, Reuther GW. JAK2-V617F-mediated signalling is dependent on lipid rafts and statins inhibit JAK2-V617F-dependent cell growth. *Br J Haematol.* 2013;160(2):177–187.
201. Hoos A. Development of immuno-oncology drugs—from CTLA4 to PD1 to the next generations. *Nat Rev Drug Discov.* 2016;15(4):235–247.
202. Wyspianska BS, Bannister AJ, Barbieri I, et al. BET protein inhibition shows efficacy against JAK2V617F-driven neoplasms. *Leukemia.* 2014;28(1):88–97.
203. Kleppe M, Kwak M, Koppikar P, et al. JAK-STAT pathway activation in malignant and nonmalignant cells contributes to MPN pathogenesis and therapeutic response. *Cancer Discov.* 2015;5(3):316–331.

PART VI

Breast Cancer: Endocrine Signaling, Resistance to Endocrine Therapy and Novel Contributors to Pathology

CHAPTER ELEVEN

Dysregulation of Aromatase in Breast, Endometrial, and Ovarian Cancers: An Overview of Therapeutic Strategies

P.R. Manna[1], D. Molehin, A.U. Ahmed
Texas Tech University Health Sciences Center School of Medicine, Lubbock, TX, United States
[1]Corresponding author: e-mail address: pulak.manna@ttuhsc.edu

Contents

Abstract

Aromatase is the rate-limiting enzyme in the biosynthesis of estrogens, which play crucial roles on a spectrum of developmental and physiological processes. The biological actions of estrogens are classically mediated by binding to two estrogen receptors (ERs), ERα and ERβ. Encoded by the cytochrome P450, family 19, subfamily A, polypeptide 1 (*CYP19A1*) gene, aromatase is expressed in a wide variety of tissues, as well as benign and malignant tumors, and is regulated in a pathway- and tissue-specific manner. Overexpression of aromatase, leading to elevated systemic levels of estrogen, is unequivocally linked to the pathogenesis and growth of a number malignancies, including breast, endometrium, and ovarian cancers. Aromatase inhibitors (AIs) are routinely used to treat estrogen-dependent breast cancers in postmenopausal women; however, their roles in endometrial and ovarian cancers remain obscure. While AI therapy is effective in hormone sensitive cancers, they diminish estrogen production throughout the body and, thus, generate undesirable side effects. Despite the effectiveness of AI therapy, resistance to endocrine therapy remains a major concern and is the leading cause of

Progress in Molecular Biology and Translational Science, Volume 144
ISSN 1877-1173
http://dx.doi.org/10.1016/bs.pmbts.2016.10.002

cancer death. Considerable advances, toward mitigating these issues, have evolved in conjunction with a number of histone deacetylase (HDAC) inhibitors for countering an assortment of diseases and cancers, including the aforesaid malignancies. HDACs are a family of enzymes that are frequently dysregulated in human tumors. This chapter will discuss the current understanding of aberrant regulation and expression of aromatase in breast, endometrial, and ovarian cancers, and potential therapeutic strategies for prevention and treatment of these life-threatening diseases.

1. INTRODUCTION

Aromatase is the key enzyme that catalyzes the final step in estrogen biosynthesis, i.e., aromatization of androgens to estrogens. Estrogens are an important group of steroid hormones that are essential for key physiological processes including growth, differentiation, and reproductive development and function.[1,2] The rate-limiting step in steroid biosynthesis is the transport of the substrate of all steroid hormones, cholesterol, from the outer to the inner mitochondrial membrane, a process that is primarily regulated by the steroidogenic acute regulatory protein (StAR).[3,4] At the inner membrane, cytochrome P450 cholesterol side chain cleavage enzyme cleaves the cholesterol side chain to form the first steroid, pregnenolone, which is then converted to various steroid hormones by a series of enzymes in specific tissues (Fig. 1).

Aromatase is expressed in a wide variety of tissues, including ovary, testis, placenta, bone, skin, brain, and adipose tissue in humans.[1,2,5,6] In premenopausal women, estrogens are synthesized by ovarian granulosa and corpus luteal cells via the classical steroidogenic pathway (Fig. 1). However, in postmenopausal women, they are synthesized in many extraovarian tissues such as adipose tissue, brain, bone, and skin. Estrogens typically exert their effects by binding to the nuclear estrogen receptors (ERs), ERα and ERβ, in which the majority of estrogenic responses is mediated by ERα.[7,8]

Expression of the *CYP19A1* (cytochrome P450, family 19, subfamily A, polypeptide 1) gene encoding aromatase is controlled by 10 tissue-specific and alternatively spliced promoters via diverse signaling pathways.[5,6,9,10] Epidemiological and experimental evidence indicate that a number of most common women's cancers, including breast, endometrium, and ovary, express high levels of both aromatase and ERα, concomitant with large amounts of estrogens.[10–12] Elevated levels of estrogens in these malignant tumors have been demonstrated to be strikingly higher than their normal counterparts and

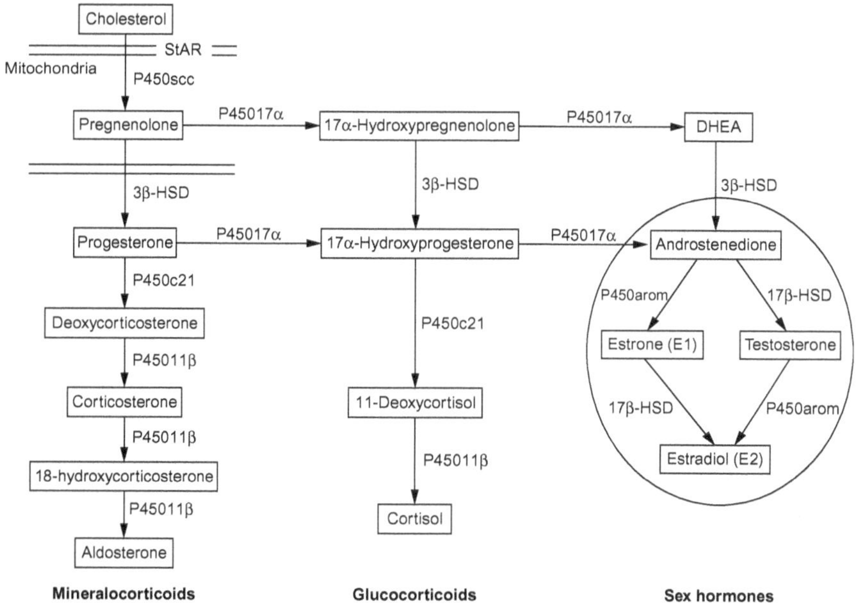

Fig. 1 Steroid biosynthetic pathway. The StAR protein primarily controls the intramitochondrial transport of cholesterol, the rate-limiting step in steroid biosynthesis. At the mitochondria, the steroid formed is pregnenolone by the action of P450scc. Pregnenolone is then converted to various steroid hormones by tissue-specific enzymes. *Circle* represents schematic of estrogen biosynthesis. Aromatase is the key enzyme in the conversion of androstenedione to estrogens (estradiol-17β is biologically active estrogen). The 17β-HSD enzyme converts androstenedione to testosterone.

circulating levels. Therefore, it is unequivocal that increased expression and/or activity of aromatase are one of the key regulatory events for elevated intratumoral production of estrogens in these malignant tissues. As a consequence, this enzyme is a molecular target for therapeutic approaches for many estrogen-dependent diseases.[13–16] It is noteworthy that the discovery of antiestrogen tamoxifen (which blocks estrogen binding to ER) in early 1970s significantly improved breast cancer survival rates.[17,18] Later, aromatase inhibitors (AIs), which effectively decrease estrogen production by inhibiting aromatase activity, are often used for treatment and prevention of hormone responsive cancers in postmenopausal women.[13,17] While AIs are more effective and well tolerated than tamoxifen, by the patients, they inhibit estrogen production throughout the body and result in many unwanted side effects.[13,16,19] Despite, considerable progress has been made for the treatment and prevention of estrogen-dependent cancers by targeting aromatase and estrogens or their signaling pathways.

Advances in genetic engineering and proteomic technologies have demonstrated a link between histone deacetylases (HDACs) and aromatase regulation. HDACs are a family (18 members classified into 4 classes) of enzymes (that remove acetyl moiety from lysine residues at histone tails) with pleiotropic activities that regulate chromatin remodeling, cell signaling, genomic stability through the dynamic process of acetylation and deacetylation of core histones.[20–27] While classes I, II, and IV HDACs are Zn^{2+} dependent, class III HDACs or sirtuins (homologous to the yeast Sir2 family of proteins) require NAD^{+} for their catalytic activity (Table 1). These epigenetic enzymes are frequently dysregulated in various tumors, and HDAC inhibitors have been shown to have multiple targets in cancer cells. To date, the United States (US) Food and Drug Administration (FDA) has approved four Zn^{2+}, but not NAD^{+}, connected HDAC inhibitors for the treatment and management of a variety of tumors and/or cancers and favorable outcomes have been reported.[26,28] Therefore, the relevance of classes I, II, and IV HDACs are appropriately discussed in conjunction with a variety of cancers, including the afore-mentioned malignancies. Given the high expression of aromatase in breast, endometrial, and ovarian malignant tumors, associated with elevated intratumoral production of estrogens, inhibition of this enzyme and, thus, estrogen synthesis, is of major therapeutic interest. Indeed, the precise mechanism involved in aromatase regulation and function in various pathological conditions, including breast, endometrial, and ovarian cancers, is the key for the development of therapies for these deadly diseases. In this chapter, we will summarize the current understanding of the regulation and expression of aromatase (the key enzyme in

Table 1 The HDAC Family and Their Classification

HDAC Classes	**HDAC Members**
Class I	HDAC1, HDAC2, HDAC3, HDAC8
Class IIA	HDAC4, HDAC5, HDAC7, HDAC9
Class IIB	HDAC6, HDAC10
Class III	SIRT1–SIRT7
Class IV	HDAC11

The HDAC family includes 18 members in four different classes. Classes I, II, and IV HDACs contain Zn^{2+} in their active site, whereas class III HDACs (SIRT1–SIRT7) require NAD^{+} for their catalytic activity.

estrogen biosynthesis) in the pathogenesis of breast, endometrial, and ovarian malignant tumors, the pharmacology and clinical applications of antiestrogens and AIs, and novel therapeutic strategies in the treatment and management of these cancers.

2. AROMATASE AND ESTROGEN BIOSYNTHESIS

Regulation of aromatase is crucial to proper functioning of estrogen responsive physiological activities. Aromatase consists of a heme group and polypeptide chain of 503 amino acid residues and is the key enzyme in the conversion of androgens to estrogens. Fig. 2 shows a ribbon diagram of the crystal structure of human aromatase that comprises of 12 major α-helices (labeled A through L) and 10 β-strands (numbered 1 through 10) distributed into 1 major and 3 minor sheets.[29,30] The active site of

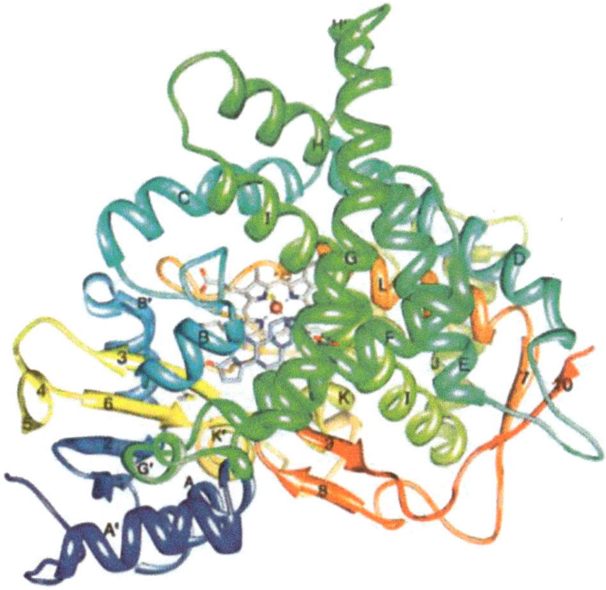

Fig. 2 A ribbon diagram showing the crystal structure of aromatase. The amino terminus, starting at residue 45, is colored in *dark blue* and the carboxyl terminus ending at residue 496 is colored *red*. The α-helices are labeled from A to L and β-strands are numbered from 1 to 10. The heme group and the bound androstenedione molecule at the active site are shown. *Revised and represented with permission from Ghosh D, Griswold J, Erman M, Pangborn W. Structural basis for androgen specificity and oestrogen synthesis in human aromatase.* Nature. *2009;457(7226):219–223.*

aromatase (that is buried in deep within the spherical molecule) is the distal cavity of the heme-binding pocket and the heme iron is the reaction center of the enzyme for androstenedione. The crystal structure of human aromatase enzyme provides a finely tuned structural and mechanistic basis for the specificity to androgens and, thus, biosynthesis of estrogens.[29,30] The structure of human aromatase has opened up the opportunity to specifically design structure-based drugs for a variety of applications. The aromatase enzyme is essentially an estrogen synthetase and is located in the endoplasmic reticulum of estrogen producing cells. In humans, aromatase is expressed in a variety of tissues, including ovary, testis, placenta, bone, skin, brain, and adipose tissue. Accordingly, estrogens are primarily produced in ovarian granulosa cells and corpus luteum, testicular Leydig cells, placental syncytiotrophoblast, various brain regions, adipose tissue, and in several cancerous tissues. Estrogens, the primary female sex hormones, include estrone (E1), the biologically active 17β-estradiol (E2), and estriol (E3). E2 is a major form of estrogens in reproductive age women. E3 is generally produced during pregnancy, whereas E1 is a predominant estrogen in postmenopausal women. Estrogens play key roles in the physiology of reproductive processes, and they also have significant effects on bone mineralization, vascular biology, lipid metabolism, and cognitive function.[2,5,9,31,32] In premenopausal women, estrogen is majorly produced in the ovary, whereas in postmenopausal women, extraovarian sites such as skin and adipose tissues are main sources of estrogen biosynthesis.[2,9,33] Noteworthy, estrogen is involved in the pathogenesis of a number of hormone responsive diseases, including breast, endometrial, and ovarian cancers, which are more prevalent in postmenopausal women. These malignant tumors express aberrant high levels of aromatase than their nonmalignant counterparts.[11,14,34] Considering the association of aromatase in estrogen-dependent tumors/cancers, inhibitors of this enzyme has been targeted for the treatment and development of hormone responsive diseases.

Aromatase belongs to a microsomal member of the cytochrome P450 superfamily and is the product of the *CYP19A1* gene that is localized at chromosome 15, band q21.1 of the human genome.[5,9,33,35] In humans and mice, a single gene (*CYP19A1*) encodes aromatase, inhibition and/or targeted disruption of which effectively eliminates estrogen production in specific tissues. The human aromatase gene spans ~123 kb and comprises of a 93 kb 5′-untranslated region (UTR), a 30 kb coding region (that includes nine exons, II–X, with ATG translational start site at exon II), and the 3′-end (Fig. 3). The UTR contains first exons (Is) that are influenced

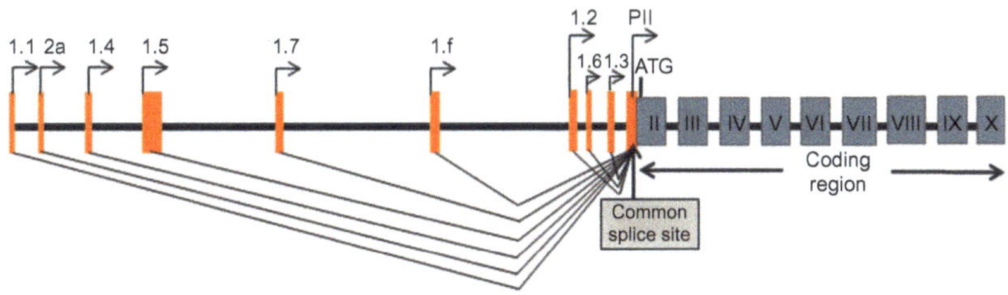

Promotors	1.1	2a	1.4	1.5	1.7	1.f	1.2	1.6	1.3	PII
Tissue	Placenta major	Placenta minor	Skin/ adipose	Fetal/ ovary	Endothelial cell/breast cancer	Brain	Placenta, ovary, brain	Bone	Adipose tissue/breast /endometrial /ovarian cancers	Adipose tissue/breast /endometrial /ovarian cancers

Fig. 3 Schematic representation of the aromatase gene. The aromatase gene contains nine coding exons (II–X). Expression of the aromatase gene is regulated by 10 tissue-specific promoters, in each case, producing aromatase gene transcripts that contain a different 5′-untranslated exon I. The exon Is are spliced onto a common 3′-splice site upstream of the translation start site ATG, resulting in the same aromatase protein regardless of the promoter used.

by 10 alternatively used tissue-specific promoters, which include I.1, I.2, and I.2a (placenta), I.4 (adipose tissue/skin), I.5 (fetal/ovary), I.7 (endothelial cells/breast cancer), I.f (brain), I.6 (bone), I.3 (adipose tissue/endothelial cells/breast and ovarian cancers), and PII (gonads/adipose tissue/endothelial cells/breast and ovarian cancers).[9,10,33,36–38] The 5′-UTR exon Is in both humans and mice are spliced with the 3′-coding exon II, thus, producing the same aromatase protein across various tissues. It is important to note that each untranslated tissue-specific exon I is regulated by distinct hormones, cytokines, growth factors, and second messenger pathways that mediate aromatase expression, and, thus, estrogen biosynthesis, under physiological and/or pathophysiological conditions.[9,33]

3. EXPRESSION OF AROMATASE IN NORMAL AND PATHOLOGICAL HUMAN TISSUES

Aromatase is encoded by the *CYP19A1* gene, and it is expressed in diverse tissues, including normal and carcinogenic tissues. As such, this enzyme is critically involved in numerous physiological and pathological processes.[10,11] Expression of aromatase, characterized mostly at the transcriptional level, is regulated by a plethora of signaling pathways, which utilizes alternatively spliced promoters in various tissues.[9,14,38] Transcriptional regulation of aromatase by tissue-specific promoters has been previously reviewed in many excellent articles[5,9,10,36,39] and will not be elaborated in great detail here.

The primary site of aromatase expression is the ovarian follicle, in which estrogens are produced from androgens, via the classical cholesterol metabolic pathway, by the granulosa cells in premenopausal women (Fig. 1). Regulation of aromatase expression, thus, estrogen production, in ovarian granulosa cells is predominantly mediated by follicle stimulating hormone (FSH) via protein kinase A (PKA)/cAMP signaling pathway, through activation of promoter PII, in cyclical fashion during reproductive ages of women.[9,10,33,39] Upon cessation of ovarian function at menopause, estrogens are produced by aromatase at extraovarian tissues such as adipose tissue, skin, bone, and brain and act in paracrine and/or intracrine manner. Of note, adipose tissue is a major source of estrogens in postmenopausal women. Estrogens produced in extragonadal sites in men and women, via activation of several tissue-specific aromatase promoters, play important roles in many physiological processes, including homeostasis, bone mineralization, and the closure of bone plates.[2,5,9,40]

Aberrant high expression of aromatase in a number of malignant tissues, including breast, endometrium, uterine fibroids, and ovary, has been shown to be tightly connected with intratumoral production of estrogens, compared to their levels present in either nonmalignant tissues or circulation.[10,11,34] Noteworthy, aromatase is also overexpressed in prostrate and colorectal cancers (Manna PR *et al.*, unpublished observations). In postmenopausal women, estrogen, derived from a number of sources, is the major risk factor in the development and growth of hormone-induced malignant disorders. Differential regulation of aromatase gene expression is mediated by hormones, cytokines, and growth factors through involvement of several alternatively spliced tissue-specific promoters.

4. ESTROGEN AND ITS RECEPTORS IN HORMONE RESPONSIVE CANCERS

Considerable evidence indicates that estrogen and its receptors are involved in the etiology of many malignant tumors, including breast, endometrial, ovarian, adrenocortical, pancreatic, and prostate cancers.[8,41–43] Classically, estrogens exert their effects by binding to two ERs: ERα and

ERβ, both of which exhibit high sequence homology and share many properties.[44] ERs belong to the nuclear receptor subfamily of ligand-activated transcription factors. Two structurally related ERα and ERβ (which possess six functional domains and are encoded by two separate genes, estrogen receptor 1 (*ESR1*) and *ESR2*, respectively) form either homo- or heterodimers and bind to the 13-bp estrogen response element (ERE; consensus sequence 5′-GGTCAnnnTGACC-3′) present at the promoter region of target genes and regulates transcription.[8,45,46] Most estrogen-dependent malignant tumors express ERα (~80%) that essentially transduces the biological actions of estrogens.[41,47,48] Despite the similarity of ERα and ERβ, they exhibit differences in the ligand binding specificity. In particular, a number of compounds with affinities to ERs have been shown to act as agonists in certain tissues and antagonists in others, as a consequence, they behave as estrogens and antiestrogens in tissue-specific manners and are termed as selective ER modulators (SERMs; discussed in Section 5.1.3).[17,49,50] A large body of evidence indicates that ER is capable of binding DNA indirectly by tethering to other transcription factors, including NF-κB (nuclear factor kappa-light-chain-enhancer of activated B cells), FOXA1/HNF3α (hepatocyte nuclear factor 3α), AP-1 (activator protein 1), Sp1 (specificity protein 1), GATA, which then recruit coregulators such as CBP (CREB-binding protein)/P300, PCAF (P300/CBP-associated factor), and steroid receptor coactivator (SRC) 1, in influencing chromatin remodeling.[7,51,52] As such, regulation of ER-mediated transcription is a complex process that involves multiple coregulatory factors and crosstalk between different signaling pathways. Beside its genomic action, estrogens have extranuclear ERE-independent activity through membrane, cytoplasmic, and G protein-coupled ERs, which when bound to chaperone proteins such as HSP90 (heat shock protein 90) and HSP70, results in formation of multiprotein complexes that trigger activation of the p44/42 MAPK (mitogen-activated protein kinase), p38 MAPK or JNK, and PI3/AKT (phosphoinositide 3-kinase/protein kinase B) pathways via nongenomic pathway.[43,53,54]

The N-terminal transactivation domain of ER is phosphorylated at several residues by a variety of kinases.[55,56] In accordance with this, studies have demonstrated that both serine (Ser)167 and Ser118 are phosphorylated by P13K/AKT and Ras-MAPK signaling cascades, which activate ER and influence its binding to coregulators.[43,57–59] Activation of Ser118 has been shown to mediate SRC3 binding and enhance E2 hypersensitivity.[60] Phosphorylated ERα/β binds to DNA both directly and indirectly, recruits coregulators, and triggers transactivation potential. Studies of Ser → Alanine (Ala) mutations at positions 104, 106, 118, and 167 have been shown to decrease transcriptional activity of ER.[55,59,61] AI-resistant breast cancer cells show marked activation of Ser118 that has been proposed as predictive marker in tamoxifen therapy.[62] Increased activation of Ser167, by tamoxifen, has been demonstrated in prolonged survival of breast cancer patients.[63] Alternatively, phosphorylation of Ser305 is connected with tumor growth in clinical trials. In addition to ER phosphorylation, five lysine (Lys) residues, i.e., Lys266, Lys268, Lys299, Lys302, and Lys303, located in the hinge region of ERα, have been acetylated by CBP/p300.[41,64,65] Also, Lys302 has been reported to be ubiquitinated, sumoylated, or methylated.[41,56] These findings imply that posttranslational modifications of ERs can be therapeutically targeted in a number of estrogen responsive malignancies.

There is increasing evidence that estrogen-dependent proliferation and growth of breast, endometrial, and ovarian cancers are primarily mediated by ERα.[66,67] On the other hand, ERβ has been demonstrated to have antiproliferative properties and is proposed to act as a tumor suppressor.[44,68,69] Specifically, whereas both ERα and ERβ share overlapping properties and coexpress in a number of benign and malignant tissues, expression levels of ERα are higher than those of ERβ in estrogen-dependent malignant tissues, including breast, uterine, endometrial, and ovarian cancers.[12,41] For example, ~80% breast cancers are ER-positive, which majorly express ERα, progesterone receptor (PR), and human epidermal growth factor receptor 2/the erythroblastosis oncogene-B2 (HER2/ErbB2), or all three, where estrogen plays a major role in the progression and survival of hormone-dependent cancers.[43,48,52] As such, ERα is considered a prognostic marker in hormone-induced malignancies, which respond to endocrine therapy.[70] By contrast, tumors with nuclear-free ER cells are classified as ER-negative, which are intrinsically resistant to hormonal therapy including AIs. Estrogen levels in malignant breast tissues have been demonstrated to be strikingly higher than their normal counterparts.[71] Targeted blockade of estrogen synthesis and/or inhibition of ERα activation are considered as effective endocrine therapies for hormone receptor positive cancers.[41,70,72–74] Consequently, several AIs that decrease the levels of estrogens, or antiestrogens (that compete with ER binding), are routinely used for treatment and management of a variety of estrogen-induced malignancies including breast, endometrium, and ovary, which are summarized later.

5. ABERRANT EXPRESSION OF AROMATASE IN THREE COMMON WOMEN'S CANCERS

5.1 Breast Cancer: Clinical Features

Breast cancer is the most prevalent cancer worldwide, comprising nearly 30% of all cancers, and it is the second leading cause of mortality in women. Breast cancer is divided into a number of categories, including invasive (infiltrating), noninvasive (in situ), and inflammatory. Both microscopic and clinical analyses are necessary to specifically determine the cancer and whether it is in situ or infiltrating and other types. Following the TNM (tumor, node, and metastasis) classification of tumors, the seriousness of breast cancer can be determined by the stage (0–IV; with stages being "0" for in situ and "IV" for most advanced). Estrogens, specifically E2, acting through ERs, play important roles in the pathogenesis and progression of breast cancer, and its incidence remains the highest in the United States.[75–77] As mentioned earlier, most breast cancers requiring estrogens for tumor growth are ER-positive, which account for ~80% among women in aged 45 years and older. Conversely, a small pool of malignancies is ER-, PR-, and HER2-negative that are named triple-negative breast cancers (TNBCs).[78] The latter subgroup of breast cancers (10–20%) is more aggressive and affects generally younger women. Besides, 40–50% women diagnosed with breast cancers will eventually progress to metastatic diseases. Nonetheless, postmenopausal, in contrast to premenopausal, women are more susceptible to developing breast cancers, suggesting that peripheral aromatization of estrogens (e.g., adipose tissue and adrenal glands) trigger development and growth of the disease through paracrine and/or intracrine mechanisms (Fig. 4). In keeping with this, intratumoral production of estrogen in malignant breast tissues can be 10–50 times higher than those found in either circulation or noncancerous counterparts.[2,6,10]

Estrogen biosynthesis requires the key enzyme aromatase and its expression is abnormally high in the majority of breast carcinomas. Induction of aromatase in malignant breast epithelial cells is mediated by activation of a number of alternatively spliced promoters.[9,40,79] It is well established that estrogens play major roles in breast cancer progression and survival, even though the epithelial carcinogenesis is multifactorial since estrogen's action involve both genomic and nongenomic signaling pathways.[33,52] Endocrine therapies, by either suppressing/inhibiting estrogen production by targeting

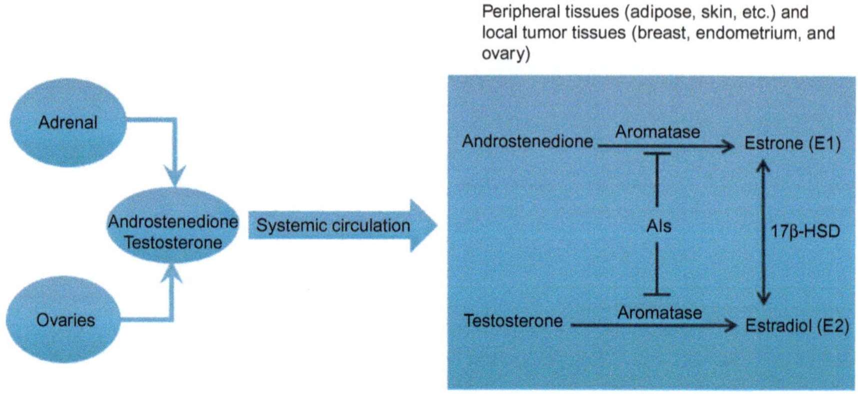

Fig. 4 Schematic representation of extraovarian estrogen synthesis. Estrogen precursors androstenedione and testosterone, which primarily originate in the adrenal in postmenopausal women, are converted to estrogen through the action of the aromatase enzyme in peripheral tissues such as adipose and skin, and locally in the breast, endometrial, and ovarian malignant tissues. AIs block the action of aromatase. 17β-HSD catalyzes the interconversion between E1 and E2.

aromatase or eliminating/downregulating ER, have proven to be effective in postmenopausal breast cancer prevention and treatment.[16,80–82] Despite various side effects, a number of selective AIs, superior to antiestrogens, are routinely used as therapeutic intervention in treating estrogen responsive breast cancers in postmenopausal women.[13,74,80,83]

5.1.1 Breast Cancer Etiology, Epidemiology, and Risk Factors

The etiology of human breast cancer is complex and is influenced by hormonal, genetic, environmental, and reproductive factors, which contribute to development of the disease. Regardless of factors involved, breast cancer is highly responsive to estrogen in promoting tumor growth which expresses high levels of aromatase and ERs.[10,11,33] Upon cessation of ovarian function, estrogens from extraovarian sources (primarily adipose tissue and skin) and produced locally in malignant breast epithelial cells, via paracrine and/or intracrine mechanisms, are the key drivers in the pathogenesis of breast cancer in postmenopausal women. Alternative views suggest the importance of circulating pool of estrogens in the progression of breast carcinogenesis.[84,85] Despite the sources, it is unambiguous that estrogen is the main culprit in the development and survival of malignant tumors; in particular, the more exposure of a woman to the hormone is more susceptible to breast cancer. Genetic alterations, involving DNA mutations, either inherited or acquired, are also involved in breast cancer development. A number of germline mutations that predispose to breast cancers include the two susceptibility genes breast cancer 1 (*BRCA1*) and *BRCA2*; and others such as *TP53/ p53* (tumor protein 53), and *PTEN* (phosphatase and tensin homolog) genes.[86–88] Recent era of molecular and genetic technological advances may identify additional factor(s) connected in the breast cancer.

With approximately 1 million new cases each year, breast cancer is a common cause of cancer death worldwide in women aged over 45. However, this life-threatening malignant disease produces typically no significant signs and symptoms at the early stage. Epidemiological evidence indicates that factors that are linked to estrogens play important roles in breast cancer progression.[76] Even so, malignant breast tissues express aberrant high levels of aromatase.[10,38,89,90] Many known breast cancer risk factors include menopause, age, family history, early menarche, obesity, oral contraceptives, and hormonal treatment during menopause (Table 2). Breast cancer is diagnosed more often in Caucasian women than other races. High breast tissue density is also a significant risk factor for the development of malignant tumors. Hormones involved in the maintenance of reproductive function are thought to influence breast cancer risk by promoting cell proliferation and growth, which are likely to involve DNA damage.[2,11,89] Studies have reported that reproductive patterns (early menstruation before age 12 or late menopause after age 55) are associated with increased risk of ER-positive breast cancers compared to ER-negative ones. As mentioned earlier, postmenopausal women with high circulating estrogen levels are more prone to developing breast cancer. There is increasing evidence to suggest that women diagnosed with breast cancer at 40 years of age or below are at increased risk of developing other cancers. Noteworthy, however, the precise causes of breast cancer are largely unknown; and many women who possess a number of risk factors never develop the disease. Conversely, women develop breast cancer without known risk factors.

5.1.2 Aromatase and Breast Cancer

The aromatase enzyme plays an indispensable role in the regulation of estrogen biosynthesis, by converting testosterone and androstenedione to E2 and E1, respectively. Estrogens act as a master regulator in a variety of physiological and pathological processes. Expression of aromatase and, thus, estrogens, is strikingly higher in malignant breast tissue compared to its normal counterpart.[5,91] In accordance with this, therapeutic approaches targeting aromatase and/or estrogen synthesis are key strategies for treatment and

Table 2 Risk Factors and Their Correlation to Breast, Endometrial, and Ovarian Cancer Incidence

Risk Factors	Breast Cancer	Endometrial Cancer	Ovarian Cancer
Late age of menarche	+	−	+
Menopause	+	+	+
Late age at last birth	+	−	?
Parity	−	−	+
Breastfeeding	+	−	−
Oral contraceptives	+	−	+
Frequent ovulatory cycles	+	+	+
Hormone replacement therapy	+	+	+
Body mass index	+	+	+
Exercise/physical activity	−	−	?
Tamoxifen use	−	+	=
Diets and diabetes	+	?	?
PCOSs	=	+	+
Lynch syndrome	−	+	+
Alcohol and smoking	+	−	+
Family history and genetic factors	+	+	+
Endogenous hormones	+	+	−
Childbearing	+	+	?
Ionizing radiation	+	?	−
Environmental estrogens	+	+	+

Factors affect tumor/cancer progression: "+", positive risk factor; "−", reduce disease incidence; "=", do not affect disease incidence; "?", clear-cut information not available.

prevention of breast cancers. Aromatase protein is tightly regulated in humans, and its expression in breast tissue is found in mesenchymal stromal and epithelial cells. In postmenopausal women, the adipose fibroblast cells express aromatase, and the majority of breast cancers is ER-/PR-positive and is responsive to estrogens. Aromatase immunoreactivity has been localized in breast cancer epithelial cells and surrounding fibroblasts.[5,9]

Long-term exposure of tissues to estrogens is well established to increase breast cancer progression and survival. Compared to noncancerous cells, malignant breast epithelial and stromal cells (adjacent to tumors) express markedly higher aromatase mRNA and protein levels. In particular, malignant breast tissues synthesize increased amounts of estrogens within the tumor, compared to their levels in circulation, in maintaining the cancer survival.

It is well established that intratumoral expression of aromatase and its activity in malignant breast tissue plays a crucial role in promoting and supporting the growth of estrogen-dependent tumors. Noteworthy, approximately 90% aromatase transcripts expressed in adipose tissues have been shown to reside in fibroblasts rather than in mature adipocytes. Expression of the human *CYP19A1* gene encoding aromatase is regulated by several alternatively spliced promoters in normal and malignant breast tissues, by a plethora of hormonal factors and signaling pathways.[11,79] A number of silenced aromatase promoters in disease-free tissues are active in cancerous breast tissue. Specifically, promoter switching from I.4 (noncancerous) to I.3, PII, and I.7 (cancerous) is a major mechanism for increased expression of aromatase in adipose tissue adjacent to breast cancer tissue.[79] These transcriptional machineries are aberrantly activated in diverse tissues including breast cancer and utilize prostaglandin E_2 (PGE_2) or PKA and PKC signaling pathways.[9–11] It is conceivable that the promiscuous nature of the splice acceptor site upstream of coding region is involved, at least in part, in transcriptional regulation of aromatase gene in a number of malignant diseases including breast cancer. Analysis of the *CYP19A1* gene in 596 breast cancer samples, obtained from the PROGgene database, indicated that overall survival of breast cancer patients (Kaplan–Meier plots) was higher with low aromatase mRNA levels,[92,93] implying higher expression of aromatase is tightly connected with advanced stage of the disease (Fig. 5).

5.1.3 Targeted Therapies for Breast Cancer

Endocrine therapy has been demonstrated to be pivotal in the treatment of hormone-dependent breast cancers, which remain the most commonly diagnosed malignancy in women. Estrogens, by binding to and activating the ERs, play central roles in the development of breast cancer, among which 80% are ERα positive, allowing the latter serves as a prognostic marker for responsiveness to hormonal therapy. Therefore, blockade/suppression of estrogen production and inhibition of ER activation are important targeted therapies in breast cancer prevention and treatment.[17,80–82,94]

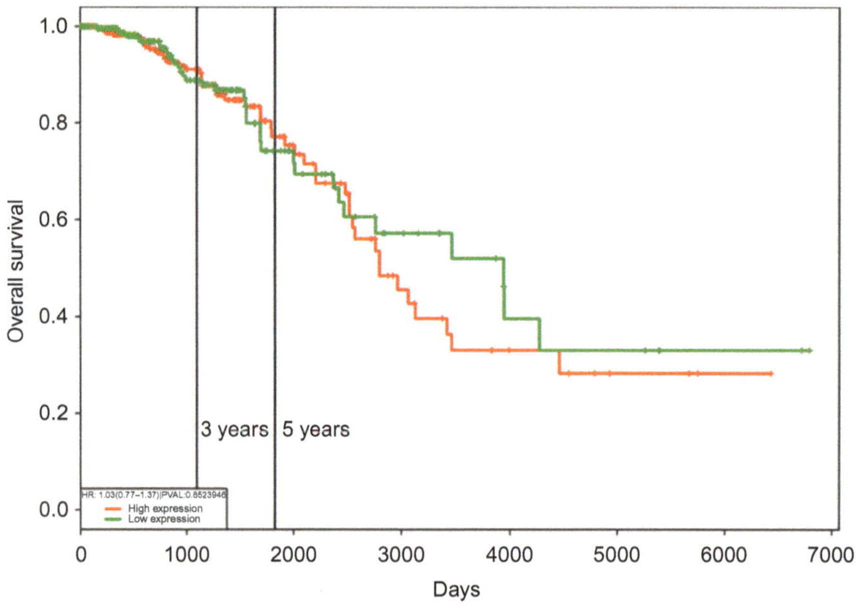

Fig. 5 Expression of the *CYP19A1* gene encoding aromatase in breast cancer and its correlation to patient survival. The overall survival of breast cancer patients was assessed in conjunction with high and low expression levels of the *CYP19A1*. The *line diagrams* generated were based on information available at the PROGgene database (analyzed 596 breast cancer samples). The *red curve* in the Kaplan–Meier plots includes all breast cancers with high expression of *CYP19A1*. The *green curve* in the Kaplan–Meier plots includes all breast cancers with low expression of *CYP19A1*. Overall survival of breast cancer patients was higher with low expression of the *CYP19A1* gene.

Two major modalities widely used for hormone responsive breast cancers include "antiestrogens" which blocks binding of estrogens to ERs and antagonize ER regulated action, and "AIs" which inhibit estrogen biosynthesis.

Antiestrogens are SERMs that antagonize the actions of estrogens, thereby block hormone-induced breast cancer proliferation and development.[17,49,50] The first-generation SERM, tamoxifen citrate, the first FDA approved nonsteroidal compound, remains the most widely used and effective antiestrogen for treatment of invasive and noninvasive breast cancers.[44,95] In both pre- and postmenopausal women, tamoxifen reduces the likelihood of breast cancer recurrence and mortality rates by 50%.[96] While tamoxifen acts as an ER antagonist to prevent estrogen responsive cell proliferation in breast cancers, it behaves as an agonist in other tissues such as endometrium, ovary, and uterus, and results in abnormal growth in these tissues. Like tamoxifen, a second-generation FDA approved SERM, raloxifene, has been shown to have both estrogen agonistic and antagonistic properties.[97] Both tamoxifen and raloxifene are used in treating breast cancers, even though they generate various unacceptable side effects. The collective results from many long-term trials indicated that tamoxifen is more effective in breast cancer prevention than raloxifene; however, the former exerts more side effects that the latter.[94,98] Side effects associated with these antiestrogen treatments include: endometrial, ovarian, and uterine cancers, blood clots, hot flushes, vaginal discharge/bleeding, and thromboembolic events. Later, a third-generation SERM, lasofoxifene, has been found to be the most potent for breast cancer prevention and treatment, resulting in an 83% decrease in ER-positive breast cancer incidence (PEARL trial, a double-blind, placebocontrolled, randomized trial with 8556 postmenopausal women).[99] In an alternative approach, a selective estrogen downregulator, fulvestrant (7α-alkylsulphinyl analog of E2), the only FDA approved steroidal pure antiestrogen, which binds to and degrades ER, is effective for the treatment of advanced and tamoxifen-resistant breast cancer.[100] Several other SERMs (for example, idoxifene, toremifene, ospemifene, ERA-923, and EM-800) have been tested for prevention and treatment of breast cancers; however, their importance has been modest to moderate in clinical settings.

The involvement of estrogens in the progression of the majority of hormone-dependent breast cancers emerged alternative therapeutic strategy for targeting estrogen signaling and ER-regulated functions. In particular, a major breakthrough in breast cancer therapy was achieved with the development of AIs, which strikingly reduce estrogen production by eliminating or inactivating the action of the aromatase enzyme to suppress/inhibit the growth of malignant breast tumors.[72–74,81] It should be noted that AIs lack estrogen agonistic activity, in comparison to SERMs. AIs are categorized as first, second, and third generation (and are grouped under steroidal and nonsteroidal compounds) depending on their efficacy and specificity in inhibiting aromatase. Whereas steroidal AIs bind covalently and irreversibly, nonsteroidal AIs bind covalently and reversibly, to the aromatase enzyme. Both first-generation (e.g., aminoglutethimide) and second-generation (e.g., formestane, fadrozole, and vorozole) AIs have been analyzed in terms of treatment of hormone-dependent breast cancer; however, they produced significant side effects that diminished their usefulness toward the disease.[17,74] The FDA approved third-generation AIs, steroidal (exemestane), and nonsteroidal (anastrozole and letrozole) compounds have been developed to combat ER-positive breast cancers. Third-generation AIs exhibit great potency and specificity for aromatase and can reduce serum estrogen levels >95%. The pharmacokinetic properties of these AIs are longer (approximately 48 h half-life for anastrozole and letrozole, and 27 h for exemestane, respectively) which allows for a limited dosing schedule. The third-generation AIs, compared with tamoxifen, are more efficacious and exhibit lower incidence of adverse side effects; however, they do lead to cardiovascular complications and osteoporosis.[13,72,81]

Whereas both AIs and antiestrogenic compounds are beneficial for treatment of ER-positive breast cancer in postmenopausal women, they have no apparent effects on TNBCs. The latter accounts for 10–20% of breast cancers and have a higher rate of recurrence and poor prognosis, all of which emphasize the identification of new druggable strategies for the treatment and prevention of these hormone receptor negative breast cancers.[78] As a consequence, a number of agents/factors, including retinoids (vitamin A and its derivatives), epidermal growth factor receptor, tyrosine kinase inhibitors, cyclooxygenase-2 (COX-2) inhibitors, and HDAC inhibitors (discussed in Section 7), have been developed and/or under investigation for targeting nonendocrine pathways.[94,101] Collectively, the third-generation AIs are superior and well-tolerated drugs, than those of tamoxifen and SERMs, and have been the mainstay of treatment for hormone receptor positive, but not receptor negative, breast cancers in postmenopausal women in both the adjuvant and metastatic settings.

5.2 Endometrial Cancer: Clinical Features

Endometrial cancer is a malignant tumor that originates in the uterine endometrium (Corpus uteri). Abnormal uterine bleeding, after menopause or between menstrual periods, is the most common symptom associated with endometrial cancer, occurring in 90% of cases. In approximately 75% of these cases, women present with early stage disease.[102] This symptomatic nature of endometrial cancer often results in its detection in early stages, especially since the incidence of endometrial cancers is much greater in postmenopausal women. The standard diagnostic strategy for endometrial cancer includes pelvic ultrasound and endometrial biopsy or dilation and curettage. Biopsy under hysteroscopy is considered most effective for endome-trial cancer diagnosis with greater accuracy compared to dilation and curettage. Imaging modalities such as magnetic resonance imaging, computed tomography, and integrated positron emission are used for evaluating disease metastasis.[103–105]

Endometrial tumors/cancers include epithelial carcinomas, mixed mesenchymal and epithelial tumors (carcinosarcomas), and mesenchymal (endometrial stromal and smooth muscle) sarcomas. Endometrial epithelial carcinomas can be further classified into several subtypes which include endometrioid, serous, clear cell, mucinous, squamous cell, transitional cell, small cell, and undifferentiated.[106–108] Overall, ~80% of endometrial cancers are of endometrioid while the other ~20% are nonendometrioid.[109] The traditional dualistic Bokhman model classifies endometrial cancers into two categories (Type I and Type II) based on histological characterization, hormone receptor expression, and grade.[110] Type I (endometrioid) endometrial cancers are generally low grade tumors that are moderately or highly differentiated, express ERs, and are dependent on estrogens. Type I tumors are associated with obesity and endometrial hyperplasia. Type II (nonendometrioid) endometrial cancers, by contrast, are high-grade tumors that are poorly differentiated and associated with an atrophic endometrium. While Type I tumors carry a favorable prognosis, Type II tumors are more aggressive and present a less favorable prognosis; indicating 5-year survival rates of 86% and 59%, respectively.[110] With the exception of hormone sensitivity and obesity, Type I and Type II endometrial cancers share many risk factors such as parity, oral contraceptive use, smoking, and age at menarche.[111] It is important to note that there is heterogeneity within each type as well as overlap between the two types of endometrial tumors.

5.2.1 Endometrial Cancer Etiology, Epidemiology, and Risk Factors

Endometrial cancer is the fourth most common cancer in women in the United States, and it is estimated that 60,050 cases of endometrial cancer will emerge in 2016 (7% of cancers in women) with 10,470 deaths (4%).[112] The lifetime probability of developing endometrial cancer in the United States has been reported to be ~3% based on analysis of data from 2010 to 2012.[113] If current trends hold, it is projected that the number of endometrial cancer cases in the United States will double to 122,000 per year by 2030.[114] The incidence of endometrial cancers in white women is twice that of black women, although the latter group tends to have more advanced disease and a less favorable tumor grade than the former.[102]

Hormone replacement therapy (HRT), especially consisting of estrogen, unopposed by progestin, has been consistently linked to various endometrial cancers. Multiple studies have linked unrestricted estrogen HRT to endometrial cancers. The addition of progesterone to HRT in the 1970s resulted in a reduction of endometrial cancer risk,[115] because progesterone antagonizes the effects of estrogens and inhibits estrogen-induced endometrial cell proliferation. Unopposed, estrogen promotes hyperproliferation and transformation of endometrial cells.[116] Long-term use of tamoxifen has been associated with endometrial thickening, dysfunctional uterine bleeding, endometrial polyps, and endometrial hyperplasia and cancers.[117]

The main risk factor for the development of endometrial cancer is exposure to estrogens, which increases with an earlier age of menarche or a later age of menopause.[115,118] Exposure to estrogens, unopposed by progesterone, as is the case in chronic anovulation that occurs in polycystic ovary syndrome (PCOS) or due to an estrogen producing tumor is also associated with an increased risk for endometrial cancer.[119] Nulliparity has also been identified as a risk factor for endometrial cancer. A number of these factors directly relate to the number of ovulatory cycles which also appears to correlate with endometrial cancer risk.[120] Age also represents a risk factor in endometrial cancer and the incidence rises sharply during the perimenopausal years, and peaks after menopause.[115] An aging population along with delays and reduction in childbearing, declining hysterectomy rates, and greater prevalence of obesity has contributed to a rise in endometrial cancer incidence.[105] Obesity is another major risk factor for developing endometrial cancer with a linear increased risk with body mass index (BMI) or weight.[121] Based on metaanalysis, an increase in BMI above 25 has also been linked to increased mortality from endometrial cancers.[122] Several familial syndromes are also associated with an increased risk in endometrial cancers, with the Lynch syndrome being the most predominant. It is an inherited, autosomal dominant, cancer susceptibility syndrome characterized by defective DNA mismatch repair and is associated with a 40–60% increased lifetime risk of endometrial cancers.

5.2.2 Aromatase and Endometrial Cancer

As mentioned earlier, aromatase is the rate-liming enzyme in the biosynthesis of estrogens. A majority of endometrial cancers are hormone sensitive and require estrogens for disease progression. Type I endometrial cancer, which accounts for 80% of cases, is associated with a high expression of ER.[109] Accumulating evidence indicates that estrogens are linked with an increased incidence of endometrial cancer.[104] Aberrant higher expression of aromatase and, thus, estrogen synthesis, has been demonstrated in malignant endometrial tissues than those present in both normal counterparts and plasma.[11,123]

There are three main sources of estrogens that influence hormone-dependent endometrial cancers. A major source of estrogens in premenopausal women is the ovarian granulosa cells. Both skin and adipose tissue convert androgens such as androstenedione and testosterone to E1 and E2, respectively, catalyzed by the aromatase enzyme (Fig. 4). E1 is also converted into E2 by 17β-hydroxysteroid dehydrogenase enzyme (17β-HSD), which reaches the tumor site through systemic circulation. Estrogens are also synthesized locally, by aromatase, within the endometrial cancer tissue. Transcriptional regulation of aromatase in endometrial cancers has been reported to be mediated by a number of alternatively spliced promoters such as 1.3 and PII[11] (Fig. 3). Ovarian estrogen synthesis is ceased in postmenopausal women. At menopause, extraovarian sources such as adipose tissue produce a majority of estrogens. It has been surmised that both skin and adipose tissue-derived estrogens play key roles in the development of a number of hormone responsive malignancies, including endometrial cancer, especially in postmenopausal women.[10,124] Several lines of evidence also indicate that synthesis of extraovarian estrogens is correlated with excess body weight, with as much as a 10-fold increase in extraovarian estrogen synthesis in morbidly obese postmenopausal women.[125] Numerous studies have highlighted the influence of obesity and, consequently, accumulation of extraovarian estrogens, in the pathogenesis of endometrial cancer. The prevalence of obesity has been rising and since endometrial cancer is predominantly a postmenopausal disease, the impact of the increase in obesity rates on endometrial cancer is yet to be fully appreciated.[121,126,127]

In postmenopausal women, while the circulating levels of estrogens decline rapidly, testosterone levels remain unchanged for many years.[128] Testosterone is one of the substrates for E2 synthesis in extraovarian tissues. Elevated levels of plasma androstenedione and testosterone have been shown to be associated with increased risks for endometrial cancers in both pre- and postmenopausal women.[129] The conversion of androstenedione and testosterone to E1 and E2, respectively, is catalyzed by aromatase (Fig. 4). Consequently, it has been reported that intratumoral expression of aromatase is considerably higher in malignant endometrial cancers than those found in disease-free tissues.[130] This suggests that production of estrogens in situ, due to aberrant high expression of aromatase, is a key factor in the development and progression of endometrial cancers.[131] Expression of the *CYP19A1* gene in 143 endometrial cancer samples, as available and analyzed from the PROGgene database, indicated that survival rate of endometrial cancer patients (Kaplan–Meier plots) is higher with low aromatase mRNA expression[92,93] (Fig. 6).

A study including 336 endometrial cancer samples (82% endometrioid and 18% nonendometrioid) has demonstrated the expression of aromatase in 65% of malignant tumors.[132] In a separate study, expression levels of ER, COX-2, and aromatase in 35 endometrioid endometrial cancer patients have been demonstrated to be ~2-fold higher than those present in non-diseased endometrial tissues.[133] Jarzabek *et al.*[134] analyzed 51 type I endometrial cancer samples and found that the majority of tumors expressed ERα (82%), aromatase (80%), and COX2 (88%) proteins. A separate study of 55 patients with endometrial cancers revealed that aromatase was expressed in epithelial (58%), stromal (36%), and myometrial (34%) cells. Whereas intratumoral expression of aromatase in stromal cells was associated with poor survival, no correlation with prognosis was found

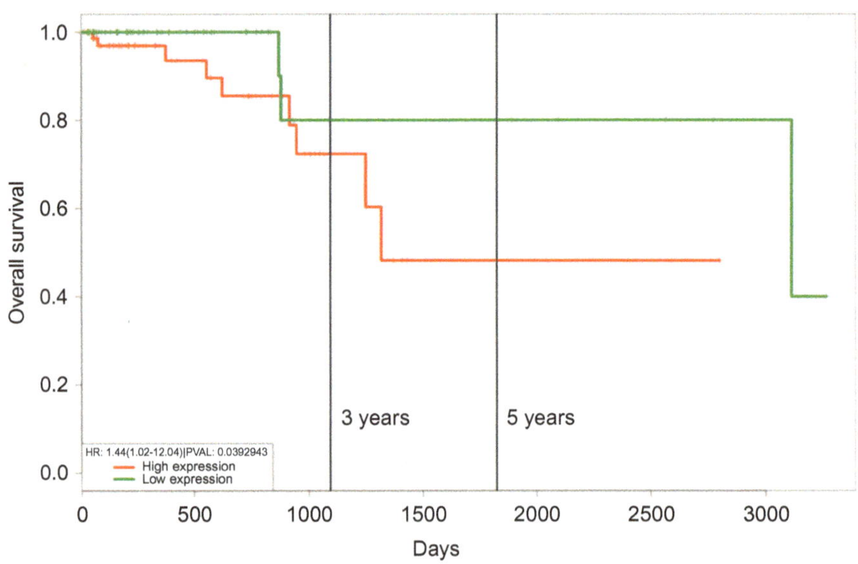

Fig. 6 Expression of the *CYP19A1* gene encoding aromatase in endometrial cancer and its correlation to patient survival. The overall survival of endometrial cancer patients was assessed in conjunction with high and low expression levels of *CYP19A1*. The *line diagrams* generated were based on information available at the PROGgene database (analyzed 143 endometrial cancer samples). The *red curve* in the Kaplan–Meier plots includes all endometrial cancers with high expression of *CYP19A1*. The *green curve* in the Kaplan–Meier plots includes all endometrial cancers with low expression of *CYP19A1*. Overall survival of endometrial cancer patients was higher with low expression of the *CYP19A1* gene.

with its expression in epithelial and myometrial cells.[135] A recent study has reported that estrogen induces IL-6 (interleukin 6) which stimulates aromatase expression in intratumoral stromal cells resulting in a positive feedback loop with in situ estrogen production in endometrial cancer cells.[20] Collectively, aromatase is overexpressed in the majority of malignant endometrial tumors, and it can be effectively targeted for the prevention and treatment of various types of endometrial cancers.

5.2.3 Targeted Therapies for Endometrial Cancer

The treatment and prevention of hormone sensitive breast cancer have relied on either blocking the production of estrogen through the use of AIs or by modulating its action on ERs through the use of SERMs such as tamoxifen. While tamoxifen is effective on ER-positive breast cancers, it promotes endometrial cancer development.[127] In a breast cancer prevention trial that involved 13,388 increased risk women, tamoxifen resulted in a 69% decrease in ER-positive breast cancer compared to placebo. However, it also resulted in an increased risk for the development of endometrial cancers.[136]

AIs have been demonstrated to be effective endocrine treatment and are superior to tamoxifen in clinical trials for the treatment of breast cancers in postmenopausal women.[9] However, AIs reduce endogenous estrogen levels thus produce many unwanted side effects. In a cohort of 17,064 women diagnosed with ER-positive breast cancers, the effect of AIs and tamoxifen on the incidence of endometrial cancers was determined in different hormone therapy groups. The findings demonstrated that endometrial cancer incidence was 48% lower in AI group compared to tamoxifen group.[137] There is increasing evidence that the use of AIs/SERMs in breast cancer therapy can modulate endometrial thickening and the risk of developing endometrial cancers.[138,139]

The efficacy of AIs in the treatment of endometrial cancers is not established, however, AIs have been shown to block estrogen production in malignant tissues.[11] In a pilot study, 16-endometrial cancer patients treated with an AI, anastrazole, for 2 weeks prior to surgery showed a marked decrease in ERα and androgen receptor (AR) expression when compared with controls. Additionally, a decrease in the expression of proliferation marker ki-67 was reported.[140] Nonetheless, treatment of advanced or recurrent endometrial cancers with AIs (letrozone or anastrozole) has been demonstrated with limited response.[141] While AIs have been shown to produce some success for treatment of endometrial cancers, further studies are required to establish their efficacy, potency, and treatment regimens.

The standard initial treatment and/or management of endometrial cancers is surgery. Treatment for stage I endometrial cancer includes hysterectomy, bilateral salpingo-oophorectomy (removal of both ovaries and fallopian tubes), and bilateral pelvic and paraaortic lymph node dissection. Surgery can be performed either laparotomically or through the use of a minimally invasive technique such as laparoscopy, which has been shown to have superior short-term safety and length of stay.[104] Patients with Federation of Gynecology and Obstetrics grade 1 or 2 endometrial carcinomas that are limited to the inner half of the endometrium do not benefit from any postsurgical therapies.[119] The most common adjuvant treatment for endometrial cancers has been radiation therapy, however, its effectiveness in stage I and II endometrial cancers is unclear.[119] Several trials have conducted to assess the benefit of external beam radiation therapy (EBRT) in patients after surgery and also in patients who received vaginal branchytherapy after surgery. While EBRT resulted in lower relapse rates at the expense of more toxic effects, no difference in overall survival was reported.

Chemotherapy is the preferred treatment for metastasis linked with endometrial cancers. The most active chemotherapeutic agents are platinum compounds, anthracyclines, and taxanes. In a Gynecologic Oncology Group study, a much higher response rate was achieved for triple therapy with doxorubicin, cisplatin, and paclitaxel (57%), compared to ~20% with a single agent.[142,143] Patients diagnosed with endometrial cancers at reproductive ages have limited fertility saving therapeutic options, with the utilization of progestins as the main treatment option. One study of 231 cases of fertility sparing therapy found an overall response rate of 68% with a recurrence rate of 12%, while 32% of patients did not respond to any treatment.[104]

5.3 Ovarian Cancer: Clinical Features

Ovarian cancer is the fourth most common gynecological cancer death in women. Abnormal neoplastic proliferation of heterogeneous cells of the ovary results in ovarian malignancy, which is categorized into different types, including epithelial, stromal, and germline ovarian cancers. Whereas epithelial carcinomas originate from the outer layer of the ovary, stromal cancer emanates from hormone producing cells of the connective tissues. Germ cell tumors develop from the egg producing cells as a result of mutations in ovarian cancer susceptibility genes.[144] Epithelial ovarian cancer is the most malignant among three types. Stromal cancer is often considered as low-grade tumors that include granulosa–theca neoplasms. Tumors originating from the germ cells are both benign and malignant types. Germ cell tumors include both mature teratomas (dermoid cyst and often benign) and immature teratomas (rare and malignant), which can be diagnosed at young age (13–25 years) and in women with reproductive capabilities.

Ovarian cancers are often diagnosed at the advanced stage, as presenting symptoms are rather unclear and frequently associated with aging and menopausal events.[145] These telltale signs include abdominal discomfort, distension or bloating, uterine bleeding, increased urinary frequency due to weakened bladder, indigestion, exhaustion, weight loss, rapid sensation of satiety, and depression. Nevertheless, persistent abdominal distension, postmenopausal bleeding, loss of appetite, and anorexia are significant indicators associated with ovarian cancers.[145] A recent study involving 97 women has presented links between ovarian tumors and abdominal mass (43%), with only 20% increased urinary frequencies, and 10% nausea, whereas 20% exhibited no symptoms. Ovarian cancer is complicated by torsion which causes abdominal pain with its acute form resulting in loss of ovarian function, tissue necrosis, and ultimately death.[146,147]

5.3.1 Ovarian Cancer Etiology, Epidemiology, and Risk Factors

Ovarian cancer is the leading cause of death from gynecologic tumors ascribed to the late diagnosis of the malignancy and emerging increase in resistance to standard chemotherapy.[148] Epithelial carcinoma is the predominant form of ovarian malignancy among postmenopausal women, and this has been attributed to protracted high levels of FSH and luteinizing hormone (LH),[149] perhaps a consequence of inadequate negative feedback control of the hypothalamic–pituitary–ovarian unit by estrogens.[149,150] Reports from the Surveillance, Epidemiology, and End Results (SEER) 18 registry (2006–12) of ovarian cancers revealed that 60% of cases had metastasized, 19% had spread to regional lymph nodes, 15% were confined to primary sites, and the rest 6% were not staged. The extent to which the cancer spreads has been linked with survival rate of approximately 92% 5-year postdiagnosis in patients with localized cancers, and 73%, 29%, and 24% in regional, distant, or unstaged ovarian cancers, respectively.[151] Despite improved medical and chemotherapeutics over the past 20 years, only a minor advancement has been made in overall survival rate for ovarian cancer patients.

Several risk factors have been implicated for the development of ovarian cancers. These include family history of cancers in breast, colorectal, endometrium, and ovary; and obesity, smoking, aging, and age-related factors.[152,153] Mutations in the familial *BRCA1/2* gene, human developmental index, HRT in postmenopausal women, and fertility treatments, all of which result in hyperestrogenic states, contribute to key risk factors for ovarian malignancies.[153,154] BRCA1/2 proteins are crucial for the maintenance of genome integrity, stability, and DNA repair via homologous recombination. According to the TCGA (The Cancer Genomic Atlas) research network, germline, and somatic mutation rates of both BRCA1 and BRCA2 are ~8% and 3.5%, respectively, in ovarian serous cystadenocarcinoma.[154,155] Emergent epidemiological data substantially implicate the use of HRT in increased risk of ovarian cancers in postmenopausal women, compared to those who never used HRT.[156] A metaanalytical study also showed that the risk of developing ovarian cancers in women using HRT is time and hormone dependent with increased risk observed with estrogen and/or progestin therapy.[157] Moreover, abnormal expression of protooncogenes, including *KRAS*, *BRAF*, *HER-2*, and *PIK3CA*, have been associated with tumorigenesis, patient survival rate, and tumor metastasis.[155,158]

5.3.2 Aromatase and Ovarian Cancer

The ovary is the major estrogen producing site in premenopausal women. Conversely, the ovary in postmenopausal women undergoes major cellular transformation, which invariably results in fewer follicles, loss of epithelium, abundant stromal content, and a reduction in estrogen biosynthesis. As such, extraovarian tissues predominantly synthesize estrogens in postmenopausal women. Epidemiological studies point to the involvement of estrogen and its metabolites in the progression of ovarian cancers expressing ERs.[159–162] However, in a recent study no association between estrogens and ovarian cancers, in postmenopausal women, not taking hormone therapy, has been demonstrated.[163] Several lines of evidence demonstrate the importance of ERα expression in the growth of high grade serous ovarian cancers (HGSOC) and that increased estrogen levels considerably impact the progression and angiogenesis of ERα-negative HGSOC xenografts in mice.[164]

Estrogens are majorly produced by the action of aromatase, and 17β-HSD as well, on androgens in specific tissues. Aromatization of androgens to estrogens is considered the rate-limiting step in estrogen biosynthesis, and abnormal high expression of aromatase has been correlated with many gynecological malignancies, including ovarian cancers.[11] Specifically, intratumoral expression of aromatase has been demonstrated to be high (33–81%) in ovarian cancers.[165,166] Expression of the *CYP19A1* gene encoding aromatase is essentially influenced by alternatively spliced tissue-specific promoters, i.e., 1.3, 1.6, and PII, among which PII is the key regulator of aromatase in the ovary.[167] PII promoter is reported to be a downstream target of FSH and LH in various phases of ovarian cycles and has been implicated in the control of aromatase in ovarian granuloma malignancies.[9,168] Expression of the *CYP19A1* gene in 578 ovarian cancer samples, as analyzed from the PROGgene database,[92,93] demonstrated that overall survival of ovarian cancer patients (Kaplan–Meier plots) was higher with low expression of *CYP19A1* (Fig. 7).

Ovarian epithelial and stromal cells express aromatase and its level has been shown to be influenced by *BRCA1* gene.[169–171] Nevertheless, increased 1.3 and PII promoter-specific transcripts have been reported in prophylactic oophorectomy samples obtained from ovaries of women carrying mutations in *BRCA1* gene.[170] Approximately 10% of all ovarian

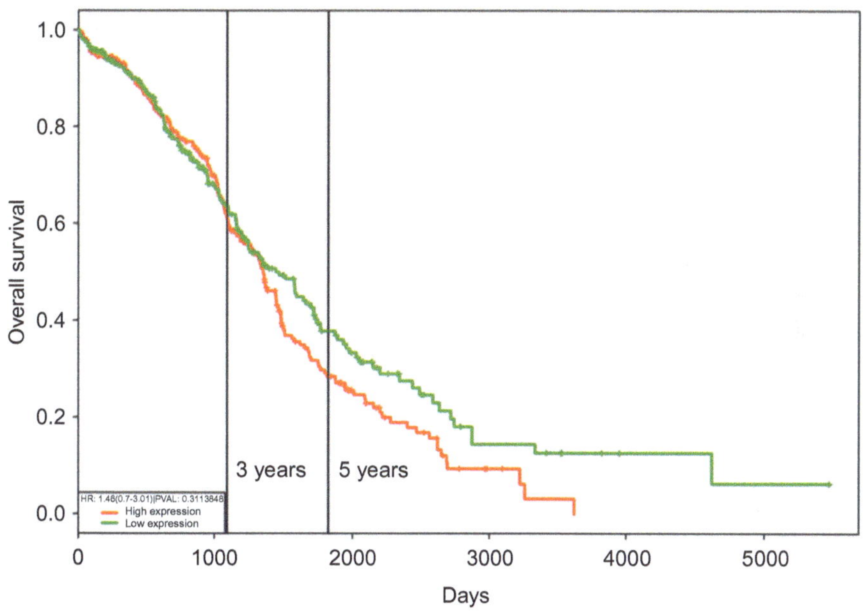

Fig. 7 Expression of the *CYP19A1* gene encoding aromatase in ovarian cancer and its correlation to patient survival. The overall survival of ovarian cancer patients was assessed in conjunction with high and low expression levels of *CYP19A1*. The *line diagrams* generated were based on information available at the PROGgene database (analyzed 578 ovarian cancer samples). The *red curve* in the Kaplan–Meier plots includes all ovarian cancers with high expression of *CYP19A1*. The *green curve* in the Kaplan–Meier plots includes all ovarian cancers with low expression of *CYP19A1*. Overall survival of ovarian cancer patients was higher with low expression of the *CYP19A1* gene.

cancers are linked with *BRCA1/2* gene mutations which enhance the chance of developing ovarian malignancy by 80%.[148,154] Loss of function in the *BRCA1* gene has been linked with elevated expression of aromatase in ovarian tissue.[172] Premenopausal women with *BRCA1/2* mutations developed a reduced tendency of ovarian cancer when their ovaries and fallopian tubes were surgically removed.[173] Therefore, a negative correlation between BRCA1 protein function and aromatase expression is a risk factor in the development of ovarian malignancy.[172] Aromatase deficiency has been linked with virilization in young females with granulosa cell tumors.[174] Estrogen levels in the ovary are also influenced by common single polymorphisms (rs749292) in the *CYP19A1* gene, suggesting that dysregulation aromatase is a risk factor in the pathogenesis of ovarian cancer in postmenopausal women.[175]

Estrogens induce ovarian cancer development by activating ER response genes and mitogenic signaling pathways, hence promoting cell proliferation.[176] A number of genes regulated by ERs include *Snail*, *hTERT* (human telomerase reverse transcriptase), and *c-Myc*, which play diverse roles such as cell proliferation, migration, invasion, and metastasis.[177] Accumulating evidence alluding to estrogen playing a key role in the progression of ovarian cancer includes a dysregulated ERα in the hypothalamic–pituitary–ovarian axis.[178] Specifically, a conditional deletion of the pituitary ERα gene in mice resulted in increased levels of LH, which consequently led to luteinization of the stromal cells, aromatase expression, and estrogen biosynthesis in ovarian epithelial cells. It is interesting to note that mice treated with letrozole have been demonstrated to markedly suppress the ovarian tumor mass,[178] suggesting the involvement of estrogen in the progression of ovarian tumorigenesis. Therefore, estrogen biosynthesis has gained the attention as a potential candidate for therapeutic target in ovarian cancers.

5.3.3 Targeted Therapies for Ovarian Cancer

The first-line treatment of ovarian malignancies includes debulking the tumor to an appreciable size, which can then be effectively treated by a chemotherapy regimen of platinum–taxane-based drugs.[179,180] Platinum-based drugs act on cancer cells by forming highly reactive platinum complexes that interact and crosslink with DNA molecules thereby preventing malignant cells from proliferation and death. Taxanes, on the other hand, are diterpenes that interact with microtubular structures of a cell to prevent reorganization, leading to inhibition of mitotic division. While these approaches help reduce tumor development, a number of patients relapse following a 6-month treatment (the platinum-resistant group) and eventually succumb to ovarian cancer. Also, hypersensitivity reactions to platinum–taxane-based therapy in ovarian cancer patients have prompted the development of other treatment options.[181] Alternative therapeutic approaches available to the platinum-resistant group of women with ovarian cancer include monoclonal antibodies, check point inhibitors and immune modulators, vaccines, adoptive T cell transfer, oncolytic viruses, and adjuvant immunotherapies.[181] Monoclonal antibodies against angiogenic and epidermal growth factors have proven to be effective in treating ovarian tumors, and vaccines raised against established malignancies elicit tumor-specific immune responses.[182] In addition, immune checkpoints and adoptive T cell transfer are developed to enhance immunity against ovarian cancer.[183]

ERs and PRs have also been the targets for ovarian cancers, as the malignant tissues express both of these receptors.[184] The use of tamoxifen is the standard adjuvant endocrine therapy for ER-positive breast cancers in postmenopausal women. However, moderate effects of tamoxifen have been reported in the treatment of ovarian cancers with overall survival rates of 13% observed from an analysis of 20 studies, in which only 4% and 9% demonstrated complete and partial responses, respectively.[176] A retrospective study evaluated the progression-free interval (PFI) for women with platinum-resistant ovarian, fallopian tube, and primary peritoneal, cancers on antiestrogen therapy (AET), and determined the relationship between PFI and ER expression.[185] It has been reported that ovarian cancer patients pretreated with at least four chemotherapeutics prior to the usage of AET displayed an average PFI of 4 months comparable to the observed PFI when patients are placed on antineoplastic drugs-like gemcitabine and liposomal doxorubicin. Tamoxifen has also been reported to be functional in ER-negative ovarian cancers,[185] suggesting that this drug may sensitize malignant tissues through estrogen-independent signaling. Adjuvant hormone therapy has recently been demonstrated to improve overall ovarian cancer survival rate, with negligible adverse effects, in women randomly administered HRT following surgical treatment of ovarian cancers.[186]

Aromatase has been targeted as a preferred therapy since estrogen biosynthesis is the final step in the steroid biosynthetic pathway (Fig. 1). AIs block aromatase and, thus, decrease estrogen biosynthesis, in all estrogen producing tissues. The first-generation AI aminoglutethimide has been shown to target cytochrome P450 enzymes and, thereby, necessitating glucocorticoid augmentation, and additionally generates considerable side effects such as lethargy, rashes, anemia, nausea, and fever. While the second-generation AIs were better than aminoglutethimide, they also produce unwanted side effects. The third-generation AIs, anastrozole, letrozole, and exemestane, are more selective for aromatase, which exhibit lesser side effects than those previous two generations. An investigative review of five studies revealed that letrozole therapy with recurrent ovarian cancers prompted clinical response rates up to ~36% and stable disease rates between 20% and 42%.[177] In a phase II clinical trial, patients with borderline or low-grade ovarian malignancies treated with letrozole demonstrated a response of 15% with stable disease rate of ~39%.[177] Following the Union for International Cancer Control criteria, a phase II trial of letrozole in ER-positive patients with relapsed ovarian cancer showed that tumors from the stable disease group expressed higher ER and PR levels than those of the progressive disease group.[187] These results reported that according to CA-125 (cancer antigen-125) test criteria 17% of patients responded and 26% achieved disease stabilization at 6 months, and that AIs exert their antitumor activity against recurrent ovarian malignancies.[187] Phase II trial studies evaluating the effect of anastrozole on recurrent ovarian cancer has been reported with only one partial response and 42% stable disease for a period of 90 days, with no correlation between hormone receptors and treatment.[188] Although a considerable success has been made with third-generation AIs in the treatment of ovarian cancers with minimal side effects, there have been incidences of resistance to the therapy and increased progression of tumors in other relevant tissues.

6 Resistance to Endocrine Therapies

Endocrine therapy resistance is a major concern, and it is broadly classified into two groups: acquired and innate (intrinsic).[43,189,190] Acquired resistance is a consequence of molecular changes and clonal selection observed after an extended period of treatment, whereas innate resistance occurs at the inception of treatment regimen. ER-negative breast cancers are examples of those who exhibit innate resistance to AI because they lack the classical ER pathway. In spite of success with third-generation AIs in the treatment of a number of estrogen-dependent cancers with minimal side effects, resistance to endocrine therapy with increased progression of tumors and poor clinical outcomes is a major concern. Acquired AI resistance is influenced by activation of growth factor receptor pathways, which constitute targets for chemotherapy at receptor–effector levels. In an effort to limit estrogen-driven tumor proliferation by AI therapy, abnormal involvement of several pathways, adaptive changes, and hypersensitivity to E2 in breast cancers, has been reported as a consequence of sustained estrogen deprivation and disruption of ER signaling.[189] These adaptive changes include dysregulation of ERs, ARs, growth factor receptor pathways, mainly epidermal growth factor (EGF) receptors 1/2 (HER1/2), as well as their downstream effector pathways such as MAP kinase, PI-3 kinase, and mTOR signaling. There is ample evidence that a number of adrenal gland-derived steroidogenic hormones that lie up stream of aromatase (Fig. 1) and are involved in the progression of breast cancers, may contribute to hormone resistance.

Preclinical and clinical studies have shown an interaction between hormone receptors and growth factor receptor pathways, which can result in activation of ER pathway in a ligand-independent manner or alteration of downstream coactivators, as well as transcriptional events. ARs have been implicated in AI resistance by the failure of anastrozole to inhibit androstenedione stimulated growth and ER transcriptional activity in MCF-7 human breast cancer cells overexpressing AR and aromatase (MCF-7-AR-arom) while inhibiting tumor cells overexpressing aromatase alone (MCF-7-arom). Activation of the insulin growth factor receptor (IGF1R) and AKT kinase has also been reported in MCF-7 cells overexpressing AR. It was proposed that cells might be evading the antitumor activity of anastrozole by the IGF1R/AKT pathway which protects cancer cells from apoptosis, as inhibition of receptors and pathways lead to restoration of anastrozole sensitivity in MCF-7-AR-arom cells.[191,192]

A quarter of invasive breast tumors undergo *HER2* gene expression, and its increased level is often associated with aberrantly cleaved forms that are characterized by enhanced tumorigenic properties.[193] These findings suggest that *HER2* might be involved in the pathogenesis of breast malignancies. Increased activation of the PI3K/Akt/mTOR pathway is important for tumor evasion of hormone-dependent regulation in breast cancer, by encouraging the continued existence of malignant cells in estrogen-deprived conditions. The PI3K/Akt/mTOR pathway has been linked with poor prognosis in breast cancer patients and also observed in AI-resistant breast tumor models.[194] Anastrozole-induced resistant MCF-7-arom cells have been shown to develop AI resistance independent of ER with decreased sensitivity to ERα antagonist and letrozole; continuous activation of the PI3K/AKT/mTOR pathway and marked overexpression of ErbB2 receptors. Sensitivity of AI resistance was reestablished with the combined treatments of anastrozole and potent inhibitors pertaining to AKT (MK-2206) and mTOR (rapamycin) pathways.[195,196] This implies that AI resistance is characterized by a transition from hormone receptor-dependent to hormone receptor-independent tumor development via growth factor signaling. Everolimus, an mTOR inhibitor has been shown a dose-dependent antitumor activity in breast cancer cells expressing aromatase and HER2-dependent resistance (BT474-AROM3), as well as in long-term estrogen deprivation induced resistant MCF-7 xenografts overexpressing HER2.[197] Moreover, increased phosphorylation of AKT and HER3 was detected in everolimus-treated cells, suggesting a role of activated Akt/HER2 in stimulating intracellular signaling that inhibits cancer cell proliferation. Everolimus has been shown to reduce ER mRNA and protein levels, as well as ER activity, and subsequently induce autophagy in AI-resistant breast cancer cells.[197,198] In postmenopausal women with recurrent breast cancers, a synergistic effect of everolimus and exemestane on progression-free survival in HR+/HER2-negative breast cancers has been reported.[199,200] These drugs have been used in the treatment of recurrent breast cancers resistant to endocrine therapies.[201]

Adaptive ER responses developed by AI-resistant cells involve a switch from steroid-dependent induction of gene expression to nonsteroid-dependent gene transcription. A typical example of elevated levels of *ER/AIB1* target gene, early growth response 3 in AI-resistant recurrent patient tumors was demonstrated by ER-chromatin immunoprecipitation sequencing when compared to primary tissues.[202] This implies that growth-related genes classically regulated by ER co-opt alternative pathways in AI resistance malignancies. AI resistance has been also linked with dependence on glycolysis in long-term estrogen-deprived MCF-7 cells displaying increased ER-related dependence on glycolytic pathway.[203] In aromatase sensitive MCF-7-2A cells, a combined therapy of letrozole and glycolysis inhibitor resulted in a decrease in cell growth. This study also revealed the role of miR-155 (micro RNA 155) in the control of metabolic plasticity of the AI-resistant cells and its expression correlated with poor response to AI therapy, indicating that miR-155 might be a prognostic biomarker for AI resistance in malignant cells.[203]

The combined administration of endocrine therapy and growth factor pathway inhibitors has been shown to improve survival of ER-positive breast malignancies; however, similar response is yet to be seen in ER-negative breast cancers that display intrinsic resistance to aromatase. In addition to lack of hormone receptors, heterogeneity in ER-negative breast cancers has been reported to play a role in poor treatment outcomes and contribute to higher tumor recurrence with reduced survival rate after relapse. Several growth factor receptor pathways are upregulated in TNBCs, and attempts to target these receptors have yielded unsatisfactory results. A combination of lapatinib, an EGFR inhibitor, and mTOR inhibitor, rapamycin, resulted in striking antitumor activity both in vivo and in vitro, consistent with substantial growth inhibition and increased cell death in TNBC patients.[198] Angiogenesis is a key event in the survival, progression, and invasion of malignant tissues, and vascular endothelial growth factor receptors have been targeted for chemotherapy against breast cancers. Antibodies and inhibitors to these receptors have also been used in clinical trials but no promising results are observed yet.[204]

Fibroblast growth factor (FGFR), a target of chemotherapy has shown encouraging results in a subcategory of TNBC preclinical tumor models.[205] The authors demonstrated that cell lines with *FGFR* gene expression responded highly to FGFR–ATP competitive inhibitor PD173074 and RNAi silencing of the FGFR2. This suggests that TNBC patients with increased expression of FGFR have a better chance of survival when administered FGFR inhibitors. Genomic instability is able to trigger the initiation of cancer, enhance its progression, and impact the overall prognosis of cancer patients. These instabilities could range from simple changes in DNA sequences to structural and chromosomal aberrations with increased gene copy numbers and rearrangements culminating in chimeric transcripts.[206] Mutations in DNA sequences or expression changes in gene products involved in ensuring proper regulation of cellular processes may support tumorigenesis and offer growth advantage to malignant cells. Somatic mutations of genes that play pivotal roles in the progression of triple negative and luminal B breast tumors have been targeted for chemotherapies in breast and other gynecological cancers. Such genes include tumor suppressor genes *BRCA*, *TP53*, glutathione *S* transferase p1, E-cadherin, retinoblastoma (*RB*), and *PTEN*.[87,88,207]

High proliferation index is the hallmark of luminal B breast cancer and blocking cell cycle checkpoints have become promising targets of chemotherapy.[208] Cyclin D1 (*CCND1*), an important oncogene in the regulation of cell cycle progression from G1 to S phase, is overexpressed in breast cancers. CCND1 works in concert with cyclin-dependent kinase (CDK) 4 and 6 to phosphorylate RB, leading to inactivation and reduction in tumor suppressive ability of RB and has been linked with AI resistance.[208,209] Tumor suppressive effects of a highly selective CDK4/6 inhibitor, PD-0332991, was demonstrated in a phase II trial on hormone-dependent refractory breast cancer model with improvement in median progression-free survival. As such, cell cycle inhibitors might be effective against ER-independent malignancies. A number of gynecological cancers share similar genetic origins, hormone receptors and may respond in similar ways to analogous anticancer agents. Whereas AIs and antiestrogens are effective and routinely used by the clinics in the prevention and treatment of estrogen-dependent malignant disorders, endocrine resistance is an important issue to hormonal therapies that may be possible to overcome by simultaneous management of multiple pathways.

7. UNDERSTANDING OF THE ROLE HDAC INHIBITORS IN COMMON WOMEN'S CANCERS

Epigenetic enzymes (HDACs) are frequently altered, dysregulated, and mutated in human tumors and/or cancers. Epigenetic erasers of acetylation marks on lysine residues of histones and nonhistone proteins have been implicated in the regulation of tumor-/cancer-related genes.[25,210–213] HDAC inhibitors interfere with deacetylase enzyme activities on biological processes, including cell proliferation, cell cycle arrest, and apoptosis in tumor cells. Due to multiple effects of HDACs in cancer cells, much attention has been placed upon the development of HDAC inhibitors targeted at various cancers. Based on their chemical structures, these compounds can be categorized into four groups: hydroxamates, cyclic peptides, benzamides, and carboxylic acids.[214–216] HDAC inhibitors are also beneficial and effective for many nonmalignant disorders, including neurodegenerative, inflammatory, and cardiovascular diseases.[27,217] HDAC inhibition results in acetylation of numerous histone and nonhistone substrates, including tumor suppressor proteins and oncogenes. To date, the US-FDA has permitted four HDAC inhibitors including Vorinostat (SAHA; trade name Zolinza; hydroxamate; approved in October 2006), Romidepsin (FK228 or depsipeptide; trade name Istodax; cyclic peptide; approved in November 2009), Belinostat (PXD101; trade name Beleodaq; hydroxamate; approved in July 2014), and Panobinostat (LBH-589; trade name Farydak; hydroxamate; approved in February 2015), which are widely used as the anticancer drugs, especially for refractory cutaneous and peripheral T cell lymphoma, and multiple melanoma. These HDAC inhibitors, either alone or in combination with other anticancer drugs, are clinically efficacious, robust, and safe; besides, they display limited toxicity, against multiple oncogenic events.[27,218] Inhibition of HDACs in tumors/cancers has been demonstrated to produce several favorable outcomes, including cell cycle arrest, antiproliferation, apoptosis, differentiation, senescence, anti-angiogenesis, and autophagy.[25,218–220] Several other HDAC inhibitors such as Valproic acid, Entinostat, MGCD0103, and PCI-24781 are on phase II/III clinical trials for a variety of tumors and/or cancers.

Whereas endocrine hormone therapies, with AIs and antiestrogens, are effective for ER-positive breast cancers, development of drug resistance is a major issue. Studies have demonstrated that inhibition of a number of HDACs promotes ubiquitin-dependent proteosomal degradation of DNA methyltransferase 1 in breast cancer cells.[221,222] It has been shown that romidepsin (HDAC 1 inhibitor) induces acetylation of histone 3 and apoptosis, and subsequently suppresses vascular epithelial growth factor and hypoxia-inducible factor 1α in breast cancer cells.[223] The HDAC6 inhibitor panobinostat synergizes with other agents and has been recognized as an effective anticancer drug against hematologic malignancies, including lymphoma, myeloma, and leukemia; and against solid tumors such as lung, thyroid, and prostrate, in numerous in vitro and preclinical trials.[224,225] Furthermore, trichostatin A and belinostat have been demonstrated to degrade ERα and altered cell proliferation in MCF-7 breast cancer cells.[226] Another HDAC inhibitor vorinostat has been shown to prevent the formation of brain metastases by 62% in TNBC cells.[227] Both vorinostat and panobinostat have been shown to acetylate Hsp90 and degrades Hsp90 client proteins such as ER, ErbB2, HDAC6, and other signaling proteins in ER-negative breast cancer cells.[228–230] Panobinostat was shown to reactivate *ER mRNA* and protein expression in TNBC cells. This inhibitor has been shown to enhance acetylation of H3 and H4 histones, induce apoptosis, decrease proliferation, and G2/M cell cycle arrest in a variety of TNBC cells.[231] Additionally, panobinostat is capable of decreasing both aromatase expression and activity, either alone or in combination with letrozole, in MCF7/human adrenocortical H295R cells.[231,232] Importantly, this HDAC inhibitor has been reported to overcome AI resistance in vitro and in vivo by suppressing the level of NF-κB1 that is usually overexpressed in AI-resistant breast cancer cells.[233]

In endometrial cancers, aberrant regulation of HDACs has also been implicated, as with other cancers and diseases.[25] The involvement of HDAC2 has been linked with more aggressive forms of endometrial cancers.[234] Class I HDACs are expressed at high levels in most endometrial cancers. Furthermore, 52% of endometrioid and 69% of high grade endometrial cancers expressed all three HDAC-1/2/3.[235] A preclinical study on HDAC inhibitors for the treatment of endometrial cancer has demonstrated their ability to reactivate tumor suppressor genes, inhibit cell cycle progression, and induce apoptosis in endometrial cancer cells.[236] It has been shown that romidepsin inhibits endometrial cancer progression and induces apoptosis through the p53/p21 pathway.[237,238] Both oxamflatin and inhibitor-1 have been shown to induce apoptosis in endometrial cancer cells.[239] Another study has demonstrated that a combined treatment of caspase-8 inhibition and vorinostat decreases endometrial cancer cell growth through the enhancement of apoptosis.[240] Other HDAC inhibitors (such as trichostatin A and valproic acid) have been shown to suppress proliferation of endometrial stromal cells by inducing cell cycle arrest and p21 expression.[241] The epigenetic reactivation of PR has been suggested as a potential approach for the treatment of advanced, PR-negative endometrial cancers with progesterone therapy.[242] HDAC inhibitors have already been approved for the treatment of other cancers, and the results reported in pertinent cancer cell models have shown promise for prevention and treatment of endometrial cancers.

HDAC inhibition has been shown to be associated with antiproliferative activity and induction of programmed cell death in ovarian cancer cells.[243] Vorinostat displayed a toxic effect and caspase-3 activity in a variety of transformed and primary ovarian cancer cell models.[244] It is conceivable that the genes encoding proteins involved in promoting apoptosis are epigenetically silenced in drug-resistant cell lines, and inhibition of HDACs results in increased acetylation and, consequently, suppress proapoptotic genes, growth inhibition, and cancer death. Clinical studies have also shown that HDAC inhibitors moderately impact the growth and progression of epithelial ovarian cancers. The findings of a phase II trial with belinostat displayed a 75% antitumor activity in patients with borderline ovarian tumor compared to epithelial ovarian cancers.[245] However, patients had allergic reactions to belinostat therapy with symptoms including thrombosis, intestinal obstruction, nausea, low levels of lymphocytes, and physical exhaustion. It has been demonstrated that panobinostat, in combination with doxorubicin (a topoisomerase inhibitor and antineoplastic agent) and carboplatin (a platinum analog used in the treatment of advanced ovarian cancer), resulted in marked synergy in ovarian cancer cell death.[246] An in vitro combination study of vorinostat and paclitaxel has been reported to synergistically reduce transcription of intracellular antiapoptotic *Bcl-2* and *c-Myc* genes, induce apoptotic *Bax* transcripts, and enhance caspase-3 and decrease angiogenic ID1 protein levels, in an OC3/P ovarian cancer cell line.[247] These combination studies attempt to exploit the diverse mechanisms of action of different therapeutic agents in a concerted effort to treat ovarian cancers. Further studies establishing the efficacy and safety of a number of HDAC inhibitors, either alone or in combination, are warranted, for the prevention and treatment of breast, endometrial, and ovarian cancers.

8. SUMMARY AND CONCLUSIONS

Aromatase is the rate-liming and final enzyme in the biosynthesis of estrogens from androgens. Estrogens, the primary female sex hormones, play pivotal roles in many important developmental and physiological processes. The major source of estrogens in premenopausal women is the ovarian granulosa cells. Upon cessation of ovarian function, extraovarian sites synthesize large amounts estrogens in postmenopausal women. Estrogens are also critically involved many pathological conditions such as breast, endometrial, ovarian, prostrate, and thyroid cancer, and uterine fibroids. The biological actions of estrogens involve both genomic (ERE-dependent) and nongenomic (ERE-independent) pathways and are dependent on cell type and context (normal vs disease state). As such, estrogens and ERs contribute to the growth and development of a number of hormone responsive tumors/cancers in pertinent tissues. In accordance, disproportionately high expression of aromatase, concomitant with increased estrogen biosynthesis, has been implicated in the etiology and epidemiology of most common women's cancers, including breast, endometrium, and ovary, which are unquestionably detrimental to reproductive function, quality of life, and lifespan. An understanding of the molecular mechanisms involved in breast, endometrial, and ovarian cancers is the key to the development of drugs and, consequently, targeted therapeutic strategies, for the prevention, treatment, and management of these deadly diseases.

Standard treatments for cancers include radiotherapy and/or chemotherapy, where the former is harmful over the more accepted chemotherapy. One of the most promising approaches for treatment of estrogen-dependent cancers in postmenopausal women is selective suppression of ER function by either antiestrogens or AIs. Targeted hormonal therapy is an important choice with recurrent/metastatic women's cancers, including breast, endometrium, and ovary. However, therapeutic approaches available at present have been effective for ER-positive, but not ER-negative, breast cancers, compared to endometrial and ovarian cancers, in postmenopausal women. There is increasing evidence that AI therapy (especially with third-generation AIs) is superior and well tolerated to antiestrogens for prevention and management of hormone-dependent breast cancers. Despite our deeper understanding on breast cancer progression and its therapeutic strategies to combat carcinogenic events, resistance to endocrine AI therapy remains a clinical challenge and is a major cause of cancer death. Additionally, AIs extensively decrease whole body estrogen levels and, thus, generates significant side effects associated with the hormone depletion (e.g., osteoporosis, aberrant lipid metabolism, etc.). Nevertheless, estrogen-dependent cancers are heterogeneous in nature and involve ER, PR, and HER2 expression at varying degrees, including distinct signaling and prognoses. Therefore, the emphasis is shifting to target signaling pathways and/or networks with combination (for example, vorinostat and tamoxifen or src inhibitor and tamoxifen) of inhibitors (that possess desired specificity, selectivity, and less toxicity) either in parallel or in series (Fig. 8), which have been shown to be beneficial against carcinogenic events in randomized clinical trials.

Deciphering the precise mechanism underlying AI resistance in conjunction with estrogen-dependent cancers has identified novel therapeutic potentials with increased specificity and efficacy. Advances in genomic and proteomic technologies have enhanced our understanding of additional signaling networks, factors, biomarkers, and coregulators, all of which play vital roles in the progression and growth of cancers. Accordingly, therapeutic strategies are targeted with markers of epigenetic modifications since dysregulation of HDACs is a primary event in tumorigenesis. HDAC inhibitors, possessing antitumor activities, represent a novel category of drugs for prevention and treatment of various malignant and nonmalignant diseases. It should be noted, however, that cancer is a multifactorial condition; as such HDAC combination therapy with other drugs targeting different aspects of the disease might be beneficial and effective (Fig. 8). Both clinical and experimental findings reveal that HDAC inhibitors (through modifications of chromatin conformation) along with other drugs can selectively and effectively act on multiple targets in cancer cells. It is plausible that the combinatorial therapy of HDAC inhibitors with AIs, and markers of DNA-damaged genes, and check point immune blockade antibodies, as well as other anticancer drugs, would be a novel approach in the treatment of estrogen responsive cancers, which requires further investigations. These strategies would limit the spread of resistant cancer cells, lead to increased sensitization to chemotherapy, and be beneficial in the treatment of advanced malignancies. Currently, four HDAC inhibitors have been

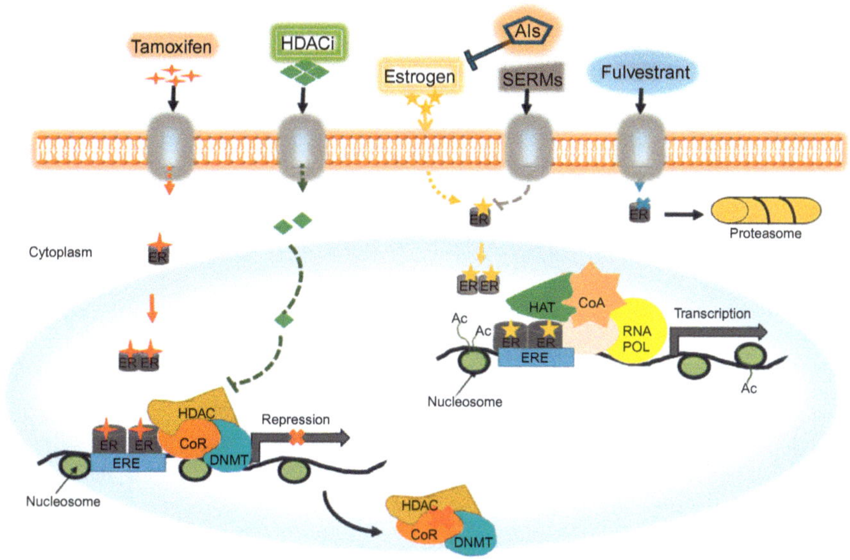

Fig. 8 A diagrammatic representation illustrating genomic and nongenomic effects of estrogen and its association to various inhibitors and/or pathways. The binding of estrogens to ERs leads to receptor dimerization that recognizes ERE present in the target gene promoters. This results in the recruitment of HAT that modifies the chromosome to enable accessibility of the transcription factors, transcription activation complexes such as CoA and RNA polymerase and initiation of transcription. AIs block estrogen synthesis and, as such, the transcription of estrogen-induced target genes. SERM blocks estrogen from binding ER, hence inhibiting estrogen-induced transcription of target genes. Tamoxifen competes with estrogen for ERs. Binding of tamoxifen to ERs results in dimerization of the receptors, recruitment of corepressors such as DNMT, CoR, as well as HDAC which deacetylates lysine residues on histone tails causing chromatin compaction, and gene expression is repressed. In the presence of HDAC inhibitors, the function of HDAC and corepressors is compromised and this leads to increased acetylation of lysine residues on the histone tails and transcription is initiated. Fulvestrant targets ER by exposing it to ubiquitination, which eventually leads to proteasomal degradation. *ERE*, estrogen receptor response element; *HAT*, histone acetyl transferases; *TF*, transcription factor; *AIs*, aromatase inhibitors; *CoR*, corepressor; *HDAC*, histone deacetylases; *HDACi*, histone deacetylase inhibitors, *DNMT*, DNA methyltransferase; *Ac*, acetylation; and *RNA POL*, RNA polymerase enzyme.

approved as anticancer drugs with clinical efficiency, either individually or in combination, and many others are in various phases of clinical trials. Characterization of selective HDAC inhibitors and their combination with biomarkers, in well-targeted trial mechanisms, is critical for increasing our understanding in optimal clinical application against a variety of malignant and nonmalignant disorders.

REFERENCES

1. Simpson ER. Sources of estrogen and their importance. *J Steroid Biochem Mol Biol.* 2003;86(3–5):225–230.
2. Simpson E, Santen RJ. Celebrating 75 years of oestradiol. *J Mol Endocrinol.* 2015;55(3): T1–T20.
3. Miller WL, Bose HS. Early steps in steroidogenesis: intracellular cholesterol trafficking. *J Lipid Res.* 2011;52(12):2111–2135.
4. Manna PR, Stetson CL, Slominski AT, Pruitt K. Role of the steroidogenic acute regulatory protein in health and disease. *Endocrine.* 2016;51(1):7–21.
5. Simpson ER, Clyne C, Rubin G, et al. Aromatase—a brief overview. *Annu Rev Physiol.* 2002;64:93–127.
6. Bulun SE, Lin Z, Zhao H, et al. Regulation of aromatase expression in breast cancer tissue. *Ann N Y Acad Sci.* 2009;1155:121–131.
7. Levin ER. Integration of the extranuclear and nuclear actions of estrogen. *Mol Endocrinol.* 2005;19(8):1951–1959.
8. Ascenzi P, Bocedi A, Marino M. Structure–function relationship of estrogen receptor alpha and beta: impact on human health. *Mol Aspects Med.* 2006;27(4):299–402.
9. Bulun SE, Lin Z, Imir G, et al. Regulation of aromatase expression in estrogen-responsive breast and uterine disease: from bench to treatment. *Pharmacol Rev.* 2005;57(3):359–383.
10. Zhao H, Zhou L, Shangguan AJ, Bulun SE. Aromatase expression and regulation in breast and endometrial cancer. *J Mol Endocrinol.* 2016;57:R19–R33.
11. Bulun SE, Chen D, Lu M, et al. Aromatase excess in cancers of breast, endometrium and ovary. *J Steroid Biochem Mol Biol.* 2007;106(1–5):81–96.
12. Bulun SE, Simpson ER. Aromatase expression in women's cancers. *Adv Exp Med Biol.* 2008;630:112–132.
13. Johnston SR, Dowsett M. Aromatase inhibitors for breast cancer: lessons from the laboratory. *Nat Rev Cancer.* 2003;3(11):821–831.
14. Hong Y, Li H, Yuan YC, Chen S. Molecular characterization of aromatase. *Ann N Y Acad Sci.* 2009;1155:112–120.
15. Dunbier AK, Hong Y, Masri S, Brown KA, Sabnis GJ, Palomares MR. Progress in aromatase research and identification of key future directions. *J Steroid Biochem Mol Biol.* 2010;118(4–5):311–315.
16. Sjoquist KM, Martyn J, Edmondson RJ, Friedlander ML. The role of hormonal therapy in gynecological cancers-current status and future directions. *Int J Gynecol Cancer.* 2011;21(7):1328–1333.
17. Santen RJ. To block estrogen's synthesis or action: that is the question. *J Clin Endocrinol Metab.* 2002;87(7):3007–3012.
18. Jordan VC. The 38th David A. Karnofsky lecture: the paradoxical actions of estrogen in breast cancer—survival or death? *J Clin Oncol.* 2008;26(18):3073–3082.
19. Bonneterre J, Buzdar A, Nabholtz JM, et al. Anastrozole is superior to tamoxifen as first-line therapy in hormone receptor positive advanced breast carcinoma. *Cancer.* 2001;92(9):2247–2258.
20. Che Q, Liu BY, Liao Y, et al. Activation of a positive feedback loop involving IL-6 and aromatase promotes intratumoral 17β-estradiol biosynthesis in endometrial carcinoma microenvironment. *Int J Cancer.* 2014;135(2):282–294.
21. Esparza-Perusquia M, Olvera-Sanchez S, Flores-Herrera O, et al. Mitochondrial proteases act on STARD3 to activate progesterone synthesis in human syncytiotrophoblast. *Biochim Biophys Acta.* 2015;1850(1):107–117.
22. Evans RM, Mangelsdorf DJ. Nuclear receptors, RXR, and the big bang. *Cell.* 2014;157(1):255–266.
23. Hubbard BA, Unger JG, Rohrich RJ. Reversal of skin aging with topical retinoids. *Plast Reconstr Surg.* 2014;133(4):481e–490e.

24. Slominski A, Zjawiony J, Wortsman J, et al. A novel pathway for sequential transformation of 7-dehydrocholesterol and expression of the P450scc system in mammalian skin. *Eur J Biochem*. 2004;271(21):4178–4188.
25. West AC, Johnstone RW. New and emerging HDAC inhibitors for cancer treatment. *J Clin Invest*. 2014;124(1):30–39.
26. Damaskos C, Karatzas T, Nikolidakis L, et al. Histone deacetylase (HDAC) inhibitors: current evidence for therapeutic activities in pancreatic cancer. *Anticancer Res*. 2015;35(6):3129–3135.
27. Yoon S, Eom GH. HDAC and HDAC inhibitor: from cancer to cardiovascular diseases. *Chonnam Med J*. 2016;52(1):1–11.
28. Chou AH, Chen SY, Yeh TH, Weng YH, Wang HL. HDAC inhibitor sodium butyrate reverses transcriptional downregulation and ameliorates ataxic symptoms in a transgenic mouse model of SCA3. *Neurobiol Dis*. 2011;41(2):481–488.
29. Ghosh D, Griswold J, Erman M, Pangborn W. Structural basis for androgen specificity and oestrogen synthesis in human aromatase. *Nature*. 2009;457(7226):219–223.
30. Ghosh D, Griswold J, Erman M, Pangborn W. X-ray structure of human aromatase reveals an androgen-specific active site. *J Steroid Biochem Mol Biol*. 2010;118(4–5): 197–202.
31. Nakamura T, Imai Y, Matsumoto T, et al. Estrogen prevents bone loss via estrogen receptor alpha and induction of Fas ligand in osteoclasts. *Cell*. 2007;130(5):811–823.
32. Arevalo MA, Azcoitia I, Garcia-Segura LM. The neuroprotective actions of oestradiol and oestrogen receptors. *Nat Rev Neurosci*. 2015;16(1):17–29.
33. Simpson ER, Misso M, Hewitt KN, et al. Estrogen—the good, the bad, and the unexpected. *Endocr Rev*. 2005;26(3):322–330.
34. Wang X, Simpson ER, Brown KA. Aromatase overexpression in dysfunctional adipose tissue links obesity to postmenopausal breast cancer. *J Steroid Biochem Mol Biol*. 2015;153:35–44.
35. Boon W, Broekgaarden R. The role of patient advocacy organisations in neuromuscular disease R&D—the case of the Dutch neuromuscular disease association VSN. *Neuromuscul Disord*. 2010;20(2):148–151.
36. Bulun SE, Sebastian S, Takayama K, Suzuki T, Sasano H, Shozu M. The human CYP19 (aromatase P450) gene: update on physiologic roles and genomic organization of promoters. *J Steroid Biochem Mol Biol*. 2003;86(3–5):219–224.
37. Golovine K, Schwerin M, Vanselow J. Three different promoters control expression of the aromatase cytochrome p450 gene (cyp19) in mouse gonads and brain. *Biol Reprod*. 2003;68(3):978–984.
38. To SQ, Knower KC, Cheung V, Simpson ER, Clyne CD. Transcriptional control of local estrogen formation by aromatase in the breast. *J Steroid Biochem Mol Biol*. 2015;145:179–186.
39. Wang X, Docanto MM, Sasano H, et al. Prostaglandin E2 inhibits p53 in human breast adipose stromal cells: a novel mechanism for the regulation of aromatase in obesity and breast cancer. *Cancer Res*. 2015;75(4):645–655.
40. Simpson E, Jones M, Misso M, et al. Estrogen, a fundamental player in energy homeostasis. *J Steroid Biochem Mol Biol*. 2005;95(1–5):3–8.
41. Renoir JM, Marsaud V, Lazennec G. Estrogen receptor signaling as a target for novel breast cancer therapeutics. *Biochem Pharmacol*. 2013;85(4):449–465.
42. Mehta A, Tripathy D. Co-targeting estrogen receptor and HER2 pathways in breast cancer. *Breast*. 2014;23(1):2–9.
43. Chan HJ, Petrossian K, Chen S. Structural and functional characterization of aromatase, estrogen receptor, and their genes in endocrine-responsive and -resistant breast cancer cells. *J Steroid Biochem Mol Biol*. 2016;161:73–83.
44. Powles TJ. Anti-oestrogenic prevention of breast cancer—the make or break point. *Nat Rev Cancer*. 2002;2(10):787–794.

45. Evans RM. The steroid and thyroid hormone receptor superfamily. *Science*. 1988;240(4854):889–895.
46. Mangelsdorf DJ, Thummel C, Beato M, et al. The nuclear receptor superfamily: the second decade. *Cell*. 1995;83(6):835–839.
47. Chen D, Riedl T, Washbrook E, et al. Activation of estrogen receptor alpha by S118 phosphorylation involves a ligand-dependent interaction with TFIIH and participation of CDK7. *Mol Cell*. 2000;6(1):127–137.
48. Yue W, Yager JD, Wang JP, Jupe ER, Santen RJ. Estrogen receptor-dependent and independent mechanisms of breast cancer carcinogenesis. *Steroids*. 2013;78(2): 161–170.
49. Jordan VC, O'Malley BW. Selective estrogen-receptor modulators and antihormonal resistance in breast cancer. *J Clin Oncol*. 2007;25(36):5815–5824.
50. Baumann CK, Castiglione-Gertsch M. Estrogen receptor modulators and down regulators: optimal use in postmenopausal women with breast cancer. *Drugs*. 2007;67(16):2335–2353.
51. Carroll JS, Meyer CA, Song J, et al. Genome-wide analysis of estrogen receptor binding sites. *Nat Genet*. 2006;38(11):1289 1297.
52. Chen GG, Zeng Q, Tse GM. Estrogen and its receptors in cancer. *Med Res Rev*. 2008;28(6):954–974.
53. Cvoro A, Tzagarakis-Foster C, Tatomer D, Paruthiyil S, Fox MS, Leitman DC. Distinct roles of unliganded and liganded estrogen receptors in transcriptional repression. *Mol Cell*. 2006;21(4):555–564.
54. Prossnitz ER, Arterburn JB, Smith HO, Oprea TI, Sklar LA, Hathaway HJ. Estrogen signaling through the transmembrane G protein-coupled receptor GPR30. *Annu Rev Physiol*. 2008;70:165–190.
55. Lannigan DA. Estrogen receptor phosphorylation. *Steroids*. 2003;68(1):1–9.
56. Le Romancer M, Poulard C, Cohen P, Sentis S, Renoir JM, Corbo L. Cracking the estrogen receptor's posttranslational code in breast tumors. *Endocr Rev*. 2011;32(5): 597–622.
57. Balthazart J, Baillien M, Charlier TD, Ball GF. Calcium-dependent phosphorylation processes control brain aromatase in quail. *Eur J Neurosci*. 2003;17(8):1591–1606.
58. Miller TW, Shin I, Kagawa N, Evans DB, Waterman MR, Arteaga CL. Aromatase is phosphorylated in situ at serine-118. *J Steroid Biochem Mol Biol*. 2008;112(1–3):95–101.
59. Campbell RA, Bhat-Nakshatri P, Patel NM, Constantinidou D, Ali S, Nakshatri H. Phosphatidylinositol 3-kinase/AKT-mediated activation of estrogen receptor alpha: a new model for anti-estrogen resistance. *J Biol Chem*. 2001;276(13):9817–9824.
60. Vendrell JA, Bieche I, Desmetz C, et al. Molecular changes associated with the agonist activity of hydroxy-tamoxifen and the hyper-response to estradiol in hydroxy-tamoxifen-resistant breast cancer cell lines. *Endocr Relat Cancer*. 2005;12(1):75–92.
61. Murphy LC, Seekallu SV, Watson PH. Clinical significance of estrogen receptor phosphorylation. *Endocr Relat Cancer*. 2011;18(1):R1–R14.
62. Weitsman GE, Li L, Skliris GP, et al. Estrogen receptor-alpha phosphorylated at Ser118 is present at the promoters of estrogen-regulated genes and is not altered due to HER-2 overexpression. *Cancer Res*. 2006;66(20):10162–10170.
63. Kirkegaard T, Witton CJ, McGlynn LM, et al. AKT activation predicts outcome in breast cancer patients treated with tamoxifen. *J Pathol*. 2005;207(2):139–146.
64. Wang C, Fu M, Angeletti RH, et al. Direct acetylation of the estrogen receptor alpha hinge region by p300 regulates transactivation and hormone sensitivity. *J Biol Chem*. 2001;276(21):18375–18383.
65. Kim SH, Kang HJ, Na H, Lee MO. Trichostatin A enhances acetylation as well as protein stability of ERalpha through induction of p300 protein. *Breast Cancer Res*. 2010;12(2):R22.

66. Ali SH, O'Donnell AL, Balu D, et al. Estrogen receptor-alpha in the inhibition of cancer growth and angiogenesis. *Cancer Res*. 2000;60(24):7094–7098.
67. Grober OM, Mutarelli M, Giurato G, et al. Global analysis of estrogen receptor beta binding to breast cancer cell genome reveals an extensive interplay with estrogen receptor alpha for target gene regulation. *BMC Genomics*. 2011;12:36.
68. Strom A, Hartman J, Foster JS, Kietz S, Wimalasena J, Gustafsson JA. Estrogen receptor beta inhibits 17beta-estradiol-stimulated proliferation of the breast cancer cell line T47D. *Proc Natl Acad Sci U S A*. 2004;101(6):1566–1571.
69. Urruticoechea A. The oestrogen-dependent biology of breast cancer. Sensitivity and resistance to aromatase inhibitors revisited: a molecular perspective. *Clin Transl Oncol*. 2007;9(12):752–759.
70. Ali S, Coombes RC. Endocrine-responsive breast cancer and strategies for combating resistance. *Nat Rev Cancer*. 2002;2(2):101–112.
71. Yager JD, Davidson NE. Estrogen carcinogenesis in breast cancer. *N Engl J Med*. 2006;354(3):270–282.
72. Choueiri TK, Alemany CA, Abou-Jawde RM, Budd GT. Role of aromatase inhibitors in the treatment of breast cancer. *Clin Ther*. 2004;26(8):1199–1214.
73. Altundag K, Ibrahim NK. Aromatase inhibitors in breast cancer: an overview. *Oncologist*. 2006;11(6):553–562.
74. Macedo LF, Sabnis G, Brodie A. Aromatase inhibitors and breast cancer. *Ann N Y Acad Sci*. 2009;1155:162–173.
75. Perou CM, Sorlie T, Eisen MB, et al. Molecular portraits of human breast tumours. *Nature*. 2000;406(6797):747–752.
76. Key TJ, Verkasalo PK, Banks E. Epidemiology of breast cancer. *Lancet Oncol*. 2001;2(3):133–140.
77. Howlader N, Chen VW, Ries LA, et al. Overview of breast cancer collaborative stage data items—their definitions, quality, usage, and clinical implications: a review of SEER data for 2004–2010. *Cancer*. 2014;120(suppl 23):3771–3780.
78. Lehmann BD, Bauer JA, Chen X, et al. Identification of human triple-negative breast cancer subtypes and preclinical models for selection of targeted therapies. *J Clin Invest*. 2011;121(7):2750–2767.
79. Chen D, Reierstad S, Lu M, Lin Z, Ishikawa H, Bulun SE. Regulation of breast cancer-associated aromatase promoters. *Cancer Lett*. 2009;273(1):15–27.
80. Chumsri S, Howes T, Bao T, Sabnis G, Brodie A. Aromatase, aromatase inhibitors, and breast cancer. *J Steroid Biochem Mol Biol*. 2011;125(1–2):13–22.
81. Palmieri C, Patten DK, Januszewski A, Zucchini G, Howell SJ. Breast cancer: current and future endocrine therapies. *Mol Cell Endocrinol*. 2014;382(1):695–723.
82. Johnston SR. Enhancing endocrine therapy for hormone receptor-positive advanced breast cancer: cotargeting signaling pathways. *J Natl Cancer Inst*. 2015;107(10):1–10.
83. Jiang Q, Zheng S, Wang G. Development of new estrogen receptor-targeting therapeutic agents for tamoxifen-resistant breast cancer. *Future Med Chem*. 2013;5(9):1023–1035.
84. Dunbier AK, Anderson H, Ghazoui Z, et al. Relationship between plasma estradiol levels and estrogen-responsive gene expression in estrogen receptor-positive breast cancer in postmenopausal women. *J Clin Oncol*. 2010;28(7):1161–1167.
85. Lonning PE, Haynes BP, Straume AH, et al. Exploring breast cancer estrogen disposition: the basis for endocrine manipulation. *Clin Cancer Res*. 2011;17(15):4948–4958.
86. Katsama A, Sourvinos G, Zachos G, Spandidos DA. Allelic loss at the BRCA1, BRCA2 and TP53 loci in human sporadic breast carcinoma. *Cancer Lett*. 2000;150(2):165–170.
87. Apostolou P, Fostira F. Hereditary breast cancer: the era of new susceptibility genes. *Biomed Res Int*. 2013;2013:747318.

88. Shiovitz S, Korde LA. Genetics of breast cancer: a topic in evolution. *Ann Oncol.* 2015;26(7):1291–1299.
89. Miller WR. Aromatase and the breast: regulation and clinical aspects. *Maturitas.* 2006;54(4):335–341.
90. Pelekanou V, Leclercq G. Recent insights into the effect of natural and environmental estrogens on mammary development and carcinogenesis. *Int J Dev Biol.* 2011;55(7–9):869–878.
91. Bulun SE, Chen D, Moy I, Brooks DC, Zhao H. Aromatase, breast cancer and obesity: a complex interaction. *Trends Endocrinol Metab.* 2012;23(2):83–89.
92. Goswami CP, Nakshatri H. PROGgene: gene expression based survival analysis web application for multiple cancers. *J Clin Bioinforma.* 2013;3(1):22.
93. Goswami CP, Nakshatri H. PROGgeneV2: enhancements on the existing database. *BMC Cancer.* 2014;14:970.
94. den Hollander P, Savage MI, Brown PH. Targeted therapy for breast cancer prevention. *Front Oncol.* 2013;3:250.
95. Fisher B, Costantino JP, Wickerham DL, et al. Tamoxifen for the prevention of breast cancer: current status of the National Surgical Adjuvant Breast and Bowel Project P-1 study. *J Natl Cancer Inst.* 2005;97(22):1652–1662.
96. Early Breast Cancer Trialists' Collaborative Group, Davies C, Godwin J, et al. Relevance of breast cancer hormone receptors and other factors to the efficacy of adjuvant tamoxifen: patient-level meta-analysis of randomised trials. *Lancet.* 2011;378(9793):771–784.
97. Jordan VC. The past, present, and future of selective estrogen receptor modulation. *Ann N Y Acad Sci.* 2001;949:72–79.
98. Vogel VG, Costantino JP, Wickerham DL, et al. Effects of tamoxifen vs raloxifene on the risk of developing invasive breast cancer and other disease outcomes: the NSABP Study of Tamoxifen and Raloxifene (STAR) P-2 trial. *JAMA.* 2006;295(23): 2727–2741.
99. Cummings SR, Ensrud K, Delmas PD, et al. Lasofoxifene in postmenopausal women with osteoporosis. *N Engl J Med.* 2010;362(8):686–696.
100. Osborne CK, Pippen J, Jones SE, et al. Double-blind, randomized trial comparing the efficacy and tolerability of fulvestrant versus anastrozole in postmenopausal women with advanced breast cancer progressing on prior endocrine therapy: results of a North American trial. *J Clin Oncol.* 2002;20(16):3386–3395.
101. Cazzaniga M, Bonanni B. Breast cancer chemoprevention: old and new approaches. *J Biomed Biotechnol.* 2012;2012:985620.
102. Sorosky JI. Endometrial cancer. *Obstet Gynecol.* 2012;120(2 pt 1):383–397.
103. Wright JD, Barrena Medel NI, Sehouli J, Fujiwara K, Herzog TJ. Contemporary management of endometrial cancer. *Lancet.* 2012;379(9823):1352–1360.
104. Burke WM, Orr J, Leitao M, et al. Endometrial cancer: a review and current management strategies: part II. *Gynecol Oncol.* 2014;134(2):393–402.
105. McAlpine JN, Temkin SM, Mackay HJ. Endometrial cancer: not your grandmother's cancer. *Cancer.* 2016;122:2787–2798.
106. Amant F, Moerman P, Neven P, Timmerman D, Van Limbergen E, Vergote I. Endometrial cancer. *Lancet.* 2005;366(9484):491–505.
107. Arora V, Quinn MA. Endometrial cancer. *Best Pract Res Clin Obstet Gynaecol.* 2012;26(3):311–324.
108. Murali R, Soslow RA, Weigelt B. Classification of endometrial carcinoma: more than two types. *Lancet Oncol.* 2014;15(7):e268–e278.
109. Westin SN, Broaddus RR. Personalized therapy in endometrial cancer: challenges and opportunities. *Cancer Biol Ther.* 2012;13(1):1–13.

110. Bokhman JV. Two pathogenetic types of endometrial carcinoma. *Gynecol Oncol.* 1983;15(1):10–17.
111. Setiawan VW, Yang HP, Pike MC, et al. Type I and II endometrial cancers: have they different risk factors? *J Clin Oncol.* 2013;31(20):2607–2618.
112. American Cancer Society. *Cancer facts and figures;* 2016.
113. Siegel RL, Miller KD, Jemal A. Cancer statistics, 2016. *CA Cancer J Clin.* 2016;66(1):7–30.
114. Sheikh MA, Althouse AD, Freese KE, et al. USA endometrial cancer projections to 2030: should we be concerned? *Future Oncol.* 2014;10(16):2561–2568.
115. Purdie DM, Green AC. Epidemiology of endometrial cancer. *Best Pract Res Clin Obstet Gynaecol.* 2001;15(3):341–354.
116. Kim JJ, Kurita T, Bulun SE. Progesterone action in endometrial cancer, endometriosis, uterine fibroids, and breast cancer. *Endocr Rev.* 2013;34(1):130–162.
117. Williams-Brown MY, Salih SM, Xu X, et al. The effect of tamoxifen and raloxifene on estrogen metabolism and endometrial cancer risk. *J Steroid Biochem Mol Biol.* 2011;126(3–5):78–86.
118. Kamal A, Tempest N, Parkes C, et al. Hormones and endometrial carcinogenesis. *Horm Mol Biol Clin Investig.* 2016;25(2):129–148.
119. Burke WM, Orr J, Leitao M, et al. Endometrial cancer: a review and current management strategies: part I. *Gynecol Oncol.* 2014;134(2):385–392.
120. McPherson CP, Sellers TA, Potter JD, Bostick RM, Folsom AR. Reproductive factors and risk of endometrial cancer. The Iowa Women's Health Study. *Am J Epidemiol.* 1996;143(12):1195–1202.
121. Calle EE, Kaaks R. Overweight, obesity and cancer: epidemiological evidence and proposed mechanisms. *Nat Rev Cancer.* 2004;4(8):579–591.
122. Secord AA, Hasselblad V, Von Gruenigen VE, et al. Body mass index and mortality in endometrial cancer: a systematic review and meta-analysis. *Gynecol Oncol.* 2016;140(1):184–190.
123. Berstein LM, Tchernobrovkina AE, Gamajunova VB, et al. Tumor estrogen content and clinico-morphological and endocrine features of endometrial cancer. *J Cancer Res Clin Oncol.* 2003;129(4):245–249.
124. Krasner C. Aromatase inhibitors in gynecologic cancers. *J Steroid Biochem Mol Biol.* 2007;106(1–5):76–80.
125. Bulun SE, Imir G, Utsunomiya H, et al. Aromatase in endometriosis and uterine leiomyomata. *J Steroid Biochem Mol Biol.* 2005;95(1–5):57–62.
126. Calle EE, Rodriguez C, Walker-Thurmond K, Thun MJ. Overweight, obesity, and mortality from cancer in a prospectively studied cohort of U.S. adults. *N Engl J Med.* 2003;348(17):1625–1638.
127. Rice LW. Hormone prevention strategies for breast, endometrial and ovarian cancers. *Gynecol Oncol.* 2010;118(2):202–207.
128. Ito K, Miki Y, Suzuki T, McNamara KM, Sasano H. In situ androgen and estrogen biosynthesis in endometrial cancer: focus on androgen actions and intratumoral production. *Endocr Relat Cancer.* 2016;23(7):R323–R335.
129. Kaaks R, Lukanova A, Kurzer MS. Obesity, endogenous hormones, and endometrial cancer risk: a synthetic review. *Cancer Epidemiol Biomarkers Prev.* 2002;11(12): 1531–1543.
130. Carlson MJ, Thiel KW, Leslie KK. Past, present, and future of hormonal therapy in recurrent endometrial cancer. *Int J Womens Health.* 2014;6:429–435.
131. Yamamoto T, Kitawaki J, Urabe M, et al. Estrogen productivity of endometrium and endometrial cancer tissue; influence of aromatase on proliferation of endometrial cancer cells. *J Steroid Biochem Mol Biol.* 1993;44(4–6):463–468.

132. Fowler JM, Ramirez N, Cohn DE, et al. Correlation of cyclooxygenase-2 (COX-2) and aromatase expression in human endometrial cancer: tissue microarray analysis. *Am J Obstet Gynecol.* 2005;192(4):1262–1271. discussion 1271–1263.
133. Knapp P, Chabowski A, Błachnio-Zabielska A, Walentowicz-Sadłecka M, Grabiec M, Knapp PA. Expression of estrogen receptors (α, β), cyclooxygenase-2 and aromatase in normal endometrium and endometrioid cancer of uterus. *Adv Med Sci.* 2013;58(1):96–103.
134. Jarzabek K, Koda M, Walentowicz-Sadlecka M, Grabiec M, Laudanski P, Wolczynski S. Altered expression of ERs, aromatase, and COX2 connected to estrogen action in type 1 endometrial cancer biology. *Tumour Biol.* 2013;34(6):4007–4016.
135. Segawa T, Shozu M, Murakami K, et al. Aromatase expression in stromal cells of endometrioid endometrial cancer correlates with poor survival. *Clin Cancer Res.* 2005;11(6):2188–2194.
136. Fisher B, Costantino JP, Redmond CK, Fisher ER, Wickerham DL, Cronin WM. Endometrial cancer in tamoxifen-treated breast cancer patients: findings from the National Surgical Adjuvant Breast and Bowel Project (NSABP) B-14. *J Natl Cancer Inst.* 1994;86(7):527–537.
137. Chlebowski RT, Schottinger JE, Shi J, Chung J, Haque R. Aromatase inhibitors, tamoxifen, and endometrial cancer in breast cancer survivors. *Cancer.* 2015;121(13):2147–2155.
138. Markovitch O, Tepper R, Fishman A, Aviram R, Cohen I. Long-term "protective" effect of aromatase inhibitors on the endometrium of postmenopausal breast cancer patients. *Breast Cancer Res Treat.* 2009;113(2):321–326.
139. Thürlimann B, Keshaviah A, Coates AS, et al. A comparison of letrozole and tamoxifen in postmenopausal women with early breast cancer. *N Engl J Med.* 2005;353(26): 2747–2757.
140. Thangavelu A, Hewitt MJ, Quinton ND, Duffy SR. Neoadjuvant treatment of endometrial cancer using anastrozole: a randomised pilot study. *Gynecol Oncol.* 2013;131(3):613–618.
141. Ma BB, Oza A, Eisenhauer E, et al. The activity of letrozole in patients with advanced or recurrent endometrial cancer and correlation with biological markers—a study of the National Cancer Institute of Canada Clinical Trials Group. *Int J Gynecol Cancer.* 2004;14(4):650–658.
142. Fleming GF, Brunetto VL, Cella D, et al. Phase III trial of doxorubicin plus cisplatin with or without paclitaxel plus filgrastim in advanced endometrial carcinoma: a Gynecologic Oncology Group Study. *J Clin Oncol.* 2004;22(11):2159–2166.
143. Leslie KK, Thiel KW, Goodheart MJ, De Geest K, Jia Y, Yang S. Endometrial cancer. *Obstet Gynecol Clin North Am.* 2012;39(2):255–268.
144. Lakhani SR, Manek S, Penault-Llorca F, et al. Pathology of ovarian cancers in BRCA1 and BRCA2 carriers. *Clin Cancer Res.* 2004;10(7):2473–2481.
145. Bankhead CR, Collins C, Stokes-Lampard H, et al. Identifying symptoms of ovarian cancer: a qualitative and quantitative study. *BJOG.* 2008;115(8):1008–1014.
146. Tsai TC, Wong LY, Wu HP. Ovarian torsion caused by teratoma masquerading as perforated appendicitis in a 5-year-old girl. *Pediatr Neonatol.* 2011;52(1):51–54.
147. Gomez-Hidalgo NR, Darin MC, Dalton H, et al. Ovarian torsion after laparoscopic ovarian transposition in patients with gynecologic cancer: a report of two cases. *J Minim Invasive Gynecol.* 2015;22(4):687–690.
148. Mungenast F, Thalhammer T. Estrogen biosynthesis and action in ovarian cancer. *Front Endocrinol (Lausanne).* 2014;5:192.
149. Mertens-Walker I, Baxter RC, Marsh DJ. Gonadotropin signalling in epithelial ovarian cancer. *Cancer Lett.* 2012;324(2):152–159.

150. Vanderhyden BC. Loss of ovarian function and the risk of ovarian cancer. *Cell Tissue Res.* 2005;322(1):117–124.
151. National Cancer Institute, SEER Cancer Statistics Factsheets. Ovarian cancer. Bethesda, MD. http://seer.cancer.gov/statfacts/html/ovary.html.
152. Hunn J, Rodriguez GC. Ovarian cancer: etiology, risk factors, and epidemiology. *Clin Obstet Gynecol.* 2012;55(1):3–23.
153. Usset JL, Raghavan R, Tyrer JP, et al. Assessment of multifactor gene-environment interactions and ovarian cancer risk: candidate genes, obesity, and hormone-related risk factors. *Cancer Epidemiol Biomarkers Prev.* 2016;25(5):780–790.
154. Arts-de Jong M, de Bock GH, van Asperen CJ, Mourits MJ, de Hullu JA, Kets CM. Germline BRCA1/2 mutation testing is indicated in every patient with epithelial ovarian cancer: a systematic review. *Eur J Cancer.* 2016;61:137–145.
155. Cancer Genome Atlas Research N. Integrated genomic analyses of ovarian carcinoma. *Nature.* 2011;474(7353):609–615.
156. Beral V, Million Women Study Collaborators, Bull D, Green J, Reeves G. Ovarian cancer and hormone replacement therapy in the Million Women Study. *Lancet.* 2007;369(9574):1703–1710.
157. Pearce CL, Chung K, Pike MC, Wu AH. Increased ovarian cancer risk associated with menopausal estrogen therapy is reduced by adding a progestin. *Cancer.* 2009;115(3):531–539.
158. Quaye L, Gayther SA, Ramus SJ, et al. The effects of common genetic variants in oncogenes on ovarian cancer survival. *Clin Cancer Res.* 2008;14(18):5833–5839.
159. Lacey Jr JV, Brinton LA, Leitzmann MF, et al. Menopausal hormone therapy and ovarian cancer risk in the National Institutes of Health-AARP Diet and Health Study Cohort. *J Natl Cancer Inst.* 2006;98(19):1397–1405.
160. Rao BR, Slotman BJ. Endocrine factors in common epithelial ovarian cancer. *Endocr Rev.* 1991;12(1):14–26.
161. Rodriguez C, Patel AV, Calle EE, Jacob EJ, Thun MJ. Estrogen replacement therapy and ovarian cancer mortality in a large prospective study of US women. *JAMA.* 2001;285(11):1460–1465.
162. Trabert B, Brinton LA, Anderson GL, et al. Circulating estrogens and postmenopausal ovarian cancer risk in the women's health initiative observational study. *Cancer Epidemiol Biomarkers Prev.* 2016;25(4):648–656.
163. Dallal CM, Lacey Jr JV, Pfeiffer RM, et al. Estrogen metabolism and risk of postmenopausal endometrial and ovarian cancer: the B approximately FIT cohort. *Horm Cancer.* 2016;7(1):49–64.
164. Ciucci A, Zannoni GF, Buttarelli M, et al. Multiple direct and indirect mechanisms drive estrogen-induced tumor growth in high grade serous ovarian cancers. *Oncotarget.* 2016;7(7):8155–8171.
165. Kitawaki J, Noguchi T, Yamamoto T, et al. Immunohistochemical localisation of aromatase and its correlation with progesterone receptors in ovarian epithelial tumours. *Anticancer Res.* 1996;16(1):91–97.
166. Slotman BJ, Kuhnel R, Rao BR, Dijkhuizen GH, de Graaff J, Stolk JG. Importance of steroid receptors and aromatase activity in the prognosis of ovarian cancer: high tumor progesterone receptor levels correlate with longer survival. *Gynecol Oncol.* 1989;33(1):76–81.
167. Bulun SE, Rosenthal IM, Brodie AM, et al. Use of tissue-specific promoters in the regulation of aromatase cytochrome P450 gene expression in human testicular and ovarian sex cord tumors, as well as in normal fetal and adult gonads. *J Clin Endocrinol Metab.* 1994;78(2):1616–1621.
168. Kaga K, Sasano H, Harada N, Ozaki M, Sato S, Yajima A. Aromatase in human common epithelial ovarian neoplasms. *Am J Pathol.* 1996;149(1):45–51.

169. Cunat S, Rabenoelina F, Daures JP, et al. Aromatase expression in ovarian epithelial cancers. *J Steroid Biochem Mol Biol.* 2005;93(1):15–24.
170. Chand AL, kConFab, Simpson ER, Clyne CD. Aromatase expression is increased in BRCA1 mutation carriers. *BMC Cancer.* 2009;9:148.
171. Kato N, Hayasaka T, Takeda J, Osakabe M, Kurachi H. Ovarian tumors with functioning stroma: a clinicopathologic study with special reference to serum estrogen level, stromal morphology, and aromatase expression. *Int J Gynecol Pathol.* 2013;32(6): 556–561.
172. Hu Y, Ghosh S, Amleh A, et al. Modulation of aromatase expression by BRCA1: a possible link to tissue-specific tumor suppression. *Oncogene.* 2005;24(56):8343–8348.
173. Kim J, Oktay K. Baseline E(2) levels are higher in BRCA2 mutation carriers: a potential target for prevention? *Cancer Causes Control.* 2013;24(3):421–426.
174. Kalfa N, Meduri G, Philibert P, et al. Unusual virilization in girls with juvenile granulosa cell tumors of the ovary is related to intratumoral aromatase deficiency. *Horm Res Paediatr.* 2010;74(2):83–91.
175. Goodman MT, Lurie G, Thompson PJ, McDuffie KE, Carney ME. Association of two common single-nucleotide polymorphisms in the CYP19A1 locus and ovarian cancer risk. *Endocr Relat Cancer.* 2008;15(4):1055–1060.
176. Simpkins F, Garcia-Soto A, Slingerland J. New insights on the role of hormonal therapy in ovarian cancer. *Steroids.* 2013;78(6):530–537.
177. Li YF, Hu W, Fu SQ, Li JD, Liu JH, Kavanagh JJ. Aromatase inhibitors in ovarian cancer: is there a role? *Int J Gynecol Cancer.* 2008;18(4):600–614.
178. Laws MJ, Kannan A, Pawar S, Haschek WM, Bagchi MK, Bagchi IC. Dysregulated estrogen receptor signaling in the hypothalamic-pituitary-ovarian axis leads to ovarian epithelial tumorigenesis in mice. *PLoS Genet.* 2014;10(3):e1004230.
179. Bookman MA. First-line chemotherapy in epithelial ovarian cancer. *Clin Obstet Gynecol.* 2012;55(1):96–113.
180. Raja FA, Chopra N, Ledermann JA. Optimal first-line treatment in ovarian cancer. *Ann Oncol.* 2012;23(suppl 10):x118–x127.
181. Boulanger J, Boursiquot JN, Cournoyer G, et al. Management of hypersensitivity to platinum- and taxane-based chemotherapy: cepo review and clinical recommendations. *Curr Oncol.* 2014;21(4):e630–e641.
182. Hu L, Hofmann J, Zaloudek C, Ferrara N, Hamilton T, Jaffe RB. Vascular endothelial growth factor immunoneutralization plus Paclitaxel markedly reduces tumor burden and ascites in athymic mouse model of ovarian cancer. *Am J Pathol.* 2002;161(5): 1917–1924.
183. Ito F, Chang AE. Cancer immunotherapy: current status and future directions. *Surg Oncol Clin N Am.* 2013;22(4):765–783.
184. Sieh W, Kobel M, Longacre TA, et al. Hormone-receptor expression and ovarian cancer survival: an ovarian tumor tissue analysis consortium study. *Lancet Oncol.* 2013;14(9):853–862.
185. Stasenko M, Plegue M, Sciallis AP, McLean K. Clinical response to antiestrogen therapy in platinum-resistant ovarian cancer patients and the role of tumor estrogen receptor expression status. *Int J Gynecol Cancer.* 2015;25(2):222–228.
186. Eeles RA, Morden JP, Gore M, et al. Adjuvant hormone therapy may improve survival in epithelial ovarian cancer: results of the AHT randomized trial. *J Clin Oncol.* 2015;33(35):4138–4144.
187. Bowman A, Gabra H, Langdon SP, et al. CA125 response is associated with estrogen receptor expression in a phase II trial of letrozole in ovarian cancer: identification of an endocrine-sensitive subgroup. *Clin Cancer Res.* 2002;8(7):2233–2239.
188. del Carmen MG, Fuller AF, Matulonis U, et al. Phase II trial of anastrozole in women with asymptomatic mullerian cancer. *Gynecol Oncol.* 2003;91(3):596–602.

189. Orlando L, Schiavone P, Fedele P, et al. Molecularly targeted endocrine therapies for breast cancer. *Cancer Treat Rev*. 2010;36(suppl 3):S67–S71.
190. Ma CX, Reinert T, Chmielewska I, Ellis MJ. Mechanisms of aromatase inhibitor resistance. *Nat Rev Cancer*. 2015;15(5):261–275.
191. Fuqua SA, Gu G, Rechoum Y. Estrogen receptor (ER) alpha mutations in breast cancer: hidden in plain sight. *Breast Cancer Res Treat*. 2014;144(1):11–19.
192. Rechoum Y, Rovito D, Iacopetta D, et al. AR collaborates with ERalpha in aromatase inhibitor-resistant breast cancer. *Breast Cancer Res Treat*. 2014;147(3):473–485.
193. Anido J, Scaltriti M, Bech Serra JJ, et al. Biosynthesis of tumorigenic HER2 C-terminal fragments by alternative initiation of translation. *EMBO J*. 2006;25(13):3234–3244.
194. Deng L, Chen J, Zhong XR, et al. Correlation between activation of PI3K/AKT/mTOR pathway and prognosis of breast cancer in Chinese women. *PLoS One*. 2015;10(3):e0120511.
195. Vilquin P, Villedieu M, Grisard E, et al. Molecular characterization of anastrozole resistance in breast cancer: pivotal role of the Akt/mTOR pathway in the emergence of de novo or acquired resistance and importance of combining the allosteric Akt inhibitor MK-2206 with an aromatase inhibitor. *Int J Cancer*. 2013;133(7):1589–1602.
196. Gadducci A, Biglia N, Tana R, Cosio S, Gallo M. Metformin use and gynecological cancers: a novel treatment option emerging from drug repositioning. *Crit Rev Oncol Hematol*. 2016;105:73–83.
197. Martin LA, Pancholi S, Farmer I, et al. Effectiveness and molecular interactions of the clinically active mTORC1 inhibitor everolimus in combination with tamoxifen or letrozole in vitro and in vivo. *Breast Cancer Res*. 2012;14(5):R132.
198. Liu T, Yacoub R, Taliaferro-Smith LD, et al. Combinatorial effects of lapatinib and rapamycin in triple-negative breast cancer cells. *Mol Cancer Ther*. 2011;10(8):1460–1469.
199. Beck JT, Hortobagyi GN, Campone M, et al. Everolimus plus exemestane as first-line therapy in HR(+), HER2(−) advanced breast cancer in BOLERO-2. *Breast Cancer Res Treat*. 2014;143(3):459–467.
200. Piccart M, Hortobagyi GN, Campone M, et al. Everolimus plus exemestane for hormone-receptor-positive, human epidermal growth factor receptor-2-negative advanced breast cancer: overall survival results from BOLERO-2dagger. *Ann Oncol*. 2014;25(12):2357–2362.
201. Steelman LS, Martelli AM, Cocco L, et al. The therapeutic potential of mTOR inhibitors in breast cancer. *Br J Clin Pharmacol*. 2016;82(5):1189–1212.
202. Vareslija D, McBryan J, Fagan A, et al. Adaptation to AI therapy in breast cancer can induce dynamic alterations in ER activity resulting in estrogen-independent metastatic tumors. *Clin Cancer Res*. 2016;22(11):2765–2777.
203. Bacci M, Giannoni E, Fearns A, et al. miR-155 Drives metabolic reprogramming of ER+ breast cancer cells following long-term estrogen deprivation and predicts clinical response to aromatase inhibitors. *Cancer Res*. 2016;76(6):1615–1626.
204. Fakhrejahani E, Toi M. Antiangiogenesis therapy for breast cancer: an update and perspectives from clinical trials. *Jpn J Clin Oncol*. 2014;44(3):197–207.
205. Turner N, Lambros MB, Horlings HM, et al. Integrative molecular profiling of triple negative breast cancers identifies amplicon drivers and potential therapeutic targets. *Oncogene*. 2010;29(14):2013–2023.
206. Ferguson LR, Chen H, Collins AR, et al. Genomic instability in human cancer: molecular insights and opportunities for therapeutic attack and prevention through diet and nutrition. *Semin Cancer Biol*. 2015;35(suppl):S5–S24.
207. Hu X, Stern HM, Ge L, et al. Genetic alterations and oncogenic pathways associated with breast cancer subtypes. *Mol Cancer Res*. 2009;7(4):511–522.
208. Chumsri S, Schech A, Chakkabat C, Sabnis G, Brodie A. Advances in mechanisms of resistance to aromatase inhibitors. *Expert Rev Anticancer Ther*. 2014;14(4):381–393.

209. Thangavel C, Dean JL, Ertel A, et al. Therapeutically activating RB: reestablishing cell cycle control in endocrine therapy-resistant breast cancer. *Endocr Relat Cancer.* 2011;18(3):333–345.
210. Ropero S, Fraga MF, Ballestar E, et al. A truncating mutation of HDAC2 in human cancers confers resistance to histone deacetylase inhibition. *Nat Genet.* 2006;38(5):566–569.
211. New M, Olzscha H, La Thangue NB. HDAC inhibitor-based therapies: can we interpret the code? *Mol Oncol.* 2012;6(6):637–656.
212. Holloway KR, Barbieri A, Malyarchuk S, et al. SIRT1 positively regulates breast cancer associated human aromatase (CYP19A1) expression. *Mol Endocrinol.* 2013;27(3):480–490.
213. Simmons Jr GE, Pruitt WM, Pruitt K. Diverse roles of SIRT1 in cancer biology and lipid metabolism. *Int J Mol Sci.* 2015;16(1):950–965.
214. Gray SG, Ekstrom TJ. The human histone deacetylase family. *Exp Cell Res.* 2001;262(2):75–83.
215. Drummond DC, Noble CO, Kirpotin DB, Guo Z, Scott GK, Benz CC. Clinical development of histone deacetylase inhibitors as anticancer agents. *Annu Rev Pharmacol Toxicol.* 2005;45:495–528.
216. Lin HY, Chen CS, Lin SP, Weng JR, Chen CS. Targeting histone deacetylase in cancer therapy. *Med Res Rev.* 2006;26(4):397–413.
217. Eom GH, Kook H. Posttranslational modifications of histone deacetylases: implications for cardiovascular diseases. *Pharmacol Ther.* 2014;143(2):168–180.
218. Smith KT, Workman JL. Histone deacetylase inhibitors: anticancer compounds. *Int J Biochem Cell Biol.* 2009;41(1):21–25.
219. Xu WS, Parmigiani RB, Marks PA. Histone deacetylase inhibitors: molecular mechanisms of action. *Oncogene.* 2007;26(37):5541–5552.
220. Kaypee S, Sudarshan D, Shanmugam MK, Mukherjee D, Sethi G, Kundu TK. Aberrant lysine acetylation in tumorigenesis: implications in the development of therapeutics. *Pharmacol Ther.* 2016;162:98–119.
221. Zhou Q, Agoston AT, Atadja P, Nelson WG, Davidson NE. Inhibition of histone deacetylases promotes ubiquitin-dependent proteasomal degradation of DNA methyltransferase 1 in human breast cancer cells. *Mol Cancer Res.* 2008;6(5):873–883.
222. Zhou Q, Shaw PG, Davidson NE. Inhibition of histone deacetylase suppresses EGF signaling pathways by destabilizing EGFR mRNA in ER-negative human breast cancer cells. *Breast Cancer Res Treat.* 2009;117(2):443–451.
223. Robertson FM, Chu K, Boley KM, et al. The class I HDAC inhibitor Romidepsin targets inflammatory breast cancer tumor emboli and synergizes with paclitaxel to inhibit metastasis. *J Exp Ther Oncol.* 2013;10(3):219–233.
224. Scuto A, Kirschbaum M, Kowolik C, et al. The novel histone deacetylase inhibitor, LBH589, induces expression of DNA damage response genes and apoptosis in Ph-acute lymphoblastic leukemia cells. *Blood.* 2008;111(10):5093–5100.
225. Ellis L, Pan Y, Smyth GK, et al. Histone deacetylase inhibitor panobinostat induces clinical responses with associated alterations in gene expression profiles in cutaneous T-cell lymphoma. *Clin Cancer Res.* 2008;14(14):4500–4510.
226. Urbinati G, Marsaud V, Plassat V, Fattal E, Lesieur S, Renoir JM. Liposomes loaded with histone deacetylase inhibitors for breast cancer therapy. *Int J Pharm.* 2010;397(1–2):184–193.
227. Palmieri D, Lockman PR, Thomas FC, et al. Vorinostat inhibits brain metastatic colonization in a model of triple-negative breast cancer and induces DNA double-strand breaks. *Clin Cancer Res.* 2009;15(19):6148–6157.
228. Zhou Q, Atadja P, Davidson NE. Histone deacetylase inhibitor LBH589 reactivates silenced estrogen receptor alpha (ER) gene expression without loss of DNA hypermethylation. *Cancer Biol Ther.* 2007;6(1):64–69.

229. Edwards A, Li J, Atadja P, Bhalla K, Haura EB. Effect of the histone deacetylase inhibitor LBH589 against epidermal growth factor receptor-dependent human lung cancer cells. *Mol Cancer Ther.* 2007;6(9):2515–2524.
230. Elsheikh SE, Green AR, Rakha EA, et al. Global histone modifications in breast cancer correlate with tumor phenotypes, prognostic factors, and patient outcome. *Cancer Res.* 2009;69(9):3802–3809.
231. Tate CR, Rhodes LV, Segar HC, et al. Targeting triple-negative breast cancer cells with the histone deacetylase inhibitor panobinostat. *Breast Cancer Res.* 2012;14(3):R79.
232. Chen S, Ye J, Kijima I, Evans D. The HDAC inhibitor LBH589 (panobinostat) is an inhibitory modulator of aromatase gene expression. *Proc Natl Acad Sci U S A.* 2010;107(24):11032–11037.
233. Kubo M, Kanaya N, Petrossian K, et al. Inhibition of the proliferation of acquired aromatase inhibitor-resistant breast cancer cells by histone deacetylase inhibitor LBH589 (panobinostat). *Breast Cancer Res Treat.* 2013;137(1):93–107.
234. Fakhry H, Miyamoto T, Kashima H, et al. Immunohistochemical detection of histone deacetylases in endometrial carcinoma: involvement of histone deacetylase 2 in the proliferation of endometrial carcinoma cells. *Hum Pathol.* 2010;41(6):848–858.
235. Weichert W, Denkert C, Noske A, et al. Expression of class I histone deacetylases indicates poor prognosis in endometrioid subtypes of ovarian and endometrial carcinomas. *Neoplasia.* 2008;10(9):1021–1027.
236. Singh BN, Zhou H, Li J, et al. Preclinical studies on histone deacetylase inhibitors as therapeutic reagents for endometrial and ovarian cancers. *Future Oncol.* 2011;7(12):1415–1428.
237. Li L-H. Histone deacetylase inhibitor, Romidepsin (FK228) inhibits endometrial cancer cell growth through augmentation of p53-p21 pathway. *Biomed Pharmacother.* 2016;82:161–166.
238. Imesch P, Fink D, Fedier A. Romidepsin reduces histone deacetylase activity, induces acetylation of histones, inhibits proliferation, and activates apoptosis in immortalized epithelial endometriotic cells. *Fertil Steril.* 2010;94(7):2838–2842.
239. Jiang S, Dowdy SC, Meng XW, et al. Histone deacetylase inhibitors induce apoptosis in both Type I and Type II endometrial cancer cells. *Gynecol Oncol.* 2007;105(2):493–500.
240. Bergadà L, Sorolla A, Yeramian A, et al. Combination of Vorinostat and caspase-8 inhibition exhibits high anti-tumoral activity on endometrial cancer cells. *Mol Oncol.* 2013;7(4):763–775.
241. Wu Y, Guo SW. Histone deacetylase inhibitors trichostatin A and valproic acid induce cell cycle arrest and p21 expression in immortalized human endometrial stromal cells. *Eur J Obstet Gynecol Reprod Biol.* 2008;137(2):198–203.
242. Zhou XC, Dowdy SC, Podratz KC, Jiang SW. Epigenetic considerations for endometrial cancer prevention, diagnosis and treatment. *Gynecol Oncol.* 2007;107(1):143–153.
243. Takai N, Narahara H. Histone deacetylase inhibitor therapy in epithelial ovarian cancer. *J Oncol.* 2010;2010:458431.
244. Sonnemann J, Gange J, Pilz S, et al. Comparative evaluation of the treatment efficacy of suberoylanilide hydroxamic acid (SAHA) and paclitaxel in ovarian cancer cell lines and primary ovarian cancer cells from patients. *BMC Cancer.* 2006;6:183.
245. Mackay HJ, Hirte H, Colgan T, et al. Phase II trial of the histone deacetylase inhibitor belinostat in women with platinum resistant epithelial ovarian cancer and micropapillary (LMP) ovarian tumours. *Eur J Cancer.* 2010;46(9):1573–1579.
246. Budman DR, Tai J, Calabro A, John V. The histone deacetylase inhibitor panobinostat demonstrates marked synergy with conventional chemotherapeutic agents in human ovarian cancer cell lines. *Invest New Drugs.* 2011;29(6):1224–1229.
247. Liu Z, Tong Y, Liu Y, et al. Effects of suberoylanilide hydroxamic acid (SAHA) combined with paclitaxel (PTX) on paclitaxel-resistant ovarian cancer cells and insights into the underlying mechanisms. *Cancer Cell Int.* 2014;14(1):112.

CHAPTER TWELVE

Molecular Changes During Breast Cancer and Mechanisms of Endocrine Therapy Resistance

S. Radhi[1]
Texas Tech University Health Science Center, Lubbock, TX, United States
[1]Corresponding author: e-mail address: saba.radhi@ttuhsc.edu

Contents

Abstract

Estrogen receptors (ERs) are expressed in 75% of breast cancers. ERs and their estrogen ligands play a key role in the development and progression of breast cancer. ERs have a genomic activity involving direct modulation of expression of genes vital to cell growth and survival by their classic nuclear receptors. The nongenomic activity is mediated by membrane receptor tyrosine kinases that activate signaling pathways resulting in activation of ER pathway modulators.

Endocrine therapies inhibit the growth promoting activity of estrogen. ERs-positive breast cancers can exhibit de novo or acquired endocrine resistance. The mechanisms of endocrine therapy resistance are complex include deregulation of ER pathway, growth factor receptor signaling, cell cycle machinery, and tumor microenvironment.

In this chapter, we will review the literature on the biology of ERs, the postulated mechanisms of endocrine therapy resistance, and their clinical implications.

Progress in Molecular Biology and Translational Science, Volume 144
ISSN 1877-1173
http://dx.doi.org/10.1016/bs.pmbts.2016.09.009

1. INTRODUCTION

Breast cancer is the most common cancer among women in the United States.[1] Majority of breast cancers are estrogen receptor (ER)-positive.[2] It is heterogeneous disease, and gene expression profiling by microarray has shed light into the complexity of this cancer. ER-positive tumors are stratified into luminal A and luminal B subtypes by molecular profiling studies. Luminal A subtype is associated with more indolent disease with high ER expression and low proliferative rate, while luminal B tumors tend to have more aggressive behavior with high proliferative rate and resistance to endocrine therapy.[3–5]

Estrogen is a mitogenic hormone that plays a vital role in normal breast development and its carcinogenesis.[6] It is predominantly synthesized in the ovaries in premenopausal women and to a lesser extent in peripheral tissue, including breast tissue. In postmenopausal women it is produced in the extragonadal, peripheral tissue.[7]

Prolonged exposure to endogenous or exogenous estrogen has been associated with increased breast cancer risk. The mechanism of breast carcinogenesis through estrogen signaling pathway is complex and not completely understood. It is thought that cancer results from genetic mutations that result in overexpression of growth factors or receptors responsible for cell growth and proliferation.[6] Animal studies suggest metabolism of estrogen to genotoxic, mutagenic metabolites with carcinogenic potential.[7]

Deprivation of estrogen signaling through endocrine-targeted therapy has become the mainstay of treatment in ER-positive disease.[8] Endocrine therapies work through different mechanisms to antagonize the effect of estrogen. Selective ER modulators (SERMs)-like tamoxifen binds ER and produce antiestrogenic activity by recruiting corepressors rather than coactivators to target promoters. Tamoxifen also has some estrogen-agonist activity depending on the concentrations of coactivators and corepressors.[9]

It also has agonist effects on certain tissue like the bone and the cardiovascular system.[10,11] Selective ER downregulators (fulvestrant) on the other hand have pure antiestrogenic activity. Aromatase inhibitors (AIs) include anastrozole, letrozole, and exemestane. They reduce the levels of estrogen by blocking the conversion of androgens of extra gonadal origin to estrogen.[12]

Although endocrine therapy is the mainstay of treatment for ER-positive breast cancer, its effectiveness is limited by de novo or acquired resistance during treatment. Understanding the mechanisms of resistance is crucial to develop targeted therapy to overcome the resistance.

In this chapter we will review the structure of the ER receptor and its mechanism of action and the mechanisms of resistance to endocrine therapy.

2. ESTROGEN RECEPTORS

There are two types of ERs, ERα and ERβ. They belong to the steroid hormone superfamily of nuclear receptors. They are encoded by two different genes, ESR1 and ESR2, respectively. Expression of ERα is increased in premalignant and many malignant cells compared to normal mammary cells.[13,14] ERα plays a vital role in breast cancer pathogenesis, while the role of ERβ is controversial. They are encoded by two separate genes but have similar structure. ERs are composed of six structural domains (domains A–F) that form three main functional domains.[15] The central DNA-binding domain (DBD), it is associated with domain C and is involved in DNA recognition and binding of the receptor to promoters of estrogen-regulated genes.[14] Two main transcription-regulator domains, AF1 is located in the amino terminal and is ligand dependent, associated with domain A/B. AF2 is in the ER carboxyl terminal and is ligand independent. Ligand-binding domain (LBD) partly overlaps with AF2, associated with domain E[16] (Fig. 1).

2.1 ER Function

The molecular mechanism of estrogen signaling is mediated through nuclear genomic and cell surface (nongenomic) functions. The genomic function includes the classic ligand dependent, ligand independent, and DNA binding independent.

2.2 Nuclear Genomic Function

The classical mechanism of action of estrogen is through ligand-dependent transcription factor (TF). It regulates expression of genes that regulate cell

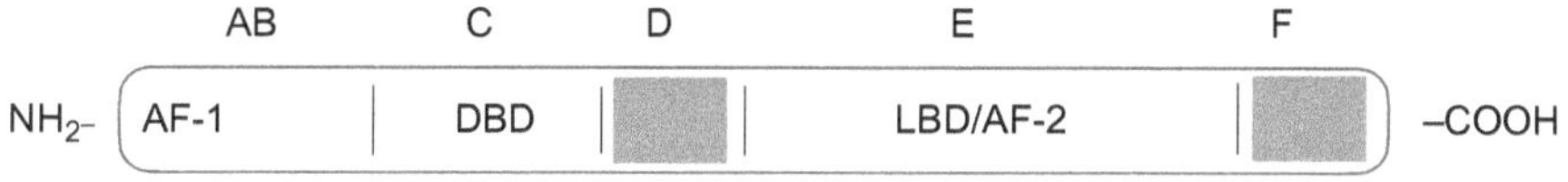

Fig. 1 The nuclear ERα consists of domains (A) through (F). The main functional domains are shown. DNA-binding domain (*DBD*), ligand-binding domain (*LBD*), and two major transcriptional activation function (*AF*) domains, AF-1 and AF-2.

proliferation, enhance invasion, metastasis, and angiogenesis.[15] The receptor is bound to protein inhibitory complex.[17] Binding of estrogen to ERα activates the receptor by phosphorylation and unbinding of the inhibitory proteins. The estrogen-bound receptor forms a homodimer with another receptor which then binds to specific DNA sequence at the promoter region of the target gene known as estrogen responsive element (ERE)[17,15] (Fig. 1A). The promoter-bound homodimer binds coregulatory proteins which regulates the transcription of various genes involved in cell growth and proliferation including receptor tyrosine kinases (RTKs). Coregulatory protein includes coactivators such as SRC family and corepressors that inhibits the transcription of the target genes.[18] AIB-1 (amplified in breast cancer-1) is a member of SRC family and is overexpressed at protein level in up to 60% of breast cancer, and the AIB-1 gene is amplified in 5–10% of breast cancer.[19] It plays a role in breast cancer development and metastasis.[20] It was also thought to be associated with tamoxifen resistance as evidenced by lower disease-free survival in patients on tamoxifen with High AIB-1 expression.[9]

In addition, ERα modulates gene expression in the absence of estrogen via ligand-independent activity through cross talk between ERα and growth factor like IGF-1, epidermal growth factor (EGF) and, cyclic AMP leading to activation of ER at ERE containing promoter as well as TFs involved in ER pathway that regulate gene expression on other response elements (REs)[17,21] (Fig. 2A).

ERα can modulate gene expression through tethering to TF protein such as c-jun or c-fos and functions as a coactivator protein by stabilizing the DNA transcription protein-binding complex and attracting other coactivator proteins. In this model the TFs bind promoter regions other than ERE, such as activating protein (AP)-1 and specificity protein-1 (SP-1)-binding sites. AP-1 TFs are suggested to be important elements in estrogen-mediated tumor growth[22] (Fig. 1A).

2.3 Nongenomic Function

In addition to the transcription function of ER, it also influences rapid stimulatory effects on various signal transduction pathways. This activity is referred to the membrane-initiated steroid signaling. This nongenomic action is thought to complement the genomic ER function and augment it.[15]

Membranous and cytoplasmic ERα, in response to estrogen, interact with tyrosine kinase receptors (RTKs) such as epithelial growth factor receptor (EGFR), insulin-like growth factor (IGF-R1), and human epidermal growth

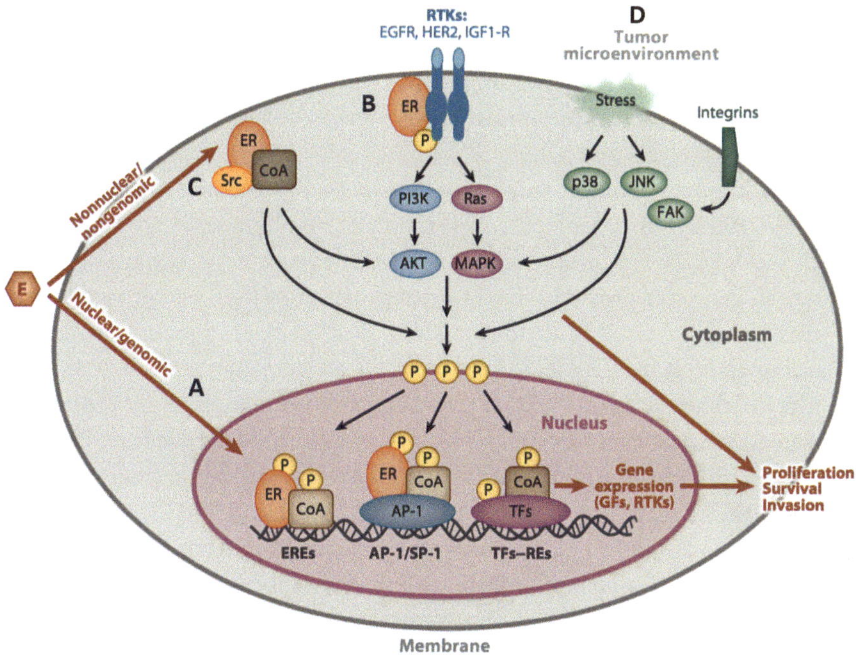

Fig. 2 Estrogen (E)-bound ER, acting as a transcription factor in the nucleus (nuclear/genomic activity), binds to DNA sequences in promoter regions of target genes either directly (at estrogen receptor elements; EREs) or indirectly via protein–protein interaction with other transcription factors at their cognate DNA responsive sites (e.g., members of the AP-1 or the SP-1 transcription complexes at AP-1 or SP-1 sites). Upon estrogen binding, ER generally recruits coactivator complexes (CoA) to induce or modulate gene transcription including genes coding growth factors (GFs) and receptor tyrosine kinases (RTKs) (A). A small subset of the cellular pool of ER localized outside the nucleus and/or at the cell membrane associates in response to estrogen with growth factor RTKs (e.g., EGFR, HER2, and IGF1-R) (B) and with additional signaling and coactivator molecules (e.g., the Src kinase) (C). This interaction, similar to GF activation of these pathways, activates multiple downstream kinase pathways (e.g., Src, PI3K/AKT, and Ras/p42/44 MAPK) which in turn phosphorylate various TFs and coregulators, including components of the ER pathway that enhance gene expression on EREs and other REs. The nonnuclear/nongenomic activity, which can also be activated by tamoxifen, is enhanced in the presence of overexpression and hyperactivation of RTKs and can contribute to endocrine therapy resistance. Overall, the nuclear/genomic and nonnuclear/nongenomic ER activities work in concert to provide breast tumor cells with proliferation, survival, and invasion stimuli. Signaling from the microenvironment activates stress-related pathways and members of the integrin family. These pathways then trigger downstream kinase pathways (e.g., FAK, JNK, and p38 MAPK) that can further modulate components of the transcriptional machinery, including ER (D). Alterations in each of these transcriptional and signaling elements can mediate resistance to endocrine therapy either by modulating ER activity or by acting as escape pathways to provide alternative proliferation and survival stimuli.

factor receptor 2 (Her2) along with various membrane-signaling molecules, including insulin receptor substrate-1 (IRS-1), the p85 subunit of phosphatidylinositol-3-OH kinase (PI3K), Src, and Shc. This in turn triggers downstream kinases including mitogen-activated protein kinase (MAPK) or PI3K/AKT[8](Fig. 2B and C).

ERα activates EGFRs indirectly acting as G proteins-coupled receptor.[23,24] ERα binds to caveolin-1 in the cell membrane, where it activates c-Src and matrix metalloproteinases, which then cleave heparin-binding epidermal growth factor (Hb-EGF) from the membrane. This growth factor in turns binds to EGFRs and activated downstream signaling pathways including MAPK and AKT.[15,23]

The microenvironment via extracellular stimuli such as inflammatory cytokines activates stress-related pathways and members of the integrin family. This stimulates downstream kinase pathways including focal adhesion kinase (FAK), c-Jun N-terminal kinase (JNK), and p38 MAPK that in turn phosphorylate ER and its coregulatory proteins[21,25,26] (Fig. 2D).

These kinases enhance tumor growth and survival, in addition they translocate to the nucleus and phosphorylate nuclear ER and coregulatory proteins resulting in activation of transcriptional components and synergize nuclear ER signaling[14,17,23] (Fig. 2).

3. MECHANISMS OF ENDOCRINE THERAPY RESISTANCE

The genomic and nongenomic ER activities act in synergy activating signaling pathways that stimulates tumor cells growth and invasion. This along with the microenvironment-activated stress-related pathways then trigger downstream kinase pathways including AKT and MAPK that can further modulate components of the transcriptional machinery, including ER, coregulators, and corepressors.[15] Resistance to endocrine therapy can occur from changes in any component of these pathways including altering ER activity or by developing alternative proliferation pathways through increased signaling of growth factors or cellular kinase pathways. The inhibitory effect of endocrine therapy can be bypassed by the bidirectional cross talk between ER and growth factor receptor signaling pathways.[21]

3.1 ESR1 Mutation

Mutation in ESR1, the gene encoding ERα, was first identified in metastatic breast cancer in 1997.[27] ESR1 mutations are rare in primary breast cancer[28] but have a high prevalence in metastatic breast cancers (11–55%) after endocrine therapy exposure more so after prolonged treatment with AIs.[29,30] Evidence support the presence of polyclonal mutations in cancer.[28] Majority of mutations take place in a small region that encodes the LBD which induce constitutive, ligand-independent effect of ER.[31,32]

Preclinical data showed that mutant ER is resistant to estrogen deprivation and is much less responsive to tamoxifen or fulvestrant, a selective ER degrader, and higher doses are needed to be as effective as in the wild-type receptors.[29] These data suggest ESR1 mutation is a mechanism of endocrine therapy resistance and the development of newer generation SERMs, and selective ER degraders are crucial for more effective therapy.

Clinical data show patients with ESR1 mutations had poor progression-free survival (PFS) on subsequent AI-based therapy but relatively better PFS on Fulvestrant,[33] suggesting ESR1 might become a predictor of resistance to certain endocrine therapies.

3.2 Epigenetic Pathways

Transcription activation by ER is known to involve several coactivators. These coactivators have acetyltransferase activity and are capable of acetylating histones, which destabilizes nucleosomes and promotes transcription. Acetylation is regulated in a steady state by the enzymes histone acetyltransferase (HAT) and histone deacetylases (HDACs) activity.[34] Expression of ER can be lost by epigenetic modifications such as histone deacetylation, which can cause endocrine resistance.

Pfister *et al.* demonstrated a correlation between histone modification and tumor phenotypes as well as patient outcomes. Higher levels of acetylation and methylation were associated with better outcomes compared with moderate or low levels.[35]

There is also some evidence of partial restoration of functional ER in cells that have lost ER expression as a result of acquired resistance to endocrine therapy.[36]

HDAC inhibitors (HDACi) were shown to induce growth arrest, differentiation, and cell death in breast cancer cells. HDACi inhibit ERα expression and cell proliferation. In breast cancer cells, TSA, which is a potent and reversible HDACi, led to a significant reduction in ERα accumulation independent of ER ligands. In human cancer cells, inhibition of HDACs regulates the expression of the ERα gene and the transcriptional activity in response to partial antiestrogens. Inhibition of HDAC enzymatic activity and modulation of ERα levels affect the relative agonist activity of partial antiestrogens on a stably integrated reporter transgene.[34]

Clinically the combination of vorinostat, a HDACi, and tamoxifen was shown to reverse hormone resistance, with a response rate of 19% and a clinical benefit rate of 40%.[37]

Epigenetic pathways therefore play a significant role in endocrine resistance and provide a potential therapeutic target.

3.3 Growth Factor Receptor Pathways

Pathways, such as the tyrosine kinase receptor family and receptors for insulin/IGF1, fibroblast growth factor (FGF), and vascular endothelial growth factor (VEGF), as well as cellular Src, AKT, and stress-related kinases, have been implicated.

3.3.1 HER2 Gene Amplification

Human epidermal receptor family members, including HER1 (EGFR), HER2, HER3, and HER4, are tyrosine kinase receptors involved in cell proliferation and survival.[6] HER2 is amplified in about 20–30% of breast cancer and in 10% of ER+ breast cancers.[38] Clinical evidence suggest that activation of the HER2 receptor pathway leads to endocrine resistance, especially to tamoxifen (SERMs).[9] In an experimental model, the agonist effect of tamoxifen was increased in breast cancer cells that expressed high levels of HER2 leading to enhanced cell growth in the presence of tamoxifen.[39] This effect was observed in breast cancer cells with high levels of AIB1 in addition to HER2.[39] AIB1 is an ER coactivator that is essential for cell growth. Clinically, patients treated with tamoxifen whose tumors expressed high levels of both AIB1 and HER-2 had worse prognosis compared to those not treated with tamoxifen.[9] Overexpression of HER2 has also been shown to potentiate the rapid membrane ER activity in response to tamoxifen in addition to estrogen.

The bidirectional cross talk between the ER and growth factor receptors could explain this observation. EGFR or HER2 activates MAPk/Akt which in turns phosphorylates both ER and the ER coactivator AIB1 leading to increased transcriptional activity of ER which further enhances the activity of the components of the pathway.[15] This cross talk was less apparent in the presence of selective EGFR tyrosine kinase inhibitor gefitinib, and the tamoxifen antagonist activity was revived.[39] This association was not observed with AIs. AIs may be more effective than tamoxifen in ER-positive breast cancer with Her2 overexpression.[40]

3.3.2 Insulin-Like Growth Factor and Type 1 Insulin-Like Growth Factor

The IGF ligands promote cell replication and survival mainly by binding to IGF1R and the subsequently activating the PI3K/Akt and RAS/MAPK pathways.[41] Up to 50% of breast cancer expresses the activated form of IGF1R. It is more expressed in ER-positive luminal A and luminal B breast cancer up to 84% and 76%, respectively.[41,42] There is bidirectional regulation between ERα and IGF1R signaling. ERα regulates expression and activity of IGFIR.[43] On the other hand, IGF1R upregulates the transcription of ERα and stimulates ER phosphorylation through activation of mTOR/S6K signaling.[44]

Hyperactivation of IGF1R and downstream signaling has been linked to the development of endocrine resistance in ER-positive breast cancer.[45,46] Trials evaluating IGFR inhibitors are underway; results so far are not encouraging. Ganitumab is a monoclonal antibody that blocks IGF-1R. It was evaluated in a phase II clinical trial in combination with endocrine therapy with fulvestrant or exemestane compared to endocrine therapy with placebo with no difference in PFS between ganitumab compared to placebo.[47] These results suggest that this pathway is complex, and further studies are needed to understand the effects of blocking this pathway and its link to other pathways.

3.3.3 FGF Receptors

FGF family signals through transmembrane tyrosine kinase FGF receptors and plays a role in regulating cell proliferation, migrations, and survival.[48]

Fibroblast growth factor receptor 1 (FGFR1) amplification occurs in about 10% of breast cancers. FGFR1 activity is required for the survival of a FGFR1-amplified breast cancer cell line.[49] FGFR1 amplification is more prevalent in luminal B subtype of breast cancer. It is found in 16–27% of luminal B breast cancers and is associated with ER-positive PR-negative phenotype, and high proliferative rate (Ki67) and may contribute to the poor prognosis in this subtype. It plays a role in resistance to endocrine therapy by stimulating both aberrant ligand-dependent and ligand-independent signaling pathway.[50] Other genomic alterations of the FGF have also been found less often in breast cancer including FGFR2 and FGFR3 amplification.[48,51] Dovitinib is an oral tyrosine kinase inhibitor of FGFR1, FGFR2, and FGFR3 which have shown efficacy in preclinical and clinical studies in breast cancers with FGF pathway amplification.[51] Other FGFR inhibitors are being evaluated in metastatic ER-positive breast cancer to overcome endocrine resistance.[38]

3.3.4 *Vascular Endothelial Growth Factor*

Angiogenesis is crucial for cell growth of the primary tumor and the development of metastases. Overexpression of VEGF and the VEGF receptor were observed in breast cancers, and high levels of VEGF have been associated with resistance to endocrine therapy.[52,53] These findings provided the basis for investigating combining VEGF inhibitors with endocrine therapy to overcome endocrine resistance. Two phase III randomized trials investigated adding bevacizumab to first-line endocrine therapy with letrozole or fulvestrant as first-line in advance ER-positive breast cancer with better PFS that was significant in one trial.[54,55] Identifying biomarkers to select the group of patients who will benefit the most will be helpful in designing further trials.

3.4 PI3K/Akt/mTOR Pathway

The PI3Ks are a family of lipid kinases. Cell stimulation by growth factors and RTK leads to phosphorylation of phosphatidylinositol lipids at the D-3 position of the inositol ring which then coordinate a sequence of events that stimulates cell growth and survival.[56] Class Ia enzymes are primarily responsible for production of D-3 phosphoinositides in response to growth factors and are deregulated in cancer. They are heterodimers of a regulatory subunit (referred to as p85) and a catalytic subunit (p110). When activated, PI3K heterodimer interacts with the intracellular portion via p85 subunit, which release the inhibitory function of p85 from p110. The activated kinase induces the phosphorylation of phosphatidylinositol bisphosphate (PIP2) to phosphatidylinositol triphosphate (PIP3). PIP3 translocates proteins from the cytoplasm to plasma membrane and activates them.[57] The serine/threonine kinase protein kinase B (Akt) is the most important to note. It is the major mediator in PI3k pathway.[58] Akt is phosphorylated and subsequently phosphorylates various proteins including the mammalian target of rapamycin (mTOR) which is pivotal in the cell cycle progression from the G_1 to the S phase.[59] MTOR has two multiprotein complexes, mTOR complexes 1 and 2 (mTORC1 and mTORC2). mTORC1 consists of mTOR that is associated with raptor (regulatory-associated protein of mTOR) and is downstream of AKT. In contrast, mTORC2 is associated with rictor (rapamycin-insensitive companion of mTOR) and phosphorylates AKT.[44] AKT phosphorylates mTORC1, which in turn phosphorylates its effectors, including S6 kinase 1 (S6K1) which directly phosphorylates the activation domain of ERs (Fig. 3).

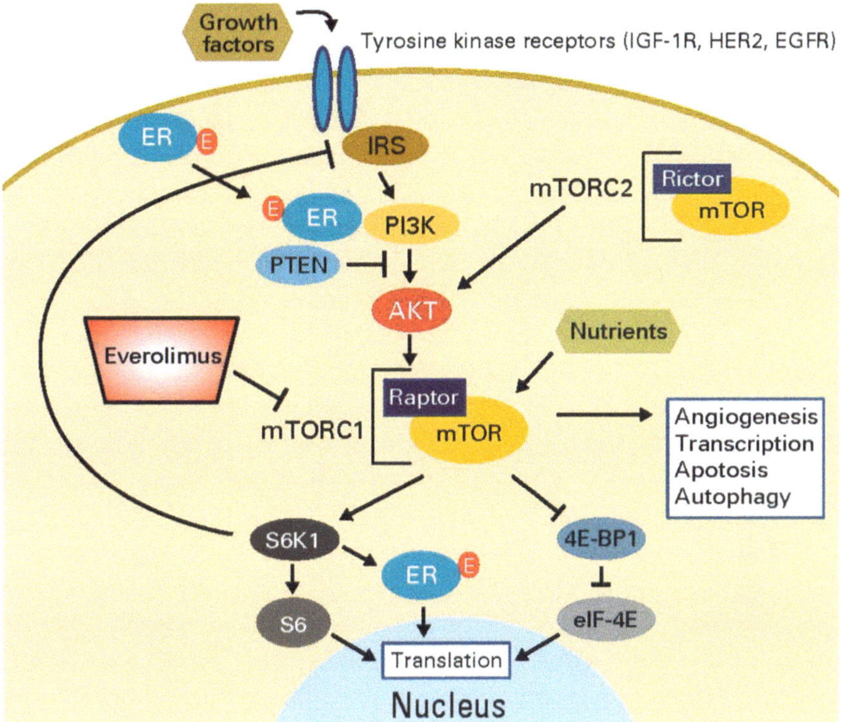

Fig. 3 Schematic of the phosphatidylinositol 3-kinase (PI3K)/AKT/mammalian target of rapamycin (mTOR) pathway. Reciprocal cross talk exists between the estrogen receptor (ER) and growth factor receptor (GFR) signaling pathways. ER can induce transcription of genes important to GFR pathways, and PI3K can phosphorylate ER to modulate this transcriptional activity. *4E-BP1*, 4E-binding protein-1; *E*, estrogen; *EGFR*, epidermal growth factor receptor; *eIF-4E*, eukaryotic initiation factor-4E; *FKBP-12*, FK506-binding protein-12; *HER2*, human epidermal growth factor receptor 2; *IGF-1R*, insulin-like growth factor-1 receptor; *IRS*, insulin receptor substrate; *mTORC1*, mTOR complex 1; *mTORC2*, mTOR complex 2; *PTEN*, phosphatase and tensin homolog deleted on chromosome 10; *S6*, 40S ribosomal protein; *S6K1*, S6 kinase 1.

The phosphatase and tensin homolog (PTEN) has an inhibitory effect. It dephosphorylates PIP3 to PIP2.

The PI3K/Akt/mTOR pathway regulates various normal cellular functions including cell proliferation, metabolism, and survival. Activation of this pathway is implicated in tumorigenesis. It is the most frequently altered pathway in breast cancer, up to 70%.[60] High PI3K signaling is found to be associated with luminal B subtype of ER-positive breast cancer and is associated with poor prognosis.[61] Activation of the PI3K pathway has been associated with resistance to endocrine therapy. It was found that direct inhibition of PI3K suppressed the hormone-independent growth of cancer cells, whereas inhibition of components upstream (RTKs) and downstream (mTOR) of PI3K only partially blocked cell growth[62]inhibition of mTOR has been linked to restoration of hormone activity.[60,63]

The majority of mutations are activating mutations in *PIK3CA*, encoding the alpha catalytic p110 subunit, and are localized in three "hot spots": E545K, E542K, and H1047R. These mutations occur in 30% of ER-positive breast cancers and are shown in experimental models to enhance downstream signaling elements including Akt and stimulate oncogenic transformation.[64–66] These mutations are more prevalent in metastatic than primary breast cancer.[67] Gene expression and protein analysis however showed PIK3CA mutation to be associated with luminal A subtype and better prognosis.[68,69] These mutations probably result in only weak activation of the PI3K pathway may be due to subsequent negative feedback regulation.[68,70]

Mutation in or loss of at least one copy of the *PTEN* gene occurs in triple-negative basal subtype breast cancer. Decreased levels are found in 50% of ER-positive breast cancer and are more prevalent in luminal B compared to luminal A subtype.[71] This results in the activation of PI3K signaling and is associated with endocrine therapy resistance and worse outcome in patients treated with tamoxifen.[72,73]

The earliest compounds used to block the PI3K pathway were mTOR inhibitors. Everolimus (RAD001) is a rapamycin analogue that inhibits mTORC1. Preclinical data demonstrated that inhibition of mTOR has been linked to restoration of hormone activity.[60,63,74] In phase II trial neoadjuvant combination of everolimus and letrozole showed better response rate and reduced cell proliferation compared to letrozole alone in ER positive untreated locally advanced postmenopausal women.[75] The phase III breast cancer trial of oral everolimus 2 (BOLERO-2) compared exemestane and everolimus to exemestane and placebo in women with ER-positive metastatic breast cancer who previously progressed on nonsteroidal AIs and demonstrated superior PFS with everolimus combination.[76] Similar patient population was studied in phase II clinical trial (TAMRAD), where a combination of tamoxifen and everolimus was superior to tamoxifen alone.[77] Based on these results, everolimus is being used in practice today in combination with endocrine therapy in metastatic ER-positive Her2-negative patients that has progressed on or recurred after AIs.

It was observed that on inhibition of mTORC1 there were increased levels of activated Akt, suggesting upregulation of Akt.[78] Normally, regulation of PI3K/AKT signaling is controlled by mTORC1-dependent phosphorylation and downregulation of IRS-1 and, IRS-1-mediated IGF-1 receptor (IGF-1R) which is suppressed by S6K downstream of mTORC1.[66,79] Inhibition of mTORC1 may upregulate Akt by relieving the negative feedback loop between the S6K and the activators of PI3K. This may potentially lower the efficacy of mTORC1 inhibitors[80] (Fig. 3). Ongoing clinical trials are evaluating the combination of everolimus and monoclonal antibody to insulin-like growth factor (NCT02123823).

Future research is exploring various methods to reverse endocrine resistance including PI3K inhibitors, Akt inhibitors, and various combinations. Pan-class I PI3K inhibitors (buparlisib [BKM120] and pictilisib [GDC0941]) and the alpha-specific inhibitors (alpelisib [BYL719] and taselisib [GDC0032]) are being studied in ongoing clinical trials.[38]

3.5 Cyclin D/CDK4/6/Rb Pathway

Normal cell replication progresses through four main sequential phases in a cell cycle: G1 (pre-DNA synthesis), S (DNA synthesis), G2 (predivision), and M (mitosis).[81] Regulation of this process is maintained by various complex proteins. The cyclin-dependent kinases (CDKs) are key regulators. They are a group of serine/threonine kinases consist of catalytic kinase subunit and a regulatory cyclin subunit.[82] This group is divided based on their role in the cell cycle and includes three interphase CDKs (CDK2, CDK4, CDK6), one mitotic CDK (CDK1), and a number of regulatory CDKs, such as CDK7, and transcriptional CDKs (CDK8, CDK9).[83] CDK4/6 facilitates G1-S transition.[84]

The retinoblastoma susceptibility protein, Rb, plays a vital role in the G_1–S transition. Hypophosphorylated (active) Rb prevents progression from G_1 to S through binding and inactivating E2F transcription family member.[85] CDK4/6 forms a complex with cyclin D which phosphorylates Rb which in turn releases E2F enabling it to stimulate gene transcription and cell cycle progression into late G_1 and S phase.[86] One of these transcription genes encodes cyclin E, which binds CDK2 and further phosphorylates pRb in addition to other key mediators of the G1–S transition (Fig. 4).[87]

CDK4/6 activity is negatively regulated by INK4 (p16, p15, p18, p19) and CIP/KIP (p21, p27, p57) protein families.[87]

Within the INK4 family of proteins, p16 (INKa) is CDK4/6 inhibitor. It prevents binding of cyclin D1 leading to cell cycle arrest (Fig. 4).[88]

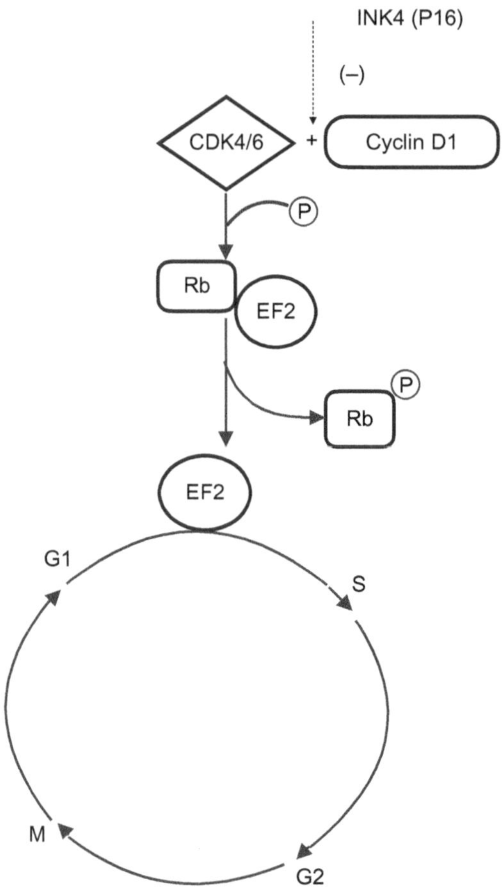

Fig. 4 The cell cycle and cyclin-dependent kinase (CDK4/6). CDK4/6 forms a complex with cyclin D1, which phosphorylates the retinoblastoma protein (Rb)—a tumor suppressor protein. This leads to Rb inactivation and release of E2F, enabling cell cycle progression.

In cancer cells, disruption of cell cycle regulation leads to uncontrolled cell growth. This happens by upregulation of cyclins, the aberrant activation of CDKs, or the inactivation of cellular CDK inhibitors and tumor suppressors proteins.[89] In breast cancer, cyclin D1 is shown to play an important role in the pathogenesis. Overexpression of cyclin D1 is observed early in breast cancer and maintained in metastatic lesions.[90] Amplification of CDK4 was reported in 15% of breast cancers, overexpression of CDK4[91] and overexpression of cyclinD1 in more than half.[87] In ER-positive breast cancer, amplification of both cyclin D1 and CDK4 is more prevalent in luminal

B (58% and 25%, respectively) compared to luminal A (29% and 14%, respectively). Lower Rb expression is more common in triple-negative basal subtype.[69]

ER signaling upregulates cyclin D1 levels and stimulates signaling pathways resulting in upregulation of CDK4/6 activity.[92] Acquired resistance could occur by activation of alternative signaling pathways (for example, PI3K/AKT) that can stimulate cyclin D1:CDK4/6 signaling independent of ER activity.[87] Also in addition to cyclin D1 function as a regulatory subunit of CDK4/4, evidence demonstrated that cyclin D1 can activate ER-mediated transcription independent of CDK binding by direct interaction with ER and stimulation of ER binding to ERE. It can also activate ER independent of estrogen ligand binding.[93] CDK gene mutations were found among potentially actionable genomic alterations in breast cancer genomic profiling.[94]

Inhibitors of these key cell cycle regulators are attractive targets for cancer therapy. First attempts at targeting CDKs involved nonspecific pan CDK inhibitors with limited efficacy.[87] More recently selective CDK4/6 inhibitors have been developed.[95] Gene expression profiles demonstrated that luminal ER-positive breast cancer cell lines were more sensitive to CDK4/6 inhibition. Palbociclib is an oral selective CDK4/6 inhibitor. It blocks Rb phosphorylation resulting in G1 arrest.[96] PALOMA-1 is a phase II trial that demonstrated superior PFS for palbociclib and letrozole compared to letrozole alone in postmenopausal women with previously untreated ER-positive advanced breast cancer.[97] PALOMA-3 is a phase 3 trial that demonstrated superior PFS for palbociclib in combination with fulvestrant compared to fulvestrant alone in patients with advanced breast cancer previously progressed on prior endocrine therapy.[98] Palbociclib is FDA approved and is being used in practice. Other CDK4/6 inhibitors are evaluated in ongoing clinical trials including ribociclib (LEE011) and abemaciclib (LY2835219).[38,99]

Preclinical data demonstrated increased Rb1 and cyclin D1 as well as decreased CDKN2A (p16) were associated with sensitivity to palbociclib on growth inhibition[96] and Rb loss, and high p16 is associated with the bypass of CDK4/6-mediated therapies.[100] Loss of Rb releases the suppression of E2F transcription leading to increased CDK2 which enhances cell cycle progression; therefore, lack of Rb may be used as a biomarker of resistance to these agents. Cyclin D1b is a splice variant of cyclin D1 was identified in breast cancer tissue and found to overcome cell cycle arrest caused by antiestrogen therapy through CDK4 and may contribute to treatment resistance.[101]

3.6 Tumor Microenvironment

Components of the microenvironment implicated in endocrine resistance include stromal cells (e.g., fibroblasts, endothelial, and immune system cells), structural elements of the extracellular matrix (ECM), and soluble factors (e.g., growth factors and cytokines), as well as additional microenvironmental conditions such as hypoxia and acidity.[21,102]

ECM serves as substrate to cells to adhere and as a reservoir for growth factors and plays a role in cell growth and metastasis. Specific ECM gene expression is found to be associated with prognosis and sensitivity to endocrine therapy. High expression of tenascin C (TNC) was associated with resistance to tamoxifen possibly via interacting with integrin and activating EGFR signaling.[26,103]

The interaction between the immune cells and cancer cells is complex. The presence of immune cells within the tumor has been evaluated. The composition of the immune cells within the tumor is heterogeneous. Tumor-infiltrating T lymphocytes is associated with poor prognostic features in breast cancer and is found to be a predictor of good response to cytotoxic chemotherapy in triple-negative breast cancer,[104] while it was linked to poor response to neoadjuvant endocrine therapy with anastrozole in a retrospective analysis.[105] In this analysis, an exploratory gene expression profiling of the inflammatory signature was most closely suggestive of dendritic cells (DCs).[105] DCs play a role in breast cancer tumorigenesis, and DCs metagene is shown to be associated with poor prognosis in tamoxifen-treated high proliferative tumors with high estrogen-related gene expression compared to trend to good prognosis in untreated patients, suggesting endocrine therapy resistance.[106]

The interaction between T cells and antigen-presenting cells involves the T cell receptor and multiple costimulatory and inhibitory receptors, programmed cell death 1 (PD-1) is an inhibitory receptor expressed by T cells and results in negative regulation when it binds the programmed death ligand 1 (PD-L1) expressed on inflamed tissue and tumor cells.[107] The immune checkpoints PD-1 and PD-L1 have been implicated in cancer immune evasion.[108] PD-L1 overexpression was observed breast cancer and stroma cells, more in basal subtype. About 4–20% of ER-positive breast cancer showed PD-L1 overexpression.[38,109–111] PD-L1 upregulation is associated with poor prognostic features such as higher grade, larger tumor size, and lack of ER receptors when all subtypes were evaluated. It is found to have favorable prognosis and predicts better response to cytotoxic chemotherapy in basal subtype.[109]

PD-1-and PD-L1 antibodies are being evaluated in clinical trials with promising preliminary results. Pembrolizumab, a PD-1 antibody, is being evaluated in phase IB KEYNOTE-28 trial that included ER-positive breast cancer and several other solid malignancies. The preliminary results showed a clinical benefit of 20% in ER-positive advanced breast cancer patients who failed or were ineligible to receive standard therapy.[112]

3.7 Apoptosis and Stem Cells

The apoptotic pathway is a regulated process of programmed cell death in human cells. Apoptosis is dysregulated in cancer. High levels of anti-apoptotic molecules in cancer cells cause resistance to treatment leading to cancer progression.[113] The B-cell lymphoma 2 (BCL2) gene family encodes regulatory proteins including both anti- and proapoptosis proteins.[114] BCL2 is an antiapoptotic protein. Increased expression in pretreated ER-positive breast cancer correlates with ER expression and good prognosis. The levels decrease after treatment with tamoxifen and are increased in residual disease which correlates with endocrine therapy resistance. Inhibition of BCl2 was found to restore antiestrogen sensitivity in cell lines.[115,116] Nuclear factor kappa B (NF-κB) is TF that plays a major role in apoptosis resistance. Inhibition of NF- κB activity restored tamoxifen sensitivity to the resistant cell line.[117] Bortezomib, a proteasome inhibitor that blocks the NF-κB pathway, is under investigation in endocrine resistant breast cancer.[118]

It has been suggested that ER-positive breast cancer has cancer stem cell-like cells, which is resistant to treatment. Gene analysis identified increased $CD44^{+}/CD24^{-/low}$ expression along with mesenchymal features in resistant cancer cells after treatment with neoadjuvant letrozole. Inhibitors of stem cell renewal pathways such as Notch and Hedgehog are potential target to overcome resistance to endocrine therapy.

REFERENCES

1. Siegel R, Ma J, Zou Z, Jemal A. Cancer statistics, 2014. *CA Cancer J Clin.* 2014;64(1):9–29. http://dx.doi.org/10.3322/caac.21208.
2. Dixon JM. Endocrine resistance in breast cancer. *New J Sci.* 2014;2014(7885):1–27. http://dx.doi.org/10.1073/pnas.1115188108.
3. Parker JS, Mullins M, Cheang MCU, et al. Supervised risk predictor of breast cancer based on intrinsic subtypes. *J Clin Oncol.* 2009;27(8):1160–1167. http://dx.doi.org/10.1200/JCO.2008.18.1370.
4. Prat A, Perou CM. Deconstructing the molecular portraits of breast cancer. *Mol Oncol.* 2011;5(1):5–23. http://dx.doi.org/10.1016/j.molonc.2010.11.003.

5. Prat A, Ellis MJ, Perou CM. Practical implications of gene-expression-based assays for breast oncologists. *Nat Rev Clin Oncol*. 2011;9(1):48–57. http://dx.doi.org/10.1038/nrclinonc.2011.178.
6. Suter R, Marcum JA. The molecular genetics of breast cancer and targeted therapy. *Biologics*. 2007;1(3):241–258.
7. Yager JD, Davidson NE. Estrogen carcinogenesis in breast cancer. *N Engl J Med*. 2006;354(3):270–282. http://dx.doi.org/10.1056/NEJMra050776.
8. Osborne CK. Estrogen-receptor biology: continuing progress and therapeutic implications. *J Clin Oncol*. 2005;23(8):1616–1622. http://dx.doi.org/10.1200/JCO.2005.10.036.
9. Osborne CK, Bardou V, Hopp TA. Role of the estrogen receptor coactivator AIB1 (SRC-3) and HER-2/neu in tamoxifen resistance in breast cancer. *J Natl Cancer Inst*. 2003;95(5):353–361.
10. Jordan VC, O'Malley BW. Selective estrogen-receptor modulators and antihormonal resistance in breast cancer. *J Clin Oncol*. 2007;25(36):5815–5824. http://dx.doi.org/10.1200/JCO.2007.11.3886.
11. Jordan VC. The science of selective estrogen receptor modulators: concept to clinical practice. *Clin Cancer Res*. 2006;12(17):5010–5013. http://dx.doi.org/10.1158/1078-0432.CCR-06-1136.
12. Lonning PE. Aromatase inhibitors in breast cancer. *Endocr Relat Cancer*. 2004;11(2):179–189. http://dx.doi.org/10.1677/erc.0.0110179.
13. Allred DC, Mohsin SK, Fuqua SA. Histological and biological evolution of human premalignant breast disease. *Endocr Relat Cancer*. 2001;8(1):47–61.
14. Schiff R, Osborne CK. Endocrinology and hormone therapy in breast cancer: new insight into estrogen receptor-alpha function and its implication for endocrine therapy resistance in breast cancer. *Breast Cancer Res*. 2005;7(5):205–211. http://dx.doi.org/10.1186/bcr1287.
15. Schiff R, Massarweh SA, Shou J, et al. Advanced concepts in estrogen receptor biology and breast cancer endocrine resistance: implicated role of growth factor signaling and estrogen receptor coregulators. *Cancer Chemother Pharmacol*. 2005;56(S1):10–20. http://dx.doi.org/10.1007/s00280-005-0108-2.
16. Shao W, Brown M. Advances in estrogen receptor biology: prospects for improvements in targeted breast cancer therapy. *Breast Cancer Res*. 2004;6(1):39–52. http://dx.doi.org/10.1186/bcr742.
17. Hall JM, Couse JF, Korach KS. The multifaceted mechanisms of estradiol and estrogen receptor signaling. *J Biol Chem*. 2001;276(40):36869–36872. http://dx.doi.org/10.1074/jbc.R100029200.
18. McKenna NJ, Lanz RB, O'Malley BW. Nuclear receptor coregulators: cellular and molecular biology. *Endocr Rev*. 1999;20(3):321–344. http://dx.doi.org/10.1210/edrv.20.3.0366.
19. Bautista S, Valles H, Walker RL, Anzick S. In breast cancer, amplification of the steroid receptor coactivator gene AIB1 is correlated with estrogen and progesterone receptor positivity. *Clin Cancer Res*. 1998;4:2925–2929.
20. Qin L, Liao L, Redmond A, et al. The AIB1 oncogene promotes breast cancer metastasis by activation of PEA3-mediated matrix metalloproteinase 2 (MMP2) and MMP9 expression. *Mol Cell Biol*. 2008;28(19):5937–5950. http://dx.doi.org/10.1128/MCB.00579-08.
21. Osborne CK, Schiff R. Mechanisms of endocrine resistance in breast cancer. *Annu Rev Med*. 2011;62(1):233–247. http://dx.doi.org/10.1146/annurev-med-070909-182917.
22. Liu Y, Ludes-Meyers J, Zhang Y, et al. Inhibition of AP-1 transcription factor causes blockade of multiple signal transduction pathways and inhibits breast cancer growth. *Oncogene*. 2002;21(50):7680–7689. http://dx.doi.org/10.1038/sj.onc.1205883.

23. Levin ER. Bidirectional signaling between the estrogen receptor and the epidermal growth factor receptor. *Mol Endocrinol.* 2003;17(3):309–317. http://dx.doi.org/10.1210/me.2002-0368.
24. Filardo EJ. Epidermal growth factor receptor (EGFR) transactivation by estrogen via the G-protein-coupled receptor, GPR30: a novel signaling pathway with potential significance for breast cancer. *J Steroid Biochem Mol Biol.* 2002;80(2):231–238.
25. Osborne CK, Schiff R, Fuqua S, Shou J. Estrogen receptor: current understanding of its activation and modulation. *Clin Cancer Res.* 2001;7:4338s–4342s.
26. Yamada KM, Even-Ram S. Integrin regulation of growth factor receptors. *Nature Cell Biol.* 2002;4:E75–E76.
27. Zhang QX, Borg A, Wolf DM, Oesterreich S, Fuqua SA. An estrogen receptor mutant with strong hormone-independent activity from a metastatic breast cancer. *Cancer Res.* 1997;57(7):1244–1249.
28. Wang P, Bahreini A, Gyanchandani R, et al. Sensitive detection of mono- and polyclonal esr1 mutations in primary tumors, metastatic lesions, and cell-free DNA of breast cancer patients. *Clin Cancer Res.* 2016;22(5):1130–1137. http://dx.doi.org/10.1158/1078-0432.CCR-15-1534.
29. Jeselsohn R, Yelensky R, Buchwalter G, et al. Emergence of constitutively active estrogen receptor-α mutations in pretreated advanced estrogen receptor-positive breast cancer. *Clin Cancer Res.* 2014;20(7):1757–1767. http://dx.doi.org/10.1158/1078-0432.CCR-13-2332.
30. Lauring J, Wolff AC. Evolving role of the estrogen receptor as a predictive biomarker: ESR1 mutational status and endocrine resistance in breast cancer. *J Clin Oncol.* 2016;34:2950–2952. http://dx.doi.org/10.1200/JCO.2016.68.4720.
31. Toy W, Shen Y, Won H, et al. ESR1 ligand-binding domain mutations in hormone-resistant breast cancer. *Nat Genet.* 2013;45(12):1439–1445. http://dx.doi.org/10.1038/ng.2822.
32. Robinson DR, Wu Y-M, Vats P, et al. Activating ESR1 mutations in hormone-resistant metastatic breast cancer. *Nat Genet.* 2013;45(12):1446–1451. http://dx.doi.org/10.1038/ng.2823.
33. Fribbens C, OLeary B, Kilburn L, et al. Plasma ESR1 mutations and the treatment of estrogen receptor-positive advanced breast cancer. *J Clin Oncol.* 2016;34:2961–2968. http://dx.doi.org/10.1200/JCO.2016.67.3061.
34. Margueron R, Duong V, Castet A, Cavaillès V. Histone deacetylase inhibition and estrogen signalling in human breast cancer cells. *Biochem Pharmacol.* 2004;68(6):1239–1246. http://dx.doi.org/10.1016/j.bcp.2004.04.031.
35. Pfister S, Rea S, Taipale M, et al. The histone acetyltransferase hMOF is frequently downregulated in primary breast carcinoma and medulloblastoma and constitutes a biomarker for clinical outcome in medulloblastoma. *Int J Cancer.* 2007;122(6):1207–1213. http://dx.doi.org/10.1002/ijc.23283.
36. Pathiraja TN, Stearns V, Oesterreich S. Epigenetic regulation in estrogen receptor positive breast cancer—role in treatment response. *J Mammary Gland Biol Neoplasia.* 2010;15(1):35–47. http://dx.doi.org/10.1007/s10911-010-9166-0.
37. Munster PN, Thurn KT, Thomas S, et al. A phase II study of the histone deacetylase inhibitor vorinostat combined with tamoxifen for the treatment of patients with hormone therapy-resistant breast cancer. *Br J Cancer.* 2011;104(12):1828–1835. http://dx.doi.org/10.1038/bjc.2011.156.
38. Ma C. Improving response to hormone therapy in breast cancer: new targets, new therapeutic options. *Am Soc Clin Oncol Educ Book.* 2016;35:e40–e54.
39. Shou J, Massarweh S, Osborne CK, et al. Mechanisms of tamoxifen resistance: increased estrogen receptor-HER2/neu cross-talk in ER/HER2-positive breast cancer. *J Natl Cancer Inst.* 2004;96(12):926–935. http://dx.doi.org/10.1093/jnci/djh166.

40. Ellis MJ, Coop A, Singh B, et al. Letrozole is more effective neoadjuvant endocrine therapy than tamoxifen for ErbB-1- and/or ErbB-2-positive, estrogen receptor-positive primary breast cancer: evidence from a phase III randomized trial. *J Clin Oncol.* 2001;19(18):3808–3816.
41. Farabaugh SM, Boone DN, Lee AV. Role of IGF1R in breast cancer subtypes, stemness, and lineage differentiation. *Front Endocrinol.* 2015;6 (article number 59). http://dx.doi.org/10.3389/fendo.2015.00059.
42. Yerushalmi R, Gelmon KA, Leung S, et al. Insulin-like growth factor receptor (IGF-1R) in breast cancer subtypes. *Breast Cancer Res Treat.* 2011;132(1):131–142. http://dx.doi.org/10.1007/s10549-011-1529-8.
43. Maor S, Mayer D, Yarden RI, et al. Estrogen receptor regulates insulin-like growth factor-I receptor gene expression in breast tumor cells: involvement of transcription factor Sp1. *J Endocrinol.* 2006;191(3):605–612. http://dx.doi.org/10.1677/joe.1.07016.
44. Rugo HS, Keck S. Reversing hormone resistance: have we found the golden key? *J Clin Oncol.* 2012;30(22):2707–2709. http://dx.doi.org/10.1200/JCO.2012.42.1271.
45. Fox EM, Miller TW, Balko JM, et al. A kinome-wide screen identifies the insulin/IGF-I receptor pathway as a mechanism of escape from hormone dependence in breast cancer. *Cancer Res.* 2011;71(21):6773–6784. http://dx.doi.org/10.1158/0008-5472.CAN-11-1295.
46. Knowlden JM, Hutcheson IR, Barrow D, Gee JMW, Nicholson RI. Insulin-like growth factor-I receptor signaling in tamoxifen-resistant breast cancer: a supporting role to the epidermal growth factor receptor. *Endocrinology.* 2005;146(11):4609–4618. http://dx.doi.org/10.1210/en.2005-0247.
47. Robertson JF, Ferrero JM, Bourgeois H, et al. Ganitumab with either exemestane or fulvestrant for postmenopausal women with advanced, hormone- receptor-positive breast cancer: a randomised, controlled, double-blind, phase 2 trial. *Lancet Oncol.* 2013;14(3):228–235. http://dx.doi.org/10.1016/S1470-2045(13)70026-3.
48. Turner N, Grose R. Fibroblast growth factor signalling: from development to cancer. *Nat Rev Cancer.* 2010;10(2):116–129. http://dx.doi.org/10.1038/nrc2780.
49. Elbauomy Elsheikh S, Green AR, Lambros MB, et al. FGFR1 amplification in breast carcinomas: a chromogenic in situ hybridisation analysis. *Breast Cancer Res.* 2007;9(2):R23. http://dx.doi.org/10.1186/bcr1665.
50. Turner N, Pearson A, Sharpe R, et al. FGFR1 amplification drives endocrine therapy resistance and is a therapeutic target in breast cancer. *Cancer Res.* 2010;70(5):2085–2094. http://dx.doi.org/10.1158/0008-5472.CAN-09-3746.
51. Andre F, Bachelot T, Campone M, et al. Targeting FGFR with dovitinib (TKI258): preclinical and clinical data in breast cancer. *Clin Cancer Res.* 2013;19(13):3693–3702. http://dx.doi.org/10.1158/1078-0432.CCR-13-0190.
52. Foekens JA, Peters HA, Grebenchtchikov N, et al. High tumor levels of vascular endothelial growth factor predict poor response to systemic therapy in advanced breast cancer. *Cancer Res.* 2001;61(14):5407–5414.
53. Manders P, Beex LVAM, Tjan-Heijnen VCG, Span PN, Sweep CFGJ. Vascular endothelial growth factor is associated with the efficacy of endocrine therapy in patients with advanced breast carcinoma. *Cancer.* 2003;98(10):2125–2132. http://dx.doi.org/10.1002/cncr.11764.
54. Dickler MN, Barry WT, Cirrincione CT, et al. Phase III trial evaluating letrozole as first-line endocrine therapy with or without bevacizumab for the treatment of postmenopausal women with hormone receptor-positive advanced-stage breast cancer: CALGB 40503 (Alliance). *J Clin Oncol.* 2016;34(22):2602–2609. http://dx.doi.org/10.1200/JCO.2015.66.1595.
55. Martin M, Loibl S, Minckwitz von G, et al. Phase III trial evaluating the addition of bevacizumab to endocrine therapy as first-line treatment for advanced breast cancer:

the letrozole/fulvestrant and avastin (LEA) study. *J Clin Oncol*. 2015;33(9): 1045–1052. http://dx.doi.org/10.1200/JCO.2014.57.2388.
56. Cantley LC. The phosphoinositide 3-kinase pathway. *Science*. 2002;296:1655–1657.
57. Stephens L, Anderson K, Stokoe D, et al. Protein kinase B kinases that mediate phosphatidylinositol 3,4,5-trisphosphate-dependent activation of protein kinase B. *Science*. 1998;279(5351):710–714.
58. Klempner SJ, Myers AP, Cantley LC. What a tangled web we weave: emerging resistance mechanisms to inhibition of the phosphoinositide 3-kinase pathway. *Cancer Discov*. 2013;3(12):1345–1354. http://dx.doi.org/10.1158/2159-8290.CD-13-0063.
59. Morgensztern D, McLeod HL. PI3K/Akt/mTOR pathway as a target for cancer therapy. *Anticancer Drugs*. 2005;16(8):797–803.
60. Chia S, Gandhi S, Joy AA, et al. Novel agents and associated toxicities of inhibitors of the pi3k/Akt/mtor pathway for the treatment of breast cancer. *Curr Oncol*. 2015;22(1):33. http://dx.doi.org/10.3747/co.22.2393.
61. Creighton CJ, Fu X, Hennessy BT, et al. Proteomic and transcriptomic profiling reveals a link between the PI3K pathway and lower estrogen-receptor (ER) levels and activity in ER+ breast cancer. *Breast Cancer Res*. 2010;12(3):R40. http://dx.doi.org/10.1186/bcr2594.
62. Miller TW, Hennessy BT, González-Angulo AM, et al. Hyperactivation of phosphatidylinositol-3 kinase promotes escape from hormone dependence in estrogen receptor-positive human breast cancer. *J Clin Invest*. 2010;120(7):2406–2413. http://dx.doi.org/10.1172/JCI41680DS1.
63. deGraffenried LA, Friedrichs WE, Russell DH, et al. Inhibition of mTOR activity restores tamoxifen response in breast cancer cells with aberrant Akt activity. *Clin Cancer Res*. 2004;10(23):8059–8067. http://dx.doi.org/10.1158/1078-0432.CCR-04-0035.
64. Bader AG, Kang S, Vogt PK. Cancer-specific mutations in PIK3CA are oncogenic in vivo. *Proc Natl Acad Sci USA*. 2006;103(5):1475–1479. http://dx.doi.org/10.1073/pnas.0510857103.
65. Saal LH, Holm K, Maurer M, et al. PIK3CA mutations correlate with hormone receptors, node metastasis, and ERBB2, and are mutually exclusive with PTEN loss in human breast carcinoma. *Cancer Res*. 2005;65(7):2554–2559. http://dx.doi.org/10.1158/0008-5472-CAN-04-3913.
66. Baselga J. Targeting the phosphoinositide-3 (PI3) kinase pathway in breast cancer. *Oncologist*. 2011;16(suppl 1):12–19. http://dx.doi.org/10.1634/theoncologist.2011-S1-12.
67. Sanchez CG, Ma CX, Crowder RJ, et al. Preclinical modeling of combined phosphatidylinositol-3-kinase inhibition with endocrine therapy for estrogen receptor-positive breast cancer. *Breast Cancer Res*. 2011;13(2):R21. http://dx.doi.org/10.1186/bcr2833.
68. Loi S, Haibe-Kains B, Majjaj S. PIK3CA mutations associated with gene signature of low mTORC1 signaling and better outcomes in estrogen receptor-positive breast cancer. *Proc Natl Acad Sci USA*. 2010;107:10208–10213. http://dx.doi.org/10.1073/pnas.0907011107/-/DCSupplemental.
69. Koboldt DC, Fulton RS, McLellan MD, et al. Comprehensive molecular portraits of human breast tumours. *Nature*. 2012;490(7418):61–70. http://dx.doi.org/10.1038/nature11412.
70. Li J, DeFea K, Roth RA. Modulation of insulin receptor substrate-1 tyrosine phosphorylation by an Akt/phosphatidylinositol 3-kinase pathway. *J Biol Chem*. 1999;274(14):9351–9356.
71. Stemke-Hale K, Gonzalez-Angulo AM, Lluch A, et al. An integrative genomic and proteomic analysis of PIK3CA, PTEN, and AKT mutations in breast cancer. *Cancer Res*. 2008;68(15):6084–6091. http://dx.doi.org/10.1158/0008-5472.CAN-07-6854.

72. Tanic N, Milovanovic Z, Tanic N, et al. The impact of PTEN tumor suppressor gene on acquiring resistance to tamoxifen treatment in breast cancer patients. *Cancer Biol Ther* 2014;13(12):1165–1174. http://dx.doi.org/10.4161/cbt.21346.
73. Pandolfi PP. Breast cancer—loss of PTEN predicts resistance to treatment. *N Engl J Med.* 2004;351(22):2337–2338. http://dx.doi.org/10.1056/NEJMcibr043143.
74. Beeram M, Tan Q-T, Tekmal R, Russell D, Middleton A, deGraffenried L. Akt-induced endocrine therapy resistance is reversed by inhibition of mTOR signaling. *Ann Oncol.* 2007;18(8):1323–1328. http://dx.doi.org/10.1093/annonc/mdm170.
75. Baselga J, Semiglazov V, van Dam P, et al. Phase II randomized study of neoadjuvant everolimus plus letrozole compared with placebo plus letrozole in patients with estrogen receptor-positive breast cancer. *J Clin Oncol.* 2009;27(16):2630–2637. http://dx.doi.org/10.1200/JCO.2008.18.8391.
76. Baselga J, Campone M, Piccart M, et al. Everolimus in postmenopausal hormone-receptor-positive advanced breast cancer. *N Engl J Med.* 2012;366(6):520–529. http://dx.doi.org/10.1056/NEJMoa1109653.
77. Bachelot T, Bourgier C, Cropet C, et al. Randomized phase II trial of everolimus in combination with tamoxifen in patients with hormone receptor-positive, human epidermal growth factor receptor 2-negative metastatic breast cancer with prior exposure to aromatase inhibitors: a GINECO study. *J Clin Oncol.* 2012;30(22):2718–2724. http://dx.doi.org/10.1200/JCO.2011.39.0708.
78. O'Reilly KE. mTOR inhibition induces upstream receptor tyrosine kinase signaling and activates Akt. *Cancer Res.* 2006;66(3):1500–1508. http://dx.doi.org/10.1158/0008-5472.CAN-05-2925.
79. Fu X, Osborne CK, Schiff R. Biology and therapeutic potential of PI3K signaling inER. *Breast.* 2013;22(S2):S12–S18. http://dx.doi.org/10.1016/j.breast.2013.08.001.
80. Wan X, Harkavy B, Shen N, Grohar P, Helman LJ. Rapamycin induces feedback activation of Akt signaling through an IGF-1R-dependent mechanism. *Oncogene.* 2006;26(13):1932–1940. http://dx.doi.org/10.1038/sj.onc.1209990.
81. Meyerson M, Enders GH, Wu CL, et al. A family of human cdc2-related protein kinases. *EMBO J.* 1992;11(8):2909–2917.
82. Shapiro GI. Cyclin-dependent kinase pathways as targets for cancer treatment. *J Clin Oncol.* 2006;24(11):1770–1783. http://dx.doi.org/10.1200/JCO.2005.03.7689.
83. Malumbres M, Barbacid M. Cell cycle, CDKs and cancer: a changing paradigm. *Nat Rev Cancer.* 2009;9(3):153–166. http://dx.doi.org/10.1038/nrc2602.
84. Sherr CJ. Gl phase progression: cycling on cue. *Cell.* 1994;79:551–555.
85. Lundberg AS, Weinberg RA. Functional inactivation of the retinoblastoma protein requires sequential modification by at least two distinct cyclin-cdk complexes. *Mol Cell Biol.* 1998;18(2):753–761.
86. Verona R, Moberg K, Estes S, Starz M. E2F activity is regulated by cell cycle-dependent changes in subcellular localization. *Mol Cell Biol.* 1997;17:7268–7282.
87. Finn RS, Aleshin A, Slamon DJ. Targeting the cyclin-dependent kinases (CDK) 4/6 in estrogen receptor-positive breast cancers. *Breast Cancer Res.* 2016;18:17. http://dx.doi.org/10.1186/s13058-015-0661-5.
88. Sherr CJ, Roberts JM. CDK inhibitors: positive and negative regulators of G. *Genes Dev.* 1999;13(12):1501–1512.
89. Choi YJ, Anders L. Signaling through cyclin D-dependent kinases. *Oncogene.* 2013;33(15):1890–1903. http://dx.doi.org/10.1038/onc.2013.137.
90. Arnold A. Cyclin D1 in breast cancer pathogenesis. *J Clin Oncol.* 2005;23(18):4215–4224. http://dx.doi.org/10.1200/JCO.2005.05.064.
91. An H-X, Beckmann MW, Reifenberger G, Bender HG, Niederacher D. Gene amplification and overexpression of CDK4 in sporadic breast carcinomas is associated with

high tumor cell proliferation. *Am J Pathol.* 1999;154(1):113–118. http://dx.doi.org/10.1016/S0002-9440(10)65257-1.

92. Foster JS, Henley DC, Bukovsky A, Seth P, Wimalasena J. Multifaceted regulation of cell cycle progression by estrogen: regulation of Cdk inhibitors and Cdc25A independent of cyclin D1-Cdk4 function. *Mol Cell Biol.* 2001;21(3):794–810. http://dx.doi.org/10.1128/MCB.21.3.794-810.2001.
93. Zwijsen RM, Wientjens E, Klompmaker R, van der Sman J, Bernards R, Michalides RJ. CDK-independent activation of estrogen receptor by cyclin D1. *Cell.* 1997;88(3):405–415.
94. Vasan N, Yelensky R, Wang K, et al. A targeted next-generation sequencing assay detects a high frequency of therapeutically targetable alterations in primary and metastatic breast cancers: implications for clinical practice. *Oncologist.* 2014;19(5):453–458. http://dx.doi.org/10.1634/theoncologist.2013-0377.
95. Ferrer J-L, Dupuy J, Borel F, Jacquamet L, Noel JP, Dulic V. Structural basis for the modulation of CDK-dependent/independent activity of cyclin D1. *Cell Cycle.* 2006;5(23):2760–2768. http://dx.doi.org/10.4161/cc.5.23.3506.
96. Finn RS, Dering J, Conklin D, et al. PD 0332991, a selective cyclin D kinase 4/6 inhibitor, preferentially inhibits proliferation of luminal estrogen receptor-positive human breast cancer cell lines in vitro. *Breast Cancer Res.* 2009;11(5):R77. http://dx.doi.org/10.1186/bcr2419.
97. Finn RS, Crown JP, Lang I, et al. The cyclin-dependent kinase 4/6 inhibitor palbociclib in combination with letrozole versus letrozole alone as first-line treatment of oestrogen receptor-positive, HER2-negative, advanced breast cancer (PALOMA-1/TRIO-18): a randomised phase 2 study. *Lancet Oncol.* 2014;16(1):25–35. http://dx.doi.org/10.1016/S1470-2045(14)71159-3.
98. Verma S, Bartlett CH, Schnell P, et al. Palbociclib in combination with fulvestrant in women with hormone receptor-positive/HER2-negative advanced metastatic breast cancer: detailed safety analysis from a multicenter, randomized, placebo-controlled, phase III study (PALOMA-3). *Oncologist.* 2016;21:1–11. http://dx.doi.org/10.1634/theoncologist.2016-0097.
99. Vidula N, Rugo HS. Cyclin-dependent kinase 4/6 inhibitors for the treatment of breast cancer: a review of preclinical and clinical data. *Clin Breast Cancer.* 2016;16(1):8–17. http://dx.doi.org/10.1016/j.clbc.2015.07.005.
100. Dean JL, McClendon AK, Hickey TE, et al. Therapeutic response to CDK4/6 inhibition in breast cancer defined by ex vivo analyses of human tumors. *Cell Cycle.* 2014;11(14):2756–2761. http://dx.doi.org/10.4161/cc.21195.
101. Wang Y, Dean JL, Millar EKA, et al. Cyclin D1b is aberrantly regulated in response to therapeutic challenge and promotes resistance to estrogen antagonists. *Cancer Res.* 2008;68(14):5628–5638. http://dx.doi.org/10.1158/0008-5472.CAN-07-3170.
102. Bissell MJ, Radisky DC, Rizki A, Weaver VM, Petersen OW. The organizing principle: microenvironmental influences in the normal and malignant breast. *Differentiation.* 2002;70(9–10):537–546. http://dx.doi.org/10.1046/j.1432-0436.2002.700907.x.
103. Helleman J, Jansen MPHM, Ruigrok-Ritstier K, et al. Association of an extracellular matrix gene cluster with breast cancer prognosis and endocrine therapy response. *Clin Cancer Res.* 2008;14(17):5555–5564. http://dx.doi.org/10.1158/1078-0432.CCR-08-0555.
104. Mahmoud SMA, Paish EC, Powe DG, et al. Tumor-infiltrating CD8+ lymphocytes predict clinical outcome in breast cancer. *J Clin Oncol.* 2011;29(15):1949–1955. http://dx.doi.org/10.1200/JCO.2010.30.5037.
105. Dunbier AK, Ghazoui Z, Anderson H, et al. Molecular profiling of aromatase inhibitor-treated postmenopausal breast tumors identifies immune-related correlates

of resistance. *Clin Cancer Res*. 2013;19(10):2775–2786. http://dx.doi.org/10.1158/1078-0432.CCR-12-1000.

106. Giampaolo B, Pusztai L, Qi Y. A dendritic metagene that predicts prognosis and endocrine resistance in breast cancer. In: *2012 ASCO Annual Meeting*; 2012.
107. Ribas A. Tumor immunotherapy directed at PD-1. *N Engl J Med*. 2012;366(26):2517–2519. http://dx.doi.org/10.1056/NEJMe1205943.
108. Pardoll DM. The blockade of immune checkpoints in cancer immunotherapy. *Nat Rev Cancer*. 2012;12(4):252–264. http://dx.doi.org/10.1038/nrc3239.
109. Sabatier R, Finetti P, Mamessier E, et al. Prognostic and predictive value of PDL1 expression in breast cancer. *Oncotarget*. 2015;6(7):5449–5464. http://dx.doi.org/10.18632/oncotarget.3216.
110. Ghebeh H, Mohammed S, Al-Omair A, et al. The B7-H1 (PD-L1) T lymphocyte-inhibitory molecule is expressed in breast cancer patients with infiltrating ductal carcinoma: correlation with important high-risk prognostic factors. *Neoplasia*. 2006;8(3):190–198. http://dx.doi.org/10.1593/neo.05733.
111. Muenst S, Schaerli AR, Gao F, et al. Expression of programmed death ligand 1 (PD-L1) is associated with poor prognosis in human breast cancer. *Breast Cancer Res Treat*. 2014;146(1):15–24. http://dx.doi.org/10.1007/s10549-014-2988-5.
112. Rugo HS, Delord JP, Im SA, Ott PA, Piha-Paul SA. Abstract S5-07: Preliminary efficacy and safety of pembrolizumab (MK-3475) in patients with PD-L1–positive, estrogen receptor-positive (ER+)/HER2-negative advanced breast cancer enrolled in KEYNOTE-028. *Cancer Res*. 2016;76(suppl 4).
113. Signore M, Ricci-Vitiani L, De Maria R. Targeting apoptosis pathways in cancer stem cells. *Cancer Lett*. 2013;332(2):374–382. http://dx.doi.org/10.1016/j.canlet.2011.01.013.
114. Yip KW, Reed JC. Bcl-2 family proteins and cancer. *Oncogene*. 2008;27(50):6398–6406. http://dx.doi.org/10.1038/onc.2008.307.
115. Schwartz-Roberts JL, Shajahan AN, Cook KL, Warri A, Abu-Asab M, Clarke R. GX15-070 (obatoclax) induces apoptosis and inhibits cathepsin D- and L-mediated autophagosomal lysis in antiestrogen-resistant breast cancer cells. *Mol Cancer Ther*. 2013;12(4):448–459. http://dx.doi.org/10.1158/1535-7163.MCT-12-0617.
116. Crawford AC, Riggins RB, Shajahan AN, Zwart A, Clarke R. Co-inhibition of BCL-W and BCL2 restores antiestrogen sensitivity through becn1 and promotes an autophagy-associated necrosis. *PLoS One*. 2010;5(1):e8604–e8611. http://dx.doi.org/10.1371/journal.pone.0008604.
117. deGraffenried LA. NF-B inhibition markedly enhances sensitivity of resistant breast cancer tumor cells to tamoxifen. *Ann Oncol*. 2004;15(6):885–890. http://dx.doi.org/10.1093/annonc/mdh232.
118. Zhou Y. The NF B pathway and endocrine-resistant breast cancer. *Endocr Relat Cancer*. 2005;12(suppl 1):S37–S46. http://dx.doi.org/10.1677/erc.1.00977.

CHAPTER THIRTEEN

Molecular and Cellular Changes in Breast Cancer and New Roles of lncRNAs in Breast Cancer Initiation and Progression

M. Kumar, R.S. DeVaux, J.I. Herschkowitz[1]
Cancer Research Center, University at Albany, Rensselaer, NY, United States
[1]Corresponding author: e-mail address: jherschkowitz@albany.edu

Contents

Abstract

Breast cancer is not just one disease but many variations on a theme, comprising a variety of molecular subtypes with distinct etiologies, cellular origins, treatment strategies, and prognoses. Like mRNAs and microRNAs (miRNAs), long noncoding RNAs (lncRNAs) differ dramatically in expression across breast cancer subtypes and can be used for classification. While there has been considerable emphasis on miRNAs, our knowledge is still in its infancy about the role of lncRNAs that comprise the majority of the mammalian transcriptome. In this chapter, we will review the critical functions that lncRNAs play in

Progress in Molecular Biology and Translational Science, Volume 144
ISSN 1877-1173
http://dx.doi.org/10.1016/bs.pmbts.2016.09.011

breast cancer development and metastatic progression. We will conclude with a discussion of current and future approaches for RNA-targeted therapeutic intervention.

1. INTRODUCTION

With the advent of genome sequencing technologies coupled with an increased ability to catalog the transcriptome with greater depth, it has become apparent that the majority of the mammalian genome is transcribed. However, less than 2% of the genome is protein coding leaving the rest of the transcriptome as noncoding. These RNAs are capable of carrying out biological functions beyond a classical role as intermediate carriers of genetic information. There has initially been considerable emphasis on the study of small noncoding RNAs referred to as microRNAs (miRNAs); however, our knowledge about the role of long noncoding RNAs (lncRNAs) that comprise the vast majority of the mammalian transcriptome is still in its infancy. Recently, a comprehensive analysis on 7256 RNA-sequencing libraries comprising the tumor, normal, and cell line data showed that 68% of the total transcribed genes are represented by lncRNAs.[1] While the ultimate function of most lncRNAs is to modulate gene expression, this can be accomplished through a variety of biological functions. This repertiore includes regulation of mRNA processing, stability, and protein synthesis, acting as a competitive RNA target or sponge for miRNAs, interactions with transcription factors, and various methods of epigenetic regulation. Through these various functions, lncRNAs are being discovered to hold critical functions that support all of the hallmarks of cancer.[2] Breast cancer, the second leading cause of cancer-related mortality in women in the United States, is a heterogeneous disease classified as a variety of subtypes with distinct molecular changes, etiologies, cellular origins, treatment strategies, and prognoses. As had been illustrated for mRNAs and miRNAs, lncRNAs differ dramatically in expression across these breast cancer subtypes. lncRNAs have a great potential to serve as useful biomarkers and, since they have been shown to have critical functions, represent new potential targets for breast cancer therapy.

2. IDENTIFYING lncRNA SIGNALS IN THE CANCER TRANSCRIPTOME

Previously discarded as sporadic cases of transcription from undesired locations in the genome, lncRNAs are now considered fundamental molecules contributing to basic processes. Classically identified lncRNAs were found through screening cDNA libraries for genes that would functionally influence a phenotype of interest. While these functional screens were assumed at the time to identify coding transcripts, critical noncoding transcripts such as XIST[3] and H19[4] were also identified. Finding lncRNAs that carried important functional roles invigorated the scientific community to annotate the noncoding RNAs that comprise the vast majority of the mammalian transcriptome. Brute force sequencing approaches followed by filtering out cDNAs with open reading frames began to annotate the noncoding compartment.[5–7] Over the past few years' advancements in sequencing technologies has enriched the catalog of lncRNAs. Deep sequencing, DNA tiling arrays, and next-generation RNA-sequencing methods have now enabled us to have better transcriptome-wide views regarding expression of lncRNAs, the various cellular pathways they are potentially associated with, and their impact on gene regulation. Applying these techniques to profile human cancers, it has become clear that lncRNAs become dysregulated in multiple cancer types at the epigenetic, genomic, and transcriptional level.[6] Intriguingly, although some lncRNAs may be altered in multiple tumor types, their dysregulation appears largely cancer-type specific.[6] Diermeier *et al.*[7a] conducted an intensive study to delineate breast tumor-specific lncRNAs form its surrounding tissue and have identified several candidate lncRNAs that could be used as biomarkers for different breast cancer subtypes. A number of next-generation sequencing studies aim to understand the lncRNA contribution to breast cancer initiation, progression, and metastasis (Table 1).

3. FUNCTIONS OF lncRNAs IN BREAST CANCER

Classically, only proteins were considered as drivers and effectors for the multistep process that leads to cellular transformation and thence development of cancers. Some of these proteins were classified as oncogenic, contributing to cancer development, while some inhibit tumor formation and are known as tumor suppressors. Interestingly, several of the classical oncogenic proteins have been found to function through lncRNAs to drive cellular transformation. The classical oncogene c-myc activates HOTAIR, an lncRNA whose overexpression drives breast cancer development and metastasis and can serve as an independent prognostic marker.[18–20] Likewise, lncRNAs can function as both oncogenes and tumor suppressors and the dysregulation of even extremely low expressing lncRNAs can lead to developing malignancy.[15] lncRNAs also function as effector molecules downstream to protein-coding oncogenes and help to maintain or advance malignancy. Recently, a lncRNA LINC00520 was reported to act downstream to the oncogenes src, PI3K, and STAT3 pathway. Although the exact

Table 1 Next-Generation Sequencing Studies for lncRNAs in Breast Cancer

Study Purpose	References
Identification of novel lncRNAs and their landscape in tumor and normal tissues	1
Expression profile of HER2-enriched breast cancer	8
Genomic characterization of long noncoding RNAs	9
Finding unbiased underrepresented noncoding transcripts	10
Prediction of lncRNA–RNA interaction	11
Prediction of noncoding RNA interacting RNAs and chromatin regions	12
Prediction of chromatin-associated lncRNAs	13
Prediction of prognostic lncRNA signature in breast cancer	14
Prediction of lncRNAs involved in cell cycle control and proliferation in breast cancer	15
Estrogen-regulated lncRNAs in breast cancer	16
lncRNA–miRNA interaction prediction based on lncRNA-competing triplets	17

molecular mechanism opted by this lncRNA has not been elucidated yet, through both gain-of-function and loss-of-function studies, it was found to be associated with cancer cell invasion and migration in breast cancer.[21] Involvement of lncRNAs in oncogenesis was further corroborated by a recent study by Niknafs *et al.*[21a] the many lncRNAs act in concert with estrogen receptor (ER) and drive tumor progression. Particularly, lncRNA *DSCAM-AS1* was found to be highly upregulated in ER-positive breast cancer and bestow invasiveness. At molecular levels, lncRNAs adopt several different mechanisms to promote or suppress tumor growth. They affect transcription, RNA splicing, mRNA stability, translation, chromatin remodeling, and many other critical cellular processes through canonical or non-canonical routes.

3.1 Regulation of mRNA Splicing by lncRNAs

lncRNAs are a diverse class of RNA molecules that can regulate other transcripts through various mechanisms. They regulate processing of mRNAs at several stages including splicing, affecting mRNA stability, and can control the translation of an mRNA through various approaches. Interestingly, the processing event of some of lncRNAs themselves leads to products that in turn help process other RNAs.

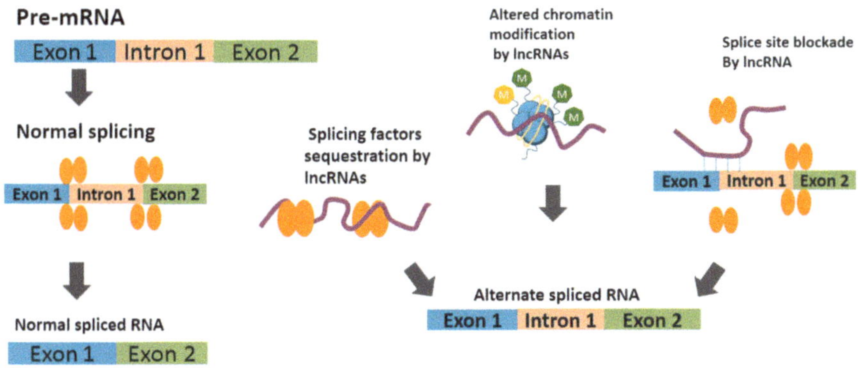

Fig. 1 lncRNAs affect different aspects of mRNA splicing. Architecture of a nascent primary transcript and alternate splicing of primary transcript orchestrated through lncRNAs.

The primary eukaryotic transcripts, when freshly transcribed, often are composed of exons interspersed with nonessential introns that must be removed to form a complete mature mRNA (Fig. 1). This splicing of pre-mRNAs involves an interplay between multiprotein complexes for effective recognition and excising of introns (Fig. 1). lncRNAs regulate the splicing of many RNAs through different mechanisms (Fig. 1). One example is the lncRNA MALAT1 which serves as a biomarker for several aggressive cancers including breast cancer. MALAT1 not only regulates the expression of alternative splicing-associated proteins but also interacts with splicing factors and modulates their distribution into nuclear speckles[22] (Fig. 1). Nuclear speckles are nuclear domains where pre-mRNA splicing factors are assembled, modified, and stored before being recruited to active transcription sites. Thus, dysregulation of MALAT1 impacts splicing of RNAs at the global level. Knockdown of MALAT1 in breast cancer cells attenuates cell invasion and proliferation and can induce apoptosis ascribing it as oncogenic lncRNA.[23]

An indirect mechanism by which lncRNAs interfere with the splicing of primary transcripts is through recruiting chromatin-modifying complexes. In both breast cancer and prostate cancer cell line models, it is reported that an lncRNA expressed in the antisense orientation from the FGFR2 locus recruits chromatin modifiers polycomb group (PcG) proteins and histone demethylase KDM2a. This conglomerate of lncRNA and chromatin modifiers precludes the binding of another repressive adapter splicing complex (Fig. 1). This results in epithelial-specific splicing of the growth factor FGFR2.[24] Similarly, an lncRNA expressed antisense to the mRNA of FAS interacts with the splicing factor 45 (SPF45) to regulate the splicing of FAS mRNA.[25]

Besides interacting directly with splicing factors, lncRNAs can also affect alternative splicing of RNAs by base pairing with a target transcript. Although the exact mechanism of regulation of splicing through this RNA–RNA duplex formation is not yet clear, the association may hide splicing signals precluding binding of splicing factors. A natural antisense transcript to Neuroblastoma myc (N-myc) binds to the first intron of N-myc and interferes with its splicing (Fig. 1).[26] Another lncRNA annotated as a natural antisense transcript prevents the splicing of an intron having an internal ribosome binding site and inhibits its splicing in Zeb2 transcript's 5′ UTR.[27]

3.2 Regulation of mRNA Stability by lncRNAs

The longevity of a protein-coding mRNA depends on several parameters including the sequence itself, splicing of the transcript, accessibility to other RNA stabilizing or degrading proteins, duplex formation and eliciting staufen1-mediated messenger RNA decay (SMD) or RNA interference (RNAi) response, etc. lncRNAs are now appreciated to be decisive players in this whole RNA stability gamut by both decreasing and increasing half-life of a transcript (Fig. 2A and B).

lncRNAs can affect the stability of a protein-coding mRNA by binding to the 3′ UTR region via Alu elements. Gong and Maquat have reported that many mRNAs have Alu elements in their 3′ UTR.[28] Based on the sequence homology, they base pair at different regions of the 3′ UTR short interspersed nuclear elements and form an RNA–RNA duplex. This RNA–RNA duplex recruits the staufen-1 (stau1) protein and leads to degradation of the target mRNA known as SMD. The lncRNAs facilitating this pathway are called

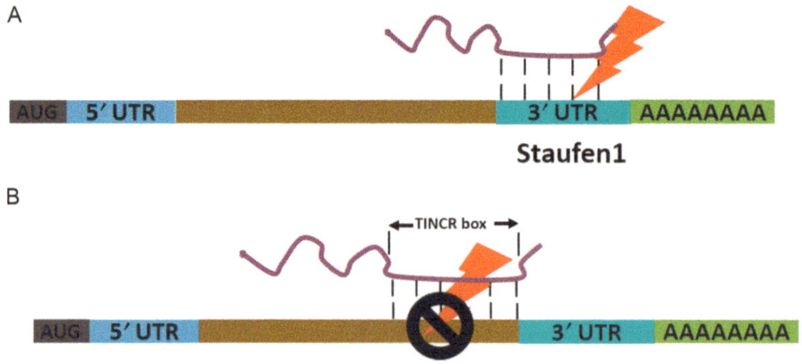

Fig. 2 Regulation of mRNA stability by lncRNA. (A) lncRNA-facilitated inhibition of staufen-1-mediated mRNA decay. (B) Stabilization of mRNA through binding of mRNA at the TINCR box of lncRNA TINCR.

half-STAU1-binding site RNAs (1/2-sbsRNAs) (Fig. 2A). lncRNA AF087999 was found to interact with the 3′ UTR of SERPINE1 mRNA and subsequently SERPINE1 mRNA was degraded via the SMD pathway.[28]

Beta-secretase 1 (BACE1) is a protein highly upregulated in Alzheimer's disease. It was shown that BACE1 is a target of miRNA miR-485-5p.[29] An antisense transcript of BACE1 masks the binding site for miR-485-5p in the sense transcript (BACE1 mRNA) and hence protects it from being cleaved by the RNA-induced silencing complex.

Kretz *et al.* have shown that a lncRNA terminal differentiation-induced noncoding RNA (TINCR) regulates the stability of many differentiation-related protein-coding mRNAs. TINCR binds to these target RNAs through a 25-nucleotide conserved sequence referred to as a TINCR box.[30] Some of the lncRNAs act via precluding binding of miRNAs to their target and increase the stability of the target mRNA (Fig. 2B).

lncRNAs also regulate the stability of mRNA indirectly by sequestering RNA-destabilizing proteins. Gadd7 is a lncRNA that binds to TDP-43, a stress-induced protein. Cdk6 mRNA is a direct target of TDP-43. By sequestering TDP-43, the lncRNA gadd7 increases the stability of cdk6.[31]

Besides regulating the processing of the protein coding mRNAs, lncRNAs can potentially regulate processing of other lncRNAs as well. The well-known lncRNA MALAT1 is devoid of polyadenylation and yet is as stable as the transcripts from other housekeeping genes. The 3′ end of MALAT1 is folded into a cloverleaf structure similar to tRNAs. The processing of the primary MALAT1 transcript is necessary for its stability. An antisense noncoding RNA at the MALAT1 locus, named TALAM1, interacts at the 3′ region of MALAT1 and helps in its processing. Further, it has been shown that both MALAT1 and TALAM1 follow the principles of mutualism. While benefiting from the presence of TALAM1, MALAT1 positively regulates expression of TALAM1 and also increases its stability.[32]

3.3 Long Noncoding RNA Acting as miRNA Sponges or Decoys

lncRNA species outnumber the mRNAs classically coding for expressing various proteins in the cell. Although many lncRNAs were found to have canonical functions including interacting with protein molecules, interacting with chromatin, binding to coding mRNAs at transcriptional and translational levels, and finding the direct targets of many lncRNAs remain elusive. With the discovery of competing endogenous RNAs (ceRNAs), the shadow over the function of the erstwhile errant lncRNAs started becoming clear. It has been hypothesized that ceRNAs might play a role in regulation of gene expression at larger scale. They have been proposed as "The Rosetta stone of a hidden RNA language".[33] This class of lncRNAs has been proposed to act as sponge sites for miRNAs. miRNAs are small RNA species and regulate the expression of various genes at transcriptional and translational levels. These miRNAs have binding sites in their target transcripts known as miRNA response elements (MREs) (Fig. 3A). Indeed, MREs have been found in several lncRNAs through bioinformatics algorithms and many of them were validated through wet lab experimental approaches. Some lncRNAs are multifunctional in nature and control the expression of several genes through more than one means. One of the most well-studied lncRNAs, HOTAIR, mitigates the expression of miR-331-3p in gastric cancer (Fig. 3B).[34] Interestingly, in triple-negative breast cancer, the MRE for miR-148-a was found to be present in HOTAIR and regulated expression of the lncRNA.[35] In another study Yuan *et al.* have shown that lncRNA-ATB acts as a sponge for the miR-200 family and increases the expression of Zeb1 and Zeb2.[36] The lncRNA GAS5 has been demonstrated as being a sponge to negate the gene regulatory role of miR-21 in triple breast cancer[37] (Fig. 3C). In triple-negative breast cancer, the lncRNA-ROR was reported to act as a sponge for miR-145.[38] lncRNA-ROR mitigates the effects of miR-145 and regulates the invasive phenotype in breast cancer by upregulating the expression of the small GTPase ADP-ribosylation factor 6 (Arf6).

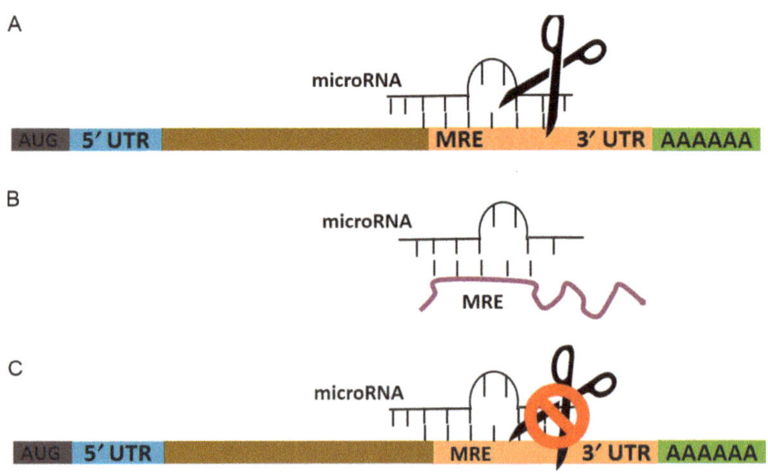

Fig. 3 lncRNAs act as microRNA sponge. (A) microRNA-mediated cleavage of mRNA. (B) Binding of microRNA at the MRE of lncRNA acts as a microRNA sponge. (C) Inhibition of microRNA-mediated mRNA cleavage due to lncRNA.

This tug of war between two distinct RNA species (lncRNA and mRNA) for binding to a common miRNA leaves the miRNA orphan without having a target to act upon. This sponging activity by lncRNAs leads to undeterred expression of the target gene. The information on the interaction between the three players, lncRNA, miRNA, and the interacting gene, could be an important resource to be harnessed for cancer prognosis. In fact, Wang *et al.* have demonstrated that this miRNA sponging mechanism has a global pattern in breast cancer.[17] The level of the association between the triplets of lncRNA–miRNA–coding gene interaction enriched for cancer type and they clustered for regulating a common cellular pathway.

3.4 Transcriptional Regulation by lncRNAs

lncRNAs adopt several mechanisms to regulate the expression of a particular gene. As described in the previous section, lncRNAs can act post-transcriptionally and snuggle into the miRNA-mediated regulation of gene expression pathways. However, increasing evidence now suggests the involvement of lncRNAs at the transcriptional level by acting as a decoy for conventional transcriptional factors. Similar to the function of transcription factors, lncRNAs can interact with the promoter region directly, base pair with the mRNA being transcribed or via interacting with other transcription factor proteins (Fig. 4A–C). Examples for this type of lncRNAs

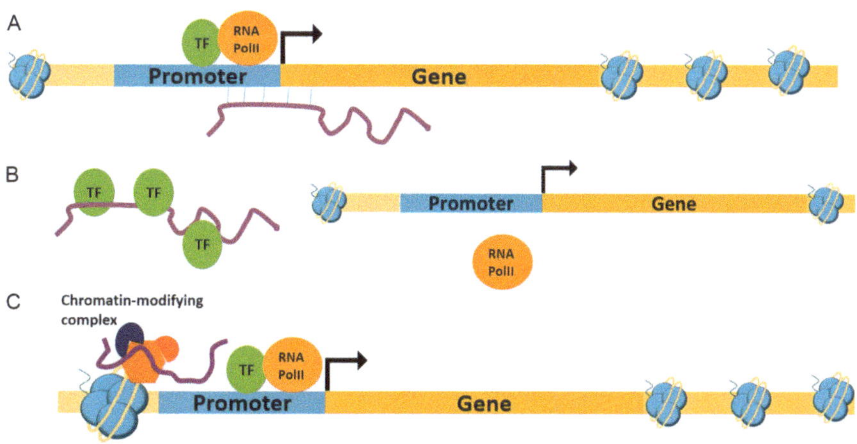

Fig. 4 Transcriptional control of mRNA expression by lncRNAs. (A) lncRNA forms a triplex at the promoter region and tethers RNA polymerase. (B) lncRNAs act as decoys for transcription factors. (C) Recruitment of chromatin-modifying complex by lncRNAs.

include HOTAIR having around 800 direct binding sites and the lncRNA Paupar found to interact with around 2800 different locations on the genome.[39] Binding sites for many of the transcription-affecting lncRNAs are located far apart and are even often found on different chromosomes. Another example for lncRNA acting in *trans* is XIST known to interact with the distant regions of the chromosome in a three-dimensional way and bring about silencing of distal loci.[40]

3.5 LncRNAs Can Directly Interact With Chromatin

lncRNAs can bind to the promoter region of a gene and form a triplet structure. This triplet structure could form by the direct binding of a lncRNA with the double-stranded DNA and also can potentially bind to a nascent transcript still undergoing transcription. Another method of regulation of gene expression through lncRNAs is recruitment of chromatin-modifying complexes (discussed in Section 3.7).

Like their protein counterparts, lncRNAs can act as transcription-associated factors and directly bind with the promoter region of the gene they regulate. This base pairing between the double-stranded DNA locus and RNA results in the formation of triplex of RNA–DNA–DNA. It has been hypothesized that formation of this type of triplex structure between nucleic acids results from non-Watson–Crick base pairing and occurs via Hoogsteen base pairing.[41] These promoter-associated lncRNAs are referred to as pRNAs. pRNAs base pair with the promoter region of the rRNA gene, form the nucleic acid triplex, and were found to effectively regulate the transcription of rRNA gene[42] (Fig. 4A).

3.6 lncRNAs Interacting With Conventional Transcription Factors

lncRNAs can target chromatin indirectly through binding with transcription factor proteins and regulate the expression of target genes. The classical example for this class of lncRNAs is XIST. lncRNA XIST helps in dosage compensation for X-chromosome-associated genes through widespread silencing of one of the two X-chromosomes. The localization of the lncRNA XIST to the chromatin is mediated by a bivalent transcription factor YY1. YY1 interacts with both the lncRNA and the DNA target on the X-chromosome, thereby tethering XIST to the chromatin[43] (Fig. 4C).

Similarly, lncRNA GAS5 controls the expression of glucocorticoid receptors under starvation conditions.[44] GAS5 acts a decoy and competes with glucocorticoid response elements for binding to glucocorticoid receptors, thus suppressing glucocorticoid-mediated transcriptional activity in growth-arrested cells (Fig. 4B). In breast cancer, expression of GAS5 is reduced relative to adjacent normal tissue and its reexpression can induce apoptosis and growth arrest, thus labeling GAS5 as a tumor suppressive lncRNA.[45]

lncRNAs might act via feedback loops to control the expression of their own gene. In the case of dihydrofolate reductase (DHFR), a noncoding transcript is expressed from the minor promoter located upstream to the gene. This noncoding transcript not only interacts with the major DHFR promoter region but has also been shown to interact with the general transcription factor TFIIb. Thus, the lncRNA tethers the general transcription factor TFIIb to the promoter region and stalls transcription[46] (Fig. 4C).

Other examples of lncRNAs interacting with transcription factors are RRMST-interacting with stem cell marker SOX2 and regulating cellular fate,[47] and lncRNAs panda, lethe, and Jpx interacting with transcription factors NFYA, NF-kB, and CCTF respectively.[48–50]

3.7 Chromatin Modification and Epigenetic Regulation

In addition to regulating genes posttranscriptionally, lncRNAs also influence breast cancer initiation and progression through epigenetic programs. Indeed, lncRNAs are found to act at every level of epigenetic regulation and can impact DNA methylation, histone modification, nucleosome positioning, etc. Several in-depth reviews are available portraying the involvement of lncRNAs in the aforementioned epigenetic events.[51,52] Although the majority of lncRNAs have undefined mechanisms of action, a number of lncRNAs have been identified to interact with PcG proteins. PcG proteins function to maintain a closed chromatin configuration through depositing repressive histone modifications. While this function is critical during development and for cells to maintain differentiation state and cell identity, this function is often co-opted in cancer, driving aberrant gene expression. PcGs assemble into multiprotein complexes, primarily represented by Polycomb Repressive Complex 1 and 2 (PRC1, PRC2). PRC1 monoubiquitinates histone H2A (H2AK119ub), while PRC2 methylates histone H3 (H3K27me3), both modifications encouraging closed chromatin structure and repressing transcriptional activation.[51] lncRNAs have been identified that interact with PcGs during normal mammary gland development. Pregnancy-induced noncoding RNA (PINC) is induced during pregnancy in progenitor cells to inhibit differentiation of alveolar cells. This process is important for preventing milk production until birth. PINC functions through interacting with PRC2 and targeting it to critical gene locations.[53] While targeting PRC2 is a critical process for normal development for cells to maintain appropriate differentiation state and cell identity, this function is often co-opted in cancer, driving aberrant gene expression.

The classic example of a lncRNA influencing PcG proteins is that of HOTAIR; however the PcG proteins have been found to interact with hundreds of additional lncRNAs, suggesting a general mechanism for a number of lncRNAs with global transcriptome consequences.[54] HOTAIR is a noncoding RNA transcribed from the HOXC locus, a locus critical to embryonic development. HOTAIR was found to function in *trans* to target PRC2 to silence HOXD located on a different chromosome.[55] In addition to PRC2, the HOX locus is regulated by the LSD1/CoREST/REST demethylase complex, an additional repressive complex. HOTAIR binds both PRC2 and LSD1 acting as a molecular scaffold and coordinating repressive gene transcription to hundreds of sites in vivo (Fig. 5A).[56] HOTAIR

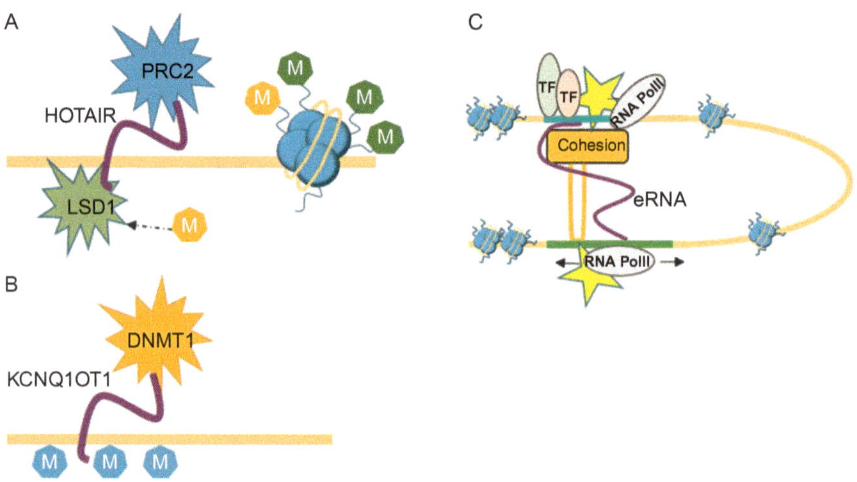

Fig. 5 lncRNAs modify chromatin structure. (A) lncRNAs can associate with and direct repressive complexes to modify histone modifications and chromatin structure. PRC2 methylates H3K27me3 (*green*), while LSD1 demethylates H3K4me2 (*yellow*). Both of these modifications contribute to gene repression. (B) lncRNAs can impact DNA methylation (*blue*) which represses gene transcription. (C) eRNAs can interact with chromatin looping factors (cohesion complex) altering interactions between enhancer (indicated in *green*) and promoter (indicated in *blue*) regions of target genes to alter gene expression. eRNAs may facilitate loading of RNA PolII and the transcription initiation complex (indicated by a *star*).

redirecting two repressive complexes results in altered repression of critical proteins including JAM2, a junctional adhesion molecule whose depletion is associated with metastasis and disease progression.[57,58] HOTAIR expression will induce expression of erstwhile repressed oncogenes such as SNAIL, an EMT and metastasis-associated transcription factor, and ABL2, metastasis-promoting nonreceptor tyrosine kinase. HOTAIR has been found to be enriched or overexpressed in approximately 25% of breast cancers leading to altered gene expression from up to 800 discreet loci.[57] This genome-wide reprogramming drives breast cancer progression and metastasis in mouse models and its overexpression is an independent prognostic indicator for metastasis and overall survival.[57,59]

In addition to HOTAIR, ANRIL (antisense noncoding RNA in the INK4 locus) is overexpressed in ~20% of invasive breast cancers[60] and found to recruit PRC1 and PRC2 to the tumor suppressor gene cluster containing INK4b–ARF–INK4A.[61,62] This gene cluster codes for critical regulators of oncogene-induced senescence and is frequently deleted or mutated in cancer initiation and progression. ANRIL has also been identified as an independent prognostic indicator of overall survival in serous ovarian cancer.[63]

Additional lncRNAs have been identified that hold critical functions in basic developmental imprinting, but frequently become co-opted in breast cancer. H19 is an imprinting RNA that functions through numerous mechanisms, is critical to early development, and is downregulated at birth; however, it has been found to reemerge in tumors.[64] H19 holds a myriad of functions, as described earlier, as a miRNA host gene (miR-675), as well as serves as a miRNA sponge. Similar to HOTAIR and ANRIL, H19 RNA can also interact with the PRC2 complex, through the EZH2 component, to direct gene repression. In breast cancer cells, H19 regulates E2F1, a G1/S-promoting transcription factor, and overexpression can drive tumor cell cycle progression, while H19 inhibition can attenuate cell growth.[65] H19 has also been implicated in driving EMT and promoting metastasis.[64] Together, H19 fills an oncogenic role in breast cancer; however, this may not serve as a good therapeutic target because in some contexts H19 expression is tumor suppressive.[66,67]

In addition to directing histone modifications, lncRNAs can influence epigenetic patterns by directly modifying imprinting proteins affecting DNA methylation. KCNQ1OT1 is a noncoding RNA expressed from the paternal allele and, similar to HOTAIR and ANRIL, can interact with repressive complexes (PRC2 and G9a) resulting in gene silencing. However, KCNQ1OT1 was also identified in mouse models to interact with

DNMT1 (DNA-methyltransferase 1) and recruit to sites for differential DNA methylation (Fig. 5B).[68] KCNQ1OT1 also targets histone (H3K9 and H3K27) methyltransferases.[69,70] In breast cancer, estrogen-dependent KCNQ1OT1 expression has been linked to repressive histone modification and DNA hypermethylation of cyclin-dependent kinase inhibitor 1C (CDKN1C), silencing the tumor suppressor and contributing to breast cancer initiation.[70] lncRNAs also control other important aspects of chromatin structure, such as chromosome looping—largely mediated by enhancer-associated RNAs (eRNAs).

3.8 Enhancer-Associated RNAs

Enhancers are genomic regions of critical importance to coordinating cell-type-specific gene expression. Transcription factors associate with enhancer regions to drive specific cell lineage determination genes as well as response to stimuli.[71] Currently, over 1 million enhancers are predicted across the genome, and recently, they have been found to be highly transcribed, resulting in eRNAs.[71–74] eRNAs may be the result of pervasive transcription initiating from the enhancer and may be rapidly degraded; however, eRNAs have also been demonstrated to functionally alter histone methylation and recruitment of transcriptional machinery to enhancer regions and may also alter chromatin looping surrounding enhancers resulting in aberrant gene expression programs.

Due to their association with enhancers, eRNAs are associated with genomic regions with enriched H3K27ac, a mark of active chromatin, negatively correlated with the repressive H3K27me3 modification, and are responsive to a broad spectrum of stimuli through signal-dependent transcription factors. Li *et al.* investigated the functional importance of eRNAs expressed from enhancers associated with estrogen-induced coding genes in breast cancer.[75] Depletion of eRNAs attenuated enhancer–promoter association and coding gene transcription. It was subsequently demonstrated that the eRNAs interact with the chromatin looping cohesion complex altering chromatin loop formation around estrogen-induced promoters (Fig. 5C).[71,75] In addition to modifying estrogen-induced enhancers, eRNAs are also critical for enhancing p53 target gene transcription[76] and can also recruit PolII to a promoter of target genes.[77]

Taken together, eRNAs are emerging as critical regulators of enhancer regions. Following comprehensive analysis of lncRNAs in breast cancer, Su *et al.* found that nearly two-thirds of lncRNAs expressed in invasive breast cancer were associated with enhancer regions (1038/1623).[78] Given that enhancers are pivotal to driving cell identity, differentiation state, and core signaling programs, corruption of these signaling nodes may contribute to cancer initiation and progression.

3.9 lncRNAs as Scaffolds

lncRNAs are often found to function within RNA–protein complexes and can serve as a scaffold holding these proteins together. In the case of LINP1 (lncRNA in NHEJ pathway), the lncRNA functions to coordinate the non-homologous end joining pathway by serving as a scaffold for Ku80 and DNA-PKcs, two critical DNA repair enzymes. LINP1 is overexpressed in triple-negative breast cancer, and when its expression is attenuated, it sensitizes breast cancer cells to DNA damaging radiation therapy.[79] Thus, LINP1 may represent a new therapeutic entry point.

4. APPROACHES FOR RNA-TARGETED THERAPEUTIC INTERVENTION

lncRNAs control a myriad of cellular pathways in breast cancer and other diseases. Some lncRNAs could be considered as master regulators of cancer phenotypes from cellular transformation to metastasis. They offer an unfathomable potential for cancer therapy as dependable targets supported by the data obtained in vitro through studies performed in different cancer cell lines as well as in animal models. As the research field of lncRNAs itself is still in its infancy, so too are the explorations regarding their therapeutic potential. Many novel and conventional therapeutic agents are being explored for effective targeting of lncRNAs as a treatment for breast and other cancer types. The following section gives an account of some of the representative drug types and their current state.

4.1 Small Interfering RNA

Small interfering RNAs (siRNAs) are artificially synthesized 19–23 nucleotide long double-stranded RNA molecules. They are routinely used in molecular biology for transient silencing of gene of interest. They elicit RNAi response upon binding to their target transcript based on the sequence complementarity. They have been rightly used to study the effect of various oncogenic lncRNAs through the loss of function. Using siRNAs, a high degree of silencing was observed against the lncRNA HOTAIR. A strong anticancer phenotype was observed both in vitro and in vivo.[57,80] Similarly, MALAT1 was targeted using siRNAs and a markedly reduced cell proliferation was observed in hepatocellular carcinoma.[81] In fact, many siRNA drugs against protein coding mRNAs are already in phase I and phase II clinical trials for testing against cancer.[82] For example, siRNA–EphA2–DOPC against EphA2, TKM-080301 targeting PLK1, and CALAA-01 inhibiting RRM2 are in phase I clinical trial. Further, siRNA drug siG12D LODER and Atu027 are being explored as anticancer drugs in combination of conventional chemotherapeutic substances.[82] The performance of these siRNA drugs will pave the way for more clinical trials for targeting oncogenic/onco-promoting lncRNAs. However, there are certain unavoidable caveats for developing siRNAs as drug molecules in cancer. The machinery for RNAi, the mechanism behind siRNAs function, is located in the cytoplasm. Therefore, it will be difficult to target nuclear-restricted lncRNAs. Another obstacle for using siRNA is the lack of availability of a suitable delivery system. Most of the in vitro studies with siRNAs are conducted using a transfection agent that cannot be used for in vivo delivery. However, many studies are in progress for conjugating the siRNA drug with nanoparticles that seem to be an effective vehicle for carrying siRNAs.

4.2 Antisense Oligonucleotides

Antisense oligonucleotide (ASOs) are small-sized single-stranded nucleic acids and offer some advantage over siRNAs in terms of targeting both nuclear and cytoplasmic located lncRNAs. Based on their sequence homology, ASOs bind to their target RNA sequence inside the cells and bring about gene silencing. Several improvements have been made with ASOs including using locked nucleic acids for better stability, lesser off-target effects, free uptake by the cell, lesser cytotoxicity in nontarget cells, etc. ASOs have been developed targeting lncRNA MALAT1 and have shown promising results to be used as an effective anticancer drug.[83] However, similar to siRNA drug, ASO drugs have similar shortcomings in terms of delivery to the target tissues. ASO drugs are also being investigated to be used in combination with nanoparticles.

4.3 Aptamers

Aptamers are mostly RNA or DNA molecules, which upon folding attain a unique three-dimensional structure and can interact with virtually any ligand having a complementary structure. In terms of binding properties and specificity, they are akin to antibodies.[84] lncRNAs have strong secondary structures that may mask the binding sites for the conventional siRNA or ASOs. Aptamers seem to be a better fit in this scenario. By virtue of secondary structures formed by the aptamers, they can find a matching complementary structure in the folded lncRNA. Many of the aptamers act by blocking the vital proteins on the cell membrane, extra cellular matrix, and the cytoplasmic proteins of the cancer cell.[85] In this case, an aptamer could disrupt the interaction between a lncRNA and a protein or other binding partner.

4.4 Small Molecules

Nucleic acids are folded into their most stable three-dimensional structure. It has been observed that the structural small domains are considerably stable and conserved. Small molecules have been developed targeting these structural elements.[86] This knowledge can be harnessed to selectively target lncRNAs using small molecules. The drawback for this approach is gaining the complete sequence information for the lncRNA first. Due to pervasive transcriptional nature of some of the larger lncRNA locus, obtaining full information regarding the functional isoforms is difficult.

5. CONCLUDING REMARKS

Significant progress has been made in analyzing and curating the data pertaining to lncRNAs obtained through next-generation RNA sequencing by using improved bioinformatics tools. Several databases are now available to assist in dissemination of information regarding a single or group of lncRNAs in various etiological context of breast cancers and several other cancer types. These databases provide a leading point to researchers studying different aspects of lncRNAs: e.g., knowledge about cellular pathways lncRNAs potentially regulate, the landscape of the lncRNA genomic locus, possible interacting partners, etc. The various databases dedicated to lncRNAs in the context of breast cancer have been compiled in Tables 2 and 3.

lncRNAs are a diverse class of molecules that function through interaction with DNA, RNA, and proteins to impact every aspect of gene regulation. With increasing advances in bioinformatics, lncRNAs now vastly outnumber annotated coding genes. A major challenge moving forward will be to distinguish functional lncRNAs from those that may simply represent byproducts of transcription. Already, many lncRNAs have been found to be critical regulators during normal processes of development and response to exogenous stimuli and threats such as DNA damage. Aberrant expression, or

Table 2 Databases for Annotation of lncRNAs

Name of Database	URL	Application	References
LNCat	http://biocc.hrbmu.edu.cn/LNCat/	A genome browser of lncRNA structures, visualization of different resources	87
LncDisease	http://www.cuilab.cn/lncrnadisease	Prediction of lncRNA disease association	88
MiTranscriptome	http://mitranscriptome.org	Polyadenylated transcript expression data	1
starBase v2.0	http://starbase.sysu.edu.cn/	Prediction of miRNA–ceRNA, miRNA–ncRNA, and protein–RNA interaction networks	89
ChIPBase	http://deepbase.sysu.edu.cn/chipbase/	Transcriptional regulation of lncRNA genes	90
lncRNA2Function	http://mlg.hit.edu.cn/lncrna2function	Functional analysis of lncRNAs	91
lnCeDB	http://gyanxet-beta.com/lncedb/	Database for lncRNAs acting as microRNA sponge	92
GeneFriends	http://www.genefriends.org/	Database for coexpression network for transcripts	93
RNA–RNA interaction database (no specific name given to the database)	http://rtools.cbrc.jp/cgi-bin/RNARNA/index.pl	Database for mining RNA–RNA interactions	11
RAID	http://www.rna-society.org/raid/index.html	Prediction and analysis of RNA–RNA and RNA–protein interactions	94
NONCODE 2016	http://www.noncode.org	A compendium of information on noncoding RNAs	95

Table 3 Bioinformatics Tools for Predicting the lncRNA–MicroRNA–Target Gene Interaction

Bioinformatics Tool	URL	References
starBase v2.0	http://starbase.sysu.edu.cn/	89
miRSponge	http://www.bio-bigdata.net/miRSponge	96
Cupid	http://cupidtool.sourceforge.net/	97
LncACTdb	http://www.bio-bigdata.net/LncACTdb/	17
lnCeDB	http://gyanxet-beta.com/lncedb/	98
miRcode	http://www.mircode.org	99

loss, of lncRNAs has been demonstrated to impact breast cancer initiation and progression. With only a handful of lncRNAs with defined functions, they represent a largely untapped source of novel therapeutic entry points in breast cancer and other diseases. Further insight into the structure and function of lncRNAs will be critical to taking full advantage of their therapeutic potential.

ACKNOWLEDGMENTS

This work was supported by NCI Grant CA166815, Breast Cancer Alliance Young Investigator Grant (J.I.H.), and DoD award (W81XWH-15-1-0495) (R.S.D.).

REFERENCES

1. Iyer MK, Niknafs YS, Malik R, et al. The landscape of long noncoding RNAs in the human transcriptome. *Nat Genet.* 2015;47(3):199–208.
2. Gutschner T, Diederichs S. The hallmarks of cancer: a long non-coding RNA point of view. *RNA Biol.* 2012;9(6):703–719.
3. Brown CJ, Ballabio A, Rupert JL, et al. A gene from the region of the human X inactivation centre is expressed exclusively from the inactive X chromosome. *Nature.* 1991;349(6304):38–44.
4. Bartolomei MS, Zemel S, Tilghman SM. Parental imprinting of the mouse H19 gene. *Nature.* 1991;351(6322):153–155.
5. Carninci P, Kasukawa T, Katayama S, et al. The transcriptional landscape of the mammalian genome. *Science.* 2005;309(5740):1559–1563.
6. Forrest AR, Abdelhamid RF, Carninci P. Annotating non-coding transcription using functional genomics strategies. *Brief Funct Genomic Proteomic.* 2009;8(6):437–443.
7. Numata K, Kanai A, Saito R, et al. Identification of putative noncoding RNAs among the RIKEN mouse full-length cDNA collection. *Genome Res.* 2003;13(6B):1301–1306.

7a. Diermeier SD, Chang KC, Freier SM, et al. Mammary tumor-associated RNAs impact tumor cell proliferation, invasion, and migration. *Cell Rep.* 2016;17(1):261–274. http://dx.doi.org/10.1016/j.celrep.2016.08.081.

8. Yang F, Lyu S, Dong S, Liu Y, Zhang X, Wang O. Expression profile analysis of long noncoding RNA in HER-2-enriched subtype breast cancer by next-generation sequencing and bioinformatics. *Onco Targets Ther.* 2016;9:761–772.
9. Yan X, Hu Z, Feng Y, et al. Comprehensive genomic characterization of long non-coding RNAs across human cancers. *Cancer Cell.* 2015;28(4):529–540.
10. Guffanti A, Iacono M, Pelucchi P, et al. A transcriptional sketch of a primary human breast cancer by 454 deep sequencing. *BMC Genomics.* 2009;10:163.
11. Terai G, Iwakiri J, Kameda T, Hamada M, Asai K. Comprehensive prediction of lncRNA-RNA interactions in human transcriptome. 2016;17:12.
12. Engreitz JM, Sirokman K, McDonel P, et al. RNA-RNA interactions enable specific targeting of noncoding RNAs to nascent pre-mRNAs and chromatin sites. *Cell.* 2014;159(1):188–199.
13. Chu C, Qu K, Zhong FL, Artandi SE, Chang HY. Genomic maps of long noncoding RNA occupancy reveal principles of RNA-chromatin interactions. *Mol Cell.* 2011;44(4):667–678.
14. Guo W, Wang Q, Zhan Y, et al. Transcriptome sequencing uncovers a three-long noncoding RNA signature in predicting breast cancer survival. *Sci Rep.* 2016;6:27931.
15. Sun M, Gadad SS, Kim DS, Kraus WL. Discovery, annotation, and functional analysis of long noncoding RNAs controlling cell-cycle gene expression and proliferation in breast cancer cells. *Mol Cell.* 2015;59(4):698–711.
16. Jonsson P, Coarfa C, Mesmar F, et al. Single-molecule sequencing reveals estrogen-regulated clinically relevant lncRNAs in breast cancer. *Mol Endocrinol.* 2015;29(11):1634–1645.
17. Wang P, Ning S, Zhang Y, et al. Identification of lncRNA-associated competing triplets reveals global patterns and prognostic markers for cancer. *Nucleic Acids Res.* 2015;43(7):3478–3489.
18. Li Y, Wang Z, Shi H, et al. HBXIP and LSD1 scaffolded by lncRNA Hotair mediate transcriptional activation by c-Myc. *Cancer Res.* 2016;76(2):293–304.
19. Ma MZ, Li CX, Zhang Y, et al. Long non-coding RNA HOTAIR, a c-Myc activated driver of malignancy, negatively regulates miRNA-130a in gallbladder cancer. *Mol Cancer.* 2014;13:156.
20. Sorensen KP, Thomassen M, Tan Q, et al. Long non-coding RNA HOTAIR is an independent prognostic marker of metastasis in estrogen receptor-positive primary breast cancer. *Breast Cancer Res Treat.* 2013;142(3):529–536.
21. Henry WS, Hendrickson DG, Beca F, et al. LINC00520 is induced by Src, STAT3, and PI3K and plays a functional role in breast cancer. *Oncotarget.* 2016.
21a. Niknafs YS, Han S, Ma T, et al. The lncRNA landscape of breast cancer reveals a role for DSCAM-AS1 in breast cancer progression. *Nat Commun.* 2016;7:12791. http://dx.doi.org/10.1038/ncomms12791.
22. Tripathi V, Ellis JD, Shen Z, et al. The nuclear-retained noncoding RNA MALAT1 regulates alternative splicing by modulating SR splicing factor phosphorylation. *Mol Cell.* 2010;39(6):925–938.
23. Miao Y, Fan R, Chen L, Qian H. Clinical significance of long non-coding RNA MALAT1 expression in tissue and serum of breast cancer. *Ann Clin Lab Sci.* Jul 2016;46(4):418–424.
24. Gonzalez I, Munita R, Agirre E, et al. A lncRNA regulates alternative splicing via establishment of a splicing-specific chromatin signature. *Nat Struct Mol Biol.* 2015;22(5):370–376.
25. Villamizar O, Chambers CB, Riberdy JM, Persons DA, Wilber A. Long noncoding RNA Saf and splicing factor 45 increase soluble Fas and resistance to apoptosis. *Oncotarget.* 2016;7(12):13810–13826.

26. Krystal GW, Armstrong BC, Battey JF. N-myc mRNA forms an RNA-RNA duplex with endogenous antisense transcripts. *Mol Cell Biol.* 1990;10(8):4180–4191.
27. Beltran M, Puig I, Pena C, et al. A natural antisense transcript regulates Zeb2/Sip1 gene expression during Snail1-induced epithelial-mesenchymal transition. *Genes Dev.* 2008;22(6):756–769.
28. Gong C, Maquat LE. lncRNAs transactivate STAU1-mediated mRNA decay by duplexing with 3′ UTRs via Alu elements. *Nature.* 2011;470(7333):284–288.
29. Faghihi MA, Zhang M, Huang J, et al. Evidence for natural antisense transcript-mediated inhibition of microRNA function. *Genome Biol.* 2010;11(5):R56.
30. Kretz M, Siprashvili Z, Chu C, et al. Control of somatic tissue differentiation by the long non-coding RNA TINCR. *Nature.* 2013;493(7431):231–235.
31. Liu X, Li D, Zhang W, Guo M, Zhan Q. Long non-coding RNA gadd7 interacts with TDP-43 and regulates Cdk6 mRNA decay. *EMBO J.* 2012;31(23):4415–4427.
32. Zong X, Nakagawa S, Freier SM, et al. Natural antisense RNA promotes 3′ end processing and maturation of MALAT1 lncRNA. *Nucleic Acids Res.* 2016;44(6):2898–2908.
33. Salmena L, Poliseno L, Tay Y, Kats L, Pandolfi PP. A ceRNA hypothesis: the Rosetta Stone of a hidden RNA language? *Cell.* 2011;146(3):353–358.
34. Liu XH, Sun M, Nie FQ, et al. Lnc RNA HOTAIR functions as a competing endogenous RNA to regulate HER2 expression by sponging miR-331-3p in gastric cancer. *Mol Cancer.* 2014;13:92.
35. Tao S, He H, Chen Q. Estradiol induces HOTAIR levels via GPER-mediated miR-148a inhibition in breast cancer. *J Transl Med.* 2015;13:131.
36. Yuan JH, Yang F, Wang F, et al. A long noncoding RNA activated by TGF-beta promotes the invasion-metastasis cascade in hepatocellular carcinoma. *Cancer Cell.* 2014;25(5):666–681.
37. Zhang Z, Zhu Z, Watabe K, et al. Negative regulation of lncRNA GAS5 by miR-21. *Cell Death Differ.* 2013;20(11):1558–1568.
38. Eades G, Wolfson B, Zhang Y, Li Q, Yao Y, Zhou Q. lincRNA-RoR and miR-145 regulate invasion in triple-negative breast cancer via targeting ARF6. *Mol Cancer Res.* 2015;13(2):330–338.
39. Vance KW, Ponting CP. Transcriptional regulatory functions of nuclear long noncoding RNAs. *Trends Genet.* 2014;30(8):348–355.
40. Engreitz JM, Pandya-Jones A, McDonel P, et al. The Xist lncRNA exploits three-dimensional genome architecture to spread across the X chromosome. *Science.* 2013;341(6147):1237973.
41. Buske FA, Mattick JS, Bailey TL. Potential in vivo roles of nucleic acid triple-helices. *RNA Biol.* 2011;8(3):427–439.
42. Schmitz KM, Mayer C, Postepska A, Grummt I. Interaction of noncoding RNA with the rDNA promoter mediates recruitment of DNMT3b and silencing of rRNA genes. *Genes Dev.* 2010;24(20):2264–2269.
43. Jeon Y, Lee JT. YY1 tethers Xist RNA to the inactive X nucleation center. *Cell.* 2011;146(1):119–133.
44. Kino T, Hurt DE, Ichijo T, Nader N, Chrousos GP. Noncoding RNA gas5 is a growth arrest- and starvation-associated repressor of the glucocorticoid receptor. *Sci Signal.* 2010;3(107):ra8.
45. Mourtada-Maarabouni M, Pickard MR, Hedge VL, Farzaneh F, Williams GT. GAS5, a non-protein-coding RNA, controls apoptosis and is downregulated in breast cancer. *Oncogene.* 2009;28(2):195–208.
46. Martianov I, Ramadass A, Serra Barros A, Chow N, Akoulitchev A. Repression of the human dihydrofolate reductase gene by a non-coding interfering transcript. *Nature.* 2007;445(7128):666–670.

47. Ng SY, Bogu GK, Soh BS, Stanton LW. The long noncoding RNA RMST interacts with SOX2 to regulate neurogenesis. *Mol Cell.* 2013;51(3):349–359.
48. Hung T, Wang Y, Lin MF, et al. Extensive and coordinated transcription of noncoding RNAs within cell-cycle promoters. *Nat Genet.* 2011;43(7):621–629.
49. Rapicavoli NA, Qu K, Zhang J, Mikhail M, Laberge RM, Chang HY. A mammalian pseudogene lncRNA at the interface of inflammation and anti-inflammatory therapeutics. *eLife.* 2013;2:e00762.
50. Sun S, Del Rosario BC, Szanto A, Ogawa Y, Jeon Y, Lee JT. Jpx RNA activates Xist by evicting CTCF. *Cell.* 2013;153(7):1537–1551.
51. Betancur JG. Pervasive lncRNA binding by epigenetic modifying complexes—the challenges ahead. *Biochim Biophys Acta.* 2016;1859(1):93–101.
52. Bohmdorfer G, Wierzbicki AT. Control of chromatin structure by long noncoding RNA. *Trends Cell Biol.* 2015;25(10):623–632.
53. Shore AN, Kabotyanski EB, Roarty K, et al. Pregnancy-induced noncoding RNA (PINC) associates with polycomb repressive complex 2 and regulates mammary epithelial differentiation. *PLoS Genet.* 2012;8(7):e1002840.
54. Zhao J, Ohsumi TK, Kung JT, et al. Genome-wide identification of polycomb-associated RNAs by RIP-seq. *Mol Cell.* 2010;40(6):939–953.
55. Rinn JL, Kertesz M, Wang JK, et al. Functional demarcation of active and silent chromatin domains in human HOX loci by noncoding RNAs. *Cell.* 2007;129(7):1311–1323.
56. Tsai MC, Manor O, Wan Y, et al. Long noncoding RNA as modular scaffold of histone modification complexes. *Science.* 2010;329(5992):689–693.
57. Gupta RA, Shah N, Wang KC, et al. Long non-coding RNA HOTAIR reprograms chromatin state to promote cancer metastasis. *Nature.* 2010;464(7291):1071–1076.
58. Zhao H, Yu H, Martin TA, Zhang Y, Chen G, Jiang WG. Effect of junctional adhesion molecule-2 expression on cell growth, invasion and migration in human colorectal cancer. *Int J Oncol.* 2016;48(3):929–936.
59. Hung T, Chang HY. Long noncoding RNA in genome regulation: prospects and mechanisms. *RNA Biol.* 2010;7(5):582–585.
60. Meseure D, Vacher S, Alsibai KD, et al. Expression of ANRIL-polycomb complexes-CDKN2A/B/ARF genes in breast tumors: identification of a two-gene (EZH2/CBX7) signature with independent prognostic value. *Mol Cancer Res.* 2016;14(7):623–633.
61. Kotake Y, Nakagawa T, Kitagawa K, et al. Long non-coding RNA ANRIL is required for the PRC2 recruitment to and silencing of p15(INK4B) tumor suppressor gene. *Oncogene.* 2011;30(16):1956–1962.
62. Li Z, Yu X, Shen J. ANRIL: a pivotal tumor suppressor long non-coding RNA in human cancers. *Tumour Biol.* 2016;37(5):5657–5661.
63. Qiu JJ, Lin YY, Ding JX, Feng WW, Jin HY, Hua KQ. Long non-coding RNA ANRIL predicts poor prognosis and promotes invasion/metastasis in serous ovarian cancer. *Int J Oncol.* 2015;46(6):2497–2505.
64. Raveh E, Matouk IJ, Gilon M, Hochberg A. The H19 Long non-coding RNA in cancer initiation, progression and metastasis—a proposed unifying theory. *Mol Cancer.* 2015;14:184.
65. Berteaux N, Lottin S, Monte D, et al. H19 mRNA-like noncoding RNA promotes breast cancer cell proliferation through positive control by E2F1. *J Biol Chem.* 2005;280(33):29625–29636.
66. Gabory A, Jammes H, Dandolo L. The H19 locus: role of an imprinted non-coding RNA in growth and development. *Bioessays.* 2010;32(6):473–480.
67. Yoshimizu T, Miroglio A, Ripoche MA, et al. The H19 locus acts in vivo as a tumor suppressor. *Proc Natl Acad Sci USA.* 2008;105(34):12417–12422.

68. Mohammad F, Mondal T, Guseva N, Pandey GK, Kanduri C. Kcnq1ot1 noncoding RNA mediates transcriptional gene silencing by interacting with Dnmt1. *Development.* 2010;137(15):2493–2499.
69. Pandey RR, Mondal T, Mohammad F, et al. Kcnq1ot1 antisense noncoding RNA mediates lineage-specific transcriptional silencing through chromatin-level regulation. *Mol Cell.* 2008;32(2):232–246.
70. Rodriguez BA, Weng YI, Liu TM, et al. Estrogen-mediated epigenetic repression of the imprinted gene cyclin-dependent kinase inhibitor 1C in breast cancer cells. *Carcinogenesis.* Jun 2011;32(6):812–821.
71. Lam MT, Li W, Rosenfeld MG, Glass CK. Enhancer RNAs and regulated transcriptional programs. *Trends Biochem Sci.* 2014;39(4):170–182.
72. Buecker C, Wysocka J. Enhancers as information integration hubs in development: lessons from genomics. *Trends Genet.* 2012;28(6):276–284.
73. Kim TK, Hemberg M, Gray JM, et al. Widespread transcription at neuronal activity-regulated enhancers. *Nature.* 2010;465(7295):182–187.
74. Xie W, Ren B. Developmental biology. Enhancing pluripotency and lineage specification. *Science.* 2013;341(6143):245–247.
75. Li W, Notani D, Ma Q, et al. Functional roles of enhancer RNAs for oestrogen-dependent transcriptional activation. *Nature.* 2013;498(7455):516–520.
76. Melo CA, Drost J, Wijchers PJ, et al. eRNAs are required for p53-dependent enhancer activity and gene transcription. *Mol Cell.* 2013;49(3):524–535.
77. Mousavi K, Zare H, Dell'orso S, et al. eRNAs promote transcription by establishing chromatin accessibility at defined genomic loci. *Mol Cell.* 2013;51(5):606–617.
78. Su X, Malouf GG, Chen Y, et al. Comprehensive analysis of long non-coding RNAs in human breast cancer clinical subtypes. *Oncotarget.* 2014;5(20):9864–9876.
79. Zhang Y, He Q, Hu Z, et al. Long noncoding RNA LINP1 regulates repair of DNA double-strand breaks in triple-negative breast cancer. *Nat Struct Mol Biol.* 2016;23(6): 522–530.
80. Kim K, Jutooru I, Chadalapaka G, et al. HOTAIR is a negative prognostic factor and exhibits pro-oncogenic activity in pancreatic cancer. *Oncogene.* 2013;32(13): 1616–1625.
81. Du Y, Kong G, You X, et al. Elevation of highly up-regulated in liver cancer (HULC) by hepatitis B virus X protein promotes hepatoma cell proliferation via down-regulating p18. *J Biol Chem.* 2012;287(31):26302–26311.
82. Li CH, Chen Y. Targeting long non-coding RNAs in cancers: progress and prospects. *Int J Biochem Cell Biol.* 2013;45(8):1895–1910.
83. Arun G, Diermeier S, Akerman M, et al. Differentiation of mammary tumors and reduction in metastasis upon Malat1 lncRNA loss. *Genes Dev.* 2016;30(1):34–51.
84. Shi H, Hoffman BE, Lis JT. RNA aptamers as effective protein antagonists in a multicellular organism. *Proc Natl Acad Sci USA.* 1999;96(18):10033–10038.
85. Lee YJ, Kim IS, Park SA, et al. Periostin-binding DNA aptamer inhibits breast cancer growth and metastasis. *Mol Ther.* 2013;21(5):1004–1013.
86. Childs-Disney JL, Disney MD. Approaches to validate and manipulate RNA targets with small molecules in cells. *Annu Rev Pharmacol Toxicol.* 2016;56:123–140.
87. Xu J, Bai J, Zhang X, et al. A comprehensive overview of lncRNA annotation resources. *Brief Bioinform.* 2016. http://dx.doi.org/10.1093/bib/bbw015.
88. Wang J, Ma R, Ma W, et al. LncDisease: a sequence based bioinformatics tool for predicting lncRNA-disease associations. *Nucleic Acids Res.* 2016;44(9):e90.
89. Li JH, Liu S, Zhou H, Qu LH, Yang JH. starBase v2.0: decoding miRNA-ceRNA, miRNA-ncRNA and protein-RNA interaction networks from large-scale CLIP-Seq data. *Nucleic Acids Res.* 2014;42(Database issue):D92–97.

90. Yang JH, Li JH, Jiang S, Zhou H, Qu LH. ChIPBase: a database for decoding the transcriptional regulation of long non-coding RNA and microRNA genes from ChIP-Seq data. *Nucleic Acids Res*. 2013;41(Database issue):D177–187.
91. Jiang Q, Ma R, Wang J, et al. LncRNA2Function: a comprehensive resource for functional investigation of human lncRNAs based on RNA-seq data. *BMC Genomics* 2015;16(Suppl. 3):S2.
92. Das S, Ghosal S, Sen R, Chakrabarti J. lnCeDB: database of human long noncoding RNA acting as competing endogenous RNA. *PLoS One*. 2014;9(6):e98965.
93. van Dam S, Craig T, de Magalhaes JP. GeneFriends: a human RNA-seq-based gene and transcript co-expression database. *Nucleic Acids Res*. 2015;43(Database issue): D1124–1132.
94. Zhang X, Wu D, Chen L, et al. RAID: a comprehensive resource for human RNA-associated (RNA-RNA/RNA-protein) interaction. *RNA*. 2014;20(7):989–993.
95. Zhao Y, Li H, Fang S, et al. NONCODE 2016: an informative and valuable data source of long non-coding RNAs. *Nucleic Acids Res*. 2016;44(D1):D203–208.
96. Wang P, Zhi H, Zhang Y, et al. miRSponge: a manually curated database for experimentally supported miRNA sponges and ceRNAs. *Database (Oxford)*. 2015;2015: bav098.
97. Chiu HS, Llobet-Navas D, Yang X, et al. Cupid: simultaneous reconstruction of microRNA-target and ceRNA networks. *Genome Res*. 2015;25(2):257–267.
98. Le TD, Zhang J, Liu L, Li J. Computational methods for identifying miRNA sponge interactions. *Brief Bioinform*. 2016. http://dx.doi.org/10.1093/bib/bbw042.
99. Jeggari A, Marks DS, Larsson E. miRcode: a map of putative microRNA target sites in the long non-coding transcriptome. *Bioinformatics*. 2012;28(15):2062–2063.

INDEX

Note: Page numbers followed by "*f*" indicate figures, and "*t*" indicate tables.

B

H

I

Q

R

W

X

FSC
www.fsc.org
MIX
Papier | Fördert
gute Waldnutzung
FSC® C083411

Kolibri 360 GmbH
Ferdinand-Jühlke-Straße 7
99095 Erfurt, Deutschland
produktsicherheit@kolibri360.de